D1046747

Kinematics and Dynamics of Machinery

An industrial robot. (*Source:* Cincinnati Milacron.)

Kinematics and Dynamics of Machinery

SECOND EDITION

Charles E. Wilson
Department of Mechanical and Industrial Engineering,
New Jersey Institute of Technology

J. Peter Sadler
Department of Mechanical Engineering,
The University of Kentucky

HarperCollins*College*Publishers

Sponsoring Editor: Don Childress
Project Editor: Janet Tilden
Assistant Art Director: Julie Anderson
Text and Cover Design: Jess Schaal
Cover Illustration: ITT Pneumotive
Production Administrator: Kathy Donnelly
Compositor: Science Typographers, Inc.
Printer and Binder: R. R. Donnelley & Sons Company
Cover Printer: New England Book Components

Kinematics and Dynamics of Machinery, Second Edition
Copyright © 1993 by HarperCollins College Publishers.

All rights reserved. Printed in the United States of America. No part of
this book may be used or reproduced in any manner whatsoever without
written permission, except in the case of brief quotations embodied in
critical articles and reviews. For information address HarperCollins College Publishers, 10 East 53rd Street, New York, NY 10022.

Library of Congress Cataloging-in-Publication Data

Wilson, Charles E.
    Kinematics and dynamics of machinery/Charles E. Wilson, J. Peter
Sadler.—2nd ed.
        p.   cm.
    Includes bibliographical references and index.
    ISBN 0-06-044474-6
    1. Machinery, Kinematics of.   2. Machinery, Dynamics of.
I. Sadler, J. Peter.   II. Title.
TJ175.W758   1991   91-24052
621.8'11—dc20   CIP

92   93   94   95   9   8   7   6   5   4   3   2   1

There is no expedient to which a man will not resort to avoid the real labor of thinking.

SIR JOSHUA REYNOLDS

It is the authors' sincere hope that users of this book will be encouraged to think and to explore designs and methods not explicitly described herein. Although worked-out examples have been used extensively to illustrate applications to engineering problems, the intent is to build a capacity for solving new problems that one may encounter in the future.

Dedication

To the memory of Walter Michels, who inspired a generation of students to achievements beyond their own expectations.

Contents

Preface

The Accreditation Board for Engineering and Technology states that "Engineering is that profession in which knowledge of the mathematical and natural sciences gained by study, experience, and practice is applied with judgment to develop ways to utilize, economically, the materials and forces of nature for the benefit of mankind. ... Engineering design is the process of devising a system, component, or process to meet desired needs. It is a decision-making process (often iterative), in which the basic sciences, mathematics, and engineering sciences are applied to convert resources optimally to meet a stated objective. Among the fundamental objectives of the design process are the establishment of objectives and criteria, synthesis, analysis, construction, testing, and evaluation. The engineering design component of a curriculum must include at least some of the following features: development of student creativity, use of open-ended problems, development and use of design methodology, formulation of design problem statements and specifications, consideration of alternative solutions, feasibility considerations, and detailed system descriptions. ... Appropriate computer-based experience must be included in the program of each student. Students must demonstrate knowledge of the application and use of digital computation techniques to specific engineering problems."* A course in kinematics and dynamics of machinery provides an opportunity to introduce many of the above objectives.

The coverage of this text includes mechanisms and machines, basic concepts; motion in machinery; velocity and acceleration analysis of mechanisms; design and analysis of cams, gears, and drive trains; static and dynamic force analysis; synthesis; and an introduction to robotic manipulators. Practical applications are considered throughout the text. Example problems and homework problems involve engineering design and provide a basis for design courses to follow. Analytical and graphical vector methods are illustrated, as well as complex number methods.

This text illustrates the design and analysis of mechanisms with the aid of mathematics software, user-written computer programs, and spreadsheets. Computer graphics and dedicated kinematics and dynamics software are discussed. Many of the example and homework problems involve calculations and plotting of results that can be done most efficiently using a computer.

---

*Accreditation Board for Science and Technology, *Criteria for Accrediting Programs in Engineering in the United States*, New York, 1990.

There are also many problems that can be solved by "hand calculations," however, using only a scientific calculator and/or simple drafting tools. The latter group may be useful as short practice problems and examination problems.

Most of the topics in this text can be covered in a three-credit-hour course given to engineering students who have completed a course in statics and dynamics of rigid bodies. Instructors may decide to cover the parts of the text that they deem essential, and then select additional topics and solution methods according to goals set for their students. For example, either analytical vector methods or complex number methods may be used as a basis for writing computer programs to solve planar linkages. However, if analysis of spatial linkages is to follow analysis of planar linkages, then vector methods might be used for both.

The authors have attempted to provide sufficient rigor and advanced material to challenge the student and provide a basis for further study. Student creativity may be fostered by the demands of the task, particularly if a few homework problems are expanded into open-ended design-type projects. Additional projects requiring creativity may be suggested by articles in technical publications or by an instructor's current research and consulting.

Kinematics and dynamics of machinery and the design of mechanisms involve modeling of physical systems. Relationships so developed have limits of applicability. The user of this text is urged to interpret the results of calculations, rather than simply obtain problem solutions. It is the reader's responsibility to assess formulas and methods to determine their applicability to a particular situation. Although the publisher, the reviewers, and the authors have made every attempt to ensure accuracy, errors invariably creep in. Suggestions and corrections are most welcome.

Acknowledgments

We wish to acknowledge the suggestions and comments of our colleagues and students at the New Jersey Institute of Technology and the University of Kentucky. We are grateful to the New Jersey Institute of Technology for granting a sabbatical leave to Charles Wilson, enabling him to increase and update computer applications in the text. Users of the first edition and reviewers of the manuscript of this edition made many worthwhile suggestions based on their extensive teaching and engineering experience. Those who reviewed all or part of the manuscript include Charles C. Adams, *Dordt College*; Ara Arabyan, *University of Arizona—Tucson*; Leo R. Maier, *Ohio Northern University*; and Saeed B. Niku, *California Polytechnic State University—San Luis Obispo*. Their careful analysis resulted in many changes in the text. Finally, we wish to thank our editors and others at HarperCollins for their careful, professional work.

Symbols

Vectors and matrices are shown in **boldface**, scalar magnitudes in lightface.

A^{-1}	Inverse of matrix A
$A \cdot B$	Dot (scalar) product of vectors A and B
$A \times B$	Cross (vector) product of vectors A and B
a	Gear tooth addendum
a, a	Acceleration
a^c, a^c	Coriolis acceleration
a^n, a^n	Normal acceleration
a^t, a^t	Tangential acceleration
bc	Velocity of C relative to B (velocity difference)
C	Cylinder pair; planet carrier
C	Force couple
C_i	Inertia couple or inertia torque
c	Center distance
CAD	Computer-aided design
d	Diameter of pitch circle
d_b	Diameter of base circle
d_c	Pitch diameter of equivalent spur gear
DF	Degrees of freedom
e	Instantaneous efficiency; cam-follower offset; piston offset; eccentricity
$e^{j\theta}$	Polar form of a complex number
F	Force
F_a	Axial or thrust gear tooth force component
F_e	External force
F_i	Inertia force
F_{ij}	Force exerted by a member i on member j
F_n	Normal gear tooth force
F_r	Radial gear tooth force component

F_s	Shaking force
F_t	Tangential gear tooth force component
f_i	Joint connectivity
G	Center of mass
H	Helix pair
h	Cam follower lift
hp	Horsepower
I	Mass moment of inertia
i, j, k	Cartesian unit vectors
j	Cam follower jerk; $\sqrt{-1}$, the imaginary unit used to represent quantities on the complex plane
L	Link length
L_d	Length of diagonal (of linkage polygon)
l	Lead of worm
l_i	Length of link i
M_s	Shaking moment
m	Mass; module; slope
m^n	Normal module
N	Number of gear teeth
N	Normal force
N_f	Formative number of gear teeth
n	Rotational speed (revolutions per minute)
n_c	Number of constraints
n_J	Number of joints
n_L	Number of links
O_1	Fixed bearing on link 1
ob	Absolute velocity of point B
P	Prism pair; planet gear; power; diametral pitch
P	Piston force
P^n	Normal diametral pitch
$^i\{P\}$	Position vector in frame i
p	Transverse circular pitch
p_b	Base pitch
p^n	Normal circular pitch
p_w	Axial pitch of worm
R	Revolute pair; ring gear; length of crank
R	Position vector

$_j^i[R]$	Rotational transformation matrix
r	Radius of pitch circle
$\boldsymbol{r}$	Position vector; vector representing a link
$\dot{\boldsymbol{r}}$	Derivative of $\boldsymbol{r}$ with respect to time
r°	Train value (speed ratio) for a planetary train relative to the carrier
r_a	Length of cam-follower arm; radius of addendum circle
r_b	Base circle radius; radius of back cone element
r_c	Center distance between cam and follower pivots; pitch radius of equivalent spur gear
r_f	Radius of cam-follower roller; radius of friction circle
r_m	Mean pitch radius
r_v	Velocity ratio
$\boldsymbol{r}^u$	Unit vector
$\boldsymbol{r}_x$	x component of vector $\boldsymbol{r}$
S	Sphere pair; sun gear
s	Displacement
s	Cam-follower displacement
$\dot{s}$	Slider velocity
s_n	Axial spacing of engine cranks and cylinders; joint offset along axis n
$\boldsymbol{T}, T$	Torque
$\boldsymbol{T}_e$	External torque
$_j^i[T]$	Transformation matrix from frame j to frame i
t	Time; gear tooth thickness
v	Velocity
v_p	Pitch line velocity
W	Work
w	Gear tooth width; weight
x, y, z	Cartesian coordinates
$\times$	Cross product
$\alpha, \boldsymbol{\alpha}$	Angular acceleration
α	Cam rotation angle; angle of approach
β	Angle of recess
Γ	Pitch angle
γ	Cam follower rotation; pitch angle
$\delta\phi, \delta x$	Virtual displacements

θ	Angular position of link; cam angle; angle of action; connecting rod angle
θ_i	Angular position of link i
θ_n	Angular spacing of engine cylinders; joint angle about axis n
λ	Lead angle of worm
μ	Coefficient of sliding friction
ρ	Radius of curvature
ρ_p	Radius of curvature of pitch curve
Σ	Angle between shafts
τ_i	Link twist of member i
ϕ	Transmission angle; pressure angle; transverse pressure angle; friction angle
ϕ_i	Angular position of link i
$\phi^{\cdot\cdot}$	Normal pressure angle
ψ	Involute angle; helix angle
ψ_n	Angular spacing of engine cranks
$\omega, \boldsymbol{\omega}$	Angular velocity

Mechanisms and Machines: Basic Concepts

1.1 INTRODUCTION

Kinematics and dynamics of machinery involve design of machines on the basis of motion requirements. A combination of interrelated parts having definite motions and capable of performing useful work may be called a **machine**. A **mechanism** is a component of a machine consisting of two or more bodies arranged so that the motion of one compels the motion of the others. The design of an automotive power train (a machine) is concerned with several mechanisms, including slider-crank linkages, cam and follower linkages, and gear trains. Many mechanisms undergo **planar motion**, motion in a single plane or in a set of parallel planes. The more general case, **spatial motion**, applies to mechanisms in which the motion must be described in three dimensions.

    Kinematics is the study of motion in mechanisms without reference to the forces that act on the system. **Dynamics** is the study of the motion of individual bodies and mechanisms under the influence of forces and torques. The study of forces and torques in stationary systems (and systems with negligible inertial effects) is called **statics**.

    Synthesis is a procedure by which a product (a mechanism, for example) is developed to satisfy a set of performance requirements. If a product configuration is tentatively specified and then examined to determine whether the performance requirements are met, the process is called **analysis**. The design of mechanisms involves both synthesis and analysis.

The design process begins with recognition of a need. A set of requirements is then listed. Creativity and inventiveness are required in selecting the connectivity and form of a mechanism or machine to satisfy the need. Specification sets and corresponding decision sets and design variables must be identified. The designer then prepares an adequacy assessment procedure, formulating a figure of merit and optimization strategy.

Detailed analyses of displacements, velocities, and accelerations are usually required. This part of the design process is then followed by analysis of forces and torques. The design process may continue long after first models have been produced and include redesigns of components that affect velocities, accelerations, forces, and torques. In order to successfully compete from year to year, most manufacturers must continuously modify their product and their methods of production. Increases in production rate, upgrading of product performance, redesign for cost and weight reduction, and motion analysis of new product lines are frequently required. Success may hinge on the accuracy of the kinematic and dynamic analysis.

1.2 TOOLS AVAILABLE TO THE DESIGNER OF LINKAGES AND OTHER MECHANISMS

A designer will ordinarily begin by making various design decisions based on experience and creative ability. These decisions may be verified and modified through analytical, graphical, numerical, and empirical methods. If a linkage is to be analyzed in only one position, graphical vector methods may provide the quickest solution. Complex number methods provide a convenient method for analyzing planar linkages. Analytical vector methods are used for solving planar and spatial linkages. While a calculator is adequate for solving a linkage problem for a single position, it is worthwhile to write a computer program when a solution is required over a range of values. Computer solutions are also effective for analysis and synthesis when it is necessary to evaluate several alternatives.

When selecting a programming language (BASIC, FORTRAN, etc.), one should consider the availability of direct and inverse trigonometric functions that appear in kinematics. Mathematics scratchpad software (e.g., MathCAD™), which handles vector and complex equations, solution of simultaneous equations, and other mathematics useful in kinematic analysis and synthesis, is a significant timesaver. Spreadsheet programs (e.g., Lotus 1-2-3™) are useful for analysis of gear trains, analysis and synthesis of linkages, and other problems in kinematics and dynamics. An important spreadsheet feature is that when one cell is changed, all related cells are updated to reflect the change. Thus a designer may try various design changes with a minimum of effort. When selecting a spreadsheet program, consider the availability of a plotting routine and the availability of functions such as the two-argument arctangent ($ARCTAN_2$), which is useful in the solution of four-bar linkages. Problems in kinematics and dynamics of machinery may also be solved with the aid of commercially available computer software written specifically for this purpose, such as ADAMS™ (Automatic Dynamic Analysis of Mechanical Systems) and IMP (Integrated Mechanisms Program), which use matrices to analyze linkages. I-DEAS™ (Integrated Design Engineering Analysis Software) includes mechanism design capabilities as well as solid modeling and finite element analysis features.

1.3 SYSTEMS OF UNITS

Any set of units may be used in the study of kinematics and dynamics of machinery, as long as consistency is maintained. Preferred systems are the International System of Units (SI), a modernized version of the meter-kilogram-second (mks) system, and the customary United States inch-pound-second system. The following basic and derived units are suggested:

SI ($m - kg_m - s$) SYSTEM

Quantity	Unit	Symbol	Relationship
Acceleration			m/s^2
Energy and work	joule	J	$N \cdot m$
Force	newton	N	$kg \cdot m/s^2$
Length	meter	m	
Mass	kilogram	kg	$N \cdot s^2/m$
Mass of moment of inertia			$kg \cdot m^2$
Power	watt	W	J/s or $N \cdot m/s$
Pressure and stress	pascal	Pa	N/m^2
Torque and moment			$N \cdot m \, (N \cdot m/rad)$
Velocity			m/s

CUSTOMARY U.S. ($in - lb_f - s$) SYSTEM

Quantity	Unit	Symbol	Relationship
Acceleration			in/s^2
Energy and work			$lb \cdot in$
Force	pound	lb or lb_f	
Length	inch	in	
Mass			$lb \cdot s^2/in$
Mass moment of inertia			$lb \cdot s^2 \cdot in$
Power	horsepower	hp	
Pressure and stress		psi	lb/in^2
Torque and moment			$lb \cdot in \, (lb \cdot in/rad)$
Velocity			in/s

COMMON TO BOTH SI AND CUSTOMARY U.S. SYSTEMS

Quantity	Unit	Symbol	Relationship
Angular acceleration			rad/s^2
Angular velocity			rad/s
Frequency	hertz	Hz	(cycles)/s
Plane angle	radian	rad	
Time	second	s	

SI prefixes may be used to eliminate nonsignificant digits and leading zeros. The following are in most common use:

Multiplication Factor	Prefix	Symbol
$1,000,000 = 10^6$	mega-	M
$1,000 = 10^3$	kilo-	k
$0.01 = 10^{-2}$	centi-	c
$0.001 = 10^{-3}$	milli-	m
$0.000001 = 10^{-6}$	micro-	μ

Although prefixes representing powers of 1000 are preferred, the *centi-* prefix is also used (e.g., centimeters, cubic centimeters).

It is generally most convenient to perform calculations by using scientific notation (powers of ten) or engineering notation (10^{-6}, 10^{-3}, 10^3, 10^6, etc.). A suitable unit and prefix should be chosen to express the results of calculations so that the numerical value falls between 0.1 and 1000 where convenient. An exception to this suggestion is in engineering drawings where, for consistency, linear dimensions would all be expressed in millimeters (mm). When a number of values are tabulated or discussed, consistent units and prefixes are preferred (e.g., a velocity range given as 0.09 m/s to 1100 m/s would be preferred to 90 mm/s to 1.1 km/s).

The advantage of SI as a coherent system may be lost if it is used with units from different systems. However, convenience and common usage suggest the use of the degree (and decimal parts of the degree) for measurement of plane angles. Obviously, time expressed in minutes, hours, and days will often be more practical than the use of seconds for some applications (e.g., vehicle velocity is commonly expressed in kilometers per hour, km/h, and the kilowatthour, kW · h, is used as a measure of energy).

Like the pound (lb), the kilogram (kg) is sometimes used as a unit of force as well as a unit of mass. The accepted force unit, however, is the newton (N). Torque may be expressed in newton-meters (N · m). Although 1 N · m equals 1 joule (J), the term *joule* should be reserved for work and energy.

1.3.1 Conversion Factors

A few of the conversion factors useful in kinematics and dynamics of machinery are listed below. An extensive list of conversion factors is given inside the front cover.

$1\ g$ (gravitational constant) $= 386.09$ in/s$^2 = 9.80665$ m/s^2

1 horsepower (hp) $= 6600$ lb $\cdot$ in/s $= 745.7$ W

1 in $= 25.4$ mm   1 m $= 39.37$ in

1 lb $= 4.4482$ N (force)

1 lb $= 0.45359$ kg (mass)

1 mi/h $= 0.44704$ m/s

1 psi $= 6894.8$ Pa   1 MPa $= 145.04$ psi

1 rad $= 180°/\pi = 57.2958°$

1 revolution per minute (rev/min) $= \pi$ rad/30 s
$$= 0.10472 \text{ rad/s}$$

1.4 TERMINOLOGY AND DEFINITIONS

Many of the basic linkage configurations have been incorporated into machines designed centuries ago, and the terms we use to describe them have changed over the years. Thus, definitions and terminology are not consistent throughout the technical literature. In most cases, however, meanings will be clear from the context of the descriptive matter. A few terms of particular interest to the study of kinematics and dynamics of machines are defined below.

Link

A *link* is one of the rigid bodies or members joined together to form a kinematic chain. The term *rigid link*, or sometimes simply *link*, is an idealization used in the study of mechanisms that does not consider small deflections due to strains in machine members. A perfectly rigid or inextensible link can exist only as a textbook type of model of a real machine member. For typical machine parts, maximum dimension changes are of the order of only one-thousandth of the part length.

Frame

The fixed or stationary link in a mechanism is called the *frame*. When there is no link that is actually fixed, we may consider one as being fixed and determine the motion of the other links relative to it. In an automotive engine, for example, the engine block is considered the frame, even though the automobile may be moving.

Joint or Kinematic Pair

The connections between links that permit constrained relative motion are called *joints*. The joint between a crank and connecting rod, for instance, may be called a *revolute joint* or a *pin joint*. The revolute joint has one degree of freedom in that if one element is fixed, the revolute joint allows the other only to rotate in a plane. (Degrees of freedom are discussed in more detail in a section that follows.) A *sphere joint* (*ball joint*) has three degrees of freedom; it allows relative motion in three angular directions. A number of common joint types are idealized in Figure 1.1. Some of the practical joints that they represent are made up of several elements. Examples include the universal joint; ball and roller bearings that are represented by the revolute joint; ball slides represented by the spline joint; and ball screws represented by the helix.

Lower and Higher Pairs

Connections between rigid bodies consist of lower and higher pairs of elements. The two elements of a *lower pair* have theoretical surface contact with one another, while the two elements of a *higher pair* have theoretical point or line contact (if we disregard deflections). Lower pairs include revolutes or pin connections—for example, a shaft in a bearing or the wrist pin joining a piston and connecting rod. Both elements joined by the pin may be considered to have the same motion at the pin center if clearance is neglected. Other basic lower pairs include the sphere, cylinder, prism, helix, and plane (Figure 1.1). Waldron[14] shows that these six are the only basic lower pairs possible.

Examples of higher pairs include a pair of gears or a disk cam and follower. The Hook-type universal joint is a combination of two lower pairs. A Bendix-Weiss type of constant-velocity universal joint includes higher pairs (see illustrations later in this chapter).

Lower pairs are desirable from a design standpoint since the load at the joint and the resultant wear is spread over the contact surface. Thus, geometric changes or failure due to high contact stresses and excessive wear may be prevented. In actual practice, we may utilize a ball or roller bearing as a revolute pair to reduce friction. In so doing, the advantages of contact over a large surface are sacrificed.

Closed-Loop Kinematic Chains

A kinematic chain is an assembly of links and pairs (joints). Each link in a closed-loop kinematic chain is connected to two or more other links. Consider, for example, the slider-crank mechanism, a component of the vertical compressor shown in Figure 1.2. Bearings (represented by a revolute joint) connect

Type of joint (pair)	Lower pair (L) or higher pair (H)	Symbol	Degrees-of-freedom (connectivity) of the joint in a spatial linkage	Schematic representation	Possible configuration	Descriptive example
Revolute	L	R	1 θ	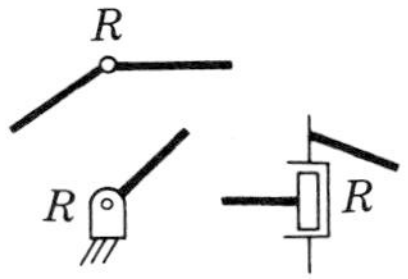	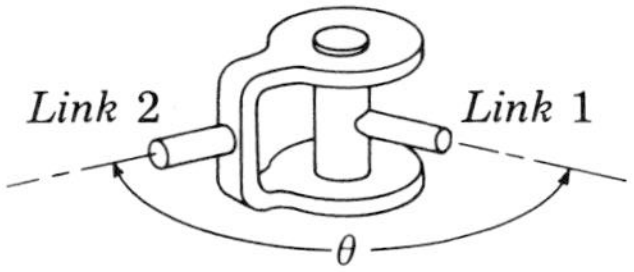	A pin joint that permits rotation only
Prism	L	P	1 x	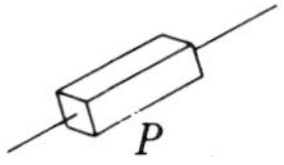	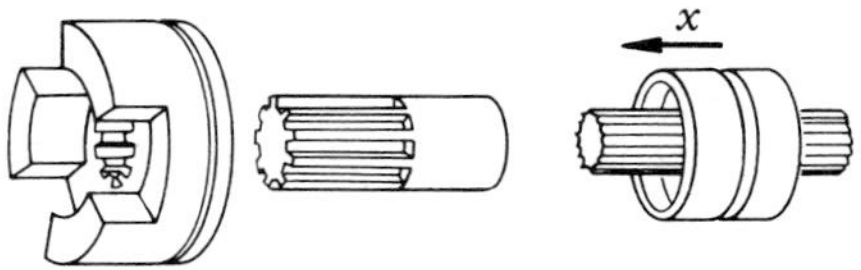	A straight spline that permits sliding only
Helix	L	H	1 x or θ	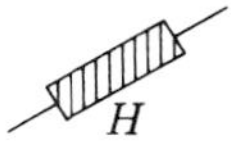	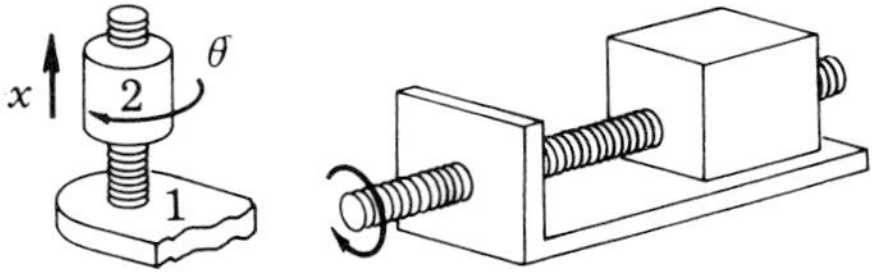	Power screw or helical spline
Cylinder	L	C	2 x, θ	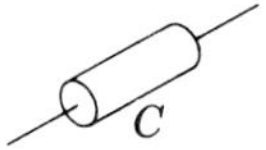	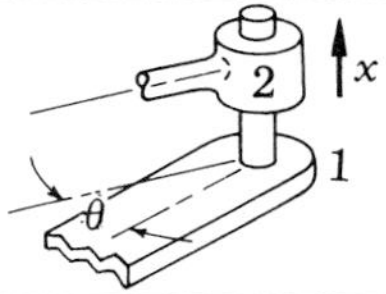	A sleeve that permits both rotation and sliding

Figure 1.1 Common linkage joints (pairs).

7

Sphere	L	S	3 θ, ϕ, γ	 S	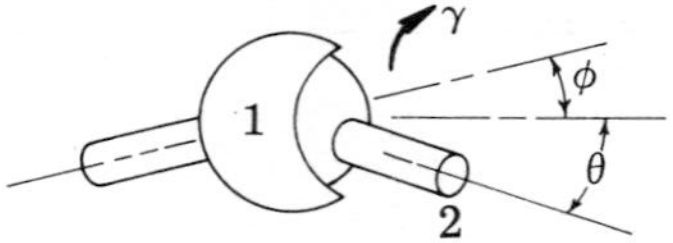	A ball (and socket) joint permitting rotation in three angular directions
Plane	L	P_L	3 x, y, θ	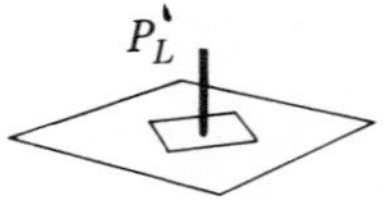 P_L	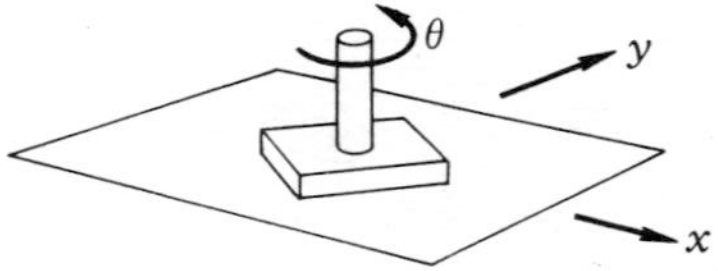	A surface restraint permitting rotation and motion parallel to the plane of the surface
Universal joint	L	U	2 θ, ϕ	 U	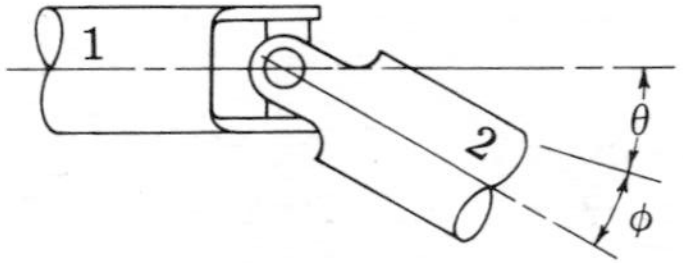	The Hooke-type universal joint that combines two revolute pairs
Spur gear pair	H	G	2 (rolling and sliding)	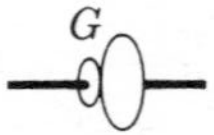 G	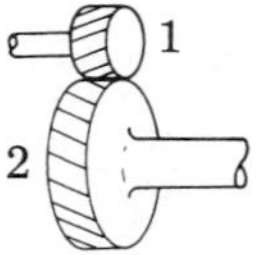	Spur gears, helical gears, and other gears
Cam pair	H	*	2 (rolling and sliding)	*	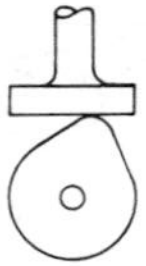	Disk cam and follower

Figure 1.1 Continued.

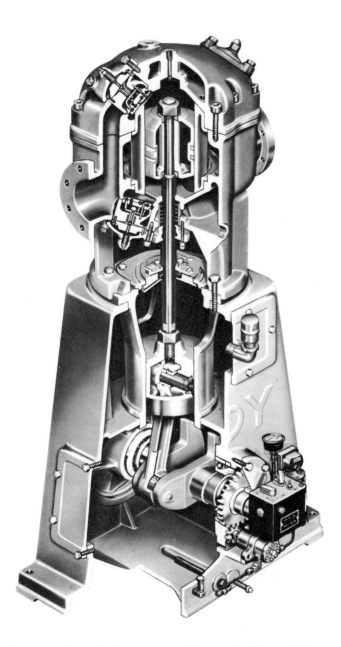

Figure 1.2 A vertical compressor. The crank (*bottom*) drives the connecting rod, which moves the crosshead within a guide. The compressor is designed with a crosshead and piston rod so that the piston may be double-acting; air is compressed as the piston moves upward and as it moves downward. (*Source:* Joy Manufacturing Company.)

the casing (frame) and crank; the crankpin (another revolute joint) connects the crank and connecting rod; the connecting rod and crosshead are joined at the wrist pin (a third revolute joint); finally, the piston and cylinder (frame) constitute a sliding pair (cylinder pair), closing the loop.

Open-Loop Kinematic Chains

A linkage failing to meet the closed-loop criterion is an open-loop kinematic chain. In this case, one (or more) of the links is connected to only one other link.

Manipulators

Manipulators designed to simulate human arm and hand motion are an example of open kinematic chains. A typical manipulator consists of a supporting base with rigid links connected in series, the final link containing a tool or "hand." Ordinarily, the rigid links are joined by revolute joints or prismatic pairs, although the hand may include a screw pair. Early systems of this type included master-slave-type manipulators for handling of radioactive materials. The slave manipulator duplicates the hand-arm motion of a human operator controlling the master manipulator.

Robots

Programmable manipulators, called robots, can follow a sequence of steps directed by a computer program. Unlike machines dedicated to a single task, robots can be retooled and reprogrammed for a variety of tasks. Typical robot tasks include spray painting, part assembly, and welding. The open-chain configuration of robots results in problems with positional accuracy. This problem is sometimes overcome by using jigs and compliant tooling systems.

It is also possible to achieve accurate positioning by incorporating a sensing system and a feedback system into the robot control system. Internal-state sensors can detect variables such as joint positions. External-state sensors may measure proximity, touch, force, and torque. Machine vision and hearing are external-state sensory capabilities available in some robot systems. Sensory-function feedback systems permit adaptive behavior of the robot. A force transducer incorporated in a robot hand may feed back a signal to the control system, which then alters the grasp pattern. Tasks requiring high precision are more often accomplished by numerically controlled machinery designed for specific operations.

Figure 1.3 shows an industrial robot with six revolute joints. This robot has a jointed-arm form, a common robot configuration. Other robot configurations are shown in Figure 1.4. Part a of the figure is a schematic representing a robot with four revolute joints and one prismatic pair. Part b represents a robot with two revolute joints, a prismatic joint, and a cylinder pair. Note that the cylinder can be replaced by a prismatic joint and a revolute joint. The

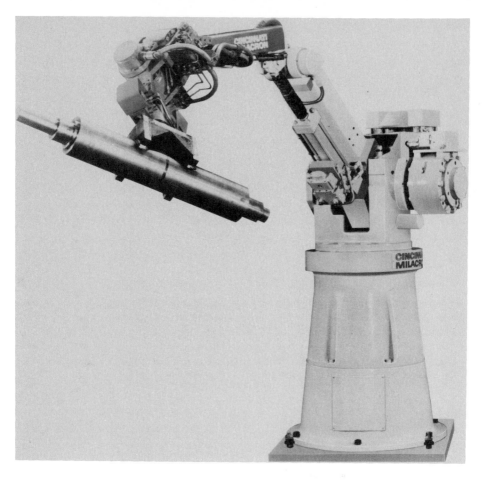

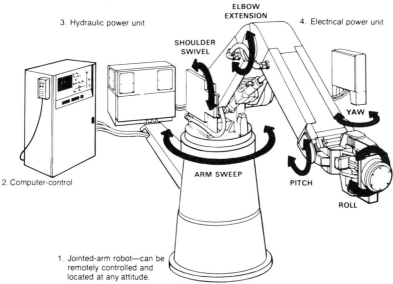

3. Hydraulic power unit

ELBOW EXTENSION

SHOULDER SWIVEL

4. Electrical power unit

YAW

ARM SWEEP

2. Computer-control

PITCH

ROLL

1. Jointed-arm robot—can be remotely controlled and located at any attitude.

Figure 1.3 Industrial robot. (*Source:* Cincinnati Milacron.)

11

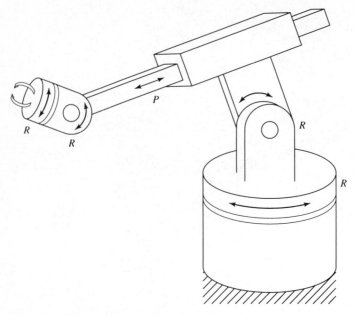

Figure 1.4 Schematic diagrams representing various robot configurations. (a) Spherical configuration.

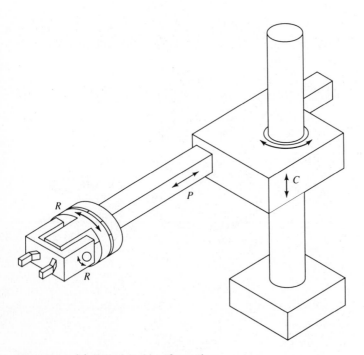

Figure 1.4 (b) Cylindrical configuration.

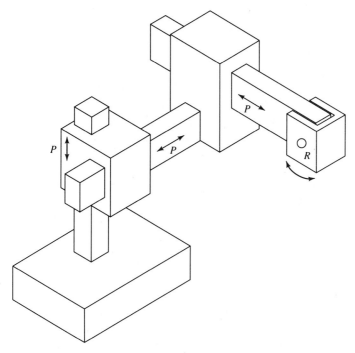

Figure 1.4 (c) Rectangular configuration.

robot schematic of part c shows three prism joints and one revolute joint. The end effectors may consist of additional links and joints. The **work envelope** or workspace is defined by all of the points that the end effector can reach. It can be seen that the type of joints in each of these robot configurations affects the shape of the work envelope. Robots are discussed extensively in Chapter 12.

Linkage

Although some references define *linkages* as kinematic chains joined by only lower pairs, the term is commonly used to identify any assemblage of rigid bodies connected by kinematic joints. The same linkage configuration may serve as a component of a mechanism, machine, or engine. Thus, the terms *linkage*, *mechanism*, *machine*, and *engine* are often used interchangeably.

Planar Motion, Planar Linkages

If all points in a linkage move in parallel planes, the system undergoes *planar motion* and the linkage may be described as a *planar linkage*. The portable drafting instrument shown in Figure 1.5a is a planar linkage. A skeleton diagram of a planar linkage is formed by projecting all of the link centerlines on one of the planes of motion, as in part b of the figure. The plane of motion

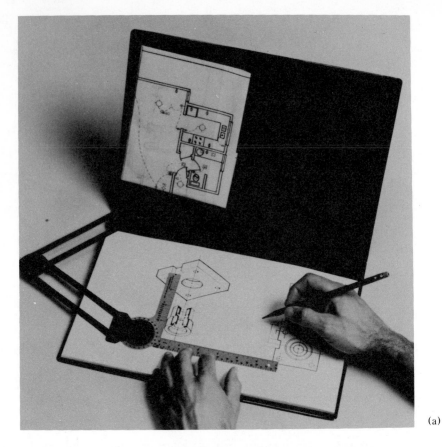

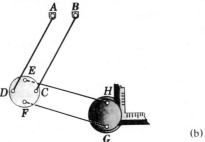

(a)

(b)

Figure 1.5 A portable drafting instrument. (*Source:* PMC Industries, Inc.)
(a) Portable drafting instrument. (b) The linkage consists of two parallelo-
grams, which permit translation of the straightedge in any direction without
rotation.

of parallelogram linkage *ABCD*, the plane of motion of parallelogram linkage *EFGH*, and the plane of motion of the straightedges are all parallel. In this linkage, planar motion is assured because the axes of revolute joints *A*, *B*, *C*, *D*, *E*, *F*, *G*, and *H* are all parallel (i.e., all perpendicular to the plane of the drawing board).

Spatial Motion, Spatial Linkages

The more general case in which motion cannot be described as taking place in parallel planes is called *spatial* motion, and the linkage may be described as a *spatial* or *three-dimensional* (3*D*) *linkage*. The industrial robot of Figure 1.3 is a spatial linkage. In order to achieve the desired range of motion, the axes of the revolute pairs in the manipulator are not all parallel.

Inversion

The absolute motion of a linkage depends on which link is fixed, that is, which link is selected as the frame. If two otherwise identical linkages have different fixed links, then each is an *inversion* of the other.

Cycle and Period

A *cycle* represents the complete sequence of positions of the links in a mechanism (from some initial position back to that initial position). In a four-stroke-cycle engine, one thermodynamic cycle corresponds to two revolutions or cycles of the crankshaft but one revolution of the camshaft and, thus, one cycle of motion of the cam followers and valves. The time required to complete a cycle of motion is called the *period*.

1.5 DEGREES OF FREEDOM (MOBILITY)

The number of degrees of freedom of a linkage is the number of independent parameters required to specify the position of every link relative to the frame or fixed link. The number of degrees of freedom of a linkage may also be called the mobility. If the instantaneous configuration of a system may be completely defined by specifying one independent variable, that system has one degree of freedom. Most practical mechanisms have one degree of freedom.

An unconstrained rigid body has six degrees of freedom: translation in three coordinate directions and rotation about three coordinate axes. If the body is restricted to motion in a plane, there are three degrees of freedom: translation in two coordinate directions and rotation within the plane.

1.5.1 Constraints Due to Joints

Each joint reduces the mobility of a system. It can be seen that a fixed, one-degree-of-freedom joint (e.g., a revolute joint) reduces a link to one degree of freedom. In general, each one-degree-of-freedom joint reduces system mobility by providing five constraints; each two-degree-of-freedom joint provides four constraints, and so on. That is, in general, each joint reduces system mobility by $(6 - f_i)$, where f_i is the number of degrees of freedom (connectivity) of the joint.

Thus, for a general spatial mechanism with n_L links (including one fixed link with zero degrees of freedom), the number of degrees of freedom of the linkage is given by

$$DF_{spatial} = 6(n_L - 1) - n_c \tag{1.1}$$

where n_c is the total number of constraints as defined above. For n_J joints with individual joint connectivity f_i, we note that $n_c = 6n_J - \sum_{i=1}^{n_J} f_i$, from which

$$DF_{spatial} = 6(n_L - n_J - 1) + \sum_{i=1}^{n_J} f_i \tag{1.2}$$

Examining the industrial robot of Figure 1.3, we see that there are seven links and six revolute joints, each joint having one degree of freedom and introducing five constraints. Using Eq. 1.1, we find

$$DF_{spatial} = 6(7 - 1) - 5 - 5 - 5 - 5 - 5 - 5 = 6$$

Alternatively, Eq. 1.2 may be used to obtain the same result:

$$DF_{spatial} = 6(7 - 6 - 1) + 1 + 1 + 1 + 1 + 1 + 1 = 6$$

Next, consider the closed-loop kinematic chain of Figure 1.6a, a general closed-loop RSSR mechanism. The mechanism is identified by its joint configuration: it has four links, two revolute joints, and two spherical joints, as shown in Figure 1.6a. Each revolute joint has one degree of freedom and introduces five constraints, while each spherical joint has three degrees of freedom and introduces three constraints. Thus, using Eq. 1.1, the RSSR mechanism has $6(4 - 1) - 5 - 3 - 3 - 5 = 2$ degrees of freedom. This particular linkage acts, for practical purposes, as a one-degree-of-freedom linkage if we ignore the degree of freedom that represents rotation of link 2 about its own axis. If the angular position of link 1 is given, the entire linkage configuration may be determined. Note, however, that the above statement assumes that applied forces or inertia effects are present to ensure a prescribed pattern of motion as the linkage passes through limiting positions.

Let one of the spherical joints in the RSSR mechanism considered above be replaced by a universal joint with two degrees of freedom (four constraints). We then form an RSUR mechanism, and the number of degrees of freedom is

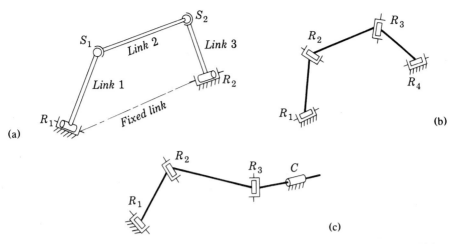

Figure 1.6 (a) An RSSR mechanism (two degrees of freedom). (b) An RRRR linkage. (c) An RRRC linkage. In the general case, no relative motion is possible in closed-loop RRRR and RRRC linkages.

given by

$$DF = 6(4 - 1) - 5 - 3 - 4 - 5 = 1$$

In a general RRRR linkage, as shown in Figure 1.6b, each joint provides five constraints, and the number of degrees of freedom is given by $DF = 6(4 - 1) - 5 - 5 - 5 - 5 = -2$. In a general RRRC linkage, the cylinder joint has two degrees of freedom, providing four constraints. The number of degrees of freedom for the RRRC linkage shown in Figure 1.6c is given by $DF = 6(4 - 1) - 5 - 5 - 5 - 4 = -1$. In both of these linkages, no relative motion would be possible for the general case. They are equivalent to statically indeterminate structures.

The actual number of degrees of freedom of a mechanism depends on the joint orientation as well as the number of links and joints. The actual constraint due to the joints may be less than given by the above equations. Thus we may write

$$DF_{spatial} \geq 6(n_L - n_J - 1) + \sum_{i=1}^{n_J} f_i \qquad (1.3)$$

to include special joint orientations. For example, the revolute joints of an RRRR linkage can be oriented so that the linkage has planar motion, with one degree of freedom (not -2 degrees of freedom as indicated for the general RRRR linkage).

When analyzing the motion of a one-degree-of-freedom linkage, the motion of one of the links may be specified as a function of time. If analyzing a robot, the rotation and translation of each joint could be specified as a function of time. In both cases, the system could then be considered a kinematically determinate (zero-degree-of-freedom) system.

1.5.2 Planar Linkages

Planar linkages, of course, represent a special case. Consider, for example, a mechanism made up of rigid links joined by three revolute joints and a cylinder joint. If the links and joints are oriented so that the links move in parallel planes, this RRRC linkage becomes a *slider-crank linkage*, the planar linkage that represents a major component of piston engines, pumps, compressors, and other common machines. Figure 1.2 shows a vertical compressor, the major components of which may be represented by an RRRC linkage. All points on the crank, connecting rod, and crosshead of the compressor move in parallel planes; the axes of the revolute joints are parallel. Thus, we have a slider-crank linkage.

A planar RRRR linkage may be called a four-bar linkage. The four-bar linkage and other planar link systems are shown in Figure 1.7.

The joints or pairs that apply to planar linkages are as follows:

	Lower or higher pair	Connectivity (Degrees of freedom of pair in plane motion)
Revolute or pin joint	Lower pair	1
Prism or sliding pair	Lower pair	1
Cam pair	Higher pair	2
Gear pair	Higher pair	2

Actual joints may sometimes be different. For example, the slider-crank linkage may be made up of three revolute joints (crankshaft bearings, crankpin, and wrist pin) and a cylinder pair. The spline-type constraint of a prism pair is unnecessary since the revolute joints prevent rotation of the piston. If the actual number of degrees of freedom is greater than would be determined by using the equation for spatial linkages, the linkage is overconstrained. Overconstraint tends to strengthen a linkage. However, overconstraint can be a disadvantage if manufacturing tolerances are poor or if the elements of a linkage undergo significant elastic deflections.

1.5.3 Determination of Degrees of Freedom for a Planar Linkage

We note that each unconstrained rigid link has three degrees of freedom in plane motion. A fixed link has zero degrees of freedom. A pin joint connecting two links produces two constraints since the motion of both links must be equal at the joint (in two coordinate directions). Thus, the number of degrees of freedom for a planar linkage made up of n_L links and n'_J one-degree-of-

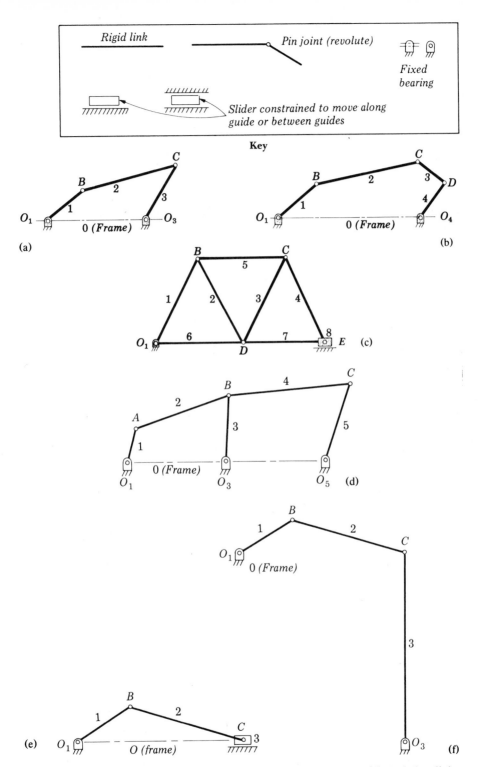

Figure 1.7 (a) Four-bar linkage. (b) Five-bar linkage. (c) Structure. (d) A six-bar linkage with one degree of freedom. (e) A slider-crank mechanism. (f) A four-bar linkage with motion approximating that of a slider-crank mechanism.

19

freedom pairs is given by

$$DF_{planar} = 3(n_L - 1) - 2n'_J \tag{1.4}$$

or, for n_J joints with individual connectivity f_i,

$$DF_{planar} = 3(n_L - n_J - 1) + \sum_{i=1}^{n_J} f_i \tag{1.5}$$

We see that for the four-bar linkage in Figure 1.7a, $n_L = 4$, $n'_J = 4$, and DF = 1. For the five-bar linkage in Figure 1.7b, $n_L = 5$, $n'_J = 5$, and DF = 2. The linkage of Figure 1.7d has a double pin at B. Thus, $n_L = 6$, $n'_J = 7$, and DF = 1.

Figure 1.7e shows a slider-crank mechanism. This is a representation of a piston engine or a piston pump, where O_1 represents the crankshaft; link 1, the crank; link 2, the connecting rod; point C, the wrist pin; and link 3, the slider or piston that is constrained by the cylinder. The figure represents the common special case, an in-line slider crank, where the extended path of wrist pin C goes through crankshaft center O_1. There are four links including the slider and frame and four lower pairs including the sliding pair. An alternative analysis uses an equivalent linkage. Figure 1.7f shows a four-bar linkage in which point C moves through an arc of radius O_3C. If link 3 is made very long, the motion of the four-bar linkage will approximate that of a slider-crank linkage. If we could construct a linkage by replacing the slider of the slider-crank mechanism with a link of infinite length and perpendicular to the path of C, then that linkage would be equivalent to the slider-crank linkage. Applying Eq. 1.4 to our equivalent linkage, with $n_L = 4$ and $n'_J = 4$, we find that DF = 3(4 − 1) − 2(4) = 1. The equivalent four-bar linkage (and, thus, the slider-crank linkage) has one degree of freedom.

Analyzing the structure in Figure 1.7c in a similar manner, and noting the double pins at O_1, B, C and E and the triple pin at D, we find $n'_J = 12$, $n_L = 9$, and DF = 0.

If a planar linkage is made up of n'_J one-degree-of-freedom pairs and n''_J two-degree-of-freedom pairs, the number of degrees of freedom of the linkage is given by

$$DF_{planar} = 3(n_L - 1) - 2n'_J - n''_J \tag{1.6}$$

A single gear mesh or the contact point of a cam and follower each represents a two-degree-of-freedom higher pair (if the two bodies do not separate). Consider the spur gear differential shown schematically in Figure 1.8. There are six links: the frame, gears S_1 and S_2 called sun gears, gears P_1 and P_2 called planet gears, and link C, the planet carrier. There are five revolute joints: the bearing between the planet carrier shaft and the frame, the bearings between the sun gear shafts and the frame, and the bearings between the planet gear shafts and the planet carrier. There are three gear pairs: S_1P_1, P_1P_2, and P_2S_2. It can be seen that the bearing axes are all parallel and that

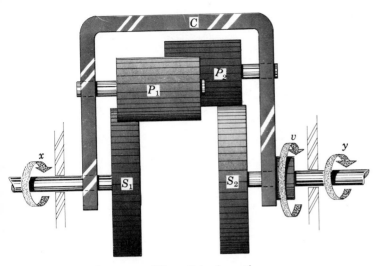

Figure 1.8 Spur gear differential schematic.

the spur gear differential is a planar linkage. Using Eq. 1.6, we find

$$\text{DF}_{\text{planar}} = 3(6 - 1) - 2 \times 5 - 3 = 2 \text{ degrees of freedom}$$

Thus there are two independent variables. For example, if we specified the motion of both sun gear shafts, the planet carrier shaft motion could be determined.

In order to achieve balance and reduce gear tooth loading, practical spur gear differentials ordinarily include two to four equally-spaced pairs of planet gears. If the gear sizes were arbitrarily chosen, the number of degrees of freedom would be reduced. However, one planet gear pair is identical to another in a practical differential, and the differential will have two degrees of freedom.

1.5.4 One-Degree-of-Freedom Configurations

Planar mechanisms with one degree of freedom are of considerable practical importance. One-degree-of-freedom planar mechanisms made up of lower pairs satisfy Grübler's criterion:

$$2n'_J - 3n_L + 4 = 0 \tag{1.7}$$

Noting that the number of links and one-degree-of-freedom pairs n_L and n'_J must be positive integers, we see that n_L must be an even number. For $n_L = 2$, we obtain $n'_J = 1$, the trivial solution that could represent two bars joined by a pin joint. Next, trying four links, we see that the number of joints must be $n'_J = 4$. This solution could represent the four-bar linkage or slider-crank linkage of Figure 1.7a and e. Inversions of the slider-crank linkage are

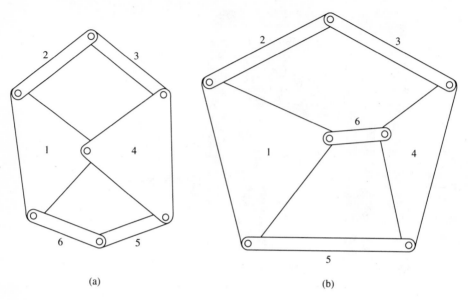

(a)   (b)

Figure 1.9 One-degree-of-freedom six-bar linkages. (a) Watt linkage. (b) Stephenson linkage.

also possible. If there are six links, then seven one-degree-of-freedom lower pairs are required to produce one degree of freedom. For pin joints only, there are two distinctly different configurations possible, as shown in Figure 1.9a and b. Any one of the links may be designated as the frame in each solution. It can be seen that the six-bar linkage of Figure 1.7d may be considered a special case of either of the linkages of Figure 1.9. Eight-bar linkage configurations will be left to an exercise.

Spatial linkages will be analyzed further in a later section. However, since planar linkages are used most frequently, the word *planar* will be used in the pages that follow only when it is necessary to compare planar and spatial linkages.

♦ **EXAMPLE PROBLEM 1.1** *Degrees of Freedom of a Variable-Stroke Pump*
(a) Determine the number of degrees of freedom for the variable-stroke pump shown in Figure 1.10. Let the adjustment cylinder be fixed in a position that results in an intermediate stroke length. (b) Suppose the adjustment cylinder position is not fixed. Find the number of degrees of freedom.

Solution. (a) It can be seen from the figure that motion takes place in a set of parallel planes. Therefore, we have a planar linkage. The number of degrees of freedom can be determined by examining the actual linkage or by examining a schematic diagram of an equivalent linkage. Selecting the latter course, the curved stroke transformer is replaced by rigid link 4 with a fixed revolute joint

(O_4) located at the center of curvature of the stroke transformer. See part b of the figure. We then have six links counting the slider and the frame. There are six revolute pairs and a sliding pair, making a total of seven (one-degree-of-freedom) lower pairs. Thus we have

$$DF_{planar} = 3(n_L - 1) - 2n_{J'} = 3(6 - 1) - 2(7) = 1$$

(b) In this case we refer to the actual pump configuration. A careful examination shows nine links including three sliders and the frame. There are eight revolute pairs and three sliding pairs, a total of eleven one-degree-of-freedom pairs. The number of degrees of freedom is given by

$$DF_{planar} = 3(n_L - 1) - 2n_{J'} = 3(9 - 1) = 2(11) = 2$$

The implication is that we must specify two variables to define the instantaneous position of the entire linkage. Ordinarily, these variables would be the position of the piston in the pump control cylinder, and the instantaneous angular position of link 1, the drive crank. ◆

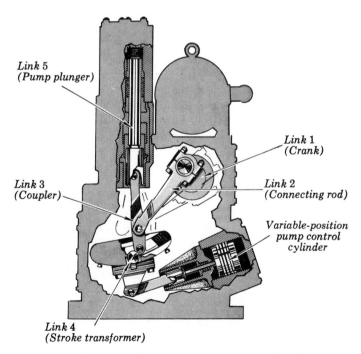

Link 5
(Pump plunger)

Link 1
(Crank)

Link 3
(Coupler)

Link 2
(Connecting rod)

Variable-position
pump control
cylinder

Link 4
(Stroke transformer)

Figure 1.10 (a) A variable-stroke pump. A curved-track *stroke transformer* allows a plunger stroke variation of 0 to 2 in on some models and of 0 to 6 in on other models. The stroke transformer is shown set (by the adjustment cylinder) for an intermediate stroke. (*Source:* Ingersoll-Rand Company.)

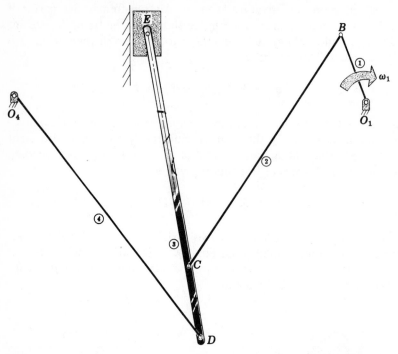

Figure 1.10 (b) In this linkage diagram, link 4 replaces the curved track of the stroke transformer. As O_4 is brought closer to E, the movement of the plunger decreases. When O_4 coincides with E, the plunger will become stationary.

1.6 CLASSIFICATION OF CLOSED PLANAR FOUR-BAR LINKAGES— THE GRASHOF CRITERION

Closed planar linkages consisting of four pin-connected rigid links are usually identified simply as **four-bar linkages**. If one of the links can perform a full rotation relative to the other three links, the linkage is called a **Grashof mechanism**.

Let the length of each link be defined as the distance between the axes of its revolute joints (the centers of its pin joints). Links are characterized by their lengths, where L_{max} is the longest link, L_{min} is the shortest link, and L_a and L_b are links of intermediate length. We may immediately eliminate combinations for which

$$L_{max} \geq L_{min} + L_a + L_b$$

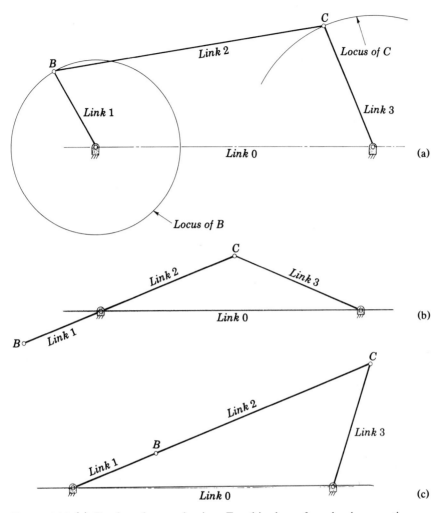

Figure 1.11 (a) Crank-rocker mechanism. For this class of mechanism, continuous rotation of the driver results in oscillation of the follower. (b) A limiting position of the crank-rocker mechanism (flexed). (c) A limiting position of the crank-rocker mechanism (extended).

since it is obvious that these links could not be assembled to form a closed four-bar linkage.

Suppose we wish to design a **crank rocker mechanism**, a linkage with a drive crank that rotates continuously, causing a driven crank (rocker) to oscillate through a limited range. Referring to Figure 1.11 we note that limiting positions of link 3, the rocker, occur when the crank (link 1) and the coupler (link 2, the link opposite the fixed link) are colinear. The triangle inequality requires that the length of one side of a triangle be less than the sum of the lengths of the other two slides. Applying the triangle inequality to parts b and c of the figure, we obtain the following relationships:

$$L_0 < L_2 - L_1 + L_3 \tag{1.8a}$$

$$L_2 - L_1 < L_3 + L_0 \tag{1.8b}$$

$$L_3 < L_0 + L_2 - L_1 \tag{1.8c}$$

$$L_0 < L_1 + L_2 + L_3 \tag{1.8d}$$

$$L_1 + L_2 < L_0 + L_3 \tag{1.8e}$$

$$L_3 < L_0 + L_1 + L_2 \tag{1.8f}$$

Adding inequalities 1.8a and 1.8c, the result is

$$L_0 + L_3 < L_2 - L_1 + L_3 + L_0 + L_2 - L_1$$

from which

$$0 < 2L_2 - 2L_1$$

or

$$L_1 < L_2$$

Similarly, combining inequalities 1.8a and 1.8e, we obtain

$$L_1 < L_3$$

and by combining inequalities 1.8c and 1.8e, we obtain

$$L_1 < L_0$$

Thus the drive crank must be the shortest link in a crank rocker mechanism. If the fixed link (link O) is the longest link, then, from inequality (1.8a),

$$L_{\max} + L_{\min} < L_a + L_b \tag{1.9}$$

If the driven crank (link 3, the rocker) is longest, then, applying inequality 1.8c, we again obtain Eq. 1.9. Finally, if the coupler (link 2) is longest, by applying inequality 1.8e, we see that Eq. 1.9 still applies.

We now consider inversions of the crank rocker mechanism. Inequality 1.9 is satisfied in each case and the shortest link can rotate continuously **relative to** the other links. If the fixed link is shortest, the other links can rotate about it as in Figure 1.12. This configuration is called a **drag link mechanism** (or double-crank mechanism). If the coupler is shortest, this inversion of the crank rocker mechanism is called a **double-rocker mechanism** (see Figure 1.13). The coupler of a double rocker can rotate continuously while the adjacent links oscillate through a limited range.

A **change-point** or **crossover-position mechanism** results when

$$L_{\max} + L_{\min} = L_a + L_b \tag{1.10}$$

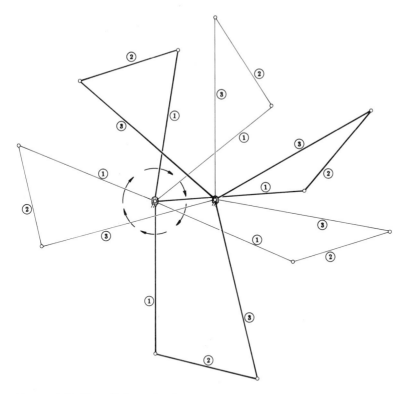

Figure 1.12 Drag link mechanism.

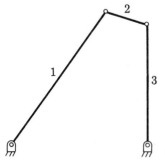

Figure 1.13 A double-rocker mechanism.

Figure 1.14 shows a change-point mechanism in which $L_0 = L_{max}$, $L_1 = L_{min}$, and Eq. 1.10 is satisfied. Relative motion of a change-point mechanism may depend on inertia, spring forces, or other forces when the links (in the skeleton diagram) become colinear. Of course, the links in an actual machine operate in parallel planes, not in a single plane. In this example of a change-point mechanism, all of the links become colinear in position b. If links 1 and 3 are rotating counterclockwise at this instant, link 3 may continue

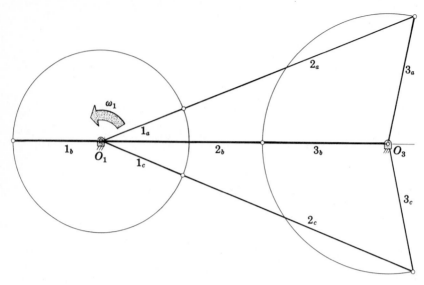

Figure 1.14 A crossover-position or change-point linkage.

rotating counterclockwise through the change point because of inertia effects. Alternatively, other forces may cause link 3 to reverse direction, resulting in a "bow-tie" configuration. A parallelogram linkage that has opposite links of equal length is another example of a change-point mechanism.

Crank rocker, drag link, double-rocker, and change-point mechanisms are called **Grashof mechanisms**, and their geometry satisfies the following relationship:

$$L_{max} + L_{min} \leq L_a + L_b \tag{1.11}$$

Any of the above classes of linkages may be driven by rotation of the coupler (the link opposite the fixed link), although the range of coupler rotation may be very limited in some. The coupler effectively provides a hinge with a moving center. Coupler-driven linkages may be called *polycentric*. Examples include polycentric door hinges and prosthetic knee joints.

Four-bar linkages that do not satisfy the Grashof criterion are called double-rocker mechanisms of the second kind or *triple-rocker mechanisms*. If $L_{max} + L_{min} > L_a + L_b$, no link can rotate through 360°. Table 1.1 summarizes the motion criteria for four-bar linkages.

♦ **EXAMPLE PROBLEM 1.2** *The Grashof Criterion*
This example concerns classification of four-bar linkages. Link lengths: L_0, fixed link; L_1, driver crank; L_2, coupler; L_3, follower crank; $L_1 = 100$ mm, $L_2 = 200$ mm, $L_3 = 300$ mm. Find the ranges of values for L_0 if the linkage is

TABLE 1.1   SUMMARY OF THE CRITERIA OF MOTION FOR EACH CLASS
OF FOUR-BAR LINKAGES

L_{min}: shortest link
L_{max}: longest link
L_a and L_b: links of intermediate length

Type of mechanism	Shortest link	Relationship between link lengths
Grashof	Any	$L_{max} + L_{min} \leq L_a + L_b$
Crank rocker	Driver crank*	$L_{max} + L_{min} < L_a + L_b$
Drag link	Fixed link	$L_{max} + L_{min} < L_a + L_b$
Double rocker	Coupler	$L_{max} + L_{min} < L_a + L_b$
Crossover-position change point	Any	$L_{max} + L_{min} = L_a + L_b$
Non-Grashof Double rocker of the second kind (triple rocker)	Any	$L_{max} + L_{min} > L_a + L_b{}^\dagger$

*If the driven crank is the shortest link, its direction of motion is uncertain when the driver crank is at a limiting position.

$^\dagger L_{max} < L_{min} + L_a + L_b$ (otherwise, a four-bar linkage cannot be constructed).

as follows:

    (a) Grashof mechanism
    (b) Crank rocker mechanism
    (c) Drag link mechanism
    (d) Double-rocker mechanism
    (e) Change-point mechanism
    (f) Triple-rocker mechanism

Solution.   Using the inequalities tabulated above, and noting that Grashof mechanisms are the crank rocker, drag link, double-rocker, and change-point mechanisms, we begin with the test for the crank rocker mechanism (b). The condition that the driver crank L_1 be shortest is satisfied by $L_0 > 100$ mm. The inequality $L_{max} + L_{min} < L_a + L_b$ yields $L_0 + 100 < 200 + 300$, from which $L_0 < 400$ if link 0 is longest. If link 0 is of intermediate length, $300 + 100 < L_0 + 200$, from which $200 < L_0$. Thus 200 mm $< L_0 < 400$ mm for the crank rocker (answer b).

The drag link criterion (c) requires that link 0 be shortest, that is, $L_0 < 100$. Also, $300 + L_0 < 100 + 200$ is required by the Grashof inequality, from which $L_0 < 0$. Thus, no drag link can be formed (answer c).

The double-rocker test (d) requires that the coupler, L_2, be shortest. Thus a double-rocker cannot be formed (answer d).

A change-point mechanism (e) exists if

$$L_{max} + L_{min} = L_a + L_b$$

If L_0 is largest,

$$L_0 + 100 = 200 + 300$$

or $L_0 = 400$.

If L_0 is smallest, $300 + L_0 = 100 + 200$, from which $L_0 = 0$ (not a four-bar mechanism). If L_0 is of intermediate length, $300 + 100 = L_0 + 200$, from which $L_0 = 200$. Thus we have a change-point mechanism if $L_0 = 200$ or 400 mm (answer e).

As noted above, any mechanism that meets the criteria in b, c, d, and e is a Grashof mechanism. Combining the above results, we have 200 mm $\leq L_0$ ≤ 400 mm (answer a).

A triple-rocker mechanism (f) exists when

$$L_{max} + L_{min} > L_a + L_b$$

If the fixed link L_0 is largest, we have $L_0 + 100 > 200 + 300$, or $L_0 > 400$. If the fixed link L_0 is shortest, we have $300 + L_0 > 200 + 100$, or $L_0 > 0$. If L_0 is of intermediate length, $300 + 100 > L_0 + 200$, or $200 > L_0$. Noting that no link length may exceed the sum of the lengths of the other three, we have $L_0 < 100 + 200 + 300$, or $L_0 < 600$. Combining the above results, we obtain either $0 < L_0 < 200$ mm or 400 mm $< L_0 < 600$ mm (answer f). ◆

1.7   TRANSMISSION ANGLE

Of course, the inequalities that classify four-bar linkages give the extreme theoretical limits of each mechanism class. Practical limitations including that of transmission angle reduce that range. The *transmission angle* is the angle between the coupler centerline and the driven crank centerline. The optimum transmission angle for a planar linkage is 90°. A transmission angle no less than 40° or 45° and no greater than 135° or 140° is usually satisfactory. Depending on the type of bearing and lubrication, values outside this range may result in binding of the linkage.

Figure 1.15 shows a linkage that satisfies the crank rocker criteria. However, if link 1 drives, the transmission angle ϕ reaches extreme values, which

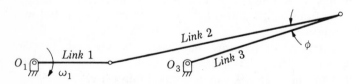

Figure 1.15 A mechanism that may fail to operate because of an unsatisfactory transmission angle.

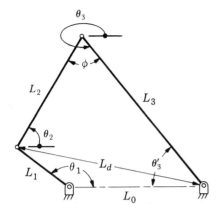

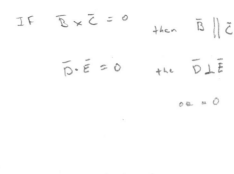

IF $\vec{B} \times \vec{C} = 0$ then $\vec{B} \parallel \vec{C}$

$\vec{D} \cdot \vec{E} = 0$ the $\vec{D} \perp \vec{E}$

$\theta = 0$

Figure 1.16 Determination of transmission angle.

may prevent satisfactory operation. As link 1 tends to rotate through the position shown in the figure, the direction of force transmitted along link 2 to link 3 results in very little torque on link 3 but a high bearing force at O_3. Wear would probably be excessive. If friction torque exceeded driving torque, the mechanism would jam and could cause the driven crank to buckle. Dimensional tolerances including looseness at pins and bearings often tend to worsen the situation. In most cases, then, it is advisable to provide a reasonable "margin of safety" in satisfying the inequalities that determine linkage motion.

Consider the four-bar linkage whose links form a quadrilateral as in Figure 1.16. For crank angle θ_1, the length of the diagonal L_d of the quadrilateral can be determined by using the cosine law. For the triangle formed by links 0 and 1 and the diagonal,

$$L_d^2 = L_0^2 + L_1^2 - 2L_0 L_2 \cos \theta_1 \qquad (1.12)$$

(with L_1 noted beneath)

Using the cosine law for the triangle formed by the diagonal and links 2 and 3, we have

$$L_d^2 = L_2^2 + L_3^2 - 2L_2 L_3 \cos \phi \qquad L_d^2 = (L_0 - L_1)^2 \qquad (1.13)$$

or

$$\cos \phi = \frac{L_2^2 + L_3^2 - L_d^2}{2L_2 L_3} \qquad L_{max} + L_{min} < L_a + L_b \qquad (1.14)$$

Thus, we may obtain transmission angle ϕ at any instant.

Of course, we are most interested in extreme values of transmission angle ϕ. For the crank rocker mechanism, maximum and minimum transmission angles occur when the driver crank and fixed link are colinear. Transmission angle ϕ_{max} corresponds to $L_{d(max)} = L_1 + L_0$ and ϕ_{min} corresponds to $L_{d(min)} = L_0 - L_1$.

◆ EXAMPLE PROBLEM 1.3 *Transmission Angle*

Given: driver crank length $L_1 = 100$ mm, coupler length $L_2 = 200$ mm, and follower length $L_3 = 300$ mm. Find the range of values for the fixed link L_0 if the linkage is to be a crank rocker, considering transmission angle. In a

$$\cos C = \frac{a^2 + b^2 - c^2}{2ab}$$

previous example, we determined that this mechanism theoretically acts as a crank rocker for 200 mm $< L_0 <$ 400 mm if we put no limit on transmission angle. Design decision: transmission angle range is limited to $45° \leq \phi \leq 135°$.

ure   for   limiting   position   and   trans ϕ

Solution.   Setting $\phi_{min} = 45°$ and using the cosine law, we have

$$(L_0 - L_1)^2 = L_2^2 + L_3^2 - 2L_2L_3 \cos \phi_{min}$$

$$(L_0 - 100)^2 = 200^2 + 300^2 - 2 \times 200 \times 300 \cos 45°$$

$$L_0 = 312.48 \text{ mm}$$

Using this value, we have

$$\cos \phi_{max} = \frac{L_2^2 + L_3^2 - (L_0 + L_1)^2}{2L_2L_3} = \frac{200^2 + 300^2 - (312.48 + 100)^2}{2 \times 200 \times 300}$$

or $\phi_{max} = 109.54°$, which is within the accepted range.

The results (which could have been determined graphically) are sketched in Figure 1.17a and b. Part c of the figure shows coupler angle θ_2, follower crank angle θ_3, and transmission angle ϕ plotted against crank position θ_1. Note that if we had set $\phi_{max} = 135°$ to obtain $L_0 = 363.52$, the value of ϕ_{min} would have been $59.69°$. Thus $312.48 \leq L_0 \leq 363.52$ is the acceptable range.

Both the follower crank (rocker) and the coupler in a crank rocker mechanism have a limited range of motion (considerably less than 180° if we require reasonable values of transmission angle). For the linkage designed in the above example, with $L_0 = 312.48$ mm, the limiting positions of the follower crank correspond to the two positions where links 1 and 2 are colinear. Using a form of the cosine law for links 1 and 2 extended, we have

$$\cos \theta_3' = \frac{L_0^2 + L_3^2 - (L_1 + L_2)^2}{2L_0L_3} = \frac{312.48^2 + 300^2 - (100 + 200)^2}{2 \times 312.48 \times 300}$$

from which $\theta_{3(max)}' = 58.61°$.

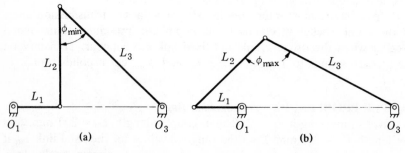

Figure 1.17 (a) Minimum value of transmission angle. (b) Maximum value of transmission angle.

$$V_{max} = r\omega$$
$$a_{max} = r\omega^2$$

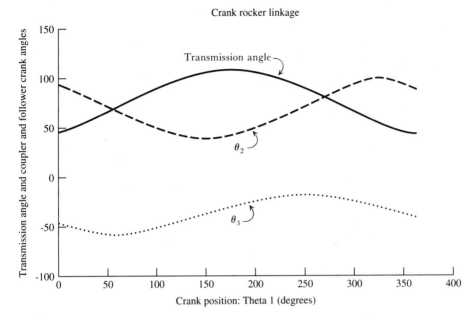

Figure 1.17 (c) Transmission angle and link positions vs. crank position.

Figure 1.17 (c) Transmission angle and link positions vs. crank position.

For the other limiting position with links 1 and 2 colinear,

$$\cos \theta_3' = \frac{L_0^2 + L_3^2 - (L_2 - L_1)^2}{2L_0 L_3}$$

from which $\theta_{3(min)}' = 18.65°$, for a range of only $39.96°$. Of course, the range of the follower crank can be increased by changing the ratios of the link lengths.

♦

We may think of the slider-crank linkage as a four-bar linkage if the slider is replaced by an infinitely long link perpendicular to the sliding path. Then, transmission angle is defined as the angle between the connecting rod and a perpendicular to the slider path. The extreme values of transmission angle occur when the crank is perpendicular to the slider path. (The proof of this statement is left as an exercise.)

Transmission angle criteria are important for spatial linkages as well. Consider, for example, an RCCC linkage made up of rigid links connected by one revolute joint and three cylindrical joints. Let the drive link (1) join the fixed link (0) at a revolute joint. Optimum transmission conditions occur when the drive link (1), coupler (2), and follower (3) are mutually perpendicular. Thus the optimum transmission angle criterion is that

$$\left| r_1^u \cdot (r_2^u \times r_3^u) \right|$$

be near unity, where r_1^u is the unit vector representing link 1, and so on. The rules of vector algebra are reviewed briefly in Chapter 2.

1.8 LIMITING POSITIONS

Consider a crank rocker mechanism driven continuously by the shortest link. The instantaneous configuration of the mechanism when the follower crank reaches one of its extreme positions is called a limiting position. A number of other mechanisms exhibit limiting positions as well. Limiting positions of the slider-crank linkage occur when the slider is at either extreme of its travel.

Limiting positions are of interest for several reasons. The limiting positions of a slider-crank mechanism define the stroke of the piston (slider). The piston has zero velocity at the instant it reaches one of the limiting positions. However, the acceleration of the piston, and consequently the inertial force, is high at that instant. When at a limiting position, a slider-crank mechanism cannot be driven by applying a force on the piston. If a single piston serves as a driver, the linkage may be driven through the limiting position by inertia of the crank. Likewise, the limiting positions define the range of the oscillating crank of a crank rocker mechanism. The oscillating crank has zero angular velocity and a high value of angular acceleration at the limiting positions.

$45° < \theta < 135°$ *values must be before that range*

1.8.1 Slider-Crank Mechanisms

A slider-crank mechanism with the usual proportions (connecting rod longer than crank) has two limiting positions, both occurring when the crank and connecting rod are colinear (in the skeleton diagram). See Figure 1.18a and b. When reciprocating steam engines were in common use, these positions were called *dead-center positions*, *crank dead center* referring to the position with the piston nearest the crankshaft (Figure 1.18, *left*), and *head dead center* referring to the position with the piston farthest from the crankshaft (Figure 1.18, *right*). Piston direction reverses at these two points.

When the extended path of the wrist pin C goes through the center of the crankshaft O_1 (as in Figure 1.18a), the linkage is then called an *in-line slider-crank mechanism*. The stroke, referred to as *piston travel*, equals $2R$, twice the crank length. The crank turns through 180° as the piston moves from left to right, and 180° as it returns to the left. If the crank turns at constant angular velocity ω, the piston takes the same time to move from left to right as it takes to return to the left.

1.8.2 Offset Slider-Crank Mechanisms

The wrist-pin path of the *offset slider-crank mechanism* (see Figure 1.18b) does not extend through the crankshaft center. The limiting positions shown represent positions of zero piston velocity, but the angles through which the crank turns between the limiting positions are not equal. If the crank turns counterclockwise, it turns through an angle greater than 180° as the piston moves from left to right, and through less than 180° as the piston moves back to the

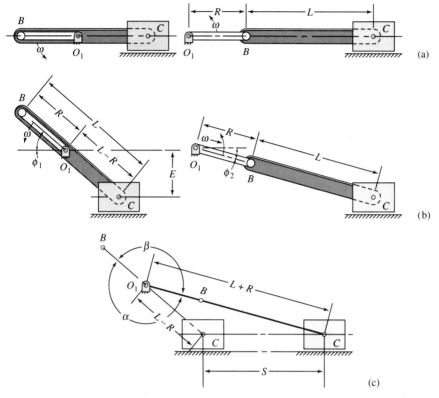

Figure 1.18 (a) The two limiting positions of an *in-line* slider-crank mechanism. (b) The two limiting positions of an offset slider-crank mechanism. (c) The limiting positions of an offset slider-crank mechanism superimposed to find the time ratio of forward to return strokes.

left. If the crank turns counterclockwise at constant angular velocity, the piston takes longer in its stroke to the right than it takes to return to the left. From its limiting position in Figure 1.18b *left*, to its limiting position in Figure 1.18b *right*, the crank turns through the angle

$$\alpha = 180° + \phi_1 - \phi_2$$

as shown in Figure 1.18c. During the return stroke, the crank turns through the angle

$$\beta = 180° - \phi_1 + \phi_2$$

where

$$\phi_1 = \sin^{-1} \frac{E}{L - R} \quad ,$$

$$\phi_2 = \sin^{-1} \frac{E}{L + R}$$

for crank length R, connecting rod length L, and offset distance E less than $L - R$.

When the crank turns at a constant angular velocity ω, the time ratio of forward to return stroke is given by α/β. The length of the stroke is

$$S = \sqrt{(L + R)^2 - E^2} - \sqrt{(L - R)^2 - E^2}$$

The limiting positions of the linkage may be superimposed to form a triangle, as in Figure 1.18c. Using the triangle inequality (the sum of the lengths of any two sides exceeds the length of the remaining side), we obtain

$$L - R + S > L + R$$

from which we see that the stroke length will always exceed $2R$ when the wrist-pin path is offset from the crankshaft. The above relationships are valid when both of the following conditions are met: offset E less than $L - R$, and R less than L. Of course, angles α and β and stroke S can be found simply by superimposing the limiting positions of the linkage as in Figure 1.18c. This graphical solution is used to check the analytical solution, or, as is often the case, the accuracy of the graphical solution alone is adequate.

1.9 QUICK-RETURN MECHANISMS

Quick-return mechanisms include an oscillating link or reciprocating slider that moves forward slowly and returns quickly (with constant speed input). The forward and return directions are arbitrarily assigned as above to correspond with machine tool usage where a working stroke would have high force capability at low speed and the return stroke could be rapid with no load.

The designation *quick return* has as much to do with the function of a mechanism as with its mode of operation. If there is an intentional difference between the time required for the forward and return strokes, the linkage may be called a *quick-return mechanism*. Most crank rocker mechanisms exhibit unequal forward and return times for the rocker. If we take advantage of the unequal strokes in designing a piece of machinery, we call the linkage a quick-return mechanism.

Forward and return strokes for the in-line slider-crank mechanism take equal time, but the offset slider crank acts as a quick-return mechanism. The time ratio of the strokes is the same as the ratio of angles given by equations $\alpha = 180° + \phi_1 - \phi_2$ and $\beta = 180° - \phi_1 + \phi_2$ (see Figure 1.18a through c).

Other linkage combinations offer considerably more flexibility for quick-return design than the offset slider crank. The *drag link*, for example, may form part of a mechanism designed for large forward-to-return time ratios. Figure 1.19 shows four-bar linkage O_1BCO_3 that appears to satisfy the criteria for a drag link mechanism. Slider D will represent a machine element that is to have different average velocities for its forward and return strokes while driving crank 1 turns at constant angular velocity. The two extreme positions of

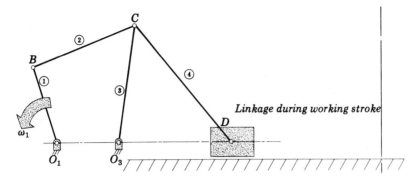

Figure 1.19 A drag link mechanism is combined with a slider to form a quick-return mechanism.

the slider occur when follower link 3 lies along the line of centers O_1O_3. Since link 4 is also colinear with the line of centers at both of the extreme positions, we see that the slider *stroke* is twice the length of link 3:

$$S = 2L_3 \tag{1.15}$$

The time for the slider to travel between limiting positions is proportional to the angle between corresponding positions of the driving crank as long as the driving crank angular velocity is constant.

◆ **EXAMPLE PROBLEM 1.4**   *Quick-Return Mechanism*
Design a *quick-return mechanism* with a 7-in stroke; the working stroke is to take 0.3 s and the return stroke 0.1 s.

Solution.

Step 1. There are many linkage configurations that will satisfy the above requirements. Because of the large ratio of working stroke to return stroke, we will try a mechanism based on the drag link. Using a mechanism suggested by Figure 1.19, a 7-in stroke will be obtained if the length of the follower link

$$L_3 = 3.5 \text{ in}$$

from Eq. 1.15. The remaining links must be selected by using only the drag link criteria and the stroke time ratio. As is the case for many practical problems we may use a trial-and-error procedure, making an educated guess and then checking our results, repeating the process until a reasonable design is obtained. Let us select the following lengths:

Fixed link: $L_0 = 2$ in (shortest link)
Driving crank: $L_1 = 2.5$ in

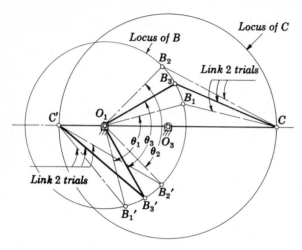

Figure 1.20 A sketch of a drag link quick-return mechanism to determine the length of link 2 needed for a 3-to-1 time ratio. The mechanism is drawn with link 3 in its critical positions.

Then, using the drag link inequality, we have

$$|2.5 - L_2| + 2 < 3.5 < 2.5 + L_2 - 2$$

from which   see   notes   for

$$3 < L_2 < 4$$ simplier   method

The mechanism will theoretically act as a drag link if L_2 is in this range. Values of transmission angle, however, may fall outside of the generally accepted range, necessitating special design provisions to ensure that the linkage operates properly.

Step 2.   The design is completed graphically in Figure 1.20. Link 3 is shown in its critical positions, O_3C and O_3C', and a circle of 2.5-in radius about O_1 represents the locus of B. For our first approximation, we try a coupler length $L_2 = 3.1$ in. In its critical positions, the linkage is described by $O_1B_1CO_3$ and $O_1B_1'C'O_3$. The slider moves from the extreme right to extreme left, as link 3 moves counterclockwise from position O_3C to O_3C'. For this stroke, the working stroke, the driver rotates counterclockwise from position O_1B_1 to O_1B_1' (through an angle $360° - \theta_1$). Then, the slider returns to the right as the driver turns through angle θ_1. The working stroke to return stroke time ratio given by $(360° - \theta_1)/\theta_1$ is less than the required value.

The value of θ for a working stroke to return stroke time ratio of 3 is found by setting

$$\frac{360 - \theta}{\theta} = 3 \quad \text{or} \quad \theta = 90°$$

Angle θ_1 obtained in our first approximation is, by observation, too large.

Step 3. As a second approximation, we try a larger coupler length, $L_2 = 3.7$. In a redrawing of the linkage, the critical positions of the linkage are $O_1B_2CO_3$ and $O_1B_2'C'O_3$. This time θ is too small and the resulting time ratio too large.

Step 4. The third approximation, $L_2 = 3.5$, results in critical positions $O_1B_3CO_3$ and $O_1B_3'C'O_3$, with θ_3 near $90°$. The required working stroke–to–return stroke time ratio of 3 is obtained (approximately) with this linkage having dimensions

$$L_0 = 2 \text{ in}$$

$$L_1 = 2.5 \text{ in}$$

$$L_2 = 3.5 \text{ in}$$

$$L_3 = 3.5 \text{ in}$$

L_4 may be any reasonable length greater than L_3. For this problem, we let

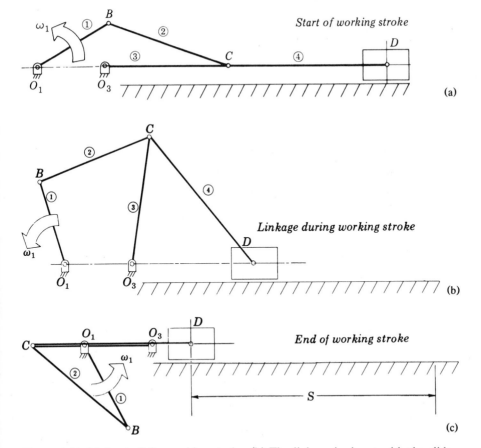

Figure 1.21 (a) Start of the working stroke. (b) The linkage is shown with the slider approximately halfway through the working stroke. (c) End of the working stroke.

$L_4 = 4.5$ in. The total time for a working stroke and a return stroke is to be 0.4 s; accordingly, crank 1 will be given one rotation in 0.4 s or $60/0.4 = 150$ rev/min. The final design is sketched in Figure 1.21 showing the linkage at the beginning and the end of the working stroke and at one intermediate position.

♦

Sliding contact linkages also form a basis for quick-return mechanisms. Figure 1.22 shows quick-return mechanism, which may be used to drive the cutting tool in a mechanical shaper. Crank 1 is the driver turning at essentially constant angular velocity and slider D represents the tool holder. Limiting positions occur when links 1 and 2 are perpendicular. The time ratio of working stroke to return stroke is equal to the ratio of angles between corresponding positions of link 1.

♦ **EXAMPLE PROBLEM 1.5** *Variable-Stroke Quick-Return Mechanism*
Design a mechanism with a stroke that may be varied from 3 to 8 in, having a working stroke to return stroke time ratio of 2 to 1 at maximum stroke length.

Solution. The 2-to-1 ratio is obtained if the angle θ between limiting positions is given by

$$\frac{360 - \theta}{\theta} = 2 \quad \text{or} \quad \theta = 120°$$

as in Figure 1.22b. Often, the key to determining link lengths is to *assign* a reasonable value to *one* or more of the unknown links. The geometric relationships in the linkage are next observed when the linkage is drawn in its limiting positions. The lengths of the remaining unknown links are then obtained. A satisfactory design may be obtained in the first trial. If not, this trial is used as a basis for improving the design.

If distance O_1O_2 is taken as 4 in, then the maximum length of the drive crank is

$$L_{1(\text{max})} = 4\sin\left(90° - \frac{\theta}{2}\right) = 2 \text{ in}$$

Since link 3 lies at the same angle at both limiting positions (the path of D is perpendicular to O_2O_1), the maximum stroke length is

$$S_{\text{max}} = D'D = C'C = 8 \text{ in}$$

from which we obtain the length of link 2:

$$O_2C = \frac{S_{\text{max}}/2}{\sin(90° - \theta/2)} = 8 \text{ in}$$

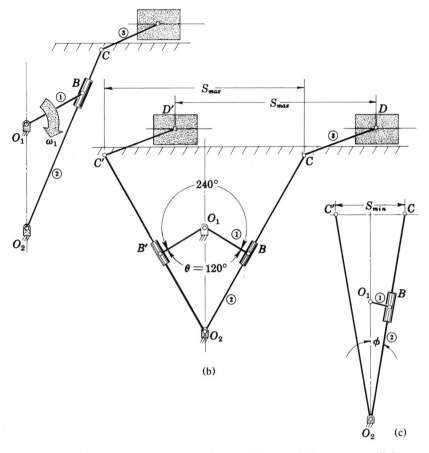

Figure 1.22 (a) Quick-return mechanism utilizing a sliding contact linkage. (b) The linkage is shown in its limiting positions (link 1 perpendicular to link 2). The stroke S of the slider is adjusted by changing the length of link 1. (c) Link 1 is shown adjusted to provide minimum slider stroke.

The length of link 3 is arbitrarily taken as 3 in, and the distance from O_1 to the path of D as 3.5 in. For the minimum stroke $S_{min} = 3$ in, the crank must be adjusted to a length of

$$L_{1(min)} = O_1O_2 \sin \phi = O_1O_2 \frac{S_{min}/2}{O_2C} = 0.75 \text{ in}$$

as shown in Figure 1.22c. The actual mechanism may differ considerably from the schematic as long as the motion characteristics are unchanged. Link 1 may be part of a large gear driven by a pinion, in which case the crankpin (B on link 1) will be moved in or out along an adjusting screw. Link 2 may be slotted so that the crankpin rides within it.   ♦

1.10 LINKAGE INTERFERENCE

For convenience in illustrating the motion of plane mechanisms, the mechanisms are shown as if they move within a single plane. Consider a crank rocker linkage as sketched in Figure 1.11. To avoid interference, the drive crank (link 1) and the coupler (link 2) must operate in two parallel planes. The plane of the drive crank should lie between the plane of the coupler and the plane of the fixed link.

The interference problem encountered in drag link mechanism design is more severe since drive crank, coupler, and follower rotate through 360°. So that link interference is avoided, the plane of the coupler should be between the planes of the cranks. The fixed bearings of the cranks must be placed on opposite sides of the linkage, with clear space for the linkage to pass between the bearings. Figure 1.23 illustrates one possible configuration schematically. It can be seen that if the plane of the coupler (link 2) is not clear, the linkage would not be able to operate through a complete rotation.

Sometimes a four-bar linkage forms part of a more complicated linkage. Motion may be transferred from the coupler of a crank rocker mechanism without much difficulty. Due to the problem of interference associated with the drag link mechanism and the requirement that the coupler plane lie between the cranks, motion transfer from the coupler of a drag link mechanism may require complicated arrangements.

The lamination-type impulse drive (described in the section that follows) illustrates motion transfer from one four-bar linkage to another. Considering the equivalent linkage, the rocker of a crank rocker mechanism acts as the

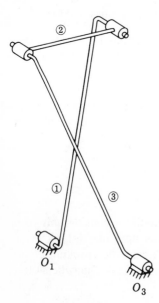

Figure 1.23 Configuration for drag link mechanism to avoid interference.

driving link of a second four-bar linkage (which oscillates due to the limited range of input motion). The drive is made up of several such mechanism combinations. Because of space limitations, eccentrics or cams are used instead of conventional cranks.

1.11 MECHANISMS FOR SPECIFIC APPLICATIONS

Before we begin detailed analysis and synthesis of mechanisms, it is worth-while to consider the basic motion characteristics of some of the commonly available linkages. In the design of a machine it may be practical to combine simple linkages and other components to obtain the required output-to-input motion relationship. The designer may wish to become familiar with many of the linkage configurations that are in the public domain and should become aware of the proprietary packaged drive trains and other machine components that are available. Then skill and ingenuity can combine these components for optimum results without a need to "reinvent the wheel." Some familiarity with various classes of available mechanisms will be obtained by leafing through this and other design-oriented books and by using manufacturers' catalogs and engineering periodicals.

Of course, the probable cost advantages of using commercially available components should not prevent the designer from exploring entirely new solutions, even though they may represent significant departures from traditional designs.

1.11.1 Drafting Instruments

The *drafting instrument* using rigid links and pins shown in Figure 1.5 is proportioned so that distances $AB = CD$ and $AD = BC$ form a parallelogram. If the line between fixed centers A and B is horizontal, then DC is also horizontal at all times. Since a straightedge attached at DC would not allow sufficient freedom of movement, another parallelogram linkage is added. A parallelogram linkage can also be used to restrain independently suspended automobile wheels to a vertical plane, reducing "tucking under" during turns.

Another drafting system, shown in Figure 1.24, uses tight steel bands (belts) on pair of disks with equal diameters. Disk 1 is not permitted to rotate and, as the arm between disks 1 and 2 is moved, disks 2 and 3 translate without rotating. The bands between disks 3 and 4 prevent rotation of disk 4 and the attached straightedges. Applications of this type were more common before the general availability of computer-aided drafting systems.

1.11.2 Pantograph Linkages

The parallelogram also forms the basis for *pantograph linkages*. At one time, pantograph linkages were used to reproduce and *change* the scale of drawings and patterns. The pantographs of Figure 1.25 are made up of rigid links AC,

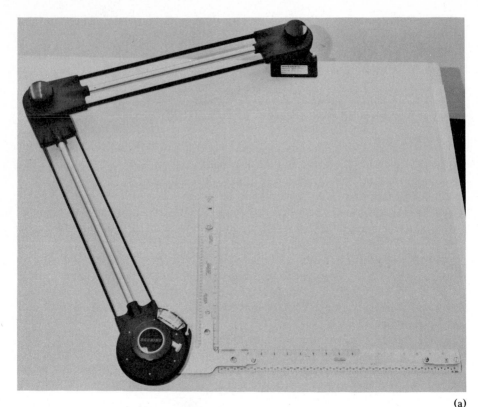

(a)

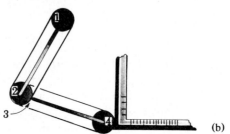

(b)

Figure 1.24 (a) A band-type drafting machine. (*Source:* A. M. Bruning.) (b) The belts arranged as shown permit translation of the straightedge but prevent rotation.

CD, *DE*, and *EB* with pin connections. Lengths *BC* = *DE* and *BE* = *CD* form a parallelogram. Link *BE* is parallel to *CD* at all times for both linkages and *F* is located on a line between *A* and *D*, making triangles *ABF* and *ACD* similar. Thus, in Figure 1.25a the ratio *DA/DF* is a constant for all positions of the linkage, and if a tracing point located at *F* is used to trace a pattern, a drawing tool at *A* will reproduce the pattern enlarged by the factor *DA/DF*. If the actual part is to be smaller than the pattern, then the tracing point can be located at *A* and the drawing tool at *F*. The result will be a size reduction of the ratio of *DF/DA*. The pantograph may be made adjustable to produce

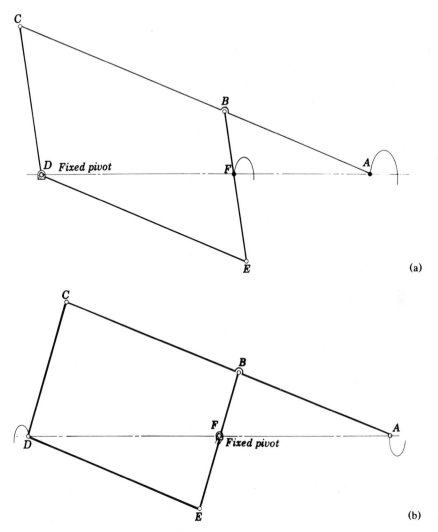

Figure 1.25 (a) A pantograph with fixed point D. The pattern can be traced-enlarged by the ratio of DA/DF if the tracing point is located at point F. Interchanging the tracing point and drawing tool produces a reduced tracing. (b) The pantograph with point F used as the fixed point will produce a tracing approximately the same size as the pattern.

various enlargement or reduction ratios provided the key features are main-tained: the linkage must form a parallelogram and points A, F, and D must lie on a straight line.

If the pattern is to be reproduced full size or nearly full size, point F will serve as the pivot, with D the tracer point and A the tool holder, as in Figure 1.25b. The pattern will be faithfully reproduced with a part-to-pattern size ratio of AF/DF but the orientation will be changed in this case.

The operation may be automated by using a sensing device to drive the tracing point over the pattern. A number of other linkages are used for similar purposes, including *engine indicators*, which reproduce a pressure signal. Engine or compressor pressure is measured by a small piston operating against a spring in the indicator. The indicator linkage, which resembles a pantograph, magnifies and records the motion of the indicator piston producing approximately straight-line motion.

◆ **EXAMPLE PROBLEM 1.6**   *Pantograph Design*

Proportion a linkage to guide an oxyacetylene torch in rough-cutting parts from steel plate. Part dimensions are to be approximately 6 in by 6 in; patterns will be 1.5 times full size.

Solution.   A pantograph of the type sketched in Figure 1.25b will be used so that the pattern will not be too near the cutting torch. The pattern dimensions will be approximately 9 in by 9 in and the linkage must be designed so that tracing point D moves freely over at least that area. It can be seen that by dimensioning the links so that $CB = DE = 12$ in and $CD = BE = 10$ in, the tracing point will cover the required area without nearing its limiting (extreme) positions. So that the size reduction factor of $1/1.5$ is obtained, $AB/CB = 1/1.5$, or $AB = 8$ in. Points A, F, and D must form a straight line, from which $BF/CD = AB/AC$, or $BF = 4$ in, locating the fixed pivot.

For a practical design, it might be necessary to allow the tracing point position A to be adjusted to various positions along the link so that several ratios of pattern to part size could be accommodated. For each position of A, a new point F, the fixed point on link BE, would have to be established to maintain the straight-line relationship between D, F, and A.   ◆

1.11.3 Slider-Crank Mechanism

The *slider-crank mechanism* is probably the most common of all mechanisms because of its simplicity and versatility. We are familiar with it in the reciprocating pump and compressor in which the input rotation is changed to reciprocating motion of the piston. Figure 1.26 shows an air-conditioning and heat pump compressor. In the piston engine, the situation is reversed and the piston is the driver. Of course, if there are several cylinders, the several pistons alternate as driver, and if the engine is a single-cylinder engine, the energy stored in the flywheel and other components actually drives the piston between power strokes. A single slider-crank mechanism and the associated cam and valve train typical of a multicylinder internal-combustion engine are shown in Figure 1.27. Figure 1.28 shows the piston and connecting rod of a small one-cylinder gasoline-powered engine.

1.11.4 Rotating Combustion Engine

The *rotating combustion (Wankel) engine* in Figure 1.29, is another solution to the same problem with little kinematic resemblance to the conventional

Figure 1.26 An air-conditioning and heat pump compressor. Capacity: 46,000 to 68,000 Btu/h. (*Source:* Tecumseh Products Company.)

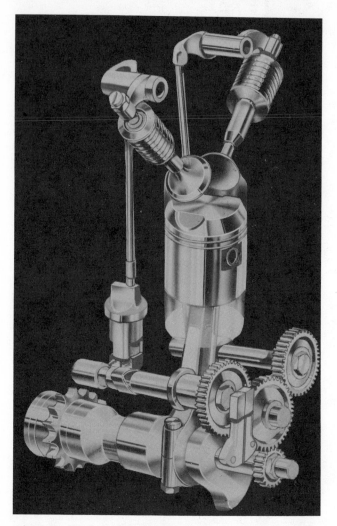

Figure 1.27 One cylinder of a reciprocating engine is illustrated, showing the basic slider-crank mechanism, camshaft, and valve train. (*Source:* Curtiss-Wright Corporation.)

piston-type engine. The three-sided rotor moves eccentrically within a two-lobed engine block. These two parts (the rotor and the shaped block) are equivalent to the pistons, cylinders, combustion chambers, and valve train of an ordinary reciprocating engine. An internal gear, part of the three-lobed rotor, actually acts as a planet gear as it meshes with a smaller fixed sun gear. Many other possible configurations of the rotating combustion engine with various numbers of rotor sides and engine block lobes were examined before this design was chosen.

The combustion cycle of a rotating combustion engine is illustrated in Figure 1.29b. At intake, an intake port is uncovered by the rotor. A mixture of

Figure 1.28 The piston and connecting rod for a one-cylinder engine. The rod was made as short as possible to reduce engine size. (*Source:* O&R Engines.)

air and fuel is drawn into the increasing space between the rotor and block. The eccentric rotor then seals the intake port and compresses the mixture in the now-decreasing space between rotor and block. The mixture is ignited when the space is very small, increasing the pressure and driving the rotor around (the expansion phase). Finally, an exhaust port is uncovered and the products of combustion are discharged. The cycle is then repeated.

In the above discussion, we traced only one charge of air and fuel through a complete cycle. The three-sided eccentric rotor and two-lobed engine block, however, correspond to three sets of pistons and cylinders. At the time of ignition of the first charge of air and fuel, the intake process is occurring in another chamber. When the first chamber is in the exhaust position, a third chamber is in the intake position. The figure shows the combustion cycle for only one chamber, but at any time a different phase of the process is occurring in each of the other chambers.

The fixed sun gear and the larger ring gear are shown as circles in Figure 1.29b. Rotation of the crankshaft and eccentric rotor carrier is seen by observing the point of contact between the fixed sun gear and the planetlike internal gear. In observing one thermodynamic cycle of this engine repre-

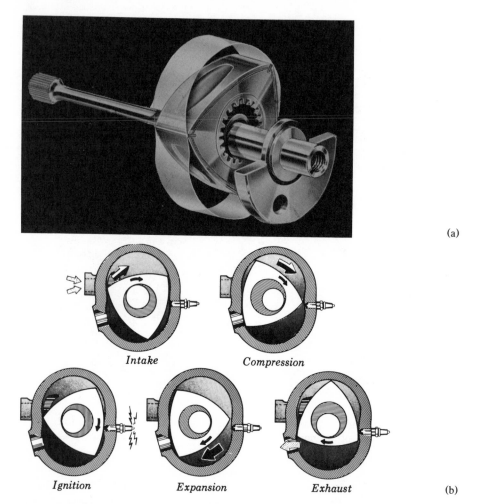

(a)

(b)

Figure 1.29 (a) The rotating combustion engine, showing the three-sided eccentric rotor and internal gear. A major advantage of this engine is its basic simplicity. The rotor and eccentric shaft are the only rotating parts; cams and valves are not required. (b) A complete combustion cycle of the rotating combustion engine. The rotor speed is one-third the eccentric crankshaft speed, maintaining one power impulse for each crankshaft revolution. (*Source:* Curtiss-Wright Corporation.)

sented by one rotor rotation, we see that the "crankshaft" (represented by the eccentric carrying the rotor) is given more than one rotation. Actually, the "crankshaft" is given three rotations, where the ratio of internal gear teeth on the rotor to teeth on the fixed gear meshing with it is 1.5 to 1. This result may be determined (with difficulty) by making successive sketches or may be calculated by using principles to be discussed in Chapter 8. The solution is left as an exercise in that chapter.

1.11.5 Fluid Links

Mechanical systems frequently include *fluid links* utilizing hydraulic or pneumatic cylinders or fluid drive transmissions. The backhoe shown in Figure 1.30 uses hydraulic cylinders arranged to give it a wide range of operating positions. Hydraulic feeds are also used for machine tools. By means of a variable delivery pump or a relief valve for control, speed and thrust may be regulated precisely by the operator. In the case of machine tools, the fluid system may be programmed to go through a complete cycle of operations automatically. For kinematic analysis, a hydraulic cylinder linkage of the type shown in Figure 1.31a is usually represented as shown in Figure 1.31b and c.

1.11.6 Swash Plate

Converting rotational motion to reciprocating rectilinear motion is a common problem, and many mechanisms have been devised for this purpose. In the *swash plate* type of mechanism, shown in Figure 1.32, a camlike swash plate is rotated about an axis that is not perpendicular to its face. The plate drives plungers in a cylinder block, the stroke of the plungers being equal to $d \tan \phi$, where the several parallel cylinders are arranged in a circle of diameter d as shown in the figure. The angle ϕ is measured between the swash plate face and a plane perpendicular to the cylinder axes. For 100 percent volumetric efficiency, the volume of liquid pumped per revolution of the swash plate is $Q = ANd \tan \phi$, where A is the cross-sectional area of one cylinder and N is the number of cylinders.

Figure 1.30 The motion of this hydraulic backhoe is determined by several independently controlled mechanisms. (*Source:* Insley Manufacturing Corporation.)

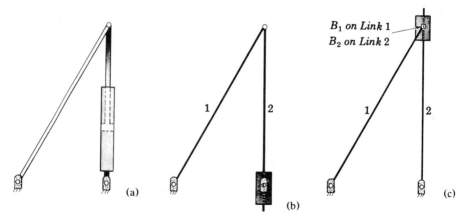

Figure 1.31 (a) A linkage that includes a hydraulic cylinder. (b) A kinematic representation of the linkage shown in part a; link 2 slides within a sleeve pinned to the frame. (c) Alternative representation of linkage shown in part a; link 2 slides within a sleeve pinned to link 1 at point B_1. Point B_2 is taken as the identical point on link 2 at this instant.

When it is operated as a hydraulic motor, fluid pressure is applied to the plungers that drive the swash plate. Each cylinder is alternately connected to the fluid supply and then to the exhaust by a distribution system operated by the swash plate shaft.

As noted earlier, an *inversion* of a mechanism exhibits the same relative motion but the links do not have the same absolute motion. The link that is fixed in the original mechanism is not fixed in the inversion. The cylinder block is fixed in the swash plate mechanism of Figure 1.32. If, instead, the cylinder block is rotated and the swash plate fixed, the motion of the plungers relative to the cylinders will not change. Hydraulic fluid will be pumped at the same rate. Figure 1.33 shows an inversion of the basic swash plate mechanism. In

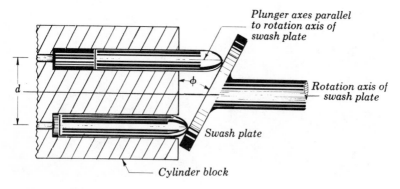

Figure 1.32 The swash plate mechanism is but one of a large number of mechanisms designed to convert rotational motion to rectilinear motion.

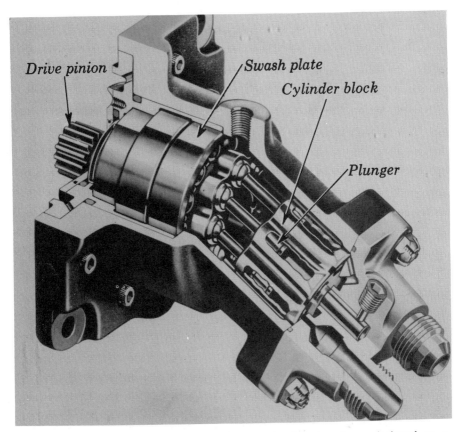

Figure 1.33 This *inversion* of a swash plate mechanism has been designed as a *fixed-displacement, piston-type hydraulic pump*. The cylinder block, drive shaft, and the nine pistons all rotate as a unit. The pump is available with the cylinder block axis offset relative to the drive shaft by 15 to 30°. This offset determines the stroke of the pistons and therefore the flow rate. (*Source:* Sperry Rand Corporation.)

this case, the link that acts as swash plate actually rotates, but its rotation is in a plane and of no significance to the relative motion. This arrangement is kinematically equivalent to the plunger ends riding on a fixed disk.

A *volume control* is effected by designing the pump so that the angle between the cylinder axes and the plane of the swash plate may be varied. The volume control may be actuated manually or automatically by a mechanical, electrical, or fluid control. When the mechanism is used as a motor, a similar control of offset angle may be used to change displacement for speed control. An *adjustable-speed transmission* may be assembled from two variable swash plates, one used as a variable-offset *pump*, and the other used as a variable-offset *hydraulic motor* (Figure 1.34). Speed is continuously variable over a wide range, with fine control and high torque capabilities. The fluid link

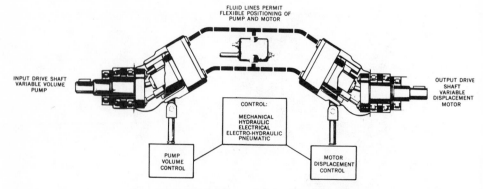

Figure 1.34 Two *variable-offset* swash plate mechanisms—one used as a pump and the other used as a motor—are combined to create an adjustable-speed transmission. In the pump, the piston stroke can be varied by changing the angle of offset. Fluid is pumped to the hydraulic motor, operating the pistons that drive the output shaft. (*Source:* Sperry Rand Corporation.)

between the two components allows considerable flexibility in positioning input and output.

1.11.7 Gear Trains

Gear trains are particularly suitable for use at high speeds and in drives with high power ratings. Since gears offer precise speed ratios, they are also used in mechanical computers, machine tools, and other applications where precision is required. Differential gears are used to distribute power in automobiles but may also be used to add or subtract inputs for control of certain processes. If two machines are to perform a production line function in a certain sequence, one machine may drive the other through a differential so that phase adjustment is possible between the operations. Figure 1.35 shows a differential transmission. The differential itself is made up of four bevel gears; the other gears in the transmission are helical gears.

Gearing is often combined with other mechanical components. A worm and worm wheel drive a power screw in the *linear actuator* of Figure 1.36. For reduction of friction, a ball screw is used. Because the translational motion of the screw is proportional to rotation of the worm, the actuator may be used as a precision jack or locating device. Gears will be discussed in detail in later chapters.

1.11.8 Lamination-Type Impulse Drive

Figure 1.37 shows a *lamination-type impulse drive* made up of several linkages. Power is transmitted from an eccentric through an adjustable linkage directly to a one-way clutch on the output shaft. There are several linkage and clutch assemblies, and each assembly operates on its own eccentric. The eccentrics

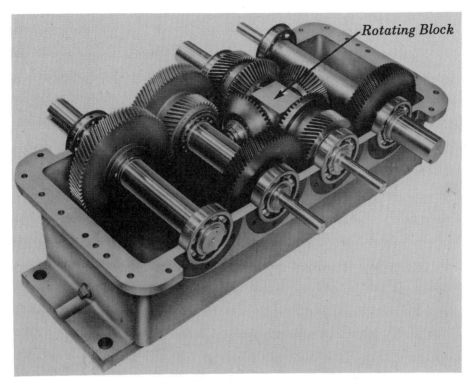

Rotating Block

Figure 1.35 A differential transmission. Differential gears are most commonly used to distribute power to the rear wheels of an automobile. Other uses of differentials and differential transmissions include phase shifting, ratio changing, and computation. (*Source:* Fairchild Industrial Products Division.)

operate in sequence throughout the entire input cycle to ensure continuous output motion, each linkage driving during its fraction of the input cycle.

The location of O_3, the control link axis, is adjustable. If it is moved toward O'_3, then link 5 oscillates through a smaller angle for each input rotation. When the control link axis is adjustable to fall on O'_3, the output shaft is stationary.

Reversing input speed direction does not change the output direction. However, output rotation may be reversed if the transmission is equipped with a reversible one-way clutch. When the clutch mechanism is reversed, the magnitude of output-to-input speed ratio also changes.

The average speed of link 5 of a given linkage assembly during the time that it is driving the output shaft clockwise is not the same as when it drives counterclockwise. If the clutch mechanism and the input rotation direction are reversed simultaneously, speed ratio does not change. Alternatively, the transmission may employ a reversing gear train so that no speed ratio change results. This transmission is designed for input speeds up to 2000 rev/min and output-to-input speed ratio may be adjusted from zero to 1/4.

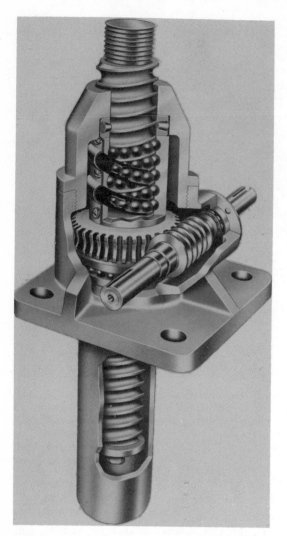

Figure 1.36 A ball-screw actuator. Rotating motion is efficiently transformed into translation through the use of a worm and worm wheel directly driving the nut of the ball screw. (*Source:* Duff-Norton Company.)

The impulse drive considered above allows for stepless variation of speed ratio, but pulsations (fluctuations in output torque or speed) do occur. If inertia of the load is relatively high, the one-way clutches permit the load to overrun the driving links, smoothing out pulsations. Linkage flexibility also aids in absorbing transmission pulsations so that their full effect is not transmitted to the driven machinery.

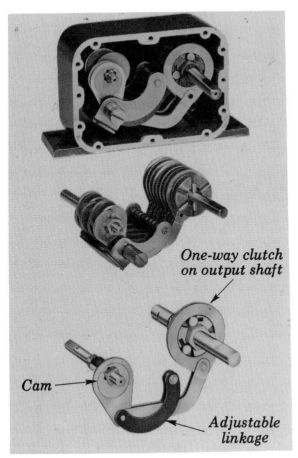

One-way clutch
on output shaft

Cam

Adjustable
linkage

Figure 1.37 (a) *Top*: Assembled view of a lamination-type impulse drive with cover plate removed. *Center*: The heart of the unit, which is a set of laminations phased to provide continuous driving. *Bottom*: A single lamination is shown with the important features identified. (*Source*: Zero-Max Ind. Inc., a unit of Barry Wright.)

Figures 1.37c and d show the equivalent mechanism representing the lamination-type impulse drive in two extreme positions. Links 1, 2, and 3 and the frame (O_1O_3) constitute a four-bar linkage driven by crank 1. Oscillating link 3 also forms part of a second four-bar linkage along with links 4 and 5 and the frame (O_3O_5).

1.11.9 Variable-Stroke Pump

There are other methods of adjusting output characteristics (speed, stroke length, stroke time ratios, etc.). Some linkages are designed so that mechanical

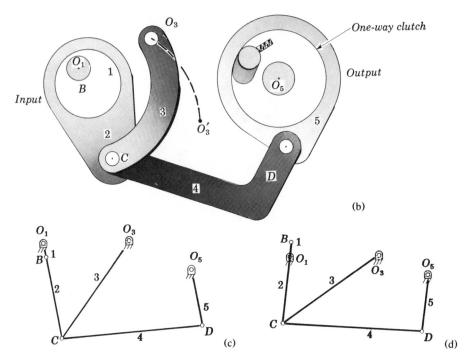

Figure 1.37 (b) A single lamination of the lamination-type impulse drive is shown in detail. The location of control axis O_3 governs output rotation. When the control link axis is adjusted to position O'_3, the output shaft becomes stationary. (c) The equivalent linkage. The linkage is shown in one of its limiting positions. (d) The other limiting position.

adjustments within the driving linkage itself can be made while the system is operating, often automatically in response to some demand on the system. The *variable-stroke pump* is a mechanism of this type. Pump stroke length is controlled by a "stroke transformer" driven by a pressure-actuated adjustment piston. Figure 1.10a is a cutaway drawing of the pump, identifying the crank, connecting rod, coupler link, stroke transformer, and plunger. Figure 1.38 shows the stroke transformer positioned for zero capacity. In Figure 1.10b the stroke transformer has been replaced by an equivalent link, where the center of rotation of the equivalent link corresponds with the center of curvature of the curved track of the stroke transformer.

Link 1, the driver, rotates at constant angular velocity. It drives connecting rod 2, which, in turn, drives links 3 and 4, moving the plunger E up and down in its cylinder. The position of O_4, the center of curvature of the stroke transformer, is adjustable. When O_4 is farthest from the wrist-pin E, we have maximum plunger stroke (maximum pump capacity). Moving O_4 closer to E reduces plunger stroke until, when O_4 and E coincide, the plunger does not move and there is no pump discharge. Velocities in the variable-stroke pump are analyzed in Chapter 3.

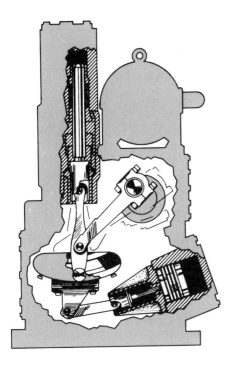

Figure 1.38 A variable-stroke pump. The adjustment cylinder piston is set so that the center of curvature of the stroke transformer coincides with the wrist pin. Thus the plunger remains stationary. (*Source:* Ingersoll-Rand Company.)

1.11.10 Power Screws

There are many ways to convert rotational motion into rectilinear motion. Cams, linkages, rack and pinion combinations, and a number of other devices are used. Power screws, one of the most common and precise methods, are frequently used as machine tool drives in conjunction with gear trains. If a screw with a single thread engages a nut that is not permitted to rotate, the nut will move relative to the screw a distance equal to the pitch for each screw rotation. (The pitch is the axial distance between adjacent corresponding thread elements.) With a double-thread screw, the nut motion is two pitches, and, in general, the motion of the nut per screw rotation will be the lead (the pitch times the number of threads). When a right-hand screw turns clockwise, the relative motion of the nut is toward the observer; for a left-hand screw turning clockwise, the relative motion of the nut is away from the observer. The nut may be split through an axial plane if it is to be engaged and disengaged from the screw as in a lathe. A split nut also permits adjustment to compensate for wear and eliminate backlash.

Differential Screws

When high-thrust, low-speed linear motion is required, a *differential screw* may be used. Figure 1.39 shows a power screw with a lead L_1 for the left half and L_2 for the right half, both right-hand threads. The motion of the slider equals the axial motion of the screw plus the axial motion of the slider with respect to

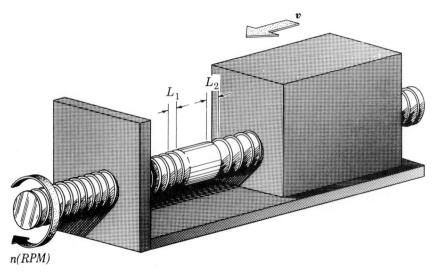

$n(RPM)$

Figure 1.39 Differential screw.

the screw:

$$v = \frac{n}{60}(L_2 - L_1)$$

where v = slider velocity

   n = clockwise revolutions per minute of the screw

   L_1 = screw lead at the frame

   L_2 = screw lead in the slider

For example, a single-thread screw may be cut with 11 threads per inch at the left end and 10 threads per inch at the right end. At 10 rev/min, the slider velocity is

$$v = \frac{10}{60}\left(\frac{1}{10} - \frac{1}{11}\right) = \frac{1}{660} \text{ in/s}$$

It is more common for power jacks, linear actuators, and other machinery controls to employ a worm drive for low-speed operation. In some cases, the outside of the nut has enveloping worm wheel teeth cut on it and it is restrained from axial motion by thrust bearings while the screw moves axially. See Figure 1.40a. Hundreds of jacks of this type were used in a single installation, a linear electron accelerator with a 4-in diameter by 2-mi-long waveguide that must be kept straight to within 1 mm. Figure 1.40b illustrates a more conventional application of the power jack.

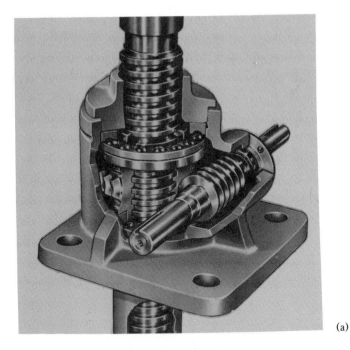

(a)

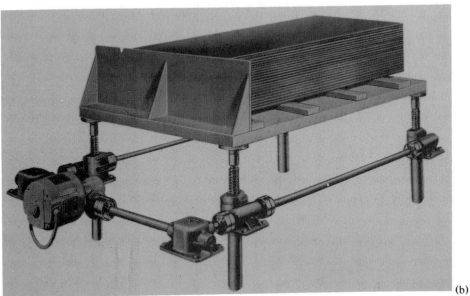

(b)

Figure 1.40 (a) Machine screw actuator; a worm-gear-driven screw, which may act as an actuator, a precision jack, or a leveling device. The ratio of worm to worm wheel teeth may be from 1:6 to 1:36. Actuators are available with up to 20 ft of screw travel. (*Source:* Duff-Norton Company.) (b) Four machine screw actuators are shown driven by a gear motor through bevel gear boxes. The actuators are self-locking. (*Source:* Duff-Norton Company.)

Ball Screws

Ball screws are used when friction must be reduced. The thrust load is carried by balls circulating in helical races, reducing typical friction losses to about 10 percent of transmitted power. Ball screws must include a ball return to provide a continuous supply to the nut. Preloading of the nut to eliminate backlash is possible if the ball race in the nut is divided into two sections. A ball-screw actuator is shown in Figure 1.36.

1.11.11 Special-Use Clutches

Special-use clutches that are self-actuating include centrifugal, torque-limiting, and one-way or overrunning types. Centrifugal clutches are actuated by a mass that locks the clutch parts together at a predetermined speed. Torque-limiting clutches, as the name implies, release at a predetermined torque. The ball-detent type has a set of steel balls that are held in detents by means of a spring force that determines the limiting torque. Any friction clutch may act as a torque limiter if the contact force is maintained by springs so that slipping occurs at torques above the limiting value.

Sprag-Type Reverse-Locking Clutches

Certain applications require that an input shaft drive the load in either direction but that the output shaft be prevented from driving the input shaft. This function is performed by the reverse-locking clutch (see Figure 1.41), through specially formed locking members called *sprags*.

Referring to the section view in Figure 1.41, assume that the input shaft (which drives the control member) turns counterclockwise. The control member contacts sprag *A* near the top, pivots it slightly counterclockwise, and thereby frees it from the outer race. The inner race is then driven by sprag *A*. (Sprag *B* performs no function during counterclockwise rotation.) Suppose, now, that the output tends to drive counterclockwise with *no* power applied to the input side. Then, the inner race slightly rotates sprag *A*, forcing it clockwise and jamming it against the fixed outer race, thus locking the system. The identical function is performed by sprag *B* for clockwise rotation of the clutch.

One-Way Clutches

One-way clutches drive in one direction only but permit freewheeling if the driven side overspeeds the driver. The clutch operation depends on balls or sprags that roll or slide when relative motion is in one direction but jam if the direction of relative motion tends to reverse.

Ratchet and pawl drives perform a similar function except that the pawl may engage the ratchet between teeth only. Either a one-way clutch or a ratchet-pawl drive may be used to change oscillation into intermittent one-way rotation. Some machinery-feed mechanisms operate in this manner.

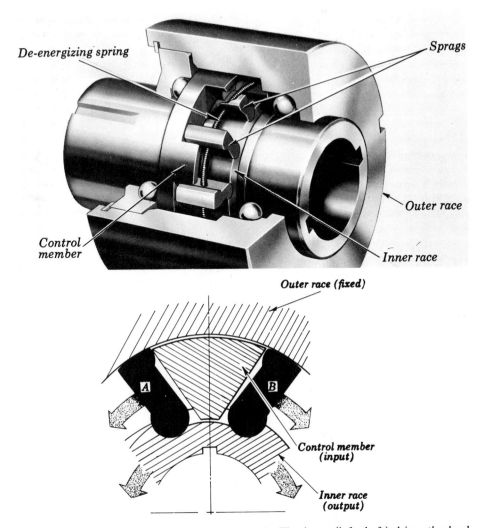

De-energizing spring

Sprags

Control member

Outer race

Inner race

Outer race (fixed)

A

B

Control member (input)

Inner race (output)

Figure 1.41 (a) Sprag-type reverse-locking clutch. The input (left shaft) drives the load (right shaft) in either direction. When the output shaft tends to drive, the sprags lock it to the outer race. Other sprag configurations are available that permit operation with freewheeling. (*Source:* Dana Corporation, Formsprag Division.) (b) Section view of the sprag-type clutch. As the input begins to rotate counterclockwise, it contacts sprag *A*. The sprag pivots slightly counterclockwise in its detent, separating from the outer race. The input pushes against the sprag, forcing the inner race (output shaft) to rotate. If the output begins to rotate faster than the input, the sprag is thereby given a slight clockwise motion, jamming the sprag against the fixed outer race and, in turn, locking the output shaft.

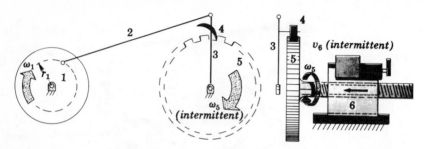

Figure 1.42 Ratchet-pawl mechanism (applied to an intermittent drive). As link 1 rotates at constant angular velocity, link 3 oscillates. A pawl (link 4) on link 3 drives the ratchet (5) during the clockwise motion of link 3. The right-hand screw drives the worktable intermittently to the left.

Figure 1.42 illustrates, in principle, the table-feed mechanisms of a mechanical shaper. The lengths of the links are such that link 3 oscillates as link 1; the driver rotates (i.e., the linkage is a crank rocker mechanism). A spring-held pawl drives the ratchet during only the clockwise motion of link 3 (approximately, but not exactly, half of each cycle). The workpiece table is intermittently fed to the left by the power screw driven by the ratchet. The cutting tool (not shown) moves perpendicular to the direction of table motion, but only during the part of the cycle when the table is stationary, to ensure straight cuts.

By an increase of the driving crank radius r_1, the angle through which link 3 oscillates is increased. Feed is increased in discrete steps; that is, the rotation of the ratchet per cycle will be an integer multiple of the pitch angle, where the pitch angle is $360°/N$ for N ratchet teeth. Feed is reversed by turning the pawl so that the counterclockwise motion of link 3 rotates the ratchet. Although this results in a change in instantaneous velocities, the feed per cycle is unchanged.

♦ **EXAMPLE PROBLEM 1.7**   *Intermittent Feed Mechanism Design*
Design an intermittent feed mechanism to provide rates of feed from 0.010 to 0.024 inches per cycle in increments of 0.002.

Solution. (There are several solutions to this design problem, each one involving many hours of work. We will take the first steps toward a practical design.)

1. A ratchet-pawl mechanism driving a power screw will be selected for this design. The required steps between minimum and maximum feed correspond to ratchet rotations of one pitch angle. For screw lead L inches, the feed per pitch angle is L/N inches. If we use a single-thread power screw with 5 threads per inch and a 100-tooth ratchet, the required 0.002-in/c feed increments are obtained.

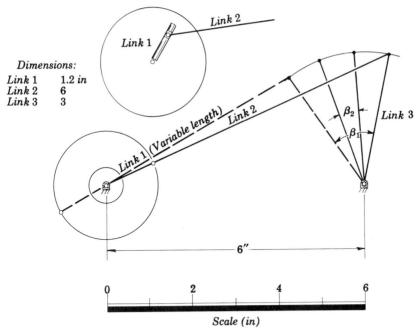

Figure 1.43 Example problem for a feed mechanism. Link 3 oscillates through angle β_1 with link 1 adjusted to 1.2 in, and through angle β_2 with link adjusted to 0.4 in.

2. A linkage with rotating driver crank and oscillating driven crank (similar to Figure 1.42) will be used to drive the pawl. The dimensions shown in Figure 1.43 will be selected tentatively, where link 1, the driving crank, is of variable length. Since the feed is 0.200 in per rotation of the screw, the screw must turn through 1/20 rotation (18°) for the 0.010-in/c feed. The 0.024-in/c feed is obtained by a 43.2° rotation of the screw.

3. For the required range of feeds, the oscillation of link 3 must be at least 43.2° when link 1 is adjusted to maximum length, and about 18° when link 1 is adjusted to minimum length. As a trial solution, we might design link 1 so that its length can be adjusted between 0.4 and 1.2 in. The mechanism is shown with link 1 adjusted to 1.2 in, at which setting link 3 oscillates through angle β_1. Angle β_2, the oscillation corresponding to a 0.4-in length of link 1, is also shown. The trial design has a wider range of feeds than required and is therefore acceptable from that standpoint. ◆

If we were to actually manufacture the mechanism, the next steps in the design process would be to find velocities and accelerations in the mechanism and to specify the members' cross sections. An investigation of tolerances and of stresses and deflections would then be required.

1.11.12 Universal Joints

When the angular relationship between the axes of two drive train elements is variable, the elements may be joined by a flexible coupling, a flexible shaft, or a universal joint. Most flexible couplings are intended only for small amounts of misalignment, and flexible shafts have very limited torque capacity. Where high torque and large misalignments occur, a *universal joint*, shown in Figure 1.44, is the typical solution. The Hooke-type universal joint has a variable output speed ω_2 for misalignment ϕ unequal to zero when the input ω_1 is constant. For the position shown in Figure 1.45, the velocity of point A is $v_A = \omega_1 r$, where ω is given in radians per second.

The angular velocity of shaft 2 is maximum at this time and equal to

$$\omega_2 = \frac{v_A}{r \cos \phi} = \frac{\omega_1}{\cos \phi}$$

In Figure 1.45b, the cross-link of the universal joint is shown as it rotates through 90°, the last position representing the minimum velocity of shaft 2: $\omega_2 = \omega_1 \cos \phi$.

At high shaft velocities, speed variations may be objectionable since acceleration and deceleration of the load can cause serious vibration and fatigue. Figure 1.46 shows two Hooke-type universal joints used to join shafts with a total misalignment of ϕ. If the shafts are in the same plane and each joint has a misalignment of $\phi/2$, as in Figure 1.46, the input shaft, 1, and output shaft, 3, travel at the same speed. The intermediate shaft, 2, turns at variable velocity, but if it is a short section, its mass will be low enough that serious vibration will not result.

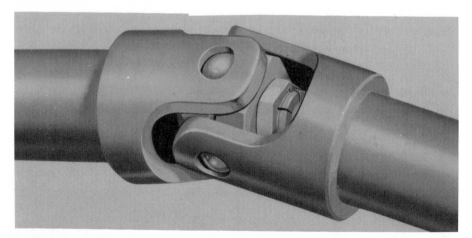

Figure 1.44 Universal joint. A preloaded universal joint is shown; it is designed for use in a steering-column-tilting mechanism and similar applications where backlash is undesirable. The recommended maximum operating angle for this type of universal joint is 18° (*Source:* Bendix Corporation.)

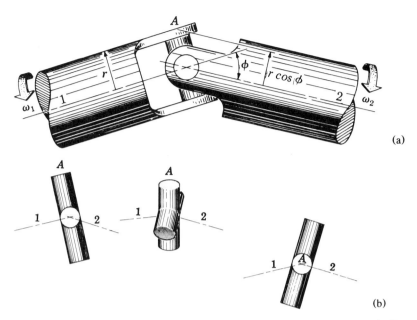

(a)

(b)

Figure 1.45 (a) The Hooke-type universal joint. The misalignment is indicated by the angle ϕ. Velocity ratio ω_2/ω_1 varies instantaneously as the joint rotates. (b) The cross-link of the universal joint is shown as it rotates through 90°.

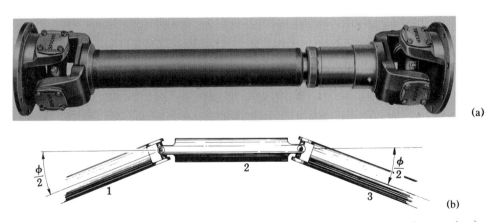

(a)

(b)

Figure 1.46 (a) A universal joint for high-torque applications. (*Source:* Dana Corporation.) (b) When two universal joints are used, input and output speeds are equal if each of the universal joints takes half of the misalignment, as shown.

An alternate method of avoiding acceleration and deceleration is through use of a constant-velocity universal joint. A constant-velocity ball joint, shown disassembled in Figure 1.47, is shown in the plane of the misaligned shafts in part b of the figure. Each half of the joint has ball grooves, the pairs of ball grooves intersecting in a plane that bisects the obtuse angle formed by the shafts. Thus, if all ball-groove center radii equal r, the velocity of the center of

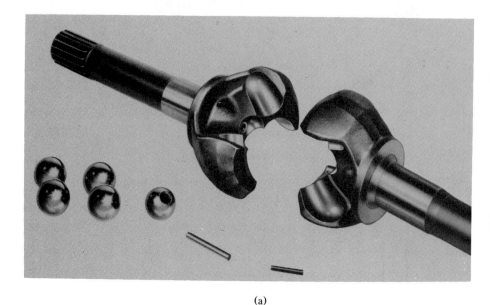

(a)

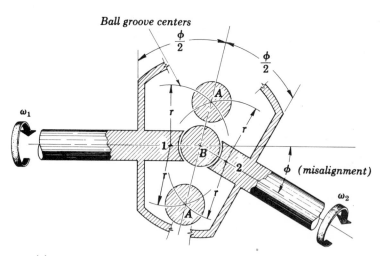

Figure 1.47 (a) A constant-velocity universal joint (Bendix-Weiss type) shown disassembled. (*Source:* Dana Corporation.) (b) Balls designated by A are held in intersecting ball grooves. The ball grooves in the input half of the joint intersect with the grooves in the output half so that the balls travel in a plane at angle $\phi/2$ to perpendiculars to either shaft. The pinned center ball is designated by B.

ball A is given by

$$v_A = \omega_1 r \cos \frac{\phi}{2}$$

and

$$\omega_2 = \frac{v_A}{r \cos \dfrac{\phi}{2}} = \omega_1$$

This constant-velocity relationship holds at all times, even as misalignment ϕ changes.

1.11.13 Automotive Steering Linkage

The Ackerman-type steering linkage is sketched in Figure 1.48. It incorporates a parallelogram linkage made up of the Pitman arm $O_1 B$, the relay rod BE, the idler arm EO_2, and the frame. Tie rods CF and DG are connected to the relay rod and to the steering arms FO_3 and GO_4. The steering arms turn the front wheels about pivoted knuckles O_3 and O_4 when the Pitman arm is rotated by a gear at O_1.

For avoidance of unnecessary tire wear when the vehicle turns, the centerlines of the four wheels should, as closely as possible, meet at a single point, the center of rotation of the vehicle.

Thus the Ackerman system is designed so that the wheels do not turn equal amounts. The wheel on the inside of the turn must be rotated through a

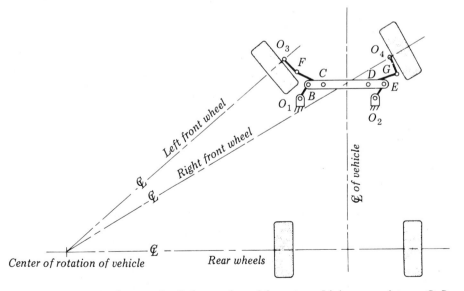

Figure 1.48 Automotive steering linkage oriented for a turn. Link nomenclature: $O_1 B$ is the Pitman arm (driven by a gear at O_1); BE is the relay rod; EO_2 is the idler arm; CF and DG are tie rods; FO_3 and GO_4 are steering arms.

greater angle about its steering knuckle than the wheel at the outside of the turn in order that the condition on the center of rotation be met.

This problem accounts, in part, for the complicated linkage design. Another problem is that the steering linkage is not strictly a planar linkage in that the wheels must follow road contours and are not rigidly pivoted to the chassis. Ball studs (ball-and-socket joints) are used at points C, D, F, and G to permit multiaxis rotation. In an alternate design, the steering linkage may be mounted forward of the centerline of the front wheels.

1.11.14 Computer-Controlled Industrial Robots

Demands for increased productivity have led to the development of computer-controlled robots. Figure 1.3 (page 11) shows an industrial robot with a highly maneuverable, six-axis jointed arm. The robot is controlled by a

Figure 1.49 An application of robotics. (*Source:* General Dynamics.)

flexible minicomputer program and may be interfaced to peripheral equipment or to an external computer. Although robots of this type have a fixed base, they may be employed in manufacturing operations involving a continuously moving production line. In one such application, the robot arm tracks moving automobile bodies and automatically spot-welds them without stopping the production line. It works on the front, middle, and rear of the auto body as the body moves through the robot's baseline station. "Abort" and "utility" sequences are included in the robot's computer control. The abort sequence directs the robot to exit from the moving part along a pretaught safe path relative to the part. The utility sequence is initiated by an external signal from malfunctioning peripheral equipment so that the robot can take corrective action. For example, the tip of a welding gun may stick to the part, requiring a twisting motion to break it free.

Figure 1.49 shows an application of robotics to drilling and perimeter routing of aircraft panels. A combination of positive-location part fixtures and compliant tooling systems was used to overcome the problem of positional inaccuracy due to joint tolerances and elastic deflection of the multi-degree-of-freedom robot manipulator.

1.12 COMPUTER-AIDED LINKAGE DESIGN

The design of linkages for specific applications has always relied heavily on human judgment and ingenuity. This design process may be illustrated by the flowchart given by Sheth and Uicker.[11] See Figure 1.50. This process of "human interaction" includes creativity and possibly lengthy periods of mathematical analysis and computation. While it is unlikely that human creativity can ever be completely replaced, computer-aided design (CAD) can be employed to relieve the designer of many of the routine processes that would otherwise be necessary. Since three-dimensional information can be stored and retrieved in various views by using CAD programs, construction of tentative physical models can often be eliminated in the design and development stage.

Currently, many organizations are integrating engineering design and drafting processes with manufacturing, administration and other functions. This approach, called *concurrent engineering*, relies heavily on computers, and is intended to reduce the time interval between the formulation of a design concept and production of the final product.

The suspension system shown in Figure 1.51 is an example of the type of problem that may be treated by CAD methods. Most CAD systems can handle open-loop (tree) systems like robots as well as closed-loop systems like four-bar and slider-crank linkages. Features of interest available in one or more CAD programs include kinematic analysis and synthesis of planar and spatial linkages, static and dynamic analysis, and interactive graphics capability. Additional features of some software include a check for redundant constraints, the capability to handle nonlinear equations; and a zero- and multi-degree-of-freedom capability as well as a capability to model one-degree-of-freedom systems.

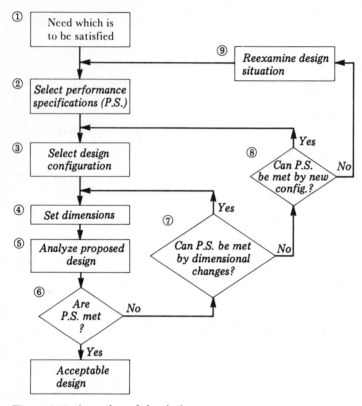

Figure 1.50 A portion of the design process.

Systems in which the motion is completely specified as a function of time are called *kinematically determinate* or zero-degree-of-freedom systems for purposes of analysis. Examples of kinematically determinate models include a robot with all joint motions specified, or a slider-crank mechanism with crank position specified.

When modeling a mechanism design, links are usually considered to be rigid. In some systems, however, link flexibility influences performance. In satellite design, for example, stringent weight requirements may result in a system that undergoes significant structural distortion. Figure 1.52, produced by ADAMS™ design software, shows a satellite deploying a flexible antenna extended with a screw jack. In this system, feedback control is employed on the momentum wheels for attitude control. Failure to consider the interaction between the flexible antenna and the feedback control system could have resulted in a satellite system that was unstable to the point of self-destruction.

A few of the software packages used for analyzing motion and forces in mechanical systems are briefly noted below.

IMP (integrated mechanisms program) is maintained by Structural Dynamics Research Corporation, Milford, Ohio. Figure 1.53 shows a flowchart

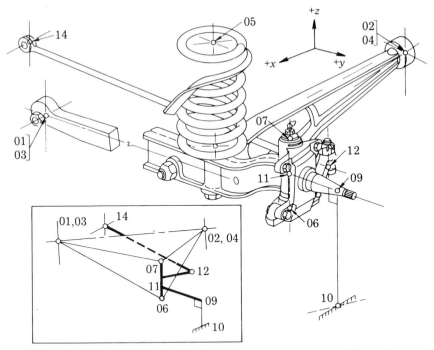

Figure 1.51 Suspension system. (*Source:* Chevrolet Motor Division, General Motors Corporation.)

for IMP processing. Capabilities of IMP include kinematic design, determination of the linkage geometry required to produce the intended motion; dynamic performance, determining the forces necessary for the mechanism to perform its intended task; and design for strength.

ADAMS (automatic dynamic analysis of mechanical systems) is a proprietary CAD program maintained by Mechanical Dynamics, Inc., Ann Arbor, Michigan. It is used to solve multiple-degree-of-freedom, three-dimensional mechanical systems. Its applications include vehicle dynamics, production machinery, high-speed mechanisms, control linkages, and robotics. It accepts spherical, universal, revolute, translational, cylindrical, and screw and rack and pinion joints.

DRAM (dynamic response of articulated machinery) is another CAD program maintained by Mechanical Dynamics, Inc. It is used to determine static equilibrium and time response (displacements, velocities, accelerations, and reaction forces) of planar multiple-degree-of-freedom, rigid-body mechanical systems that operate through large displacements. Figure 1.54 shows a simulation of a spring-reset plow that was designed to relieve shock loading when an embedded rock is struck. The operation was analyzed and refined with the DRAM program. Another CAD example, a front-end loader, is shown in Figure 1.55.

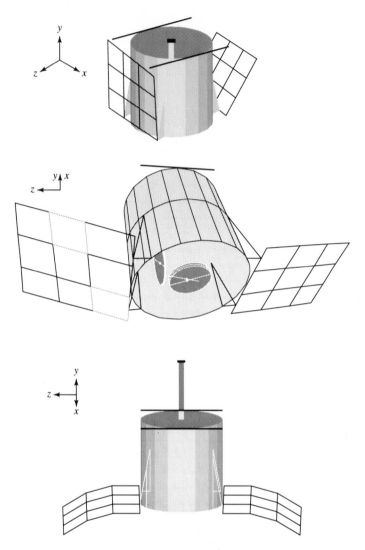

Figure 1.52 A satellite system deploying a flexible antenna (modeled by ADAMS™ software). (*Source:* Mechanical Dynamics Inc.)

RECSYN (synthesis with rectification) is an interactive linkage synthesis program that uses the precision position approach.[15] The designer specifies two, three, four, or five positions, and the solution linkage moves precisely through those positions. The rectification feature eliminates spurious solutions and solutions that pass through the design positions in the wrong order. RECSYN will output the Grashof type of each selected solution linkage (crank rocker, drag link, etc.) and the maximum and minimum transmission angle. Although basically intended for synthesis of four-bar linkages, RECSYN will also design slider-crank linkages and other four-link mechanisms with sliders.

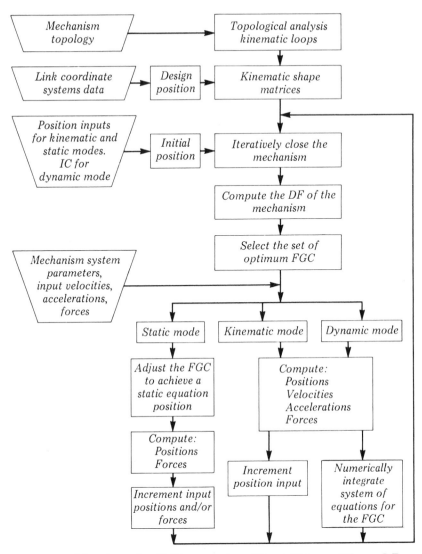

Figure 1.53 Flowchart for IMP processing; IC = initial conditions; DF = degrees-of-freedom; FGC = free generalized coordinates. (*Source:* Structural Dynamics Research Corporation.)

RECSYN can be used for function generation, motion generation, and path-angle generation. Path-angle generation refers to moving a point through specified positions while the driving crank rotates through specified angles. The motion generation or link guidance problem is that of designing a four-bar linkage so that a body attached to its coupler moves through a number of specified positions. Additional information on computer-aided design of mechanisms is given by Kaufman[6] and Rubel and Kaufman.[9]

Software packages that plot a linkage in a series of positions aid engineers in visualizing linkage motion and aid in making design changes to improve

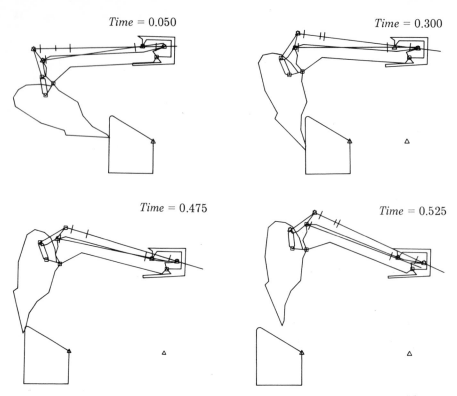

Figure 1.54 Analysis of a spring-reset plow, designed with the help of the DRAM program. (*Source:* Mechanical Dynamics, Inc.)

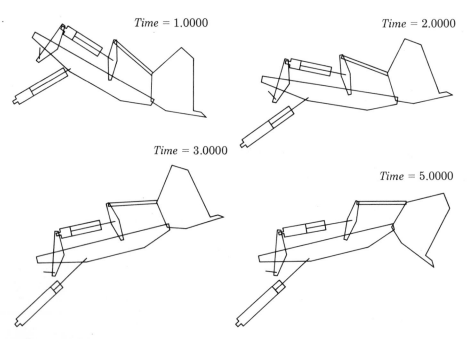

Figure 1.55 Computer-aided design applied to construction machinery. (*Source:* Mechanical Dynamics, Inc.)

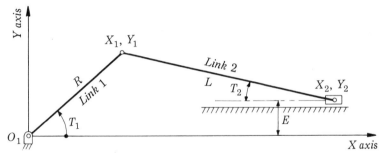

Figure 1.56 (a) Key to in-line and offset slider-crank linkage plots. Nomenclature for computer graphics displacement analysis.

performance. Some programs simulate animation by replotting successive linkage positions on a graphic display to give the appearance of continuous motion.

When examining basic linkages, it may be most convenient to write a simple computer program to plot skeleton diagrams of the linkage. Figure 1.56 shows a slider-crank linkage with offset E. Part b of the figure (page 78) is a flowchart of a sample program for calculating and plotting positions of the crank and connecting rod when the crank angle changes in 20° increments. Parts c, d, and e of the figure (page 79) show, respectively, linkage configurations for offset ratios $E/R = 0$, 0.25, and 0.40. On examining part e of the figure, we see that the connecting rod becomes nearly perpendicular to the slider path at certain crank angles. If the crank drives the connecting rod, which in turn drives the slider, then the normal force would be relatively large on the slider, possibly causing jamming. Thus we would probably reject a linkage with the proportions shown.

1.12.1 Research in Engineering Design Theory and Methodology

Design of kinematic systems, as with most engineering design, is a blend of art and science. Some investigators are studying design theory and methodology, attempting to obtain a deeper and more fundamental understanding of the design process. Finger and Dixon[2] reviewed research in descriptive, prescriptive, and computer-based models of design processes. They summarized studies of how humans create mechanical designs, processes, strategies, and problem-solving methods. They reviewed computer-based models that emulate human problem solving; morphological analysis, a methodology to generate and select alternatives; and configuration design in which a physical concept is transformed into a configuration with a set of attributes.

1.13 COMPUTER-IMPLEMENTED NUMERICAL METHODS

Some engineering problems do not have a simple closed-form solution. Or a closed-form solution may exist for a particular problem, but the solution is not

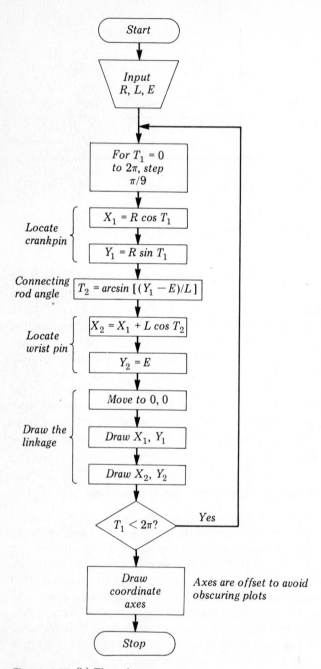

Figure 1.56 (b) Flow chart.

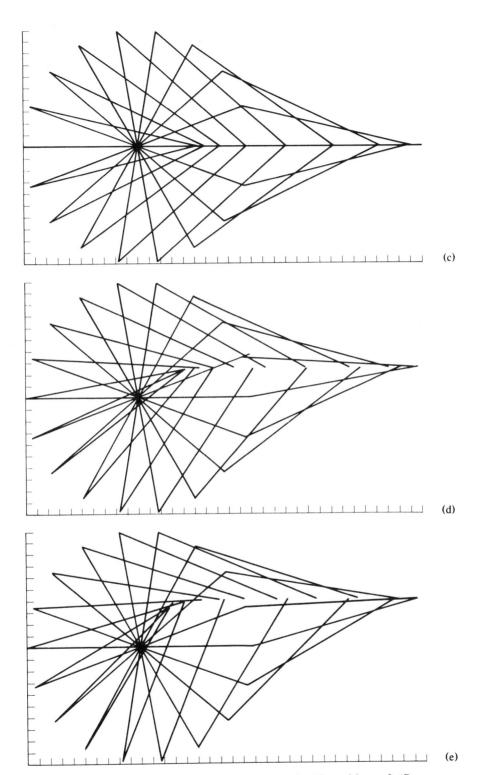

Figure 1.56 (c) In-line slider-crank linkage shown in 18 positions; $L/R = 1.5$, $E = 0$. (d) Offset slider-crank linkage shown in 18 positions; $L/R = 1.5$, $E/R = 0.25$. (e) Offset slider-crank linkage shown in 18 positions; $L/R = 1.5$, $E/R = 0.4$; key in a.

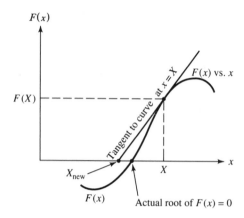

Figure 1.57 The Newton-Raphson method.

immediately obvious. We are then likely to try a numerical method of solution. Numerical methods differ from "trial and error" in that each successive approximation in a numerical method is guided by the previous result. Many of the numerical methods in current use were developed long before computers became available. With computers, however, we may make many iterations to obtain a high degree of accuracy while avoiding hours of tedious calculation.

1.13.1 The Newton-Raphson Method

Newton's method, also called the Newton-Raphson method, can be introduced by considering the root of a nonlinear equation in a single variable:

$$F(x) = 0$$

Figure 1.57 shows a curve that could resemble a plot of $F(x)$ vs. x [we do not actually plot $F(x)$ vs. x]. Let the first approximation of the root be $x = X$. Unless we were lucky enough to make a perfect guess, $F(X) \neq 0$. It can be seen from the figure that the first approximation can be improved by considering the error $F(X)$ and the slope of the curve $G(X)$. The second approximation of the root is given by

$$X_{new} = X - F(X)/G(X)$$

where $G(X) = dF/dx$ evaluated at $x = X$. The next approximation can be obtained by using the above equation, except that the value of X is replaced by the value of X_{new} obtained in the previous step. The process is repeated until $F(X_{new}) = 0 \pm$ a predetermined tolerance.

Success in finding a root depends on the behavior of $F(x)$ in the region of interest. It can be seen that the Newton-Raphson method fails if $G(X) = 0$ at any step. If $F(x) = 0$ has multiple roots, a poor first approximation may result in convergence at a root other than the desired one. Figure 1.58 shows a flowchart outlining the Newton-Raphson method. When using a computer for iterative processes, a limit may be placed on the number of iterations.

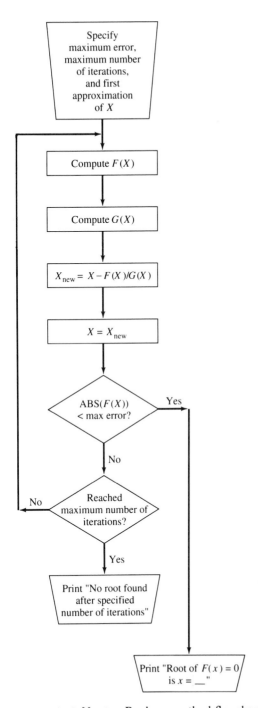

Figure 1.58 Newton-Raphson method flowchart.

◆ EXAMPLE PROBLEM 1.8   *An Iterative Solution*

Consider an offset slider-crank linkage for which the ratio of connecting rod to crank length is to be 3. Find the eccentricity for which the ratio of stroke to crank length will be 2.7.

Solution.   If we superpose sketches of the linkage in its two limiting positions, it can be seen that the resulting triangle can be measured or solved by the cosine law, leading to the answer directly. Instead, to illustrate a numerical method, it will be assumed that we fail to notice a closed-form solution and resort to the Newton-Raphson method. The problem is described by

$$S = \left[(L + R)^2 - E^2\right]^{1/2} - \left[(L - R)^2 - E^2\right]^{1/2}$$

Dividing by R, inserting the given values, and rearranging, the above equation becomes

$$F(x) = [16 - x^2]^{1/2} - [4 - x^2]^{1/2} - 2.7 = 0$$

where $x = E/R$. Differentiating, we obtain

$$G(x) = dF(x)/dx = -x/(16 - x^2)^{1/2} + x/(4 - x^2)^{1/2}$$

A computer program was written, based on the Newton-Raphson method flowchart. Using a tolerance of 10^{-6} and an initial approximation of $x = 1.9$, the program produced the following successive values: $x = 1.821906$, 1.800705, 1.799787, and 1.799786, the final value being the root of $f(x) = 0$ ($\pm 10^{-6}$). The root is then checked by substituting it in the initial equation.   ◆

When using numerical methods, the physical problem should be considered before making a first approximation of a root. In the above problem, an initial guess of $x > 2$ will result in one term that is the square root of a negative number. An initial guess of $x = 2$ produces an infinite value of $G(x)$ and a message that "no root is found after the specified maximum number of iterations." In this problem 20 iterations were allowed before the program "gives up" trying to find a root. On observing a sketch of the linkage, it is obvious that values of $x \geq 2$ are not valid.

1.13.2 Other Numerical Methods

Numerical methods are commonly used for solving complicated nonlinear equations. One disadvantage of the Newton-Raphson method is the necessity of obtaining the derivative $dF(x)/dx$.

The **secant method** uses a difference quotient instead of a derivative. However, we are required to supply two initial approximations of the root. We begin by making two approximations, X_1 and X_2, for the root of $F(x) = 0$. A new (hopefully improved) approximation of the root is given by

$$X_{new} = X_2 - F(X_2)/G(X_2)$$

where the difference quotient is given by

$$G(X_2) = [F(X_2) - F(X_1)]/[X_2 - X_1]$$

The procedure continues with X_1 assuming the old value of X_2, and X_2 assuming the value of X_{new}. The iteration continues until the root X_{new} is found, satisfying the equation $F(x) = 0$ within an acceptable tolerance. Otherwise the calculation is stopped after, say, 20 or 30 iterations, noting that the process does not converge.

The secant method is available in some mathematical software packages. One software package, MathCAD™, requires the user to input only one estimate of the root after $F(x)$ is defined. The second estimate of the root is taken as the tolerance if the first estimate is zero. Otherwise the second estimate is given by $X_2 = X_1 + $ tolerance.

Other sections of this text include kinematics problems involving more than one variable. Some of these are used as example problems and are solved by numerical methods, with the aid of mathematical software packages. References 1, 3, 5, 8, 10, 12, and 13 may be helpful in developing user-written numerical analysis computer programs or in understanding commercially available mathematical software.

1.14 MECHANISM DESIGN CONSIDERATIONS

The material in this chapter is intended to form a basis for the analytical work to follow and to introduce some of the analytical tools and approaches used in mechanism. In addition, the more common terminology is brought to the reader's attention to form a common ground for communication.

Design and manufacture of a product by one person working alone is seldom possible and rarely practical. Consider, for example, the complexity of the fully automated machine tool, or the case of a relatively simple mechanism to be mass-produced. In either case, many people are involved due to the interaction of one linkage with the machine as a whole and the relationship between design and manufacture. The designer must transmit ideas to others through mathematical equations and graphical representations, as well as through clear written and oral descriptions.

Past and even present practice relies heavily on ingenious designers taking advantage of their own inventiveness and years of practical experience. But the trend is toward more kinematic analysis and synthesis including computer-aided optimization. One automobile manufacturer investigated about 8000 linkage combinations in a computer-aided study of four-bar window regulator mechanisms. From those satisfying all design requirements (less than 500 did), the computer proceeded to select the one "best" linkage based on a set of predetermined criteria. Computer-aided design cannot replace inventiveness and human judgment, but it can extend the capabilities of an engineer and reduce the tedium of repetitive tasks.

Practical considerations often make it necessary to "freeze" a product design at some stage, thus preventing significant changes. However, a designer should investigate many possible design configurations early in the design process. Suppose, for example, a quick-return mechanism is required for a particular application. It was noted earlier in this chapter that offset slider-crank linkages, drag-link–slider-crank combinations, and sliding-contact–slider-crank combinations may be utilized as quick-return mechanisms. There are many other possibilities. We could examine cam-controlled and numerically controlled mechanisms and other combinations that include two or more sliders, in both planar and spatial configurations.

♦ **EXAMPLE PROBLEM 1.9**   *Linkage Design*
Design a quick-return mechanism with maximum stroke of 170 mm, 40 working strokes/minute, and a forward-to-return ratio of 4 to 1. Use a configuration not previously illustrated in this chapter.

Solution.   There are many possible configurations that could be used as quick-return mechanisms. A one-degree-of-freedom linkage is desired. Using Grubler's criterion, for mechanisms made up of lower pairs,

$$2n'_J - 3n_L + 4 = 0$$

or

$$n'_J = 3n_L/2 - 2$$

The following combinations produce an integer number of lower pairs:

n_L	n'_J	n_L	n'_J
2	1	8	10
4	4	10	13
6	7	12	16

One possible combination that includes four revolute pairs, three sliding pairs, and six links is shown in Figure 1.59. The designer may investigate various link lengths and locations for fixed pivot 0_1 relative to the slider paths in order to produce a given stroke length and forward-to-return ratio. Slider link 4 or 5 may be used as a toolholder. For some link-length ratios, one slider will undergo two oscillations per rotation of link 1, while the other slider undergoes one oscillation.

If the intended application of the quick-return mechanism calls for variable stroke or variable forward-to-return ratio, the design must include a means to adjust effective link lengths. A screw adjustment of distance O_1C is one possibility. Movement of point O_1 perpendicular to the plane of motion of

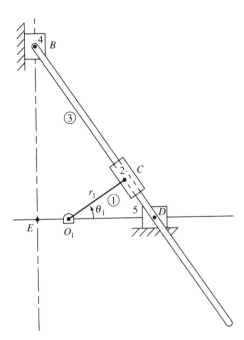

Figure 1.59 Three-slider quick-return mechanism.

link 3 provides an alternative means of adjustment. This second option would require a redesign, possibly including spherical pairs. A clever design would allow adjustment of motion characteristics while the mechanism was operating. Designs of this type require careful placement of links and bearings to avoid interference between moving parts. Design software such as I-DEAS™ (Lawry[7]) with solid modeling and assembly modeling capabilities may be used to aid the task of interference checking. ◆

References

1. Anderson, R. B., *The Student Edition of MathCAD*, Addison-Wesley, Reading, MA, 1989.
2. Finger, S., and J. R. Dixon, "A Review of Research in Mechanical Engineering Design. Part I: Descriptive, Prescriptive, and Computer-Based Models of Design Processes," *Research in Engineering Design*, vol. 1, 1989, pp. 51–67.
3. Frank, W. H., L. Kloepping, M. St. Norddahl, K. Olaffson, and D. Shook, *TK Solver Plus*, Universal Technical Systems, Rockford, IL, 1988.
4. Grashoff, F., *Theoretische Maschinenlehre*, L. Voss, Leipzig, 1883.
5. Hornbeck, R. W., *Numerical Methods*, Prentice-Hall, Englewood Cliffs, NJ, 1975.
6. Kaufman, R. E., "Mechanism Design by Computer," *Machine Design*, Oct. 26, 1978, pp. 94–100.

7. Lawry, M. H., *I-DEAS Student Guide*, Structural Dynamics Research Corp., Milford, OH, 1990.

8. Pipes, L. A., *Applied Mathematics for Engineers and Physicists*, McGraw-Hill, New York, 1958.

9. Rubel, A. J., and R. E. Kaufman, "KINSYN III: A New Human Engineered System for Interactive Computer-Aided Design of Planar Linkages," *Transactions of the American Society of Mechanical Engineers, Journal of Engineering for Industry*, vol. 99, no. 2, May 1977.

10. Scheid, F., *Numerical Analysis*, McGraw-Hill, New York, 1968.

11. Sheth, P. N., and J. J. Uicker, "IMP (Integrated Mechanisms Program), a Computer-Aided Design Analysis System for Mechanisms and Linkages," *Transactions of the American Society of Mechanical Engineers, Journal of Engineering for Industry*, May 1972.

12. Stark, P. A., *Introduction to Numerical Methods*, Macmillan, New York, 1970.

13. Stoer, J., and R. Bulirsch, *Introduction to Numerical Analysis*, Springer, New York, 1980.

14. Waldron, K. J., "A Method of Studying Joint Geometry," in *Mechanisms and Machine Theory*, vol. 7, Pergamon Press, Elmsford, NY, 1972, pp. 347–353.

15. Waldron, K. J., *RECSYN: NSF User's Manual*, National Science Foundation Grants ENG 77-22745, SER 78-00500, CME 8002122, 1977.

Other Publications of Interest

Mechanism and Machine Theory, Pergamon, Oxford and New York (the scientific journal of the International Federation for the Theory of Machines and Mechanisms). The subject matter includes:
Dynamics of machine systems
Gears and power transmission
Robots and manipulator systems
Computer-aided design methods

Machine Design, Penton Publishing, Cleveland, OH (a periodical covering the practice of machine design). The subject matter includes:
Recent developments related to the design of machines and mechanisms
Evaluation of software for computer-aided design
The designer may also find useful information about machine components in the advertisements.

PROBLEMS

Figures that accompany the problems are indicated with the prefix P. Otherwise, references apply to figures in the text proper.

* **1.1 (a)** Find the number of degrees of freedom for the spatial linkage of Figure P1.1. This open-loop kinematic chain includes cylindrical, prismatic, and spherical pairs.

  (b) Figure P1.2 is a schematic representation of a piece of construction machinery. It has two hydraulic cylinders (links 1 and 2, and links 6 and 7) which may be considered cylindrical pairs. The other pairs are revolute joints, including a double revolute joint where links 2, 3, and 8 meet. Find the

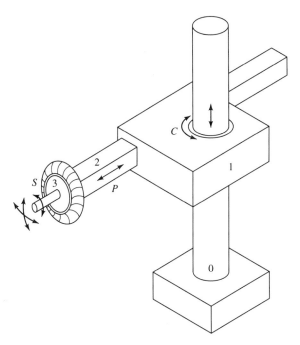

Figure P1.1

number of degrees of freedom if this linkage is treated as a spatial linkage.

(c) Find the number of degrees of freedom if the linkage of Figure P1.2 is treated as a planar linkage.

(d) Referring to Figure P1.2, indicate the conditions that must be met for the planar linkage assumption to be valid.

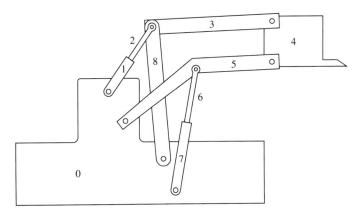

Figure P1.2

1.2 (a) Change the structure shown in Figure 1.7c into a (one-degree-of-freedom) mechanism by removing one link. (There are several solutions to this problem).

(b) Change the two-degrees-of-freedom linkage in Figure 1.7b into a one-de-gree-of-freedom mechanism by adding a link.

• **1.3** Find the average piston velocity (between limiting positions) for an in-line slider-crank mechanism. Crank length is 2 in, and the crank rotates at 3000 rev/min.

1.4 Find the average piston velocity in each direction (between limiting positions) for an offset slider-crank mechanism. Crank length is 2 in, connecting rod length is 4 in, and offset is 1 in. Crank speed is 3000 rev/min clockwise.

1.5 Repeat Problem 1.4 for eccentricity of 1.5.

1.6 Repeat Problem 1.4 for 100-mm crank length, 200-mm connecting rod length, and 50-mm offset.

1.7 Referring to Figure 1.56a, let $L/R = 1.5$ and $E/R = 0.2$. Show linkage position for crank angle $T_1 = 0, 20, 40°$, and so on. If computer graphics facilities are used, show 18 positions. Otherwise, show only 4.

1.8 In Problem 1.7, let crank length be 100 mm and crank speed be 5000 rev/min (constant) counterclockwise (ccw). Find average piston velocity during the 0.002-s interval beginning with zero crank angle.

1.9 Repeat Problem 1.7 for $L/R = 1.8$ and $E/R = 0.4$.

1.10 In Problem 1.9, let $R = 50$ mm and $\omega = 500$ rad/s. Find average piston velocity as the crank angle goes from (a) 0 to 40°, (b) 0 to 180°, (c) 0 to 360°.

• **1.11** Given an in-line slider-crank linkage with $R = 2$ and $L = 4$, (a) plot displacement x versus crank angle θ for $\theta = 30, 60,$ and 90°. If the crank rotates counterclockwise at 100 rev/min, find (b) the average piston velocity for θ between 30 and 60°, and (c) the average piston velocity for θ between 60 and 90°.

Problems 1.12 Through 1.22 Refer to Figure 1.11a

For each of the following problems: (a) determine whether the links can actually form a mechanism with the dimensions given; (b) if a mechanism exists, identity the motion relationship (crank rocker, drag link, double-rocker, triple-rocker, or change-point); (c) show the limiting positions if the mechanism is a crank rocker, double rocker, or triple-rocker mechanism.

1.12 $L_0 = 5, L_1 = 2, L_2 = 1.5, L_3 = 2$

• **1.13** $L_0 = 1, L_1 = 2, L_2 = 1.5, L_3 = 3.5$

1.14 $L_0 = 1, L_1 = 3.25, L_2 = 1.5, L_3 = 3.5$

1.15 $L_0 = 1.5, L_1 = 4, L_2 = 2, L_3 = 3.5$

1.16 $L_0 = 1, L_1 = 4, L_2 = 1.5, L_3 = 4.5$

1.17 $L_0 = 2, L_1 = 1.25, L_2 = 2, L_3 = 3$

1.18 $L_0 = 4, L_1 = 1, L_2 = 2, L_3 = 1.5$

1.19 $L_0 = 3, L_1 = 2, L_2 = 3, L_3 = 2$

1.20 $L_0 = 2.5, L_1 = 1, L_2 = 2.5, L_3 = 2$

1.21 $L_0 = 5, L_1 = 10, L_2 = 15, L_3 = 15$

1.22 $L_0 = 16, L_1 = 7, L_2 = 10, L_3 = 15$

• **1.23** Refer to Figure 1.11a. $L_0 = 300$ mm, $L_1 = 100$ mm, $L_3 = 280$ mm. Find the range of coupler length (L_2) for which the mechanism will theoretically act as a crank rocker.

Problems 1.24 Through 1.30 Refer to Figure 1.11a

For each of these problems, find the range of values for the unknown link if the linkage will theoretically act as a (a) crank rocker mechanism (where link 1 rotates through 360°); (b) drag link mechanism; (c) double-rocker mechanism; (d) change-point mechanism; (e) Grashof mechanism; (f) triple-rocker mechanism.

1.24 Find L_0 for $L_1 = 100$ mm, $L_2 = 280$ mm, $L_3 = 360$ mm.

1.25 Find L_1 for $L_0 = 40$, $L_2 = 60$, $L_3 = 40$.

1.26 Find L_2 for $L_0 = 50$, $L_1 = 200$, $L_3 = 250$.

1.27 Find L_3 for $L_0 = 120$, $L_1 = 220$, $L_2 = 80$.

1.28 Find L_0/L_1 for $L_2/L_1 = 1.5$, $L_3/L_1 = 1.2$.

1.29 Find L_2/L_1 for $L_0/L_1 = 2.2$, $L_3/L_1 = 1.7$.

1.30 Find L_0 for $L_1 = 400$, $L_2 = 600$, $L_3 = 750$.

1.31 In Figure 1.11a, let $L_0/L_1 = 2$. Plot the range of permissible values of L_3/L_1 versus L_2/L_1 if the linkage is to be a crank rocker mechanism.

1.32 Repeat Problem 1.31 for $L_0/L_1 = 1.5$.

1.33 Sketch a flowchart to classify four-bar linkages according to motion (crank rocker, etc.).

1.34 Write a computer or calculator program to classify four-bar linkages according to motion. Test the program with values corresponding to a crank rocker linkage, and so on.

Problems 1.35 Through 1.41 Refer to Figures 1.16 and 1.17

Find the length of the unknown link so that the linkage forms a crank rocker mechanism and transmission angle falls between 45 and 135°.

1.35 $L_1 = 100$ mm, $L_2 = 140$ mm, $L_0 = 120$ mm; find L_3.

• **1.36** $L_1 = 50$ mm, $L_2 = 200$ mm, $L_0 = 210$ mm; find L_3.

1.37 $L_1 = 110$ mm, $L_2 = 150$ mm, $L_0 = 150$ mm; find L_3.

1.38 $L_2/L_1 = 1.5$, $L_3/L_1 = 1.2$; find L_0/L_1.

1.39 $L_0/L_1 = 3.2$, $L_3/L_1 = 1.7$; find L_2/L_1.

1.40 Plot the range of permissible values of L_3/L_1 versus L_2/L_1 (if any) for $L_0/L_1 = 2$.

1.41 Repeat Problem 1.40 for $L_0/L_1 = 4$.

Problems 1.42 Through 1.47 Refer to a Slider-Crank Mechanism

The transmission angle is limited to the range between 45 and 135°, i.e., the angle between the connecting rod and the slider path must fall between -45 and $+45°$.

1.42 Crank length is 100 mm; find the minimum connecting rod length for zero offset.

1.43 Crank length is 500 mm, offset is 100 mm; find the connecting rod length (minimum).

1.44 Crank length is 400 mm, offset is 200 mm; find the minimum connecting rod length.

1.45 Crank length is 300 mm, offset is 50 mm; find the minimum connecting rod length.

1.46 Crank length to connecting rod length is $R/L = 0.5$; find the maximum possible offset.

1.47 Repeat Problem 1.46 for $R/L = 0.25$.

1.48 Sketch the flowchart for a program to design a crank rocker mechanism with maximum follower crank range if transmission angle is limited.

1.49 Write and test a calculator or computer program as designed in Problem 1.48.

1.50 Prove that extreme values of transmission angle occur when the crank is colinear with the fixed link in a four-bar linkage.

1.51 Prove that the extreme values of transmission angle occur when the crank is perpendicular to the slider path in a slider-crank linkage.

1.52 Referring to Figures 1.16 and 1.17, let $L_0 = 18$, $L_1 = 7$, $L_2 = 9$, and $L_3 = 17$.
(a) Classify the linkage according to its theoretical motion.
(b) Find extreme values of transmission angle.
(c) Will the linkage operate as determined in part a?

1.53 Sketch a double-rocker mechanism, showing bearing locations to avoid interference.

1.54 Will a drafting machine of the type shown in Figure 1.24 operate properly if the pulleys are arbitrarily selected of different diameters? Show the motion of the straightedges if pulley diameters and $d_1 = 50$ mm, $d_2 = d_3 = d_4 = 100$ mm.

1.55 Design and dimension a pantograph that may be used to double the size of the pattern.

1.56 Design and dimension a pantograph that may be used to increase pattern dimensions by 10 percent. Let the fixed pivot lie between the tracing point and the marking point or tool holder.

1.57 Design and dimension a pantograph that will decrease pattern dimensions by 40 percent.

• **1.58** Referring to the swash plate pump shown in Figure 1.32, (a) determine the dimensions of a pump with a capacity of 120 ft^3/h at 600 rev/ min (assume 100 percent volumetric efficiency), and (b) find average plunger velocity. (There are many possible solutions to this problem.)

1.59 Repeat Problem 1.58 for a capacity of 0.01 m^3/s at 300 rad/s.

1.60 Design a differential power screw for a linear velocity of approximately 0.1 mm/s when the angular velocity is 65 rad/s.

1.61 Design a differential power screw for a linear velocity of approximately 0.0005 in/s when the screw rotates at 60 rev/min.

1.62 Design a ratchet-pawl mechanism to provide feed rates of 0.005 to 0.012 in/c in increments of 0.001 in/c.

1.63 Repeat Problem 1.62 for feed rates of 1 to 3 mm/c in increments of 100 microns per cycle (μm/c).

1.64 A Hooke-type universal joint has 20° misalignment. Find output shaft speed range if the input shaft rotates at a constant 1000 rev/min.

1.65 Repeat Problem 1.64 for 15° misalignment.

1.66 Find permissible misalignment of a Hooke-type universal joint if speed variation is limited to ±2 percent.

1.67 Repeat Problem 1.66 for ±3 percent variation.

Problems 1.68 Through 1.75 Are Based on a Constant-Speed Crank

It is suggested that a drag link mechanism be incorporated in the design.

1.68 Design a mechanism with a 10-in stroke and a forward-to-return stroke time ratio of 2 to 1 (approximately).

1.69 Repeat Problem 1.68 for a time ratio of 2.5 to 1 (approximately).

1.70 Design a mechanism with a 150-mm stroke and a forward-to-return stroke time ratio of approximately 2 to 1.

1.71 Design a mechanism with a 100-mm stroke and a forward-to-return time ratio of approximately 2 to 1.

1.72 Design a mechanism with a stroke length that may be varied between 5 and 10 in. Forward-to-return stroke time ratio will be 1.5 to 1 at maximum stroke. Utilize a sliding contact linkage.

1.73 Repeat Problem 1.72 for a time ratio of 2.5 to 1.

1.74 Repeat Problem 1.72 for a stroke of 100 to 200 mm.

1.75 Repeat Problem 1.72 for a stroke of 180 to 280 mm.

1.76 Describe the motion of each link in Figure P1.3. Show the linkage in its limiting positions (corresponding to extreme positions of the slider). Measure the angle through which link 1 turns as the slider moves from extreme left to extreme right. Compare this with the corresponding angle as the slider moves to the left. Find the stroke of the slider (distance between limiting positions).

1.77 Repeat Problem 1.76 for Figure P1.4.

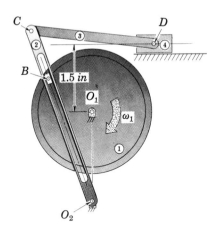

$O_1O_2 = 2$ in
$O_1B = 1.25$ in
$O_2C = 4$ in
$CD = 3$ in
Pin B is part
of link 1

Figure P1.3

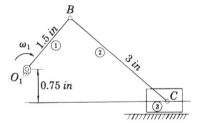

Figure P1.4

1.78 Repeat Problem 1.77 for $O_1B = 60$ mm, $BC = 120$ mm, and offset (O_1 to slider path) $= 20$ mm.

Problems 1.79 Through 1.83 Refer to Figures 1.18a, 1.18b, and 1.18c

1.79 $R = 2$, $L = 5$, and $E = 0.4$.
(a) Find the forward-to-return stroke ratio.
(b) Find the stroke length S.

1.80 Repeat Problem 1.79 for offset $E = 0.8$.

1.81 Repeat Problem 1.79 for offset $E = 0.65$.

1.82 $R = 1$, $L = 3$, and $E = 1$. Plot the path of the midpoint of the connecting rod.

1.83 $R = 150$ mm, $L = 450$ mm, $E = 150$ mm. Plot the path of the midpoint of the connecting rod.

1.84 Find the number of degrees of freedom of the lamination drive in Figure 1.37, when the control linkage position is given.

Problems 1.85 Through 1.92 Refer to a Linkage That Is to Act as a Crank Rocker Mechanism

Link lengths are fixed link L_0, drive crank L_1, coupler L_2, and driven crank L_3.

1.85 $L_0 = 200$ mm, $L_1 = 50$ mm, $L_3 = 150$ mm; find the theoretical range of L_2.

1.86 $L_0 = 210$ mm, $L_1 = 50$ mm, $L_3 = 150$ mm; find the theoretical range of L_2.

1.87 $L_0 = 200$ mm, $L_1 = 45$ mm, $L_3 = 150$ mm; find the theoretical range of L_2.

1.88 $L_0 = 200$ mm, $L_1 = 50$ mm, $L_3 = 160$ mm; find the theoretical range of L_2.

1.89 (a) Find L_2 so that minimum transmission angle $\phi_{min} = 45°$ for the data of Problem 1.85.
(b) Find the maximum transmission angle for this linkage.

1.90 (a) Find L_2 so that minimum transmission angle $\phi_{min} = 45°$ for the data of Problem 1.86.
(b) Find the maximum transmission angle for this linkage.

1.91 (a) Find L_2 so that minimum transmission angle $\phi_{min} = 45°$ for the data of Problem 1.87.
(b) Find the maximum transmission angle for this linkage.

1.92 (a) Find L_2 so that minimum transmission angle $\phi_{min} = 45°$ for the data of Problem 1.88.
(b) Find the maximum transmission angle for this linkage.

1.93 Find average piston velocity in each direction between limiting positions for a slider-crank linkage with 150-mm crank length, 350-mm connecting rod length, and 100-mm offset. The crank rotates at 240 rad/s cw.

1.94 Referring to Figure 1.6a, let spherical pair S_1 be replaced by a cylindrical pair. Find the number of degrees of freedom.

1.95 Referring to Figure 1.6a, let revolute joint R_1 be replaced by a spherical (ball) joint. Find the number of degrees of freedom.

1.96 How many degrees of freedom will a four-bar spatial CCCC linkage have?

1.97 How many degrees of freedom will a four-bar spatial SSSS linkage have?

1.98 Identify five or more spatial four-bar linkages having one degree of freedom. Select linkages that include revolute, prism, helix, cylinder, and sphere joints.

1.99 Show as many one-degree-of-freedom, planar, pin-connected linkage configurations as you can. Use up to eight links.

1.100 Consider an offset slider-crank linkage for which the ratio of connecting rod to crank length is to be 3.
   (a) Write a computer program utilizing the Newton-Raphson method or another numerical method to determine eccentricity for a given value of stroke length.
   (b) Find the eccentricity for which the ratio of stroke to crank length will be 2.5.

1.101 Consider an offset slider-crank linkage for which the ratio of connecting rod to crank length is to be 3.
   (a) Write a computer program utilizing the Newton-Raphson method or another numerical method to determine eccentricity for a given value of stroke length.
   (b) Find the eccentricity for which the ratio of stroke to crank length will be 2.2.

1.102 Consider an offset slider-crank linkage for which the ratio of connecting rod to crank length is to be 3. Use the Newton-Raphson method or another numerical method to determine eccentricity for which the ratio of stroke to crank length will be 3.

PROJECTS

Project topics will often be suggested by the instructor's research interests and by current publications. The following topics may also be used as projects.

1.1 Aircraft landing gear sometimes fail to lower into position.
   (a) Consider a system that will remedy this problem.
   (b) Investigate the feasibility of a redundant landing gear system that can be deployed if the primary system fails to operate.

1.2 Investigate the possibility of an innovative system to transport skiers to the top of a slope. Try to avoid conventional chair lifts, T-bars, rope tows, gondolas, etc.

1.3 Operators of power tools are sometimes injured because of inadequate guarding or removal of guards that prevent efficient use of the tool. Design a system to feed wood into a circular saw in such a way that the operator's hands cannot contact the blade.

1.4 Back-country skiing and Telemark skiing combine aspects of both alpine and Nordic skiing. Back-country skiers utilize free-heel bindings on skis with metal edges. Some users of this equipment ski on steep slopes.
   (a) Consider the design of an innovative release binding for free-heel skiing.
   (b) Design a ski brake to stop a released ski.

1.5 Aircraft accidents have been attributed to an improperly latched cargo-hold door. In one or more instances, this problem resulted in loss of pressure in the cargo hold. The then-greater cabin pressure caused the floor between the cabin and cargo hold to compress. This, in turn, pinched cables needed to steer and control the aircraft. Investigate a new cargo-hold door and latch design.

1.6 Environmental concerns make recycling a necessity. Investigate the design of a system to sort glass and plastic bottles, as well as steel and aluminum cans.

Suggestion: Projects may be assigned to an individual or a group, depending on the instructor's goals. Most mechanical devices in current use are the result of many person-years of effort by experienced engineers. However, student creativity may be stimulated by the demands of the task. It is expected that the mechanical engineering aspects of the design will be given priority. Detailed analysis may be limited to one aspect of the design project, if the scope of the project is too large for the time available. The degree of success can be measured by the quality of innovative thinking, analysis, and interpretation rather than by comparing a proposed design with a commercially available product. As the project progresses, it is expected that it will be necessary to consult several sections in this text as well as other information sources.

Motion in Machinery

Some machine elements are constrained to rectilinear motion and some to planar motion; others undergo more general spatial or three-dimensional motion. Both graphical and analytical methods are used to analyze linkage motion. Methods of vector analysis and complex number methods are important analytical tools used for mechanism analysis and mechanism synthesis. When velocities and accelerations in a linkage are to be determined for a single linkage position, graphical methods are often most convenient. Analytical methods are practical when a sequence of positions of a mechanism must be analyzed. The use of a computer permits a detailed study of a full cycle of motion. Once the initial programming is completed, little effort is required to examine the effect of design changes. On the other hand, if we were to use graphical methods, each linkage position would require a separate plot and each change in length of a link would require a new sequence of plots.

2.1 MOTION

When the motion of the elements of a linkage is *not* restricted to a single plane or to a set of parallel planes, that linkage undergoes spatial motion. Three independent coordinates are required to specify the location of a point

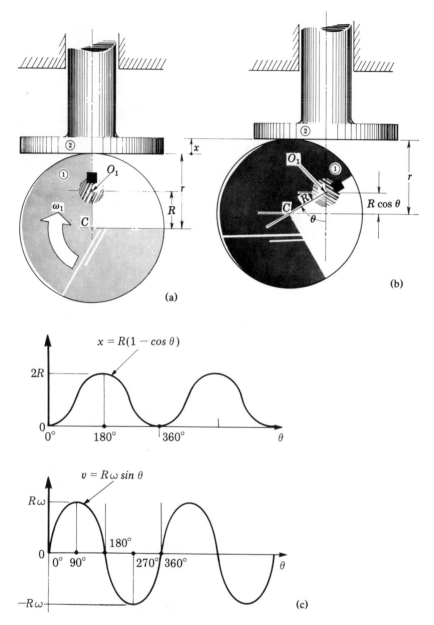

Figure 2.1 (a) The cam, link 1, is formed by a circular disk of radius r and eccentricity R. (b) The follower, link 2, undergoes displacement x as the cam rotates through an angle θ. (c) The displacement and velocity of the follower are plotted against cam rotation.

in spatial motion. A rigid body in spatial motion has six degrees of freedom subject to restrictions imposed by joints. Spatial or three-dimensional linkages often include joints with two or three degrees of freedom, such as cylinder pairs and sphere pairs. A spatial linkage would be required if we were to attempt to duplicate the motion of the human arm.

If the motion of all points in a linkage system is restricted to a plane or to a set of parallel planes, then the motion is *planar*. A point in planar motion is located by two independent coordinates. An unconstrained rigid link has three degrees of freedom in planar motion. We may, for example, identify the x and y coordinates of a point on the link and the angular position of the link centerline. Planar motion may be characterized by two-dimensional vectors. Plane mechanisms, a special case of the more general spatial mechanisms, are of particular interest. They include major components of the internal-combustion engine, spur gear trains, and most cams, as well as a variety of other mechanisms.

The plane motion of a rigid link may be pure translation (also called rectilinear motion), in which case all points in the link move in the same direction at the same speed. For example, the cam follower in Figure 2.1 undergoes rectilinear motion (i.e., translation along a straight line). Translation of a rigid body in general implies motion in space such that any line connecting two points in the body remains parallel to its original position.

In another special case, the plane motion of a body is described by pure rotation, in which case a point in the link is fixed (e.g., a cam). Oscillation refers either to a back-and-forth rotation (e.g., the motion of a pendulum) or a back-and-forth translation (the motion of a piston). In the study of mechanisms, oscillation is commonly used in the first sense, and the motion of a piston is described instead as reciprocating motion. In every case, the meaning should be clear in context. Rectilinear motion and rotation of a rigid body about a point may be described by one independent variable (e.g., x and θ, respectively).

2.1.1 Examples of Rectilinear Motion: The Eccentric Cam and Scotch Yoke

The eccentric cam with flat-face follower and the Scotch yoke are examples of mechanisms having simple mathematical representations. The *eccentric cam* of Figure 2.1a is circular in form, but the center of the circle, C, is offset a distance R from the center of the camshaft O_1. For circle radius r, the distance from camshaft center to follower face is $r - R$ when the follower is at its lowest position. After the cam turns through an angle θ (see Figure 2.1b), the distance becomes $r - R \cos \theta$. Therefore, follower displacement during the interval is

$$x = r - R \cos \theta - (r - R) = R(1 - \cos \theta)$$

for any value of θ, where x is measured from the lowest position.

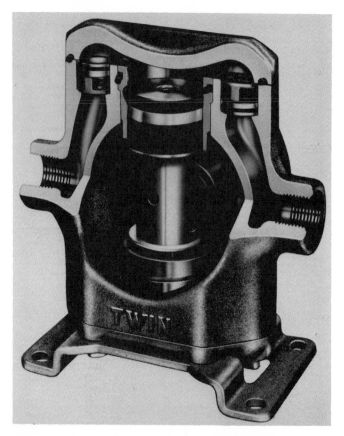

Figure 2.2 A two-cylinder cam-type piston pump. An eccentric cam (in the form of a sealed-roller-type bearing) transmits power from crankshaft to follower (the piston drive). Piston location and velocity are the same as given for the eccentric cam except that the roller bearing eliminates sliding. (*Source:* Hypro, a division of Lear Siegler, Inc.)

If this expression is differentiated with respect to time, we obtain follower velocity

$$v = R\omega \sin \theta$$

where angular velocity $\omega = d\theta/dt$ (rad/s). For constant angular velocity, follower acceleration is

$$a = R\omega^2 \cos \theta$$

Displacement and velocity are plotted in Figure 2.1c for constant ω. Note that velocity is proportional to eccentricity R but independent of cam circle radius r.

In the two-cylinder piston pump of Figure 2.2, the cam raises and lowers a shaft with a piston at each end by acting alternately on two separate follower

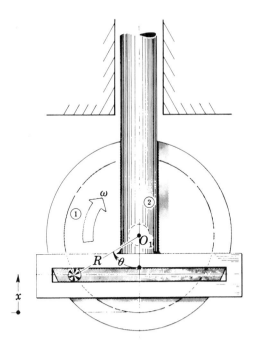

Figure 2.3 A Scotch yoke is kinematically equivalent to the eccentric cam.

faces. The cam pushes against the upper follower face, lifting the shaft (upper piston) through 180° of cam rotation. Through the remaining 180° of cam rotation the cam acts against the lower follower face, forcing the shaft (lower piston) downward.

The *Scotch yoke* (see Figure 2.3) is kinematically equivalent to the eccentric cam considered above. In this case, link 1, the driver, has a pin on which the slotted follower, link 2, rides. The pin is a distance R from O_1, the axis of the driver. Measuring from the lowest position of the follower, displacement, velocity, and acceleration are given by the equations that were used to describe the eccentric cam.

2.2 VECTORS

Vectors are an important part of the language of mechanism and the other branches of mechanics. They provide us with a graphical and analytical means to represent motion or force. A quantity described by its magnitude, direction, and sense can be considered a vector, and can be represented by an arrow. Consider a vector representing a point on a piston that is constrained to move vertically. The vector direction is vertical; with further information, we may determine the vector sense (upward or downward) and vector magnitude.

Graphical and analytical vector methods may be applied to linear displacements, velocities, accelerations, and forces, and to torques and angular velocities and accelerations. Although finite angular displacements possess magnitude and direction, they are not generally considered vectors because they do not follow the rules of vector addition. Vectors are usually identified by **boldface** type to distinguish them from scalar quantities. A line above or below the letter symbol may be used as an alternative identification for a vector.

Some rules for vector addition and multiplication by a scalar follow, where A, B, and C are vectors and m is a scalar quantity.

$$A + B = B + A \quad \text{(commutative law for addition)}$$
$$A + (B + C) = (A + B) + C \text{ (associative law for addition)}$$
$$mA = Am \text{ (commutative law for multiplication by a scalar)} \quad \text{(2.1)}$$
$$m(A + B) = mA + mB \text{ (distributive law for multiplication by a scalar)}$$

As noted above, finite angular displacements are not generally treated as vectors (or scalars). Consider the motion of an aircraft, denoting yaw by the angle θ and pitch by the angle ϕ, where both θ and ϕ are moving coordinates, referred to the axes of the aircraft. Let the aircraft make a 90° right turn ($\theta = 90°$ cw) and then pitch downward by 90° ($\phi = 90°$). Using any rigid body to represent the aircraft, we see that if the order of the above maneuvers is reversed, then the result is different. Thus,

$$\theta + \phi \neq \phi + \theta$$

showing that finite rotations about nonparallel axes do not follow the commutative law for addition.

2.2.1 Unit Vectors

In general, a vector of unit magnitude can be called a unit vector. Thus, $A^u = A/A$ is a unit vector, where $A = |A|$ is the magnitude of vector A. Unit vectors i, j, k (or I, J, K) parallel to the x, y, z (or X, Y, Z) coordinate axes, respectively, are particularly useful. These unit vectors are also called rectangular unit vectors. A right-hand system of mutually perpendicular coordinates is shown in Figure 2.4. A vector may be described in terms of its components along each coordinate axis.

2.2.2 Vector Components

Let the location of a point in space be described by a vector extending from the origin of a coordinate system to the point. Point P in Figure 2.4 is located

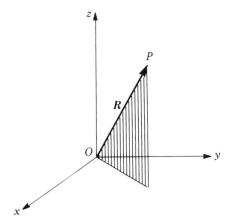

Figure 2.4 Vector R locates the position of point P in the xyz coordinate system.

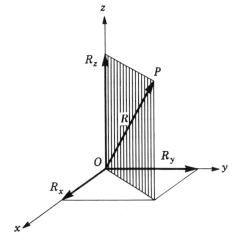

Figure 2.5 The vector R can be resolved into vector components along the x, y, and z axes.

by the vector R. The x, y, and z coordinate axes in that figure are mutually perpendicular. Any motion of P will result in a change in the vector R, either in its magnitude or direction or both.

A line through P perpendicular to the x axis intersects the x axis at a distance R_x from the origin O. (See Figure 2.5). The distance R_x is called the *projection* of the vector R on the x axis, or the x component of R. The projections of R on the y axis and z axis are labeled R_y and R_z. Vectors i, j, and k are *unit vectors* in the x, y, and z directions, respectively, as shown in Figure 2.6. That is, each has unit length and is used only to assign a direction. The scalar length R_x multiplied by the unit vector i gives us the vector iR_x of length R_x parallel to the x axis. Forming the vector sum of the three components times their corresponding unit vectors, we obtain the original vector:

$$R = iR_x + jR_y + kR_z \tag{2.2}$$

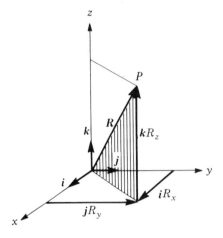

Figure 2.6 Each component vector can be considered the product of a unit vector times the scalar magnitude of the component.

2.2.3 Vector Equations

If two vectors are equal, then each component of the first is equal to the corresponding component of the second. Let

$$A = iA_x + jA_y + kA_z$$

$$B = iB_x + jB_y + kB_z$$

$$A = B \quad \text{implies that}$$

$$A_x = B_x \qquad A_y = B_y \qquad \text{and} \qquad A_z = B_z$$

2.2.4 Vector Addition and Subtraction

Addition of vectors simply involves adding the x-, y-, and z-direction components (i.e., the i, j, and k components) individually. Let

$$C = A + B$$

where vectors A, B, and C are represented in terms of their scalar components and rectangular unit vectors. Then

$$C = i(A_x + B_x) + j(A_y + B_y) + k(A_z + B_z)$$

Vector subtraction is similar. If

$$D = A - B$$

then

$$D = i(A_x - B_x) + j(A_y - B_y) + k(A_z - B_z)$$

Graphical Addition and Subtraction of Planar Vectors

Vectors may be added graphically by joining them head to tail. Although graphical procedures may be used to treat both planar and spatial linkages, graphical solutions are most commonly used with planar problems. Consider the vector equation

$$A + B + D + E = F$$

where all vectors lie in the same plane as in Figure 2.7a. Beginning at an arbitrary point o, we draw vector A, then successively add vectors B, D, and E, the tail of each added vector beginning at the head of the vector drawn last.

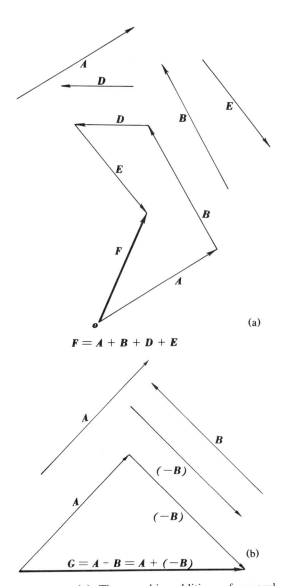

(a)

$$F = A + B + D + E$$

(b)

$$G = A - B = A + (-B)$$

Figure 2.7 (a) The graphic addition of several vectors. (b) Vector subtraction. Simply *reverse* the sense of the vector being subtracted and add it like any other vector.

The vector sum is given by the vector F with tail at o and head drawn to the head of the last vector of the series to be added, vector E. The reader may verify that the addition of vectors is commutative; that is, the above vectors may be added in any order to obtain the same result: $A + B = B + A$, and so on.

Vector subtraction is sometimes required when we consider relative motion. Thus, if the vector G is given by vector A minus vector B, we write

$$G = A - B \quad \text{or} \quad G = A + (-B)$$

where vector $(-B)$ is identical with vector B except that the sense is reversed as shown in Figure 2.7b. The second form of the expression for the difference between two vectors is preferred, particularly when many vectors are to be combined.

2.2.5 The Vector Cross Product

In addition to the product of a scalar and a vector, two types of products involving vectors are defined: the vector or cross product and the scalar or dot product. The *vector cross product*, or simply *cross product*, of two vectors is a vector perpendicular to the plane of the said vectors. The cross product of vectors A and B separated by angle θ is $A \times B = C$, where the magnitude is $C = AB \sin \theta$ and the direction of C is given by the right-hand rule, as follows: the thumb and index finger of the right hand are extented in the direction of vectors A and B, respectively. The middle finger is bent perpendicular to A and B; it points in the direction of vector C. $A \times B$ is read "A cross B." Thus $A \times A = 0$.

Note that the cross product does not follow the commutative law. Interchanging the order of the vectors changes the sign of the vector product (i.e., $B \times A = -A \times B$). It can be seen that the vector products of unit vectors i, j, and k are as follows:

$$\begin{aligned}
i \times j &= k & j \times i &= -k \\
j \times k &= i & k \times j &= -i \\
k \times i &= j & i \times k &= -j \\
i \times i &= j \times j &= k \times k = 0
\end{aligned} \tag{2.3}$$

Vectors A and B may be expressed in terms of their scalar components and unit vectors as

$$A = iA_x + jA_y + kA_z$$

$$B = iB_x + jB_y + kB_z$$

Then

$$A \times B = i(A_y B_z - A_z B_y) + j(A_z B_x - A_x B_z) + k(A_x B_y - A_y B_x)$$

which may be written more concisely in determinant form as

$$A \times B = \begin{vmatrix} i & j & k \\ A_x & A_y & A_z \\ B_x & B_y & B_z \end{vmatrix} \tag{2.4}$$

One application of the vector cross product is the velocity of a point on a link rotating about a fixed center at angular velocity

$$\omega = i\omega_x + j\omega_y + k\omega_z$$

If the vector from the fixed center to the point in question is given by

$$r = ir_x + jr_y + kr_z$$

then the velocity of the point is given by

$$v = \omega \times r = \begin{vmatrix} i & j & k \\ \omega_x & \omega_y & \omega_z \\ r_x & r_y & r_z \end{vmatrix} \tag{2.5}$$

◆ **EXAMPLE PROBLEM 2.1** *Rotating Link*

Suppose a link that is held in a ball joint (spherical pair) at one end rotates with an instantaneous angular velocity

$$\omega = i2 + j(-1) + k4 \ (\text{rad}/\text{s})$$

Find the instantaneous velocity of P, a point on the link defined by the radius vector

$$r = i(-1) + j10 + k2 \ (\text{mm})$$

measured from the ball joint (see Figure 2.8a).

Solution. The velocity of the point is given by

$$v = \omega \times r = \begin{vmatrix} i & j & k \\ \omega_x & \omega_y & \omega_z \\ r_x & r_y & r_z \end{vmatrix} = \begin{vmatrix} i & j & k \\ 2 & -1 & 4 \\ -1 & 10 & 2 \end{vmatrix}$$

$$= i(-2 - 40) + j(-4 - 4) + k(20 - 1)$$

$$= i(-42) + j(-8) + k19 \ (\text{mm}/\text{s})$$

The velocity is perpendicular to the plane of ω and r; its magnitude is given by the sum of the squares of the velocity components

$$v = \sqrt{v_x^2 + v_y^2 + v_z^2} = \sqrt{(-42)^2 + (-8)^2 + 19^2} = 46.79 \ \text{mm}/\text{s} \quad ◆$$

Another application of the vector cross product involves torque. For example, in Figure 2.8b let force F be applied to a rigid link. The resultant

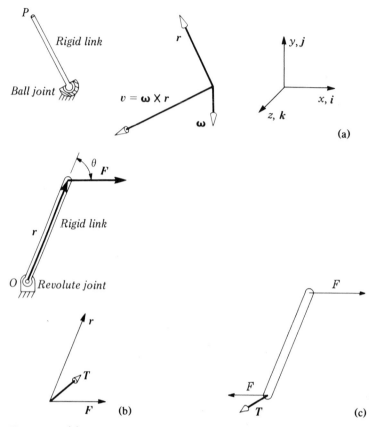

Figure 2.8 (a) Velocity of a point on a link in terms of the cross product. (b) The torque of force F about point O is $T = r \times F$. (c) A link in equilibrium.

torque of force F about point O may be represented by the vector product

$$T = r \times F = \begin{vmatrix} i & j & k \\ r_x & r_y & r_z \\ F_x & F_y & F_z \end{vmatrix}$$ (2.6)

The magnitude of T is $rF \sin \theta$, and the direction is perpendicular to the plane of r and F, as shown in the figure. Note that applying the right-hand rule to $T = r \times F$ gives the direction of the resultant torque. If both r and F lie in the xy plane, then T is represented by a vector in the $\pm z$ direction. The link could be in static equilibrium if a force and torque were applied as in Figure 2.8c.

2.2.6 The Dot or Scalar Product

The dot product of two vectors is a scalar equal to the product of the vector magnitude times the cosine of the angle between them. The dot product of vectors A and B separated by angle θ is given by

$$A \cdot B = AB \cos \theta \tag{2.7}$$

Thus,

$$i \cdot i = j \cdot j = k \cdot k = 1$$
$$i \cdot j = j \cdot k = k \cdot i = 0 \tag{2.8}$$

Let A and B be expressed in terms of their scalar components and unit vectors as

$$A = iA_x + jA_y + kA_z$$
$$B = iB_x + jB_y + kB_z$$

Then

$$A \cdot A = A_x^2 + A_y^2 + A_z^2 = A^2 \tag{2.9}$$
$$A \cdot B = A_x B_x + A_y B_y + A_z B_z \tag{2.10}$$

The dot product of two perpendicular vectors is zero. Thus, since the cross product of two vectors is perpendicular to the plane of the two vectors, we have

$$A \cdot (A \times B) = B \cdot (A \times B) = 0$$

This relationship may be used as a partial check on the calculation of velocity of a point on a rigid link.

◆ EXAMPLE PROBLEM 2.2   *The Dot Product*
Use the dot product to check the results of a previous example in which:

$$\omega = i2 + j(-1) + k4$$
$$r = i(-1) + j10 + k2$$
$$v = \omega \times r = i(-42) + j(-8) + k19$$

Solution.

$$\omega \cdot v = (2)(-42) + (-1)(-8) + (4)(19) = 0$$
$$r \cdot v = (-1)(-42) + (10)(-8) + (2)(19) = 0$$

Thus, v is shown to be perpendicular to the plane of ω and r.   ◆

As another example of an application of the dot product, consider a spatial linkage where we require that the rotation of a certain link about its own axis be zero. Designating the link by the vector

$$r = ir_x + jr_y + kr_z$$

and its angular velocity

$$\boldsymbol{\omega} = i\omega_x + j\omega_y + k\omega_z$$

the requirement will be satisfied if $\boldsymbol{\omega} \cdot \boldsymbol{r} = 0$, from which

$$(i\omega_x + j\omega_y + k\omega_z) \cdot (ir_x + jr_y + kr_z) = \omega_x r_x + \omega_y r_y + \omega_z r_z = 0$$

Some additional laws relating to the dot product and vector cross product are noted below.

$$A \cdot B = B \cdot A \qquad \text{(the commutative law holds for the dot product, but not the cross product)}$$

$$A \cdot (B + C) = A \cdot B + A \cdot C$$
$$A \times (B + C) = A \times B + A \times C \qquad \text{(distributive law)} \tag{2.11}$$

$$A \cdot (B \times C) = B \cdot (C \times A) = C \cdot (A \times B) = \begin{vmatrix} A_x & A_y & A_z \\ B_x & B_y & B_z \\ C_x & C_y & C_z \end{vmatrix}$$

$$A \times (B \times C) = (A \cdot C)B - (A \cdot B)C$$
$$(A \times B) \times C = (A \cdot C)B - (B \cdot C)A$$

When solving kinematics and dynamics problems, it is often convenient to use mathematics software with built-in vector functions. Some types of software (e.g. MathCAD™) express vectors in column form:

$$A = \begin{bmatrix} A_x \\ A_y \\ A_z \end{bmatrix}$$

instead of the form $iA_x + jA_y + kA_z$

◆ **EXAMPLE PROBLEM 2.3**   *Vector Operations*
The following vectors are given (in column format):

$$A = \begin{bmatrix} 4 \\ 3 \\ 1 \end{bmatrix} \qquad B = \begin{bmatrix} 2 \\ -1 \\ -5 \end{bmatrix} \qquad C = \begin{bmatrix} 16 \\ -8 \\ -10 \end{bmatrix} \qquad D = \begin{bmatrix} -9 \\ -5 \\ 7 \end{bmatrix}$$

$$E = \begin{bmatrix} 16 \\ -8 \\ -10 \end{bmatrix}$$

Find: $A + B + C + D + E$, $\quad A - B + C - D + E$, $\quad A \cdot B$, $\quad A \cdot C$,
$|A|$, $\quad |B|$, $\quad (|A|)^2$, $\quad A \cdot A$, $\quad A \times A$, $\quad C \times E$, $\quad A \times B$,
$|A \times B|$, $\quad |A| \cdot |B|$, $\quad A \times C$, $\quad C \times A$, $\quad A \cdot (A \times C)$,
$C \cdot (A \times C)$.

Solution. Software with vector capabilities was used to obtain the following results:

$$A + B + C + D + E = \begin{bmatrix} 29 \\ -19 \\ -17 \end{bmatrix}, \quad A - B + C - D + E = \begin{bmatrix} 43 \\ -7 \\ -21 \end{bmatrix},$$

$$A \cdot B = 0, \quad A \cdot C = 30, \quad |A| = 5.099, \quad |B| = 5.477,$$

$$(|A|)^2 = 26,$$

$$A \cdot A = 26, \quad A \times A = \begin{bmatrix} 0 \\ 0 \\ 0 \end{bmatrix}, \quad C \times E = \begin{bmatrix} 0 \\ 0 \\ 0 \end{bmatrix}, \quad A \times B = \begin{bmatrix} -14 \\ 22 \\ -10 \end{bmatrix},$$

$$|A \times B| = 27.928, \quad |A| \cdot |B| = 27.928, \quad A \times C = \begin{bmatrix} -22 \\ 56 \\ -80 \end{bmatrix},$$

$$C \times A = \begin{bmatrix} 22 \\ -56 \\ 80 \end{bmatrix}, \quad A \cdot (A \times C) = 0, \quad C \cdot (A \times C) = 0$$

The result: $A \cdot B = 0$ may be unexpected. Since

$$A \cdot B = AB \cos \theta$$

where θ = the angle separating the vectors, and neither A nor B is zero, then $\cos \theta = 0$. This indicates that vectors A and B are perpendicular to one another. The magnitude of the cross product is given by

$$|A \times B| = AB \sin \theta$$

For this special case, with A and B perpendicular, the cross product equals the product of the magnitudes. The calculations show this.

We note that $A \cdot C = C \cdot A$ and that $A \times C = -C \times A$. These relationships are true in general. Of course, $A \cdot A$, the dot product of a vector with itself, is the square of the magnitude of A. $A \times A$, the cross product of a vector with itself, is always zero. Since vectors C and E are identical, $C \times E$ is zero as well. The cross product $A \times C$ is perpendicular to both vectors A and C. Thus both dot products $A \cdot (A \times C)$ and $C \cdot (A \times C)$ are zero. ◆

2.2.7 Differentiation of a Vector

If vector A varies in magnitude and direction with time, then the derivative of A with respect to time is given by

$$\frac{dA}{dt} = \lim_{\Delta t \to 0} \frac{A(t + \Delta t) - A(t)}{\Delta t} \tag{2.12}$$

where Δt represents an increment in time. Note that dA/dt is a vector and that the numerator of the fraction on the right involves changes in both the magnitude and direction of A.

If A, B, and C are vector functions of time and m is a scalar function of time, then the following rules hold:

$$\frac{d}{dt}(A + B) = \frac{dA}{dt} + \frac{dB}{dt}$$

$$\frac{d}{dt}(A \times B) = A \times \frac{dB}{dt} + \frac{dA}{dt} \times B$$

$$\frac{d}{dt}(A \cdot B) = A \cdot \frac{dB}{dt} + \frac{dA}{dt} \cdot B \qquad (2.13)$$

$$\frac{d}{dt}(mA) = m\frac{dA}{dt} + \frac{dm}{dt}A$$

$$\frac{d}{dt}[A \times (B \times C)] = A \times \left(B \times \frac{dC}{dt}\right) + A \times \left(\frac{dB}{dt} \times C\right) + \frac{dA}{dt} \times (B \times C)$$

Recall that $A \times B = -B \times A$. Thus, the order of the vectors in the vector cross products should be retained when we apply the chain rule of differentiation.

The time differential of a unit vector fixed within a fixed coordinate system is, of course, zero since neither magnitude nor direction changes. The time differential of a unit vector in a moving coordinate system is not zero. In applications of vectors to velocity and acceleration, we will consider both fixed and moving coordinate systems.

2.2.8 Solution of Vector Equations

Let a vector equation in the form

$$r_0 + r_1 + r_2 + r_3 = 0 \qquad (2.14)$$

represent a spatial linkage where the null vector

$$0 = i0 + j0 + k0$$

Henceforth, we simply use a scalar zero to represent the null vector. Each vector may be expressed in terms of its x, y, and z components and the unit vectors i, j, and k such as

$$r_1 = r_{1x}i + r_{1y}j + r_{1z}k$$

and so on. Since the sum of the components in each coordinate direction must equal zero, we have three independent scalar equations:

$$r_{0x} + r_{1x} + r_{2x} + r_{3x} = 0$$

and so on. Thus, in general, the vector equation $r_0 + r_1 + r_2 + r_3 = 0$ may be solved for three unknown components. For example, we may solve for r_{3x}, r_{3y}, and r_{3z} if the other components are known. Alternatively, each vector could be expressed in terms of its magnitude and two independent angular coordinates. Vector Eq. 2.14 could then be solved for any combination of three magnitudes and directions. Planar linkages can be represented by two independent scalar equations. Only two unknown components are permitted if Eq. 2.14 is applied to planar linkages.

2.2.9 Solution of Planar Vector Equations

Consider the planar vector equation

$$A + B + C = 0 \tag{2.15}$$

or in terms of unit vectors (A^u, etc.) and magnitudes (A, etc.),

$$A^u A + B^u B + C^u C = 0 \tag{2.16}$$

If the magnitude and direction of the same vector are unknown, then the solution is easily obtained. If C is unknown, we use

$$C = -(A_x + B_x)i - (A_y + B_y)j \tag{2.17}$$

or

$$C = -(A \cdot i + B \cdot i)i - (A \cdot j + B \cdot j)j \tag{2.18}$$

If the magnitudes of two different vectors are unknown, a vector elimination method suggested by Chace[3] may be used. Suppose, for example, magnitudes $A = |A|$ and $B = |B|$ are unknown in the vector equation $A^u A + B^u B + C^u C = 0$. We take the dot product of each term with $B^u \times k$, noting that $B^u \cdot (B^u \times k) = 0$ since vector B^u is perpendicular to vector $B^u \times k$. Thus, we obtain

$$A^u A \cdot (B^u \times k) + C \cdot (B^u \times k) = 0$$

from which the magnitude of vector A is given by

$$A = \frac{-C \cdot (B^u \times k)}{A^u \cdot (B^u \times k)} \tag{2.19}$$

Similarly, the magnitude of B is given by

$$B = \frac{-C \cdot (A^u \times k)}{B^u \cdot (A^u \times K)} \tag{2.20}$$

If the vector directions A^u and B^u are unknown but all vector magnitudes are known, the solution to the equation $A + B + C = 0$ is more difficult. The results in this case are

$$A = \mp \left[B^2 - \left(\frac{C^2 + B^2 - A^2}{2C} \right)^2 \right]^{1/2} (C^u \times k)$$

$$+ \left[\frac{C^2 + B^2 - A^2}{2C} - C \right] C^u \tag{2.21}$$

$$B = \pm \left[B^2 - \left(\frac{C^2 + B^2 - A^2}{2C} \right)^2 \right]^{1/2} (C^u \times k) - \left[\frac{C^2 + B^2 - A^2}{2C} \right] C^u \tag{2.22}$$

The significance of the signs before the radical are illustrated in the section that follows.

When the magnitude of A and the direction of B are unknown, A and B may be found by the following equations:

$$A = \left[-C \cdot A^u \mp \sqrt{B^2 - [C \cdot (A^u \times k)]^2} \right] A^u \tag{2.23}$$

$$B = -[C \cdot (A^u \times k)](A^u \times k) \pm \sqrt{B^2 - [C \cdot (A^u \times k)]^2} A^u \tag{2.24}$$

The above approach uses vector notation throughout, unlike alternative methods that use vector analysis to derive scalar equations. If the above method is to be used for computer-aided analysis and design of mechanisms, it is desirable to incorporate subroutines to handle vectors.

2.3 DISPLACEMENT ANALYSIS OF PLANAR LINKAGES: ANALYTICAL VECTOR METHODS

If a mechanism is to be examined for only one or two positions, graphical methods of analysis are convenient. For most planar linkages, position analysis requires only that we locate the intersection of plane curves. Referring to Figure 2.9a for example, let the crank position be given. A line is drawn representing the path of the wrist pin. An arc of radius equal to the connecting rod length is drawn with the center at the crankpin. The wrist pin is located at the intersection of the arc and line. For this linkage, there are, in general, two solutions, depending on how the links were assembled. Spatial linkages may also be treated graphically. Methods of descriptive geometry applicable to spatial linkages are illustrated later in this chapter. If many linkage positions are to be analyzed, then analytical methods are preferred.

In Chapter 1, a slider-crank linkage was described analytically by trigonometric functions, and linkage positions were plotted. Alternatively, a closed-loop linkage may be modeled as a closed vector chain, i.e., as a null vector. Unambiguous vector representations are often used as a basis for computer-aided analysis and design. If we are given the position of the crank for the linkage of Figure 2.9b, then the unknowns will be the connecting rod orientation and slider position. These may be found by using the equations of the previous section to solve the vector equation

$$r_0 + r_1 + r_2 = 0 \tag{2.25}$$

♦ **EXAMPLE PROBLEM 2.4**   *Slider-Crank Linkage*
Suppose as in-line slider-crank linkage (see Figure 2.9a) has a crank described by the vector $r_1 = i + j2$ (at the instant shown) and connecting rod length $r_2 = 4$. Find the connecting rod orientation r_2^u and slider position r_0. (See Figure 2.9a and b.)

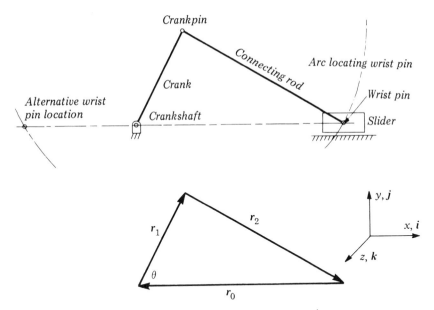

Figure 2.9 (a) Slider-crank linkage. (b) Vector representation.

Solution.   Slider position is given by the vector

$$r_0 = \left[-r_1 \cdot r_0^u \mp \sqrt{r_2^2 - [r_1 \cdot (r_0^u \times k)]^2} \right] r_0^u$$

scalar

$$= \left[-(i + j2) \cdot (-i) \mp \sqrt{4^2 - [(i + j2) \cdot (-i \times k)]^2} \right] (-i)$$

where the sign of the root depends on the initial assembly configuration. The positive root applies to the configuration in the figure, yielding $r_0 = -i4.464$. The alternative wrist pin location is determined by taking the negative root. It is defined by the vector $r_0 = +i2.464$.

The vector representing the connecting rod is

$$r_2 = -[(i + j2) \cdot (-i \times k)](-i \times k)$$

$$\pm \sqrt{4^2 - [(i + j2) \cdot (-i \times k)]^2} (-i)$$

$$= -2j \pm \sqrt{12} (-i)$$

The negative root applies to the configuration shown, yielding

$$r_2 = i3.464 - j2$$

and

$$r_2^u = \frac{(i3.464 - j2)}{4} = i0.866 - j0.5$$

For the alternative wrist pin location, the positive root applies, yielding

$$r_2 = -2j + \sqrt{12}(-i) = -i3.464 - j2$$

and

$$r_2^u = -i0.866 - j0.5 \quad \blacklozenge$$

2.3.1 The Four-Bar Linkage

A graphical layout of a four-bar linkage is easily constructed. We only require that the position of one link be given relative to the frame and that the link lengths be known. Then, the linkage may be drawn with the aid of a compass. It can be quite time-consuming, however, to develop the analytical formulas for link positions that are to be used to write a computer program.

2.3.2 Position Analysis Using the Vector Cross Product

Eqs. 2.21 and 2.22 may be used to find linkage displacements. These equations apply when the directions of vectors A and B are unknown. The four-bar planar linkage of Figure 2.10a is described by the vector equation

$$r_0 + r_1 + r_2 + r_3 = 0 \qquad (2.26)$$

Suppose we have already decided the length of each of the links in a tentative design, where link 1 is to be the driver. For a given angular position of link 1, the diagonal vector (also called the **prime diagonal**) is given by

$$r_d = r_0 + r_1$$

and the triangle formed by links 2 and 3 and the diagonal is described by

$$r_d = -(r_2 + r_3) \qquad (2.27)$$

As the lengths of the links are specified and the orientation of link 1 is given, then the following substitution may be made in Eqs. 2.21 and 2.22:

$$A = r_2$$
$$B = r_3$$
$$C = r_d = r_0 + r_1$$

yielding

$$r_2 = \mp \sqrt{r_3^2 - \left(\frac{r_3^2 - r_2^2 + r_d^2}{2r_d}\right)^2} \, (r_d^u \times k) + \left(\frac{r_3^2 - r_2^2 + r_d^2}{2r_d} - r_d\right) r_d^u \qquad (2.28)$$

$$r_3 = \pm \sqrt{r_3^2 - \left(\frac{r_3^2 - r_2^2 + r_d^2}{2r_d}\right)^2} \, (r_d^u \times k) - \left(\frac{r_3^2 - r_2^2 + r_d^2}{2r_d}\right) r_d^u \qquad (2.29)$$

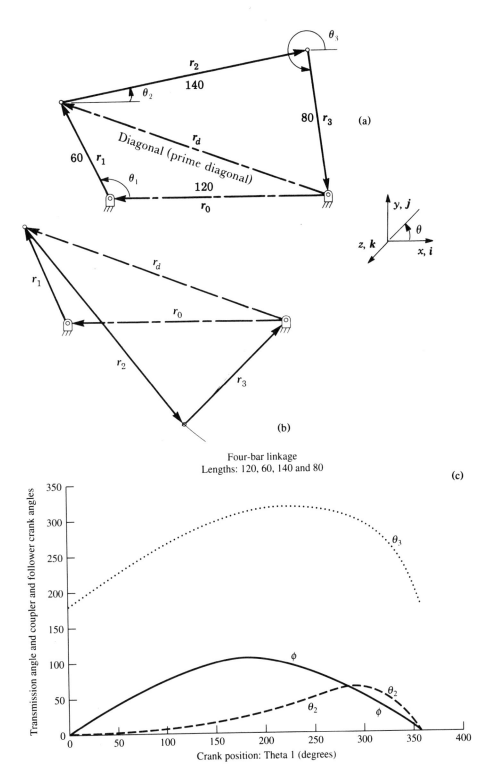

Figure 2.10 (a) Four-bar planar linkage. (b) Four-bar linkage (alternative mode of assembly). (c) Coupler angle θ_2, follower crank angle θ_3, and transmission angle ϕ plotted against crank angle θ_1.

If the linkage is assembled so that the vector loop $r_2 r_3 r_d$ is clockwise, then the lower set of signs in the above equation applies (see Figure 2.10a). If the loop is counterclockwise, the upper set of signs applies (Figure 2.10b).

◆ **EXAMPLE PROBLEM 2.5** *Position Analysis Using the Vector Cross Product*
A planar mechanism (see Figure 2.10a) has the following link lengths:

> Link 0, fixed: 120 mm
>
> Link 1, crank: 60 mm
>
> Link 2, coupler: 140 mm
>
> Link 3, follower: 80 mm

(a) Find the orientation of links 2 and 3 at the instant when the internal angle between the crank and fixed link is 116.56°. (b) Plot θ_2, θ_3, and transmission angle against θ_1.

Solution. (a) If the links are drawn to scale, the joint between links 2 and 3 can be located by using a compass. Alternatively, an analytical vector solution may be obtained as follows. Select the coordinate system so that the fixed link lies in the $-x$ direction. (as in Figure 2.10a.) Then, the known values are

$$r_0 = -i120$$

$$r_1 = 60(i \cos 116.56° + j \sin 116.56°) = -i26.83 + j53.67$$

$$r_2 = 140$$

$$r_3 = 80$$

The diagonal is given by the vector

$$r_d = (r_0 + r_1) = -i146.83 + j53.67$$

which has a magnitude $r_d^{\text{scalar}} = 156.34$ and unit vector

$$r_d^u = -i0.9392 + j0.3433$$

The direction of the cross product $r_d^u \times k$ is given by the right-hand rule. It lies in the xy plane and is perpendicular to r_d^u. Expressing the cross product in determinant form, we have

$$r_d^u \times k = \begin{vmatrix} i & j & k \\ -0.9392 & 0.3433 & 0 \\ 0 & 0 & 1 \end{vmatrix} = i0.3433 + j0.9392$$

Using Eqs. 2.28 and 2.29, we define

$$a = \frac{r_3^2 - r_2^2 + r_d^2}{2r_d} = \frac{80^2 - 140^2 + 156.324^2}{2 \times 156.34} = 35.95$$

When the vector loop $r_2 r_3 r_d$ is clockwise, the lower set of signs in Eqs. 2.28 and 2.29 applies. The vector representing the coupler is given by

$$r_2 = +\sqrt{r_3^2 - a^2}\,(r_d^u \times k) + r_d^u(a - r_d)$$

$$= \sqrt{80^2 - 35.95^2}\,(i0.3433 + j0.9392)$$

$$+ (35.95 - 156.34)(-i0.9392 + j0.3433)$$

$$= i137.60 + j25.79 = 140\angle 10.6°$$

The vector representing the follower is given by

$$r_3 = -\sqrt{r_3^2 - a^2}\,(r_d^u \times k) - ar_d^u$$

$$= -\sqrt{80^2 - 35.95^2}\,(i0.3433 + j0.9392)$$

$$- 39.95(-i0.9392 + j0.3433)$$

$$= i9.24 - j79.46 = 80\angle - 83.4°$$

(Note that the magnitudes of vectors r_2 and r_3 agree with the given data.)

In some linkage configurations, it is impossible to go from one mode to another without disassembling the links. Applying Grashof-type criteria to check this linkage (see Chapter 1), we find

$$L_{max} + L_{min} = L_a + L_b$$

$$(140 + 60 = 120 + 80)$$

This is a crossover-position or change-point mechanism. The mechanism my go from one mode to another, depending on inertia forces, spring forces, or other forces. Using Eqs. 2.28 and 2.29 with the upper set of signs, for the position where r_2, r_3, and r_d form a counterclockwise loop, we obtain the vector representing the coupler:

$$r_2 = i88.54 - j108.44 = 140\angle - 50.8°$$

The vector representing the follower is

$$r_3 = i58.30 + j54.68 = 80\angle 43°$$

(b) Figure 2.10c shows coupler angle θ_2, follower crank angle θ_3, and transmission angle ϕ plotted against crank angle θ_1. Note that $\phi = 0$ at $\theta_1 = 0$. If this linkage was stopped in the $\theta_1 = 0$ position, it could not be restarted by driving link 1. Special consideration has to be given to problems of driving change-point linkages. ◆

2.3.3 Position Analysis Using the Dot Product

As an alternative to using the vector cross product, we may analyze linkage displacement by using the dot product. Referring to Figure 2.11, a vector

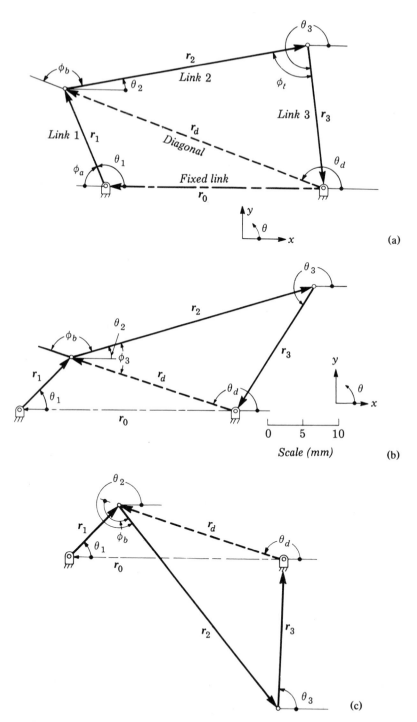

Figure 2.11 (a) Four-bar planar linkage; position analysis by dot product method. (b) Position analysis example problem. (c) Alternate assembly mode.

representation of a planar four-bar linkage, the diagonal vector is given by the vector loop equation:

$$r_d = r_0 + r_1 \tag{2.30}$$

Suppose we are to determine linkage position for a given angle θ_1, if all of the link lengths are known. Taking the dot product of each side of the above equation with itself, we have

$$r_d \cdot r_d = (r_0 + r_1) \cdot (r_0 + r_1) \tag{2.31}$$

from which

$$r_d^2 = r_0^2 + 2r_0 r_1 \cos \phi_a + r_1^2 \tag{2.32}$$

where the angle between vectors r_0 and r_1 is $\phi_a = 180° - \theta_1$. Thus, Eq. 2.32 is equivalent to the cosine law:

$$r_d^2 = r_0^2 - 2r_0 r_1 \cos \theta_1 + r_1^2$$

The direction of the diagonal is given in terms of the x and y components of r_0 and r_1:

$$\tan \theta_d = \frac{r_{0y} + r_{1y}}{r_{0x} + r_{1x}} \tag{2.33}$$

where $r_{0y} = 0$ if the x axis is selected parallel to the fixed link. For some four-bar linkage proportions, θ_d can be in any of the four quadrants. If this equation is programmed for machine calculations, it is important that θ_d be located in its proper quadrant. This may be possible if a two-argument arctangent function is available wherein both numerator and denominator become inputs and the quadrant of the angle is determined by the signs of the numerator and denominator. As an alternative, we note that

$$\sin \theta_d = \frac{r_{0y} + r_{1y}}{r_d} \tag{2.34}$$

$$\cos \theta_d = \frac{r_{0x} + r_{1x}}{r_d} \tag{2.35}$$

It can be shown that

$$\tan\left(\frac{\theta_d}{2}\right) = \frac{1 - \cos \theta_d}{\sin \theta_d} \tag{2.36}$$

Using the last three expressions instead of Eq. 2.33, we are able to determine θ_d in its correct quadrant.

We continue the analysis with the loop closure equation

$$r_d + r_2 + r_3 = 0 \tag{2.37}$$

from which

$$-r_3 = r_d + r_2$$

Taking the dot product of each side of this equation with itself, we have

$$r_3 \cdot r_3 = (r_d + r_2) \cdot (r_d + r_2) \tag{2.38}$$

from which

$$r_3^2 = r_d^2 + 2r_d r_2 \cos \phi_b + r_2^2 \tag{2.39}$$

or

$$\cos \phi_b = \frac{r_3^2 - r_2^2 - r_d^2}{2r_d r_2} \qquad \text{where} \qquad 0 \le \phi_b \le 180° \tag{2.40}$$

Then,

$$\theta_2 = \theta_d \mp \phi_b \tag{2.41}$$

The sign before ϕ_b depends on the mode of assembly of the linkage. If the vector loop $r_2 r_3 r_d$ is clockwise, as in Figure 2.11a, the negative sign applies; if counterclockwise, the positive sign applies. In most cases, once the linkage is assembled, the mode will not change.

The position of link 3 may now be found by using the dot product in a similar manner, or by using the law of sines. So that errors of angle quadrant are avoided, however, the x and y components of r_3 will be determined from the loop equation for the entire linkage, written as follows:

$$r_3 = -(r_0 + r_1 + r_2) \tag{2.42}$$

Then,

$$\sin \theta_3 = \frac{r_{3y}}{r_3} \tag{2.43a}$$

$$\cos \theta_3 = \frac{r_{3x}}{r_3} \tag{2.43b}$$

and θ_3 is found from the equation

$$\tan\left(\frac{\theta_3}{2}\right) = \frac{1 - \cos \theta_3}{\sin \theta_3} \tag{2.44}$$

Angle ϕ_t is the transmission angle at the instant shown. Extreme values of ϕ_t occur when link 1 and the fixed link are colinear. If the linkage satisfies the crank rocker criteria, then limiting positions of link 3 will occur when links 1 and 2 are colinear. Transmission angle, linkage classification, and limiting positions were discussed in more detail in Chapter 1.

◆ **EXAMPLE PROBLEM 2.6** *Position Analysis Using the Dot Product*
A four-bar linkage has the following link lengths:

Fixed link:	$r_0 = 30$ mm
Drive Crank:	$r_1 = 10$ mm
Coupler:	$r_2 = 35$ mm
Follower:	$r_3 = 20$ mm

Find the position of links 2 and 3 when $\theta_1 = 45°$, as shown in Figure 2.11b.
    This problem can be solved quickly if we simply draw the drive crank and fixed link and then strike arcs of radius r_2 and r_3 to locate the pin joining the coupler and follower. A practical application, however, could require a solution for a series of values of θ_1. Therefore, analytical vector or complex number methods might be selected so that a computer or programmable calculator could be used to generate a large number of solutions. Furthermore, a machine solution would readily allow redesign by adjustment of link lengths.

Solution. We will use the equations developed from the position analysis using the dot product, which could easily be programmed if required. From Eq. 2.32,

$$r_d^2 = 30^2 + 2 \times 30 \times 10 \cos(180° - 45°) + 10^2$$

$$r_d = 23.99 \text{ mm}$$

From Eqs. 2.34 through 2.36,

$$\sin \theta_d = \frac{0 + 10 \sin 45°}{23.99} = 0.2947$$

$$\cos \theta_d = \frac{-30 + 10 \cos 45°}{23.99} = -0.9558$$

$$\tan\left(\frac{\theta_d}{2}\right) = \frac{(1 + 0.9558)}{0.2947}$$

$$\theta_d = 162.86°$$

Note that Eq. 2.33 yields

$$\theta_d = \arctan\frac{0 + 10 \sin 45°}{-30 + 10 \cos 45°} = \arctan(-0.3084)$$

which will usually be evaluated as $-17.14°$. The error would be obvious if we used a sketch but might be undetected in machine calculations. From Eq. 2.40,

$$\cos \phi_b = \frac{20^2 - 35^2 - 23.99^2}{2 \times 23.99 \times 35}$$

$$\phi_b = 146.51°$$

For the assembly mode shown in Figure 2.11b (clockwise vector loop),

$$\theta_2 = \theta_d - \phi_b = 162.86 - 146.51 = 16.35°$$

and for the alternative mode in Figure 2.11c

$$\theta_2 = \theta_d = \phi_b = 162.86 + 146.51 = 309.37°$$

Using the loop closure equation, Eq. 2.42, for the x and y components of r_3, we have, for the assembly mode of Figure 2.11b,

$$r_{3x} = -(30 + 10\cos 45° + 35\cos 16.35°) = -10.66 \text{ mm}$$

$$r_{3y} = -(0 + 10\sin 45° + 35\sin 16.35°) = -16.92 \text{ mm}$$

Then, the position of link 3 is found by using an $xy:r\theta$ conversion, from which $r_3 = 20\angle -122.21°$. If this conversion is not available, then we may use Eqs. 2.43a through 2.44, from which

$$\tan\left(\frac{\theta_3}{2}\right) = \frac{1 + 10.66/20}{-16.92/20}$$

or

$$\theta_3 = -122.2° \ (237.8°)$$

For the alternative mode, Figure 2.11c

$$r_{3x} = -(-30 + 10\cos 45° + 35\cos 309.37°) = 0.73 \text{ mm}$$

$$r_{3y} = -(0 + 10\sin 45° + 35\sin 309.37°) = 19.99 \text{ mm}$$

$$r_3 = 20 \text{ mm at } \theta_3 = 87.91°$$

For a given value of θ_1, the diagonal vector r_d is the same for both assembly modes. Thus, the triangle formed by the diagonal and links 2 and 3 for one mode is congruent to the corresponding triangle for the other (but reflected about r_d).

For the dimensions given in this example, link 1, the drive crank, is shortest, and $L_{max} + L_{min} < L_a + L_b$ (35 + 10 < 30 + 20). Thus, we have a crank rocker linkage as defined in Chapter 1. If it is assembled in one mode (Figure 2.11b), it cannot, without reassembly, assume the other mode.

The vector methods of position analysis using the dot product and cross product are not limited to one type of linkage. The four-bar linkage was only used as an illustration; other configurations can be solved by using the same principles and a bit of ingenuity.

2.4 COMPLEX NUMBERS

Complex numbers, each made up of a real and an imaginary part, provide an alternative representation for vectors that lie in a plane. Once the rules for handling complex numbers have been mastered, it is fairly easy to apply

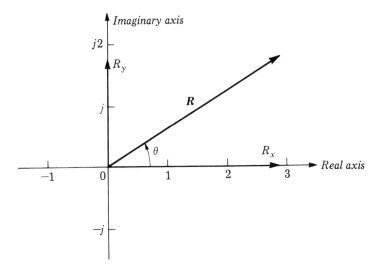

Figure 2.12 A vector in the complex plane.

complex numbers to the analysis of planar linkages. Note that the imaginary quantity is a mathematical artifice. When used in the analysis of mechanisms, the real and imaginary parts of complex numbers represent components of actual dimensions, velocities, and accelerations.

2.4.1 Rectangular Form

A complex number may be written in the rectangular form

$$z = x + jy$$

where j is the imaginary unit defined by $j = \sqrt{-1}$. The term x represents a real number called the real part of the complex number z. The term y represents a real number called the imaginary part of z.

The *complex plane* shown in Figure 2.12 permits graphical representation of vectors as complex numbers. Consider a vector R of magnitude R with components R_x and R_y along the real and imaginary axes, respectively. Using angle $\theta = \arctan(R_y/R_x)$ in the complex plane and magnitude $R = \sqrt{R_x^2 + R_y^2}$, vector R may be identified by its real part $R_x = R \cos \theta$ and its imaginary part $R_y = R \sin \theta$ and written in the form

$$R = R_x + jR_y \tag{2.45}$$

2.4.2 Polar Form

It can be shown that

$$e^{j\theta} = \cos \theta + j \sin \theta \tag{2.46}$$

(an identity called the Euler formula). Using this and Eq. 2.45, the vector can be expressed in *polar form* in terms of its magnitude and the complex exponential:

$$R = R(\cos\theta + j\sin\theta) = Re^{j\theta} \tag{2.47}$$

where θ is expressed in radians.

2.4.3 Complex Arithmetic

The rectangular form of a complex number is convenient for addition and subtraction. For example, if

$$R_1 = R_{1x} + jR_{1y}$$

and

$$R_2 = R_{2x} + jR_{2y}$$

then

$$R_1 + R_2 = (R_{1x} + R_{2x}) + j(R_{1y} + R_{2y}) \tag{2.48}$$

Two complex numbers are equal if and only if the real and imaginary parts of the first are respectively equal to the real and imaginary parts of the second.

2.4.4 Multiplication, Division, and Differentiation

The polar form of a complex number may be more convenient for multiplication, division, and differentiation. Following the rules of algebra and calculus, if

$$R_1 = R_1 e^{j\theta_1} \qquad \leftarrow \text{magnitude}$$

and

$$R_2 = R_2 e^{j\theta_2}$$

$$V_{max} = r\omega$$
$$a_{max} = r\omega^2$$

then

$$R_1 \times R_2 = R_1 R_2 e^{j(\theta_1 + \theta_2)} \tag{2.49}$$

$$\frac{R_1}{R_2} = \left(\frac{R_1}{R_2}\right) e^{j(\theta_1 - \theta_2)} \tag{2.50}$$

$$\frac{dR}{dt} = j\omega Re^{j\theta} + e^{j\theta}\frac{dR}{dt} \qquad \text{where } \omega = \frac{d\theta}{dt} \tag{2.51}$$

If R represents a rigid link of fixed length R, then

$$\frac{dR}{dt} = 0$$

and

$$\frac{dR}{dt} = j\omega Re^{j\theta} = j\omega R \tag{2.52}$$

Using the last expression for constant R, noting that j can be expressed in polar form as $j = e^{j\pi/2}$, we obtain

$$\frac{dR}{dt} = \omega R e^{j(\theta + \pi/2)} \tag{2.53}$$

This result is useful in velocity analysis of linkages.

When one is using a calculator, the rectangular-polar ($xy : R\theta$) conversion feature may be employed to put a complex number in a form convenient for addition, differentiation, and so on.

When we are solving problems involving planar linkages, vector analysis and complex number methods as well as other methods are at our disposal. In some ways, the imaginary unit j resembles a unit vector in the y direction, similar to the unit vector $\boldsymbol{j}$. However, operations with the imaginary unit are different. For example, multiplication of a vector by the unit vector $\boldsymbol{j}$ is only defined by the dot and cross products, neither of which has the same meaning as multiplication by the imaginary unit j.

Multiplication of a complex number by the imaginary unit j represents a counterclockwise rotation of $\pi/2$ rad (90°) in the complex plane. Recalling that $j = e^{j\pi/2}$, we have the following results:

$$je^{j\theta} = e^{j(\theta + \pi/2)}$$

$$j^2 = e^{j\pi} = -1$$

$$j^3 = e^{j3\pi/2} = -j \tag{2.54}$$

$$j^4 = e^{j2\pi} = +1$$

and so on.

2.5 COMPLEX NUMBER METHODS APPLIED TO DISPLACEMENT ANALYSIS OF LINKAGES

The displacement, velocity, and acceleration of planar linkages may be analyzed by using complex number methods. Consider, for example, the sliding contact linkage shown in Figure 2.13a, where link 1 rotates and the slider moves relative to link 2, causing link 2 to oscillate. The linkage can be described at any instant by the vector equation

$$R_2 = R_0 + R_1 \tag{2.55}$$

(see Figure 2.13b), where R_0 represents the fixed link O_2O_1, R_1 represents the crank O_1B, and R_2 represents the portion of link 2 between O_2 and pin B.

If link lengths R_0 and R_1 are given, Eq. 2.55 may be solved for any given crank angle θ_1. For convenience, we select the real axis in the direction of the fixed link. Then, expressing Eq. 2.55 in complex form, we have

$$R_2 = R_{2x} + jR_{2y} = R_0 + R_{1x} + jR_{1y} \qquad (2.56)$$

where

$$R_{1x} = R_1 \cos \theta_1$$

and

$$R_{1y} = R_1 \sin \theta_1$$

Equating the real parts of Eq. 2.56, we obtain

$$R_{2x} = R_0 + R_{1x} = R_0 + R_1 \cos \theta_1$$

Equating the imaginary parts, we have

$$R_{2y} = R_{1y} = R_1 \sin \theta_1$$

from which the magnitude of R_2 is

$$R_2 = \sqrt{R_{2x}^2 + R_{2y}^2}$$

$$= \sqrt{R_0^2 + 2R_0 R_1 \cos \theta_1 + R_1^2 \cos^2 \theta_1 + R_1^2 \sin^2 \theta_1}$$

$$= \sqrt{R_0^2 + 2R_0 R_1 \cos \theta_1 + R_1^2} \qquad (2.57)$$

The last equation is identical with the cosine law. However, the cosine law is generally written in terms of the internal angle $180° - \theta_1$. Angle θ_2 may be found from

$$\tan \theta_2 = \frac{R_{2y}}{R_{2x}} = \frac{R_1 \sin \theta_1}{R_0 + R_1 \cos \theta_1} \qquad (2.58)$$

Although the arctangent is actually multivalued, most calculators and computers will give values of θ between $0°$ and $+90°$ (0 to $\pi/2$ rad) for positive arctan arguments and θ between $0°$ and $-90°$ (0 to $-\pi/2$ rad) for negative arctan arguments. Therefore, it may be preferable to use a rectangular-polar conversion routine or the two-argument arctangent, $\arctan_2(x, y)$. Otherwise, if a computer program is used and the computer produces arctangents in the first and the fourth quadrants, we might introduce the following test:

$$\theta_2 = \arctan\left(\frac{R_{2y}}{R_{2x}}\right)$$

If R_{2x} is negative, $\theta_2 = 180° + \theta_2$. Then, the computed arctangent will be replaced by the angle θ_2 in the correct quadrant. Computation of $\tan(\theta_2/2)$ as in Section 2.3.3 produces the same result.

◆ **EXAMPLE PROBLEM 2.7**   *Sliding Contact Linkage Solved by the Complex Number Method*

Refer to Figures 2.13a and 2.13b. Given link lengths $O_2O_1 = 320$ mm and $O_1B = 170$ mm. Locate slider pin B when $\theta_1 = 50°$.

Solution.   Pin B is located by Eq. 2.56, from which

$$|R|\, e^{+\tan^{-1}\left(\frac{y}{x}\right)}$$

$$R_2 = R_{2x} + jR_{2y} = R_0 + R_{1x} + jR_{1y}$$

$$= 320 + 170\cos 50° + j170\sin 50° = 429.3 + j130.2$$

The magnitude and direction of R_2 are given immediately by using a rectangular-polar $(xy : r\theta)$ conversion available on many calculators. This conversion may be programmed to produce the correct angle quadrant:

$$R_2 = 448.6 \text{ mm} \quad \text{at} \quad \theta_2 = 16.9° \quad \times \frac{\pi}{180} = 0.294$$

or in complex exponential form:

$$R_2 = 448.6e^{j0.294}$$

These values can be checked with Eqs. 2.57 and 2.58, which were avoided by using the $xy : r\theta$ conversion. The dimensions of this particular linkage $(R_1 < R_0)$ permit values of θ_2 in only the first and fourth quadrants. In a similar linkage, but with $R_1 > R_0$, then θ_2 can fall in any quadrant.   ◆

Using the principles illustrated above, complex number methods may be used to analyze many other types of linkages. However, since the complex plane is two-dimensional, complex numbers methods are not applied to spatial linkages.

2.5.1 Limiting Positions

Limiting positions for some linkages were discussed in Chapter 1. For the sliding contact linkage of Figure 2.13, link 2 oscillates as link 1 rotates continuously. The locus of B_1 (point B on link 1) is a circle of radius R_1. We may find the limiting positions of link 2 graphically by drawing link 2 tangent to that circle. The tangent (representing link 2) will be perpendicular to the radius (representing link 1).

As an alternative, we may use the calculus. Noting that extreme values of θ_2 correspond to extreme values of $\tan\theta_2$ for this linkage, we differentiate Eq. 2.58 with respect to θ_1 and set the result equal to zero:

$$\frac{d\tan\theta_2}{d\theta_1} = 0 = R_1\cos\theta_1(R_0 + R_1\cos\theta_1)^{-1}$$

$$-R_1\sin\theta_1(R_0 + R_1\cos\theta_1)^{-2}(-R_1\sin\theta_1)$$

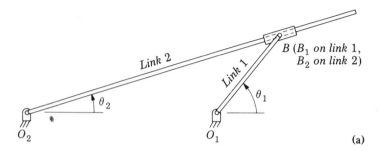

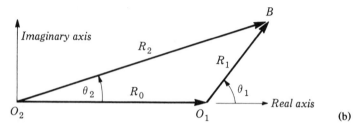

Figure 2.13 (a) Sliding contact linkage. (b) Complex plane representation.

from which

$$\cos \theta_1 = \frac{-R_1}{R_0}$$

Thus the three links form a right triangle with hypotenuse R_0. And as determined above graphically, links 1 and 2 are perpendicular when link 2 is at a limiting position.

2.5.2 The Geneva Mechanism

The Geneva mechanism (Figure 2.14) provides intermittent motion of the driven link while the driver rotates continuously. It is equivalent to the sliding contact mechanism (Figure 2.13a) during part of its cycle. For the position shown, pin B on the driver is entering the slot on the driven member. Thus, O_1B is equivalent to link 1 in Figure 2.13a, while the slot is equivalent to link 2 for the next $\frac{1}{4}$ rotation of the driven member. Then, the driven member remains stationary until pin B enters the next slot.

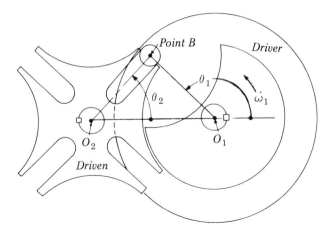

Figure 2.14 Geneva mechanism; a mechanism equivalent to the sliding contact linkage during part of its cycle.

2.6 SPATIAL (THREE-DIMENSIONAL) LINKAGES

The motion of all points in a planar linkage is restricted to a single plane or a set of parallel planes. The motion of points in a spatial linkage is more general. For example, the motion of a given point in a spatial linkage may describe a curve that does not lie in a plane, or the motion of two points in the same spatial linkage may lie in two nonparallel planes.

2.6.1 Kinematic Pairs (Joints)

Spatial linkages employ single-degree-of-freedom joints (e.g., pin joints) and multiple-degree-of-freedom joints (e.g., ball joints). Some common joints were identified in Chapter 1. Many others are possible.

2.6.2 Types of Spatial Mechanisms

Spatial linkages are identified by their joint configuration symbols. For example, a PRCR mechanism consists of a prism (spline), revolute, cylinder, and revolute. A variety of spatial mechanism configurations are shown in Figure 2.15a through 2.15i. Limiting positions occur when motion about or along a joint stops and then changes direction. A joint's limiting position defines its range of motion. Absence of a limiting position indicates that continuous motion is possible with respect to that joint. In general, limiting positions are not as obvious as in simple planar mechanisms.

Baker[2] describes a method for determining limit positions by utilizing the transient mobility of the remainder of the linkage. Examples include determination of limiting positions of PRCR, PHPH, HCCH, CCCR, and PCSR

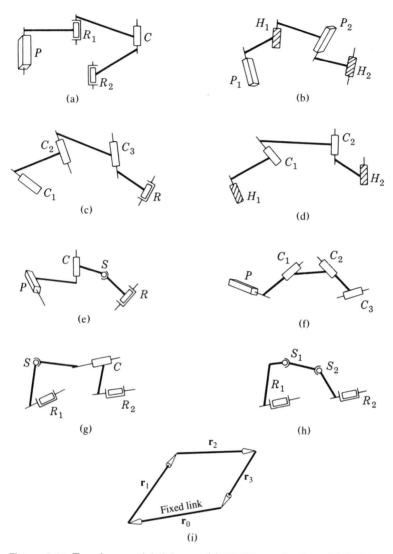

Figure 2.15 Four-bar spatial linkages. (a) PRCR mechanism. (b) PHPH mechanism. (c) CCCR mechanism. (d) HCCH mechanism. (e) PCSR mechanism. (f) PCCC mechanism. (g) RSCR mechanism. (h) RSSR mechanism. (i) Typical vector representation of a four-bar closed-loop spatial linkage.

mechanisms. Only a limited number of spatial linkages can be solved analytically. Denavit et al.[4] describe an algebraic method using 4×4 matrices for the analysis of velocities, accelerations, and static forces. The method applies to one-degree-of-freedom, single-loop spatial linkages consisting of revolute and prismatic joints. The joints may be combined; for example, combining three revolute joints produces a spherical joint. Velocities and accelerations are obtained by differentiation of the matrix loop or position equations. This method is suitable for computer calculations, using an iteration program.

A single-loop, one-degree-of-freedom kinematic chain with only revolute and prismatic joints generally requires seven links and seven joints. There are exceptions, however, including planar linkages and the Hooke-type universal joint.

2.6.3 Analysis of Four-Link Spatial Linkages

Vector methods may be used in the analysis of spatial linkages. For example, if there are four links (including the frame) that form a closed loop as in Figure 2.15i, then the vector equation

$$r_0 + r_1 + r_2 + r_3 = 0 \qquad (2.59)$$

may be used for displacement analysis. If each vector is written in terms of its components and the unit vectors i, j, and k, that is, $r_0 = r_{0x}i + r_{0y}j + r_{0z}k$, and so on, then the x, y, and z components of the vectors in the closed loop must each sum to zero. Thus we have three scalar equations:

$$r_{0x} + r_{1x} + r_{2x} + r_{3x} = 0$$
$$r_{0y} + r_{1y} + r_{2y} + r_{3y} = 0 \qquad (2.60)$$
$$r_{0z} + r_{1z} + r_{2z} + r_{3z} = 0$$

Considering the restraints imposed by the joints and by the link lengths, it may be possible to determine the position of the links analytically. One input link position variable is required to solve for displacements in a one-degree-of-freedom spatial linkage. For a spatial linkage with more than one degree of freedom, more than one input position variable would be required to solve for displacements.

Methods of descriptive geometry may be used as an alternative to an analytical solution for linkage displacements. Consider, for example, an RSSR linkage as represented in Figure 2.15h. Let the position of the left-hand crank, R_1S_1, be given. The locus of possible positions of point S_2 lies on a sphere of radius S_1S_2 and center at S_1. However, revolute R_2 restricts point S_2 to move in a circle. The actual position of S_2 may be determined graphically by constructing the intersection of the sphere and plane loci of S_2 and using the radius of the circle R_2S_2.

For given input values of velocity and acceleration, it may be possible to determine the velocity and acceleration of every point and the angular velocity and acceleration of every link in a spatial linkage. Vector methods may be used as with planar linkages, but all three coordinate directions must be considered. Straightforward solutions can be obtained for some spatial linkages. Others require ingenuity as well as considerable time and effort. An example of the latter type, displacement analysis of spatial mechanism with seven revolute joints, is given by Duffy and Derby[5]. In analyzing this special case of a 7R (RRRRRRR) mechanism, the authors derived an input-output equation of degree 24, making a major step toward solution of the general 7R mechanism.

2.6.4 Analysis of a Spatial Linkage Made Up of Two Revolute Pairs and Two Cylindrical Pairs (an RSSR Linkage)

RSSR linkages are shown in Figures 1.6a and 2.15h. The number of degrees of freedom for an RSSR linkage is given by

$$DF_{spatial} \geq 6(n_L - n_J - 1) + \Sigma f_1$$

$$= 6(4 - 4 - 1) + 1 + 3 + 3 + 1$$

$$DF \geq 2$$

Inspecting the linkage configuration shown in the figures, we see that one of the degrees of freedom corresponds to rotation of link 2 (link $S_1 S_2$) about its own axis. If this motion is not relevant to the intended application of the mechanism, the RSSR linkage acts essentially as a one-degree-of-freedom linkage.

Linkage Displacements

Referring to Figure 2.15h, the RSSR linkage may be described by the vector equation

$$r_0 + r_1 + r_2 + r_3 = 0 \qquad\qquad (2.59)$$

where the vectors form a closed loop as in Figure 2.15i. Let the position of links 0 and 1 be specified. Then, the positions of links 2 and 3 may be identified by three components each, resulting in six unknowns. There are six equations: three from Eq. 2.60 considering the x, y, and z directions; two equations based on the lengths of links 2 and 3; and one equation based on the plane of rotation of link 3.

◆ **EXAMPLE PROBLEM 2.8** *An RSSR Spatial Linkage*

Consider an RSSR linkage similar to that in Figure 1.6a, where link lengths are $r_0 = 3$, $r_1 = 1$, $r_2 = 3$, and $r_3 = 2$. Link 0 lies on the x axis, link 1 rotates in the xy plane, and link 3 rotates in the xz plane. Plot the vector components representing the position of links 2 and 3 against angular position θ of link 1.

Solution. Define a vector

$$c = r_0 + r_1$$

where c is known for any given value of θ. Then,

$$c + r_2 + r_3 = 0$$

and, considering the plane of rotation of link 3, there are five unknowns: r_{2x}, r_{2y}, r_{2z}, r_{3x}, and r_{3z}. Equating the vector components in each coordinate direction and noting the link lengths, we have the following five equations:

$$c_x + r_{2x} + r_{3x} = 0$$

$$c_y + r_{2y} = 0$$

$$r_{2z} + r_{3z} = 0$$

$$r_{2x}^2 + r_{2y}^2 + r_{2z}^2 = r_2^2$$

$$r_{3x}^2 + r_{3z}^2 = r_3^2$$

The five equations can be reduced to a single equation for r_{2x} in terms of known values. The solution and plots of vector components of r_2 and r_3, which were obtained using mathematics software, are shown in Figure 2.16.

♦

2.6.5 Analysis of a Spatial Linkage Made Up of a Revolute Pair, Two Spherical Pairs, and a Cylinder Pair

Figure 2.17 shows an RSSC spatial linkage where link RS_1 acts as a crank. The path of the sliding link intersects the plane of the crank at point A, making an angle γ with the plane of the crank. Joints C, S_2, and R of this linkage will be in the xy plane, and link 1 will move within the yz plane. Link lengths are r_1 (crank RS_1), r_2 (coupler S_1S_2), r_3 (instantaneous distance S_2A on the sliding link), and r_0 (fixed distance AR).

Analytical Solution for Displacements

The links could be identified as vectors and the vector equation

$$r_0 + r_1 + r_2 + r_3 = 0$$

could be used along with constraint equations to solve for displacements. For this configuration, however, it is convenient to express the length of the coupler link $r_2 = S_1S_2$ in terms of its components in three mutually perpendicular directions.

$$r_2^2 = r_{2x}^2 + r_{2y}^2 + r_{2\mathbf{z}}^2$$

$$= (r_3 \sin \gamma)^2 + (r_0 - r_1 \cos \theta + r_3 \cos \gamma)^2 + (r_1 \sin \theta)^2 \qquad (2.61)$$

If crank angle θ is given along with link lengths r_0, r_1, and r_2 and path angle γ, the result is a quadratic equation in r_3:

$$r_3^2 + 2 \cos \gamma (r_0 - r_1 \cos \theta) r_3 + r_0^2 + r_1^2 - r_2^2 - 2r_0 r_1 \cos \theta = 0 \qquad (2.62)$$

The solution gives the sliding link position, locating ball joint S_2. The two

RSSR linkage. Link 1 rotates in xy plane; link 3 rotates in xz plane

$$r_0 = 3 \qquad r_1 = 1 \qquad r_2 = 3 \qquad r_3 = 2 \qquad \theta = 0, \frac{\pi}{6} \cdots 2 \cdot \pi$$

$$c_x(\theta) = -r_0 + r_1 \cdot \cos(\theta) \qquad c_y(\theta) = r_1 \cdot \sin(\theta)$$

$$r_{2x}(\theta) = \frac{-r_2^2 + c_y(\theta)^2 + r_3^2 - c_x(\theta)^2}{2 \cdot c_x(\theta)} \qquad r_{3x}(\theta) = -c_x(\theta) - r_{2x}(\theta)$$

$$r_{2y}(\theta) = -c_y(\theta) \qquad r_{2z}(\theta) = \left[r_3^2 - r_{3x}(\theta)^2 \right]^{.5} \qquad r_{3z}(\theta) = -r_{2z}(\theta)$$

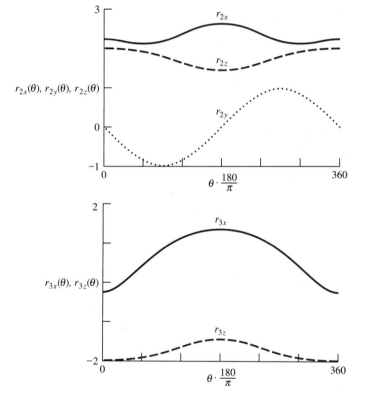

Figure 2.16 Solution and plots of vector components of r_2 and r_3.

roots of the quadratic equation are

$$r_3 = \frac{-b \pm \sqrt{b^2 - 4ac}}{2a}$$

where $a = 1$

$$b = 2 \cos \gamma (r_0 - r_1 \cos \theta)$$

$$c = r_0^2 + r_1^2 - r_2^2 - 2r_0 r_1 \cos \theta$$

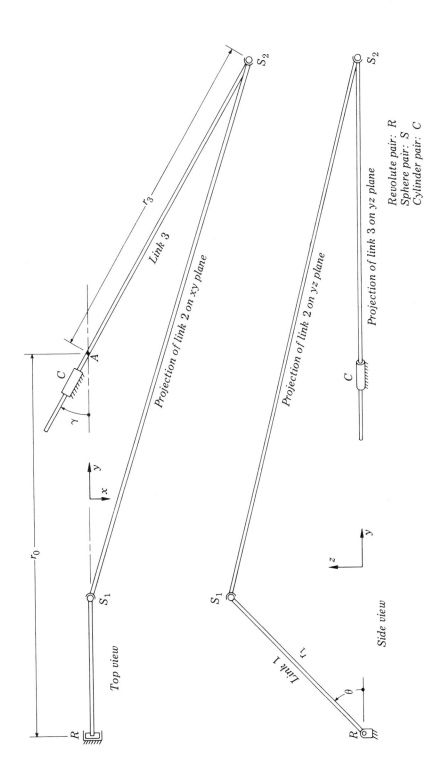

Figure 2.17 RSSC spatial linkage.

Revolute pair: R
Sphere pair: S
Cylinder pair: C

Top view

Side view

Link 1

Link 3

Projection of link 2 on xy plane

Projection of link 2 on yz plane

Projection of link 3 on yz plane

135

◆ EXAMPLE PROBLEM 2.9   *RSSC Spatial Linkage*

Displacement analysis of a three-dimensional RSSC linkage is the subject here. Let dimensions r_0, r_1, and r_2 be given for the RSSC linkage described above. Sketch a flowchart for finding the displacement of the sliding link for every 15° of crank angle θ at various path angles γ.

Solution.   See Figure 2.18. The mechanism is equivalent to a planar slider-crank linkage when the path angle is zero. For nonzero values of path angle, we then have a spatial linkage. For path angle $\gamma = 90°$ for the configuration of this example, the coupler link $S_1 S_2$ is perpendicular to the sliding link when $\theta = 180°$. This results in an unacceptable transmission angle, analogous to a transmission angle of 0° or 180° in a planar four-bar linkage. Thus, for values of $\gamma = 90°$ or near that value, if the crank RS_1 drives, the linkage is likely to jam.   ◆

More complicated spatial linkages are discussed in the technical literature. Lee and Liang[7] describe a new vector theory for the analysis of spatial linkages. They include displacement, velocity, and acceleration equations for open-chain and closed-loop mechanisms. The same authors[8] also analyze displacements of a general spatial seven-link 7-R mechanism. The analysis involves a sixteenth-degree polynomial input-output equation for displacement in the form of an 8 by 8 determinant. Fanghella[6] describes the kinematics of spatial linkages by group algebra.

2.6.6 Alternate Analysis of a Spatial Linkage Using Graphical Methods

Using descriptive geometry methods, we can find the position of the sliding link of the RSSC spatial mechanism described above for path angle $\gamma = 30°$ and crank angle $\theta = 45°$. Let crank length $r_1 = 100$ mm, coupler length $r_2 = 300$ mm, and let point A lie a distance $r_0 = 200$ mm away from the revolute joint R, as in Figure 2.19.

A side view (yz plane) and top view (xy plane) are used. Link 1, the crank, which lies in the yz plane, is drawn to scale in that plane and projected to the xy plane. The path of S_2 is located in the xy plane. Link 2 will not, in general, lie in either the xy or yz planes. If link 2 were constrained only at S_1, the locus of all possible points S_2 would lie on a sphere of radius r_2 with center at S_1. A 300-mm-radius circle with its center at S_1 is drawn in the yz plane to represent the outline of the sphere. The intersection of the sphere and the horizontal plane containing the sliding link is a circle. Its projection on the yz plane is a line segment ending at point B, as marked in the figure. Point B is projected upward to the xy plane and a circular arc with its center at S_1 is drawn tangent to the projection line. The intersection of the circle and the path of S_2 in the xy plane locates S_2 and determines the value of r_3. While the graphical solution is useful for providing an independent check of the analytical solution, if many values of θ and γ were to be studied, a graphical solution for each value would be impractical.

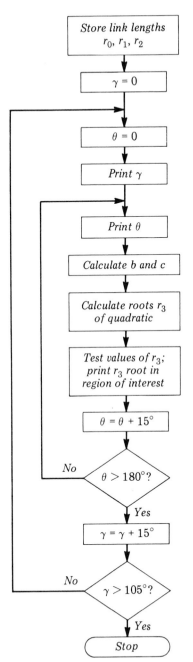

Figure 2.18 Flowchart. Displacement of an RSSC spatial linkage.

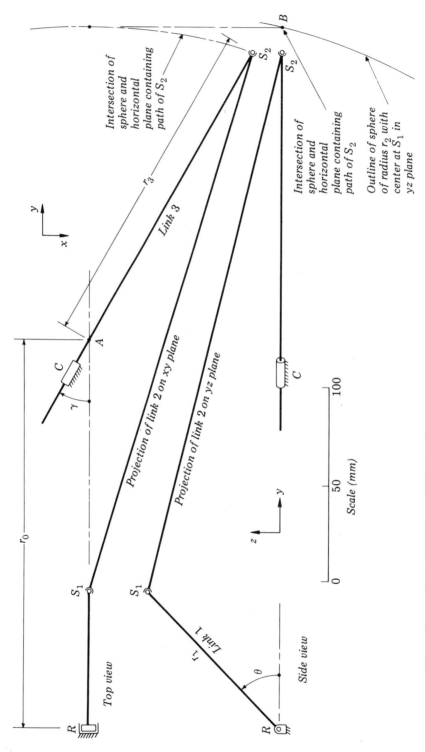

Figure 2.19 Alternative analysis of RSSC spatial linkage by graphical methods.

2.7 COMPUTER-IMPLEMENTED NUMERICAL METHODS OF POSITION ANALYSIS

Linkage displacement relationships tend to be nonlinear. The angular position of the driven crank of a four-bar linkage, for example, is not proportional to the input position. Simple closed-form solutions are available for some mechanisms, while other mechanisms, particularly multiloop linkages, are best solved using iterative numerical methods.

2.7.1 The Newton-Raphson Method for Two or More Variables

The Newton-Raphson method was utilized to solve a problem in a single variable in Chapter 1. Using the same concept, we may solve linkage problems involving two or more variables. Unfortunately, the Newton-Raphson methods for n variables involves an n by n matrix of partial derivatives. Those preferring less mathematical complexity may seek closed-form solutions or use preprogrammed numerical routines such as those found in MathCAD™ or other mathematical software.

Suppose a problem is represented by a set of simultaneous equations as follows:

$$F_1(x_1, x_2, \ldots, x_n) = 0$$

$$F_2(x_1, x_2, \ldots, x_n) = 0$$

$$\cdots \cdots \cdots$$
$$\cdots \cdots \cdots$$

$$F_n(x_1, x_2, \ldots, x_n) = 0$$

(2.63)

or, in vector form,

$$F(x_1, x_2, \ldots, x_n) = [F_1, F_2, \ldots, F_n] = 0 \tag{2.64}$$

where 0 is the null vector.

In order to find the unknown variables, the (state) vector

$$x = [x_1, x_2, \ldots, x_n]$$

we begin by making a first approximation of each of the variables:

$$X = [X_1, X_2, \ldots, X_n]$$

Then we compute the vector F at $x = X$. Unless we are very fortunate, the first approximations will not be correct, i.e., $F(X)$ is unequal to 0. As with the Newton-Raphson method in one variable, we make a linear adjustment to arrive at what we hope to be a better approximation of x.

The second approximation is computed from

$$X_{new} = X - G^{-1}F(X) \tag{2.65}$$

where

$$G = \begin{bmatrix} \partial F_1/\partial x_1 & \partial F_1/\partial x_2 & \cdots & \partial F_1/\partial x_n \\ \partial F_2/\partial x_1 & \partial F_2/\partial x_2 & \cdots & \partial F_2/\partial x_n \\ \cdots\cdots\cdots\cdots\cdots\cdots\cdots\cdots\cdots \\ \cdots\cdots\cdots\cdots\cdots\cdots\cdots\cdots\cdots \\ \partial F_n/\partial x_1 & \cdots\cdots & \cdots & \partial F_n/\partial x_n \end{bmatrix} \tag{2.66}$$

The process is repeated with X replaced by X_{new} for as many iterations as necessary, i.e., until each component of $F(X) = 0 \pm$ the tolerance. Otherwise the process is stopped after a set number of iterations (say 20) with a message that "the process does not converge in 20 iterations."

Success may depend on initial guesses of the values. If there is more than one set of roots, a poor first approximation of X may lead to a solution other than the desired one. The determinant of the matrix G, $J = |G|$ is called the Jacobian of the system of simultaneous equations. The Jacobian must not vanish during any of the iterations.

The above discussion is based on an extension of the single-variable Newton-Raphson method. Those desiring a more rigorous approach may refer to Taylor[12] or Stark.[11]

♦ EXAMPLE PROBLEM 2.10   *A Numerical Method Applied*
to the Four-Bar Linkage

Consider the four-bar linkage of Figure 2.10 where link lengths r_0, r_1, r_2, and r_3 are given, and angular positions $\theta_0 = \pi$ rad and $\theta_1 = \pi/3$ rad. Find θ_2 and θ_3.

Solution. In this case, a graphical solution or a closed-form mathematical solution is possible. However, this problem will be solved by the Newton-Raphson method in order to illustrate the numerical procedures involved. Referring to the figure, we see that the x and y components of diagonal vector r_d are determined, respectively, by

$$r_{dx} = r_0 \cos \theta_0 + r_1 \cos \theta_1 \tag{2.67}$$

and

$$r_{dy} = r_0 \sin \theta_0 + r_1 \sin \theta_1 \tag{2.68}$$

The problem of locating links 2 and 3 can be expressed as two simultaneous equations describing the horizontal and vertical projections of the triangle formed by vectors r_d, r_2, and r_3:

$$r_2 \cos \theta_2 + r_3 \cos \theta_3 = -r_{dx} \tag{2.69}$$

and

$$r_2 \sin \theta_2 + r_3 \sin \theta_3 = -r_{dy} \tag{2.70}$$

We will utilize Eqs. 2.63 to 2.66 where x_1 becomes θ_2 and x_2 becomes θ_3. Thus we have

$$F_1 = r_2 \cos \theta_2 + r_3 \cos \theta_3 + r_{dx} = 0$$

$$F_2 = r_2 \sin \theta_2 + r_3 \sin \theta_3 + r_{dy} = 0$$

or, in vector form,

$$F = \begin{bmatrix} r_2 \cos \theta_2 + r_3 \cos \theta_3 + r_{dx} \\ r_2 \sin \theta_2 + r_3 \sin \theta_3 + r_{dy} \end{bmatrix}$$

In order to form the matrix G, we find $\partial F_1/\partial\theta_2 = -r_2 \sin \theta_2$, etc., from which

$$G = \begin{bmatrix} -r_2 \sin \theta_2 & -r_3 \sin \theta_3 \\ r_2 \cos \theta_2 & r_3 \cos \theta_3 \end{bmatrix}$$

Suppose we are interested in the linkage configuration for which r_d, r_2, and r_3 form a clockwise loop. Then a guess (first approximation) of

$$x = X = \begin{bmatrix} \theta_2 \\ \theta_3 \end{bmatrix} = \begin{bmatrix} 1 \\ 3 \end{bmatrix}$$

seems reasonable. The result is

$$F = \begin{bmatrix} F_1 \\ F_2 \end{bmatrix}_{(x=X)} = \begin{bmatrix} -93.56 \\ 181.06 \end{bmatrix}$$

(instead of the desired value, $F = 0$), indicating that the first approximation was not very accurate.

A second approximation of the roots is found by computing

$$X_{new} = X - G^{-1}F = \begin{bmatrix} 0.072 \\ 4.400 \end{bmatrix}$$

Successively replacing X by X_{new}, the third through sixth approximations are, respectively:

$$\begin{bmatrix} 0.116 \\ 4.077 \end{bmatrix} \quad \begin{bmatrix} 0.077 \\ 4.043 \end{bmatrix} \quad \begin{bmatrix} 0.077 \\ 4.044 \end{bmatrix} \quad \text{and} \quad \begin{bmatrix} 0.077 \\ 4.044 \end{bmatrix}$$

where, for the final value, $F = 0 \pm$ a small tolerance. Note that there is no change (to three decimal places) between the fifth and sixth approximations. Thus $\theta_2 = 0.077$ rad and $\theta_3 = 4.044$ rad for the desired linkage configuration.

$\blacklozenge$

2.7.2 Multiloop Linkages

The link positions of a planar **single-loop linkage** can be described by a single vector equation or two scalar equations describing a skeleton diagram that forms a single closed polygon. The slider-crank linkage and four-bar linkage

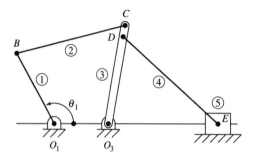

Figure 2.20 Drag-link–slider-crank linkage.

are examples of single-loop linkages. Planar **multiloop linkages** require one vector equation or two scalar equations for each loop. Quick-return mechanisms including the drag link–slider crank linkage and the sliding contact–slider crank linkage described in Chapter 1 are two-loop linkages.

The degree of difficulty of an analytical solution for displacements of a multiloop linkage depends on the linkage configuration and the given data. Consider the drag link–slider crank linkage shown in Figure 2.20. The skeleton diagram forms two loops, $O_1 BCO_3 O_1$ and $O_3 DEO_3$. If the angular position of link 1 is given, the orientation of links 2 and 3 in four-bar linkage $O_1 BCO_3$ can be determined without considering links 4 and 5. Then, using the orientation of link 3, the slider-crank linkage is easily solved. Closed-form solutions for both the four-bar linkage and slider-crank linkage are given earlier in this chapter.

If the angular position of link 1 is given for the double-slider two-loop linkage of Figure 2.21 the solution cannot be uncoupled as above. The two kinematic loops are represented by four simultaneous equations. Numerical solutions are suggested for the determination of displacements in linkages of this type.

♦ **EXAMPLE PROBLEM 2.11**   *Displacement Analysis of a Multiloop Linkage*
Consider the double-slider two-loop linkage of Figure 2.21 where $r_1 = 1$, $r_2 = 5.2$, $r_F = 5$, $r_{DE} = 4$, $r_{CD} = 1.5$, and $\theta_1 = \pi/3$ rad. (Distance DE is identified as r_{DE}, etc.) Find θ_2, θ_3, r_D, and r_E.

Solution.   The equations describing the horizontal and vertical components of loop, $O_1 BCEO_1$ are, respectively:

$$r_1 \cos \theta_1 + r_2 \cos \theta_2 + r_{CE} \cos \theta_3 - r_E = 0$$
$$r_1 \sin \theta_1 + r_2 \sin \theta_2 + r_{CE} \sin \theta_3 = 0$$

and the equations describing the horizontal and vertical components of loop *FDEF* are, respectively:

$$r_{DE} \cos \theta_3 - r_E + r_F = 0$$
$$r_D + r_{DE} \sin \theta_3 = 0$$

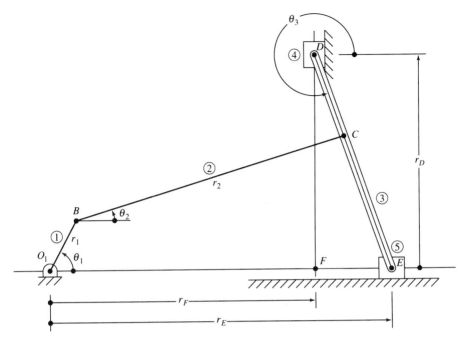

Figure 2.21 A double-slider two-loop linkage.

After substituting the given values, the four above equations may be solved by the Newton-Raphson method as used in the previous section. However, in this case, the Newton-Raphson method requires a 4×4 Jacobian matrix. Another alternative will be chosen, the Levenberg-Marquart method, which is a quasi-Newtonian method (a variation of the gradient method). This method is in the public domain (see More et al.[10]) and is available on software such as MathCAD™ (see Anderson[1]).

We begin by approximating the unknowns. A trial-and-error graphical solution may be used to generate these approximations for one position. In this case, the approximations are

$$\theta_2 = 0.3 \qquad \theta_3 = 5 \qquad r_D = 3.8 \qquad \text{and} \qquad r_E = 6.3$$

After a few iterations, the program yields

$$\theta_2 = 0.293 \qquad \theta_3 = 5.037 \qquad r_D = 3.792 \qquad \text{and} \qquad r_E = 6.274$$

The graphical solution was more accurate than necessary; a less accurate set of approximations would have yielded the same results. If the linkage must be solved for a number of successive positions, it may not be necessary to make additional graphical approximations. The results of one solution are likely to be satisfactory as a first approximation of the unknowns after θ_1 is incremented. ◆

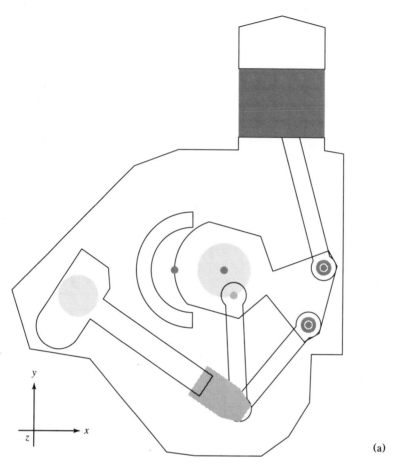

(a)

Figure 2.22 A variable-stroke engine (modeled by ADAMS™ software). (*Source:* Mechanical Dynamics, Inc.)

Mechanical Systems Software Packages

Mechanical systems software packages are sometimes used to avoid the extensive programming required to solve complicated multiloop linkages and spatial linkages. Most of these packages use matrix methods and other common mathematical procedures, including Newton-Raphson-type iteration methods. However, these packages incorporate many person-years of development with a goal toward increased speed and efficiency. Furthermore, once the system configuration is described, both the kinematic and the dynamic capabilities of the software may be utilized.

A proposed variable-stroke engine is shown in Figure 2.22a. The crankshaft is linked to a pivoted slide anchored in the crankcase. Stroke may be adjusted according to power demand. Parts b and c of the figure were obtained using ADAMS™ software. Part a shows slide and piston relationships for approxi-

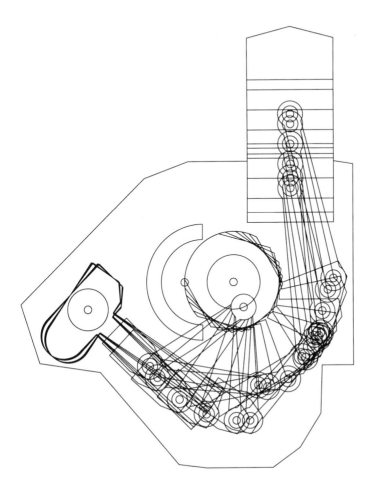

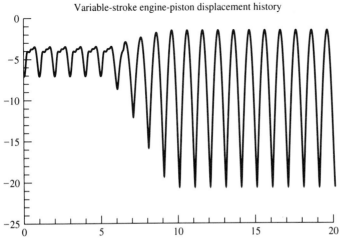

Variable-stroke engine-piston displacement history

Figure 2.22 *Continued.*

mately the same crankshaft position. A piston travel history during transition from short to long stroke is shown in part c.

References

1. Anderson, R. B., *The Student Edition of MathCAD*, Version 2.0, Addison-Wesley, Reading, MA, 1988, pp. 166–168, 221–222.
2. Baker, J. E., "Limit Positions of Spatial Linkages via Connectivity Sum Reduction," *Transactions of the American Society of Mechanical Engineers, Journal of Mechanical Design*, vol. 101, July 1979, pp. 504–508.
3. Chace, M. A., "Vector Analysis of Linkages," *Transactions of the American Society of Mechanical Engineers, Journal of Engineering for Industry, Series B 55*, no. 3, August 1963, pp. 289–297.
4. Denavit, J., R. S. Hartenberg, R. Razi, J. J. Uicker, "Velocity, Acceleration and Static-Force Analyses of Spatial Linkages," *Transactions of the American Society of Mechanical Engineers, Journal of Applied Mechanics*, December 1965, pp. 903–910.
5. Duffy, J., and S. Derby, "Displacement Analysis of a Spatial 7R Mechanism—A Generalized Lobster's Arm," *Transactions of the American Society of Mechanical Engineers, Journal of Mechanical Design*, vol. 101, April 1979, pp. 224–228.
6. Fanghella, P., "Kinematics of Spatial Linkages by Group Algebra: A Structure-Based Approach," *Mechanism and Machine Theory*, Penton, Cambridge, vol. 23, no. 3, 1988, pp. 171–184.
7. Lee, H.-Y. and C.-G. Liang, "A New Vector Theory for the Analysis of Spatial Mechanisms," *Mechanism and Machine Theory*, Penton, Cambridge, vol. 23, no. 3, 1988, pp. 209–218.
8. Lee, H.-Y. and C.-G. Liang, "Displacement Analysis of the General Seven-Link 7-R Mechanism," *Mechanism and Machine Theory*, Penton, Cambridge, vol. 23, no. 3, 1988, pp. 219–226.
9. Mischke, C. R., *Elements of Mechanical Analysis*, Addison-Wesley, Reading, MA, 1963.
10. More, J., B. Garbow, and K. Hillstrom, *Users Guide to MINPACK I*, ANL-80-74, Argonne National Laboratory, 1980.
11. Stark, P. A., *Introduction to Numerical Methods*, Macmillan, New York, 1970, pp. 115–147.
12. Taylor, A. E., *Advanced Calculus*, Ginn, Boston, 1955.

PROBLEMS

2.1 List three to five machine components in each of the following categories: motion described by (a) pure translation, (b) pure rotation, (c) combined rotation and translation.

• **2.2** The stroke of a Scotch yoke is 60 mm. The driver rotates at 1740 rev/min (constant). Find the following:
   (a) Maximum velocity
   (b) Maximum acceleration
   (c) Maximum jerk (time rate of change of acceleration)

2.3 Repeat Problem 2.2 except that the driver is to rotate at 3000 rev/min.

2.4 Repeat Problem 2.2 except that the stroke is to be 45 mm.

2.5 Design a two-cylinder, cam-type piston pump to deliver a flow rate of 0.5 m³/s at 90 rad/s. Let stroke equal piston diameter. Assume 80 percent volumetric efficiency.

2.6 Repeat Problem 2.5 except that the flow rate is to be 0.25 m³/s at 880 rev/min.

2.7 Repeat Problem 2.5 except that the flow rate is to be 120 gal/min at 1760 rev/min.

2.8 Repeat Problem 2.5 except that the flow rate is to be 12 ft³/min at 700 rev/min.

2.9 A circular cam with 20-mm eccentricity drives a flat-face follower. Plot displacement, velocity, and acceleration versus time for $\omega = 200$ rad/s.

2.10 Repeat Problem 2.9 for 28-mm eccentricity and a cam speed of 1500 rev/min.

Problems 2.11 Through 2.14 Refer to a Circular Cam with a Flat-face Follower. Eccentricity Is R and Angular Velocity ω.

2.11 Find the maximum angular velocity if $R = 5$ mm and the follower acceleration cannot exceed one g (the acceleration of gravity).

2.12 Find the maximum eccentricity in millimeters if the follower acceleration cannot exceed 1 g (the acceleration of gravity). The cam rotates at 4000 rev/min.

2.13 Find follower velocity and acceleration amplitude for $R = 0.25$ in at a cam speed of 1200 rev/min.

2.14 Find follower velocity and acceleration amplitude for $\omega = 300$ and $R = 0.5$ in.

Problems 2.15 Through 2.27 Refer to Vector Angles Measured Counterclockwise from the Horizontal Axis

For the following problems, $A = 1.0$ at $30°$; $B = 2$ at $60°$; $C = 1.5$ at $180°$; and $D = 3$ at $225°$. All vectors lie in a plane.

2.15 Find $A + B$.
   (a) Add the vectors by using trigonometric functions.
   (b) Solve graphically.

2.16 Find $A + B + D$. Solve graphically and analytically.

2.17 Find $A + B + C + D$. Solve graphically.

2.18 Find $A + B + C - D$, that is, $A + B + C + (-D)$. Solve graphically and analytically.

2.19 Find the vector product $A \times B$.

2.20 Find the vector product $C \times D$.

2.21 Find the vector product $C \times (D \times E)$, where $E = 2.5$ pointing inward in a direction perpendicular to the plane of C and D.

2.22 Repeat Problem 2.21 but find $D \times (C \times E)$.

2.23 Find $A \cdot B$.

2.24 Find $A \cdot (B + C)$.

2.25 Find $A \cdot (B \times C)$.

2.26 Find $B \cdot (C \times A)$.

2.27 Find $C \cdot (A \times B)$.

Problems 2.28 Through 2.30

Let $\omega = (5 + 3t)k$ and $r = ir_x(t) + jr_y(t)$, where i, j, and k are unit vectors in a fixed coordinate system.

2.28 Find $\omega \times r$.

• **2.29** Find $d/dt(\omega \times r)$.

2.30 Find $d/dt(\omega \times r)$ if i and j are vectors in a moving coordinate system.

2.31 The Immelman turn was a World War I aircraft maneuver used to gain altitude while turning to fly in the opposite direction. It consists of a half loop followed by a half roll to resume normal level position. Describe the maneuver in terms of pitch and roll coordinates referred to the aircraft. Consider reversing the order of the rotations. Does the commutative law of vector addition apply?

Problems 2.32 Through 2.34

$r_1 + r_2 + r_3 = 0$

2.32 $r_1 = 10i + 25j$, $r_2 = 40i - 20j + 10k$; find r_3.

2.33 Vectors r_1, r_2, and r_3 represent a planar linkage.

$$\frac{r_{1y}}{r_{1x}} = 2 \qquad \frac{r_{2y}}{r_{2x}} = 0.5 \qquad r_3 = i50 + j75$$

Find r_1 and r_2 by using the vector elimination method suggested by Chace (Section 2.2.9).

2.34 Repeat Problem 2.33 except that $r_3 = i5 - j8$.

2.35 A force $F = 22$ N acts on a bar of length $r = 180$ mm (see Figure 2.8b), where $\theta = 68°$. Find torque about point O.

2.36 Repeat Problem 2.35 for $\theta = 2$ rad.

• **2.37** An in-line slider-crank linkage has a crank length r_1 and connecting rod length $r_2 = 1.5r_1$ (see Figure 2.9). Find connecting rod and slider position when $\theta = 40°$ by using analytical vector methods.

2.38 Repeat Problem 2.37 for $\theta = 140°$.

2.39 Plot slider position versus θ for an in-line slider-crank linkage where the ratio of connecting rod to crank length is 1.5. Let $\theta = 0$, $\pi/9$, $2\pi/9$, $\pi/3$ rad, and so on.

• **2.40** The link lengths of a planar four-bar mechanism are $r_0 = 120$, $r_1 = 60$, $r_2 = 140$, and $r_3 = 80$. Find the orientation of links 2 and 3 when the internal angle between the crank and fixed link is 30° and the linkage is in the open phase (i.e., the coupler does not cross the fixed link). Use the vector cross product method.

2.41 Repeat Problem 2.40 for a 60° internal angle.

2.42 Repeat Problem 2.40 by using the dot product method.

2.43 Repeat Problem 2.41 by using the dot product method.

2.44 Repeat Problem 2.40 for the crossed phase.

2.45 Repeat Problem 2.40 for the crossed phase, using the dot product method.

Problems 2.46 and 2.47

$$R_1 = 200 \text{ mm at } \theta_1 = \frac{\pi}{3} \text{ rad}$$

$$R_2 = 150 \text{ mm at } \theta_2 = \frac{5\pi}{3} \text{ rad}$$

2.46 (a) Express R_1 and R_2 in complex rectangular form.
(b) Find R_0, where $R_0 + R_1 = R_2$.
(c) Express R_0 in polar form.

2.47 Find dR_1/dt if R_1 is constant in magnitude and $d\theta_1/dt = 120$ rad/s.

2.48 Repeat the sliding contact linkage example, Example Problem 2.7, for $\theta_1 = 80°$.

2.49 Repeat Problem 2.48 for $\theta_1 = 110°$.

2.50 Repeat Problem 2.37, using complex number methods.

2.51 Repeat Problem 2.37 with $\theta = 140°$, using complex number methods.

2.52 For the RSSC linkage described in Example Problem 2.9, let path angle $\gamma = 40°$. Find displacement of the sliding link for crank angle $\theta = 60°$.

2.53 Repeat Problem 2.52 for $\theta = 0, 15, 30°$, and so on. Use a computer.

2.54 Solve Problem 2.52, using methods of descriptive geometry.

2.55 Use vectors A, B, C, D, and E given in Example Problem 2.3: Vector Operations. Determine $A + B + C + D$, $A - C - D - E$, $B \cdot C$, $C \cdot B$, $|B|$, $|C|$, $(|B|)^2$, $B \cdot B$, $B \times B$, $C \times D$, $D \times C$, $C \cdot (C \times D)$.

2.56 Use vectors A, B, C, D, and E given in Example Problem 2.3: Vector Operations. Determine $A + B + C + E$, $A + B - C - D - E$, $D \cdot E$, $E \cdot D$, $|D|$, $|E|$, $(|D|)^2$, $D \cdot D$, $D \times D$, $E \times D$, $D \times E$, $E \cdot (E \times D)$.

2.57 Use vectors A, B, C, D, and E given in Example Problem 2.3: Vector Operations. Determine $D + B + C + E$, $B + B - C - D - E - A$, $B \cdot D$, $D \cdot B$, $|D|$, $|B|$, $(|E|)^2$, $E \cdot E$, $E \times E$, $D \cdot D$, $E \cdot (E \times B)$.

2.58 Consider an RSSR linkage similar to that in Figure 1.6a, where link lengths are $r_0 = 32$, $r_1 = 10$, $r_2 = 28$, and $r_3 = 20$. Link 0 lies on the x axis, link 1 rotates in the xy plane, and link 3 rotates in the xz plane. Plot the vector components representing the position of links 2 and 3 against angular position θ of link 1.

2.59 Consider an RSSR linkage similar to that in Figure 1.6a, where link lengths are $r_0 = 62$, $r_1 = 20$, $r_2 = 55$, and $r_3 = 45$. Link 0 lies on the x axis, link 1 rotates in the xy plane, and link 3 rotates in the xz plane. Plot the vector components representing the position of links 2 and 3 against angular position θ of link 1.

2.60 Consider a four-bar linkage for the assembly configuration with r_d, r_2, and r_3 forming a counterclockwise loop. Link lengths are to be $r_0 = 120$, $r_1 = 60$, $r_2 = 140$, and $r_3 = 80$. Let $\theta_0 = \pi$ rad. Find θ_2 and θ_3 at the instant that $\theta_1 = \pi/3$ rad. Use the Newton-Raphson method. A first approximation may be obtained by sketching the linkage.

2.61 Consider a four-bar linkage for the assembly configuration with r_d, r_2, and r_3 forming a clockwise loop. Link lengths are to be $r_0 = 3$, $r_1 = 1$, $r_2 = 3.6$, and $r_3 = 2.1$. Let $\theta_0 = \pi$ rad. Find θ_2 and θ_3 at the instant that $\theta_1 = \pi/4$ rad. Use the Newton-Raphson method. A first approximation may be obtained by sketching the linkage.

2.62 Consider the double-slider two-loop linkage illustrated in Section 2.7, where $r_1 = 1$, $r_2 = 5.2$, $r_F = 5$, $r_{DE} = 4$, $r_{CD} = 1.5$, and $\theta_1 = \pi/6$ rad. (Distance DE is identified as r_{DE}, etc.) Find θ_2, θ_3, r_D, and r_E. Use a numerical method.

2.63 Consider the double-slider two-loop linkage illustrated in Section 2.7, where $r_1 = 1$, $r_2 = 5.2$, $r_F = 5$, $r_{DE} = 4$, $r_{CD} = 1.5$, and $\theta_1 = \pi/2$ rad. (Distance DE is identified as r_{DE}, etc.) Find θ_2, θ_3, r_D, and r_E. Use a numerical method.

2.64 Consider the double-slider two-loop linkage illustrated in Section 2.7, where $r_1 = 1$, $r_2 = 5.2$, $r_F = 5$, $r_{DE} = 4$, $r_{CD} = 1.5$, and $\theta_1 = 2\pi/3$ rad. (Distance DE is identified as r_{DE}, etc.) Find θ_2, θ_3, r_D, and r_E. Use a numerical method.

2.65 Consider the double-slider two-loop linkage illustrated in Section 2.7, where $r_1 = 20$, $r_2 = 115$, $r_F = 100$, $r_{DE} = 85$, $r_{CD} = 40$, and $\theta_1 = \pi/4$ rad. (Distance DE is identified as r_{DE}, etc.) Find θ_2, θ_3, r_D, and r_E. Use a numerical method.

2.66 Consider the double-slider two-loop linkage illustrated in Section 2.7, where $r_1 = 20$, $r_2 = 115$, $r_F = 100$, $r_{DE} = 85$, $r_{CD} = 35$, and $\theta_1 = 2\pi/3$ rad. (Distance DE is identified as r_{DE}, etc.) Find θ_2, θ_3, r_D, and r_E. Use a numerical method.

PROJECTS

See Projects 1.1 to 1.6 and suggestions in Chapter 1. Select a project at this time, or continue with the previously selected project. Establish a set of performance requirements for the project. Examine linkages involved in the chosen project. Describe and plot motion characteristics of the linkages. Make use of computer software wherever practical. Evaluate the linkage in terms of the performance requirements.

Velocity Analysis of Mechanisms

3.1 BASIC CONCEPTS

Velocity is a vector representing the change in position of a moving point divided by the time interval. If the time interval is finite, the result is **average velocity**:

$$v_{\text{average}} = \frac{\Delta s}{\Delta t}$$

where Δs = the change in the x, y, and z coordinates of the point. If the time interval is infinitesimal, the result is **instantaneous velocity**:

$$v = \text{limit}_{\Delta t \to 0} \frac{\Delta s}{\Delta t} = \frac{ds}{dt}$$

Since we are largely concerned with instantaneous values, the term *velocity* will refer to instantaneous velocity unless otherwise noted.

A vector representing the change in angular position of a body divided by the time internal is called the **angular velocity**:

$$\omega = \text{limit}_{\Delta t \to 0} \frac{\Delta \theta}{\Delta t} = \frac{d\theta}{dt}$$

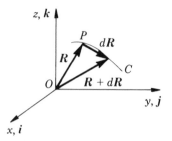

Figure 3.1 Velocity of a point.

Angular velocity is sometimes treated as a scalar when dealing with planar linkages. Analytical and graphical vector methods, including representation of vectors in complex form, are useful in velocity studies related to linkage design.

3.1.1 Velocity of a Point

Let the location of a point be described by a vector R. Consider point P in Figure 3.1, which moves along curve C through a displacement dR during a time interval dt. The new position vector is then $R + dR$ representing a change in the direction of R or a change in the magnitude of R or both. If we allow the time interval dt to become infinitesimal, the corresponding infinitesimal displacement of dR lies on the curve C. Then, the instantaneous velocity of point P is given by

$$v = \frac{dR}{dt} \tag{3.1}$$

where the direction of v is given by a tangent to curve C at P. A dot above a vector or scalar quantity is sometimes used to indicate differentiation with respect to time; thus dR/dt becomes $\dot{R}$.

   In general, the vector dR represents a change in the x, y, and z components of R. The velocity of point P may be expressed in terms of the components R_x, R_y, and R_z and the unit vectors i, j, and k parallel to the coordinate axes, as noted in Chapter 2. Then,

$$v = \dot{R} = i\dot{R}_x + j\dot{R}_y + k\dot{R}_z \tag{3.2}$$

if the x, y, z coordinate system is stationary.

3.1.2 Angular Velocity

Angular velocity may be treated as a vector quantity. Consider a link whose angular position changes at a rate of

$$\omega = \omega_x i + \omega_y j + \omega_z k \text{ radians per second}$$

(where $\omega = 2\pi n/60$ for rotation speed n revolutions per minute). The direc-

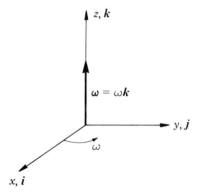

Figure 3.2 Rotation in the *xy* plane.

tion of vector $\boldsymbol{\omega}$ is perpendicular to the plane of rotation, and its sense is found by curving the fingers of the right hand in the direction of rotation. The thumb, then, points in the direction of vector $\boldsymbol{\omega}$. Alternatively, consider a right-hand screw rotating clockwise. The direction of the vector $\boldsymbol{\omega}$ is the direction of advance along the screw axis. For a body that rotates in the *xy* plane, $\boldsymbol{\omega}$ will be in the $\pm z$ direction. In Figure 3.2 for example, $\boldsymbol{\omega} = \omega \boldsymbol{k}$.

3.1.3 Motion of a Rigid Body About a Fixed Axis (Without Translation)

Consider a rigid body rotating about an axis that is fixed in a stationary coordinate system (see Figure 3.3). Then, angular velocity $\boldsymbol{\omega}$ has a fixed direction (along that axis). If a point P, which is fixed in the body, is identified by vector $\boldsymbol{R}$ (measured from the rotation axis), point P moves in a curved path of radius $R \sin \theta$ about the rotation axis, where θ is the angle between the rotation axis and $\boldsymbol{R}$. The speed (velocity magnitude) of P is given by

$$v = \omega R \sin \theta \tag{3.3}$$

This is identical to the vector cross product identified in Chapter 2. Thus, we

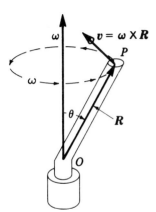

Figure 3.3 Rigid-body motion about a fixed axis.

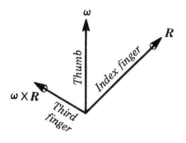

Figure 3.4 Using the right-hand rule to find the direction of the vector cross product.

may write

$$\dot{R} = v = \omega \times R \qquad (3.4)$$

since the velocity direction is perpendicular to the plane of ω and R and is given by the right-hand rule. The thumb of the right hand is pointed in the ω direction, and the index finger in the R direction (see Figure 3.4). Velocity is then in the direction of the third finger.

If ω varies (with time) in magnitude or direction, velocity is still given by the above equations provided that $d\omega/dt$ is finite. Thus, the instantaneous velocity of a point P in a rigid body that rotates about a fixed point (e.g., a ball joint) with instantaneous angular velocity ω is given by

$$v = \omega \times R$$

where R is measured from any point on the instantaneous rotation axis. Thus, for a link that moves in a plane about a stationary revolute joint, ω and R are perpendicular, and velocity magnitude of a point in the link is given by the product ωR. If we consider sliding along a rotating link or if we consider links that are not fixed at a point, additional terms enter into our analysis.

♦ **EXAMPLE PROBLEM 3.1** *Surface Speed*
Surface speeds of from 800 to 2000 ft/min (4064 to 10,150 mm/s) are recommended for milling aluminum. Find the *corresponding speeds (angular velocities) in revolutions per minute* for a 4-in- (101.5-mm-) diameter milling cutter.

Solution. We will let the radius of the cutter be represented by $R = 2$ in. The lowest surface speed is

$$v = 800 \text{ ft/min} \times 12 \text{ in/ft} \times 1 \text{ min/60 s} = 160 \text{ in/s}$$

The angular velocity vector and radius vector are perpendicular. Thus, $v = \omega \times R = \omega R$ tangent to the surface, and

$$\omega = \frac{v}{R} = \frac{160 \text{ in/s}}{2 \text{ in}} = 80 \text{ rad/s}$$

We divide 80 rad/s by 0.1047 (rad/s)/(1 rev/min) to obtain 764 rev/min, which is the minimum value. Similarly, 2000 ft/min gives us a maximum value of 1910 rev/min. ♦

◆ **EXAMPLE PROBLEM 3.2**   *An In-Line Slider-Crank Mechanism*
(a) Determine the velocity of the piston in a pump modeled as an in-line slider-crank linkage.
(b) Let $R = 2$ in, $L = 3.76$ in, $\theta = 70°$, and $\omega = 10$ rad/s. Find the slider velocity analytically.
(c) An in-line slider-crank mechanism has a crank length of 200 mm and a connecting rod length of 560 mm. The crank rotates at a constant angular velocity of 50 rad/s counterclockwise. Find the average slider velocity for one stroke.

Solution.   **(a)** An analytical examination of the in-line slider-crank mechanism has some resemblance to the Scotch yoke considered in Chapter 2. In fact, the Scotch yoke can be considered a special case of the slider crank, a slider-crank linkage with an infinite connecting rod. It is the connecting rod and the angle ϕ it forms with the slider path that complicates our analytical solution. Figure 3.5 shows the in-line slider crank first in its extreme extended position (top dead center), and then in a general position with angular displacement θ of the crank.

Measuring piston displacement x from the original position, we have

$$x = R + L - (R \cos \theta + L \cos \phi)$$

$$= R(1 - \cos \theta) + L(1 - \cos \phi)$$

We may express ϕ in terms of θ by dropping a perpendicular from B to line O_1C, forming two right triangles. The length of the perpendicular is

$$a = R \sin \theta = L \sin \phi$$

Using this equation and the identity $\sin^2 \phi + \cos^2 \phi = 1$, we obtain exact slider

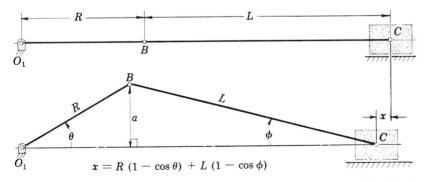

Figure 3.5 In-line slider-crank mechanism, shown in its extended position (*top*) and shown an instant later when the crank has moved through an angle θ (*bottom*).

displacement in terms of θ only:

$$x = R(1 - \cos\theta) + L\left[1 - \sqrt{1 - \left(\frac{R}{L}\right)^2 \sin^2\theta}\right]$$

Slider velocity is obtained by differentiating x with respect to time.

Exact slider velocity is given by the following equation where angular velocity ω is the time rate of change in θ

$$v = R\omega\sin\theta\left[1 + \left(\frac{R}{L}\right)\frac{\cos\theta}{\sqrt{1 - (R/L)^2\sin^2\theta}}\right]$$

If the slider-crank mechanism considered is a piston engine or piston pump for which the ratio of L to R is fairly large, say 3 or more, the following simplification is in order. The displacement equation is expanded by the binomial theorem, retaining only the terms

$$x = R\left[1 - \cos\theta + \left(\frac{1}{2}\right)\left(\frac{R}{L}\right)\sin^2\theta\right]$$

Using this or simplifying the velocity equation directly, we obtain the approximate slider velocity

$$v = R\omega\sin\theta\left[1 + \left(\frac{R}{L}\right)\cos\theta\right]$$

Using the trigonometric identity $\sin\theta \cdot \cos\theta = \frac{1}{2}\sin(2\theta)$, the approximate velocity equation may be written in the form

$$v = R\omega\left[\sin\theta + \frac{1}{2}(R/L)\sin(2\theta)\right]$$

or, for constant angular velocity of the crank,

$$v = R\omega\left[\sin(\omega t) + \frac{1}{2}(R/L)\sin(2\omega t)\right]$$

The latter form may be used if we wish to determine accelerations, and vibration force amplitudes.

(b) Using these values in the exact equation, we have

$$v = 2(10)\sin 70°\left[1 + \left(\frac{2}{3.76}\right)\frac{\cos 70°}{\sqrt{1 - (2/3.76)^2(\sin 70°)^2}}\right]$$

$$= 22.8 \text{ in/s to the left.}$$

From the approximate velocity equation, we obtain $v = 22.2$ in/s, which is a fairly good approximation when one notes that the L/R ratio is *not* within the recommended range for the approximate equation.

(c) We might be tempted to integrate an expression for velocity or to average values over an entire plot. The exact solution, however, is simply the stroke, $2R$, divided by the time for half of 1 cycle. For any in-line slider-crank mechanism with constant crank speed, this becomes

$$v_{av} = \frac{2R}{\pi/\omega}$$

Thus average slider velocity is given by

$$v_{av} = \frac{2 \times 200 \times 50}{\pi} = 6366 \text{ mm/s} \quad \blacklozenge$$

Parameter Studies

Parameter studies are an aid in selecting optimum linkage dimensions and speeds. We might examine velocities in a particular class of linkages, the slider crank, for example, without specifying actual dimensions or crank speed. To be as general as possible, a family of velocity-versus-crank-angle curves can be plotted, each curve for a different ratio of connecting rod length L to crank length R. See Figure 3.6a. For normalization of results, the product of crank

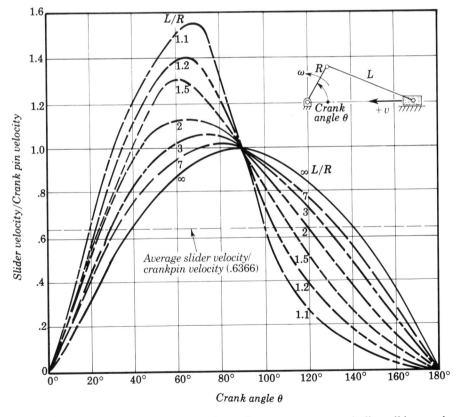

Figure 3.6 (a) The family of curves shown here represents an in-line slider-crank mechanism *parameter study*. The ratio of slider velocity to crankpin velocity is plotted against the crank angle for various L/R ratios. Slider velocity at any position of the mechanism is found by multiplying the ordinate (slider velocity/crankpin velocity) by the actual $R\omega$.

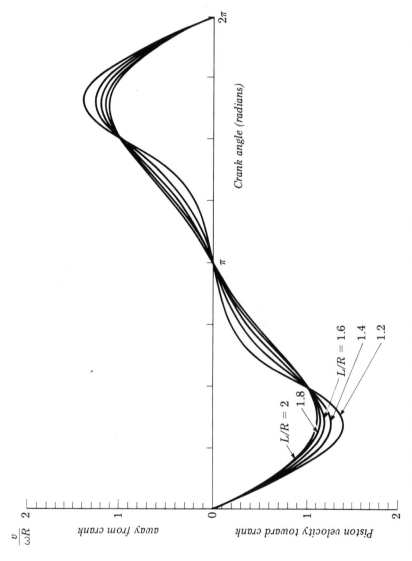

Figure 3.6 (b) Normalized piston velocity vs. crank position for various L/R ratios (in-line slider-crank mechanism).

angular velocity and crank length may be assigned a value of 1. Then, all velocities will later be multiplied by the actual ωR value to obtain actual velocity.

Figure 3.6a is a family of curves of slider velocity versus crank angle for an in-line slider-crank mechanism. The $L/R = \infty$ curve is a sine wave, representing the actual velocity of a Scotch yoke mechanism, or the limiting velocity relationship for the connecting rod length many times greater than crank length. Note how closely the curve $L/R = 7$ resembles the sine curve. To obtain curves for which the ratio of connecting rod length to crank length is near 1, the exact analytical solution is preferred.

Figure 3.6b shows normalized piston velocity $v/(\omega R)$ versus crank angle for an in-line slider-crank mechanism over a full cycle of motion. In this plot, L/R ratios range from 1.2 to 2, and piston velocity toward the crank is shown below the axis. It can be seen that all the plots for the in-line slider-crank mechanism are antisymmetric about a crank angle of π radians.

◆ EXAMPLE PROBLEM 3.3 *Specification of a Linkage to Satisfy Velocity Conditions*

Let us specify the dimensions of a linkage to meet a simple set of requirements. Suppose a certain process requires rectilinear motion with a velocity between 75 and 100 in/s in one direction during at least 15 percent of each cycle. The linkage is to be driven by a shaft that turns at 600 rev/min.

Solution. The requirements are not very rigid and therefore several mechanisms would be satisfactory, but we will consider the in-line crank mechanism that has already been examined in detail. Checking the curve representing $L/R = 3$ in Figure 3.6a, we see that this connecting rod-to-crank length ratio may be satisfactory. For that curve, $v/(R\omega)$ ranges from about 0.8 to 1.05 during the interval from $\theta = 40°$ to $\theta = 110°$ (which is greater than 15 percent of one cycle). If we let $v/(R\omega) = 0.8$ correspond to $v = 75$ in/s, when

$$\omega = \frac{2\pi}{60} \times 600 \text{ rev/min} = 62.8 \text{ rad/s}$$

then

$$R = \frac{v}{0.8\omega} = \frac{75}{0.8 \times 62.8} = 1.5$$

At maximum velocity, $v/(R\omega) = 1.05$, or $v = 99$ in/s (approximately), which is within the required range. The tentative solution, then, is an in-line slider-crank mechanism with crank length $R = 1.5$ and connecting rod length $L = 3R = 4.5$, and the conditions are satisfied between crank angles of $\theta = 40$ and 110° (approximately). Since the curves are only approximate, the next step is an accurate determination of velocity during the interval chosen.

For almost every practical design situation, the first step involves sketching of as many linkages as possible that might be suitable. The only limits are the designer's creativity and experience. Then, the motion characteristics of the linkages are analyzed, first to see that the displacement pattern meets all requirements, then to check velocity and acceleration. ◆

3.2 MOVING COORDINATE SYSTEMS

It is sometimes convenient to establish a coordinate system that translates and/or rotates along with a moving link. In most cases, we then refer velocities and accelerations back to a fixed coordinate system.

Consider the two coordinate systems of Figure 3.7. Coordinate axes X, Y, and Z and the corresponding unit vectors I, J, and K are fixed. (For most work with mechanisms, this would mean that the XYZ coordinate system is an inertial reference frame; it does not move with respect to the earth.) The origin o of coordinate system xyz is defined by the position vector R_o. Unit vectors i, j, and k for this set of moving axes lie along, and move with, the x, y, and z axes, respectively. The xyz-ijk system may translate and/or rotate in any direction. A point P in a linkage is described by the vector r (the position vector oP) in the moving coordinate system xyz. The total *position vector* of P is

$$R = R_o + r \tag{3.5}$$

measured from the origin of the fixed coordinates.

Expressing vectors R_o and r in terms of their components and corresponding unit vectors, the radius vector to point P is given by

$$R = R_0 + r = R_{0X}I + R_{0Y}J + R_{0Z}K + r_x i + r_y j + r_z k \tag{3.6}$$

Note that in Eq. 3.6, R_o is written in terms of the fixed coordinate system, while r is written in terms of the moving coordinate system. The *velocity* of

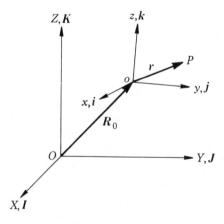

Figure 3.7 A moving coordinate system. System xyz moves within fixed system XYZ. Point P moves within xyz. The absolute position of point P is given by the vector sum $R_0 + r$, where r is the position of P with respect to the moving system and R_0 locates the origin of the moving system.

point P is given by the time rate of change in R:

$$\dot{R} = \dot{R}_{0X}I + \dot{R}_{0Y}J + \dot{R}_{0Z}K + \dot{r}_x i + \dot{r}_y j + \dot{r}_z k + r_x \dot{i} + r_y \dot{j} + r_z \dot{k} \qquad (3.7)$$

The first vector on the right side of the equation is the rate of change of R_o in the X direction, the X component of the velocity of o. Since the X, Y, and Z coordinate frame is fixed, unit vectors I, J, and K do not change and the velocity of o is given completely by the first three vectors on the right in Eq. 3.7. The sum of the first three will be identified by the symbol $\dot{R}_o$. The next three vectors, $\dot{r}_x i$ and so on, represent the rate of change in the r vector *with respect to the moving coordinates*, or velocity of P relative to the moving coordinate system *xyz*. Their sum will be identified by the symbol $\dot{r}_r$. (The velocity of P relative to the *xyz* system will be subscripted with the letter r.)

The last three vectors of Eq. 3.7 represent the effect of the rotating coordinate system (*xyz*) in any expression for the absolute velocity of P. Unit vectors i, j, and k are fixed relative to the moving *xyz* system (i.e., i, j, k move *with* the *xyz* system). Relative to the fixed *XYZ* system, however, unit vectors i, j, and k rotate; thus their positions relative to fixed system *XYZ* are functions of time.

The first derivative of a vector of constant magnitude is the cross product of the angular velocity of the vector (i.e., the angular velocity of the moving coordinate system) and the vector itself. Thus, for the last three vectors in Eq. 3.7,

$$r_x \dot{i} = r_x (\omega \times i)$$

$$r_y \dot{j} = r_y (\omega \times j) \qquad (3.8)$$

$$r_z \dot{k} = r_z (\omega \times k)$$

so that

$$r_x \dot{i} + r_y \dot{j} + r_z \dot{k} = r_x (\omega \times i) + r_y (\omega \times j) + r_z (\omega \times k)$$

$$= \omega \times (r_x i + r_y j + r_z k)$$

$$= \omega \times r \qquad (3.9)$$

Thus, the last three vectors of Eq. 3.7 can be replaced by the vector product $\omega \times r$, where ω is the angular velocity of the *xyz* coordinate system and r is the position vector of P in the *xyz* system.

The velocity may be expressed more concisely as

$$\dot{R} = \dot{R}_o + \dot{r}_r + \omega \times r \qquad (3.10)$$

where $\dot{R}$ = absolute velocity of point P (relative to XYZ)

$\dot{R}_o$ = velocity of the origin o of the xyz system

$\dot{r}_r$ = velocity of point P relative to the xyz system

$\boldsymbol{\omega} \times \boldsymbol{r}$ = cross product of the angular velocity of the moving system xyz in the XYZ system and the position vector $\boldsymbol{r}$

3.3 RELATIVE VELOCITY

In the preceding section, we referred to absolute velocity, that is, a velocity measured in a fixed coordinate system (an inertial reference frame). In addition, a relative velocity was identified; in that case, the velocity of a point with respect to a moving coordinate system. In the study of mechanisms, it is sometimes useful to describe the velocity of a point by referring to another moving point.

Consider nonstationary points B and C. The term v_{CB} is defined as *the absolute vector velocity of C minus the absolute vector velocity of B*; that is,

$$v_{CB} = v_C - v_B \tag{3.11}$$

Frequently, v_{CB} is referred to as the velocity of C *relative to* B or the velocity of C *with respect to* B. Other terms include *velocity difference* and the velocity of C *about* B. Some references use a different notation, such as $v_{C/B}$ instead of v_{CB}. When the terms "relative to," "with respect to," and "about" are used, it is understood that motion is viewed from an inertial or nonrotating reference frame. An observer in a rotating reference frame would not, in general, detect the correct relative velocity as defined above. In the study of mechanisms, the earth is most often selected as a "stationary reference frame." However, problems of space flight and even problems of long-range ballistics on the earth require that the earth's motion be considered.

Eq. 3.11 may be written in the equivalent form

$$v_C = v_B + v_{CB} \tag{3.12}$$

where the addition sign indicates the *vector* sum. For a simple example of relative velocity, let an aircraft carrier B move northward with a velocity $v_B = 15$ kn (7.72 m/s). Suppose an aircraft C on the flight deck has a velocity relative to the carrier $v_{CB} = 25$ kn (12.86 m/s). The direction of the path of the aircraft across the flight deck differs from aircraft carrier velocity direction by $20°$, as shown in Figure 3.8. The vector v_B, which represents the carrier velocity, is drawn to a convenient scale starting at an arbitrary point o. Then the vector v_{CB}, which represents the velocity of the aircraft *relative to the carrier*, is drawn, beginning at the head of vector v_B. The vector sum

$$v_B + v_{CB} = v_C$$

is the vector beginning at o with its head at the head of v_{CB}. Measuring it

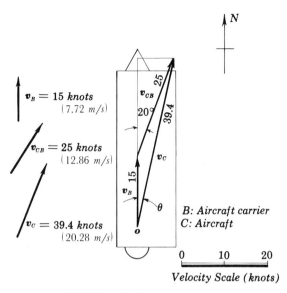

Figure 3.8 Relative velocity.

against the scale, we find the absolute aircraft velocity to be 39.4 kn (20.28 m/s) at $\theta = 12.5°$.

The concept of relative motion also applies to machine operations. For example, consider the motion of a robot manipulator on a fixed base, where the manipulator is to interact with an assembly on a moving production line. As another example, consider a lathe tool that moves axially to cut a helical thread in a rotating workpiece.

3.4 APPLICATION OF ANALYTICAL VECTOR METHODS TO LINKAGES

Analytical vector methods may be used to find velocities in planar and spatial linkages. Basically, it is necessary to determine the link orientations by solving the position equation and then differentiate the position equation with respect to time. If the linkage may be described by the vector polygon

$$r_0 + r_1 + r_2 + r_3 = 0$$

then the differentiation with respect to time yields the velocity equation

$$\dot{r}_0 + \dot{r}_1 + \dot{r}_2 + \dot{r}_3 = 0 \tag{3.13}$$

The form of solution depends on the given data and the type of linkage. The solution for a linkage with sliding pairs is somewhat different from the solution for a linkage with only revolute joints.

Four-Bar Linkage

Consider a four-bar planar linkage represented by vectors, as shown in Figure 3.9. All the links are rigid and there is no sliding contact. Velocities depend only on the rotation of links 1, 2, and 3. For the frame, $\dot{r}_0 = 0$. Thus, Eq. 3.13 becomes

$$\boldsymbol{\omega}_1 \times \boldsymbol{r}_1 + \boldsymbol{\omega}_2 \times \boldsymbol{r}_2 + \boldsymbol{\omega}_3 \times \boldsymbol{r}_3 = 0 \tag{3.14}$$

For this planar linkage, the coordinate axes are selected so that vectors $\boldsymbol{r}_1$, $\boldsymbol{r}_2$, and $\boldsymbol{r}_3$ have components in the x and y directions and angular velocities $\boldsymbol{\omega}$ in the z direction. Thus, examining the first term in Eq. 3.14 as typical, we have

$$\boldsymbol{\omega}_1 = \boldsymbol{k}\omega_1$$

and

$$\boldsymbol{r}_1 = \boldsymbol{i} r_{1x} + \boldsymbol{j} r_{1y}$$

from which

$$\boldsymbol{\omega}_1 \times \boldsymbol{r}_1 = \begin{vmatrix} \boldsymbol{i} & \boldsymbol{j} & \boldsymbol{k} \\ 0 & 0 & \omega_1 \\ r_{1x} & r_{1y} & 0 \end{vmatrix} = -\boldsymbol{i}\omega_1 r_{1y} + \boldsymbol{j}\omega_1 r_{1x} \tag{3.15}$$

Since the remaining terms in Eq. 3.14 are similar in form, that equation may be written as

$$-\boldsymbol{i}(\omega_1 r_{1y} + \omega_2 r_{2y} + \omega_3 r_{3y}) + \boldsymbol{j}(\omega_1 r_{1x} + \omega_2 r_{2x} + \omega_3 r_{3x}) = 0 \tag{3.16}$$

The $\boldsymbol{i}$ component and the $\boldsymbol{j}$ component of the above equation must each

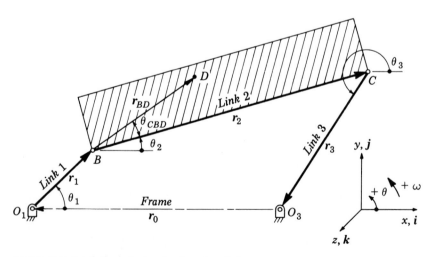

Figure 3.9 Analytical study of a four-bar linkage.

separately equal zero, yielding two simultaneous equations

$$\omega_1 r_{1y} + \omega_2 r_{2y} + \omega_3 r_{3y} = 0$$

and

$$\omega_1 r_{1x} + \omega_2 r_{2x} + \omega_3 r_{3x} = 0 \tag{3.17}$$

Handwritten: $V_{CD} = \omega_2 DC$ $bc \perp DC$ $V_{CD} = \omega_2 DC$ $= \omega_1 \theta_1 \theta$ $V_D = \omega_2 DC$ $V_C = \omega_2 DC$

A number of methods are commonly used for solving simultaneous equations. For example, if ω_1 is known and we wish to determine the other angular velocities, we may write the above equations in matrix form:

$$\begin{bmatrix} r_{2y} & r_{3y} \\ r_{2x} & r_{3x} \end{bmatrix} \begin{bmatrix} \omega_2 \\ \omega_3 \end{bmatrix} = -\omega_1 \begin{bmatrix} r_{1y} \\ r_{1x} \end{bmatrix} \tag{3.18}$$

Angular velocity of link 2, the coupler, is given by the determinant expression

$$\omega_2 = \frac{-\omega_1}{D} \begin{vmatrix} r_{1y} & r_{3y} \\ r_{1x} & r_{3x} \end{vmatrix}$$

where

$$D = \begin{vmatrix} r_{2y} & r_{3y} \\ r_{2x} & r_{3x} \end{vmatrix}$$

Handwritten: diagram

and the angular velocity of link 3 by

$$\omega_3 = \frac{-\omega_1}{D} \begin{vmatrix} r_{2y} & r_{1y} \\ r_{2x} & r_{1x} \end{vmatrix}$$

Expanding the expressions, we have

$$\omega_2 = \frac{-\omega_1(r_{1y}r_{3x} - r_{3y}r_{1x})}{r_{2y}r_{3x} - r_{3y}r_{2x}} \tag{3.19}$$

and

$$\omega_3 = \frac{-\omega_1(r_{2y}r_{1x} - r_{1y}r_{2x})}{r_{2y}r_{3x} - r_{3y}r_{2x}} \tag{3.20}$$

Handwritten margin notes:

$r_0 + r_1 + r_2 + r_3 = 0$

$r_1 = r_1 e^{j\theta_1}$ ect

$\dot{r}_1 = \frac{dr_1}{dt} = \dot{r}_1 e^{j\theta} + j r_1 \omega_1 e^{j\theta_1}$

$0 = j\omega_1 r_1 e^{j\theta_1} + j\omega_2 r_2 e^{j\theta_2} + j\omega_3 r_3 e^{j\theta_3}$

multiply by $e^{-j\theta_2}$

use $e^{j\theta} = \cos\theta + j\sin\theta$

$\omega_1 r_1 [\cos(\theta_1 - \theta_2) + j\sin(\theta_1 - \theta_2)]$
$+ \omega_2 r_2$
$+ \omega_3 r_3 [\cos(\theta_3 - \theta_2) + j\sin(\theta_3 - \theta_2)] = 0$

separate imaginary

$j[\omega_1 r_1 \sin(\theta_1 - \theta_2) + \omega_3 r_3 \sin(\theta_3 - \theta_2)] = 0$

$\omega_2 = \frac{-\omega_1 r_1 \sin(\theta_1 - \theta_2)}{r_3 \sin(\theta_3 - \theta_2)}$

Since link 1 in Figure 3.9 rotates about fixed point O_1, the velocity of any point on link 1 is given by $\omega_1 \times r$, where r is the vector measured from O_1 to the point in question. Link 2, the coupler, has no fixed point. The velocity of an arbitrary point D on link 2 may be found by using Eq. 3.10, where link 2 is fixed in a rotating coordinate system with origin at B. Then,

$$\dot{R}_0 = \omega_1 \times r_1$$

$$\dot{r}_r = 0$$

and

$$\omega \times r = \omega_2 \times r_{BD}$$

Handwritten: $\omega = \frac{d\theta}{dt}$ $\alpha = \frac{d\omega}{dt}$

$j\omega_1 r_1 e^{j\theta_1} + \cdots$

$j\alpha_1 r_1 e^{j\theta_1} + j\omega_1 r_1 e^{j\theta_1}(j\omega_1) + \cdots$

$j\alpha_1 r_1 e^{j\theta_1} - \omega_1^2 r_1 e^{j\theta_1}$
$+ j\alpha_2 r_2 e^{j\theta_2} - \omega_2^2 r_2 e^{j\theta_2}$
$+ j\alpha_3 r_3 e^{j\theta_3} - \omega_3^2 r_3 e^{j\theta_3} = 0$

and the velocity of point D is given by

$$v_D = \omega_1 \times r_1 + \omega_2 \times r_{BD} \tag{3.21}$$

where the first term on the right represents the velocity of point B and the last term the velocity of D with respect to B. The same result could be obtained by noting that the position of point D could be described by the equation

$$r_D = r_1 + r_{BD}$$

where vectors r_1 and r_{BD} have fixed magnitude. Note that point D does not move relative to link 2.

Point C in Figure 3.9 represents the revolute (pin) joint between links 2 and 3. The velocity of C can be found by the equation

$$v_C = \omega_1 \times r_1 + \omega_2 \times r_2 \tag{3.22}$$

where the first term on the right represents the velocity of B and the last term the velocity of C with respect to B (compare v_D in Eq. 3.21). An alternative expression is

$$v_C = \omega_3 \times (-r_3) \tag{3.23}$$

since link 3 rotates about fixed center O_3 and $-r_3$ represents the radius vector O_3C. If we subtract Eq. 3.23 from Eq. 3.22, the result is Eq. 3.14.

Eqs. 3.22 and 3.23 form the basis of the graphical relative velocity and velocity polygon methods for plane linkages of this type. For example, if the velocity of point B (Figure 3.9) is known, we use the fact that the cross products are perpendicular to the link vectors to find the velocity of point C.

Although the equations developed above are general for the four-bar planar linkage (Figure 3.9), the solutions are actually *instantaneous velocities* and *instantaneous angular velocities*. Even if ω_1 is constant in magnitude, only the magnitude of v_B will be constant. Due to the changing position, ω_2 and ω_3 will, in general, vary as will the velocities of points on links 2 and 3.

♦ **EXAMPLE PROBLEM 3.4** *Linkage Velocities by Analytical Vector Methods*
Referring to Figure 3.9, let $\omega_1 = 100$ rad/s ccw, $\theta_1 = 45°$, $\theta_{CBD} = 20°$ (constant), $r_0 = 30$ mm, $r_1 = 10$ mm, $r_2 = 35$ mm, $r_3 = 20$ mm, and $r_{BD} = 15$ mm. Find ω_2, ω_3, v_C, and v_D.

Solution. For the assembly configuration shown, we determined from the position analysis (Chapter 2) that $\theta_2 = 16.35°$ and $\theta_3 = 237.81°$. The components of the link vectors are given by

$$r_x = r \cos \theta \quad \text{and} \quad r_y = r \sin \theta$$

Programming a polar:rectangular/rectangular:polar conversion subroutine or using a preprogrammed one saves time. For this linkage at the instant considered,

$$r_{1x} = 7.0711 \qquad r_{1y} = 7.0711$$

$$r_{2x} = 33.590 \qquad r_{2y} = 9.835$$

$$r_{3x} = -10.660 \quad r_{3y} = -16.922$$

Using these values in Eqs. 3.18 and 3.19, we find

$$\omega_2 = \frac{-100(-7.0711 \times 10.660 + 16.922 \times 7.0711)}{-9.835 \times 10.660 + 16.992 \times 33.590}$$

$$= -9.567 \text{ rad/s} \quad (9.567 \text{ rad/s cw})$$

$$\omega_3 = \frac{-100(9.835 \times 7.0711 - 7.0711 \times 33.590)}{-9.835 \times 10.660 + 16.992 \times 33.590}$$

$$= 36.208 \text{ rad/s ccw}$$

The velocity of point C is given by the vector sum of the velocity of B and the velocity of C with respect to B (Eq. 3.22). Noting that $\boldsymbol{\omega}_1 = 100\boldsymbol{k}$ and $\boldsymbol{\omega}_2 = -9.567\boldsymbol{k}$, we have

$$v_C = \begin{vmatrix} i & j & k \\ 0 & 0 & 100 \\ 7.0711 & 7.0711 & 0 \end{vmatrix} + \begin{vmatrix} i & j & k \\ 0 & 0 & -9.567 \\ 33.590 & 9.835 & 0 \end{vmatrix}$$

$$= -i612.83 + j385.80 = 724.16 \text{ mm/s} \angle 147.8°$$

Note that the velocity vector is perpendicular to the link vector.

Using, instead, the angular velocity and length of link 3 (Eq. 3.23), we obtain

$$v_C = \begin{vmatrix} i & j & k \\ 0 & 0 & 36.208 \\ 10.660 & 16.922 & 0 \end{vmatrix}$$

which differs from the previous solution only due to round-off error. This provides a partial check of our arithmetic operations or of our coding of the program.

In finding the velocity of point D, which lies in link 2, we note that the components of the vector location are defined by the angle $\theta_2 + \theta_{CBD} = 16.35 + 20°$, from which

$$r_{BDx} = 12.081 \quad \text{and} \quad r_{BDy} = 8.891$$

Adding the vector velocities of B, and D with respect to B, (Eq. 3.21) we obtain

$$v_D = \begin{vmatrix} i & j & k \\ 0 & 0 & 100 \\ 7.0711 & 7.0711 & 0 \end{vmatrix} + \begin{vmatrix} i & j & k \\ 0 & 0 & -9.567 \\ 12.081 & 8.891 & 0 \end{vmatrix}$$

$$= -i622.04 + j591.53 = 858.4 \text{ mm/s} \angle 136.4° \quad \blacklozenge$$

TABLE 3.1 SPREADSHEET

	A	B	C	D	E	F	G
1		Four-bar linkage.					
2	Position analysis based on the vector cross product method.						
3	If Rd-R2-R3 loop clockwise, config = 1; -1 if ccw.						
4	R0	deg	R1	R2	R3	config	
5	30	180	10	35	20	1	
6							
7	Theta1	Rdx	Rdy	Rd	Rdux	Rduy	a
8	deg						
9	0	-20.000	.000	20.000	-1.000	.000	-10.625
10	5	-20.038	.872	20.057	-.999	.043	-10.538
11	10	-20.152	1.736	20.227	-.996	.086	-10.281
12	15	-20.341	2.588	20.505	-.992	.126	-9.865
13	20	-20.603	3.420	20.885	-.986	.164	-9.308
14	25	-20.937	4.226	21.359	-.980	.198	-8.633
15	30	-21.340	5.000	21.918	-.974	.228	-7.862
16	35	-21.808	5.736	22.550	-.967	.254	-7.018
17	40	-22.340	6.428	23.246	-.961	.277	-6.122
18	45	-22.929	7.071	23.994	-.956	.295	-5.194
19	50	-23.572	7.660	24.786	-.951	.309	-4.250
20	55	-24.264	8.192	25.610	-.947	.320	-3.302

	O	P	Q	R	S
1					
2					
3					
4					
5					
6					
7	Trans	r1x	r1y	Omega2	Omega3
8	angle				
9	28.955	10.000	.000	-50.000	-50.000
10	29.147	9.962	.872	-45.403	-37.734
11	29.716	9.848	1.736	-40.141	-25.128
12	30.639	9.659	2.588	-34.666	-12.900
13	31.880	9.397	3.420	-29.334	-1.581
14	33.400	9.063	4.226	-24.379	8.523
15	35.156	8.660	5.000	-19.918	17.296
16	37.109	8.192	5.736	-15.982	24.762
17	39.219	7.660	6.428	-12.549	31.020
18	41.455	7.071	7.071	-9.567	36.208
19	43.788	6.428	7.660	-6.977	40.467
20	46.192	5.736	8.192	-4.716	43.931
21	48.646	5.000	8.660	-2.728	46.717

Velocity Analysis

$$\dot{r}_0 = j\,\omega_1 r_1 e^{j\theta_1} + j\,\omega_2 r_2 e^{j\theta_2}$$

Euler

$$\dot{r}_0 = j\,\omega_1 r_1 (j\cos\theta_1 + j\sin\theta_1) + j\,\omega_2 r_2 (j\cos\theta_2 + j\sin\theta_2)$$

Imaginary $\rightarrow 0 = j(\omega_1 r_1 \cos\theta_1 + \omega_2 r_2 \cos\theta_2)$

$$\omega_2 = \frac{-\omega_1 r_1 \cos\theta_1}{r_2 \cos\theta_2}$$

$o'c' = o'b' + b'c'$

$a^n_c + a^t_c = (a^n_B + a^t_B) + (a^n_{CB} + a^t_{CB})$

$\omega_1^2\, O_1B \quad \alpha_1\, O_1B \quad \omega_2^2\, BC$

tangential $\quad 0\, \bot O_1B \quad \frac{bc}{BC}$

O_1

ACCel

o For suben CnorL

168

TABLE 3.1 *CONTINUED*

H	I	J	K	L	M	N
Omega1 100	Alpha1 0	rBD 15	ThetaCBD 20			
r2x	r2y	theta2 deg	r3x	r3y	theta3 deg	theta3 deg
30.625	16.944	28.955	-10.625	-16.944	-122.090	237.910
31.305	15.653	26.566	-11.267	-16.525	-124.286	235.714
31.867	14.473	24.426	-11.715	-16.209	-125.858	234.142
32.323	13.425	22.556	-11.982	-16.013	-126.806	233.194
32.685	12.518	20.957	-12.082	-15.938	-127.163	232.837
32.969	11.750	19.616	-12.032	-15.976	-126.984	233.016
33.189	11.112	18.510	-11.850	-16.112	-126.333	233.667
33.359	10.592	17.615	-11.550	-16.328	-125.276	234.724
33.488	10.177	16.904	-11.148	-16.605	-123.877	236.123
33.584	9.854	16.353	-10.655	-16.925	-122.192	237.808
33.654	9.612	15.941	-10.082	-17.273	-120.272	239.728
33.703	9.441	15.650	-9.438	-17.633	-118.159	241.841

T	U	V	W	X	Y

rBDx	rBDy	vDx	vDy	vD	vD dir deg
9.850	11.313	565.646	507.512	759.949	41.899
10.313	10.893	407.400	527.963	666.873	52.345
10.712	10.500	247.826	554.800	607.635	65.930
11.049	10.145	92.852	582.891	590.240	80.949
11.328	9.832	-53.594	607.390	609.750	95.043
11.555	9.565	-189.440	624.601	652.698	106.872
11.737	9.340	-313.966	632.237	705.903	116.409
11.882	9.155	-427.255	629.254	760.597	124.176
11.995	9.007	-529.761	615.529	812.110	130.717
12.081	8.891	-622.042	591.528	858.395	136.440
12.144	8.804	-704.619	558.058	898.841	141.621
12.189	8.742	-777.922	516.092	933.549	146.439
12.217	8.703	-842.283	466.670	962.923	151.011

Solution slider crank

$$r_0 = r_1 + r_2$$

$$r_0 = r_1 e^{j\theta} + r_2 e^{j\theta_2}$$

$$r_0 = r_1 (\cos\theta_1 + j\sin\theta_1) + r_2 (\cos\theta_2 + j\sin\theta_2)$$

Imaginary $\rightarrow 0 = j(r_1 \sin\theta_1 + r_2 \sin\theta_2)$

$$\sin\theta_2 = \frac{-r_1 \sin\theta_1}{r_2} \qquad \theta_2 =$$

$$r_0 = r_1 \cos\theta_1 + r_2 \cos\theta_2 =$$

169

Inverse Matrix Solution Using Mathematics Software

Equation 3.18 may be written in the form

$$AX = B \qquad (3.24)$$

Let the numerical values of the elements of matrix A and state vector B be specified. Then state vector X, representing the angular velocities of links 2 and 3, is the only unknown. Multiplying the above equation by A^{-1}, the inverse of matrix A, we obtain

$$A^{-1}AX = A^{-1}B$$

or

$$X = A^{-1}B \qquad (3.25)$$

The solution is easily obtained using mathematics software to evaluate Eq. 3.25 directly.

3.5 USING A SPREADSHEET TO SOLVE PROBLEMS IN KINEMATICS

An electronic spreadsheet program is a convenient tool for analyzing a linkage in a series of positions. Spreadsheets allow for rapid evaluation of potential design changes, and most spreadsheet programs include built-in plotting routines. One disadvantage is that spreadsheet formulas are written in terms of cell references, while we are accustomed to writing equations in terms of physical parameters.

Spreadsheets allow copying of formulas in a given range of cells to an additional range of cells. If the cell reference is not to change when a formula is copied to another cell, then an **absolute cell reference** is used. A **relative cell reference** is the default form. The column letter and/or row number of a relative cell reference change according to the cell into which a formula is copied. Example Problem 3.5 is intended to illustrate an application of spreadsheets to kinematics. Complete instructions for manipulating spreadsheets are found in the manuals that accompany the software.

♦ **EXAMPLE PROBLEM 3.5**   *Utilizing a Spreadsheet to Plot Velocities in a Four-Bar Linkage*

Let crank angle θ_1 vary from $0°$ to $360°$ in $5°$ increments in the four-bar linkage described in Example Problem 3.4. Plot the angular velocities of the coupler and follower crank and the velocity of point D vs. θ_1.

Solution.   We begin by listing the given data in cells that are identified by their column letter and row number. Referring to Table 3.1 (pages 168–169), fixed link length $r_0 = 30$ (cell A5) at an orientation of $180°$ (cell B5). Other data are given in cells C5 through K5. Initial crank position $\theta_1 = 0$ is given in cell A9. In cell A10, we type the formula: $5 + A9$ to obtain the second crank position,

TABLE 3.2 SPREADSHEET FORMULAS

```
A9:  0
B9:  +$A$5*@COS(@RAD($B$5))+$C$5*@COS(@RAD(A9))
C9:  +$A$5*@SIN(@RAD($B$5))+$C$5*@SIN(@RAD(A9))
D9:  @SQRT(B9^2+C9^2)
E9:  +B9/D9
F9:  +C9/D9
G9:  ((($E$5^2-$D$5^2+D9^2)/(2*D9))
H9:  +$F$5*@SQRT($E$5^2-G9^2)*F9+(G9-D9)*E9
I9:  +$F$5*@SQRT($E$5^2-G9^2)*(-E9)+(G9-D9)*F9
J9:  @DEG(@ATAN2(H9,I9))
K9:  -$F$5*@SQRT($E$5^2-G9^2)*F9-G9*E9
L9:  -$F$5*@SQRT($E$5^2-G9^2)*(-E9)-G9*F9
M9:  @DEG(@ATAN2(K9,L9))
N9:  @IF(M9<0,360+M9,M9)
O9:  @ABS(N9-J9-180)
P9:  +$C$5*@COS(@RAD($A9))
Q9:  +$C$5*@SIN(@RAD($A9))
R9:  -$H$5*(Q9*K9-L9*P9)/(I9*K9-L9*H9)
S9:  -$H$5*(I9*P9-Q9*H9)/(I9*K9-L9*H9)
T9:  +$J$5*@COS(@RAD($K$5+J9))
U9:  +$J$5*@SIN(@RAD($K$5+J9))
V9:  -$H$5*Q9-R9*U9
W9:  +$H$5*P9+R9*T9
X9:  @SQRT(V9^2+W9^2)
Y9:  @DEG(@ATAN2(V9,W9))
A10: 5+A9
B10: +$A$5*@COS(@RAD($B$5))+$C$5*@COS(@RAD(A10))
C10: +$A$5*@SIN(@RAD($B$5))+$C$5*@SIN(@RAD(A10))
D10: @SQRT(B10^2+C10^2)
E10: +B10/D10
F10: +C10/D10
G10: (($E$5^2-$D$5^2+D10^2)/(2*D10))
H10: +$F$5*@SQRT($E$5^2-G10^2)*F10+(G10-D10)*E10
I10: +$F$5*@SQRT($E$5^2-G10^2)*(-E10)+(G10-D10)*F10
J10: @DEG(@ATAN2(H10,I10))
K10: -$F$5*@SQRT($E$5^2-G10^2)*F10-G10*E10
L10: -$F$5*@SQRT($E$5^2-G10^2)*(-E10)-G10*F10
M10: @DEG(@ATAN2(K10,L10))
N10: @IF(M10<0,360+M10,M10)
O10: @ABS(N10-J10-180)
P10: +$C$5*@COS(@RAD($A10))
Q10: +$C$5*@SIN(@RAD($A10))
R10: -$H$5*(Q10*K10-L10*P10)/(I10*K10-L10*H10)
S10: -$H$5*(I10*P10-Q10*H10)/(I10*K10-L10*H10)
T10: +$J$5*@COS(@RAD($K$5+J10))
U10: +$J$5*@SIN(@RAD($K$5+J10))
V10: -$H$5*Q10-R10*U10
W10: +$H$5*P10+R10*T10
X10: @SQRT(V10^2+W10^2)
Y10: @DEG(@ATAN2(V10,W10))
```

5 (degrees). We then direct that the formula in cell A10 be copied into the next 71 cells (A11 through A81). Cell reference A9 is a relative reference. Thus the formula becomes 5 + A10 in cell A11, 5 + A11 in cell A12, and so on. Values of θ_1 from 0° to 360° in 5° increments appear on the spreadsheet (only the first few rows are shown in the figure).

The diagonal vector

$$r_d = r_0 + r_1$$

as defined in Chapter 2. The x component of r_d is given by

$$r_{dx} = r_0 \cos \theta_0 + r_1 \cos \theta_1$$

This equation is typed in cell B9 as the formula

$$\$A\$5 * @ \cos[@ \operatorname{rad}(\$B\$5)] + \$C\$5 * @ \cos[@ \operatorname{rad}(A9)].$$

The column letters and row numbers preceded by a $ sign are absolute references. For example, $A$5 refers to r_0, which is constant. The @ sign precedes functions (@ cos, @ sin, etc.) where the argument of trigonometric functions is in radians. The function @ rad changes angles from degrees to radians. When the formula in cell B9 is copied into cell B10, it changes to $A$5 * @ cos[@ rad($B$5)] + $C$5 * @ cos[@ rad(A10)]. The absolute references are unchanged, but θ_1 changes from 0° to 5° (cell A9 to A10).

The relevant equations from Chapters 2 and 3 are changed to cell form and typed into row 9 of the spreadsheet (corresponding to $\theta_1 = 0$). They

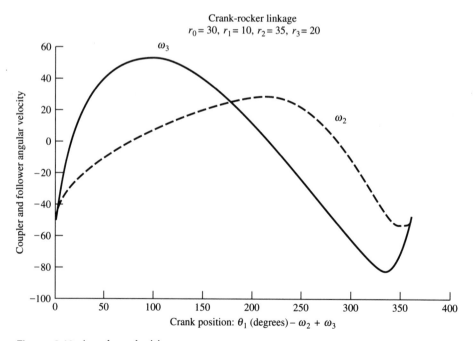

Crank-rocker linkage
$r_0 = 30, \ r_1 = 10, \ r_2 = 35, \ r_3 = 20$

Figure 3.10 Angular velocities.

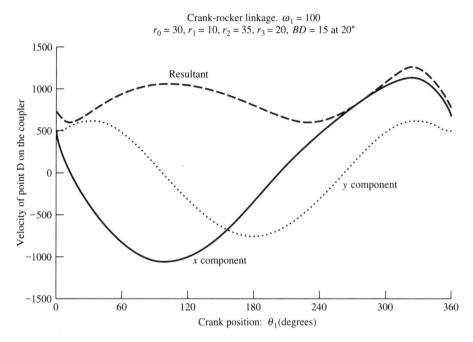

Figure 3.11 Velocity of a point on the coupler of a four-bar linkage.

compute the values needed for position analysis of the linkage using the vector cross product method. We find coupler and follower crank positions, the transmission angle, the angular velocities of the coupler and follower crank, and the velocity of point D on the coupler. These equations are represented by formulas B9 through Y9 in Table 3.2 (page 171). Using a single spreadsheet command, the 24 formulas (B9 through Y9) are copied to the 1704 remaining empty cells in the spreadsheet (cells C9 through Y81) to complete the analysis for a full crank rotation. Formulas A10 through Y10 are also shown in the table of spreadsheet formulas. Figure 3.10 shows coupler and follower angular velocity plotted against crank position. Figure 3.11 shows the velocity of point D on the coupler of the linkage. The x and y components are shown, as well as the resultant velocity. ◆

3.6 MATHEMATICS SOFTWARE APPLIED TO VECTOR SOLUTIONS OF KINEMATICS PROBLEMS

If we choose to solve kinematics problems in vector form, we may write vector-manipulation routines or use commercially available mathematics software. Consider the offset slider-crank linkage sketched in Figure 3.12. It can be described by the vector equation

$$e + r_1 + r_2 + r_0 = 0 \tag{3.26}$$

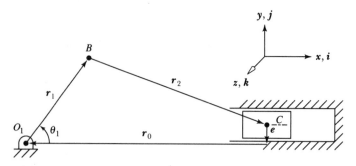

Figure 3.12 Offset slider crank linkage.

where the magnitude of e is the offset, the distance from the centerline of the slider path to crank bearing O_1.

Position Analysis

If offset e, crank length R_1, connecting rod length R_2, and crank position θ_1 are given, then it is convenient to rewrite Eq. 3.26 as

$$r_3 + r_2 + r_0 = 0 \tag{3.27}$$

where $r_3 = e + r_1$. Noting that the magnitude of r_0 is unknown and the direction of r_2 is unknown, and referring to Section 2.2.9, we see the solution is given by Eqs. 2.23 and 2.24 where $r_0 = A$, $r_2 = B$, and $r_3 = e + r_1 = C$.

Velocity Analysis

We note that the magnitudes of r_1 and r_2 are constant, that r_0 has a fixed direction, and that e is constant in direction and magnitude. Differentiating Eq. 3.26, the result is

$$\omega_1 \times r_1 + \omega_2 \times r_2 - v_c = 0 \tag{3.28}$$

where $dr_0/dt = -v_c$. Slider velocity v_c is in the x direction and the angular velocity vectors are in the z direction. To eliminate v_c in Eq. 3.28, we take the dot product of each term with j from which

$$\omega_1(k \times r_1) \cdot j + \omega_2(k \times r_2) \cdot j = 0 \tag{3.29}$$

Noting that

$$(k \times r) \cdot j = r \cdot (j \times k)$$

and

$$j \times k = i$$

Eq. 3.29 is equivalent to

$$\omega_1 r_1 \cdot i + \omega_2 r_2 \cdot i = 0$$

from which

$$\omega_2 = -\omega_1 r_1 \cdot i / (r_2 \cdot i) \tag{3.30}$$

Equation 3.28 may now be solved for v_c.

◆ **EXAMPLE PROBLEM 3.6** *Velocity Analysis of an Offset Slider-Crank Linkage Using Vector Methods*
Referring to Figure 3.12, let $R_2/R_1 = 1.8$ and $e/R_1 = -0.5j$ where $R_1 =$ crank length and $R_2 =$ connecting rod length. The angular velocity of the crank is constant. Plot slider position, slider velocity, and angular velocity of the connecting rod vs. crank angle θ_1.

Solution. Commercially available mathematics software was used to obtain the solution shown in Figure 3.13. In this solution we set $R_1 = 1$ and $\omega_1 = 1$. Then the results apply to other slider-crank linkages with the same proportions if r_0 is multiplied by the actual R_1, ω_2 by the actual ω_1, and v_c by the actual $\omega_1 R_1$. Crank angle θ_1 is to range from 0 to 2π rad in steps of $\pi/9$. Known vectors e, r_0^u, r_1, and r_3 are expressed in column form. Rectangular unit vectors i, j, and k are also defined in column form. Equations 2.23 and 2.24 are then used to determine r_0 and r_2. The magnitude of r_0 is tabulated and plotted against θ_1 (which is converted to degrees). Connecting rod angular

Slider-crank linkage $c_0 = 1$ (clockwise configuration positive)

$$\theta_1 = 0, \frac{\pi}{9} \cdot \cdot 2 \cdot \pi \qquad R_1 = 1 \qquad R_2 = 1.8 \qquad e = \begin{bmatrix} 0 \\ -0.5 \\ 0 \end{bmatrix} \qquad r_{0u} = \begin{bmatrix} -1 \\ 0 \\ 0 \end{bmatrix}$$

$$r_1[\theta_1] = \begin{bmatrix} R_1 & \cdot & \cos[\theta_1] \\ R_1 & \cdot & \sin[\theta_1] \\ & 0 & \end{bmatrix} \qquad i = \begin{bmatrix} 1 \\ 0 \\ 0 \end{bmatrix} \qquad j = \begin{bmatrix} 0 \\ 1 \\ 0 \end{bmatrix} \qquad k = \begin{bmatrix} 0 \\ 0 \\ 1 \end{bmatrix}$$

$$r_3[\theta_1] = r_1[\theta_1] + e$$

$$r_0[\theta_1] = \left[-r_3[\theta_1] \cdot r_{0u} + c_0 \cdot \left[R_2^2 - [r_3[\theta_1] \cdot [r_{0u} \times k]]^2 \right]^{.5} \right] \cdot r_{0u}$$

$$a[\theta_1] = -[r_3[\theta_1] \cdot [r_{0u} \times k]] \cdot [r_{0u} \times k]$$

$$r_2[\theta_1] = a[\theta_1] - c_0 \cdot \left[R_2^2 - [r_3[\theta_1] \cdot [r_{0u} \times k]]^2 \right]^{.5} \cdot r_{0u}$$

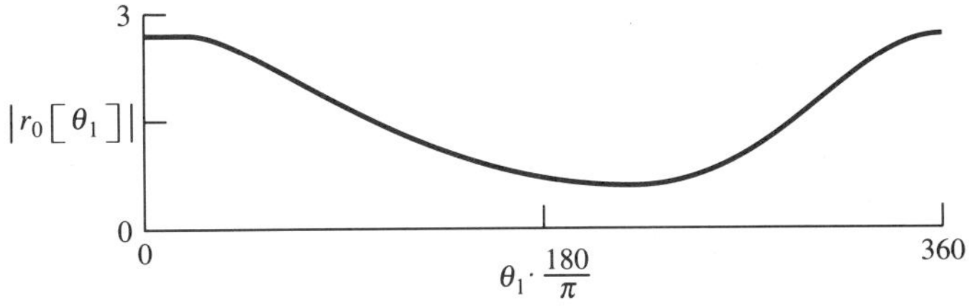

Figure 3.13 Example solution using mathematics software (*continued on page 176*).

$$\omega_1 = 1 \qquad \omega_2[\theta_1] = -\omega_1 \cdot \frac{r_1[\theta_1] \cdot i}{r_2[\theta_1] \cdot i}$$

$$v_c[\theta_1] = \omega_1 \cdot k \times r_1[\theta_1] + \omega_2[\theta_1] \cdot k \times r_2[\theta_1] \qquad \mathbf{V}_c[\theta_1] = v_c[\theta_1] \cdot i$$

$\theta_1 \cdot \dfrac{180}{\pi}$	$|r_0[\theta_1]|$	$\omega_2[\theta_1]$	$V_c[\theta_1]$
0	2.729	− 0.578	0.289
20	2.733	− 0.524	− 0.259
40	2.56	− 0.427	− 0.704
60	2.262	− 0.284	− 0.97
80	1.907	− 0.1	− 1.033
100	1.56	0.1	− 0.936
120	1.262	0.284	− 0.762
140	1.028	0.427	− 0.582
160	0.853	0.524	− 0.425
180	0.729	0.578	− 0.289
200	0.651	0.591	− 0.155
220	0.625	0.551	0.013
240	0.672	0.427	0.283
260	0.844	0.171	0.731
280	1.191	− 0.171	1.238
300	1.672	− 0.427	1.449
320	2.157	− 0.551	1.272
340	2.531	− 0.591	0.839
360	2.729	− 0.578	0.289

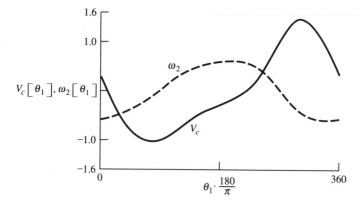

Figure 3.13 *Continued.*

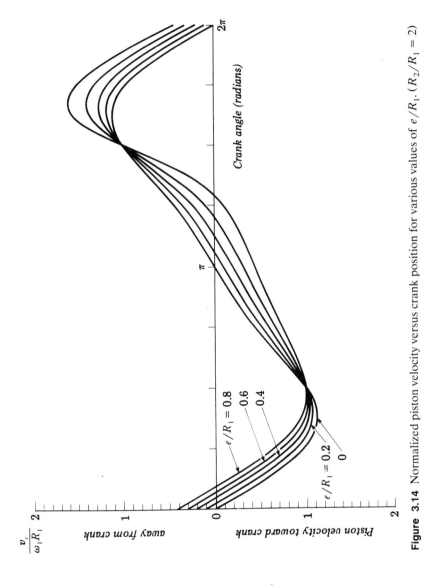

Figure 3.14 Normalized piston velocity versus crank position for various values of e/R_1. ($R_2/R_1 = 2$)

velocity ω_2 and slider velocity v_c are then calculated by Eqs. 3.30 and 3.28, tabulated, and plotted. Even though v_c lies only in the $\pm x$ direction, it has been stored as a vector. In order to tabulate and plot its magnitude and sense, we compute

$$V_c = v_c \cdot i \quad \blacklozenge$$

Figure 3.14 shows the effect of varying offset e. In this plot, slider velocity v_c is plotted against crank angle θ_1 for a slider-crank linkage where $R_2/R_1 = 2$.

3.7 COMPLEX NUMBER METHODS APPLIED TO VELOCITY ANALYSIS

Complex numbers are a convenient form for representing vectors. They may be used to develop analytical solutions to linkage velocity problems. With complex number methods, we are limited, of course, to planar linkages. First, the loop closure (displacement) equation is solved for unknown directions and magnitudes. Then, the displacement equation is differentiated with respect to time to obtain the velocity equation.

Consider the sliding contact linkage shown in Figure 3.15a and b, where the slider moves along link 2. The linkage is described by the equation

$$R_2 = R_0 + R_1 \tag{3.31}$$

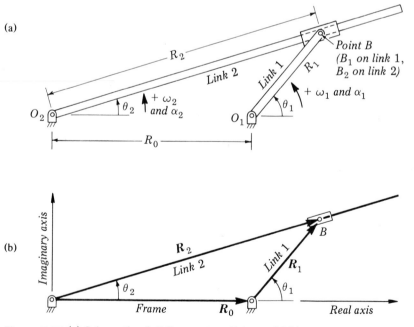

Figure 3.15 (a) Schematic of sliding contact linkage. (b) Vector representation.

where R_0, representing the frame, is fixed in magnitude and direction. Vector R_1, representing the rotating crank, has constant magnitude, but the magnitude of R_2 changes with time. If the real axis is selected parallel to the fixed link, we may then write

$$R_0 + R_1 e^{j\theta_1} = R_2 e^{j\theta_2} \tag{3.32}$$

Differentiating with respect to time, we have

$$j\omega_1 R_1 e^{j\theta_1} = j\omega_2 R_2 e^{j\theta_2} + v_{B_1 B_2} e^{j\theta_2} \tag{3.33}$$

where

$$\omega = \frac{d\theta}{dt}$$

is the angular velocity,

$$v_{B_1 B_2} = \frac{dR_2}{dt}$$

is the relative (sliding) velocity of B_1 with respect to B_2 (positive if R_2 is increasing),

$$\omega_1 R_1 = v_{B_1}$$

$$\omega_2 R_2 = v_{B_2}$$

Thus, we have the vector equation

$$v_{B_1} = v_{B_2} + v_{B_1 B_2} \tag{3.34}$$

with the complex exponentials indicating the directions. Note that the order of the subscripts is critical ($v_{B_2 B_1} = -v_{B_1 B_2}$).

Referring to Eq. 3.33, suppose link lengths R_0 and R_1 are given along with the orientation and angular velocity of link 1. Then R_2 and θ_2 may be found by solving the displacement equation (see Chapter 2). Two unknowns remain, the angular velocity of link 2 and the sliding velocity, both of them part of complex expressions. We can now take full advantage of the complex number method by multiplying both sides of Eq. 3.33 by $e^{-j\theta_2}$. The result is

$$j\omega_1 R_1 e^{j(\theta_1 - \theta_2)} = j\omega_2 R_2 + v_{B_1 B_2} \tag{3.35}$$

Expressing the exponential in rectangular form (using the Euler formula), we obtain

$$j\omega_1 R_1 \cos(\theta_1 - \theta_2) - \omega_1 R_1 \sin(\theta_1 - \theta_2) = j\omega_2 R_2 + v_{B_1 B_2} \tag{3.36}$$

The imaginary parts of the above equation yield

$$\omega_1 R_1 \cos(\theta_1 - \theta_2) = \omega_2 R_2$$

from which we obtain the angular velocity of link 2:

$$\omega_2 = \omega_1 R_1 \cos(\theta_1 - \theta_2)/R_2 \tag{3.37}$$

Equating the real parts, we obtain the sliding velocity, the relative velocity of B_1 with respect to B_2:

$$v_{B_1 B_2} = -\omega_1 R_1 \sin(\theta_1 - \theta_2) \tag{3.38}$$

Instead of the above solution, we could have separated Eq. 3.33 into its real and imaginary parts. However, it would then be necessary to solve a pair of simultaneous equations, making the solution a bit more difficult.

For $R_1 < R_0$, link 2 oscillates while link 1 rotates. We see that $\omega_2 = 0$ when $\cos(\theta_1 - \theta_2) = 0$, or $\theta_1 - \theta_2 = \pi/2, 3\pi/2, 5\pi/2$, and so on. Alternatively, these limiting positions (link 1 perpendicular to link 2) may be obtained by sketching the linkage. The magnitude of sliding velocity $v_{B_1 B_2}$ is maximum at the limiting positions.

♦ **EXAMPLE PROBLEM 3.7**   *Analysis of a Sliding Contact Linkage*
  Using Complex Numbers
We will examine velocities in a sliding contact linkage, utilizing the results of complex number analysis. Referring to Figure 3.15, let $\omega_1 = 20$ rad/s counterclockwise (constant), $R_0 = 40$ mm, and $R_1 = 20$ mm. Find R_2, θ_2, ω_2, and $v_{B_1 B_2}$ for the instant when $\theta_1 = 75°$.

Solution.   Using the displacement equations from Chapter 2, we proceed as follows. Slider position:

$$R_2 = \sqrt{R_0^2 + R_1^2 + 2R_0 R_1 \cos \theta_1}$$

$$= \sqrt{40^2 + 20^2 + 2 \times 40 \times 20 \times \cos 75°} = 49.13 \text{ mm}$$

Link 2 position:

$$\theta_2 = \arcsin\left(\frac{R_1 \sin \theta_1}{R_2}\right) = \arcsin\left(\frac{20 \sin 75°}{49.13}\right) = 23.15°$$

For this linkage, the proportions are such that θ_2 is limited to the first and fourth quadrants, and the above result is correct. A safer procedure utilizes the function $\arctan_2$; alternatively, following Section 2.3.3, compute

$$\cos \theta_2 = \frac{R_1 \cos \theta_1 + R_0}{R_2}$$

$$\sin \theta_2 = \frac{R_1 \sin \theta_1}{R_2}$$

and

$$\tan\left(\frac{\theta_2}{2}\right) = \frac{1 - \cos\theta_2}{\sin\theta_2}$$

which yields the same value of θ_2.

Angular velocity:

$$\omega_2 = \frac{\omega_1 R_1 \cos(\theta_1 - \theta_2)}{R_2} = \frac{20 \times 20 \cos(75° - 23.15°)}{49.13}$$

$$= +5.03 \text{ rad/s} \quad (\text{ccw})$$

Sliding velocity (motion of the slider relative to link 2):

$$v_{B_1 B_2} = \frac{dR_2}{dt} = -\omega_1 R_1 \sin(\theta_1 - \theta_2)$$

$$= -20 \times 20 \sin(75° - 23.15°) = -314.6 \text{ mm/s}$$

(i.e., 314.6 mm/s along link 2 toward O_2).   ◆

3.8 SPATIAL LINKAGES

Consider a spatial linkage with $n \geq 3$ links that form a closed loop. If links are represented by vectors r_0, r_1, and so on, the three-dimensional loop closure equation

$$r_0 + r_1 + r_2 + \cdots + r_n = 0 \tag{3.39}$$

may be used to describe the linkage. Differentiating with respect to time yields the velocity equation:

$$\dot{r}_0 + \dot{r}_1 + \dot{r}_2 + \cdots + \dot{r}_n = 0 \tag{3.40}$$

For a vector of fixed magnitude, an $\dot{r}$ term may be replaced by $\omega \times r$. When sliding occurs, the relative velocity of coincident points must be considered.

The velocity vector equation for a spatial linkage is, in general, three-dimensional rather than two-dimensional, as for a planar linkage. As a result, it is often desirable to utilize a computer or programmable calculator since several simultaneous equations may be involved.

3.8.1 Analysis of an RSSC Spatial Linkage

For a given value of θ, the position of all links in the RSSC linkage of Figure 3.16 may be found. Both graphical and analytical solutions are described in Chapter 2. If the angular velocity of link 1 is given, the velocity of spherical pair S_2 is given by

$$v_{S_2} = v_{S_1} + v_{S_2 S_1} = \omega_1 \times r_1 + \omega_2 \times r_2 \tag{3.41}$$

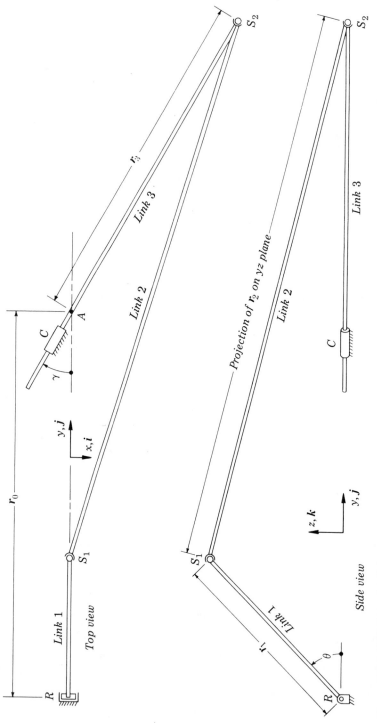

Figure 3.16 RSSC spatial linkage.

For the configuration shown, v_{s_2} lies in the xy plane and

$$\frac{v_{s_2x}}{v_{s_2y}} = \tan \gamma$$

a constant. The spherical joints permit link 2 to rotate about its own axis. This motion may be set equal to zero without affecting other kinematic aspects of the problem. Thus, we may write $\omega_2 \cdot r_2 = 0$. These relationships result in a set of simultaneous equations from which the unknown velocities are determined.

◆ EXAMPLE PROBLEM 3.8   RSSC Spatial Linkage

Here, a spatial linkage is analyzed using vector methods. Find link orientations and velocities for the RSSC spatial linkage of Figure 3.16 for $\gamma = 30°$, $\theta = 45°$, and crank angular velocity $\omega_1 = 50$ rad/s counterclockwise; $r_0 = 200$ mm, $r_1 = 100$ mm, and $r_2 = 300$ mm. Use vector notation.

Solution.   The links may be described in terms of unit vectors as follows:

$$r_0 = 0i \qquad -200j \qquad\qquad +0k$$
$$r_1 = 0i \qquad +100(\cos \theta)j \quad +100(\sin \theta)k$$
$$r_2 = r_{2x}i \quad +r_{2y}j \qquad\qquad +r_{2z}k$$
$$r_3 = r_{3x}i \quad +r_{3y}j \qquad\qquad +0k$$

The equation $r_0 + r_1 + r_2 + r_3 = 0$ applies to each vector direction. Thus, summing the i components, we have $r_{3x} = -r_{2x}$. Since $r_{3x} = r_{3y} \tan \gamma$, we have $r_{2x} = -r_{3y} \tan \gamma$. From the j components, $r_{2y} = 200 - 100 \cos \theta - r_{3y}$, and from the k components, $r_{2z} = -100 \sin \theta$. The length of link 2 is 300 mm, from which

$$r_{2x}^2 + r_{2y}^2 + r_{2x}^2 = 300^2$$

Substituting the given values of θ and γ, we have a quadratic equation in r_{3y}. The root of the equation that corresponds to the assembly configuration in the figure is

$$r_{3y} = -149.23$$

Using the above equations, we may then write

$$r_0 = 0i - 200j + 0k$$
$$r_1 = 0i + 70.71j + 70.71k$$
$$r_2 = 86.16i + 278.52j - 70.71k$$
$$r_3 = -86.16i - 149.23j + 0k$$

The velocity of link 3, the sliding link, is given by

$$v_{s_2} = v_{s_1} + v_{s_2s_1}$$

where

$$v_{s_1} = \omega_1 \times r_1$$

$$v_{s_2 s_1} = \omega_2 \times r_2$$

$$v_{s_2} = v_{s_2 x} i + v_{s_2 y} j + 0k$$

$$\omega_1 = 50i + 0j + 0k$$

$$\omega_2 = \omega_{2x} i + \omega_{2y} j + \omega_{2z} k$$

We ignore rotation of link 2 about its own axis, which does not affect the motion of the rest of the linkage. Thus we may set the component of ω_2 in the r_2 direction equal to zero, which is equivalent to stating $\omega_2 \cdot r_2 = 0$, from which

$$\omega_2 \cdot r_2 = 86.16\omega_{2x} + 278.52\omega_{2y} - 70.71\omega_{2z} = 0$$

The motion of link 3 is limited by the cylindrical joint so that

$$v_{s_2 x} = v_{s_2 y} \tan \gamma$$

This last equation may be used to reduce the problem to a system of four unknowns and four equations.

The velocity of the spherical joint on the crank is given by

$$v_{s_1} = \omega_1 \times r_1 = \begin{vmatrix} i & j & k \\ \omega_{1x} & \omega_{1y} & \omega_{1z} \\ r_{1x} & r_{1y} & r_{1z} \end{vmatrix}$$

$$= \begin{vmatrix} i & j & k \\ 50 & 0 & 0 \\ 0 & 70.71 & 70.71 \end{vmatrix}$$

$$= 0i - 3535.5j + 3535.5k$$

$$v_{s_2 s_1} = \omega_2 \times r_2 = \begin{vmatrix} i & j & k \\ \omega_{2x} & \omega_{2y} & \omega_{2z} \\ r_{2x} & r_{2y} & r_{2z} \end{vmatrix}$$

$$= \begin{vmatrix} i & j & k \\ \omega_{2x} & \omega_{2y} & \omega_{2z} \\ 86.16 & 278.52 & -70.71 \end{vmatrix}$$

$$= (-70.71\omega_{2y} - 278.52\omega_{2z})i$$

$$+ (86.16\omega_{2z} + 70.71\omega_{2x})j$$

$$+ (278.52\omega_{2x} - 86.16\omega_{2y})k$$

$$v_{s_2} = 0.5774 v_{s_2 y} i + v_{s_2 y} j + 0k$$

The velocity equation

$$v_{s_2} = v_{s_1} + v_{s_2 s_1}$$

applies in each vector direction.
From the i components,

$$0.5774 v_{s_2 y} = 0 - 70.71 \omega_{2y} - 278.52 \omega_{2z}$$

From the j components,

$$v_{s_2 y} = -3535.5 + 86.16 \omega_{2z} + 70.71 \omega_{2x}$$

From the k components,

$$0 = 3535.5 + 278.52 \omega_{2x} - 86.16 \omega_{2y}$$

Combining these with the result of $\omega_2 \cdot r_2 = 0$, we have

$$
\begin{bmatrix}
0 & 70.71 & 278.52 & 0.5774 \\
70.71 & 0 & 86.16 & -1 \\
278.52 & -86.16 & 0 & 0 \\
86.16 & 278.52 & -70.71 & 0
\end{bmatrix}
\begin{bmatrix}
\omega_{2x} \\
\omega_{2y} \\
\omega_{2z} \\
v_{s_2 y}
\end{bmatrix}
=
\begin{bmatrix}
0 \\
3535.5 \\
-3535.5 \\
0
\end{bmatrix}
$$

It is most convenient to use a calculator or computer program to solve this set of linear simultaneous equations. The results are

$$\omega_{2x} = -11.190$$

$$\omega_{2y} = 5.091$$

$$\omega_{2z} = 6.505$$

scalar

$$v_{s_2 y} = -3761$$

Noting the direction of v_{s_2}, we have $v_{s_2 x} = -2172$. The angular velocity of link 2 is given by

$$\omega_2 = -11.190 i + 5.091 j + 6.505 k$$

The velocity of the sliding link is given by

$$v_{s_2} = -2172 i - 3761 j$$

The relative velocity is

$$v_{s_2 s_1} = \boldsymbol{\omega}_2 \times r_2 = -2172i - 231j - 3535k$$

We may now check to ensure that the vector velocity equation $v_{s_2} = v_{s_1} + v_{s_2 s_1}$ is satisfied. ◆

3.8.2 Velocity Analysis of an RSSR Spatial Linkage

Displacement analysis of an RSSR spatial linkage was considered in Chapter 2. Since each link vector has a constant magnitude, differentiation of the vector position equation yields

$$\boldsymbol{\omega}_1 \times R_1 + \boldsymbol{\omega}_2 \times R_2 + \boldsymbol{\omega}_3 \times R_3 = 0 \tag{3.42}$$

If the angular velocity of link 1 is given, and all link positions are known from the displacement analysis, there are four unknowns: the three components of the angular velocity of link 2 and the angular velocity of link 3 (which is restrained to rotation in a particular plane by a revolute joint).

Recall that an RSSR linkage has two degrees of freedom, but we are not ordinarily concerned with rotation of link 2, the coupler, about its own axis. Thus we may set

$$\boldsymbol{\omega}_2 \cdot r_2 = 0 \tag{3.43}$$

Three more equations are necessary; they are obtained by considering the x, y, and z components of Eq. 3.42.

◆ **EXAMPLE PROBLEM 3.9** *Velocity Analysis of an RSSR Linkage*
Consider an RSSR linkage similar to that in Figure 1.6a, where link lengths are $r_0 = 3$, $r_1 = 1$, $r_2 = 3$, and $r_3 = 2$. Link 0 lies on the x axis. Link 1 rotates in the xy plane with an angular velocity of 1 rad/s (constant) and link 3 rotates in the xz plane. Plot the vector components representing the angular velocity of link 2 and the angular position and angular velocity of link 3 against angular position θ of link 1.

Solution. The vector components representing the link positions are determined as in Chapter 2. Then Eqs. 3.42 and 3.43 result in four simultaneous scalar equations which may be written in the matrix form

$$AX = B$$

where X is a column matrix of the four unknowns. Using mathematics software with matrix capability, the solution is given directly by

$$X = A^{-1}B$$

The equations and plotted results are shown in Figure 3.17. ◆

RSSR linkage. Link 1 rotates in xy plane; link 3 rotates in xz plane

$$r_0 = 3 \qquad r_1 = 1 \qquad r_2 = 3 \qquad r_3 = 2 \qquad \theta = 0, \frac{\pi}{12} \cdot \cdot 2 \cdot \pi$$

$$\mathbf{k} = \begin{bmatrix} 0 \\ 0 \\ 1 \end{bmatrix} \qquad \omega_1 = lk \qquad \alpha_1 = 0$$

$$c_x(\theta) = -r_0 + r_1 \cdot \cos(\theta) \qquad c_y(\theta) = r_1 \cdot \sin(\theta) \qquad r_{1x}(\theta) = r_1 \cdot \cos(\theta)$$

$$r_{2x}(\theta) = \frac{-r_2^2 + c_y(\theta)^2 + r_3^2 - c_x(\theta)^2}{2 \cdot c_x(\theta)} \qquad \begin{array}{l} r_{1y}(\theta) = r_1 \cdot \sin(\theta) \\ r_{3x}(\theta) = -c_x(\theta) - r_{2x}(\theta) \end{array}$$

$$r_{2y}(\theta) = -c_y(\theta) \qquad r_{2z}(\theta) = \left[r_3^2 - r_{3x}(\theta)^2 \right]^{.5} \qquad r_{3z}(\theta) = -r_{2z}(\theta)$$

$$\theta_3(\theta) = \text{angle}\left[r_{3z}(\theta), r_{3x}(\theta) \right]$$

$$A(\theta) = \begin{bmatrix} 0 & r_{2z}(\theta) & -r_{2y}(\theta) & r_{3z}(\theta) \\ -r_{2z}(\theta) & 0 & r_{2x}(\theta) & 0 \\ r_{2y}(\theta) & -r_{2x}(\theta) & 0 & -r_{3x}(\theta) \\ r_{2x}(\theta) & r_{2y}(\theta) & r_{2z}(\theta) & 0 \end{bmatrix} \qquad B(\theta) = |\omega_1| \cdot \begin{bmatrix} r_{1y}(\theta) \\ -r_{1x}(\theta) \\ 0 \\ 0 \end{bmatrix}$$

$$X(\theta) = A(\theta)^{-1} \cdot B(\theta)$$

$$\mathbf{I} = \begin{bmatrix} 1 \\ 0 \\ 0 \\ 0 \end{bmatrix} \qquad \mathbf{J} = \begin{bmatrix} 0 \\ 1 \\ 0 \\ 0 \end{bmatrix} \qquad \mathbf{K} = \begin{bmatrix} 0 \\ 0 \\ 1 \\ 0 \end{bmatrix} \qquad \mathbf{L} = \begin{bmatrix} 0 \\ 0 \\ 0 \\ 1 \end{bmatrix}$$

$$\omega_{2x}(\theta) = X(\theta) \cdot I \qquad \omega_{2y}(\theta) = X(\theta) \cdot J \qquad \omega_{2z}(\theta) = X(\theta) \cdot K$$

$$\omega_3(\theta) = X(\theta) \cdot L$$

$$\omega_2(\theta) = \left[\omega_{2x}(\theta)^2 + \omega_{2y}(\theta)^2 + \omega_{2z}(\theta)^2 \right]^{.5}$$

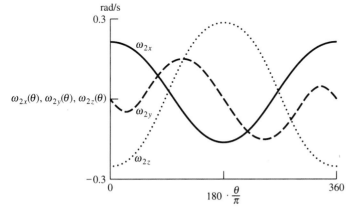

Figure 3.17 Velocity analysis of a spatial linkage (*continued on page* 188). (*continued on page* 188)

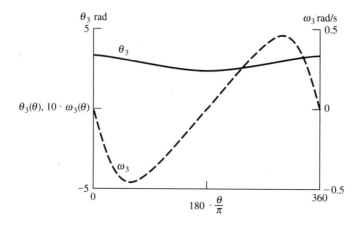

$180 \cdot \dfrac{\theta}{\pi}$	$\omega_2(\theta)$	$\theta_3(\theta)$	$\omega_3(\theta)$
0	0.333	3.267	0
30	0.297	3.165	− 0.357
60	0.178	2.94	− 0.46
90	0.144	2.712	− 0.397
120	0.24	2.533	− 0.28
150	0.31	2.422	− 0.145
180	0.333	2.384	0
210	0.31	2.422	0.145

Figure 3.17 *Continued.*

Mechanical systems software packages are useful in the solution of complicated spatial linkages. Figure 3.18 is a composite drawing of a satellite deploying panels mounted on flexible arms. It was necessary to find the speed of deployment of the arms which were driven by highly nonlinear rotary springs. This solution to this problem involves the response of the flexible system to forces which, in turn, depend on the instantaneous position of the arms. Analysis methods considered in this chapter are inadequate for solving the problem. The superimposed display of the deployment history shown in the figure was obtained with ADAMS[TM] software after inputting the results of tests of the springs and other system data. Solution methods for problems of this type are discussed in manuals furnished by Mechanical Dynamics, Inc.[7–9] Velocity analysis of spatial linkages is also treated by Angeles[1] and Lee and Liang.[6]

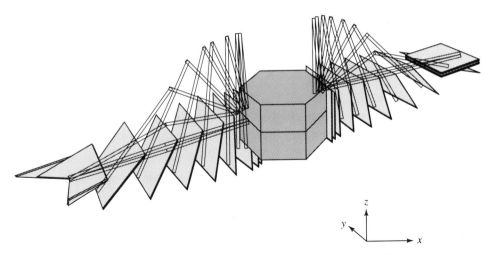

Figure 3.18 A satellite deploying flexible arms (modeled by ADAMS™ software). (*Source:* Mechanical Dynamics, Inc.)

3.9 GRAPHICAL ANALYSIS OF LINKAGE MOTION UTILIZING RELATIVE VELOCITY

It is desirable to have an independent method of solution for use in testing and debugging computing programs and programmable calculator procedures. Graphical solutions are ideal for this purpose. In addition, they provide insight into kinematic problems in a way that analytical solutions cannot.

Relative velocity was defined earlier in this chapter as the difference between velocities. For *points B and C on the same rigid link, the relative velocity v_{CB} must be perpendicular to line BC between the points.* This is demonstrated by the cross product relationship

$$\dot{r} = \boldsymbol{\omega} \times r$$

where $\boldsymbol{\omega}$ is the angular velocity of the link and vector r (of constant magnitude) represents line *BC*. If the relative velocity was not perpendicular to line *BC*, there would be a component of v_{CB} along line *BC*, representing a change in length. Obviously, real links deflect due to load, but these small strains are ordinarily negligible compared to the rigid-body motion.

When sliding occurs, we consider the motion of instantaneously coincident points. Then, for coincident points B_1 on link 1 and B_2 on link 2, relative velocity $v_{B_2B_1}$ is tangent to the relative path. To be more specific, the relative velocity of two points is tangent to the path that one point traces on the link on which the second point is defined. Both of these relationships are utilized to find velocities in linkages by graphical means.

Before a velocity analysis is performed, a position analysis must be made to determine the direction of all links. As observed in the previous sections, an analytical position analysis may be the most difficult part of the analysis. Graphical position analysis of a planar linkage is simple, since it is only necessary to draw the linkage to scale, generally using a compass, scale, protractor and straightedge.

3.9.1 Analyzing Motion of the In-Line Slider-Crank Mechanism

The slider-crank mechanism is a basic part of reciprocating engines, pumps, compressors, and other machines. See Figure 3.19. Figure 3.20a is a representation of an in-line slider crank. The sketch is further simplified in Figure 3.20b by showing only the centerlines, sizes, and angular positions of the links. Link O represents the frame, link 1 the crank, link 2 the connecting rod, and

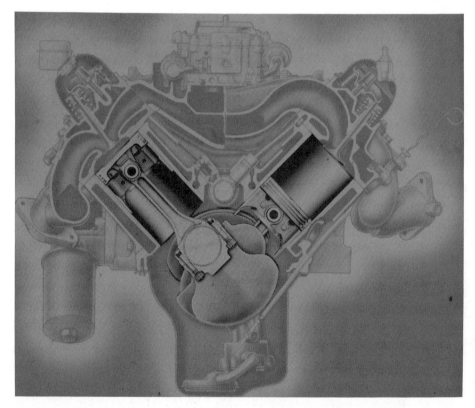

Figure 3.19 This section view of a V-8 engine shows two pistons and connecting rods (slider-crank mechanisms) at their extreme positions. The crankshaft represents the crank of the mechanisms. (*Source:* General Motors Corporation.)

link 3 the piston. The crankshaft center is point O_1, the crankpin point B and the wrist pin point C.

Suppose that point *B* has a velocity of 20 in/s as link 1 turns counter-clockwise. A velocity scale is selected that will result in vectors large enough for accurate results. Velocities v_B and v_{CB} are drawn perpendicular to lines O_1B and BC, respectively, as shown in Figure 3.20c, an exploded view of the mechanism. The direction of the velocity of C is horizontal because the piston is constrained to move within the cylinder.

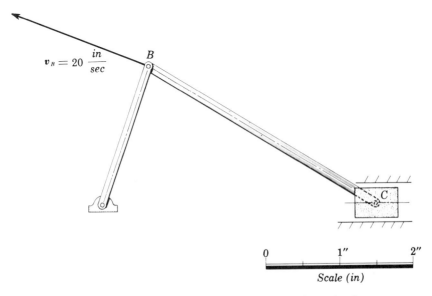

Figure 3.20 (a) Simplified sketch of an in-line slider-crank mechanism.

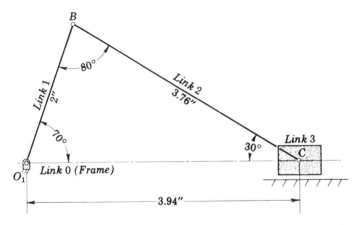

Figure 3.20 (b) Linkage diagram of the mechanism in Figure 3.21.

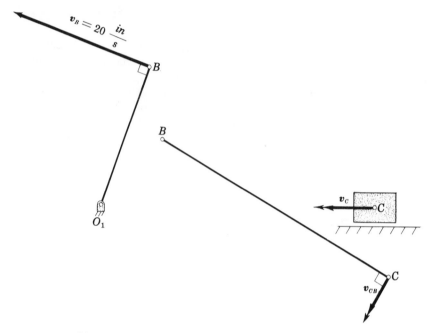

Figure 3.20 (c) Velocity vector v_B is perpendicular to link O_1B, v_{CB} is perpendicular to link BC, and v_C is constrained to a straight line as shown.

The single arrowhead of v_B in Figure 3.20c indicates that it has been drawn to scale to represent a known magnitude and direction. Vectors v_{CB} and v_c are given *double arrowheads* indicating that, while their directions are known, their magnitudes are not. When we draw a vector of unknown magnitude, we will call it a *trial vector*. The term *magnitude* will be interpreted to mean both vector length and vector sense. Thus, trial vector v_c may be to the left, as shown, or to the right. v_{CB} may be oriented as shown or it may be in exactly the opposite direction.

The solution to the problem of finding v_C, the velocity of the piston, is again based on the vector equation $v_C = v_B + v_{CB}$. Beginning at an arbitrary point o in Figure 3.20d, the vector v_B is drawn to scale. Then, trial vector v_{CB} is added to it starting at the head of v_B. Next, trial vector v_C is drawn beginning at the point o. Since $v_C = v_B + v_{CB}$, we have the equivalent of two simultaneous equations, one representing the line v_C and the other the line v_{CB} added to v_B. The solution is represented by the intersection. In Figure 3.20e, the double arrowheads have been replaced by single arrowheads since the magnitudes of the relative velocity v_{CB} and piston velocity v_C are determined by the construction. The vector lengths are measured, and, using the velocity scale, the velocities represented are written directly on the figure. We note that the piston velocity is 22.8 in/s.

Figure 3.20d has been redrawn in Figure 3.20e only to illustrate the steps in obtaining a solution; in actual practice, the construction would simply be

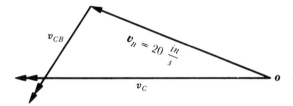

Figure 3.20 (d) The vectors can be added together to form a *velocity diagram*. Absolute velocity vectors (velocities relative to the fixed frame) are drawn from a common reference, point o. The relative velocity vector is drawn starting at the head of the known absolute velocity vector and with the correct direction ($\perp BC$). The points of intersection determine the magnitude of the unknown vectors.

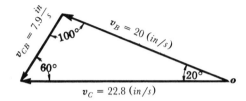

Figure 3.20 (e) The completed velocity diagram.

"cleaned up" and darkened for clarity. It can be seen that the solution does not depend on our ability to guess the correct sense of v_{CB} and v_C. If, for example, v_C were assumed to be to the right, there would be no intersection, and we would then try drawing it in the opposite direction and obtain the correct solution.

Sometimes the required accuracy is greater than can be obtained by a simple graphical solution. Or we may wish to make a velocity analysis based on freehand sketches without using drafting tools. Using the above example problem, let crank angle BO_1C be given as 70°.

The law of sines states that the ratio of the length of a side divided by the sine of the opposite angle is the same for all three sides in a triangle.

For the triangle formed by the linkage,

$$\frac{BC}{\sin(BO_1C)} = \frac{O_1B}{\sin(O_1CB)}$$

or

$$\frac{3.76}{\sin 70} = \frac{2}{\sin(O_1CB)}$$

from which angle $O_1CB = 30°$.

In the velocity diagram, v_C is horizontal. Vector v_B, which is perpendicular to crank O_1B, makes a 20° angle with v_C. Vector v_{CB}, perpendicular to link BC, makes a 60° angle with v_C. The sum of the internal angles in any triangle is 180°, from which we obtain the remaining angle in the velocity polygon: 100°. Using the law of sines once more,

$$\frac{v_B}{\sin 60°} = \frac{v_{CB}}{\sin 20°} = \frac{v_C}{\sin 100°}$$

Crankpin velocity v_B was given as 20 in/s. Substituting in the above equation, we obtain relative velocity $v_{CB} = 7.9$ in/s and piston velocity $v_C = 22.8$ in/s to the left.

3.10 THE VELOCITY POLYGON

If the relative velocity vector v_{CB} is replaced by the vector bc with c at the head of v_{CB}, we have the basis for an alternative form of notation for the method of relative velocity, the *velocity polygon*. See Figure 3.21. Absolute velocity v_B might as well have been called v_{BO}, the velocity of point B with respect to the frame O. Thus, v_B is replaced by ob and, likewise, v_C (or v_{CO}) by oc. The velocity equation $v_C = v_B + v_{CB}$ now becomes

$$oc = ob + bc \tag{3.44}$$

We identify ob with the actual velocity in millimeters per second or inches per second rather than letting ob mean a length in inches on a sketch.

The velocity polygon constructed in Figure 3.22 is based on the linkage of Figures 3.20a and b. In addition to oc, the piston velocity, we have determined bc, the velocity of the wrist pin C relative to the crankpin B. If bc is not zero,

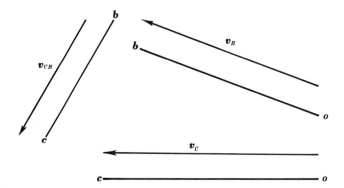

Figure 3.21 Alternative form of notation for velocity vectors. The notation v_B is replaced by ob; both indicate the velocity of point B with respect to point O. Thus v_C becomes oc. Note, however, that v_{CB}, the velocity of point C relative to point B, becomes bc (the letters are reversed).

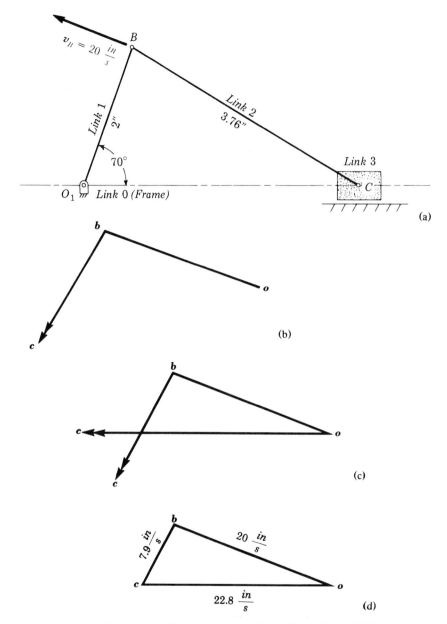

Figure 3.22 (a) Slider-crank linkage redrawn from Figure 3.20. (b) Velocity of point *B* (given) is drawn to scale as vector *ob* perpendicular to crank O_1B. From the head of *ob*, vector *bc*, the velocity of point *C* relative to point *B*, is drawn perpendicular to link *BC*. We do not know the magnitude of vector *bc*. (c) Point *C* is constrained to move horizontally. Thus vector *oc* is drawn horizontally to intersect vector *bc*. The intersection determines the lengths (magnitudes of the unknown vectors). (d) The completed velocity polygon.

the motion of the connecting rod BC includes rotation. The value of ω_2, the angular velocity of rod BC, is determined just as with links having a fixed center of rotation. It is equal to the relative velocity divided by the distance BC. Thus,

$$\omega_2 = \frac{bc}{BC} \tag{3.45}$$

where bc is the magnitude of the velocity of C relative to B and BC is the distance between pins B and C on the actual linkage, not a distance on the sketch. Substituting in Eq. 3.45, we find

$$\omega_2 = \frac{7.9 \text{ in/s}}{3.76 \text{ in}} = 2.1 \text{ rad/s}$$

Angular velocity direction of rod BC is found by locating the velocity of C relative to B at point C on the linkage sketch. The direction of the relative velocity given by the order of the letters bc is downward and to the left, as shown by the velocity diagram in Figure 3.22d. Therefore, ω_2 is clockwise as shown in Figure 3.23a. Note that while the piston velocity oc must be horizontal, there is a different restriction on the relative velocity bc; it must be perpendicular to BC.

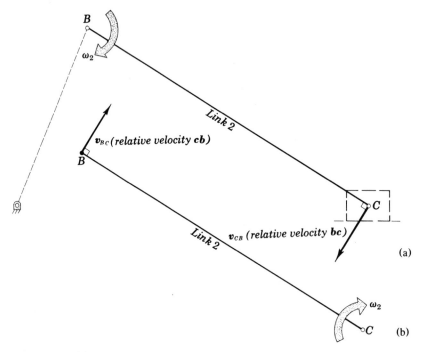

Figure 3.23 (a) Determining the angular velocity of link BC from bc, the velocity of point C relative to point B. (b) An alternative method.

Alternatively, the velocity of B relative to C may be located at point B to find the direction of ω_2. Since $cb = -bc$ (Figure 3.23b) the magnitude of ω_2 has the same value as found above.

Noting the letter order cb and inspecting Figure 3.22d, we see that cb is upward and to the right and, therefore, ω_2 is clockwise. The result is the same whether we consider bc at C or cb at B. In fact, the angular velocity of a link can be determined from the relative velocity of any two points on the link.

3.10.1 Layout Techniques

A few words about layout may be helpful at this point. In many linkage problems, the velocity polygon can be drawn with sufficient accuracy that measurements may be taken directly from it. But the care used in drawing the velocity polygon bears heavily on the results.

The mechanism to be analyzed should be sketched in skeleton form. Only link centerlines, pins, fixed centers, and sliders are shown. If a sketch must be copied, the use of tracing paper or dividers is preferred. The scale of the drawing is indicated, and the length of each link (from pin to pin) is shown directly on the link.

When one is working at a desk and using letter size paper, lines can be drawn parallel and perpendicular to one another by using two triangles, as shown in Figure 3.24. The paper should be taped in place to keep it from

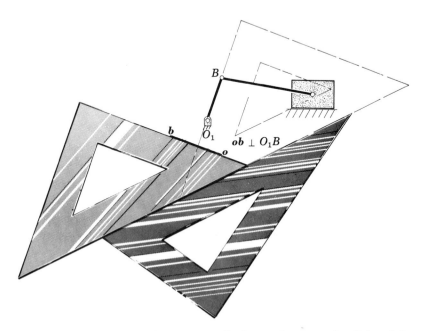

Figure 3.24 Vectors are drawn perpendicular to the respective links of the linkage diagram by using two triangles.

slipping since accuracy in both vector direction and vector length is critical. A ruler with decimal graduations is preferable to one graduated in sixteenths and thirty-seconds. The velocity polygon (and, later, the acceleration and force polygons) should be on the same sheet with the sketch of the mechanism in order to avoid vector orientation errors. A graphical solution for one link position may be used to check a computer solution.

3.10.2 Velocity Image

The utility of the velocity polygon notation is illustrated by problems where several points lie on the same link. Consider a rigid link in plane motion such as *BCD* of Figure 3.25. On any rigid link, each relative velocity is perpendicular to the line between the points considered. Thus:

bc ⊥ *BC*
bd ⊥ *BD*
cd ⊥ *CD* satisfy the conditions for similar triangles

Triangle *bcd* of the velocity polygon is similar to triangle *BCD*, the rigid link, and *bcd* is said to be the *velocity image* of rigid link *BCD*. As a result,

$$\frac{bc}{BC} = \frac{bd}{BD} = \frac{cd}{CD} \tag{3.46}$$

since corresponding sides of similar triangles are proportional. In order to draw *bcd* to the correct scale, however, we must know one of the relative velocities, for example, relative velocity *bc*. For any configuration of points on a rigid link, the velocity polygon contains the exact image except for size and orientation.

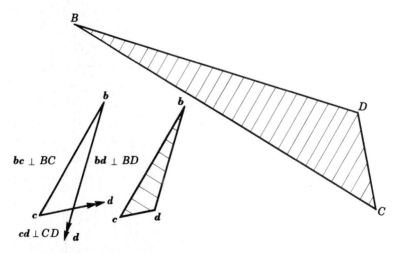

Figure 3.25 The velocity image of a link. The vectors are drawn perpendicular to the lines connecting the points on the link. The result is a similar triangle.

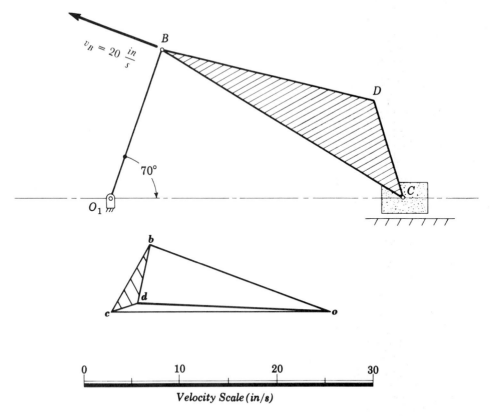

Figure 3.26 The velocity image of the connecting rod of a slider-crank mechanism.

Figure 3.26 shows a mechanism containing link BCD. Let distances O_1B, BC, and O_1C and the velocity of B be the same as in the example illustrated in Figure 3.22. Then we can take the velocity polygon *obc* directly from that figure. Drawing **bd** perpendicular to BD and **cd** perpendicular to CD, as in Figure 3.25, we obtain **bcd**, the image of BCD directly on the velocity polygon. The absolute velocity of D is found by measuring *od*, a vector of about 20-in/s magnitude to the left and slightly upward. The reader is again reminded that the velocity image principle applies only to points that lie on the same rigid link.

The path of point D is neither circular (like B) nor a straight line (like C), as can be shown by drawing the mechanism at several different crank positions. A point such as D on a mechanism may provide just the right motion required to perform a given task. In the design of machinery, it is often necessary to investigate a large number of mechanisms before the desired input-output relationship is obtained.

Let us now consider the velocity image of three points, B, C, and E lying on a straight line, all on the same rigid link as in Figure 3.27. Let the link have

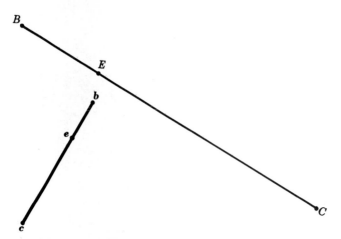

Figure 3.27 The velocity image of three points lying on a line in a rigid link is itself a line.

planar motion, which includes in general both rotation at an angular velocity ω and translation. The rotation gives us the following relative velocities:

$$\boldsymbol{bc} = BC\omega \quad \text{which is } \perp BC$$

$$\boldsymbol{be} = BE\omega \quad \text{which is } \perp BE \tag{3.47}$$

$$\boldsymbol{ec} = EC\omega \quad \text{which is } \perp EC$$

Dividing the second of Eqs. 3.47 by the first and the third by the first, we obtain

$$\frac{be}{bc} = \frac{BE}{BC} \tag{3.48}$$

and

$$\frac{ec}{bc} = \frac{EC}{BC} \tag{3.49}$$

In practice, the velocity image of points B, E, and C on one rigid link is obtained by using either Eq. 3.48 or 3.49 and the fact that the order of b, e, and c in the velocity polygon is the same as for B, E, and C on the link. One will recall the preceding discussion where the velocity image and link were similar triangles. If the points considered lie on a straight line, this becomes a special case, the triangles having angles $0°$, $0°$, and $180°$.

Let us examine the velocity of a point E lying on connecting rod BC, as in Figure 3.28a. The linkage is identical with that of Figure 3.22 (except for the addition of point E), and B will again be given a velocity of 20 in/s. The velocity polygon in Figure 3.28b may be taken directly from Figure 3.22,

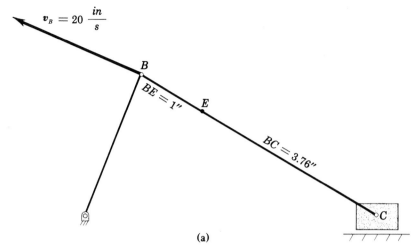

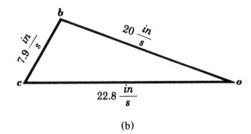

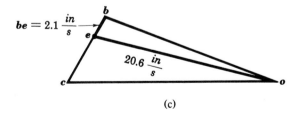

Figure 3.28 (a) The problem is to find the velocity of point E (which could be the center of gravity) at the instant when the linkage is in the position shown.

Figure 3.28 (b) The velocity polygon of the entire linkage is first drawn. (c) With the velocity image principle, point e is located on vector bc, the velocity image of link BC. The velocity of point E is found by drawing oe on the velocity polygon and measuring its length.

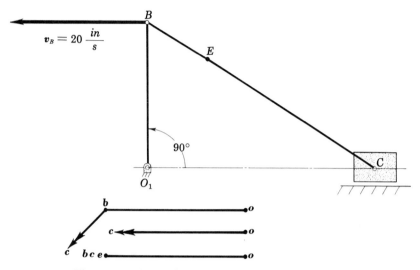

Figure 3.29 The connecting rod is shown at the instant when its angular velocity is zero. *At this instant, there is no relative velocity* $(bc = 0)$. The velocity image of the connecting rod therefore shrinks to a single point. While *ob* and *oc* are shown parallel, they are actually colinear.

leaving only point *e* to be found. From Eq. 3.48

$$\frac{be}{bc} = \frac{BE}{BC} \quad \text{or} \quad \frac{be}{7.9 \text{ in/s}} = \frac{1 \text{ in}}{3.76 \text{ in}}$$

from which $be = 2.1$ in/s.

Point *e* is located a distance from *b* corresponding to 2.1 in/s. Since E falls between B and C, *e* falls between *b* and *c*. Scaling the vector *oe*, we find the velocity of E to be approximately 20.6 in/s upward and to the left. See Figure 3.28c.

While the velocity image relationships hold in every case, the velocity image of a link moving in translation shrinks to a single point. This is true in the case of a slider that always translates and in the case of a connecting rod at the instant when its angular velocity is zero. The latter case is illustrated by the in-line slider-crank mechanism of Figure 3.29. When the crank angle is 90°, *ob*, the velocity of point B, is horizontal. The slider velocity *oc* is horizontal, and thus, *ob* and *oc* are colinear for an instant.

3.11 GRAPHICAL ANALYSIS OF BASIC LINKAGES

3.11.1 The Four-Bar Linkage

Although the slider-crank mechanism discussed in the preceding sections is a four-bar linkage, this term is ordinarily used to denote a mechanism made up

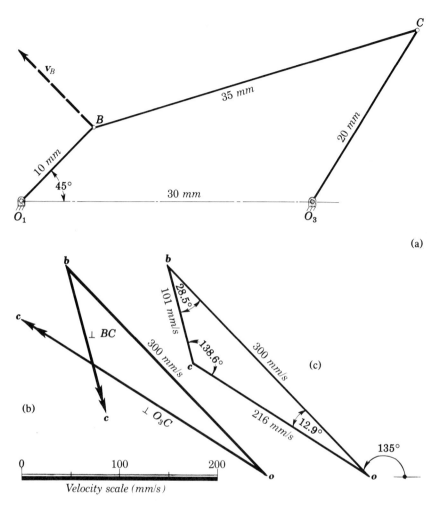

Figure 3.30 (a) A four-bar linkage. The velocity of point *B* is given. (b) Once a convenient scale is determined, the velocity polygon for the four-bar linkage is begun. While we know the magnitude of only one of the vectors, we know the directions (perpendicular to the links) for all the vectors. (c) The completed velocity polygon. The points of intersection determine the unknown vector quantities. Velocity magnitudes are finally obtained by measuring the vector lengths against the velocity scale.

of two cranks, a connecting rod, and a fixed link (the frame). A common four-bar linkage of this type has the driving crank rotating while the output crank oscillates. The velocity analysis differs little from analysis of the slider-crank mechanism. For the mechanism of Figure 3.30a, the velocity of *B* is given as 300 mm/s at the instant shown as the crank rotates counterclockwise.

In order to find all of the velocities, we select a velocity scale, Figure 3.30b, and the vector **ob** is drawn perpendicular to O_1B to represent the velocity of B. Relative velocity vector $\boldsymbol{bc}$ of unknown length is drawn perpendicular to BC, starting at $\boldsymbol{b}$. The velocity polygon is completed by drawing $\boldsymbol{oc}$ beginning at $\boldsymbol{o}$ and perpendicular to O_3C. The last two steps locate $\boldsymbol{c}$ on the velocity polygon. Figure 3.30c shows the velocity polygon with all construction lines removed and the values of the vectors shown directly on the polygon. We see that

$$\boldsymbol{bc} = \boldsymbol{v}_{CB} = 101 \text{ mm/s} \perp BC \quad \text{(downward and to the right)}$$

$$\boldsymbol{oc} = \boldsymbol{v}_C = 216 \text{ mm/s} \perp O_3C \quad \text{(upward and to the left)}$$

From these results, the angular velocities of the coupler 2 and follower 3 can be determined as follows:

$$\omega_2 = \frac{bc}{BC} = \frac{101}{35} = 2.89 \text{ rad/s cw}$$

and

$$\omega_3 = \frac{oc}{O_3C} = \frac{216}{20} = 10.8 \text{ rad/s ccw}$$

The above values can also be found by drawing line BO_3 on the mechanism to form two triangles and solving the triangles by the cosine law and the law of sines:

$$z^2 = x^2 + y^2 - 2xy \cos Z$$

and

$$\frac{\sin X}{x} = \frac{\sin Y}{y} = \frac{\sin Z}{z} \tag{3.50}$$

where X, Y, and Z represent the internal angles opposite the sides x, y, and z, respectively. In addition, we use the relation $X + Y + Z = 180°$ for any triangle and note that velocity directions differ from link directions by $90°$ to draw velocity polygon $\boldsymbol{obc}$.

3.11.2 Analyzing Sliding Contact Linkages

Sliding contact exists between slider and frame in the slider-crank mechanism. In cams, gears, and certain other mechanisms, moving links slide on one another. If a point on one link slides in a curved path on a second link, the relative velocity of the common points is tangent to the path described on the second link. In the example that follows, the relative path is straight.

The mechanism of Figure 3.31a has a slider pinned to link 1. The slider is constrained to slide along link 2. This mechanism is basic to the mechanically driven shaper and is utilized in combination with other linkages, like the jackknife boom shown in Figure 3.32. The key to solving problems of this type is the designation of a double point B. See Figure 3.31a. B_1 is a point on the slider and on link 1, and B_2 is a common point on link 2. While, at this instant,

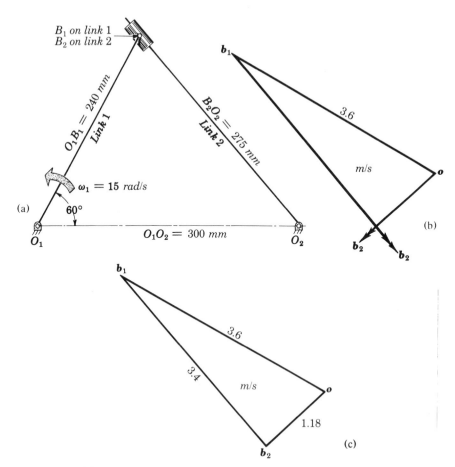

Figure 3.31 (a) A sliding contact linkage. The slider is pinned to link 1 and slides along link 2. (b) The velocity polygon for a sliding contact linkage. Note that the relative velocity vector b_2b_1 is drawn parallel to link 2, since the motion of B_1 relative to B_2 must be along link 2 at any given instant. (c) The completed polygon.

B_1 and B_2 are the same point, B_1 moves relative to B_2 by sliding along link 2. Thus the direction of relative velocity b_2b_1 (the velocity of B_1 with respect to B_2) is along link 2. Relative velocity b_1b_2 (equal and opposite b_2b_1) is therefore also along link 2.

Contrast the relative velocity of two coincident points on different links with the relative velocity of two points on the same link. In cases where two points lie on the same rigid link (considered in earlier sections), the relative velocity is perpendicular to the line between the two points.

Suppose link 1 of Figure 3.31a rotates counterclockwise at 15 rad/s, making the velocity of B_1 equal to 3.6 m/s perpendicular to O_1B_1, upward and to the left. To solve the mechanism, we use the relationship

$$v_{B2} = v_{B1} + v_{B2B1} \quad \text{or} \quad ob_2 = ob_1 + b_1b_2 \tag{3.51}$$

Figure 3.32 A jackknife boom. The boom may be represented by a combination of four-bar and sliding contact linkages. (*Source:* Dorsey Trailers, Inc./Holan Equipment.)

Vector ob_1 (representing the velocity of B_1) is drawn to scale in Figure 3.31b, beginning at an arbitrary pole point o. Sliding velocity vector b_1b_2 is added to ob_1 beginning at b_1. The vector direction of b_1b_2 is parallel to link 2 but its length and sense are unknown. Thus, we draw trial vector b_1b_2 with a double arrowhead at b_2 (not caring about vector sense because, if our original guess of vector sense is wrong, we later reverse the vector to obtain the velocity polygon). Trial vector ob_2 (also of unknown sense and magnitude) is drawn starting at pole point o and perpendicular to O_2B_2. The length (and, if necessary, the sense) of both vectors b_1b_2 and ob_2 is corrected, completing the velocity polygon ob_1b_2 in Figure 3.31c.

Scaling vector b_1b_2, we find the sliding velocity to be 3.4 m/s; at this instant, link 2 moves downward and to the right relative to the slider at 3.4 m/s. The order of the subscripts is important. Vector b_2b_1 refers to the velocity of point B on link 1 relative to the coincident point on link 2. Thus, the slider moves upward and to the left at 3.4 m/s relative to link 2. The velocity of B_2 on link 2 scales to 1.18 m/s downward and to the left, from which we obtain

$$\omega_2 = \frac{ob_2}{O_2B_2} = \frac{1.18 \text{ m/s}}{0.275 \text{ m}} = 4.29 \text{ rad/s}$$

counterclockwise. Note that ob_2 and O_2B_2 refer, respectively, to actual velocity of B_2 and length O_2B_2 on the actual link 2. The method for determining the velocities of the sliding contact linkage is essentially the same, even if links 1 and 2 are curved. Velocity vector ob_1 is perpendicular to O_1B_1 and ob_2 perpendicular to O_2B_2. Sliding velocity b_1b_2 is in the direction of the relative path, that is, b_1b_2 is *tangent* to the instantaneous path of B_1 on link 2 (at B_1). If greater accuracy is desired, the velocity polygon may be solved analytically by using the law of sines, as was done in an earlier example.

Sliding velocity is of particular interest because of friction and wear considerations. (Some references state the coefficient of friction in terms of sliding velocity.) In addition, we must find the sliding velocity in order to compute the Coriolis acceleration. This phase of the problem is treated in the chapter that follows.

Comparison of Results with an Analytical Solution

A sliding contact linkage was analyzed previously by using complex number methods (Section 3.7). With the notation of that section (see Figure 3.15), $\theta_1 = 240°$, $R_0 = 300$ mm, $R_1 = 240$ mm, and $\omega_1 = 15$ rad/s, from which we obtain

$$R_2 = \sqrt{R_0^2 + R_1^2 - 2R_0R_1 \cos\theta_1} = 275.0 \text{ mm}$$

$$\theta_2 = \arcsin\left(\frac{R_1 \sin\theta_1}{R_2}\right) = -49.1°$$

$$\omega_2 = \frac{\omega_1 R_1 \cos(\theta_1 - \theta_2)}{R_2} = 4.285 \text{ rad/s} \quad (\text{ccw})$$

$$v_{B_2} = \omega_2 R_2 = 1178 \text{ mm/s} \quad (1.178 \text{ m/s})$$

$$v_{B_1B_2} = -\omega_1 R_1 \sin(\theta_1 - \theta_2) = 3402 \text{ mm/s}$$

(3.402 m/s along link 2, away from point O_2). All of these values correspond closely to the graphical (velocity polygon) solution.

The reader should be alert for mechanisms that are kinematically equivalent to the sliding contact linkage of Figure 3.31. Examples include a variable-displacement pump, in which the plungers move within a rotating cylinder block, and the Geneva mechanism. In the Geneva mechanism, a pin on a rotating wheel (the driver) enters radial slots in the driven member, giving it intermittent rotation as the driver rotates at constant velocity.

3.11.3 Cams and Cam Followers

Almost any motion-time relationship may be generated by using one or more cams. Usually, the cam rotates at constant angular velocity, giving the follower reciprocating or oscillating motion having some predetermined sequence. *When sliding occurs between cam and follower, the key to solving for velocities is again a double point where the two make contact, and the solution proceeds as with other sliding contact mechanisms.* The velocity of the point of contact of the follower is equal to the vector sum of the velocity of the point of contact on the cam plus the sliding velocity. If B is the point of tangency, the above statement may be expressed symbolically as

$$ob_2 = ob_1 + b_1b_2$$

where subscripts 1 and 2 refer, respectively, to cam and follower.

In the case of a cam with an oscillating follower, Figure 3.33a, ob_2 is perpendicular to a line between the center of rotation of the follower and the contact point and ob_1 is perpendicular to a line between the center of rotation of the cam and the contact point. Sliding velocity b_1b_2 is parallel to the common tangent to cam and follower.

Figure 3.33a shows a cam formed by an eccentric circle. Let cam angular velocity be 20 rad/s and, at the instant shown, let the distance from the center of rotation of the cam to the point where it makes contact with the follower be 1.9 in. Thus velocity $ob_1 = 38$ in/s is drawn to scale (to the right and downward beginning at an arbitrary pole point o). Trial vector b_1b_2 is added to ob_1 and trial vector ob_2 is drawn beginning at point o. The two trial vectors are made to intersect and the intersection is labeled b_2.

Sliding velocity, scaled from the velocity polygon, is 39 in/s with B_2 on the follower sliding upward and to the left with respect to B_1 on the cam (or B_1 on the cam sliding downward and to the right with respect to B_2 on the follower). The velocity of B_2, found by scaling vector ob_2, is 8 in/s upward and to the right, from which follower angular velocity

$$\omega_2 = \frac{ob_2}{O_2B_2} = \frac{8 \text{ in/s}}{1.5 \text{ in}} = 5.3 \text{ rad/s}$$

clockwise. At this instant, then, the ratio of follower angular velocity to cam angular velocity is

$$\frac{\omega_2}{\omega_1} = \frac{5.3 \text{ rad/s}}{20 \text{ rad/s}} = +0.27$$

where a positive sign is used when both turn in the same direction. In order to analyze motion of this mechanism, we could redraw it for several cam positions and plot the results.

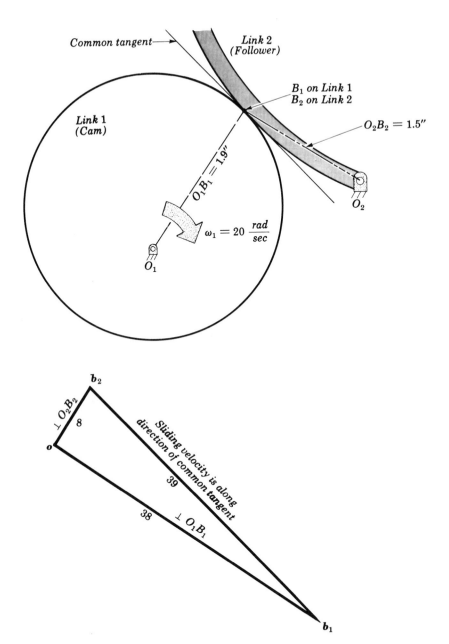

Figure 3.33 (a) A cam with an oscillating follower. To obtain the velocity polygon, $\boldsymbol{ob_2}$ is drawn perpendicular to O_2B_2 and $\boldsymbol{ob_1}$ is drawn perpendicular to O_1B_1. Sliding velocity $\boldsymbol{b_1b_2}$ is drawn parallel to the common tangent at B_1B_2.

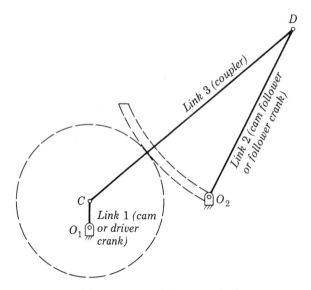

Figure 3.33 (b) A four-bar linkage, equivalent to a cam and follower.

An Equivalent Linkage for a Cam Mechanism

If, at the point of contact, the cam and follower both have finite radii, then a four-bar linkage may be used to analyze the motion. Referring to Figure 3.33a, for example, the equivalent links would form a crank rocker mechanism as sketched in Figure 3.33b. The driver crank would consist of a link from O_1 to the center of curvature of the cam at the point of contact (C, the center of the circular cam in this case). The driven crank (the rocker) would consist of a link from O_2 to D, the center of curvature of the follower at the point of contact. The connecting rod CD would extend form one center of curvature to the other.

Equivalent link CD is the common normal to the surfaces of the cam and follower at the point of contact. All points in the cam can be considered rigidly connected to link 1 and all points in the follower can be considered rigidly connected to link 2. Thus angular velocities in the cam and follower are represented by angular velocities in links 1 and 2, respectively.

In the above examples, both cam and follower have constant radii of curvature. If, instead, the cam surface consisted of a series of circular arcs, then a different four-bar linkage would be required to represent each segment of the surface.

When either the cam or follower has a straight section (infinite radius of curvature) at the point of contact, then the cam and follower are equivalent to a sliding-contact linkage. The basic analysis using the sliding velocity component (similar to that illustrated in Figure 3.33a) is still valid.

In most cases, a cam is produced to generate a certain predetermined follower displacement versus time or follower displacement versus cam angle relationship. For example, the relationship could be that follower displacement x is expressed as a fifth-order polynomial in cam angle θ. Then, this expression can be differentiated to obtain follower velocity, acceleration, and jerk (rate of change in acceleration) at any cam angle for given cam angular velocity. Cam design for various analytical follower displacement versus cam angle relationships (synthesis) is treated in Chapter 5.

The Scotch Yoke

The Scotch yoke (see Figure 3.34) is kinematically equivalent to the eccentric cam considered analytically in Chapter 2. Construction of the velocity polygon proceeds as follows.

Step 1. The velocity of the pin on link 1 is $oc_1 = R\omega$ perpendicular to O_1C.

Step 2. Trial vector oc_2 is drawn in a vertical direction to represent the velocity of link 2 (the common point C_2 on link 2). Then, the trial vector representing sliding velocity c_1c_2 is drawn in a horizontal direction, locating velocity point c_2.

Step 3. It is noted that velocity vectors oc_1 and c_1c_2 meet at an angle θ, from which

$$oc_2 = oc_1 \sin \theta \quad \text{or} \quad v = R\omega \sin \theta$$

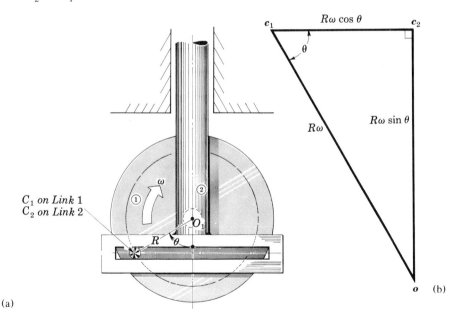

(a)

Figure 3.34 (a) Scotch yoke. Pin C is attached to wheel 1. The pin is constrained to ride in the slot, driving the follower, yoke 2. The Scotch yoke is kinematically equivalent to the eccentric cam, which is described analytically in Chapter 2. (b) The velocity polygon for the Scotch yoke. A common point (C_1 on link 1, C_2 on link 2) is the key to the solution.

The sliding velocity, incidentally, is given by

$$c_1 c_2 = oc_1 \cos \theta = R\omega \cos \theta$$

3.11.4 Friction Drives

Motion may be transmitted between two shafts by disks that roll on one another. Consider the friction drive of Figure 3.35, where P is a point common to both disks. *If they roll together without slipping, the relative velocity $p_1 p_2$ is zero, and Eq. 3.51 becomes*

$$op_2 = op_1 = \omega_1 O_1 P_1 \tag{3.52}$$

Since $\omega_2 = op_2/O_2 P_2$, but in a direction opposite ω_1, we have the angular velocity ratio

$$\frac{\omega_2}{\omega_1} = -\frac{O_1 P_1}{O_2 P_2} \tag{3.53}$$

A drive of this type is quite satisfactory for a small motion picture projector or other low-power applications. However, when large torques are involved, the designer might turn to a gear drive, sacrificing the simplicity of a friction drive to ensure that power will be transmitted under all conditions.

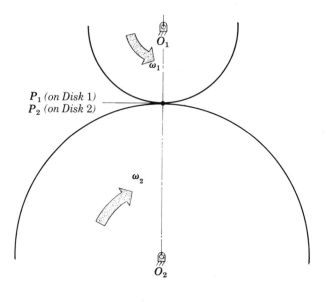

$$op_2 = op_1 = \omega_1 O_1 P_1$$

$$\frac{\omega_2}{\omega_1} = -\frac{O_1 P_1}{O_2 P_2}$$

Figure 3.35 A friction drive. There is no relative velocity ($p_1 p_2 = 0$) if the disks roll without slipping.

3.11.5 Straight-Toothed Spur Gears

Spur gear velocities may be found by examining a pair of teeth *at their point of contact*. The velocity of that point on the driven gear is the vector sum of the velocity of the same point on the driver and the relative velocity. The equation used earlier for cams and mechanisms that include sliding also applies to gears: $ob_2 = ob_1 + b_1b_2$. In this case, the common point or point of contact is B with subscripts 1 and 2 referring to driver and driven gears, respectively.

Figure 3.36a shows a pair of involute spur gears with contact occurring at point B. Velocity vector $ob_1 = \omega_1 O_1 B_1$ is drawn perpendicular to $O_1 B_1$ in Figure 3.36b. Then, trial vector ob_2 is drawn, beginning at o and perpendicular to $O_2 B_2$. Trial vector $b_1 b_2$ (sliding velocity) is drawn from b_1 (parallel to the common tangent to the teeth where they contact) until it intersects trial vector ob_2. The true location of b_2 is thus found, and the angular velocity of gear 2 is given by $\omega_2 = ob_2/O_2 B_2$. Figure 3.36c shows the velocity polygon for contact on the line of centers (the pitch point).

The above construction would not be used to find angular velocities, however, since the angular velocity ratio of a pair of gears is given simply by the inverse ratio of the numbers of teeth N_1 and N_2:

$$\frac{|\omega_2|}{|\omega_1|} = \frac{N_1}{N_2} \tag{3.54}$$

This relationship, which holds for any pair of gears except those in planetary trains, will be derived in Chapters 6 and 7.

Helical Gears

While the tooth elements of straight spur gears are parallel to the gear shaft, helical gear tooth elements are not. When a pair of helical gears are mounted on parallel shafts, the sliding velocity vector for any contact point will lie in a plane perpendicular to the shafts (as with straight spur gears). This is not the case with crossed helical gears.

Crossed Helical Gears

In the case of crossed helical gears (helical gears on nonparallel shafts), the sliding velocity at a general point of contact has a component across the face of the gears. See Figure 3.37a. Let a pair of crossed helical gears make contact at a point B, the pitch point (B_1 on gear 1, B_2 on gear 2), which lies on a common perpendicular to the two shafts. The velocity of B_1, represented by ob_1, is $\omega_1 d_1/2$, perpendicular to shaft 1. The velocity of B_2, unknown at this time, is represented by ob_2 perpendicular to shaft 2, using a double arrowhead. Velocity $ob_2 = ob_1 + b_1b_2$, where the relative velocity b_1b_2 must be the sliding velocity *parallel to the gear tooth* in a plane through B that is parallel to both shafts (see Figure 3.37b). The point b_2 is therefore located by drawing b_1b_2 *parallel to the gear tooth* until it meets the line ob_2, Figure 3.37c. The

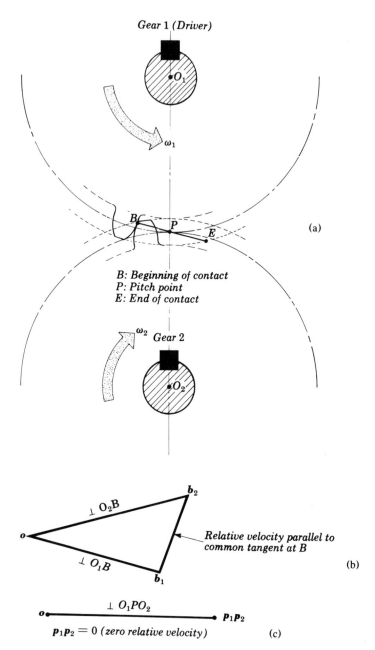

Gear 1 (Driver)

ω_1

B

P

E

(a)

B: Beginning of contact
P: Pitch point
E: End of contact

ω_2 Gear 2

b_2

$\perp O_2B$

Relative velocity parallel to
common tangent at B

(b)

$\perp O_1B$

b_1

$\perp O_1PO_2$

p_1p_2

$p_1p_2 = 0$ (zero relative velocity)   (c)

Figure 3.36 (a) A pair of spur gears is shown. For clarity, only
two teeth are shown. Point B is the point of initial contact (B_1
on gear 1, B_2 on gear 2). As the gears rotate, the point of
contact will follow line BPE. (b) The velocity polygon. Vectors
ob_1 and ob_2 are drawn perpendicular, respectively, to O_1B_1
and O_2B_2. Relative velocity vector b_1b_2 is drawn from b_1
parallel to the common tangent to the teeth at the point of
contact B. (c) The velocity polygon when contact occurs at the
pitch point P. At the pitch point, vectors op_1 and op_2 of the
vector polygon become colinear.

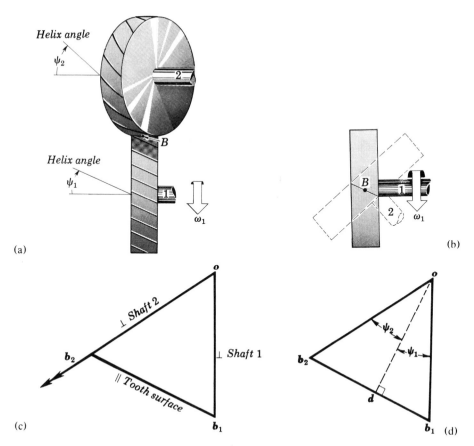

Figure 3.37 (a) Crossed helical gears. Pitch diameters of gears 1 and 2 are d_1 and d_2, respectively. (b) Section view of the crossed helical gears. Gear 2 is shown in dashed lines. (c) Velocity polygon showing pitch line velocities ob_1 and ob_2 and sliding velocity b_1b_2. (d) Line od is perpendicular to sliding velocity b_1b_2.

pitch line velocity of gear 2, ob_2, may then be scaled to find angular velocity

$$\omega_2 = \frac{ob_2}{d_2/2}$$

We see from the direction of ob_2 that gear 2 turns clockwise when viewed from the right.

Greater accuracy may be obtained analytically. In Figure 3.37d, a perpendicular, od, from o to line b_1b_2 forms angles ψ_1 and ψ_2. It can be seen that $od = ob_1 \cos \psi_1 = ob_2 \cos \psi_2$. Substituting $ob_1 = \omega_1 d_1/2$ and $ob_2 = \omega_2 d_2/2$, the velocity ratio

$$\frac{n_2}{n_1} = \frac{\omega_2}{\omega_1} = \frac{d_1 \cos \psi_1}{d_2 \cos \psi_2} \tag{3.55}$$

is obtained. Alternatively, the tooth numbers, if known, may be used to find the velocity ratio

$$\left| \frac{n_2}{n_1} \right| = \frac{N_1}{N_2} \tag{3.56}$$

A section view through the point of contact will aid in establishing the direction of rotation.

When we consider contact at a point other than the pitch point, there are additional sliding velocity components to be considered. Angular velocity ratio, however, is constant and may be found by one of the methods described above.

Gears and cams are among the most commonly used and most versatile mechanisms. The above material on velocities only touches on the problem of analysis and design of gears and cams, which is covered in considerably more detail in other sections.

3.12 ANALYZING COMBINATIONS OF BASIC LINKAGES

3.12.1 Toggle Linkage

Many practical multilink mechanisms are made up of basic linkage combinations such as slider-crank and four-bar mechanisms. The *toggle linkage* shown in Figure 3.38a is an example of a mechanism of this type; the toggle principle is applied in ore crushers and in essentially static linkages that act as clamps. The linkage analysis is made by considering the basic linkages separately. To solve for the velocities of the mechanism, we may begin by ignoring the slider and its connecting rod while solving the four-bar linkage separately. Then, the velocity polygon may be completed by finding slider velocity.

In this example, link 1 will be 100 mm long and ω_1 will be 100 rad/s clockwise. Examining links 1, 2, and 3 alone (Figure 3.38b), we draw the following vectors to some convenient scale:

Vector $ob = \omega_1 O_1 B = 10$ m/s $\perp O_1 B$ (downward and to the left)
Trial vector $oc \perp O_3 C$
Trial vector $bc \perp BC$ (Figure 3.38c)

Velocity point c is located at the intersection of trial vectors oc and bc, completing the velocity polygon for links 1, 2, and 3.

In order to illustrate the mental process involved in solving multilink mechanisms, we will now consider links 3 and 4 and the slider separately (Figure 3.38d). Beginning at a new pole point (o in Figure 3.38e), vector oc is redrawn, its direction and magnitude taken from the four-bar polygon. Then, trial vectors are drawn, $cd \perp CD$ and od in the direction of the path of the slider, from which velocity point d is located. Slider velocity at this instant is given by od, 4 m/s to the left.

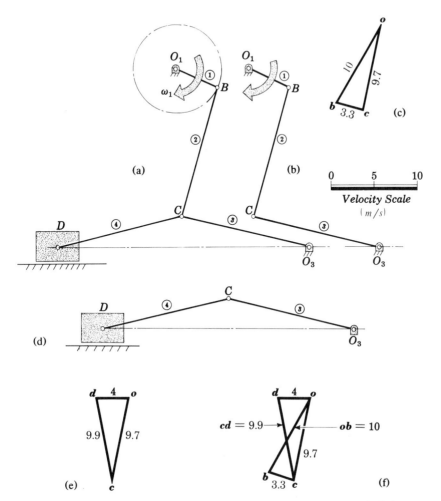

Figure 3.38 (a) This toggle mechanism is a combination of two basic linkages, the four-bar mechanism and the slider-crank mechanism. (b) The velocities for the entire toggle mechanism are found by solving for the velocities of the component mechanisms. Thus the four-bar linkage component is considered first (link 4 is not shown). (c) After a suitable velocity scale is selected, the velocity polygon for the four-bar mechanism is drawn. (d) The remainder of the toggle mechanism, the slider crank, is next considered. (e) The velocity polygon for the slider crank is drawn. Note that link O_3C is common to both mechanisms. The velocity of point C, oc, was already determined in the four-bar polygon. (f) The velocity polygon for the entire mechanism.

*The above solution in two parts is given for demonstration purposes only. After velocity polygon **obc** is completed, it is faster and slightly more accurate to continue by finding point **d** on the same polygon. Figure 3.38f shows velocity polygon **obcd** found in this manner, a compact velocity representation for the entire mechanism. An additional advantage of the complete polygon is that it is only necessary to join points **b** and **d** to find the velocity of D with respect to B if this velocity is of importance in a particular machine.*

An important feature of the toggle mechanism is its ability to produce high values of force at the slider with relatively low torque input. While the study of mechanisms is primarily concerned with motion, forces are of great importance to the designer and intimately related to motion analysis. If a rigid mechanism has a single input and single output with negligible losses, the rate of energy input equals the rate of output, from which force ratios are the inverse of velocity ratios. Specifically, in the toggle linkage at the instant shown, the horizontal force at D divided by the tangential force at B equals ob/od or $10/4$ (the mechanical advantage of the mechanism at this instant). Clockwise rotation of link 1 from the position shown toward the limiting position of the mechanism produces very high ratios of output force to input torque.

If we sketch the mechanism at the instant that links 3 and 4 are colinear, we see that slider velocity $od = 0$. Thus, the ratio ob/od, the theoretical mechanical advantage of the toggle mechanism, becomes infinite. Clamps and ore crushers using the toggle linkage principle are designed to operate near this limiting position. Actual forces at the slider are, of course, finite due to bearing clearances and elasticity of the linkage. In this exceptional case, the small amount of motion due to elastic deformation of the linkage and deformation of the workpiece is of the same order of magnitude as the slider motion. Therefore, any analysis of this problem (which assumes perfectly rigid links) must serve only as a "first approximation."

3.12.2 Shaper Mechanisms

The *mechanical shaper mechanism* in Figure 3.39 is another example of a combination of simple linkages. It is made up of a slider-crank mechanism (links 2 and 3 and the slider at D) and a sliding contact mechanism (links 1 and 2 and a slider at B moving along link 2). When the mechanism is operating, the angular velocity of link 1 is essentially constant.

Beginning with the velocity of point B at the end of the crank, we find the velocity of a common point on link 2. The velocity of point C at the end of link 2 is found by forming the velocity image of link 2. The solution is completed by examining the mechanism formed by links 2 and 3 and the slider at D (a slider crank of unusual proportions).

A detailed solution follows for the instant shown with the angular velocity of link 1 equal to 10 rad/s clockwise.

Step 1. Select a reasonable velocity scale and draw velocity vector $ob_1 = \omega_1 O_1 B_1 = 20$ in/s perpendicular to link 1 (downward and to the right), as shown in Figure 3.39.

Step 2. Draw trial vector ob_2 perpendicular to link 2 and trial vector $b_1 b_2$ parallel to link 2, locating b_2.

Step 3. Use the proportion $oc/ob_2 = O_2 C/O_2 B_2$ to locate point c on the velocity polygon.

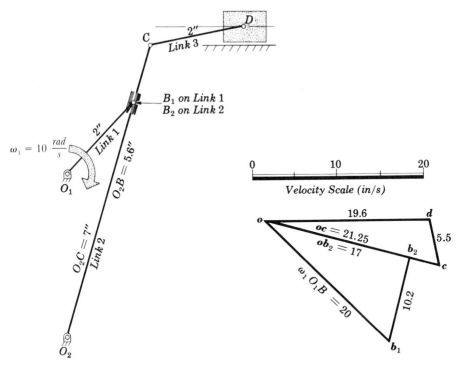

Figure 3.39 A mechanical shaper mechanism is another combination of two simple mechanisms. The velocity diagram is constructed as usual. To find the magnitude of *oc*, however, we must use velocity *ob*$_2$ and the proportionality of link 2 (O_2C to O_2B_2).

Step 4. Draw trial vector *od* from *o* parallel to the path of the slider *D*, and trial vector *cd* from *c* perpendicular to link *CD*, locating velocity point *d*. The velocity of *D* (the shaper tool velocity) is 19.6 in/s to the right at the instant shown, as given by the scaled vector *od*. When angular velocity ω_1 is constant and clockwise, the average velocity of point *D* is greater when *D* moves to the left than when *D* moves to the right. This is the quick-return feature of the shaper that ensures a slow, powerful cutting stroke and a quick return. The stroke length (distance between extreme positions of *D*) is varied by adjusting the length of link 1. Since this adjustment affects the velocity of *D*, it may be necessary to compensate by changing the angular velocity of link 1 at the same time.

3.12.3 Beam Pump

The four-bar and slider-crank mechanism combination of Figure 3.40 forms a *beam pump*. Connecting rod (4) moves practically along its own axis, and the rod may be made very long. If we know the angular velocity of crank 1, it is a simple matter to construct velocity polygon *obc*. Then velocity image *cod* is

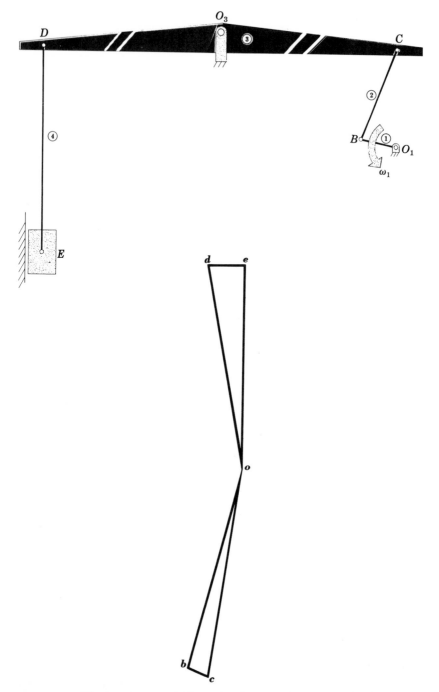

Figure 3.40 The beam pump. This mechanism may be analyzed by first solving the four-bar linkage (links 1, 2, 3, and the frame) and then drawing the velocity image of link 3. The slider-crank part of the mechanism (links 3 and 4 and the slider and frame) is then solved.

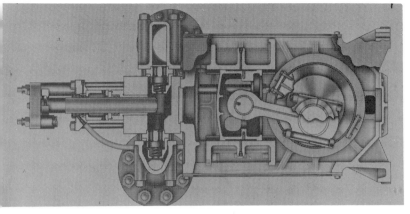

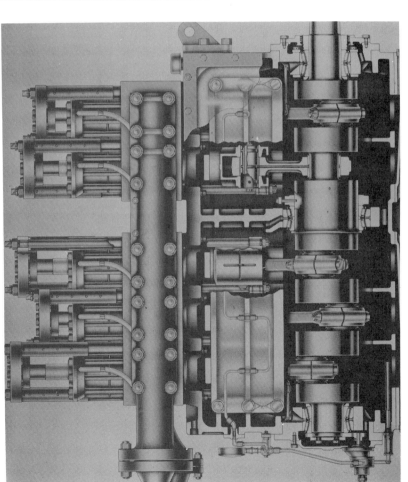

Figure 3.41 A high-pressure pump. The in-line slider-crank mechanism is basic to several types of machines. In this case, five slider-crank mechanisms are used to produce a continuous flow at up to several thousand pounds per square inch. At any instant, each linkage is at a different point in its stroke. As seen in the end view (*right*), the crank drives the connecting rod, which moves the crosshead. "Trombone" side rods connect the crossheads to the plungers, which enter the top of the cylinder. (*Source:* Cooper Energy Services.)

drawn similar to CO_3D on link 3. The velocity polygon is completed by locating point *e* corresponding to the slider. Slider velocity is given by vector *oe*.

3.12.4 Multiple Slider-Crank Mechanisms

Single-cylinder internal-combustion engines provide adequate power for lawn mowers, portable tools, and similar applications. Power output, however, is limited by cylinder size. If we were to design a single-cylinder engine with a capacity of several hundred horsepower, the piston, connecting rod, and crank might be unreasonably large. At full speed, inertia forces could be a serious problem. Furthermore, the single power stroke per revolution in the case of a two-stroke-cycle engine (or one power stroke for each two revolutions in the case of a four-stroke-cycle engine) might cause unacceptable fluctuations in speed, even when a flywheel is used. A pump or air compressor poses the same problems, particularly when high capacity or uniform output is called for. The solution is a design with several separate cylinders, which might be in-line (with parallel axes) or in a V arrangement (two separate banks of cylinders at an angle to one another).

The multicylinder high-pressure pump shown in front and end section views in Figure 3.41 is an almost trivial example of a mechanism combination

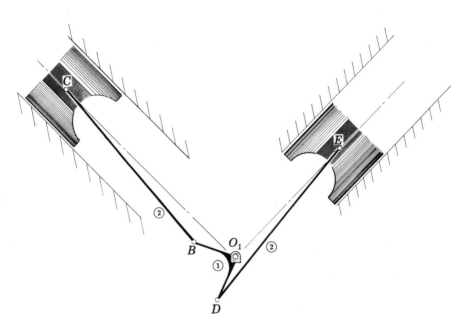

Figure 3.42 A typical multicylinder engine or compressor configuration. This V design shows two slider-crank linkages on a common crankshaft but with separate crankpins at different axial locations along the crankshaft.

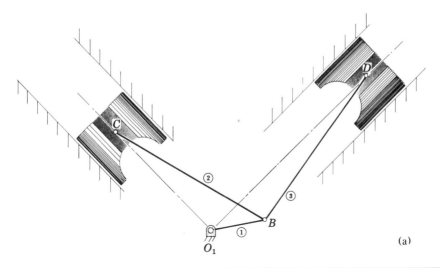

(a)

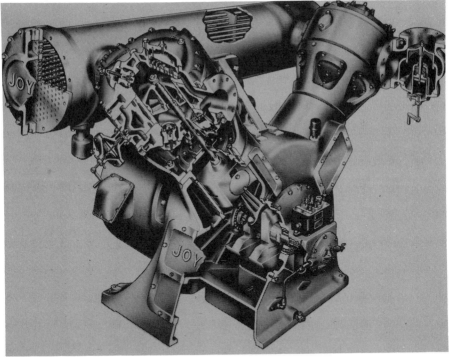

(b)

Figure 3.43 (a) Note that in this variation of the crank–connecting rod arrangement, both connecting rods are attached to a single crankpin (at different axial locations). (b) This V type of compressor is a commercial application of the single crankpin arrangement. Both linkages are similar kinematically except for instantaneous position. This unit compresses air in two stages. The large-diameter piston compresses the air in the first stage. The air is then cooled and brought to still higher pressure by the small (upper) piston in the second stage. (*Source:* Joy Manufacturing Company.)

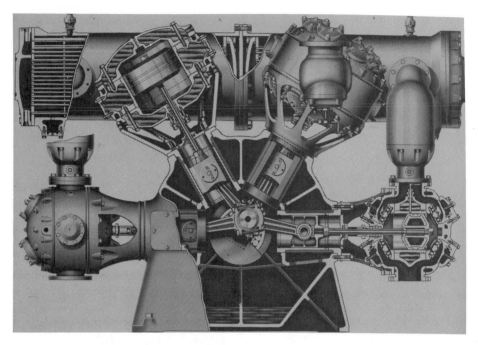

Figure 3.43 (c) This *semiradial compressor* also employs a single crankpin. (*Source:* Joy Manufacturing Company.)

(if we consider only the slider-crank linkages). Similarly, Figure 3.42 is a sketch representing two cylinders of a V block engine or pump. When we have drawn the velocity polygon for links 1, 2, and the slider at *C*, it is only necessary to note that the velocity of *D* has the same magnitude as the velocity of *B* as we complete the polygon.

Ordinarily, all of the slider-crank linkages in a multicylinder engine are identical except for instantaneous link orientation. For example, we might examine piston velocity as a function of crank position for a single piston of an eight-cylinder engine. If crankshaft speed is maintained, the results would apply equally to the other pistons, the only difference being the individual cylinder orientation and the phasing of the motion.

In Figures 3.41 and 3.42, each connecting rod is attached to a *separate* crankpin. There are, however, several variations of the crank–connecting rod arrangement. Figure 3.43a shows an alternative configuration for a two-cylinder engine, with both connecting rods (*BC* and *BD*) attached to a *single* crankpin, *B*. A practical example of this alternative arrangement is shown in the cutaway view of the two-stage, V type of compressor in Figure 3.43b. A view of a similar arrangement is seen in the cutaway view of the four-cylinder semiradial compressor shown in Figure 3.43c, where the four connecting rods are again attached to a common crankpin.

3.12.5 Articulated Connecting Rods

When several cylinders are to be arranged radially in an engine or compressor, still another arrangement, the *articulated connecting rod*, may be used. The articulated connecting rod linkage, sketched in Figure 3.44a, consists of a single crank pinned to a *master connecting rod*. The connecting rods of the remaining cylinders are in turn pinned (at different points) to the master connecting rod. A practical example of an articulated connecting rod linkage is shown in Figure 3.44b.

While the articulated connecting rod linkage and the single-crankpin linkage of the semiradial compressor of Figure 3.44b appear to be very similar, the connecting rods of the semiradial compressor must be placed in line with, or next to, each other, whereas the articulated connecting rod linkage permits the connecting rods, and the pistons, to lie in a single plane. The use of an articulated connecting rod therefore permits the design of a multicylinder engine with all the cylinder centerlines in a single plane.

Graphical methods were used to analyze several multiloop mechanisms illustrated in this section. Analytical vector methods and complex number methods may be used as well, particularly if many positions are to be considered. The key to solving these linkages is similar in each case. We begin

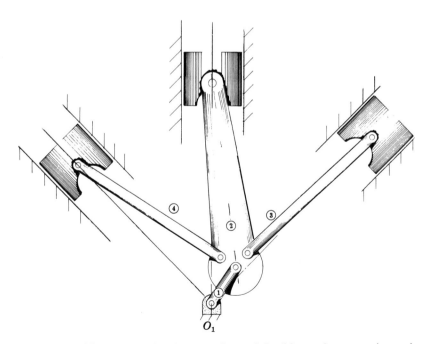

Figure 3.44 (a) The *articulated* connecting rod. In this crank–connecting rod arrangement, the crank is pinned to a master connecting rod. The rest of the connecting rods are, in turn, pinned to the master rod.

Figure 3.44 (b) Commercial application of an articulated connecting rod. The cylinder centerlines of the compressor shown are arranged radially in a single plane perpendicular to the crankshaft axis for better balance. (*Source:* Worthington Group, McGraw-Edison Company.)

by considering a single basic linkage, which is a component of the more complex system, and then proceed through a common point to solve the next component of the system.

3.13 ANALYZING LINKAGES THROUGH TRIAL SOLUTIONS AND INVERSE METHODS

For some mechanisms, the solution is not as straightforward as for the cases examined earlier. We may frequently find that step-by-step drawing of velocity vectors leads us to a point where trial solutions are suggested. Alternatively, it may be more practical to use an inverse method, that is, to begin with the solution and work the problem backward.

Figure 3.45b, which is an equivalent linkage for the variable-stroke pump shown in Figure 3.45a, is a problem of this type. Link 4 is an equivalent link; in the actual pump, point D represents the pin in the guide block riding in a curved track (of radius O_4D). Stroke length (the length of the path of E) may be varied from zero to a maximum value by tilting the curved track, which is equivalent to changing the location of point O_4. The control mechanism may be manually operated or may incorporate an air-operated plunger (not shown) to rotate the track automatically in response to a remote signal.

If we replace the slider by an infinite link (as illustrated in the centro method in a later section), then there are six links counting the frame and seven pins. Thus Grübler's criterion is satisfied (see Chapter 1) and there is one degree of freedom. If the motion of one link is specified, it should be possible to describe the motion of the entire linkage.

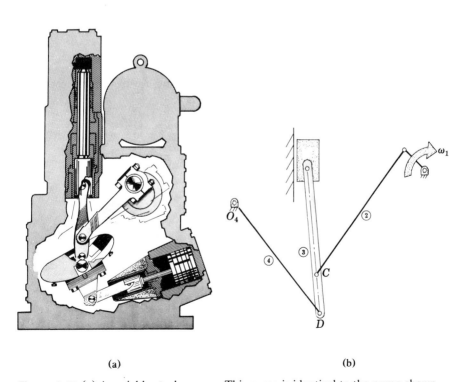

(a)   (b)

Figure 3.45 (a) A variable-stroke pump. This pump is identical to the pump shown in Chapter 1. Here the adjustment cylinder is set so that the stroke transformer provides a maximum plunger stroke. (*Source:* Ingersoll-Rand Company.) (b) The equivalent linkage for the variable-stroke pump.

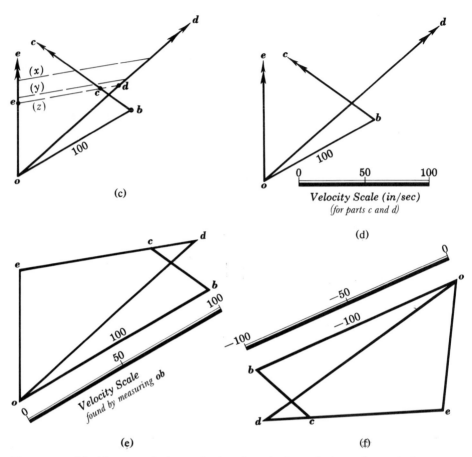

(c)

Velocity Scale (in/sec)
(for parts c and d)

(d)

(e)

(f)

Figure 3.45 (c) After a velocity scale is selected, the velocity polygon is begun. However, the location of the velocity image *dce* is not immediately obvious. (d) Trial solution method of locating line *dce*. The third trial, line *z*, satisfies the relationship $dc/de = DC/DE$, with *c* between *d* and *e* since *C* lies between *D* and *E*. (e) In the inverse method of solution, vector **oe** is given an arbitrary length, and the polygon is completed without using a velocity scale. The scale is then found by measuring **ob** (the only magnitude we are given) on the polygon. (f) If the assumed sense of vector **oe** is wrong, the vector magnitudes will still be correct, but the orientation of the polygon will be in error by 180°.

3.13.1 Trial Solution Method

Suppose it is necessary to find the velocity of the piston of the mechanism in Figure 3.45a and b when the velocity of crankpin *B* is 100 in/s with crank O_1B rotating clockwise. A *trial solution* procedure follows.

Step 1. Select a convenient velocity scale and draw **ob** = 100 in/s perpendicular to O_1B, as in Figure 3.45c.

Step 2. It is usually best to indicate all known vectors and vector directions on the sketch. Draw these trial vectors:

od perpendicular to $O_4 D$
oe in the direction of sliding (vertical)
bc perpendicular to BC (*bc* is added to *ob*)

Step 3. In a straightforward problem, we would continue the polygon by finding another velocity point, say point *c*. The following vector equations might be used:

$$oc = ob + bc \quad \text{and} \quad oc = oe + ec$$

But the magnitudes of *bc*, *oe*, and *ec* are unknown, and the magnitude and direction of *oc* are unknown; there are too many unknowns to utilize the equations. A similar problem exists with the vector equation $oc = od + dc$ since the magnitudes of *od* and *dc* are also unknown.

Step 4. The velocity image relationship is the missing tool; with it, a solution is possible. On the velocity polygon, *dce* is the (straight-line) image of *DCE*, three points on rigid link 3. The relationship is used by noting that *d*, *c*, and *e* lie on a line perpendicular to line *DCE* with *c* between *d* and *e*, proportioning the line by the equation $dc/de = DC/DE$. Now that we have the direction and relative proportions of line *dce*, we may satisfy the velocity image condition by trial and error. Line *x* in Figure 3.45d is the first approximation. It intercepts trial vectors *od*, *bc*, and *oe* in such a way that the ratio dc/de would be too large. We can see that the ratio is decreased as the trial line is moved downward. Line *y*, the second approximation, is somewhat better, and the line *z*, the third approximation, closely satisfies the required relationship $dc/de = DC/DE$. Line *z* completes the polygon, and velocity points *d*, *c*, and *e* are located where trial vectors *od*, *bc*, and *oe* intercept it. The piston velocity is given by vector *oe*.

Trial solutions are required more frequently in engineering practice than indicated by typical academic problem assignments. In the latter situation, a shortage of time favors the use of problems in which the answer is obtained directly. The reader should, however, be prepared for both types.

3.13.2 Inverse Method

Piston velocity of the variable-speed pump may also be found by a method that avoids the inconvenience and potential error of making several approximations. (We are not always so fortunate; for many types of problems, *only* trial solutions are known to exist.)

The reader may have observed in Figure 3.45c that, had piston velocity been given instead of the crankpin velocity, we could draw the velocity polygon directly. The solution would proceed from the slider-crank mechanism, the velocities of *E* and *D*, to the velocity of *C* by proportion and thence the velocity of B. Let us, then, solve the problem by an *inverse method* (i.e., we assume the answer at the beginning). The following steps constitute an entirely new solution to the problem without making use of the trial solution.

Step 1.   Represent piston velocity *oe* by a vertical vector of arbitrary length (Figure 3.45e). We cannot give *oe* an actual velocity magnitude or select a scale since we would undoubtedly guess wrong unless we had already solved the problem by another method.

Step 2.   Draw trial vectors *od* perpendicular to O_4D and *de* perpendicular to DE to locate velocity point *d*. Velocity point *c* is located between *d* and *e* on line *de* by using the proportion $dc/de = DC/DE$. Trial vectors *cb* perpendicular to CB and *ob* perpendicular to O_1B locate velocity point *b*, completing the polygon.

Step 3.   Finally, the scale of the velocity polygon must be determined from the given data. In this problem, the velocity of the crankpin *ob* = 100 in/s. The length representing *ob* on the velocity polygon becomes 100 in/s and all other vectors are scaled accordingly and labeled with their correct velocity magnitudes. Measurements taken directly from the velocity polygon give *oe/ob* = 0.58, from which piston velocity *oe* = 0.58 × 100 = 58 in/s upward at this instant.

In this particular example, the angular velocity of link 1 is clockwise, and it was clear at the start that the velocity of E would be upward. For some positions of this mechanism, it is impossible to determine the sense of piston velocity by inspection. Suppose we had guessed wrong this time and drawn *oe* downward in the velocity polygon. The result is shown in Figure 3.45f, where *ob* is drawn downward and to the left (its orientation is in error by 180°). The measured length of *ob* on the polygon is equal to − 100 in/s, and each of the vectors is given a negative value. For example, *oe* = − 58 in/s downward (i.e., 58 in/s upward). Of course, the easiest way to rectify our error is to cut out the velocity polygon, rotate it 180° about point *o*, and tape it back in place.

3.14 CENTROS

The velocity polygon method is not the only graphical method of velocity analysis in linkages. Other methods include the centro method, the instant center method, and the method of resolution (see Hinkle[3]).

3.14.1 Centros Defined

The centro method has been, traditionally, of considerable interest. It consists of locating a point, the **centro**, which is considered to lie in two links and have the same (vector) velocity in both of those links. The centro is then used to relate unknown velocities to known velocities.

The centro method does not lead directly to acceleration analysis, as does the velocity polygon method. Thus, while the centro method can provide some

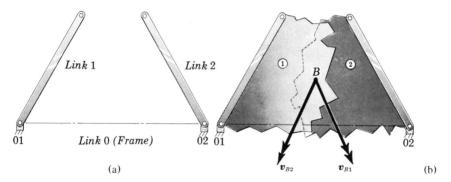

Figure 3.46 (a) A centro is a point common to two links that has the same vector velocity in each link. The pins joining links 1 and 2 to the frame (link 0) are, respectively, centros 01 and 02 (zero velocity). (b) Three links have three centros, all of which lie on a straight line. To prove this theorem, let us assume that B, common to extended links 1 and 2, is the third centro. Vectors v_{B_1} and v_{B_2} violate the requirement for a centro that the velocity vector must be *identical* for both links. To meet this basic requirement, centro 12 cannot occur anywhere except on $\overline{01\ 02}$.

useful insight into mechanism velocities, it will not be considered the best general method of analysis. However, the centro method can add to our understanding of mechanisms and can be used as a partial check of computer-generated plots.

The pin connecting two links in a mechanism is obviously the centro of the two since it has the same velocity in both. See Figure 3.46a. If two bodies (two friction disks, for instance) roll on one another without sliding, then the instantaneous point of contact is the centro. The centros just mentioned are *observed centros* and are to be located and labeled before any others are found by construction. The centro label will consist of the two link numbers, the smaller number first. The point having the same velocity in both link 0 and link 1, for example, will be labeled centro 01. The label 10 would apply to the same centro, but we will avoid duplication by always labeling this centro 01.

Since each pair of links has a centro, three links, which can form three different pairs, will have three centros. In general, given sufficient information, we should be able to find $n(n-1)/2$ centros for a linkage with n links. In this case, links 0, 1, and 2 form three centros, which we will label 01, 02, and 12. Figure 3.46b shows three links including 0, the frame. There exist centros 01 (where link 1 is pinned to the frame), 02 (where link 2 is pinned to the frame), and 12 (which we will attempt to locate).

An arbitrary point B which does not lie on (line) $\overline{01\ 02}$ is selected as a possible location for centro 12. The bar on $\overline{01\ 02}$ will be used to represent a line in the mathematical sense (extending infinitely on both sides of the line segment between 01 and 02). Let us imagine an extension of links 1 and 2 so that B may lie in both. The velocity of B in link 1, v_{B_1}, is perpendicular to $\overline{01\ B}$. Likewise, the velocity of B in link 2, v_{B_2}, is perpendicular to $\overline{02\ B}$. (The velocity magnitudes are unknown.) Since $\overline{01\ B}$ and $\overline{02\ B}$ are not parallel or colinear, v_{B_1} and v_{B_2} cannot be the same; the directions of v_{B_1} and v_{B_2} are different. Certainly B is not the centro 12.

3.14.2 Kennedy's Theorem

We can see from the above that centro 12 cannot be any point that does *not* lie on $\overline{01\ 02}$. A similar examination of the velocity direction of points lying on $\overline{01\ 02}$ shows that centro 12 may lie somewhere on $\overline{01\ 02}$. Considering the infinite extent of $\overline{01\ 02}$, there must be some point on it which has the same velocity magnitude as well as direction in both link 1 and link 2. Thus, we have Kennedy's theorem (the three-link theorem) in a nutshell: *three links have three centros that lie on a line.*

Kennedy's theorem applies except in trivial cases (e.g., a linkage without any relative motion), but it does not assure us of finding all the centros that exist in a theoretical sense. As for the original problem of actually locating centro 12, we have failed because there is a need for additional data.

3.14.3 Centros of a Four-Bar Linkage

The mechanism of Figure 3.47a has four links including the frame. There are six possible pairs formed by four numbers and, hence, six centros. They are observed centros 01, 12, 23, and 03 and the two remaining centros, 02 and 13, which must be located by construction. A procedure for finding the unobserved centros follows.

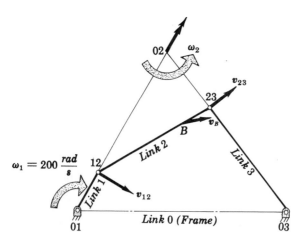

Figure 3.47 (a) Locating unknown centro 02 using the three-link (Kennedy's) theorem. Centro 02 is determined by the intersection of the extensions of $\overline{01\ 12}$ and $\overline{03\ 23}$.

Step 1. The three-link theorem will be used to draw a line on which centro 02 must lie. For the necessary three links, we must include links 0 and 2 since we are looking for centro 02; then, either of the remaining links will do. Using links 0, 2, and 1, we have centros 01 and 12 (both already labeled) and 02 (the unknown), all three on a line. Line $\overline{01\ 12}$ is drawn (of arbitrary length) and labeled 02 for our unknown centro.

Step 2. The three-link theorem is used again to draw another line to locate centro 02. Links 0 and 2 must again be included, this time along with the other remaining link, link 3. Links 0, 2, and 3 have three centros, 03 and 23 (already labeled) and the still unknown centro 02. Line $\overline{03\ 23}$ is drawn until it reaches the extension of $\overline{01\ 02}$, at which point centro 02 is located.

Step 3. Centro 02 is a point in link 0, the frame, and in link 2, the connecting rod. Centro 02 has the same velocity in both links by the definition of a centro. While it is not actually a part of either, it is considered to be "in" the links for velocity analysis. Link 0 is fixed, and thus the point common to link 0 and link 2, centro 02, is a point in link 2, which is (instantaneously) stationary.

Step 4. Since link 2 rotates about centro 02, as in Figure 3.47a, velocities of points on link 2 are proportional to their distances from 02. We may then state the ratio of velocity magnitudes of points 23 and 12:

$$\frac{v_{23}}{v_{12}} = \frac{02-23}{02-12} \tag{3.57}$$

where 02—23 and 02—12 are distances scaled from the linkage drawing.

The magnitude of velocity v_{12} is given by the length of link 1 times ω_1 and, using this, the magnitude of velocity v_{23} may be found by Eq. 3.57. If ω_1 is clockwise, v_{12} is to the right and downward. Observing the location of 02, the instant center of link 2, we see that the link must rotate counterclockwise at this instant (since v_{12} is to the right and downward). Velocity v_{23}, then, is to the right and upward (perpendicular to $\overline{03\ 23}$.)

Step 5. The magnitude of the angular velocity of link 2 is given by

$$\omega_2 = \frac{v_{12}}{02-12}$$

where 02—12 is the *actual distance* from point 12 to centro 02 of link 2 for the full size mechanism. The velocity of an arbitrary point B on link 2 of Figure 3.47a is given by

$$\frac{v_B}{v_{12}} = \frac{02-B}{02-12} \tag{3.58}$$

where lengths 02—B and 02—12 are scaled from the diagram.

Velocity v_B is perpendicular to 02—B and is to the right and upward, its sense determined in the same manner as for v_{23}.

Step 6. The angular velocity of link 3 follows immediately from the above calculations, but we will use centro 13 to complete the problem in order to

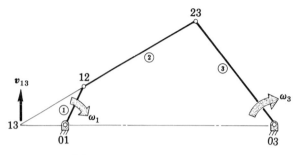

Figure 3.47 (b) Unknown centro 13 is determined by the intersection of $\overline{01\ 03}$ and $\overline{12\ 23}$.

illustrate further the centro method. Centro 13 is located in a manner similar to the procedure for locating centro 02. In this case, lines $\overline{01\ 03}$ and $\overline{12\ 23}$ both contain centro 13. Now 13 is a centro in the general sense, having the same nonzero velocity in links 1 and 3. Using that property, we have

$$v_{13} = \omega_1 01\text{—}13 \quad \text{and} \quad v_{13} = \omega_3 03\text{—}13$$

Equating these two expressions, we obtain

$$\omega_3(03\text{—}13) = \omega_1(01\text{—}13) \quad \text{or} \quad \frac{\omega_3}{\omega_1} = \frac{01\text{—}13}{03\text{—}13} \tag{3.59}$$

which might be expressed in words as follows: *For two links with fixed pivots, the angular velocities of two links are inversely proportional to distances from the respective fixed pivots to the common centro.* In Figure 3.47b, we see that v_{13} is upward when ω_1 is clockwise. Then ω_3 is seen to be clockwise also. In general, when the common centro falls between the fixed centers of a pair of links, one link turns clockwise and the other counterclockwise; otherwise, both turn clockwise or both turn counterclockwise. The reader will observe that the pitch point for a pair of gears and the tangent point for a pair of friction disks represent the common centro. For these examples, the expression boils down to this: *Angular velocities are inversely proportional to radii.*

Referring to the skeleton diagram of a planar four-bar linkage, if links 0 and 1 are colinear, then centros 02 and 03 are coincident. Note that links 2 and 3 are joined at centro 23. As a result, the angular velocities of links 2 and 3 are equal at the instant that links 0 and 1 are colinear. This condition serves as a partial check of the angular velocity plots shown earlier in this chapter.

3.14.4 Analyzing Basic Linkages Using Centros

The slider-crank mechanism in Figure 3.48 is solved by first examining the equivalent linkage shown in that figure. It is seen that the slider may be replaced by a link of infinite length perpendicular to the slider path. The slider moves in a horizontal path. Its motion can be duplicated by an equivalent link,

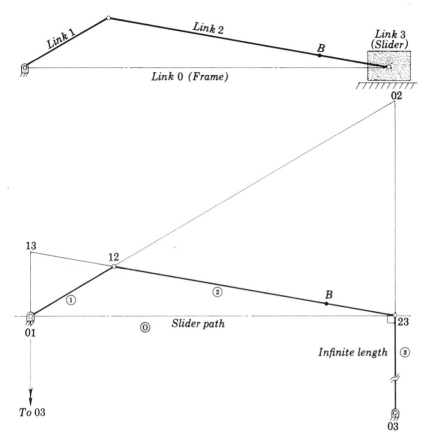

Figure 3.48 For the use of centros to find velocities in a slider-crank mechanism, an equivalent linkage must be used, where the slider is replaced by a link of infinite length. The solution then proceeds as for a four-bar linkage.

link 3, which is vertical. Centro 03, the "fixed center" of the equivalent link, is shown an infinite distance below the slider path. (Centro 03 could, as well, have been located an infinite distance above the slider path, as long as the equivalent link 3 is perpendicular to the slider path.)

Observed centros 01, 12, and 23 are labeled on the sketch along with centro 03. Centro 02 is found by construction as in the four-bar linkage, and the same equations apply. To find slider velocity in terms of crankpin velocity, we may use

$$\frac{v_{23}}{v_{12}} = \frac{02-23}{02-12}$$

For the velocity of an arbitrary point B on link 2 (e.g., the center of gravity), we can use the formula

$$\frac{v_B}{v_{12}} = \frac{02—B}{02—12}$$

and for angular velocity of link 2,

$$\omega_2 = \frac{v_{12}}{02—12}$$

In the last equation, the reader is reminded that the dimension 02—12 refers to the true distance between 02 and 12 on a full-scale drawing. Then ω_2 is given in radians per second. The dimensions in the first two equations may be to any scale as long as they are consistent.

It is clear from the equivalent linkage that the slider-crank mechanism is a special case of the four-bar linkage. In the above discussion, only five of the six centros for the linkage have been found. The remaining one, centro 13, is of interest because it gives us slider velocity directly from crank angular velocity. Using the three-link theorem with links 1, 2, and 3, we draw line $\overline{23\ 12}$ to locate centro 13. Applying the three-link theorem again, using links 0, 1, and 3 this time, centro 13 is located on line $\overline{03\ 01}$. Since 03 is located at an infinite distance from the slider, the line $\overline{03\ 01}$ must be drawn parallel to equivalent link 3 (perpendicular to the slider path); any other orientation would require that 03 be located at a finite distance from the slider. In Figure 3.48, line $\overline{03\ 01}$, a perpendicular to the slider path, is extended upward until it intersects line $\overline{23\ 12}$, an extension of the centerline of link 2. Centro 13 is located at the intersection.

By definition, centro 13 is a point "in" both links 1 and 3. Given ω_1, the angular velocity of link 1, the velocity of the centro is $v_{13} = \omega_1(01—13)$, where 01—13 is the full-scale distance between the centro and crank center. Since link 3 translates, all points in it (including centro 13) have the same velocity. Thus, v_{13} is also the slider velocity.

Of course link 3, which represents the slider, does not rotate. The angular velocity ratio (as given by Eq. 3.59 for the more general four-bar linkage) is

$$\frac{\omega_3}{\omega_1} = \frac{01—13}{03—13} = 0$$

since the numerator is finite and the denominator infinite. Note that the equation $v_{13} = \omega_3(03—13)$ is useless in this case; ω_3 is zero and distance 03—13 is infinite, therefore giving us an indeterminate product.

Rolling contact is easily treated by the centro method. In Figure 3.49, a wheel (1) rolls on the frame (0). The point having the same velocity in both (zero velocity) is the point of contact, centro 01. Centro 01 is the instant center of the wheel, and given the velocity of any point on the wheel (01 excepted), we may find the velocity of any other point. Using the relationship that

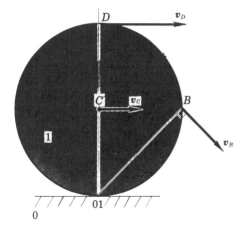

Figure 3.49 Using the centro (instant center) to analyze rolling contact. The velocity of any point on the circle is proportional to its distance from the point of contact.

velocities are proportional to distances from the instant center, we have

$$\frac{v_B}{v_C} = \frac{01-B}{01-C} \quad \text{and} \quad \frac{v_D}{v_C} = \frac{01-D}{01-C} \qquad (3.60)$$

The Cam and Follower

An additional concept allows us to locate the centros common to a cam and follower. Centros 01 and 02 are the fixed centers of the cam and follower, respectively, in Figure 3.50. Using the three-link theorem, centro 12 is known to lie somewhere on line $\overline{01\ 02}$, but more information is needed to actually locate 12. The relative motion between cam and follower at the point of contact is sliding, having some resemblance to the motion of the slider relative to the frame in a slider-crank mechanism. And in both cases, the common centro lies on a line perpendicular to the relative motion. Thus centro 12 is located at the intersection of the common normal through the point of contact and the line of centers of cam and follower. By definition, centro 12 has the same velocity "in" links 1 and 2 from which

$$v_{12} = \omega_1(01-12) = \omega_2(02-12)$$

Using this equation, we see that the angular velocity ratio of follower to cam is inversely proportional to the distances from the common centro to the fixed centers:

$$\frac{\omega_2}{\omega_1} = \frac{01-12}{02-12}$$

For the position shown, centro 12 does not fall between centros 01 and 02; therefore, cam and follower both turn clockwise or both turn counterclockwise. Measuring the ratio of distances given by the above equation, we find the same angular velocity ratio as when this problem was solved by the velocity

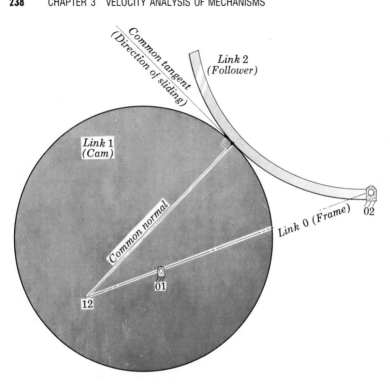

Figure 3.50 The cam and follower analyzed by the centro method. Unknown centro 12 lies on the intersection of the line of centers and the common normal through the point of contact.

polygon method in a previous section. (Slight variations may be attributed to graphical error.)

Spur Gears

A *pair of spur gears* has much in common with the cam and follower considered above. The use of centros in this case give us some insight into gear design for a constant angular velocity ratio. In Figure 3.51, centro 12 is located on the intersection of the line of centers and common normal through the point of contact of a pair of teeth. Angular velocity ratio is given by

$$\left[\frac{\omega_2}{\omega_1} \right] = \frac{01-12}{02-12}$$

and if the location of centro 12 is fixed in space (i.e., if the gears rotate on fixed shafts), the angular velocity ratio is constant. Here, of course, gears and cams differ because a cam is usually designed to transmit cyclic motion to the follower. Except in a few instances, a constant angular velocity ratio is a fundamental gear system requirement, and each gear tooth form must be cut

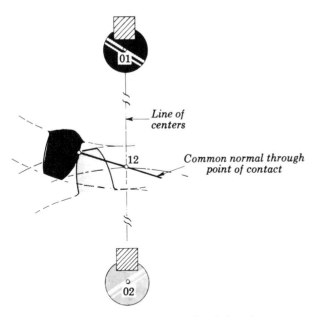

Figure 3.51 A pair of gears analyzed by the centro method. Centro 12 is located at the intersection of the line of centers $\overline{01\ 02}$ and the common normal through the point of contact. This point of intersection is referred to as the *pitch point*.

to ensure that centro 12 *is* fixed in space. An involute curve meets this requirement and is used almost exclusively. Involute curves will be discussed in detail in Chapter 6.

Centro 12 of Figure 3.51 is called the *pitch point* and defines the pitch radius of a gear. The ratio $(01—12)/(02—12)$ represents the ratio of pitch radius of gear 1 to pitch radius of gear 2. When we say angular velocities are inversely proportional to diameters,

$$\left[\frac{\omega_2}{\omega_1}\right] = \frac{d_1}{d_2}$$

d_1 and d_2 refer to pitch diameters.

Shaper Mechanisms

The *mechanical shaper* mechanism (Figure 3.52) may be considered a six-bar linkage when the frame and both sliders are counted as links. To begin, seven centros are observed (located by inspection). They are the fixed centers of links 1 and 2, three pin joints, the "fixed" center of the equivalent link 5 (located perpendicular to the path of 5 at an infinite distance), and the centro of link 2 and slider 4. By analogy to the problem of a slider moving on a fixed

Centros Found by Three Link Theorem

Three Links	Three Centros		
	Known		Unknown
012	01	02	12
124	14	24	12
014	01	14	04
024	02	24	04
023	02	23	03
035	05	35	03
025	02	05	25
235	23	35	25
234	23	24	34
034	04	03	34
125	12	23	13
134	14	34	13
045	04	05	45
245	24	25	45
015	01	05	15
145	14	45	15

Observed Centros

01
02
05
14
23
24
35

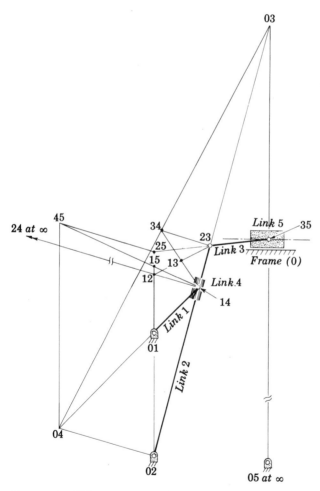

Figure 3.52 Using centros to determine the velocities of a six-bar linkage. Since keeping track of the centros may be confusing, an orderly procedure is necessary. A table like the one shown is recommended.

path, centro 24 is located perpendicular to link 2, the relative path, at an infinite distance.

Altogether the six-bar linkage has $n(n-1)/2 = 15$ centros, where n represents the number of links in the mechanism. The three-link theorem is applied sixteen times, twice for each of the remaining eight centros. Using links 0, 1, and 2, for example, we draw line $\overline{01\ 02}$ since unknown centro 12 must lie on that line. Then using links 1, 2, and 4, line $\overline{14\ 24}$ is drawn to intersect line $\overline{01\ 02}$, locating centro 12. The procedure progresses until all 15 centros are located, treating centros at infinite distances in the same manner as with the slider-crank mechanism. Figure 3.52 also shows all 15 centros of the six-bar linkage with the procedure tabulated. It is not necessary to locate the centros in the order given, however.

For a mechanism of the type just considered, we might be interested in the velocity of slider 5 in terms of the angular velocity of link 1. This is obtained directly through centro 15 since 15 may be considered a part of slider 5 and link 1. The velocity of any point in slider 5 is

$$v_5 = \omega_1(01—15)$$

where 01—15 represents the full-scale distance between centros 01 and 15.

An alternative method that requires the location of only a few centros follows.

Step 1. Let ω_1 be given. Centro 12 is located and used to obtain ω_2.

Step 2. Using ω_2, v_{23} is calculated.

Step 3. Centro 03 is then located and used to obtain v_{35}.

Since both of the above procedures fail to give us relative velocities directly, they are not recommended when accelerations of the linkage are also required. Their major practical use is to serve as an independent check on another procedure or to gain insight into a particular problem.

References

1. Angeles, J., *Spatial Kinematic Chains: Analysis-Synthesis-Optimization*, Springer, New York, 1982.
2. Barton, L. O., *Mechanism Analysis—Simplified Graphical and Analytical Techniques*, Marcel Dekker, New York, 1984.
3. Hinkle, R., *Kinematics of Machines*, 2d ed. Prentice-Hall, Englewood Cliffs, NJ, 1960.
4. Hirschhorn, J., "A Graphical Investigation of the Velocity Pattern of a Rigid Body in Three-Dimensional Motion," *Mechanism and Machine Theory*, Penton, Cambridge, vol. 23, no. 3, 1988, pp. 185–189.
5. JML Research, Inc., *Integrated Mechanisms Program*, JML Research, Inc., Madison, WI, 1988.
6. Lee, H.-Y., and C.-G. Liang, "A New Vector Theory for the Analysis of Spatial Mechanisms," *Mechanism and Machine Theory*, Penton, Cambridge, vol. 23, no. 3, 1988, pp. 209–218.

7. Mechanical Dynamics, Inc., *ADAMS Applications Manual*, Mechanical Dynamics, Inc., Ann Arbor, MI, 1987.

8. Mechanical Dynamics, Inc., *ADAMS User's Manual*, Mechanical Dynamics, Inc., Ann Arbor, MI, 1987.

9. Mechanical Dynamics, Inc., *ADAMS Introductory Tutorial* (draft release), Mechanical Dynamics, Inc., Ann Arbor, MI, 1989.

10. Mischke, C. R., *Elements of Mechanical Analysis*, Addison-Wesley, Reading, MA, 1963.

11. Stephenson, J., *VP-Planner Plus*, 2d ed., Vol. I: Tutorial, Paperback Software International, Berkeley, CA, 1987.

12. Stephenson, J., *VP-Planner Plus*, 2d ed., Vol. II: Reference, Paperback Software International, Berkeley, CA, 1987.

PROBLEMS

3.1 Find the velocity of a point P in a rigid body with angular velocity $\omega = i10 + j15 + k20$. The body rotates about fixed point O_1, and the radius vector O_1P is given by $r = -i100 + j50 + k60$.

3.2 Repeat Problem 3.1, except that $\omega = i200 + j0 - k150$.

3.3 Find the velocity of a point on the circumference of a 30-in-diameter flywheel rotating at 800 rev/min.

3.4 Low-carbon steel is turned on a lathe at typical surface speeds of 100 to 400 ft/min. Find the corresponding lathe spindle speeds in revolutions per minute for (**a**) a 2-in-diameter bar, and (**b**) a 3-in-diameter bar.

3.5 Write an equation to determine lathe spindle speed n (in revolutions per minute) when bar diameter d (in inches) and surface speed s (in feet per minute) are given. Is the same equation valid if n represents the speed of a milling cutter of diameter d?

3.6 A point P is described in terms of a fixed coordinate system XYZ with unit vectors I, J, and K and a moving coordinate system xyz with unit vectors i, j, and k. At a given instant, the location of the origin of the moving system is $12I + 5J$, and the velocity of the origin of the moving system is $80I - 90J$. The velocity of P relative to the moving system is $50i + 45j$; the radius of point P is $3i + 4j$ in the moving system, which rotates at angular velocity $\omega = 100k$. Find the velocity of point P in the fixed system if the x axis is rotated 30° counterclockwise from the X axis.

3.7 Repeat Problem 3.6, except that $\omega = -95k$.

3.8 A 500-mm-diameter wheel rolls in a straight path, rotating at 800 rev/min. Find relative velocity v_{CB}, where C is the center of the wheel and B is at the top.

3.9 Repeat Problem 3.8 for a 16-in-diameter wheel.

3.10 Velocity $v_B = 15$ in/s at 45°. Velocity $v_{CB} = 20$ in/s at 0°. Find v_C (Use $v_C = v_B + v_{CB}$.)

3.11 Velocity $v_C = 30$ in/s at 0°. Relative velocity $v_{CB} = 10$ in/s at 135°. Find v_B. [Use $v_B = v_C + (-v_{CB})$.]

3.12 Velocity $v_B = 20$ in/s at 45°. Relative velocity v_{CB} is an unknown vector at 315°. Velocity v_C is an unknown vector at 0°. Find v_{CB} and v_C. (Use $v_C = v_B + v_{CB}$.)

3.13 Referring to Figure 3.9, let $r_0 = 30$ mm, $r_1 = 10$ mm, $r_2 = 35$ mm, $r_3 = 20$ mm, and $\theta_1 = 45°$. Calculate θ_2, θ_3, v_C, ω_2, and ω_3 for $\omega_1 = 30$ rad/s ccw. Use analytical vector methods.

3.14 Repeat Problem 3.13 for $\theta_1 = 60°$.

3.15 Plot ω_3/ω_1 versus θ_1 for the linkage of Figure 3.9 if $r_0/r_1 = 3$, $r_2/r_1 = 3.5$, and $r_3/r_1 = 2$. Use a computer or programmable calculator.

3.16 Referring to Figure 3.9, let $r_0 = 30$ mm, $r_1 = 10$ mm, $r_2 = 35$ mm, $r_3 = 20$ mm, $r_{BD} = 15$ mm, $\theta_{CBD} = 20°$, and $\theta_1 = 30°$. Determine $v_D/(\omega_1 r_1)$. Use analytical vector methods.

3.17 Repeat Example Problem 3.7 by using analytical vector methods.

3.18 Refer to Figure 3.15. Find ω_2/ω_1 by using analytical vector methods. Write a computer or calculator program to determine ω_2/ω_1 for a series of values of θ_1 where ω_1 is constant and counterclockwise.

3.19 Refer to Figure 3.15. Let $R_1 = 30$ mm, $O_1O_2 = 50$ mm, $\omega_1 = 10$ rad/s ccw, and $\theta_1 = 15°$. Find **(a)** O_2B, **(b)** θ_2, **(c)** ω_2. Use complex number methods.

3.20 Repeat Problem 3.19 for $\theta_1 = 30°$.

3.21 Repeat Problem 3.19 for $\theta_1 = 45°$.

3.22 Repeat Problem 3.13 by complex number methods.

3.23 Refer to Figure 3.15. Calculate ω_2/ω_1 for $\theta_1 = 0$ to $180°$ in $15°$ steps if $R_0/R_1 = 3$. Use a computer or programmable calculator.

3.24 Refer to Figure P3.1, an RSSR spatial linkage. The x, y, and z coordinates of the joints are, respectively, as follows: using R_1 as the origin:

$$R_1: 0,0,0 \qquad S_1: -25,0,35$$

$$R_2: -20,95,0 \qquad S_2: -20,80,40$$

(all in millimeters). Link 1 rotates at $\omega_1 = 25$ rad/s (constant) in the xz plane. Sphere joint S_1 is moving away from the observer at this instant. Link 3 rotates in the yz plane. Find velocities v_{s_1}, v_{s_2} and angular velocity ω_3. Set the angular velocity of link 2 about its own axis $= 0$.

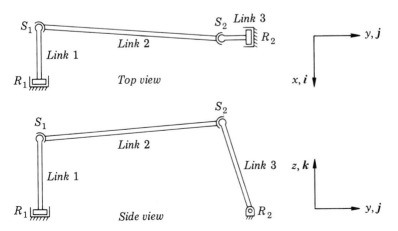

Figure P3.1 RSSR spatial linkage.

3.25 Repeat Problem 3.24, except that $\omega_3 = 10$ rad/s cw (constant). Find v_{s_1}, v_{s_2}, and ω_1.

3.26 (a) Find the crank position corresponding to maximum piston velocity for an in-line slider-crank mechanism. Crank speed ω is constant, and the ratio of connecting rod to crank length is $L/R = 2$.

(b) Find maximum piston velocity in terms of R and ω.

3.27 Repeat Problem 3.26 for $L/R = 1.5$.

Problems 3.28 Through 3.61

(a) *Write the appropriate vector equation.*

(b) *Solve graphically unless directed otherwise, using velocity polygon notation.*

(c) *Dimension all vectors of the velocity polygon.*

(d) *Express angular velocities in radians per second and indicate direction.*

● **3.28** In Figure P3.2, $\theta = 45°$ and $\omega_1 = 10$ rad/s. Draw and dimension the velocity polygon. Find ω_2.

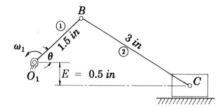

Figure P3.2

3.29 Repeat Problem 3.28 for $\theta = 120°$.

3.30 Repeat Problem 3.28 for the mechanism in the limiting position (with C to the extreme right).

3.31 In Figure P3.3, $\omega_1 = 100$ rad/s. Draw and dimension the velocity polygon. Find v_D and ω_2. Use the scale 1 in = 100 in/s.

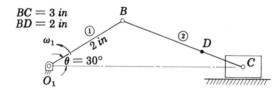

Figure P3.3

3.32 In Figure P3.4, $\omega_1 = 50$ rad/s. Draw and dimension the velocity polygon. Find v_D, ω_2, and ω_3. Use the scale 1 in = 50 in/s.

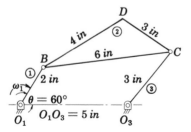

Figure P3.4

3.33 In Figure P3.5, $\omega_1 = 20$ rad/s.

    (a) Draw and dimension the velocity polygon for the limiting position shown. Find relative velocity v_{CB}. Use the scale 1 in = 10 in/s.

    (b) Repeat the problem for the other limiting position.

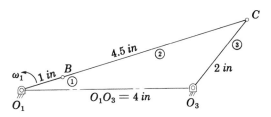

Figure P3.5

• **3.34** In Figure P3.6, $\theta = 105°$ and $\omega_2 = 20$ rad/s. Draw and dimension the velocity polygon. Identify the sliding velocity. Find ω_1. Use the scale 1 in = 20 in/s.

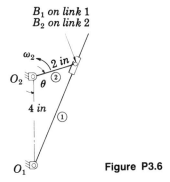

Figure P3.6

3.35 Repeat Problem 3.34 for $\theta = 30°$.

3.36 In Figure P3.3, $\omega_1 = 100$ rad/s.

    (a) Find v_C *analytically* for the position shown.

    (b) Find v_C analytically at both limiting positions.

• **3.37** In Figure P3.7, let the angular velocity of the crank be ω.

    (a) Draw the velocity polygon for the position shown. Identify relative velocity v_{CB}.

    (b) Repeat for the other limiting position.

Figure P3.7

3.38 In Figure P3.8, $\omega_1 = 100$ rad/s. Draw and dimension the velocity polygon. Use the scale 1 in = 100 in/s.

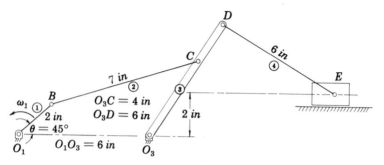

$O_3C = 4\ in$
$O_3D = 6\ in$
$O_1O_3 = 6\ in$

Figure P3.8

3.39 In Figure P3.9, $\theta = 135°$ and $\omega_1 = 10$ rad/s. Draw and dimension the velocity polygon. Identify the follower velocity and sliding velocity. Use the scale 1 in = 5 in/s.

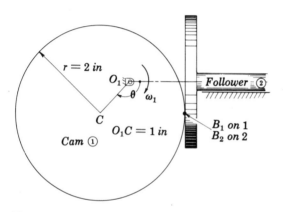

Figure P3.9

3.40 Repeat Problem 3.39 for $\theta = 30°$.

3.41 In Figure P3.10, $\omega_1 = 35$ rad/s. Draw the velocity polygon. Use the scale 1 in = 10 in/s. Find ω_2 and ω_3. Find the velocity of the midpoint of each link.

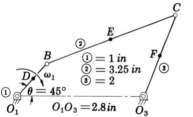

Figure P3.10

3.42 In Figure P3.11, $\omega_1 = 20$ rad/s.
   (a) Draw and dimension the velocity polygon. Find ω_2 and identify the sliding velocity.

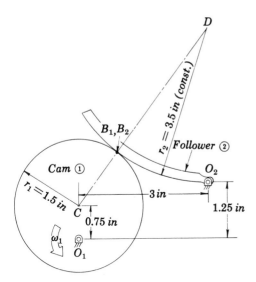

Figure P3.11

(b) Note that CD is a fixed distance. Thus, we can use the equivalent linkage O_1CDO_2. Draw the velocity polygon and again find the angular velocity of the follower (represented by DO_2).

3.43 Consider a pair of involute spur gears with a 20° pressure angle. Let the driver speed be 300 rev/min clockwise and the follower speed be 1000 rev/min counterclockwise. Find sliding velocity when contact occurs 1.2 in from the pitch point.

• **3.44** In Figure P3.12, $\theta = 105°$ and $\omega_2 = 20$ rad/s. Draw and dimension the velocity polygon, using the scale 1 in = 20 in/s.

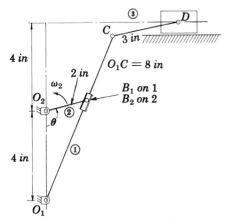

Figure P3.12

3.45 Repeat Problem 3.44 for $\theta = 30°$.

3.46 In Figure P3.9, $\omega_1 = 10$ rad/s. Find the follower velocity analytically when (a) $\theta = 135°$, (b) $\theta = 30°$.

3.47 In Figure P3.13, ω_1 (angular velocity of O_1B_1) = 30 rad/s. Draw and dimension

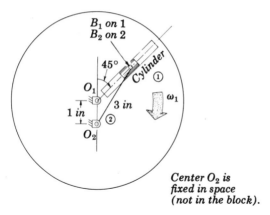

Figure P3.13

the velocity polygon, using the scale 1 in = 20 in/s. Identify the sliding velocity and find ω_2, the angular velocity of $O_2 B_2$.

3.48 For Figure P3.14, draw and dimension the velocity polygon, using the scale 1 in = 100 in/s. Find ω_2 and v_D.

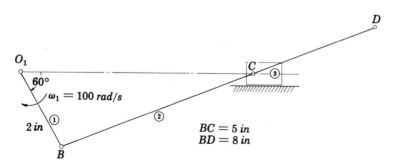

Figure P3.14

3.49 Locate all of the centros in Figure P3.14. Using centro 13, write an expression for slider velocity in terms of ω_1. Calculate v_C.

3.50 In Figure P3.14, use centro 02 in order to write an expression for (a) ω_2 in terms of v_B; (b) ω_2/ω_1; (c) v_C in terms of ω_2; (d) v_D in terms of ω_2; (e) calculate ω_2, v_C, and v_D.

3.51 In Figure P3.15, $\omega_5 = 15$ rad/s. Use a vector 3 in long to represent the velocity of *B*. Complete the velocity polygon and determine the velocity scale. Dimension the polygon and find the angular velocity of each link.

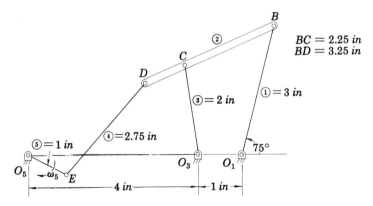

$BC = 2.25\ in$
$BD = 3.25\ in$

Figure P3.15

3.52 Refer to Figure P3.16. Solve graphically for $\theta_1 = 15°$.
  (a) Draw to $1:1$ scale.
  (b) Let 1 mm = 5 mm/s; draw velocity polygon ob_1b_2. Add point c where $O_2C = 100$ mm.
  (c) Find v_{B2}.
  (d) Find v_c.
  (e) Find ω_2.

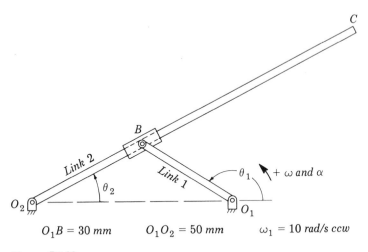

$O_1B = 30\ mm$ $O_1O_2 = 50\ mm$ $\omega_1 = 10\ rad/s\ ccw$

Figure P3.16

3.53 Repeat 3.52 for $\theta_1 = 30°$.
3.54 Repeat Problem 3.52 for $\theta_1 = 45°$.
3.55 Refer to Figure P3.16. Find θ_1 when $\omega_2 = 0$.
3.56 Refer to Figure P3.17.
  (a) Draw velocity polygon obc.
  (b) Find ω_2.

$O_1B = 120$ mm
$BC = 200$ mm
$\theta = 15°$
$\omega_1 = 500$ rad/s ccw
$BD = 40$ mm

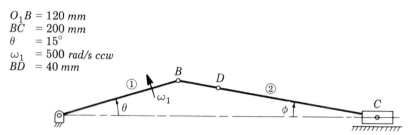

Figure P3.17

(c) Locate d on the velocity polygon. Find v_D.
(d) Identify angles θ and ϕ in the above solution. Find velocity v_c and relative velocity bc in terms of ω_1, O_1B, θ, and ϕ.

3.57 A four-bar planar linkage has the following dimensions:

r_0 (frame) $= 4$

r_1 (driver) $= 1.25$

r_2 (coupler) $= 5.25$

r_3 (driven link) $= 2.5$

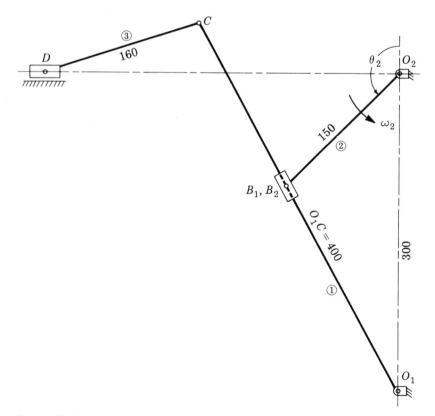

Figure P3.18

If $\omega_1 = 5.75$ rad/s constant cw, plot ω_2 and ω_3 versus θ. Use a computer or programmable calculator. Check values at $\theta = 120°$ by using a velocity polygon.

3.58 In Figure P3.18, $\theta_2 = 135°$, $\omega_2 = 500$ rad/s ccw (constant), $O_1O_2 = 300$ mm, $O_1C = 400$ mm, $O_2B = 150$ mm, and $CD = 160$ mm.
(a) Draw velocity polygon ob_1b_2.
(b) Find oc.
(c) Find ω_1.
(d) Add c to the polygon.
(e) Add d to the polygon.

3.59 In problem 3.58, let $\theta_2 = 120°$.
(a) Find v_{B_1}, v_D, and ω_1 at $\theta_2 = 120°$.
(b) For the interval $120° < \theta < 135°$, find the average angular acceleration of link 1 and average acceleration of point D.

3.60 Refer to Figure 3.15. Let $R_1 = 30$ and $O_1O_2 = 50$. Find θ_1 when $\omega_2 = 0$.

3.61 (a) Derive equations for position angles and angular velocities of a four-bar planar linkage. Use complex number methods. Refer to Figure 3.9.
(b) Given $\theta_1 = 60°$, $r_0 = 4$, $r_1 = 2$, $r_2 = 4.5$, $r_3 = 3$, and $\omega_1 = 50$ rad/s cw. Find ω_2 and ω_3, using the above equations.
(c) Check, using graphical methods.

3.62 Solve Problem 3.28 by analytical vector methods.

3.63 Solve Problem 3.29 by analytical vector methods.

3.64 Solve Problem 3.28 by complex number methods.

3.65 Solve Problem 3.29 by complex number methods.

3.66 A four-bar linkage has the following dimensions: $r_0 = 312.48$, $r_1 = 100$, $r_2 = 200$, $r_3 = 300$. The assembly is such that the vector loop $r_2r_3r_d$ is clockwise. Let the angular velocity of the crank be unity and let crank angular acceleration be zero.
(a) Tabulate link positions, transmission angle, and coupler and follower crank angular velocities for values of θ_1 from 0 to $360°$.
(b) Plot coupler and follower crank angular velocities for values of θ_1 from 0 to $360°$.

3.67 A four-bar linkage has the following dimensions: $r_0 = 312.48$, $r_1 = 100$, $r_2 = 200$, $r_3 = 300$. The assembly is such that the vector loop $r_2r_3r_d$ is counterclockwise. Let the angular velocity of the crank be unity and let crank angular acceleration be zero.
(a) Tabulate link positions, transmission angle, and coupler and follower crank angular velocities for values of θ_1 from 0 to $360°$.
(b) Plot coupler and follower crank angular velocities for values of θ_1 from 0 to $360°$.

3.68 A four-bar linkage has the following dimensions: $r_0 = 37$, $r_1 = 10$, $r_2 = 25$, $r_3 = 40$. Point D lies on the coupler at a distance of 15 from the crankpin, at an angle of $20°$. The assembly is such that the vector loop $r_2r_3r_d$ is clockwise. Let the angular velocity of the crank be 100 rad/s and let crank angular acceleration be zero.
(a) Tabulate link positions, transmission angle, coupler and follower crank angular velocities, and the velocity of point D for values of θ_1 from 0 to $360°$.
(b) Plot coupler and follower crank angular velocities for values of θ_1 from 0 to $360°$.
(c) Plot the velocity of D and its x and y components.

3.69 A four-bar linkage has the following dimensions: $r_0 = 38$, $r_1 = 13$, $r_2 = 27$, $r_3 = 41$. Point D lies on the coupler at a distance of 12 from the crankpin, at an angle of

$20°$. The assembly is such that the vector loop $r_2 r_3 r_d$ is clockwise. Let the angular velocity of the crank be 50 rad/s and let crank angular acceleration be zero.

(a) Tabulate link positions, transmission angle, coupler and follower crank angular velocities, and the velocity of point D for values of θ_1 from 0 to $360°$.

(b) Plot coupler and follower crank angular velocities for values of θ_1 from 0 to $360°$.

(c) Plot the velocity of D and its x and y components.

3.70 Consider an offset slider-crank linkage for which the connecting rod to crank length ratio is $R_2/R_1 = 2.5$ and the offset ratio $e/R_1 = -j\,0.4$. Tabulate and plot normalized slider position r_0/R_1, normalized slider velocity $v_c/(\omega_1 R_1)$, and angular velocity ratio ω_2/ω_1, all against crank position. Crank angular velocity is constant. Suggestion: Write a vector manipulation routine or use commercially available mathematics software.

3.71 Consider an offset slider-crank linkage for which the connecting rod to crank length ratio is $R_2/R_1 = 2$ and the offset ratio $e/R_1 = -j0.5$. Tabulate and plot normalized slider position r_0/R_1, normalized slider velocity $v_c/(\omega_1 R_1)$, and angular velocity ratio ω_2/ω_1, all against crank position. Crank angular velocity is constant. Suggestion: Write a vector manipulation routine or use commercially available mathematics software.

• **3.72** Consider an offset slider-crank linkage for which the connecting rod to crank length ratio is $R_2/R_1 = 3$ and the offset ratio $e/R_1 = -j0.7$. Tabulate and plot normalized slider position r_0/R_1, normalized slider velocity $v_c/(\omega_1 R_1)$, and angular velocity ratio ω_2/ω_1, all against crank position. Crank angular velocity is constant. Suggestion: Write a vector manipulation routine or use commercially available mathematics software.

3.73 Consider an RSSR linkage similar to that in Figure 1.6a where link lengths are $r_0 = 4$, $r_1 = 1$, $r_2 = 3.5$, and $r_3 = 2.5$. Link 0 lies on the x axis. Link 1 rotates in the xy plane with an angular velocity of 1 rad/s (constant) and link 3 rotates in the xz plane. Plot the vector components representing the angular velocity of link 2 and the angular position and angular velocity of link 3 against angular position θ of link 1. Tabulate resultant angular velocity of link 2, angular position of link 3, and angular velocity of link 3 against angular position of link 1.

3.74 Consider an RSSR linkage similar to that in Figure 1.6a where link lengths are $r_0 = 3.2$, $r_1 = 1$, $r_2 = 2.8$, and $r_3 = 2$. Link 0 lies on the x axis. Link 1 rotates in the xy plane with an angular velocity of 1 rad/s (constant) and link 3 rotates in the xz plane. Plot the vector components representing the angular velocity of link 2 and the angular position and angular velocity of link 3 against angular position θ of link 1. Tabulate resultant angular velocity of link 2, angular position of link 3, and angular velocity of link 3 against angular position of link 1.

PROJECTS

See Projects 1.1 to 1.6 and suggestions in Chapter 1. Examine linkages involved in the chosen project. Describe and plot velocity and angular velocity characteristics of the linkages. Make use of computer software wherever practical. Check the results by a graphical method for at least one linkage position. Evaluate the linkage in terms of the performance requirements.

Acceleration Analysis of Mechanisms

4.1 BASIC CONCEPTS

Acceleration in linkages is of particular importance because inertial forces are proportional to rectilinear acceleration and inertial torques are proportional to angular accelerations. Graphical and analytical vector techniques, including representation of vectors in complex form, are useful in determining linkage accelerations.

4.1.1 Acceleration of a Point

Consider a point moving along a curve in three-dimensional space and located by a vector R. The acceleration a of the point is given by the time rate of change in velocity:

$$a = \frac{dv}{dt} = \frac{d^2R}{dt^2} \tag{4.1}$$

The acceleration may be expressed in terms of the x, y, and z components and their respective unit vectors i, j and k in a *fixed coordinate system*:

$$a = i\ddot{R}_x + j\ddot{R}_y + k\ddot{R}_z \tag{4.2}$$

where two dots above the variable represent the second derivative with respect to time.

4.1.2 Angular Acceleration

Angular acceleration α represents the time rate of change in angular velocity ω. In general,

$$\alpha = \frac{d\omega}{dt} = \alpha_x i + \alpha_y j + \alpha_z k \tag{4.3}$$

For the special case of planar motion, the vector direction of α is perpendicular to the plane of rotation. For planar motion in the xy plane, vector α is in the $\pm z$ direction, that is, $\alpha = \alpha k$.

4.1.3 Motion of a Rigid Body About a Fixed Point

This important special case occurs frequently in linkage analysis. As noted in Chapter 3, the velocity of a point in a rigid body rotating about a fixed point is given by

$$\dot{R} = v = \omega \times R \tag{4.4}$$

where R is the vector from the fixed point to the point in question. Differentiating v with respect to time, we obtain the acceleration of the moving point:

$$\dot{v} = \dot{\omega} \times R + \omega \times \dot{R} \tag{4.5}$$

Using Eqs. 4.3 and 4.4, Eq. 4.5 may be rewritten as follows:

$$a = \alpha \times R + \omega \times (\omega \times R) \tag{4.6}$$

The first vector on the right in Eq. 4.6 is tangent to the path of the moving point and is called the *tangential acceleration*. The second term, which is normal to the path and directed toward the fixed point, is called the *normal acceleration*.

♦ **EXAMPLE PROBLEM 4.1**   *Average Angular Acceleration*
An automobile accelerates from 0 to 60 mi/h (0 to 96.56 km/h) in 15 s. Find the *average angular acceleration of the rear axle*. The tires have a 13-in. (330.2-mm) outer radius.

Solution. This problem is equivalent to a dynamometer test, where the 60-mi/h speed is the linear velocity of a point on the tread of the tire. Let us convert this speed to more manageable dimensions:

$$60 \text{ mi/h} \times \frac{5280 \text{ ft}}{1 \text{ mi}} \times \frac{1 \text{ h}}{3600 \text{ s}} \times \frac{12 \text{ in}}{1 \text{ ft}} = 1056 \text{ in/s}$$

Since we know that the point rotates about a circle of radius $R = 13$ in, we can use the formula equating the angular velocity of a rotating body ω to the linear velocity v of a point on the body at a distance R from the axis of rotation: $v = \omega \times R = R\omega$. Dividing by wheel radius, then, we find

$$\omega = \frac{1056 \text{ in/s}}{13 \text{ in}} = 81.2 \text{ rad/s}$$

Average angular acceleration is defined as the time rate of change in the angular velocity. Therefore,

$$\alpha_{av} = \frac{\omega(\text{final}) - \omega(\text{initial})}{\text{time interval}}$$

$$= \frac{81.2 - 0 \text{ rad/s}}{15 \text{ s}} = 5.41 \text{ rad/s}^2 \quad \blacklozenge$$

$\blacklozenge$ **EXAMPLE PROBLEM 4.2**   *Acceleration of a Point*

A 2-in (50.8-mm) link rotates in a plane with an angular velocity of 1000 rad/s clockwise. It is given an angular acceleration $\alpha = 750{,}000$ rad/s^2 counter-clockwise (causing the link to slow down). Find the acceleration of point B at the end of the link when it is oriented as in Figure 4.1.

Solution.   The normal acceleration of point B is independent of the angular acceleration; thus,

$$a_B^n = \omega \times (\omega \times R) = (-1000 \text{ rad/s})^2 (2 \text{ in})$$

$$= 2{,}000{,}000 \text{ in/s}^2 \angle \theta + \pi \text{ (along } BO \text{ toward } O)$$

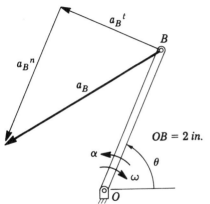

$OB = 2 \text{ in.}$

Figure 4.1 Acceleration of a point.

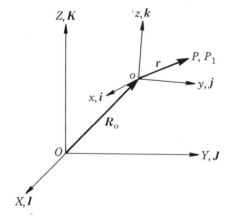

Figure 4.2 A moving coordinate system. System *xyz* moves within fixed system *XYZ*.

The tangential acceleration is given by

$$a_B^t = \dot{\omega} \times R = (750{,}000 \text{ rad/s}^2)(2 \text{ in})$$

$$= 1{,}500{,}000 \text{ in/s}^2 \angle \theta + \pi/2$$

(perpendicular to *OB* to the left since α is counterclockwise).

Adding the vectors, we obtain the total acceleration of point *B*:

$$a_B = a_B^t + a_B^n = \sqrt{(a_B^t)^2 + (a_B^n)^2}$$

$$= \sqrt{(1{,}500{,}000)^2 + (2{,}000{,}000)^2}$$

$$= 2{,}500{,}000 \text{ in/s}^2$$

to the left and downward. ◆

4.1.4 Moving Coordinate Systems

A more general case of linkage motion may be described by first considering a point within a moving coordinate system (Figure 4.2). A coordinate system *xyz* with respective unit vectors *i*, *j*, and *k* moves within a fixed system *XYZ* with respective unit vectors *I*, *J*, and *K*. The velocity of a point *P* may be described by

$$\dot{R} = \dot{R}_0 + \dot{r}_r + \omega \times r \tag{4.7}$$

(repeated from Chapter 3), where

$\dot{R}$ = absolute velocity of point *P* relative to *XYZ*

$\dot{R}_0$ = velocity of the origin *o* of the *xyz* system

$\dot{r}_r$ = velocity of point *P* relative to the *xyz* system

$\omega \times r$ = cross product of the angular velocity of the rotating system *xyz* in the *XYZ* system and the position vector *r*

Differentiating the first term on the right with respect to time, we have

$$\frac{d}{dt}\dot{R}_0 = \ddot{R}_0 \tag{4.8}$$

The next term could be written

$$\dot{r}_r = \dot{r}_{rx}i + \dot{r}_{ry}j + \dot{r}_{rz}k \tag{4.9}$$

Differentiating,

$$\frac{d\dot{r}_r}{dt} = \ddot{r}_{rx}i + \ddot{r}_{ry}j + \ddot{r}_{rz}k + \dot{r}_{rx}\dot{i} + \dot{r}_{ry}\dot{j} + \dot{r}_{rz}\dot{k} \tag{4.10}$$

The differential of a unit vector with respect to time is the cross product of the angular velocity of the moving coordinate system and the unit vector, for example, $\dot{i} = \omega \times i$. Then $\dot{r}_{rx}\dot{i} = \dot{r}_{rx}(\omega \times i) = \omega \times \dot{r}_{rx}i$, and so on. Then,

$$\frac{d\dot{r}_r}{dt} = \ddot{r}_{rx}i + \ddot{r}_{ry}j + \ddot{r}_{rz}k + \omega \times \dot{r}_{rx}i + \omega \times \dot{r}_{ry}j + \omega \times \dot{r}_{rz}k \tag{4.11}$$

which, observing Eq. 4.9, may be written

$$\frac{d\dot{r}_r}{dt} = \ddot{r}_r + \omega \times \dot{r}_r \tag{4.12}$$

The last term on the right side of Eq. 4.7 may be written

$$\omega \times r = \omega \times (r_x i + r_y j + r_z k) \tag{4.13}$$

Differentiating, we have

$$\frac{d}{dt}(\omega \times r) = \dot{\omega} \times (r_x i + r_y j + r_z k) + \omega \times (\dot{r}_{rx}i + \dot{r}_{ry}j + \dot{r}_{rz}k)$$
$$+ \omega \times (r_x \dot{i} + r_y \dot{j} + r_z \dot{k})$$

Noting that $\dot{i} = \omega \times i$ and so on, as in the above case, and combining terms, we obtain

$$\frac{d}{dt}(\omega \times r) = \dot{\omega} \times r + \omega \times \dot{r}_r + \omega \times (\omega \times r) \tag{4.14}$$

Then, using Eqs. 4.8, 4.12, and 4.14, the *total acceleration of point P* is

$$\ddot{R} = \ddot{R}_0 + \dot{\omega} \times r + \omega \times (\omega \times r) + \ddot{r}_r + 2\omega \times \dot{r}_r \tag{4.15}$$

The first term on the right of Eq. 4.15 is the total acceleration of the origin o of the moving coordinates; the next two give the acceleration of P_1 relative to o, where P_1 is a point instantaneously coincident with P having no motion relative to the moving coordinates xyz. The last two terms are the motion of P relative to P_1.

The term $\dot{\omega} \times r$ may be identified as tangential acceleration a^t; $\omega \times (\omega \times r)$ may be identified as normal acceleration a^n; and $2\omega \times \dot{r}_r$ may be identified as Coriolis acceleration a^c. The sum $\ddot{R}$ is the total acceleration of P relative to fixed coordinates *XYZ*. It is important to remember that $\boldsymbol{\omega}$ and $\dot{\boldsymbol{\omega}}$ refer, respectively, to the angular velocity and the angular acceleration of the moving coordinate system.

Problems involving spatial linkages (mechanisms involving motion that does not lie in a plane or in a set of parallel planes) require that Eq. 4.15 be applied in its general form. In plane mechanisms, the vector products take the following forms: $\dot{\boldsymbol{\omega}} \times \boldsymbol{r}$ becomes αr, the tangential acceleration; $\boldsymbol{\omega} \times (\boldsymbol{\omega} \times \boldsymbol{r})$ becomes $\omega^2 r$, the normal acceleration; and $2\boldsymbol{\omega} \times \dot{\boldsymbol{r}}_r$ becomes $2\omega v$, the Coriolis acceleration. The Coriolis acceleration term appears when sliding occurs along a rotating link. Referring to Figure 4.2, the term $v = \dot{r}_r$ is the velocity of point *P* relative to a point P_1, where P_1 is instantaneously coincident with *P* but has no motion relative to the moving coordinates. Normal acceleration, tangential acceleration, and Coriolis acceleration will appear later as we use graphical and analytical methods to investigate the motion of linkages.

4.1.5 Relative Acceleration

In Eq. 4.15, the acceleration of a point *P* is described in terms of the acceleration of the origin *o* of a set of moving coordinates and four terms representing the difference between the acceleration of *P* and the acceleration of the origin *o* of a set of moving coordinates. Consistent with the terminology used for velocities, the acceleration difference is called the acceleration of *P* *relative to o* or the acceleration of *P* *with respect to o*.

The special case involving two points on the same rigid link is frequently encountered. Consider link *BC* of Figure 4.3, which is not fixed at any point. If the acceleration of point *B* is known, we may find the acceleration of any point *C* on the link by adding *the acceleration of point C with respect to B to the acceleration of B*. Symbolically, then, the acceleration of point *C* is given by

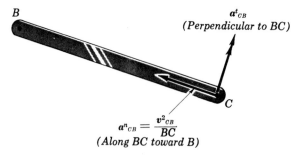

Figure 4.3 The acceleration of point *C* relative to point *B*, $\boldsymbol{a}_{CB}$, is shown broken into its normal and tangential components. The vector representing the normal component, $\boldsymbol{a}_{CB}^n$, lies along *BC* and is directed toward *B*; the vector representing the tangential component, $\boldsymbol{a}_{CB}^t$, is perpendicular to *BC*.

the expression

$$a_C = a_B + a_{CB} \tag{4.16}$$

where a_{CB}, the acceleration of point C with respect to point B, may be broken into its normal and tangential components:

$$a_{CB} = a_{CB}^n + a_{CB}^t \tag{4.17}$$

The normal acceleration of C with respect to B is given by

$$a_{CB}^n = \omega_{BC} \times (\omega_{BC} \times BC) \tag{4.18}$$

and the tangential acceleration is given by

$$a_{CB}^t = \alpha_{BC} \times BC \tag{4.19}$$

Using Eqs. 4.18 and 4.19, we have

$$a_C = a_B + \omega_{BC} \times (\omega_{BC} \times BC) + \alpha_{BC} \times BC \tag{4.20}$$

which may be solved analytically or graphically if sufficient data are given.

If link BC moves in a plane, the magnitude of the angular velocity of link BC is given by

$$\omega_{BC} = \frac{v_{CB}}{CB}$$

The magnitude of the normal acceleration is

$$a_{CB}^n = \omega_{BC}^2 BC = \frac{v_{CB}^2}{BC}$$

and the tangential acceleration magnitude is

$$a_{CB}^t = \alpha_{BC} BC$$

4.2   ANALYSIS OF A FOUR-BAR LINKAGE USING ANALYTICAL VECTOR METHODS

The vector equations developed in the preceding section may be applied to analysis of linkages. Consider the four-bar planar linkage of Figure 4.4. The loop equation for the linkage,

$$r_0 + r_1 + r_2 + r_3 = 0$$

was solved in Chapter 2 to determine relative link positions. Differentiating the loop equation, and making the substitutions indicated in Chapter 3, we obtain the velocity equation

$$\omega_1 \times r_1 + \omega_2 \times r_2 + \omega_3 \times r_3 = 0$$

(Equation 3.14), repeated). Solutions of this equation were obtained in Chapter 3.

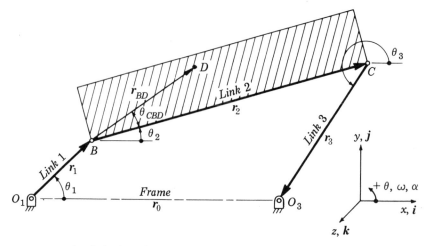

Figure 4.4 Analytical study of a four-bar linkage.

Differentiating the velocity equation, while noting that the links are fixed in length, we obtain the acceleration equation

$$\alpha_1 \times r_1 + \omega_1 \times (\omega_1 \times r_1) + \alpha_2 \times r_2 + \omega_2 \times (\omega_2 \times r_2)$$

$$+ \alpha_3 \times r_3 + \omega_3 \times (\omega_3 \times r_3) = 0 \tag{4.21}$$

The sense of vector r_3 is such that the last two terms of the above equation represent $-a_C$. Thus, the equation is equivalent to writing

$$a_B^t + a_B^n + a_{CB}^t + a_{CB}^n = a_C^t + a_C^n \tag{4.22}$$

We may orient the coordinate axes so that the linkage lies in the xy plane, with angular velocity and angular accelerations given in the form $\omega = \omega k$ and $\alpha = \alpha k$. Then, typical terms in Eq. 4.21 have the form

$$\alpha \times r = \begin{vmatrix} i & j & k \\ 0 & 0 & \alpha \\ r_x & r_y & 0 \end{vmatrix} = \alpha(jr_x - ir_y)$$

$$\omega \times r = \begin{vmatrix} i & j & k \\ 0 & 0 & \omega \\ r_x & r_y & 0 \end{vmatrix} = \omega(jr_x - ir_y)$$

$$\omega \times (\omega \times r) = \begin{vmatrix} i & j & k \\ 0 & 0 & \omega \\ -\omega r_y & \omega r_x & 0 \end{vmatrix} = -\omega^2(ir_x + jr_y)$$

Making the indicated substitutions in Eq. 4.21, the result is

$$\alpha_1(-ir_{1y} + jr_{1x}) - \omega_1^2(ir_{1x} + jr_{1y}) + \alpha_2(-ir_{2y} + jr_{2x}) - \omega_2^2(ir_{2x} + jr_{2y})$$

$$+ \alpha_3(-ir_{3y} + jr_{3x}) - \omega_3^2(ir_{3x} + jr_{3y}) = 0 \qquad (4.23)$$

The above equation must be satisfied separately for the coefficients of unit vectors i and j. Assume that the position, angular velocity, and angular acceleration of driving crank, link 1, are given. If we have already solved the displacement and velocity equations (Chapters 2 and 3), the remaining unknowns are α_2 and α_3. Separating the components of vector i, we have

$$\alpha_2 r_{2y} + \alpha_3 r_{3y} = -\alpha_1 r_{1y} - \omega_1^2 r_{1x} - \omega_2^2 r_{2x} - \omega_3^2 r_{3x} \qquad (4.24)$$

Separating the components of vector j, we have

$$\alpha_2 r_{2x} + \alpha_3 r_{3x} = -\alpha_1 r_{1x} + \omega_1^2 r_{1y} + \omega_2^2 r_{2y} + \omega_3^2 r_{3y} \qquad (4.25)$$

The simultaneous equations may be solved by elimination or another convenient form. For example, we may use the matrix form: $AX = B$, that is,

$$\begin{bmatrix} r_{2y} & r_{3y} \\ r_{2x} & r_{3x} \end{bmatrix} \begin{bmatrix} \alpha_2 \\ \alpha_3 \end{bmatrix} = \begin{bmatrix} a \\ b \end{bmatrix} \qquad (4.26)$$

where a represents the right side of Eq. 4.24 and b the right side of Eq. 4.25. Then the solution is given by

$$X = \begin{bmatrix} \alpha_2 \\ \alpha_3 \end{bmatrix} = A^{-1}B$$

Alternatively, using determinants, angular acceleration of the coupler is given by

$$\alpha_2 = \frac{1}{D} \begin{vmatrix} a & r_{3y} \\ b & r_{3x} \end{vmatrix} = \frac{ar_{3x} - br_{3y}}{D}$$

and the angular acceleration of the follower crank by

$$\alpha_3 = \frac{1}{D} \begin{vmatrix} r_{2y} & a \\ r_{2x} & b \end{vmatrix} = \frac{br_{2y} - ar_{2x}}{D}$$

where

$$D = \begin{vmatrix} r_{2y} & r_{3y} \\ r_{2x} & r_{3x} \end{vmatrix} = r_{2y}r_{3x} - r_{2x}r_{3y} \qquad (4.27)$$

◆ **EXAMPLE PROBLEM 4.3** *Accelerations in a Four-Bar Linkage*
Referring to Figure 4.4, let $\omega_1 = 100$ rad/s (constant), $\theta_1 = 45°$, $\theta_{CBD} = 20°$, (constant), $r_0 = 30$ mm, $r_1 = 10$ mm, $r_2 = 35$ mm, $r_3 = 20$ mm, and $r_{BD} = 15$ mm. Find α_2, α_3, a_B, a_C, and a_D for the assembly mode shown.

Solution. As determined in Chapters 2 and 3 using the same data,

$$r_{1x} = 7.0711 \qquad r_{1y} = 7.0711$$

$$r_{2x} = 33.590 \qquad r_{2y} = 9.835$$

$$r_{3x} = -10.660 \quad r_{3y} = -16.929$$

$$\omega_2 = -9.567 \qquad \omega_3 = 36.208$$

Using these values, we have

$$a = -\alpha_1 r_{1y} - \omega_1^2 r_{1x} - \omega_2^2 r_{2x} - \omega_3^2 r_{3x}$$

$$= 0 - 100^2(7.0711) - 9.567^2(33.590) - 36.208^2(-10.660)$$

$$= -59,810 \text{ mm/s}^2$$

$$b = -\alpha_1 r_{1x} + \omega_1^2 r_{1y} + \omega_2^2 r_{2y} + \omega_3^2 r_{3y}$$

$$= 0 + 100^2(7.0711) + 9.567^2(9.835) + 36.208^2(-16.929)$$

$$= 49,381 \text{ mm/s}^2$$

$$D = r_{2y}r_{3x} - r_{2x}r_{3y} = 9.835(-10.660) - 33.590(-16.929)$$

$$= 463.80$$

$$\alpha_2 = \frac{ar_{3x} - br_{3y}}{D} = \frac{(-59,810)(-10.660) - (49,381)(-16.929)}{463.80}$$

$$= 3180 \text{ rad/s}^2$$

$$\alpha_3 = \frac{br_{2y} - ar_{2x}}{D} = \frac{(49,381)(9.835) - (-58,600)(33.590)}{463.80}$$

$$= 5386 \text{ rad/s}^2$$

Once the angular accelerations are determined, we may find the acceleration of any point on the linkage. The acceleration of point B is given by

$$a_B = \alpha_1 \times r_1 + \omega_1 \times (\omega_1 \times r_1) = -\omega_1^2(ir_{1x} + jr_{1y})$$

$$= -100^2(i7.0711 + j7.0711) = -i70711 - j70711$$

$$= 100,000 \text{ mm/s}^2 \angle -135°$$

The acceleration of point C, as determined directly from its position on link 3, is

$$a_C = \alpha_3 \times (-r_3) + \omega_3 \times (\omega_3 \times (-r_3))$$

$$= \alpha_3(-jr_{3x} + ir_{3y}) - \omega_3^2(-ir_{3x} - jr_{3y})$$

$$= 5386(j10.660 - i16.929) - 36.208^2(i10.660 + j16.929)$$

$$= -i105,100 + j35,200 = 110,900 \text{ mm/s}^2 \angle 161.5°$$

As an alternative, we may calculate the acceleration of point C in terms of the acceleration of B and the acceleration of C with respect to B.

$$a_C = a_B + a_{CB}$$

where

$$a_{CB} = \alpha_2 \times r_2 + \omega_2 \times (\omega_2 \times r_2) = \alpha_2(jr_{2x} - ir_{2y}) - \omega_2^2(ir_{2x} + jr_{2y})$$

Substituting the values obtained above, we have

$$a_C = -i70{,}711 - j70{,}711 + 3180(j33.590 - i9.835)$$

$$- 9.567^2(i33.590 + j9.835)$$

which differs from the previous result only due to rounding errors.

The acceleration of point D in the coupler is given by

$$a_D = \alpha_1 \times r_1 + \omega_1 \times (\omega_1 \times r_1) + \alpha_2 \times r_{BD} + \omega_2 \times (\omega_2 \times r_{BD})$$

where the radius vector extending from B to D is

$$r_{BD} = r_{BD} \angle \theta_2 + \theta_{CBD} = 15\text{mm} \angle 16.35 + 20° = i12.081 + j8.891$$

$$a_D = -100^2(i7.0711 + j7.0711) + 3180(j12.081 - i8.891)$$

$$- 9.567^2(i12.081 + j8.891)$$

$$= -i100{,}100 - j33{,}100 = 105{,}400 \text{ mm/s}^2 \angle -161.7° \quad \blacklozenge$$

4.3 ACCELERATION ANALYSIS USING A SPREADSHEET

If the displacement and velocity formulas for a given type of linkage are already programmed on a spreadsheet, then the acceleration formulas can be added with little difficulty. For analysis of a four-bar linkage, we may use the equations in the section immediately preceding, converting them to spreadsheet form. The results of the analysis may be plotted, using the spreadsheet plotting routines. The graphical results aid in checking for programming errors, since inconsistencies are more easily detected in plotted results than in tabulated results.

♦ **EXAMPLE PROBLEM 4.4** *Utilizing a Spreadsheet to Determine Angular Accelerations in a Four-Bar Linkage*

For the four-bar linkage described in Example Problem 4.3, let angular velocity of the crank be 100 rad/s (constant and counterclockwise). Tabulate and plot angular velocities and angular accelerations of links 2 and 3 against crank angle.

Solution. The equations used in Example Problem 4.3 were converted into spreadsheet formulas. Crank position was changed in 5° increments and the

TABLE 4.1 ACCELERATION ANALYSIS OF A FOUR-BAR LINKAGE

	A	B	C	D	E	F	G
1	Four-bar linkage.			Acceleration analysis.			
2	Position analysis based on the vector cross product method.						
3	If Rd-R2-R3 loop clockwise, config = 1; -1 if ccw.						
4	R0	deg	R1	R2	R3	config	
5	30	190	10	35	20	1	
6							
7	Theta1	Rdx	Rdy	Rd	Rdux	Rduy	a
8	deg						
9	0	-20	0	20	-1	0	-10.625
10	20	-20.6031	3.420201	20.88503	-.9865	.1637633	-9.30848
11	40	-22.3396	6.427876	23.24593	-.961009	.2765162	-6.12207
12	60	-25	8.660254	26.45751	-.944911	.3273268	-2.36228
13	80	-28.2635	9.848078	29.9301	-.944317	.3290359	1.182941
14	100	-31.7365	9.848078	33.22934	-.955074	.2963669	4.20094
15	120	-35	8.660254	36.05551	-.970725	.2401922	6.587065
16	140	-37.6604	6.427876	38.20506	-.985745	.1682467	8.305531
17	160	-39.3969	3.420201	39.54511	-.996253	.0864886	9.341428
18	180	-40	0	40	-1	0	9.6875
19	200	-39.3969	-3.4202	39.54511	-.996253	-.086489	9.341428
20	220	-37.6604	-6.42788	38.20506	-.985745	-.168247	8.305531
21	240	-35	-8.66025	36.05551	-.970725	-.240192	6.587065
22	260	-31.7365	-9.84808	33.22934	-.955074	-.296367	4.20094
23	280	-28.2635	-9.84808	29.9301	-.944317	-.329036	1.182941
24	300	-25	-8.66025	26.45751	-.944911	-.327327	-2.36228
25	320	-22.3396	-6.42788	23.24593	-.961009	-.276516	-6.12207
26	340	-20.6031	-3.4202	20.88503	-.9865	-.163763	-9.30848
27	360	-20	0	20	-1	0	-10.625

	O	P	Q	R	S	T	U
1							
2							
3							
4							
5							
6							
7	Trans	r1x	r1y	Omega2	Omega3	a1	b1
8	angle						
9	28.95502	10	0	-50	-50	-150000	0
10	31.88023	9.396926	3.420201	-29.3345	-1.58073	-122065	44934.23
11	39.21914	7.660444	6.427876	-12.5486	31.02024	-71150.2	49903.28
12	48.64563	5	8.660254	-2.72811	46.71703	-31191.8	47403.28
13	58.61077	1.736482	9.848078	3.393613	52.83555	-2532.98	44871.65
14	68.1605	-1.73648	9.848078	8.025953	53.49536	20307.85	42122.39
15	76.57674	-5	8.660254	12.25964	50.41742	40294.95	37675.52
16	83.21617	-7.66044	6.427876	16.56474	44.30701	57614.27	29868.79
17	87.4952	-9.39693	3.420201	20.95568	35.6277	70212.13	17402.89
18	88.97681	-10	0	25	25	75000	0
19	87.4952	-9.39693	-3.4202	27.88739	13.21537	69864.73	-21404.5
20	83.21617	-7.66044	-6.42788	28.62672	.884453	54951.1	-45472.1
21	76.57674	-5	-8.66025	26.2019	-11.9559	31912.65	-71552.2
22	68.1605	-1.73648	-9.84808	19.52268	-25.9467	2404.241	-99818.7
23	58.61077	1.736482	-9.84808	7.301845	-42.1401	-32111.1	-129759
24	48.64563	5	-8.66025	-11.5576	-61.0027	-68654	-155548
25	39.21914	7.660444	-6.42788	-35.4972	-79.066	-101666	-155959
26	31.88023	9.396926	-3.4202	-54.0742	-81.8279	-130511	-95816.9
27	28.95502	10	0	-50	-50	-150000	0

TABLE 4.1   *CONTINUED*

H	I	J	K	L	M	N

Omega1	Alpha1					
100	0	15	20			

r2x	r2y	theta2 deg	r3x	r3y	theta3 deg	theta3 deg
30.625	16.9443	28.95502	-10.625	-16.9443	-122.09	237.91
32.68478	12.51819	20.95672	-12.0817	-15.9384	-127.163	232.8369
33.48778	10.17685	16.90389	-11.1482	-16.6047	-123.877	236.123
33.73285	9.332446	15.4645	-8.73285	-17.9927	-115.89	244.1101
33.71564	9.394436	15.56982	-5.45212	-19.2425	-105.819	254.1806
33.51938	10.0723	16.72509	-1.7829	-19.9204	-95.1144	264.8856
33.14159	11.25322	18.7549	1.858408	-19.9135	-84.6684	275.3316
32.53437	12.90405	21.63467	5.126071	-19.3319	-75.1492	284.8508
31.62	15.00585	25.38754	7.776926	-18.4261	-67.1173	292.8827
30.3125	17.49721	29.99473	9.6875	-17.4972	-61.0285	298.9715
28.561	20.2304	35.3108	10.83592	-16.8102	-57.194	302.806
26.41224	22.96505	41.00646	11.2482	-16.5372	-55.7774	304.2226
24.06995	25.4094	46.55067	10.93005	-16.7491	-56.8726	303.1274
21.92916	27.27841	51.20413	9.807317	-17.4303	-60.6354	299.3646
20.57725	28.31213	53.99036	7.686268	-18.4641	-67.3989	292.6011
20.73143	28.19943	53.67771	4.268568	-19.5392	-77.6767	282.3233
22.95807	26.41831	49.00867	-.618513	-19.9904	-91.7722	268.2278
26.88698	22.40737	39.80752	-6.28391	-18.9872	-108.312	251.6877
30.625	16.9443	28.95502	-10.625	-16.9443	-122.09	237.91

U	W	X	Y	Z	AA	AB

Omega* = Omega/Omega1
Alpha* = Alpha/Omega1 2

D	Alpha2 rad/s^2	Alpha3 rad/s^2	Omega2*	Omega3*	Alpha2*	Alpha3*
338.886	4702.908	13555.44	-.5	-.5	.4702908	1.355544
369.7017	5926.207	12313.04	-.293345	-.015807	.5926207	1.231304
442.6017	3664.309	6530.75	-.125486	.3102024	.3664309	.653075
525.4462	2141.62	2844.394	-.027281	.4671703	.214162	.2844394
597.5541	1468.074	848.3667	.0339361	.5283555	.1468074	.0848367
649.7607	1235.666	-394.664	.0802595	.5349536	.1235666	-.039466
680.8772	1211.87	-1338.67	.1225964	.5041742	.121187	-.133867
695.0992	1255.585	-2142.16	.1656474	.4430701	.1255585	-.214216
699.3312	1239.328	-2801.19	.2095568	.356277	.1239328	-.280119
699.8884	1038.112	-3248.29	.25	.25	.1038112	-.324829
699.3312	568.0204	-3472.5	.2788739	.1321537	.056802	-.34725
695.0992	-192.604	-3590.35	.2862672	.0088445	-.01926	-.359035
680.8772	-1247.85	-3798.38	.262019	-.119559	-.124785	-.379838
649.7607	-2641.42	-4271.75	.1952268	-.259467	-.264142	-.427175
597.5541	-4422.53	-5042.24	.0730184	-.421401	-.442253	-.504224
525.4462	-6341.89	-5639.12	-.115576	-.610027	-.634189	-.563912
442.6017	-6901.94	-4035.5	-.354972	-.79066	-.690194	-.40355
369.7017	-2702.64	3684.196	-.540742	-.818279	-.270264	.3684196
338.886	4702.908	13555.44	-.5	-.5	.4702908	1.355544

TABLE 4.2 SPREADSHEET CODING

(The table shows the formulas for a small portion of the spreadsheet)

```
A9:   0
B9:   +$A$5*@COS(@RAD($B$5))+$C$5*@COS(@RAD(A9))
C9:   +$A$5*@SIN(@RAD($B$5))+$C$5*@SIN(@RAD(A9))
D9:   @SQRT(B9^2+C9^2)
E9:   +B9/D9
F9:   +C9/D9
G9:   (($E$5^2-$D$5^2+D9^2)/(2*D9))
H9:   +$F$5*@SQRT($E$5^2-G9^2)*F9+(G9-D9)*E9
I9:   +$F$5*@SQRT($E$5^2-G9^2)*(-E9)+(G9-D9)*F9
J9:   @DEG(@ATAN2(H9,I9))
K9:   -$F$5*@SQRT($E$5^2-G9^2)*F9-G9*E9
L9:   -$F$5*@SQRT($E$5^2-G9^2)*(-E9)-G9*F9
M9:   @DEG(@ATAN2(K9,L9))
N9:   @IF(M9<0,360+M9,M9)
O9:   @ABS(N9-J9-180)
P9:   +$C$5*@COS(@RAD($A9))
Q9:   +$C$5*@SIN(@RAD($A9))
R9:   -$H$5*(Q9*K9-L9*P9)/(I9*K9-L9*H9)
S9:   -$H$5*(I9*P9-Q9*H9)/(I9*K9-L9*H9)
T9:   -$I$5*Q9-$H$5^2*P9-R9^2*H9-S9^2*K9
U9:   -$I$5*P9+$H$5^2*Q9+R9^2*I9+S9^2*L9
V9:   +I9*K9-H9*L9
W9:   (T9*K9-U9*L9)/V9
X9:   (U9*I9-T9*H9)/V9
Y9:   +R9/$H$5
Z9:   +S9/$H$5
AA9:  +W9/$H$5^2
AB9:  +X9/$H$5^2
A10:  5+A9
B10:  +$A$5*@COS(@RAD($B$5))+$C$5*@COS(@RAD(A10))
C10:  +$A$5*@SIN(@RAD($B$5))+$C$5*@SIN(@RAD(A10))
D10:  @SQRT(B10^2+C10^2)
E10:  +B10/D10
F10:  +C10/D10
G10:  (($E$5^2-$D$5^2+D10^2)/(2*D10))
H10:  +$F$5*@SQRT($E$5^2-G10^2)*F10+(G10-D10)*E10
I10:  +$F$5*@SQRT($E$5^2-G10^2)*(-E10)+(G10-D10)*F10
J10:  @DEG(@ATAN2(H10,I10))
K10:  -$F$5*@SQRT($E$5^2-G10^2)*F10-G10*E10
L10:  -$F$5*@SQRT($E$5^2-G10^2)*(-E10)-G10*F10
M10:  @DEG(@ATAN2(K10,L10))
N10:  @IF(M10<0,360+M10,M10)
O10:  @ABS(N10-J10-180)
P10:  +$C$5*@COS(@RAD($A10))
Q10:  +$C$5*@SIN(@RAD($A10))
R10:  -$H$5*(Q10*K10-L10*P10)/(I10*K10-L10*H10)
S10:  -$H$5*(I10*P10-Q10*H10)/(I10*K10-L10*H10)
T10:  -$I$5*Q10-$H$5^2*P10-R10^2*H10-S10^2*K10
U10:  -$I$5*P10+$H$5^2*Q10+R10^2*I10+S10^2*L10
V10:  +I10*K10-H10*L10
W10:  (T10*K10-U10*L10)/V10
X10:  (U10*I10-T10*H10)/V10
Y10:  +R10/$H$5
Z10:  +S10/$H$5
AA10: +W10/$H$5^2
AB10: +X10/$H$5^2
```

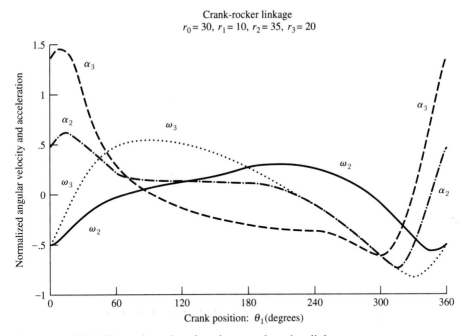

Figure 4.5 Velocities and accelerations in a crank-rocker linkage.

formulas were copied throughout the spreadsheet, changing reference cells as described in Chapter 3. Table 4.1 shows the spreadsheet in abbreviated form, with 20° crank angle increments. Spreadsheet formulas are shown in Table 4.2 to illustrate the cell format and the automatic changing of reference cells during copying. Only the formulas for 0° and 5° crank angles are shown. For nonzero values of crank angular acceleration, it would be necessary to change the formulas to increment crank angular velocity.

For plotting purposes, angular velocities are normalized by dividing by crank angular velocity; angular accelerations are normalized by dividing by the square of crank angular velocity. The plotted results are shown in Figure 4.5. Note that the angular velocities of links 2 and 3 are equal at $\theta_1 = 0°$ and also equal at $\theta_1 = 180°$. We observed that this is the case when considering the centros of a four-bar linkage. Note also that zero angular acceleration of a given link corresponds to an angular velocity extremum (a maximum or minimum).   ◆

4.4   MATHEMATICS SOFTWARE APPLIED TO ACCELERATION ANALYSIS

If we choose to work directly with vectors in the solution of linkage acceleration problems, we may write programs for vector manipulation or use commercially available mathematics software. In most cases, the vector solution will require less calculation on our part but will require more computer time than solutions in scalar form.

Consider the offset slider-crank linkage of Figure 3.12. The linkage may be described by the following position and velocity equations as given in Chapter 3:

$$e + r_1 + r_2 + r_0 = 0$$

and

$$\omega_1 \times r_1 + \omega_2 \times r_2 - v_c = 0$$

Differentiating the last equation with respect to time, the result is

$$\omega_1 \times (\omega_1 \times r_1) + \omega_2 \times (\omega_2 \times r_2) + \alpha_2 \times r_2 - a_c = 0 \qquad (4.28)$$

if the angular velocity of the crank is constant.

Noting that the acceleration of the slider lies along the x axis, we may eliminate the last term in the above equation by taking the dot product of each term with the unit vector j. The result is

$$\omega_1 \times (\omega_1 \times r_1) \cdot j + \omega_2 \times (\omega_2 \times r_2) \cdot j + \alpha_2 \times r_2 \cdot j = 0$$

from which

$$\alpha_2 = \left[-\omega_1 \times (\omega_1 \times r_1) \cdot j - \omega_2 \times (\omega_2 \times r_2) \cdot j \right] / (k \times r_2 \cdot j) \qquad (4.29)$$

where $\omega = \omega k$ and $\alpha = \alpha k$. Slider acceleration may now be obtained by rearranging Eq. 4.28 as follows:

$$a_c = \omega_1 \times (\omega_1 \times r_1) + \omega_2 \times (\omega_2 \times r_2) + \alpha_2 \times r_2 \qquad (4.30)$$

where $a_c = A_c i$.

◆ **EXAMPLE PROBLEM 4.5** *Acceleration Analysis of an Offset Slider-Crank Linkage Using Mathematics Software*

A slider-crank linkage has a connecting rod to crank length ratio of 1.8 and an offset to crank length ratio of 0.4. Tabulate and plot connecting rod angular velocity and angular acceleration, and slider velocity and acceleration, all against crank angle.

Solution. The first part of the solution is similar to that given in Section 3.6, utilizing commercially available mathematics software. Equations 4.29 and 4.30 were used to complete the solution shown in Figure 4.6. Crank length and crank angular velocity were set at unity. For other values of crank length and angular velocity, multiply the calculated values of ω_2 by ω_1, α_2 by ω_1^2, v_c by $\omega_1 r_1$, and A_c by $\omega_1^2 r_1$. Since we wish to tabulate and plot scalars, we take the dot product of the slider acceleration vector with unit vector i. Connecting rod angular velocity and angular acceleration are plotted together so we can verify that angular velocity reaches an extremum when angular acceleration is zero. Slider velocity and acceleration are also plotted together as a partial check for programming errors. ◆

Slider-crank linkage $c_0 = 1$ (clockwise configuration positive)

$$\theta_1 = 0, \frac{\pi}{9} \cdots 2 \cdot \pi \qquad R_1 = 1 \qquad R_2 = 1.8 \qquad e = \begin{bmatrix} 0 \\ -0.4 \\ 0 \end{bmatrix} \qquad r_{0u} = \begin{bmatrix} -1 \\ 0 \\ 0 \end{bmatrix}$$

$$r_1[\theta_1] = \begin{bmatrix} R_1 \cdot \cos[\theta_1] \\ R_1 \cdot \sin[\theta_1] \\ 0 \end{bmatrix} \qquad i = \begin{bmatrix} 1 \\ 0 \\ 0 \end{bmatrix} \qquad j = \begin{bmatrix} 0 \\ 1 \\ 0 \end{bmatrix} \qquad k = \begin{bmatrix} 0 \\ 0 \\ 1 \end{bmatrix} \qquad \omega_1 = 1$$
$$\alpha_1 = 0$$

$$r_3[\theta_1] = r_1[\theta_1] + e$$

$$r_0[\theta_1] = [-r_3[\theta_1] \cdot r_{0u} + c_0 \cdot [R_2^2 - [r_3[\theta_1] \cdot [r_{0u} \times k]]^2]^{.5}] \cdot r_{0u}$$

$$a[\theta_1] = -[r_3[\theta_1] \cdot [r_{0u} \times k]] \cdot [r_{0u} \times k]$$

$$r_2[\theta_1] = a[\theta_1] - c_0 \cdot [R_2^2 - [r_3[\theta_1] \cdot [r_{0u} \times k]]^2]^{.5} \cdot r_{0u}$$

$$\omega_2[\theta_1] = -\omega_1 \cdot \frac{r_1[\theta_1] \cdot i}{r_2[\theta_1] \cdot i}$$

$$v_c[\theta_1] = \omega_1 \cdot k \times r_1[\theta_1] + \omega_2[\theta_1] \cdot k \times r_2[\theta_1] \qquad V_c[\theta_1] = v_c[\theta_1] \cdot i$$

$$\alpha_2[\theta_1] = \frac{-\omega_1^2 \cdot [k \times [k \times r_1[\theta_1]]] \cdot j - \omega_2[\theta_1]^2 \cdot [k \times [k \times r_2[\theta_1]]] \cdot j}{[k \times r_2[\theta_1]] \cdot j}$$

$$b[\theta_1] = \omega_2[\theta_1]^2 \cdot [k \times [k \times r_1[\theta_1]]] + \omega_2[\theta_1]^2 \cdot [k \times [k \times r_2[\theta_1]]]$$

$$A_c[\theta_1] = [b[\theta_1] + \alpha_2[\theta_1] \cdot [k \times r_2[\theta_1]]] \cdot i$$

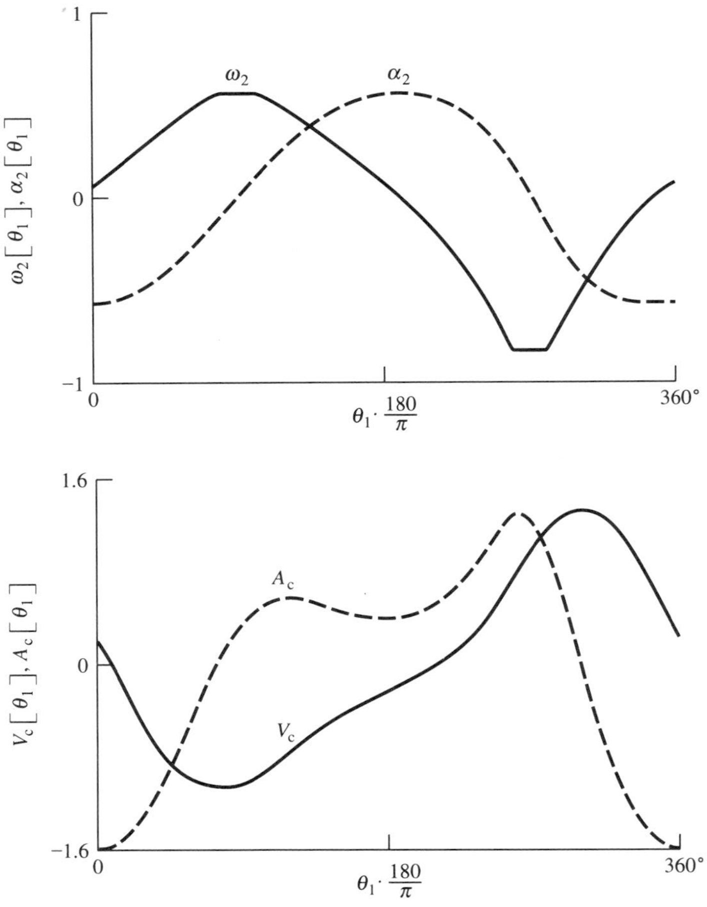

Figure 4.6 A slider-crank linkage analyzed by mathematics software. *(Continued on page 270.)*

$\theta_1 \cdot \dfrac{180}{\pi}$	$|r_0[\theta_1]|$	$\omega_2[\theta_1]$	$V_c[\theta_1]$	$\alpha_2[\theta_1]$	$A_c[\theta_1]$
0	2.755	-0.57	0.228	0.074	-1.599
20	2.739	-0.522	-0.312	0.199	-1.442
40	2.55	-0.43	-0.747	0.335	-1.014
60	2.239	-0.288	-1	0.476	-0.422
80	1.876	-0.102	-1.044	0.575	0.145
100	1.529	0.102	-0.925	0.575	0.492
120	1.239	0.288	-0.732	0.476	0.578
140	1.018	0.43	-0.539	0.335	0.518
160	0.859	0.522	-0.372	0.199	0.437
180	0.755	0.57	-0.228	0.074	0.401
200	0.7	0.573	-0.083	-0.06	0.446
220	0.701	0.522	0.098	-0.244	0.621
240	0.78	0.391	0.371	-0.526	0.97
260	0.976	0.151	0.776	-0.829	1.295
280	1.324	-0.151	1.194	-0.829	0.948
300	1.78	-0.391	1.361	-0.516	-0.03
320	2.233	-0.522	1.187	-0.244	-0.911
340	2.58	-0.573	0.767	-0.06	-1.434
360	2.755	-0.57	0.228	0.074	-1.599

Figure 4.6 Continued.

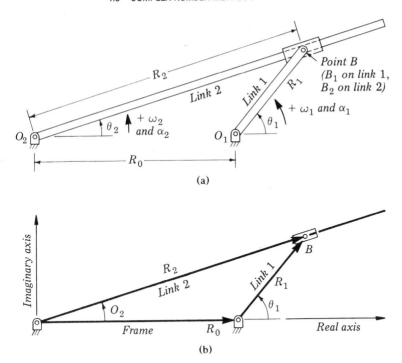

Figure 4.7 (a) Schematic for sliding contact linkage. (b) Vector representation.

4.5 COMPLEX NUMBER METHODS APPLIED TO ACCELERATION ANALYSIS

As illustrated in Chapters 2 and 3, complex numbers are a convenient form for representing the vectors that model planar linkage elements and velocities. It follows, therefore, that complex number methods can be applied to acceleration analysis of planar linkages. Consider the sliding contact linkage described in Section 3.7 and shown here in Figure 4.7a and b. The displacement equation is given by

$$R_0 + R_1 e^{j\theta_1} = R_2 e^{j\theta_2}$$

and the velocity equation by

$$j\omega_1 R_1 e^{j\theta_1} = j\omega_2 R_2 e^{j\theta_2} + v_{B_1 B_2} e^{j\theta_2}$$

as in Chapter 3.

Differentiating with respect to time, we obtain the acceleration equation:

$$-\omega_1^2 R_1 e^{j\theta_1} + j\frac{d\omega_1}{dt}R_1 e^{j\theta_1}$$

$$= -\omega_2^2 R_2 e^{j\theta_2} + j2\omega_2 v_{B_1 B_2} e^{j\theta_2} + j\frac{d\omega_2}{dt}R_2 e^{j\theta_2} + \frac{dv_{B_1 B_2}}{dt}e^{j\theta_2} \quad (4.31)$$

Thus we have the vector equation

$$a_{B_1}^n + a_{B_1}^t = a_{B_2}^n + a_{B_2}^t + a_{B_1 B_2}^c + a_{B_1 B_2}^t \quad (4.32)$$

where the vector magnitudes are as follows:

$$a_{B_1}^n = \omega_1^2 R_1 \quad \text{and} \quad a_{B_2}^n = \omega_2^2 R_2$$

(the normal accelerations),

$$\frac{d\omega}{dt} = \alpha$$

(the angular acceleration),

$$\alpha_1 R_1 = a_{B_1}^t \quad \text{and} \quad \alpha_2 R_2 = a_{B_2}^t$$

(the tangential accelerations),

$$2\omega_2 v_{B_1 B_2} = a_{B_1 B_2}^c$$

(the Coriolis acceleration) (note again that the order of the subscripts is critical),

$$\frac{dv_{B_1 B_2}}{dt} = a_{B_1 B_2}^t$$

[the relative (tangential) acceleration of B_1 with respect to B_2 (positive if dR_2/dt is increasing)].

4.5.1 Solving the Complex Acceleration Equation

In a problem of this type, it is likely that link lengths R_0 and R_1 would be specified, as well as the angular velocity and acceleration of link 1. Then R_2 and θ_2 can be found for given values of θ_1, using the displacement equations as in Chapter 2. Angular velocity ω_2 and relative velocity $v_{B_1 B_2}$ can be found as in Chapter 3. The remaining unknowns in Eq. 4.31 are $d\omega_2/dt$ and $dv_{B_1 B_2}/dt$. All of the terms in Eq. 4.31 are, in general, complex. If each term is multiplied by $e^{-j\theta_2}$, the equation then takes the form

$$\left(-\omega_1^2 + j\alpha_1\right)R_1 e^{j(\theta_1 - \theta_2)} = -\omega_2^2 R_2 + j2\omega_2 v_{B_1 B_2} + j\alpha_2 R_2 + a_{B_1 B_2}^t \quad (4.33)$$

where $\alpha_2 = d\omega_2/dt$ and $a_{B_1 B_2}^t = dv_{B_1 B_2}/dt$. The two unknowns can now be separated since the term containing $a_{B_1 B_2}^t$ is real and the term containing α_2 is imaginary. Using the Euler formula

$$e^{j(\theta_1 - \theta_2)} = \cos(\theta_1 - \theta_2) + j\sin(\theta_1 - \theta_2) \tag{4.34}$$

and noting that $j^2 = -1$, we equate the real parts of the resulting equation to obtain

$$a_{B_1 B_2}^t = -\omega_1^2 R_1 \cos(\theta_1 - \theta_2) + \alpha_1 R_1 \sin(\theta_2 - \theta_1) + \omega_2^2 R_2 \tag{4.35}$$

Equating the imaginary parts

$$\alpha_2 = \frac{1}{R_2}\left[\omega_1^2 R_1 \sin(\theta_2 - \theta_1) + \alpha_1 R_1 \cos(\theta_2 - \theta_1) - 2\omega_2 v_{B_1 B_2}\right] \tag{4.36}$$

◆ **EXAMPLE PROBLEM 4.6**   *Accelerations in a Sliding Contact Linkage*
Referring to Figure 4.7a and b, let $\omega_1 = 20$ rad/s (constant and counterclockwise), $R_0 = 40$ mm, and $R_1 = 20$ mm. Find $a_{B_1 B_2}^t$, and α_2 at $\theta_1 = 75°$.

Solution.   Using the equations of Chapters 2 and 3, we obtain $R_2 = 49.13$ mm, $\theta_2 = 23.15$, $\omega_2 = 5.03$ rad/s ccw, and $v_{B_1 B_2} = -314.6$ mm/s.
Relative acceleration is found by noting that ω_1 is constant ($\alpha_1 = 0$):

$$a_{B_1 B_2}^t = -\omega_1^2 R_1 \cos(\theta_1 - \theta_2) + \alpha_1 R_1 \sin(\theta_2 - \theta_1) + \omega_2^2 R_2$$

$$= -20^2 \times 20\cos(75° - 23.15°) + 5.03^2 \times 49.13 = -3698.7 \text{ mm/s}^2$$

(i.e., 3698.7 mm/s² along link 2 toward O_2). Angular acceleration:

$$\alpha_2 = \frac{\omega_1^2 R_1 \sin(\theta_2 - \theta_1) + \alpha_1 R_1 \cos(\theta_2 - \theta_1) - 2\omega_2 v_{B_1 B_2}}{R_2}$$

$$= \frac{20^2 \times 20\sin(23.15° - 75°) - 2 \times 5.03(-314.6)}{49.13}$$

$$= -63.63 \text{ rad/s}^2$$

(i.e., 63.63 rad/s² clockwise).   ◆

4.6 THE ACCELERATION POLYGON

4.6.1 Analysis of Slider-Crank Mechanisms

The acceleration polygon is analogous to the velocity polygon discussed in Chapter 3. The vector polygon provides us with a convenient method of finding unknown vectors through their relationship to known (easily calculated) vectors. In this case, the vectors being considered are accelerations. The acceleration polygon is simply the graphical expression of the acceleration vector equation, Eq. 4.22, where $a_C = a_C^t$. Vector $a_C^n = 0$ since the slider moves in a straight path.

◆ **EXAMPLE PROBLEM 4.7**   *Acceleration Polygon for a Slider-Crank Linkage*
Analysis of the slider-crank mechanism with constant crank velocity is shown.
Figure 4.8 shows a slider-crank linkage that was examined in the preceding
chapter. We want to find the acceleration of point C on the slider.

Solution.   The velocity polygon in Figure 4.8b is taken from Chapter 3. Since
the crank has a constant angular velocity,

$$a_B^t = \alpha_1 \times O_1 B = 0$$

Thus, the total acceleration of B is

$$a_B = a_B^n = \omega_1 \times (\omega_1 \times O_1 B)$$

with the magnitude given by

$$a_B^n = \omega_1^2 O_1 B = \frac{v_B^2}{O_1 B} = \frac{(ob)^2}{O_1 B} = \frac{(20 \text{ in/s})^2}{2 \text{ in}}$$

$$= 200 \text{ in/s}^2$$

in a direction parallel to $O_1 B$ toward O_1. See Figure 4.8c and the tabulated
values of acceleration (Table 4.3).

Before the construction of the actual acceleration polygon, it will prove
helpful to first note the acceleration vector directions, which are apparent
from the linkage drawing, Figure 4.8a. After noting the linkage orientation,
the restraints on the mechanism, and the given data, we can readily identify
the presence and the direction of the following acceleration vectors (Figure
4.8c):

a_C along the horizontal path to which the slider is constrained
a_B^n parallel to $O_1 B$ and toward fixed point O_1
a_B^t $= 0$, since crank $O_1 B$ rotates with constant angular velocity
a_{CB}^n parallel to link BC and directed toward B
a_{CB}^t perpendicular to link BC

The procedure for constructing the acceleration polygon is similar in some
ways to the velocity polygon procedure. The first step, Figure 4.8c, includes
selection of an acceleration scale that will result in an acceleration polygon of
reasonable size. Of course, the accelerations are not all known at this time, but
it may be assumed that they are of the same order of magnitude as a_B, the
acceleration of the crankpin.

In Figure 4.8d, the vector a_B^n has already been drawn. The tangential
acceleration of the crankpin is zero in the special case under consideration,
eliminating the term a_B^t from Eq. 4.22. (*Note:* It does not necessarily follow
that the tangential acceleration of C with respect to B is likewise zero; in fact,
it will be shown that a_{CB}^t is quite large in this example.) Since $a_B^t = 0$, we see
that a_{CB}^n, the normal acceleration of C with respect to B, must be evaluated

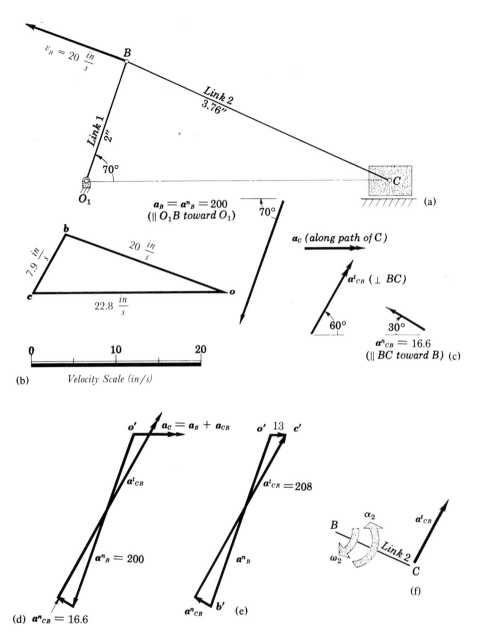

Figure 4.8 (a) The slider-crank mechanism is drawn to scale. (b) After a suitable velocity scale is selected, the velocity polygon for the slider-crank mechanism is drawn. (c) After a suitable acceleration scale is selected, the directions of the various acceleration components are identified by inspecting the linkage orientation. Where vector magnitudes can be determined (with the aid of the velocity polygon and link lengths), the vectors are drawn to scale. Where the vector magnitudes are unknown, they are drawn with double arrowheads. (d) The acceleration polygon is begun. (e) Accelerations a_C and a_{CB}^t are scaled directly from the acceleration polygon. (f) Knowing the tangential acceleration of C relative to B, a_{CB}^t, and knowing the length of the link BC, we can find the angular acceleration of the link.

next. The normal acceleration of C with respect to B is given by

$$a_{CB}^n = \omega_2 \times (\omega_2 \times BC)$$

The magnitude is

$$a_{CB}^n = \omega_2^2 BC = \frac{v_{CB}^2}{BC} = \frac{(bc)^2}{BC} = \frac{7.9^2}{3.76} = 16.6 \text{ in/s}^2$$

and the direction is opposite the vector BC. That is, it will lie along the connecting rod BC, directed toward B. Vector a_{CB}^n is then drawn to scale and added to the head of vector a_B^n (Figure 4.8d), parallel to BC, and is directed toward point B. The final vector, a_{CB}^t, is added at the head of and perpendicular to a_{CB}^n to complete the vector sum of Eq. 4.22. A double arrowhead is used in Figure 4.8d to indicate that the length of a_{CB}^t is not yet known. The sum $a_B^n + a_{CB}^n + a_{CB}^t$ represents the total acceleration of C; the true direction of the acceleration of C is horizontal. (The direction of C was obvious at the onset and was drawn as a horizontal vector in Figure 4.8c. We therefore draw a_C horizontally (to the right to close the polygon) from pole point o' in the acceleration polygon of Figure 4.8d, again using a double arrowhead since the magnitude of a_C is unknown. Both vectors a_C and a_{CB}^t end where they intersect, a point which we label c'. Measuring the lengths of each on the acceleration scale we find that

$$a_{CB}^t = 208 \text{ in/s}^2 \text{ (upward and to the right)}$$

$$a_C = 13 \text{ in/s}^2 \text{ (to the right)}$$

as shown in Figure 4.8e. If greater accuracy is required we may compute the exact angles and use trigonometric functions or use analytical or computer methods from the start.

Knowing the length BC and the tangential acceleration of C with respect to B, we can also find the angular acceleration of the connecting rod, link 2. From the formula for tangential acceleration, $a^t = r\alpha$, we obtain

$$\alpha_2 = \frac{a^t}{r} = \frac{a_{CB}^t}{BC} = \frac{208 \text{ in/s}^2}{3.76 \text{ in}} = 55 \frac{\text{rad}}{\text{s}^2}$$

The method for finding the direction of α_2 is similar to the method for finding the direction of ω. Tangential acceleration vector a_{CB}^t is placed at C on link BC as in Figure 4.8f. We see immediately that α_2 is counterclockwise (opposite the direction of ω_2 found in the preceding chapter). Thus, at this instant, the angular acceleration α_2 is opposing the angular velocity ω_2, which means that ω_2 is decreasing. The reader will recall that ω_1, the angular velocity of the crank, is constant in this example.

Let us now review the above steps for finding accelerations of the slider-crank mechanism in Figure 4.8a.

Step 1.   Draw the linkage to scale. Draw the velocity polygon **obc** representing the solution to the vector equation.

$$v_C = v_B + v_{CB} \quad \text{or} \quad oc = ob + bc$$

TABLE 4.3   VECTOR TABULATION FOR THE ACCELERATION POLYGON
OF A SLIDER-CRANK MECHANISM WITH UNIFORM CRANK VELOCITY

Vector	$a_C =$	a_B^n	$+a_B^t$	$+a_{CB}^n$	$+a_{CB}^t$
Vector magnitude	?	$\dfrac{(ob)^2}{O_1 B}$	$\alpha_1 O_1 B$	$\dfrac{(bc)^2}{BC}$	?
Vector direction	∥ path of C	∥ $O_1 B$ toward O_1	$\perp O_1 B$	∥ BC toward B	$\perp BC$
Vectors used to construct polygon	$\overset{?}{\longrightarrow}$	200 in/s² ↙	0	16.6 in/s² ↖	? ↗

Step 2. Solve the general acceleration vector equation for the slider-crank mechanism graphically as demonstrated in Table 4.3.

Step 3. The acceleration of B is labeled $o'b'$. The "prime" used with the lowercase boldface letters indicates that the vector is an acceleration and not a velocity. In this case, $a_B^t = 0$ since $\alpha_1 = 0$, from which $a_B = a_B^n$ only. To a_B, we add vectors a_{CB}^n (the magnitude is found with the aid of the velocity polygon) and a_{CB}^t (drawn perpendicular to BC and of unknown magnitude). The intersection of a_C and a_{CB}^t completes the polygon and determines the magnitude of each.

Step 4. The acceleration vectors have been identified by their components (e.g., a_B^n and a_B^t) and by an acceleration polygon notation patterned after the velocity polygon notation. For the linkage considered above:

$$a_B = o'b'$$

$$a_{CB} = b'c'$$

Although not shown in the acceleration polygon, the normal and tangential components of a_{CB} could be replaced by a single vector representing their sum and extending from b' to c' ($b'c'$). Also,

$$a_C = o'c'$$

Note the reversal of letters in acceleration polygon notation: acceleration a_{CB} becomes $b'c'$, just as velocity v_{CB} becomes bc in velocity polygon notation. The acceleration polygon will be used to advantage later when we consider the acceleration image. ◆

4.6.2 Comparison with an Analytical Solution

In Chapter 3, the velocity of the slider of an in-line slider-crank linkage was approximated by

$$v = R\omega \sin \theta \left[1 + \left(\frac{R}{L} \right) \cos \theta \right]$$

For the special case of constant angular velocity ω, we may replace θ in the above equation by ωt, setting $t = 0$ when $\theta = 0$. Slider accelerations may be obtained by differentiating the velocity equation. Differentiating the approximate equation for the velocity of the slider of a slider-crank mechanism, we obtain the approximate slider acceleration

$$a = R\omega^2 \left[\cos \theta + \left(\frac{R}{L} \right) \cos 2\theta \right] \tag{4.37}$$

The above equations give a positive value for velocity and acceleration directed toward the crankshaft and a negative value when directed away from the crankshaft. They are valid for the in-line slider crank when crank speed is constant and the ratio L/R does not approach a value of 1, say L/R greater than or equal to 3. For the data from the above example, the approximate slider acceleration is $a_C = -13.1$ (13.1 in/s² to the right), which corresponds closely to the result obtained using the acceleration polygon.

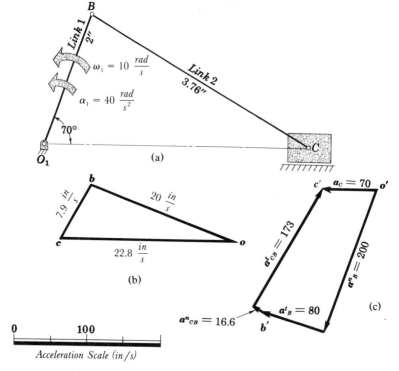

Figure 4.9 (a) The mechanism of Figure 4.8 is redrawn. This time, the crank is given an angular acceleration instead of constant angular velocity. (b) The addition of an angular acceleration does not affect the velocity polygon for the instant considered. (c) The acceleration polygon with angular acceleration of the crank. The expression for a_C is now made up of all the following vectors: $a_B^n + a_B^t + a_{CB}^n + a_{CB}^t$.

TABLE 4.4   VECTOR TABULATION FOR THE ACCELERATION POLYGON
OF A SLIDER-CRANK MECHANISM WITH ANGULAR ACCELERATION
OF THE CRANK

Vector	$a_C =$	a_B^n	$+a_B^t$	$+a_{CB}^n$	$+a_{CB}^t$
Vector magnitude	?	$\dfrac{(ob)^2}{O_1 B}$	$\alpha_1 O_1 B$	$\dfrac{(bc)^2}{BC}$	?
Vector direction	∥ path of C	$\parallel O_1 B$ toward O_1	$\perp O_1 B$	$\parallel BC$ toward B	$\perp BC$
Vectors used to construct polygon	←— ?	200 in/s²	80 in/s²	16.6 in/s²	? ↗

◆ **EXAMPLE PROBLEM 4.8**   *Linkage with Angular Acceleration of the Crank*
Here we consider acceleration analysis of the slider-crank linkage with angular
acceleration of the crank. Find the accelerations for the linkage of Figure 4.9a.
Given data are the same as those for the preceding problem except that link 1
does not have a constant angular velocity but instead accelerates at a rate
$\alpha_1 = 40$ rad/s² counterclockwise.

Solution. The addition of an angular acceleration has no effect on the
instantaneous velocity so that the velocity polygon remains unchanged. (See
Figure 4.9b). We will again use Eq. 4.22. In this case, however, the tangential
acceleration of point B does not equal zero:

$$a_B^t = \alpha_1 \times O_1 B = (40 \text{ rad/s}^2)(2 \text{ in}) = 80 \text{ in/s}^2 \perp O_1 B$$

For convenience, the terms of the acceleration equation are tabulated in Table
4.4.
   We can now construct the acceleration polygon. Beginning at an arbitrary
pole point o' in Figure 4.9c, we draw a_B^n to scale and add a_B^t at the head of
a_B^n. The sum $a_B^n + a_B^t$ forms a_B (or $o'b'$). Thus the head of a_B^t is labeled point
b'. Next, the two components of relative acceleration are added: first, a_{CB}^n
starting at b' and, to it, a_{CB}^t (of unknown length). Finally, the length of a_C is
found by drawing vector a_C horizontally (as in the previous example, the slider
is physically restricted to horizontal motion) until it intersects the vector a_{CB}^t.
The point of intersection determines the magnitudes of a_{CB}^t and a_C; it is
labeled c'.
   Scaling the vectors in Figure 4.9c, we obtain the slider acceleration

$$a_C = 70 \text{ in/s}^2 \text{ (to the left)}$$

and the tangential component of relative acceleration

$$a_{CB}^t = 173 \text{ in/s}^2 \text{ (upward and to the right)}$$

From the latter acceleration, we can also obtain the angular acceleration of the connecting rod:

$$\alpha_2 = \frac{a^t_{CB}}{CB} = 46 \text{ rad/s}^2 \text{ (counterclockwise)}. \quad \blacklozenge$$

4.6.3 Acceleration Image

In Example Problem 4.8, we determined the accelerations of the crankpin B and the slider C. We may also wish to find the acceleration of another point on the crank or connecting rod. The acceleration of the center of gravity, for example, would be used to perform a dynamic analysis of a link, or, in a more complicated linkage, an intermediate point on a link that serves as a connecting point would be investigated. We may resort to the acceleration image method (similar to the velocity image method) in order to find the acceleration of any point.

Consider any three points B, C, and D that lie on the same rigid link shown in Figure 4.10a. Let the link have an angular velocity ω_2 and an angular acceleration α_2. Then, the magnitudes of accelerations are

$$a^n_{CB} = \omega_2^2 BC \quad \text{and} \quad a^t_{CB} = \alpha_2 BC$$

The total acceleration of C with respect to B, a_{CB}, is the vector sum of its normal and tangential components. Its magnitude is given by the expression

$$a_{CB} = b'c' = \sqrt{\left(a^n_{CB}\right)^2 + \left(a^t_{CB}\right)^2}$$

$$= \sqrt{\left(\omega_2^2 BC\right)^2 + \left(\alpha_2 BC\right)^2}$$

from which

$$b'c' = BC\sqrt{\omega_2^4 + \alpha_2^2}$$

Similarly, the other relative acceleration magnitudes for the connecting rod are

$$b'd' = BD\sqrt{\omega_2^4 + \alpha_2^2}$$

$$c'd' = CD\sqrt{\omega_2^4 + \alpha_2^2}$$

from which we obtain the convenient *acceleration image* relationships:

$$\frac{b'd'}{b'c'} = \frac{BD}{BC}, \quad \frac{c'd'}{b'c'} = \frac{CD}{BC}, \quad \frac{b'd'}{c'd'} = \frac{BD}{CD} \tag{4.38}$$

Eq. 4.38 may be summarized by stating that triangle $b'c'd'$ (the acceleration image of BCD) is similar to triangle BCD for any points B, C, and D on the same rigid link. The angle relationship between a line connecting two points on a rigid link and their relative acceleration depends on the angular acceleration α and the angular velocity ω and is the same for any pair of points on the same rigid link. In the example problems that follow, we will utilize this relationship without having to calculate α and ω.

◆ **EXAMPLE PROBLEM 4.9** *Acceleration Image*
Three points, B, C, and D, lie on the rigid link in Figure 4.10a but do not lie on a straight line. Using the acceleration image method, find the acceleration of point D of the mechanism, using the data given on the illustration.

Solution. This problem and the problem of Figure 4.9a are identical except for the addition of an arbitrary point D. The velocity polygon, including the velocity image, is constructed (Figure 4.10b) as described in Chapter 3. The acceleration polygon $o'b'c'$ (Figure 4.10c) is taken directly from Figure 4.9c, but the normal and tangential components of a_B and a_{CB} have been omitted here to clarify the construction.

We observe in Figure 4.10c that the relative acceleration vector $b'c'$ (forming one leg of the required acceleration image) lies in the direction of line BC rotated approximately 95° counterclockwise. Since the acceleration image $b'c'd'$ and link BCD are similar triangles, we know that each leg of the acceleration image will make a 95° angle with its respective side in the linkage drawing. Beginning at b', we construct the acceleration image by first drawing trial vector $b'd'$, its direction determined by rotating line BD 95° counterclockwise. Then, trial vector $c'd'$ is drawn from point c'; its direction is also found by rotating line CD 95° counterclockwise. Point d' is thus determined by the intersection of trial vectors $b'd'$ and $c'd'$, as in Figure 4.10c, completing triangle $b'c'd'$, the acceleration image of link BCD on the acceleration polygon. The acceleration of D is thus given by vector $o'd'$. Using the acceleration scale, we find that $a_D = o'd' = 113$ in/s², slightly downward to the left. ◆

◆ **EXAMPLE PROBLEM 4.10** *Acceleration Image of Three Points on a Line*
In this problem we are required to find the acceleration of point E, which lies *on* the line BC in Figure 4.11a.

Solution. Again, this problem and the problem of Figure 4.9a are identical except for the addition of a point E along line (link) BC. We are again spared the necessity of constructing the velocity and acceleration polygons for the mechanism. See Figures 4.11b and c. Acceleration polygon $o'b'c'$ is again taken directly from Figure 4.9c. This problem is simpler than the preceding problem because no additional construction is necessary after the acceleration polygon is constructed. Since point E lies on line BC, we know that e' must lie somewhere on vector $b'c'$. A proportion similar to one of Eq. 4.38 gives us the desired acceleration image relationship:

$$\frac{b'e'}{b'c'} = \frac{BE}{BC}$$

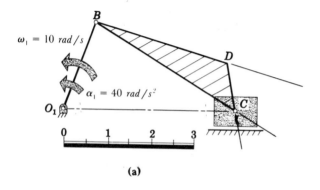

(a)

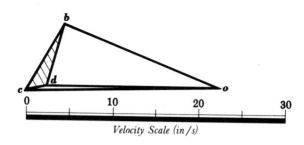

Velocity Scale (in/s)

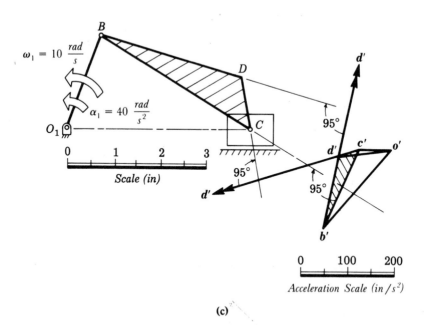

Acceleration Scale (in/s²)

(c)

Figure 4.10 The slider-crank mechanism of Figure 4.9a is repeated here. The dimensions of the linkage and the motion of the crank remain unchanged. We interested in finding the acceleration of an arbitrary point D on the connecting rod. (b) The velocity polygon for the slider-crank mechanism, showing the velocity image of the connecting rod. (c) The acceleration image of BCD is constructed.

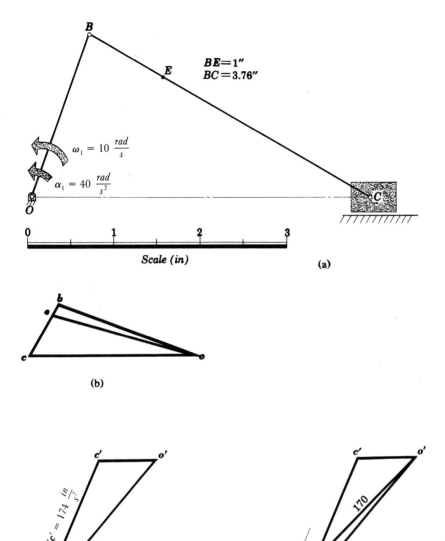

Figure 4.11 (a) Figure 4.9 is again repeated. We want to find the acceleration of a point E lying on line BC. (b) The velocity polygon for the linkage at the instant shown. (c) The acceleration polygon for the linkage at the instant shown. (d) Once the acceleration polygon is constructed, the position of e' (which must lie along $b'c'$) is determined by the proportion $b'e'/b'c' = BE/BC$. Drawing vector $o'e'$ then gives us the magnitude and direction of the acceleration of E.

from which the acceleration of E relative to B is

$$b'e' = b'c'\left(\frac{BE}{BC}\right) = (174 \text{ in/s}^2)\left(\frac{1 \text{ in}}{3.76 \text{ in}}\right) = 46.3 \text{ in/s}^2$$

This locates point e' on line $b'c'$. (Note that e' lies between b' and c' just as E lies between B and C.) Vector $o'e'$ is then drawn to obtain the acceleration of E, as shown in Figure 4.11d. Measuring $o'e'$ against the acceleration scale, we obtain

$$a_E = o'e' = 170 \text{ in/s}^2 \text{ (to the left and downward)}$$

The image principle illustrates the analogy between velocity polygons and acceleration polygons. The acceleration image principle (as well as the velocity image examined in Chapter 3) applies to a set of points on any rigid link, whether the link acts as a crank rotating about a fixed point or as a connecting rod. The only restriction is that the set of points considered must all lie on the same rigid link. ◆

4.6.4 Graphical Analysis of the Four-Bar Linkage

The acceleration analysis of a four-bar linkage requires no new concepts. Referring to Figure 4.12a, for example, we may again relate acceleration by the vector equation

$$a_C = a_B + a_{CB}$$

just as for the slider-crank mechanism, but with one additional complication. Each of the above acceleration vectors will have, in general, both a normal and tangential component and the equation takes the following form:

$$a_C = a_C^n + a_C^t = a_B^n + a_B^t + a_{CB}^n + a_{CB}^t$$

(Eq. 4.22 repeated).

The first steps of the acceleration analysis of a four-bar linkage are the construction of the skeleton drawing and the velocity polygon. The dimensions of the linkage together with the velocities taken from the velocity polygon allow us to calculate the normal components of acceleration of the links, which, as we have seen, are usually the starting point for the acceleration polygon.

◆ **EXAMPLE PROBLEM 4.11**   *Four-Bar Linkage*

Figure 4.12a shows a four-bar linkage for which an acceleration analysis will be made. The lengths of all the links are indicated on the skeleton drawing. The crank has an angular velocity $\omega_1 = 30$ rad/s and an angular acceleration $\alpha_1 = 200$ rad/s^2.

Solution.   The velocity polygon is constructed in Figure 4.12b after selection of a suitable scale. The velocities are then indicated directly on the velocity

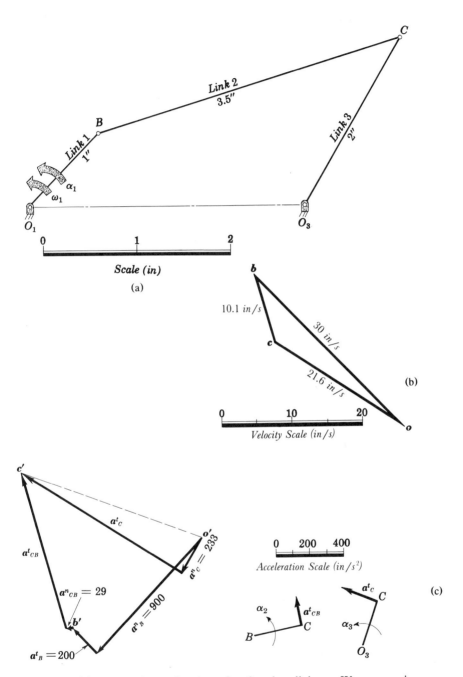

Figure 4.12 (a) The skeleton drawing of a four-bar linkage. We are again required to find the acceleration of point C, using the relationship $a_C = a_B + a_{CB}$. (b) Velocity polygon for the four-bar mechanism. (c) Acceleration polygon for the four-bar mechanism. The presence of the many vectors makes a method of tabulation of the various vectors desirable to ensure correct vector addition and orientation.

TABLE 4.5   VECTOR TABULATION FOR THE ACCELERATION ANALYSIS
OF A FOUR-BAR MECHANISM, FIGURE 4.12

Vector	a_C^n	$+a_C^t$	$= a_B^n$	$+a_B^t$	$+a_{CB}^n$	$+a_{CB}^t$
Vector magnitude	$\dfrac{(oc)^2}{O_3C}$	?	$\dfrac{(ob)^2}{O_1B}$	$\alpha_1 O_1 B$	$\dfrac{(bc)^2}{BC}$	?
Vector direction	$\parallel O_3C$ toward O_3	$\perp O_3C$	$\parallel O_1B$ toward O_1	$\perp O_1B$	$\parallel BC$ toward B	$\perp BC$
Vectors used to construct polygon	233 in/s²	?	900 in/s²	200 in/s²	29 in/s²	?

polygon. The skeleton drawing and the velocity polygon give us the information needed to determine the normal components of acceleration. Given the
angular acceleration of link 1, we can also calculate the tangential component
of acceleration for point B. We can now put together our acceleration vector
table (Table 4.5) and begin the construction of the acceleration polygon.
Velocities ob, oc, and bc are taken from the velocity polygon in Figure 4.12b.

We will now construct the acceleration polygon (see Figure 4.12c), adding
the vectors in the order indicated by Table 4.5. Beginning at the pole point o',
we draw a_C^n to the scale selected. To a_C^n, we add the trial vector a_C^t. The head
of a_C^t may be labeled c', but we do not as yet know the true magnitude of that
vector. Again, beginning at o', we add the vectors on the right side of Eq. 4.22
in the order indicated. The sum $a_B^n + a_B^t = a_B$ (or $o'b'$); thus the head of a_B^t is
labeled b'. Adding the last two of the four vectors on the right side of the
equation, which includes trial vector a_{CB}^t, we again obtain a_C (or $o'c'$). Point c'
is located at the intersection of the trial vectors a_C^t and a_{CB}^t, completing the
polygon and determining in turn the magnitude of each tangential component.
Thus, $a_C^t = 1110$ in/s² and $a_{CB}^t = 930$ in/s². Using the acceleration scale, we
can obtain the acceleration of point C:

$$a_C = a_C^n + a_C^t = o'c' = 1134 \text{ in/s}^2$$

to the left and upward.

Using the tangential acceleration of point C, we can find the angular
acceleration of link 3 (O_3C):

$$\alpha_3 = \frac{a_C^t}{O_3C} = \frac{1110 \text{ in/s}^2}{2 \text{ in}} = 555 \text{ rad/s}^2 \text{ (counterclockwise)}$$

Similarly, the angular acceleration of link 2 (BC) is given by

$$\alpha_2 = \frac{a_{CB}^t}{CB} = \frac{930 \text{ in/s}^2}{3.5 \text{ in}} = 266 \text{ rad/s}^2 \text{ (counterclockwise)} \quad \blacklozenge$$

◆ **EXAMPLE PROBLEM 4.12** *Four-Bar Linkage Described by Complex Numbers*

A linkage is described by complex numbers, as follows:

Link 1 (crank): $O_1B = -j100$

Link 2 (coupler): $\dot{BC} = -200 - j200$

Link 3 (rocker): $CO_3 = j200$

(all in millimeters). Given: angular velocity $\omega_3 = 1$ rad/s and angular acceleration $\alpha_3 = 1$ rad/s^2 at the instant shown. Find angular velocities and angular accelerations of links 1 and 2.

Solution. Converting to polar form, we have

$O_1B = 100e^{-j1.571}$ or 100 mm at $\theta_1 = -90°$

$BC = 282.8e^{-j2.356}$ or 282.8 mm at $\theta_2 = -135°$

$CO_3 = 200\,e^{j1.571}$ or 200 mm at $\theta_3 = 90°$

The linkage is sketched in Figure 4.13a.

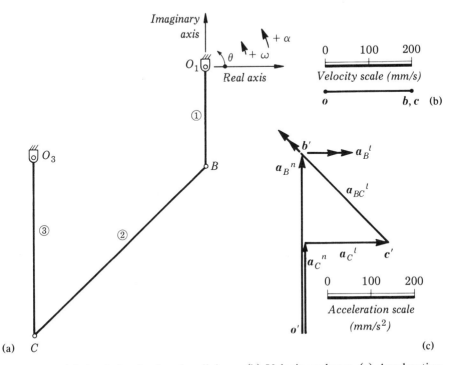

(a)   C   (c)

Figure 4.13 (a) Analysis of a four-bar linkage. (b) Velocity polygon. (c) Acceleration polygon.

The velocity of C is

$$oc = \omega_3 O_3 C = 1 \times 200 = 200 \text{ mm/s} \perp O_3 C$$

and to the right as shown in Figure 4.13b. Using $ob \perp O_1 B$ and $cb \perp CB$, we complete the velocity polygon $ob = oc + cb$ to obtain $ob = 200$ mm/s to the right, as shown, and $cb = 0$.

Alternatively, we could have expressed the velocity vectors in complex form as follows, where the real component is subscripted x and the imaginary, y:

$$\boldsymbol{v}_C = 200 + j0 \text{ (mm/s)}$$

$$\boldsymbol{v}_B = v_{B_x} + j0$$

$$\boldsymbol{v}_{BC} = v_{BC_x} + j v_{BC_y}$$

$$v_{BC_y} = -v_{BC_x}$$

Noting that the velocity equation $\boldsymbol{v}_B = \boldsymbol{v}_C + \boldsymbol{v}_{BC}$ must be satisfied by the imaginary components, we find $v_{BC_y} = 0$, from which $v_{BC_x} = 0$. Then, substituting the real parts, we find $v_{B_x} = 200$, or $\boldsymbol{v}_B = 200 + j0$ (mm/s).

Angular velocities are

$$\omega_1 = \frac{ob}{O_1 B} = \frac{200}{100} = 2 \text{ rad/s ccw}$$

$$\omega_2 = \frac{cb}{CB} = 0$$

$$\omega_3 = \frac{oc}{O_3 C} = \frac{200}{200} = 1 \text{ rad/s ccw}$$

For the acceleration analysis, we may solve the acceleration equation in complex form or graphically as follows. (See Table 4.6.) The acceleration

TABLE 4.6 ACCELERATION ANALYSIS OF A FOUR-BAR MECHANISM, FIGURE 4.13

	$o'b'$		$o'c'$		$c'b'$	
Vector	a_B^n	$+a_B^t$	$= a_C^n$	$+a_C^t$	$+a_{BC}^n$	$+a_{BC}^t$
Vector magnitude	$\frac{(ob)^2}{O_1 B}$	?	$\frac{(oc)^2}{O_3 C}$	$\alpha_3 O_3 C$	$\frac{(cb)^2}{CB}$	?
Vector direction	$\| O_1 B$ toward O_1	$\perp O_1 B$	$\| O_3 C$ toward O_3	$\perp O_3 C$	$\| CB$ toward C	$\perp CB$
Vectors used to construct polygon (mm/s^2)	400 ↑	? $\longrightarrow$	200 ↑	200 $\longrightarrow$	0	? ↖

polygon is drawn beginning at pole point o', as shown in Figure 4.13c. The intersection of vectors a_{BC}^t and a_B^t completes the polygon at a point that we label b'. For this position, we find $a_B^t = 0$ and $a_{BC}^t = 282.8$ mm/s$^2 \perp CB$, as shown. We may now calculate angular accelerations:

$$\alpha_1 = \frac{a_B^t}{O_1B} = 0 \quad \alpha_2 = \frac{a_{BC}^t}{BC} = \frac{282.8}{282.8} = 1 \text{ rad/s}^2 \text{ ccw}$$

$$\alpha_3 = \frac{a_C^t}{O_3C} = \frac{200}{200} = 1 \text{ rad/s}^2 \text{ ccw} \quad (\text{check}) \quad \blacklozenge$$

4.6.5 An Analytical Solution Based on the Acceleration Polygon

In Chapter 3, it was shown that the law of sines could be used to solve a sketched velocity polygon. A sketched acceleration polygon may be solved in the same manner, except that the angles are more difficult to compute. A slightly different procedure will be used in the illustration that follows.

Consider Example Problem 4.3, except that we will use an acceleration polygon approach. Since $\alpha_1 = 0$, the acceleration of point B is given by

$$o'b' = a_B = \omega_1^2 r_1 = 100^2 \times 10 = -100,000 \angle \theta_1$$

Also, after determining velocities, we find

$$a_{CB}^n = \frac{bc^2}{BC} = -3193 \angle \theta_2$$

$$a_C^n = \frac{oc^2}{O_3C} = 26,260 \angle \theta_3$$

An arbitrary pole point o' is selected, with coordinates $x = 0$, $y = 0$. Vector a_C^n is drawn. The coordinates of its head are identified as

$$x_1 = -a_C^n \cos \theta_3 = 26,260 \cos(-122.21°) = -13,997$$

and

$$y_1 = -a_C^n \sin \theta_3 = 26,260 \sin(-122.21°) = -22,219$$

as shown in Figure 4.14. Vector $o'b'$ is drawn, and a_{CB}^n is added to it. The coordinates of the head of a_{CB}^n are identified as

$$x_2 = -o'b' \cos \theta_1 - a_{CB}^n \cos \theta_2$$

$$= -100,000 \cos 45° - 3193 \cos 16.35° = -73,775$$

and

$$y_2 = -o'b' \sin \theta_1 - a_{CB}^n \sin \theta_2$$

$$= -100,000 \sin 45° - 3193 \sin 16.35° = -71,610$$

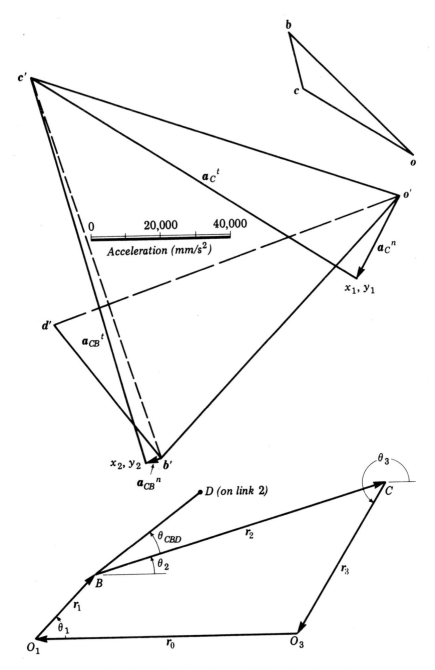

Figure 4.14 Analytical solution of a four-bar linkage based on the acceleration polygon.

At x_1, y_1, we add vector a_C^t with unknown magnitude but with slope

$$m = \tan(\theta_3 - 90°) = \tan(237.79 - 90°) = -0.62997$$

a_C^t lies on a line described by

$$y = y_1 + m(x - x_1) = -22,219 - 0.62997(x + 13,997)$$

At x_2, y_2, we add vector a_{CB}^t with unknown magnitude but with slope

$$n = \tan(\theta_2 + 90°) = \tan(16.5 + 90°) = -3.4087$$

a_{CB}^t lies on a line described by

$$y = y_2 + n(x - x_2) = -71,610 - 3.4087(x + 73,775)$$

The intersection of vectors a_C^t and a_{CB}^t locates point c' on the acceleration polygon. Equating the right sides of the two equations of the lines describing a_C^t and a_{CB}^t, we find the x coordinate of a_C:

$$x = a_{C_x} = -105,100$$

Substituting x in one of the above equations, the result is

$$y = a_{C_y} = 35,174$$

from which we find

$$a_C = o'c' = 110,800 \text{ mm/s}^2 \angle 161.5°$$

Relative acceleration is given by

$$b'c' = a_{CB} = a_C - a_B$$

from which

$$a_{CB_x} = -105,102 - 100,000 \cos 225°$$

$$a_{CB_y} = 35,174 - 100,000 \sin 225°$$

and

$$a_{CB} = 111,330 \text{ mm/s}^2 \angle 108°$$

Using the image principle, we find the relative acceleration of two points on the coupler.

$$a_{BD} = b'd' = \frac{b'c'BD}{BC} = \frac{(111,330)(15)}{35}$$

$$= 47,712 \text{ mm/s}^2 \angle 108° + 20°$$

The acceleration of point D on the coupler is given by

$$a_D = o'd' = o'b' + b'd'$$

from which

$$a_{D_x} = 100,000 \cos 225° + 47,712 \cos 128° = -100,100$$

$$a_{D_y} = 100,000 \sin 125° + 47,712 \sin 128° = -33,100$$

$$\boldsymbol{a}_D = 105,400 \text{ mm/s}^2 \angle -162°$$

4.7 EQUIVALENT LINKAGES

Equivalent linkages (which duplicate the motion of an actual mechanism) are sometimes useful in velocity and acceleration analysis. Figure 4.15a illustrates a commercially available curved-wing air pump, a mechanism which is obviously not a four-bar linkage. The pump has four evenly spaced wings, but for purposes of analysis, we need only consider the motion of one of these wings. The key to arriving at an equivalent linkage is to consider the forces acting on the mechanism and the restraints that restrict the mechanism to its specific

Figure 4.15 (a) A curved-wing air pump. Air is carried from inlet to outlet by the curved wings, which are held against the housing by acceleration forces. (*Source:* ITT Pneumotive.)

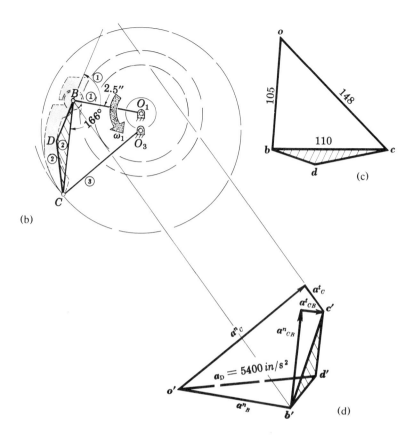

(b)

(c)

(d)

Figure 4.15 (b) The equivalent linkage for the curved-wing air pump is shown superimposed on the outline of one of the wings. The length of equivalent link 2 depends on the point of contact of the wing with the housing. (c) The velocity polygon for the four-bar (equivalent) linkage. (d) The acceleration polygon for the four-bar (equivalent) linkage.

path. These forces and restraints are replaced by links that are arranged so that the linkage duplicates the motion of the actual mechanism.

Figure 4.15b shows only one of the four wings (dashed lines) and its equivalent linkage for the instant shown (solid lines). Link 1 represents the driver crank and link 2 the wing. Point O_3 is the geometric center of the housing, and link 3 represents the distance from the center of the housing to the point of contact between wing and housing. (Actually, the wing is restrained by the rotational force of the pump to follow the curvature of the housing. At the instant being considered, this restraint can be considered a rigid link that forces point C to rotate about a circle of radius O_3C.) Link 3 does not exist on the actual pump, but it is essential to the equivalent linkage. Note that the equivalent linkage we have devised is the familiar four-bar linkage, and the mechanism will be analyzed as such. In the following example problem, we will assign dimensions to the pump, assume a reasonable rotating speed, and analyze the acceleration of this mechanism.

◆ **EXAMPLE PROBLEM 4.13** *Equivalent Linkages*

The pump shown in Figure 4.15a rotates at a constant 400 rev/min. The wing pins rotate about a circle of radius 2.5 in. The pump is shown drawn to scale. Find the angular acceleration of the wing and the acceleration of its center of gravity.

Solution. The equivalent linkage is drawn as shown in Figure 4.15b. The lengths of links 2 and 3 are obtained from the scale to which the skeleton drawing is made.

The velocity polygon in Figure 4.15c is based on the given speed of 400 rev/min and a given length of 2.5 in for link 1. The velocity polygon represents the solution to the vector equation

$$v_C = v_B + v_{CB} \quad \text{or} \quad oc = ob + bc$$

Point D of our equivalent linkage represents the center of gravity of the curved wing. Equivalent link 2 (*BCD*) forms the velocity image *bcd* on the velocity polygon. (This example is intended only to illustrate principles of acceleration analysis; the dimensions used may not correspond to an actual pump.) The velocities of B and C and the relative velocity *bc* shown on the velocity polygon are obtained in the usual manner and are used to construct the basic acceleration polygon $o'b'c'$ in the order tabulated in Table 4.7.

The magnitudes of the normal accelerations as calculated for Table 4.7 indicate the need for an acceleration scale on the order of 1 in = 2000 in/s². Since link 1 rotates at constant angular velocity (no angular acceleration), a_B^n represents the total acceleration of point B, $o'b'$. Beginning with this acceleration, we draw $o'b'$ parallel to O_1B (see Figure 4.15d). To it, we add known relative acceleration a_{CB}^n parallel to BC, and trial vector a_{CB}^t perpendicular to BC, as shown in Figure 4.15d. This completes the addition of the vectors on the right side of the equation.

TABLE 4.7   VECTOR TABULATION FOR THE ACCELERATION ANALYSIS
OF THE AIR PUMP EQUIVALENT LINKAGE

Vector	a_C^n	$+a_C^t$	$= a_B^n$	$+a_B^t$	$+a_{CB}^n$	$+a_{CB}^t$
Vector magnitude	$\dfrac{(oc)^2}{O_3C}$	?	$\dfrac{(ob)^2}{O_1B}$	$\alpha_1 O_1 B$	$\dfrac{(bc)^2}{BC}$	?
Vector direction	$\|O_3C$ toward O_3	$\perp O_3C$	$\|O_1B$ toward O_1	$\perp O_1B$ when $\alpha_1 \neq 0$	$\|BC$ toward B	$\perp BC$
Vectors used to construct polygon	6080 in/s²	?	4400 in/s²	0 for $\alpha_1 = 0$	3560 in/s²	?

Starting again at pole point o', we draw the known acceleration a_C^n parallel to O_3C. To it, we add trial vector a_C^t perpendicular to O_3C. The basic acceleration equation is satisfied when we locate c' at the intersection of the trial vectors a_C^t and a_{CB}^t. The angular acceleration of the wing is then given by

$$\alpha_2 = \frac{a_{CB}^t}{CB} = 265 \text{ rad}/\text{s}^2 \quad (\text{counterclockwise})$$

using values for a_{CB}^t and BC scaled from the illustration.

To find the acceleration of the center of gravity D, we construct the acceleration image of link 2. Points B, C, and the center of gravity D all lie on the same rigid link, permitting us to use the *acceleration image principle* to find a_D. Points b' and c' on the acceleration polygon are joined to form the image of BC. Using a protractor, we see that the orientation of $b'c'$ on the acceleration polygon is given by rotating BC on the skeleton linkage 166° counterclockwise. Similarly, BD and CD are rotated 166° counterclockwise to obtain the directions of $b'd'$ and $c'd'$, respectively. The acceleration image is thereby completed, locating d' at the intersection of $b'd'$ and $c'd'$. Vector $o'd'$ represents the total acceleration of the center of gravity of the wing, point D. Measuring vector $o'd'$ on the acceleration scale, we obtain

$$a_D = 5400 \text{ in}/\text{s}^2$$

to the right and slightly upward.

The construction used in the above problem (rotating the links to determine the position of the acceleration image) is equivalent to simply transferring angles BCD and CBD to the acceleration polygon (as angles $b'c'd'$ and $c'b'd'$, respectively). The acceleration image triangle $b'c'd'$ is similar to triangle BCD. Since we read BCD going around the link clockwise, $b'c'd'$ must appear in that same order, reading clockwise around the acceleration image.

The equivalent linkage that we have used to analyze the air pump is valid for any instant, so long as point C on the wing contacts the housing. It may not be used, however, for the portion of the cycle when point C leaves the housing (see the wing at the top of the mechanism). ◆

4.8 GRAPHICAL ANALYSIS OF SLIDING CONTACT LINKAGES

4.8.1 Coriolis Acceleration

In the preceding graphical studies, we considered the acceleration of a slider moving along a fixed path (e.g., a piston in a cylinder) and we also considered the acceleration of a point on a rotating link (e.g., either of the moving pin joints in a four-bar linkage). We will now look into the case of a link that slides along a rotating member. When these two conditions are met (a rotating path and a point that has a velocity relative to that path), there exists an additional acceleration component, Coriolis acceleration. Thus, the total acceleration of a

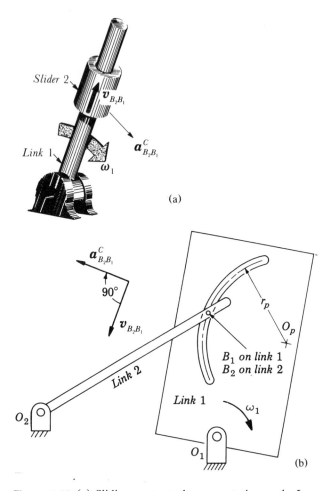

Figure 4.16 (a) Sliding contact along a rotating path. In this case, an additional component of acceleration, *Coriolis acceleration*, is required to describe the motion of the slider. (b) Coriolis acceleration direction.

point on the slider in Figure 4.16a consists of the following components:

1. Normal and tangential accelerations of a coincident point on the rotating link
2. Relative acceleration of the slider along the rotating link
3. *Coriolis acceleration*

Referring to Eq. 4.15, we may use the vector product $a^c = 2\boldsymbol{\omega} \times \dot{\boldsymbol{r}}_r$ to obtain the magnitude and direction of the Coriolis acceleration. For planar linkages, the direction is given by rotating the relative velocity vector $\dot{\boldsymbol{r}}_r$ 90° in the direction of ω of the link over or through which the slider moves.

◆ **EXAMPLE PROBLEM 4.14**   *Coriolis Acceleration*
Link 1 in Figure 4.16b has an angular velocity $\omega_1 = 20$ rad/s clockwise. The velocity of point B_2 on link 2 with respect to a coincident point on link 1 is $v_{B_2B_1} = 30$ in/s in the direction shown. Find the Coriolis acceleration.

Solution.   The Coriolis component of slider acceleration is given by

$$a^c_{B_2B_1} = 2\omega_1 \times v_{B_2B_1} = 2(20 \text{ rad/s})(30 \text{ in/s}) = 1200 \text{ in/s}^2$$

The direction of $a^c_{B_2B_1}$ is found by rotating relative velocity vector $v_{B_2B_1}$ in the direction of ω_1 (clockwise) by 90°. In this example, $a^c_{B_2B_1}$ is to the left and upward as shown. Note that the subscripts of a^c and v must correspond. We use angular velocity ω_1 and relative velocity $v_{B_2B_1}$ since link 1 guides the slider (i.e., we know the relative path of link 2 along link 1).

Note, however, that we may replace the linkage in the figure with an equivalent four-bar linkage (O_1 O_p B O_2). Acceleration analysis of the equivalent linkage would not involve determining Coriolis acceleration.   ◆

◆ **EXAMPLE PROBLEM 4.15**   *Coriolis Acceleration in a Sliding Contact Linkage*
Coriolis acceleration in a sliding contact linkage is investigated here. The slider in the mechanism of Figure 4.17a travels along link 1, which rotates. Thus Coriolis acceleration is involved in an analysis of this linkage. To find the angular acceleration of link 1 in this problem, we will find the tangential acceleration of B_1 by using the acceleration polygon method.

Solution.   Velocities were computed for the linkage in Chapter 3, but the link numbers have been changed. In this problem, the continuously rotating crank is identified as link 2. Point B_2 represents the pin (revolute pair) joining link 2 to the slider. We have from the velocity polygon (Figure 4.17b)

$$v_{B_1} = ob_1 = 1.18 \text{ m/s} \quad \text{(downward to the left)}$$

$$v_{B_2} = ob_2 = 3.6 \text{ m/s} \quad \text{(upward to the left)}$$

$$v_{B_2B_1} = b_1b_2 = 3.4 \text{ m/s} \quad \text{(upward to the left)}$$

In solving for accelerations, we will express the acceleration of a point on the slider as the vector sum of the acceleration of a coincident point on link 1 *plus* the relative acceleration of the slider on the link, which includes Coriolis acceleration.

We may apply Eq. 4.15, where point B_1 is taken as the origin of a coordinate system that rotates with link 1. Note that we have selected a

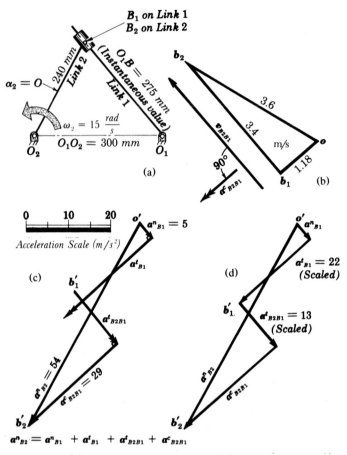

Figure 4.17 (a) A sliding contact linkage. (b) The velocity polygon for the sliding contact linkage. When the direction of the relative velocity is determined, the direction of the Coriolis acceleration is found by rotating the relative velocity vector 90° in the direction of the angular velocity of the link on which the slider rides. (c) The acceleration polygon. (d) The "cleaned up" acceleration polygon. The trial vectors are measured against the acceleration scale and noted on the polygon.

coordinate system rotating with link 1 because sliding occurs along link 1. Then

$$\ddot{R} = a^n_{B_2} + a^t_{B_2}$$

$$\ddot{R}_0 = a^n_{B_1} + a^t_{B_1}$$

$$\ddot{r}_r = a^t_{B_2 B_1}$$

$$2\omega \times \dot{r}_r = a^c_{B_2 B_1}$$

The remaining terms in Eq. 4.15 are zero since the relative path is straight

TABLE 4.8   VECTOR TABULATION FOR THE ACCELERATION ANALYSIS
OF A SLIDING CONTACT LINKAGE, FIGURE 4.17

Vector	$a_{B_2}^n$	$= a_{B_1}^n$	$+a_{B_1}^t$	$+a_{B_2B_1}^t$	$+a_{B_2B_1}^c$
Vector magnitude	$\dfrac{(ob_2)^2}{O_2B}$	$\dfrac{(ob_1)^2}{O_1B}$	?	?	$2(b_1b_2)\omega_1$
Vector direction	$\|O_2B$ toward O_2	$\|O_1B$ toward O_1	$\perp O_1B$	$\|$ link 1 (the relative slider path)	direction of b_1b_2 rotated $90°$ counterclockwise
Vectors used to construct polygon	54 m/s²	5 m/s²	?	?	29 m/s²

(along link 1). Thus we have

$$a_{B_2}^n + a_{B_2}^t = a_{B_1}^n + a_{B_1}^t + a_{B_2B_1}^t + a_{B_2B_1}^c \qquad (4.39)$$

It can be seen that when we refer to coincident points on two different links (B_1 and B_2 in the above problem) the solution differs from analysis of linkages using two different points on the same rigid link (e.g., B and C in Eq. 4.22).

For this problem, the condition $\alpha_2 = 0$ eliminates the term $a_{B_2}^t$. We will now set up the vector table (Table 4.8) and proceed to solve for the remaining terms in the acceleration equation. Since all vector directions are known and we require only two magnitudes, $a_{B_1}^t$ and $a_{B_2B_1}^t$, we are prepared to construct the acceleration polygon. Beginning at an arbitrary point o' on Figure 4.17c, we draw $a_{B_2}^n$ to a convenient scale. Vector $a_{B_2}^n$ represents the total acceleration of B_2 in this problem; the head of $a_{B_2}^n$ is labeled b_2'. Then, working with the right side of the above equation and again beginning at o', we draw $a_{B_1}^n$ and *add trial vector $a_{B_1}^t$.*

It would be convenient to continue by adding $a_{B_2B_1}^t$, the next term in the equation, to the head of $a_{B_1}^t$, but we do not know where to begin; the head of $a_{B_1}^t$ has not been located. Instead, we observe that both sides of the equation represent $o'b_2'$, the acceleration of B_2. Then, the last term $a_{B_2B_1}^c$ may be put in its logical place, the head of $a_{B_2B_1}^c$ at point b_2'. Working backward, the next to last term, trial vector $a_{B_2B_1}^t$, is placed with its head at the tail of $a_{B_2B_1}^c$. The intersection of the trial vectors locates the tail of $a_{B_2B_1}^t$ and head of $a_{B_1}^t$, and we label that point b_1'. Figure 4.17d shows the "cleaned up" acceleration polygon with scaled values of $a_{B_1}^t$ and $a_{B_2B_1}^t$. Having obtained the tangential acceleration of B_1, the solution to the problem, the angular acceleration of link 1, is given by

$$\alpha_1 = \frac{a_{B_1}^t}{O_1B} = 22/0.275 = 80 \text{ rad/s}^2$$

Transferring tangential acceleration $a_{B_1}^t$ to point B_1 on link 1, it is seen that α_1 is counterclockwise. ◆

Comparison of Results with an Analytical Solution

Example Problem 4.15 may be solved analytically, using equations derived by complex number methods in Section 4.5. Changing the equations so that link numbers and data correspond to those of Figure 4.17a-d, with $R_1 = O_1B$ and $R_2 = O_2B$, we find relative acceleration of the coincident points (using results obtained in Section 3.11.2).

$$a_{B_2B_1}^t = -\omega_2^2 R_2 \cos(\theta_2 - \theta_1) + \alpha_2 R_2 \sin(\theta_1 - \theta_2) + \omega_1^2 R_1$$

$$= -12.62 \text{ m/s}^2$$

The slider acceleration relative to link 1 is toward O_1; that is, the outward relative velocity is decreasing. Angular acceleration of link 1 is given by

$$\alpha_1 = \frac{\omega_2^2 R_2 \sin(\theta_1 - \theta_2) + \alpha_2 R_2 \cos(\theta_1 - \theta_2) - 2\omega_1 v_{B_2B_1}}{R_1}$$

$$= 79.53 \text{ rad/s}^2 \quad \text{(counterclockwise)}$$

from which tangential acceleration of B_1 on link 1, $a_{B_1}^t = \alpha_1 R_1 = 21.9 \text{ m/s}^2$, is downward and to the left. The results compare closely with those obtained in the graphical solution.

The significance of the order of the subscripts will be noted once more. In the above problem, the solution takes the form

$$a_{B_2} = a_{B_1} + a_{B_2B_1}$$

The Coriolis component is therefore $a_{B_2B_1}^c = 2\omega_1 \times v_{B_2B_1}$, where ω_1 is the angular velocity of the link along which the slider slides. The subscripts of a^c and v must correspond.

If link 1 in the above problem was curved, or if the end of link 2 was sliding in a curved slot in link 1, then a normal relative acceleration term would be added in Eq. 4.39, resulting in the following equation:

$$a_{B_2}^n + a_{B_2}^t = a_{B_1}^n + a_{B_1}^t + a_{B_2B_1}^n + a_{B_2B_1}^t + a_{B_2B_1}^c \tag{4.40}$$

In Figure 4.16b, the relative path radius is r_p. Acceleration $a_{B_2B_1}^n$ is given by $v_{B_2B_1}^2/r_p$, which is directed from B_1 toward O_p. A similar situation occurs with an oscillating roller follower and a disk cam. In cases such as these, an equivalent linkage may be used as an alternative form of solution.

◆ EXAMPLE PROBLEM 4.16 Vane Pump

The *straight-wing air pump* (Figure 4.18a) has four sliding vanes that are held against the housing by inertia forces. Let us examine a single vane (Figure 4.18b) and find the acceleration of a point B on that vane.

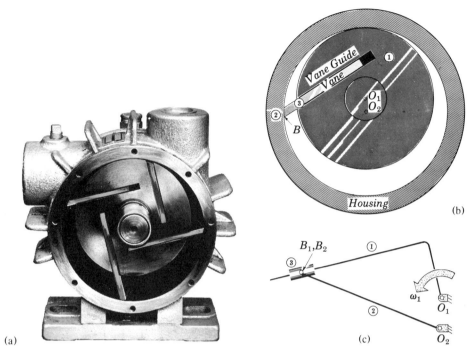

Figure 4.18 (a) This straight-wing air pump can be considered a sliding contact mechanism. This machine is used as a compressor or vacuum pump and is operated at 500 to 1500 revolutions per minute. Acceleration forces hold the four sliding vanes against the housing. (*Source:* ITT Pneumotive.) (b) A single vane of the pump is examined. (c) The equivalent linkage for the mechanism.

Solution. The housing is circular, and as long as the vane makes contact with the housing, the contact point will describe a circle. Thus we may introduce an artificial link 2 with its center at O_2, the center of the housing, and its length equal to $O_2 B$. The equivalent mechanism is shown in Figure 4.18c, where link 1 represents the vane guide with center of rotation O_1. Let the angular velocity of the vane guide be a constant value ω_1 (given).

Eq. 4.39 applies to sliding-contact problems of this type:

$$a_{B_2}^n + a_{B_2}^t = a_{B_1}^n + a_{B_1}^t + a_{B_2B_1}^t + a_{B_2B_1}^c$$

The given data for this problem differ only slightly from data given in the previous problem. In this case, angular velocity ω_1 is constant for the vane guide (link 1); angular acceleration α_1 is therefore zero and we can eliminate the term for the tangential acceleration, $a_{B_1}^t$. In this problem, the angular acceleration α_2 of equivalent link 2 is unknown and the tangential component of acceleration for B_2, $a_{B_2}^t$, will not, in general, equal zero.

Eq. 4.39 can now be rewritten as in Table 4.9. The components making up the acceleration polygon are handled most easily when added in the order indicated in the table. After we specify ω_1 and construct a velocity polygon, we

TABLE 4.9 VECTOR TABULATION FOR THE ACCELERATION
OF THE AIR PUMP EQUIVALENT LINKAGE, FIGURE 4.17c

Vector	$a^n_{B_2}$	$+a^t_{B_2}$	$=a^n_{B_1}$	$+a^c_{B_2B_1}$	$+a^t_{B_2B_1}$
Vector magnitude	$\dfrac{(ob_2)^2}{O_2B}$	?	$\dfrac{(ob_1)^2}{O_1B}$	$2(b_1b_2)\omega_1$	?
Vector direction	$\|O_2B$ toward O_2	$\perp O_2B$	$\|O_1B$ toward O_1	$\perp$ path of slider on link 1 (found by rotating b_1b_2 90° in direction of ω_1)	$\|$ path of slider on link 1

may substitute in the above equations and construct the acceleration polygon for the straight-wing air pump. ◆

4.9 CAMS AND CAM FOLLOWERS

Cam design is ordinarily a synthesis process in which we determine the required cam shape to meet a predetermined set of conditions on displacement, velocity, and acceleration. This process is treated for a number of follower displacement versus cam angle relationships in Chapter 5.

In some cases, an equivalent linkage may be used to analyze a cam and follower by graphical methods. As noted in Chapter 3, if a cam and follower both have finite radii at the point of contact, then a four-bar linkage may be used as an equivalent linkage for velocity analysis. Acceleration analysis of the equivalent linkage then follows. Angular velocity and angular acceleration of the follower are represented by angular velocity and angular acceleration of the equivalent link. Figure 3.28b shows a four-bar linkage equivalent to a cam and follower having finite radii. Figure 4.19a shows a cam and a roller follower rotating on an oscillating arm, which may be analyzed by examining an equivalent four-bar linkage, Figure 4.19b. Both equivalent linkages are crank rocker mechanisms. In Figures 4.19a and b, the driver crank consists of a link from O_1, the cam rotation axis, to C, the center of curvature of the cam at the point of contact. The driven crank (the rocker) consists of a link from the follower pivot O_2 to the center of curvature D of the follower, which is the center of the roller. The connecting rod CD has no counterpart on the actual cam and follower linkage. Since the roller follower usually includes a ball bearing, roller bearing, or needle bearing, friction will be very low.

If the cam is circular in shape and has a roller follower, the same four-bar linkage may be used to obtain angular velocity and angular acceleration of the follower arm throughout the entire cycle. If instead, the cam surface consists of a series of circular arcs, then a different four-bar linkage is needed to represent each segment of the surface. In that case, there may be jumps (instantaneous changes) in the angular acceleration versus time relationship for the follower arm. For the cam designs referred to above, the velocity of

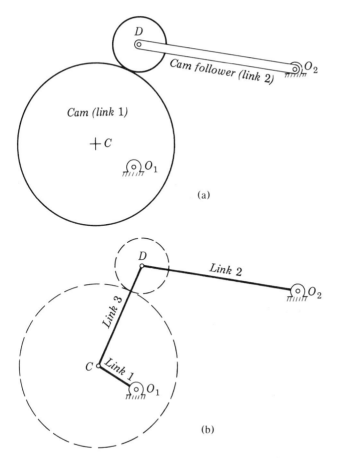

Figure 4.19 (a) Cam and roller follower on oscillating arm.
(b) Equivalent linkage.

point D is given by

$$od = oc + cd$$

from which we find the angular velocity of the follower arm

$$\omega_2 = \frac{od}{O_2D}$$

The acceleration of point D may be found from the equation

$$a_D^n + a_D^t = a_C^n + a_C^t + a_{DC}^n + a_{DC}^t$$

from which we find angular acceleration of the follower arm

$$\alpha_2 = \frac{a_D^t}{O_2D}$$

When the cam has a straight section (infinite radius of curvature) at the point of contact, then it cannot be represented by a four-bar linkage with finite

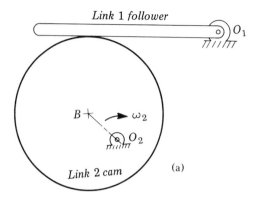

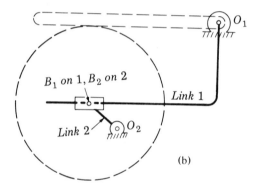

Figure 4.20 (a) Cam with flat-face oscillating follower. (b) Equivalent linkage.

links. A similar difficulty occurs with an oscillating flat-surface follower. In Figure 4.20a, for example, we observe that the cam is circular and that its center lies at point B. An equivalent sliding contact linkage is shown in Figure 4.20b. Equivalent link 1 is drawn parallel to the follower face through point B. Link 2 equivalent to the cam is represented by O_2B_2. The applicable velocity equation is

$$v_{B_1} = v_{B_2} + v_{B_1B_2}$$

where

$$v_{B_2} = \omega_2 O_2 B_2 \perp O_2 B_2$$

$$v_{B_1} \perp O_1 B_1$$

$$v_{B_1B_2} \| \text{link 1 at the slider}$$

Then, follower angular velocity $\omega_1 = v_{B_1}/O_1B_1$. The acceleration equation may be written as follows:

$$a_{B_2}^n + a_{B_2}^t = a_{B_1}^n + a_{B_1}^t + a_{B_2B_1}^t + a_{B_2B_1}^c$$

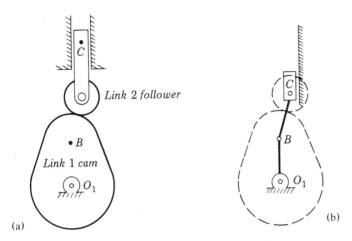

Figure 4.21 (a) Cam with reciprocating roller follower. (b) Equivalent linkage.

where

$$a_{B_2}^n = \omega_2^2 O_2 B_2 \| O_2 B_2 \text{ toward } O_2$$

$$a_{B_2}^t = \alpha_2 O_2 B_2 \perp O_2 B_2$$

$$a_{B_1}^n = \frac{v_{B_1}^2}{O_1 B_1} \| O_1 B_1 \text{ toward } O_1$$

$$a_{B_1}^t \perp O_1 B_1$$

$$a_{B_2 B_1}^t \| \text{flat-face follower}$$

$a_{B_2 B_1}^c = 2\omega_1 v_{B_2 B_1} \perp$ flat-face follower (direction given by $2\omega_1 \times v_{B_2 B_1}$). Follower angular acceleration is given by

$$\alpha_1 = \frac{a_{B_1}^t}{O_1 B_1}$$

If the cam surface may be represented by a series of circular arcs, then a different equivalent linkage applies to each segment of the cam surface.

For a cam with a reciprocating roller follower, as in Figure 4.21a, the equivalent linkage is a slider-crank mechanism, as shown in Figure 4.21b. Follower velocity and acceleration are represented by velocity and acceleration of the slider in the equivalent linkage.

4.10 ANALYZING COMBINATIONS OF BASIC LINKAGES

In most cases, mechanisms made up of more than four links may be broken down into simple basic linkages and solved in a straightforward manner for velocities and accelerations. As an example, let us find slider acceleration in

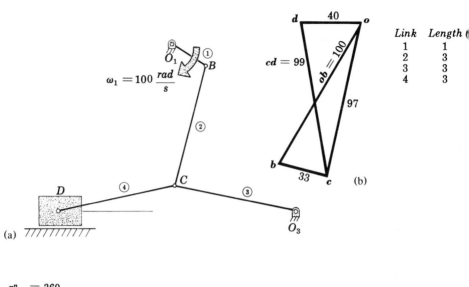

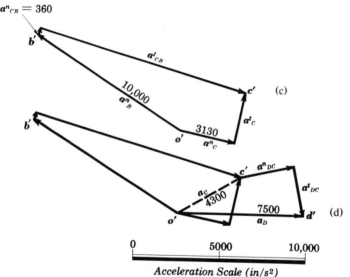

Figure 4.22 (a) The toggle mechanism is analyzed by considering it as made up of two simpler mechanisms: a four-bar linkage (1, 2, 3, and frame): and a slider-crank mechanism (3, 4, and frame). (b) The velocity polygon for the entire mechanism, easily constructed as a single polygon. (We know the length and angular velocity of link 1.) (c) The acceleration polygon for the four-bar linkage portion of the toggle mechanism. (d) The acceleration polygon for the entire toggle mechanism is completed by adding the acceleration vectors for the slider crank to the acceleration polygon for the four-bar mechanism.

TABLE 4.10 PART ONE OF THE VECTOR TABULATION
FOR THE ACCELERATION
POLYGON FOR THE TOGGLE MECHANISM
OF FIGURE 4.22

Vector	a_C^n	$+a_C^t$	$= a_B^n$	$+a_{CB}^n$	$+a_{CB}^t$
Vector magnitude	$\dfrac{(oc)^2}{O_3 C}$	?	$\dfrac{(ob)^2}{O_1 B}$	$\dfrac{(bc)^2}{BC}$	?
Vector direction	$\|O_3 C$ toward O_3	$\perp O_3 C$	$\|O_1 B$ toward O_1	$\|BC$ toward B	$\perp BC$
Vectors used to construct polygon	3130 in/s²	?	10,000 in/s²	360 in/s²	?

the *toggle mechanism* of Figure 4.22a. The velocity polygon in Figure 4.22b is constructed according to the methods given in Chapter 3. Using values from the velocity polygon, we will proceed to construct the acceleration polygon for *part* of the toggle mechanism, the four-bar linkage made up of links 1, 2, and 3 and the frame. If link 1 rotates at constant angular velocity, $a_B^t = 0$, and the acceleration equation for a four-bar mechanism reduces to the equation shown in Table 4.10. Noting the magnitudes of the accelerations as indicated in Table 4.10, a convenient scale is selected. An acceleration polygon is drawn for the four-bar linkage made up of links 1, 2, and 3 and the frame, using the above vectors. See Figure 4.22c. The two unknown tangential components, a_C^t and a_{CB}^t, may then be scaled from the completed polygon.

To complete the acceleration analysis, we must now consider the rest of the mechanism—the slider crank made up of links 3 and 4 and the frame. We can set up the general equation for the slider acceleration, a_D, as equal to the acceleration of point C plus the acceleration of D with respect to C. The formula is given in Table 4.11.

TABLE 4.11 PART TWO OF THE VECTOR TABULATION FOR THE ACCELERATION
POLYGON FOR THE TOGGLE MECHANISM OF FIGURE 4.22

Vector	$a_D =$	a_C	$+a_{DC}^n$	$+a_{DC}^t$
Vector magnitude	?	Vector $o'c'$ (already found)	$\dfrac{(cd)^2}{CD}$	?
Vector direction	Along slider	See figure	$\|CD$ toward C	$\perp CD$
Vectors used to construct polygon	?	4300 in/s²	3270 in/s²	?

We can either construct a separate acceleration polygon for the slider-crank part of the toggle mechanism to find a_D, or we can simply add to the polygon constructed in the first half of the problem. The latter method is usually more convenient. In this case, for example, the first vector of the slider-crank acceleration polygon, a_C, has already been constructed in the four-bar polygon of Figure 4.22c. Adding these vectors to the acceleration polygon in Figure 4.22c, then, we find slider acceleration $a_D = 7500 \text{ in/s}^2$ to the right (measured on the completed acceleration polygon in Figure 4.22d). Slider acceleration direction ($o'd'$ on the acceleration polygon) is opposite the slider velocity direction (od from the velocity polygon); that is, the slider is slowing down.

4.11 INVERSE AND TRIAL SOLUTION METHODS APPLIED TO LINKAGE ACCELERATION ANALYSIS

A straightforward solution may not be apparent in the analysis of some mechanisms. As with problems in velocity analysis, inverse and trial solutions can be attempted. Since we must deal with both tangential and normal acceleration components, a detailed analysis is often very time-consuming, suggesting the use of computer-aided procedures. The problem that follows illustrates a graphical procedure that will apply to certain types of problems.

Figure 4.23a shows a mechanism with six links and seven joints (if the sliders are treated as infinite links). Using Grübler's criterion (see Chapter 1), we find there is one degree of freedom. Thus, if the motion of one link is specified, we should be able to describe the motion of the entire mechanism.

Let the velocity of point B be given where link 1 rotates clockwise at the instant shown. Although a direct method would be used if v_D or v_E were given, since v_B is given in this example, we will use the inverse method to complete the velocity polygon (see Chapter 3). Using the relationship $oe = od + de$, vector od is given arbitrary length (Figure 4.23b). Then, velocity point e is located by noting that oe is horizontal and de is perpendicular to DE. Velocity point c is found from the image principle, $dc/de = DC/DE$. The vector equation $ob = oc + cb$, where $cb \perp CB$, allows us to complete the velocity polygon. The scale of the velocity polygon is obtained by comparing vector ob on the polygon with the given velocity v_B. Thus, the velocity scale is v_B (given)/ob (measured), and $v_E = oe$ (measured) $\times$ (velocity scale), $v_{CB} = bc$ (measured) $\times$ (velocity scale), and so on.

4.11.1 Inverse Method

Suppose the given data for Figure 4.23a are $v_B = 4$ m/s downward and to the right and that the acceleration magnitudes of D and E are equal with a_E to the left and a_D upward.

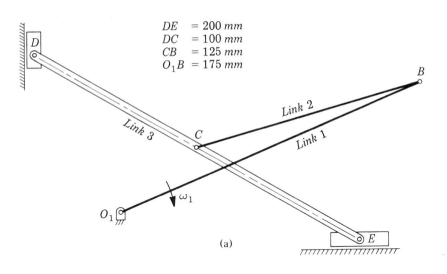

$$
\begin{aligned}
DE &= 200 \ mm \\
DC &= 100 \ mm \\
CB &= 125 \ mm \\
O_1B &= 175 \ mm
\end{aligned}
$$

(a)

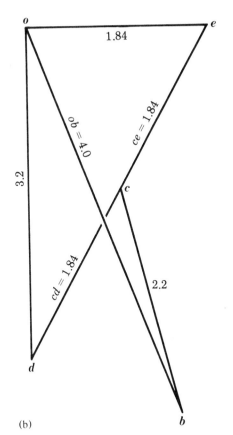

(b)

Figure 4.23 (a) A six-bar linkage examined by inverse and trial solution methods. (b) Velocity polygon; velocities in meters per second.

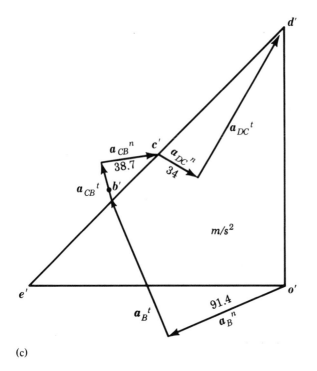

(c)

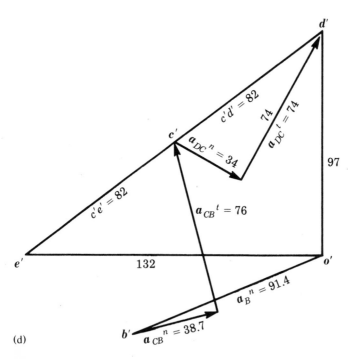

(d)

Figure 4.23 (c) Acceleration polygon for $|a_D| = |a_E|$; accelerations in meters per second squared. (d) Acceleration polygon for constant ω_1 accelerations in meters per second squared.

The velocity polygon in Figure 4.23b may be used for this problem, where the velocity scale is 4 $(m/s)/ob$ (measured). Acceleration equations

$$o'c' = o'b' + b'c'$$

$$o'e' = o'd' + d'e'$$

$$\frac{d'c'}{d'e'} = \frac{DC}{DE}$$

and so on, do not provide a straightforward graphical solution, but an inverse method may be used by giving $o'd'$ an arbitrary length, as in Figure 4.23c. Then $o'e'$ has the same length, and c' is located by the acceleration image principle:

$$\frac{c'd'}{e'd'} = \frac{CD}{ED}$$

Noting that

$$c'd' = a_{DC}^n + a_{DC}^t$$

we break this vector into components. Determining relative velocity cd from the velocity polygon and using $a_{DC}^n = cd^2/CD = 34$ m/s^2, we find the acceleration scale to be 34 $(m/s^2)/a_{DC}^n$ (measured). Now, the normal acceleration of B,

$$a_B^n = \frac{ob^2}{O_1 B} = 91.4 \text{ m/s} \| O_1 B$$

can be drawn to that scale. Writing the acceleration equation $o'c' = o'b' + b'c'$ in the following order for convenience,

$$a_C = a_B^n + a_B^t + a_{CB}^t + a_{CB}^n$$

we continue by adding trial vector $a_B^t \perp O_1 B$. Then

$$a_{CB}^n = \frac{bc^2}{BC} = 38.7 \text{ m/s} \| CB$$

toward B is drawn to scale with its head at point c'. Trial vector $a_{CB}^t \perp CB$ is added to the tail of a_{CB}^n. The intersection of trial vectors a_B^t and a_{CB}^t locates acceleration point b'. The acceleration of any point on the linkage may now be determined by using the acceleration scale found above, and the angular accelerations may be calculated.

4.11.2 Trial Solution Method

The given data in a problem may be such that neither straightforward nor inverse graphical solutions are practical. This would be the case if the velocity of point B and the angular acceleration of link 1 were specified in Figure 4.23a. We may proceed by a set of approximations of one of the vector magnitudes, attempting to satisfy the vector equations.

TABLE 4.12 ACCELERATION BY A TRIAL SOLUTION METHOD

	$o'd'$	$=$	$o'b'$	$+$	$b'c'$	$+$	$c'd'$
Vector	a_D	$= a_B^n$	$+a_B^t$	$+a_{CB}^n$	$+a_{CB}^t$	$+a_{DC}^n$	$+a_{DC}^t$
Vector magnitude	?	$\dfrac{ob^2}{O_1B}$	$\alpha_1 O_1 B_1$	$\dfrac{bc^2}{BC}$	$\alpha_2 BC$	$\dfrac{cd^2}{CD}$	$\alpha_3 CD$
Vector direction	Along slider path	$\|O_1B$ toward O_1	$\perp O_1B$	$\|BC$ toward B	$\perp BC$	$\|CD$ toward C	$\perp CD$
Vectors used to construct polygon (m/s^2)	?	91.4	0	38.7	?	34	?

For example, let the velocity of point B in Figure 4.23a be 4 m/s at the instant shown, with ω_1 constant and clockwise. The acceleration of point D may be expressed as in Table 4.12. In addition, we see that $a_E(o'e')$ is horizontal and that $d'c'e'$ is the image of DCE, that is, d', c', and e' lie on a line and $d'c'/d'e' = DC/DE$.

Using Table 4.12, we draw $a_B^n = o'b'$ to a convenient scale, add a_{CB}^n, and then add a first approximation of a_{CB}^t. Vector a_{DC}^n is then added along with trial vector a_{CD}^t, which intersects the vertical trial vector $o'd'$ at d'. Line $d'c'$ is extended to intersect horizontal trial vector $o'e'$ at e'. We then check the acceleration image proportion $d'c'/d'e' = DC/DE$. If it is not satisfied, we make a second approximation of a_{CB}^t, attempting to obtain the correct proportion. Several approximations may be required for each mechanism position we wish to examine.

Figure 4.23d shows an acceleration polygon that satisfies the equations with reasonable accuracy. For the data given in this problem, we obtain $a_E = 132$ m/s² to the left and $a_D = 97$ m/s² upward. The solution differs, of course, from that of Figure 4.23c, where $|a_D| = |a_E|$, resulting in $\alpha_1 \neq 0$.

4.12 LIMITING POSITIONS

Limiting positions of the slider-crank mechanism and the crank rocker mechanism were discussed in previous sections. As the driver crank rotates continuously, the follower link (slider or driven crank) stops and changes direction at the limiting position. At the limiting position, the follower velocity is zero, but the follower acceleration, in general, is *not* equal to zero. When analyzing linkages at their limiting positions, it is important to obtain the relative velocities for use in the acceleration equation.

Consider, for example, the in-line slider-crank mechanism of Figure 4.24 with the crank rotating at a constant angular velocity ω. In the limiting

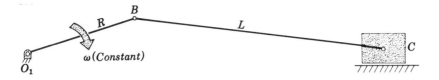

Figure 4.24 The in-line slider-crank mechanism. The crank R rotates at constant angular velocity ω.

position in Figure 4.25a, crankpin velocity is given by $ob = \omega R$, perpendicular to $O_1 B$ (downward). In forming the velocity polygon for the linkage at its limiting position in Figure 4.25b, we note that relative velocity bc is colinear with velocity ob since the crank $(O_1 B)$ and connecting rod (BC) form a straight line. Point C travels only in a horizontal path; therefore, velocity point c must lie at pole point o to correctly position bc perpendicular to BC. The velocity polygon tells us what we already knew: the velocity of the slider is zero. In addition, it gives us the relative velocity:

$$bc = \omega R \quad \text{(upward)}$$

If the crank has a constant angular velocity, the acceleration equation for the slider may be written as shown in the acceleration table, Table 4.13. But the last vector, a'_{CB}, is clearly inconsistent with the acceleration equation since there are no vertical acceleration components. Therefore, $a'_{CB} = 0$, and our

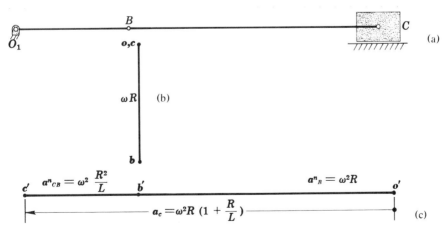

Figure 4.25 (a) The mechanism is shown in its extended limiting position, where O_1, B, and C are colinear. (b) Velocity polygon at the instant the mechanism is in its limiting position. (c) Accelerating polygon for the mechanism in its limiting position.

vector equation becomes the simple scalar equation

$$a_C = a_B^n + a_{CB}^n = \omega^2 R\left(1 + \frac{R}{L}\right) \tag{4.41}$$

for the limiting position with the mechanism extended. See Figure 4.25c. When the slider is at or near to the extreme right position, slider acceleration is to the left, as shown in the acceleration polygon. For crank angular acceleration $\alpha \neq 0$, we must include the tangential acceleration $a_B^t = \alpha R$, which is balanced by the equal and opposite vector a_{CB}^t. Therefore, Eq. 4.41 holds for the limiting position in Figure 4.25a for constant or variable crank angular velocity.

The other limiting position, with B and C at opposite sides of O_1, results in a slider acceleration

$$a_C = \omega^2 R\left(1 - \frac{R}{L}\right) \tag{4.42}$$

When the slider is at or near its extreme left position, slider acceleration is to the right.

The results of the acceleration polygon method are exact. If we substitute $\theta = 0$ and $\theta = 180°$ in the approximate acceleration equation (4.37), the results are identical to Eqs. 4.41 and 4.42 above. Thus the approximate analytical expressions are exact at $\theta = 0$ and $180°$. The reader is reminded that the above equations were derived for the in-line slider-crank mechanism in its limiting positions. The offset slider-crank mechanism and other linkages will require separate analyses.

4.13 SPATIAL LINKAGES

Many of the principles used to analyze planar linkages apply to spatial linkages as well. Although both graphical and analytical methods of spatial linkage displacement analysis were illustrated in Chapter 2, analytical methods of

TABLE 4.13   VECTOR TABULATION FOR THE ACCELERATION ANALYSIS
OF AN IN-LINE SLIDER CRANK IN A LIMITING POSITION, FIGURE 4.25a

Vector	$a_C =$	a_B^n	$+a_{CB}^n$	$+a_{CB}^t$
Vector magnitude	?	$\dfrac{(ob)^2}{R}$ or $\omega^2 R$	$\dfrac{(bc)^2}{BC}$	?
Vector direction	Along slider path	$\parallel O_1 B$ toward O_1	$\parallel BC$ toward B	$\perp BC$
Vectors used to construct polygon	$\xleftarrow{?}$	$\xleftarrow{\omega^2 R}$	$\xleftarrow{\omega^2 \frac{R^2}{L}}$	$?$ $\downarrow$

velocity and acceleration analysis will generally be found most practical. For analytical vector methods, velocity and acceleration terms in the form $\boldsymbol{\alpha} \times \boldsymbol{r}$, $\boldsymbol{\omega} \times (\boldsymbol{\omega} \times \boldsymbol{r})$, and so on, must, of course, be applied in their general, three-dimensional form:

$$\boldsymbol{\alpha} \times \boldsymbol{r} = \begin{vmatrix} \boldsymbol{i} & \boldsymbol{j} & \boldsymbol{k} \\ \alpha_x & \alpha_y & \alpha_z \\ r_x & r_y & r_z \end{vmatrix}$$

and so on. Chace[1] illustrates a method of displacement, velocity, and acceleration analysis that utilizes vector notation throughout. A spatial four-bar RCCC linkage is analyzed as an example.

4.13.1 Analysis of an RSSC Linkage

A spatial RSSC linkage was described previously (see Section 2.6.5 and Figure 4.26). For that linkage, it was found convenient to write a quadratic equation in terms of link lengths and positions:

$$r_3^2 + 2 \cos \gamma (r_0 - r_1 \cos \theta) r_3 + r_0^2 + r_1^2 - r_2^2 - 2r_0 r_1 \cos \theta = 0$$

In the above displacement equation, we note that link lengths r_0, r_1, and r_2 and path angle γ are constants. Differentiating with respect to time, we have

$$2r_3 \frac{dr_3}{dt} + 2 \cos \gamma \, r_1 r_3 \sin \theta \frac{d\theta}{dt}$$

$$+ 2 \cos \gamma (r_0 - r_1 \cos \theta) \frac{dr_3}{dt} + 2r_0 r_1 \sin \theta \frac{d\theta}{dt} = 0 \qquad (4.43)$$

from which we obtain the velocity of the sliding link:

$$\frac{dr_3}{dt} = \frac{-\omega_1 r_1 \sin \theta (r_0 + r_3 \cos \gamma)}{(r_0 - r_1 \cos \theta) \cos \gamma + r_3} \qquad (4.44)$$

where $\omega_1 = d\theta/dt$, the angular velocity of the crank. Differentiating Equation 4.43 or 4.44 with respect to time and simplifying the result, we obtain the acceleration of the sliding link:

$$\frac{d^2 r_3}{dt^2} = - \left[\frac{(dr_3/dt)[(dr_3/dt) + 2\omega_1 r_1 \cos \gamma \sin \theta]}{r_3 + (r_0 - r_1 \cos \theta) \cos \gamma} \right] \qquad (4.45)$$

where $\alpha_1 = d^2\theta/dt^2$, the angular acceleration of the crank.

For the special case where path angle $\gamma = 0$, we have, of course, an in-line planar slider-crank linkage. The displacement equation reduces to

$$r_2^2 = (r_0 - r_1 \cos \theta + r_3)^2 + (r_1 \sin \theta)^2$$

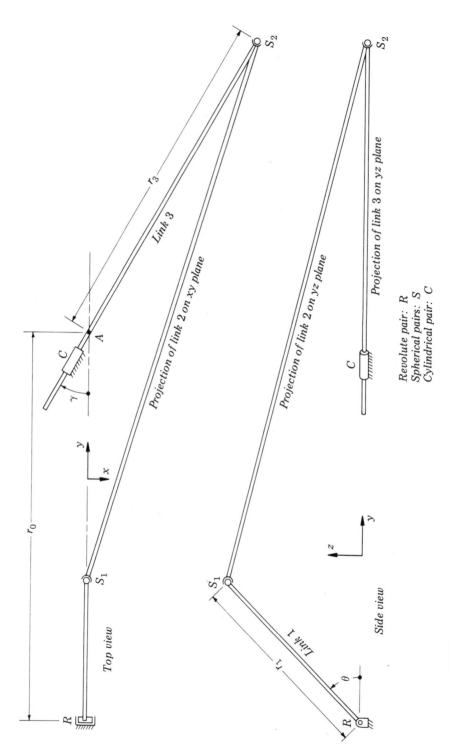

Top view

Link 3

r_3

C

A

γ

y

x

r_0

S_1

R

Projection of link 2 on xy plane

S_2

Side view

y

z

S_1

Link 1

r_1

θ

R

Projection of link 2 on yz plane

S_2

C

Projection of link 3 on yz plane

Revolute pair: R
Spherical pairs: S
Cylindrical pair: C

Figure 4.26 RSSC spatial linkage.

where point A is undefined but $r_0 + r_3$ is the distance between the revolute R and ball joint S_2. For $\gamma = 0$, the two ball joints could be replaced by revolute joints since the motion is restricted to a plane. For this special case, the velocity equation reduces to

$$\frac{dr_3}{dt} = -r_1 \sin \theta \frac{d\theta}{dt} \left[1 + \frac{r_1 \cos \theta}{r_0 + r_3 - r_1 \cos \theta} \right]$$

It can be shown that the above equation is identical (except for the sign convention) to the planar slider-crank velocity equation derived earlier for crank length R and connecting rod length L:

$$v = R\omega \sin \theta \left[1 + \left(\frac{R}{L} \right) \frac{\cos \theta}{\sqrt{1 - (R/L)^2 \sin^2 \theta}} \right]$$

◆ **EXAMPLE PROBLEM 4.17**   *Spatial Linkage*
Velocity and acceleration analysis of a spatial RSSC linkage are considered. Let the RSSC linkage of Figure 4.26 have the following proportions:

$$\frac{r_0}{r_1} = 2 \qquad \frac{r_2}{r_1} = 3$$

where crank angular velocity ω_1 is constant and the sliding link (3) and revolute joint R lie in the xy plane. Plot values of sliding link velocity and acceleration for various values of crank angle θ where path angle $\gamma = 30°$.

Solution.   The results will be plotted in dimensionless form. Substituting

$$r_0 = \frac{r_0}{r_1} = 2 \qquad r_1 = 1 \qquad r_2 = \frac{r_2}{r_1} = 3$$

and given values of θ and γ, we may solve the quadratic position equation. The root of interest represents r_3/r_1, the ratio of slider displacement to crank length.

With this value and the above substitutions along with $\omega_1 = 1$ in Eq. 4.44, the result dr_3/dt represents $(dr_3/dt)/(\omega_1 r_1)$, the dimensionless ratio of slider velocity to velocity of joint S_1.

We use this value and the above substitutions in Eq. 4.45, noting that crank angular acceleration $\alpha_1 = 0$. The result, $d^2 r_3/dt^2$, represents $(d^2 r_3/dt^2)/(\omega_1^2 r_1)$, the dimensionless ratio of slider acceleration to normal acceleration of joint S_1.

The results are plotted in Figure 4.27 for path angle $\gamma = 30°$. The plotted curves appear "jerky" since the plotter used for this example is limited to 20 lines. The actual motion of the sliding link is smooth for $\gamma = 30°$.

A partial check of results may be made by comparing average velocity over a finite time interval with instantaneous velocity. A given change in θ repre-

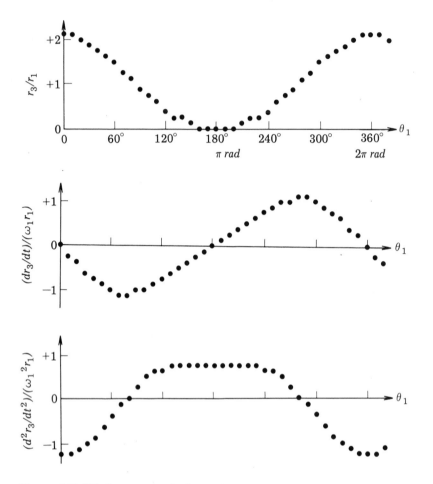

Figure 4.27 Displacement, velocity, and acceleration of an RSSC linkage.

sents a time interval of

$$\Delta t = \Delta \theta (\text{radians}) / \omega_1 \qquad (4.46)$$

Using the interval from $\theta = 30$ to $60°$ for $r_1 = 1$, $\omega = 1$, and $\gamma = 30°$, we have

$$V_{\text{av}} = \frac{R_3(60°) - R_3(30°)}{\Delta t} = \frac{1.4736 - 1.9211}{\pi / 6} = 0.855$$

This differs from the instantaneous velocity at $\theta = 45°$ (the center of the interval) by about 1.6 percent. Average acceleration over the same interval differs from instantaneous acceleration at the center of the interval by about 1.4 percent. If smaller intervals were used, we would expect average values to approach instantaneous values.

For the linkage of this example, when path angle $\gamma = 90°$ and crank angle $\theta = 180°$, velocity $V_3 = 0/0$ (indeterminate) and A_3 is infinite. The sliding link velocity cannot be determined by further examination of the limit (e.g., L'Hôpital's rule) because the defect exists in the mechanism configuration. Since an infinite slider acceleration would require an infinite force, this linkage cannot be operated near crank angle $\theta = 180°$ when path angle $\gamma = 90°$. ♦

4.13.2 Vector Methods

As an alternative to the analytical scalar method used above, an analytical vector solution could have been used to solve the RSSC linkage. The procedure is as follows, using displacements already determined (see Chapters 2 and 3).

Accelerations may be determined by using the vector equation

$$a^n_{s_2} + a^t_{s_2} = a^n_{s_1} + a^t_{s_1} + a^n_{s_2 s_1} + a^t_{s_2 s_1} \tag{4.47}$$

where

$$a^n_{s_1} = \omega_1 \times (\omega_1 \times r_1)$$

$$a^t_{s_1} = \alpha_1 \times r_1$$

$$a^n_{s_2 s_1} = \omega_2 \times (\omega_2 \times r_2)$$

$$a^t_{s_2 s_1} = \alpha_2 \times r_2$$

In the above example of an RSSC spatial linkage, $a^n_{s_2} = 0$ since S_2 travels in a straight path and the direction of $a^t_{s_2}$ is along that path. Taking angular acceleration of link 2 along its own axis to be zero, we have the dot product $\alpha_2 \cdot r_2 = 0$. If ω_1 and α_1 are given, we may solve for the unknown vectors α_2 and $a^t_{s_2}$.

4.13.3 Mechanical Systems Software Packages

Some design studies involve acceleration analysis, mechanism dynamics, and closed-loop feedback control. Software packages may be used to aid in reducing the massive programming effort that would be required to model the mechanisms involved. Figure 4.28a illustrates a design study involving man-machine interaction, combining vehicular analysis and anthropomorphic characteristics. This study utilized ADAMS/ANDROID™ and ADAMS/TIRE™ to model a driver and vehicle. Using closed-loop feedback control, the upper musculature of the driver was programmed to move based on a sensed road obstacle. Figure 4.28b shows computed accelerations at the driver mass center during a maneuver to avoid a tree. Details concerning the application of software packages are given by Mechanical Dynamics Inc.[4-6] and JML Research, Inc.[2]

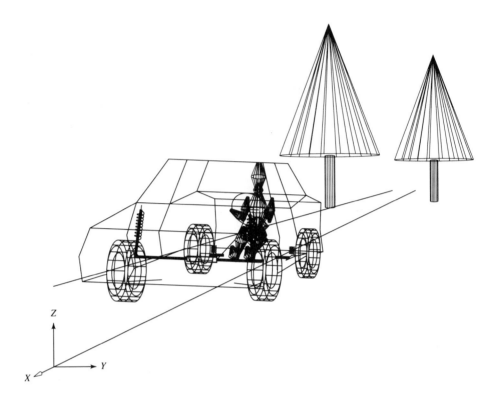

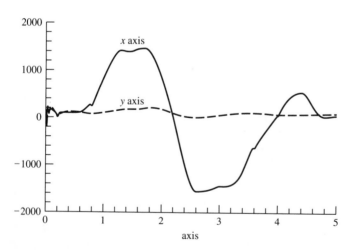

Figure 4.28 (a) Man-machine interaction. (*Source:* Mechanical Dynamics, Inc., modeled by ADAMS™ software.) (b) Acceleration at the driver mass center. (*Source:* Mechanical Dynamics, Inc., modeled by ADAMS™ software).

References

1. Chace, M. A., "Vector Analysis of Linkages," *Transactions of the American Society of Mechanical Engineers, Journal of Engineering for Industry*, vol. 55, ser. B, August 1963, pp. 289–297.
2. JML Research, Inc., *Integrated Mechanisms Program*, JML Research, Inc., Madison, WI, 1988.
3. Lawry, M. H., *I-DEAS^{TM} Student Guide*, Structural Dynamics Research Corp., Milford, OH, 1990.
4. Mechanical Dynamics Inc., *ADAMS Applications Manual*, Mechanical Dynamics Inc., Ann Arbor, MI, 1987.
5. Mechanical Dynamics Inc., *ADAMS User's Manual*, Mechanical Dynamics Inc., Ann Arbor, MI, 1987.
6. Mechanical Dynamics Inc., *ADAMS Introductory Tutorial* (draft release), Mechanical Dynamics Inc., Ann Arbor, MI, 1989.

PROBLEMS

4.1 Point P lies in a rigid body that rotates at angular velocity

$$\omega = i10 - j20 + k30$$

and angular acceleration

$$\alpha = -i20 - j40 + k80$$

The body rotates about fixed point O_1, and the radius vector O_1P is given by

$$r = i50 + j100 - k30$$

Find the acceleration of P. Unit vectors i, j, and k lie in a fixed coordinate system.

4.2 Repeat Problem 4.1 for

$$r = -i40 - j80 + k100$$

4.3 Find the average acceleration as a piston increases speed from 120 in/s to 140 in/s during a 0.01-s interval.

4.4 A body moving at 1000 mi/h is brought to a stop in 0.05 s. Find average acceleration in inches per second square.

4.5 Find average angular acceleration (in radians per second square) as a flywheel goes (a) from zero to 1000 rev/min in 20 s, and (b) from 1000 to 990 rev/min in 0.5 s.

4.6 A vehicle traveling at 50 km/h makes a turn of 12-meter radius. Find the normal acceleration of the vehicle.

4.7 Repeat Problem 4.6 for a 10-meter radius.

4.8 In Figure 4.1, let $OB = 125$ mm, $\omega = 24$ rad/s cw, and $\alpha = 30$ rad/s^2 ccw. Find the velocity and acceleration of point B.

4.9 Repeat Problem 4.8 for $\alpha = 100$ rad/s^2 cw.

4.10 Referring to Figure 4.2, let the I, J, and K axes be initially coincident with the i, j, and k axes (respectively). Let $R_0 = 40i$, $\dot{R}_0 = 10i$, $\ddot{R}_0 = -20i$, $r = 20i + 60j$, $\dot{r}_r = 5i + 15j$, $\ddot{r}_r = 10i + 30j$, $\omega = 150k$, and $\dot{\omega} = 80k$, where millimeter, radian, and second units are used. Find the location, velocity, and acceleration of point P.

4.11 Repeat Problem 4.10 with the same data except $R_0 = 0$, $\dot{R}_0 = 10k$, $\ddot{R}_0 = 20k$.

4.12 Refer to Figure 4.4 and Example Problem 4.3. Find the acceleration of point D for $\theta_{CBD} = 20°$ and $r_{BD} = 15$ mm. Solve by analytical vector methods for $\theta_1 = 120°$.

4.13 Refer to Figure 4.4 and Example Problem 4.3. Find the acceleration of C by using relative acceleration, that is, $a_C = a_B + a_{CB}$ for $\theta_1 = 120°$. Compare with the value obtained by using α_3 and ω_3. Solve by analytical vector methods.

• **4.14** Refer to Figure 4.4 and Example Problem 4.3. Using the linkage proportions given, determine the ratios ω_3/ω_1 and α_3/ω_1^2 for a range of values of θ_1 if ω_1 is constant. Use a computer or programmable calculator if available.

4.15 Repeat Problem 4.14 except that $r_0 = 10$ mm, $r_1 = 35$ mm, $r_2 = 20$ mm, and $r_3 = 30$ mm.

4.16 Plot slider acceleration versus crank angle for an in-line slider-crank linkage with connecting rod-to-crank length ratio $L/R = 3$. Crank speed is constant. Use analytical vector methods and a computer or programmable calculator.

4.17 Repeat Problem 4.16 for $L/R = 2$ and offset $E/R = 0.5$.

4.18 In Figure P3.2, $\theta = 45°$ and $\omega_1 = 10$ rad/s (constant). Draw the acceleration polygon, using the scale 1 in = 100 in/s². Find α_2.

4.19 Repeat Problem 4.18 for $\theta = 120°$.

4.20 Repeat Problem 4.18 for the mechanism in the limiting position (point C to the extreme right).

4.21 In Figure P3.3, $\theta = 30°$ and $\omega_1 = 100$ rad/s (constant). Draw the acceleration polygon, using the scale 1 in = 10,000 in/s². Find a_D and α_2.

4.22 Repeat Problem 4.21 for $\alpha_1 = 2000$ rad/s² counterclockwise.

4.23 In Figure P3.4, $\theta = 60°$, $\omega_1 = 50$ rad/s clockwise, and $\alpha_1 = 500$ rad/s² counterclockwise. Draw the acceleration polygon, using the scale 1 in = 2000 in/s². Find a_D, α_2, and α_3.

4.24 Repeat Problem 4.23 for $\alpha_1 = 0$.

4.25 In Figure P3.5, $\omega_1 = 20$ rad/s counterclockwise and $\alpha_1 = 100$ rad/s² counterclockwise. Draw the acceleration polygon, using the scale 1 in = 200 in/s². Find α_2 and α_3.

4.26 Repeat Problem 4.25 for the mechanism in the other limiting position, using the scale 1 in = 100 in/s².

4.27 In Figure P3.6, $\theta_2 = 105°$ and $\omega_2 = 20$ rad/s (constant). Draw the acceleration polygon, using the scale 1 in = 200 in/s². Find α_1.

4.28 Repeat Problem 4.27 for $\theta_2 = 30°$, using the scale 1 in = 500 in/s².

• **4.29** In Figure P3.3, $\omega_1 = 100$ rad/s (constant). Solve for a_C analytically **(a)** for $\theta = 30°$ and **(b)** for both limiting positions. (Note that the approximate analytical expression is not intended for a linkage with the dimensions of Figure P3.3). Compare the results with a graphical analysis.

4.30 In Figure P3.7, the crank is given an angular velocity ω and an angular acceleration α (both counterclockwise).
 (a) Draw the acceleration polygon for the position shown. Write an expression for a_C in terms of ω, R, and L.
 (b) Draw the acceleration polygon for the other limiting position.
 (c) Examine the effect of α on the value of a_C when this mechanism is in a limiting position.

4.31 In Figure P3.8, $\omega_1 = 100$ rad/s (constant). Draw the acceleration polygon, using the scale 1 in = 10,000 in/s^2.

4.32 Repeat Problem 4.31 for $\alpha_1 = 2000$ rad/s^2.

4.33 In Figure P3.9, $\theta = 135°$ and $\omega_1 = 10$ rad/s (constant). Note that distance B_1C on the cam is constant. Thus, we may take C as a double point—C_1 on the cam and C_2 on the follower. Using point C, draw the acceleration polygon, using the scale 1 in = 50 in/s^2.

4.34 Repeat Problem 4.33 for $\theta = 30°$.

4.35 In Figure P3.10, $\omega_1 = 35$ rad/s (constant). Draw the acceleration polygon, using the scale 1 in = 200 in/s^2. Find the acceleration of the midpoint of each link (D, E, and F).

4.36 In Figure P3.11, $\omega_1 = 20$ rad/s counterclockwise (constant). Note that CD is a fixed distance. Thus the cam and follower can be replaced by an equivalent four-bar linkage, O_1CDO_2. Draw the velocity and acceleration polygons for the equivalent linkage. Find α_2, the angular acceleration of the follower (represented by O_2D).

4.37 In Figure P3.12, $\theta = 105°$ and $\omega_2 = 20$ rad/s (constant). Draw the acceleration polygon, using the scale 1 in = 200 in/s^2. Find α_3.

4.38 Repeat Problem 4.37 for $\theta = 30°$. Use the scale 1 in = 500 in/s^2.

4.39 In Figure P3.13, $\omega_1 = 30$ rad/s (constant). Draw the velocity and acceleration polygons, using the scales 1 in = 20 in/s and 1 in = 1000 in/s^2. Find α_2.

• **4.40** In Figure P3.14, $\omega_1 = 100$ rad/s and $\alpha_1 = 2000$ rad/s^2 (both clockwise). Draw the velocity and acceleration polygons, using the scales 1 in = 100 in/s and 1 in = 10,000 in/s^2. Identify a_D and find α_2.

4.41 An in-line slider-crank linkage has a crank length of 4 in and connecting rod length of 10 in. Maximum slider acceleration occurs at the limiting position with the slider farthest from the crankshaft. At what crank velocity will slider acceleration reach 50,000 in/s^2?

4.42 In Figure P3.15, $\omega_5 = 15$ rad/s and $\alpha_1 = 0$. Draw the acceleration polygon, using the scale 1 in = 50 in/s^2. Explain the difficulty encountered in solving this problem were α_5 specified instead of α_1.

• **4.43** In Figure P4.1, draw the acceleration polygon for the limiting position shown. Let $\omega_2 = 100$ rad/s (constant). Use the scales 1 in = 100 in/s and 1 in = 10,000 in/s^2. Find α_1 and α_3.

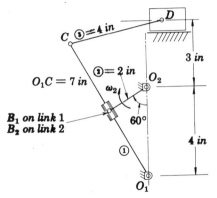

Figure P4.1

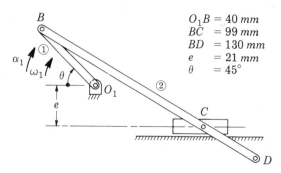

$$O_1B = 40 \text{ mm}$$
$$BC = 99 \text{ mm}$$
$$BD = 130 \text{ mm}$$
$$e = 21 \text{ mm}$$
$$\theta = 45°$$

Figure P4.2

4.44 In Figure P4.2, $\omega_1 = 200$ rad/s cw and $\alpha_1 = 4000$ rad/s^2 cw. Find the acceleration of points B, C, and D and angular acceleration of link 2.

4.45 Write an expression for slider acceleration in a slider-crank mechanism if crank angular velocity is not constant.

4.46 (a) Find the crank position corresponding to maximum piston acceleration for an in-line slider-crank mechanism. Crank speed is constant and the ratio of connecting rod to crank length $L/R = 2$.
(b) Find maximum piston acceleration in terms of R and ω.

4.47 Repeat Problem 4.46 for $L/R = 1.5$.

4.48 Refer to Figure P3.3, $O_1B = 6$ cm, $BC = 15$ cm, $\theta = 210°$, $\omega_1 = 150$ rad/s (constant). Draw and dimension the velocity and acceleration polygons.

4.49 Determine position, velocities, and accelerations for the offset sliding contact linkage of Figure P4.3. Assume r_0, r_1, e, θ_1, ω_1, and α_1 are given. Use complex number methods.

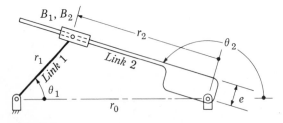

Figure P4.3 Sliding contact linkage.

4.50 Refer to Figure P4.4, the Oldham coupling. Find r_1, r_2, $\dot{r}_1$, $\dot{r}_2$, $\ddot{r}_1$ and $\ddot{r}_2$ in terms of r_0, θ_1, ω_1, α_1, and fixed angle ϕ. Use complex number methods.

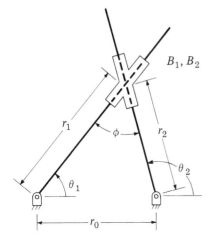

Figure P4.4 Oldham coupling.

4.51 Repeat Problem 4.50 by analytical vector methods.

4.52 In Figure P3.18, $\theta_2 = 135°$, $\omega_2 = 500$ rad/s ccw (constant), $O_1O_2 = 300$ mm, $O_1C = 400$ mm, $O_2B = 150$ mm, and $CD = 160$ mm.
  (a) Draw acceleration polygon $o'b_1'b_2'$.
  (b) Locate c'.
  (c) Complete polygon $o'b_1'b_2'c'd'$.
  (d) Find α_1 and a_D.

4.53 In Figure P3.2, let $\theta = 45°$, $O_1B = 80$ mm, $BC = 300$ mm, $E = 40$ mm, $\omega_1 = 100$ rad/s cw, and $\alpha_1 = 0$. Draw the velocity and acceleration polygons. Show all accelerations. Find α_2.

4.54 Repeat Problem 4.53 for $\omega_1 = 100$ rad/s ccw.

4.55 Refer to Figure P3.16. Let $\theta_1 = 45°$, $O_1O_2 = 100$ mm, $O_1B_1 = 40$ mm, $O_2C = 150$ mm, $\omega_1 = 10$ rad/s cc, and $\alpha_1 = 0$. Draw the velocity and acceleration polygons. Show all velocities and accelerations.

4.56 Repeat Problem 4.55. Use complex number methods.

4.57 Refer to Figure P4.5. $O_1B = 60$ mm, $BC = 150$ mm, $\theta_1 = 120°$, and $\omega_1 = 150$ rad/s (constant). Draw the velocity and acceleration polygons. Show all values.

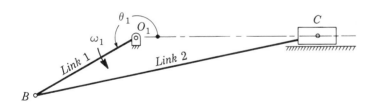

Figure P4.5

4.58 Repeat Problem 4.57 for $\theta = 180°$.

4.59 Repeat Problem 4.57 by analytical vector methods.

4.60 Refer to Figure P4.6. $O_1B_1 = 28$ mm, $O_2B_2 = 58$ mm, $O_2C = 83$ mm, and $O_1O_2 = 80$ mm. Draw the velocity and acceleration polygons. Show all values. Find α_2.

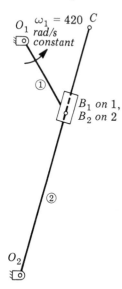

Figure P4.6

4.61 Refer to Figure 4.7. Let $R_1 = 30$ mm, $O_1O_2 = 50$ mm, $\omega_1 = 10$ rad/s ccw (constant), and $\theta_1 = 15°$. Find α_2 and $a'_{B_1B_2}$. Use complex number methods.

4.62 Repeat Problem 4.61 for $\theta_1 = 30°$.

4.63 Refer to Figure 4.7. Compute α_2/ω_1^2 for $\theta_1 = 0$ to $180°$ in $15°$ steps if $R_0/R_1 = 3$. Use a computer or programmable calculator.

4.64 Develop the equations for acceleration analysis of a four-bar linkage, using complex number methods.

4.65 Consider a four-bar linkage where drive crank length $r_1 = 10$ mm, coupler length $r_2 = 35$ mm, driven crank length $r_3 = 40$ mm, and fixed link $r_0 = 30$ mm. Find accelerations when $\theta = 45°$ and $\omega_1 = 30$ rad/s ccw (constant) and the linkage is in the open position.

4.66 Repeat Problem 4.65 for $r_1 = 15$ mm.

4.67 Refer to Figure 4.26, a spatial linkage. Let $r_0 = 2.5$, $r_1 = 1$, $r_2 = 2.5$, $\omega_1 = 100$ ccw, and $\gamma = 45°$. Find the acceleration of link 3 when $\theta = 60°$.

4.68 In Problem 4.67, find $a_3/(\omega_1^2 r_1)$ for one complete cycle of motion (in $30°$ steps). Use a computer or programmable calculator.

4.69 Refer to Figure 4.26. Let $r_0 = 6$, $r_1 = 1$, $r_2 = 9$, and $\gamma = 90°$. Find $a_3/(\omega_1^2 r_1)$ for one complete cycle of motion (in $30°$ steps). Use a computer or programmable calculator.

● **4.70** Consider a four-bar linkage for which $r_0 = 3.5$, $r_1 = 1.2$, $r_2 = 2.2$, and $r_3 = 3.1$. The angular velocity of the crank is 400 rad/s (constant and counterclockwise). Tabulate and plot angular velocities and angular accelerations of links 2 and 3 against crank angle. Suggestion: Use a spreadsheet program.

4.71 Consider a four-bar linkage for which $r_0 = 38$, $r_1 = 13$, $r_2 = 27$, and $r_3 = 41$. The angular velocity of the crank is 50 rad/s (constant and counterclockwise). Tabulate and plot angular velocities and angular accelerations of links 2 and 3 against crank angle. Suggestion: Use a spreadsheet program.

4.72 Consider a four-bar linkage for which $r_0 = 312.48$, $r_1 = 100$, $r_2 = 200$, and $r_3 = 300$. The angular velocity of the crank is 250 rad/s (constant and counterclockwise). Tabulate and plot angular velocities and angular accelerations of links 2 and 3 against crank angle. Suggestion: Use a spreadsheet program.

4.73 Consider a four-bar linkage for which $r_0 = 37$, $r_1 = 10$, $r_2 = 25$, and $r_3 = 40$. The angular velocity of the crank is 100 rad/s (constant and counterclockwise). Tabulate and plot angular velocities and angular accelerations of links 2 and 3 against crank angle. Suggestion: Use a spreadsheet program.

• **4.74** A slider-crank linkage has a connecting rod-to-crank length ratio of 2 and an offset-to-crank length ratio of 0.5. Tabulate and plot connecting rod angular velocity and angular acceleration, and slider velocity and acceleration, all against crank angle. Crank angular velocity is constant. Suggestion: Assume crank length = 1 and crank angular velocity = 1. Results are then normalized on these parameters. Consider the effect of other values of crank length and angular velocity. Solve by computer, using vector methods.

4.75 A slider-crank linkage has a connecting rod-to-crank length ratio of 3 and an offset-to-crank length ratio of 0.5. Tabulate and plot connecting rod angular velocity and angular acceleration, and slider velocity and acceleration, all against crank angle. Crank angular velocity is constant. Suggestion: Assume crank length = 1 and crank angular velocity = 1. Results are then normalized on these parameters. Consider the effect of other values of crank length and angular velocity. Solve by computer, using vector methods.

4.76 A slider-crank linkage has a connecting rod-to-crank length ratio of 2.5 and an offset-to-crank length ratio of 0.6. Tabulate and plot connecting rod angular velocity and angular acceleration, and slider velocity and acceleration, all against crank angle. Crank angular velocity is constant. Suggestion: Assume crank length = 1 and crank angular velocity = 1. Results are then normalized on these parameters. Consider the effect of other values of crank length and angular velocity. Solve by computer, using vector methods.

PROJECTS

See Projects 1.1 to 1.6 and suggestions in Chapter 1.
Describe and plot acceleration and angular acceleration characteristics of linkages in the design. Make use of computer software wherever practical. Check the results by a graphical method for at least one linkage position. Evaluate the linkage in terms of the performance requirements.

Cams: Design and Analysis

5.1 INTRODUCTION

Perhaps the simplest method by which a body can be given a certain prescribed motion is to make use of a cam and follower. A cam is a mechanism component that causes another body, known as the follower, to conform to a definite path and progress of motion.

Since the cam mechanism has its motion prescribed, it is a good example of kinematic synthesis. That is, rather than an analysis of a mechanism to determine its motion, the cam follower has a predetermined motion, and the process consists of designing the cam to give the desired motion. It is theoretically possible to obtain almost any type of follower motion by proper design of the cam. However, practical design considerations often necessitate modification of desired follower motions.

A great variety of cam types are available. Some examples are shown in Figure 5.1. Practical applications for cam mechanisms are numerous; one of the better-known applications is the timing system used in automotive engines. See Figure 5.2.

A cam mechanism usually consists of a cam (the driver), the follower (the driven element), and the frame (the support for the cam and follower). The follower normally has either translating or rotating motion.

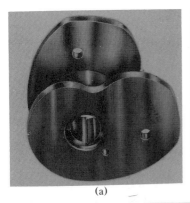

(b)

(a)

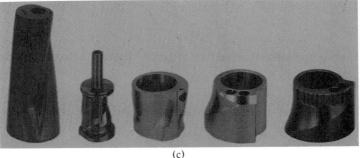

(c)

Figure 5.1 (a) A conjugate disk cam and (b) a cylindrical barrel cam. (*Source:* Commercial Cam Division, Emerson Electric Company.) (c) Three-dimensional cams. (*Source:* Parker-Hartford Corporation.)

The simplest type of cam is the *disk cam* or *plate cam*. Figure 5.3 illustrates this type of cam with four different followers: the translating, or reciprocating, roller follower (Figure 5.3a); the translating flat-faced follower (Figure 5.3b); the rotating, or oscillating, roller follower (Figure 5.3c); and the rotating flat-faced follower (Figure 5.3d). These followers can satisfy most common one-dimensional motion requirements, although other specialized follower types are sometimes used.

5.2 GRAPHICAL CAM DESIGN

In order to illustrate cam design and associated terminology, we will consider the design of a disk cam that is to give a prescribed motion to a translating roller follower. A translating roller follower (see Figure 5.3a) consists of an arm, constrained to move in a straight line, at the end of which a roller is attached by means of a pin. As the cam rotates, the roller rolls on the cam profile and causes the arm to translate. This rolling action helps to reduce wear, and, for this reason, the roller follower is often preferred over followers

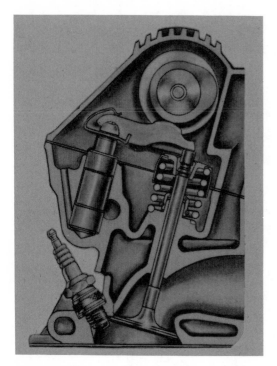

Figure 5.2 Cams are used to operate the intake and exhaust valves of automotive engines. (*Source:* General Motors Corporation.)

that have sliding contact. If the follower centerline intersects the centerline (i.e., the axis of rotation) of the cam, the follower is said to have radial motion. This will be the case in this example. (Observe that this is not the case in Figure 5.3a.)

Let us now assign a specific set of movements to the follower. The follower is to rise $1\frac{1}{2}$ in. with simple harmonic motion in 180° of clockwise cam rotation and then return to the initial position, also with simple harmonic motion, during the remaining 180° of cam rotation. The motion is to repeat this pattern with every camshaft revolution. Simple harmonic motion is one form of motion that can be imparted to a follower, and the details of this motion are described below. Later in this chapter, we will investigate various follower motions in order to determine which give the best operating characteristics.

The follower motion could have been described in terms of time rather than angles of cam rotation. For example, suppose the follower is to rise $1\frac{1}{2}$ in. in $\frac{1}{4}$ s, and return to the initial position in $\frac{1}{4}$ s. If the cam is made to rotate at a constant velocity of two revolutions per second, then one revolution of the cam will occur each half second, as required. The motions, whether described in terms of cam angle or time, are thus the same.

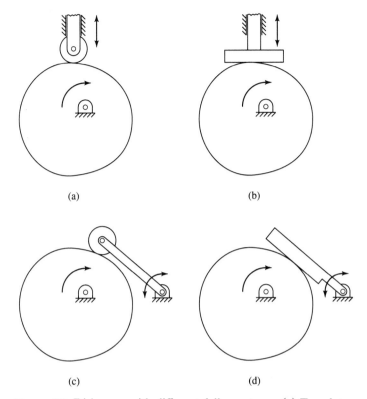

(a)   (b)

(c)   (d)

Figure 5.3 Disk cams with different follower types. (a) Translating roller follower. (b) Translating flat-faced follower. (c) Rotating roller follower. (d) Rotating flat-faced follower.

5.2.1 Follower Displacement Diagram

There are many types of follower motion used in cam design, some of which are discussed in Section 5.3. One of these motions is simple harmonic motion, which is briefly described here, and used for illustrative purposes in the following sections.

Figure 5.4 shows a point P, moving with constant speed on a circle centered at a fixed point O. For every position of its orbit, point P can be projected onto the vertical diameter AB. The projections of the successive positions of point P on AB are labeled P'. For example, when point P is at position P_1, the projection is P'_1; when it is at P_2, the projection is P'_2, and so on. The motion of the projected points on AB (the P''s) as point P rotates about O is simple harmonic motion. As point P continues to rotate about point O, the motion of P' keeps repeating itself.

Figure 5.5 shows a plot of follower displacement versus cam angle for the harmonic motion specified. Follower displacements are usually longitudinal

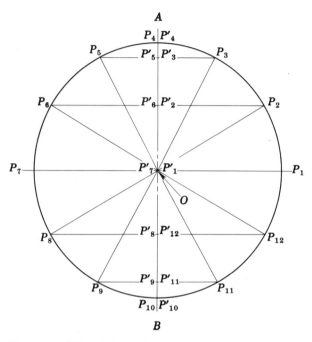

Figure 5.4 Point P is a point moving at a constant rate in a circular path about a fixed point O. The projections of point P on the vertical diameter are labeled P'. The motion of these projections is simple harmonic motion.

displacements expressed in millimeters or inches, while cam displacements are angular displacements measured in degrees or time units. The plot of follower displacement is usually drawn to scale, with follower displacement plotted along the ordinate and cam displacement plotted along the abscissa.

The displacement diagram is the starting point for the design of the cam. The diagram in Figure 5.5 is divided into a number of equal intervals. In this case, 12 intervals were chosen. This choice is convenient not only for dividing the cam into equal intervals, 30° each, but also for obtaining the proper displacements for simple harmonic motion. In an actual cam design, since more accuracy is necessary in determining the cam profile, a larger number of divisions is used.

The following procedure is used to draw the displacement diagram. The semicircle shown in the figure is drawn with a radius of $\frac{3}{4}$ in (half of the follower rise) to the left of the displacement diagram proper. It is then divided into 6 equal sectors of 30° each, that is, the same number of intervals as used on the abscissa for the rise portion of the motion. Horizontal lines are then drawn through the points 1 to 6 of the semicircle. The intersections of these lines with vertical lines drawn through points 1 to 6 on the abscissa of the displacement diagram determine points 1' to 6' on the displacement curve. A smooth curve drawn through these points, 0' to 6', completes the desired

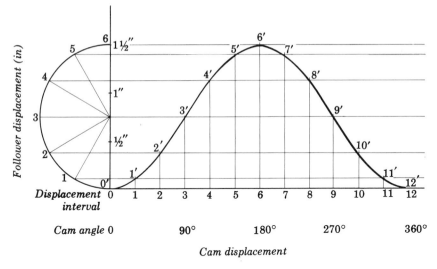

Figure 5.5 This illustration shows the plot of follower displacement versus cam displacement for simple harmonic motion. The follower rise occurs from 0 to 180°, while the return occurs from 180 to 360°. The follower lift, the cam displacement during the rise, and the cam displacement during the return are each divided into six intervals.

displacement curve for the first 180° of cam rotation. A similar procedure is used to obtain the other (symmetric) half of the displacement curve.

To sum up, if the cam profile is designed by using the displacement curve shown, the follower will rise $1\frac{1}{2}$ in, with simple harmonic motion, during the first 180° of cam rotation. The follower will then return $1\frac{1}{2}$ in, with simple harmonic motion, during the next 180° of cam rotation.

5.2.2 Graphical Cam Layout

Figure 5.6 shows the graphical layout of the corresponding cam. This graphical procedure is based on the mechanism inversion principle, wherein the relative motion of the members of a mechanism is independent of the member that is fixed. In this procedure, the cam is treated as the fixed member and the follower is drawn in the positions that it would take relative to the cam. The curve drawn tangent to these successive follower positions is then the cam profile. This process is illustrated in Figure 5.6 based on the displacement diagram of Figure 5.5.

The initial position of the follower is at point O, where the reference point on the follower is the center of the roller (see the inset in Figure 5.6). This point can be thought of as coinciding with a point O' on the extension of the cam. A circle the same size as the roller is then drawn with O' as its center. The distance from point K on this circle to the center of rotation of the cam, point C, is the radius of the *base circle* of the cam. The base circle radius,

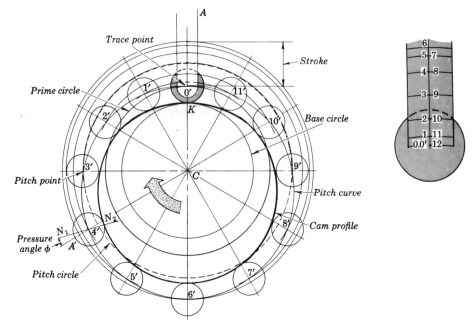

Figure 5.6 The cam layout for a roller follower and the displacement of Figure 5.5 is shown. The important terms illustrated in the figure are as follows: stroke: maximum follower rise; trace point: reference point for follower displacement, located at roller center; base circle: smallest circle, having its center at the cam center of rotation, that can be drawn tangent to cam profile; pitch curve: curve drawn through the center of the roller at the various positions around the circumference of the cam; cam profile: actual shape of the disk cam surface; pressure angle: angle between line of travel of follower and a normal drawn to pitch curve; pitch point: that point on the pitch curve having the largest pressure angle; pitch circle: circle drawn through pitch point and having its center at cam center C; and prime circle: smallest circle, having its center at cam center, that can be drawn tangent to pitch curve.

which is chosen by the designer, is directly related to the overall size of the cam. Therefore, cam size is one factor to be considered when selecting the base circle radius, especially if there are space limitations in a given machine; however, other factors must also be considered, including cam profile curvature, the stresses developed, and inertia effects, which often dictate a compromise in the final cam size.

Next, the follower displacements 1 to 12, taken from the displacement diagram, are laid off along the follower centerline, beginning at the pin, in Figure 5.6 (again shown in the enlarged inset of the follower for clarity). The cam is divided into the same number of equal parts, in this case 12 with 30° spacings between the radial lines. Arcs are swung from the numbered points on the follower centerline, using the center of the cam as the center of each circular arc. The intersections of these arcs with their respective radial lines on the cam determine points 1′ to 12′. For example, the arc from point 1 on the follower centerline intersects the 30° radial line at 1′, the arc from point 2 intersects the 60° radial line at point 2′, and so on. The curve drawn through these points is the pitch curve. A circle having a radius equal to the roller radius is then drawn at each of the primed points. The smooth curve, tangent to each of these circles, is the cam profile.

As pointed out earlier, the viewpoint adopted here is that the cam remains stationary, while the follower rotates about the cam in the opposite direction. The relative motion is the same, and the results obtained are exactly the same as those for the true situation, in which the cam rotates and the follower translates. This viewpoint makes the determination of the cam profile much simpler.

The final result of this example is that, as the cam rotates clockwise through 180° (points 0′ through 6′), the follower rises $1\frac{1}{2}$ in., and during the last 180° of rotation (points 6′ through 12′), the follower returns to its initial position.

Once the cam has been designed, there are various processes by which it can be manufactured. A common approach involves milling or grinding with the cutting tool taking the role of the follower. By indexing the cam by a fraction of a degree, a very accurate profile can be generated.

The cam layout procedure for the flat-faced translating follower, shown in Figure 5.3b, is similar to that presented above for a roller follower, except that the positions of the follower relative to the cam are represented by a series of straight lines (the follower face), and the cam profile is the curve drawn tangent to each of these straight lines. This same basic inversion strategy can also be employed for rotating follower types (Figure 5.3c and d).

5.2.3 Additional Cam Terms

Some of the important procedures and terms, illustrated by the example of Figure 5.6, should be discussed at this point. The greatest distance through which the follower moves (in this case, the distance from points 0 to 6) is known as the *total follower travel*, the *stroke*, or the *throw*. The center of the roller is known as the *trace point*. The *pitch curve* is the curve along which the trace point moves if the cam is stationary and the follower rotated.

When a disk cam is used, the follower is usually kept in contact with the cam surface by means of a spring force. Occasionally, when the follower motion is along a vertical line, the force of gravity is sufficient to maintain the contact during the return portion of the motion. However, great care must be exercised in the design of cams to ensure contact at all times between cam and follower.

A complication associated with roller followers is that the point of contact between the roller and cam does not always lie on the follower centerline. The force that exists between the follower and the cam acts along a line perpendicular to the tangent drawn to the surfaces in contact. It is therefore necessary to define the *pressure angle* of a roller follower as the *angle between the line of travel of the follower and a normal drawn to the pitch curve*.

In Figure 5.6, the pressure angle is ϕ, the normal to the pitch curve is N_1N_2, and the line of travel of the follower is $A'C$. Since the force between

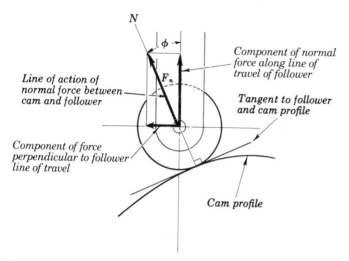

Figure 5.7 The relationship between the contact force and the pressure angle for a roller follower constrained to move vertically.

follower and cam does not act entirely along the line of travel of the follower, it will have a component acting perpendicularly to the follower centerline. See Figure 5.7. This component of force, $F_n \sin \phi$, varies directly with the pressure angle, so that the larger the pressure angle is, the larger is the force. It is quite possible for the pressure angle to be large enough to result in a substantial lateral component of force. This component of force could cause the follower to jam in its bearings. It is therefore desirable to keep the size of the pressure angle as small as possible. However, the smaller the pressure angle is, the larger is the required cam size. Since cam size is usually limited by space availability, a compromise with respect to the magnitude of the pressure angle is usually necessary. Most cams are designed with pressure angles less than 30°.

The flat-faced follower in Figure 5.3b has the advantage of a zero-degree pressure angle throughout its motion, and the side thrust present in roller followers is therefore not present in flat-faced followers.

It should be clear from Figure 5.6 that the pressure angle does not in general remain constant as the cam rotates. The *pitch point*, Figure 5.6, is that point on the pitch curve having the largest pressure angle. While exact methods are available for determining the maximum pressure angle, sufficiently accurate results can usually be obtained by graphical methods, that is, by drawing normals to the pitch curve and measuring the angle.

The *pitch circle* in Figure 5.6 is defined as the circle drawn through the pitch point with its center at the cam center.

The last term to be defined for this example is the *prime circle* of Figure 5.6, which is the smallest circle, having its center at the cam center, that can be drawn tangent to the pitch curve.

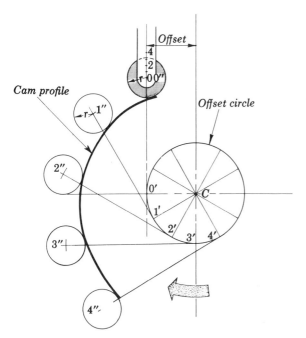

Figure 5.8 The offset, translating roller follower shown has less side thrust than an in-line or radial roller follower. Points 0″ to 4″ are determined by drawing tangents to the offset circle at points 0′ to 4′ and laying off along these lines distances 0′1 from 1′, 0′2 from 2′, and so on. Points 0″ to 4″ are then used as centers for circles having radii equal to the roller radius. A smooth curve drawn tangent to these circles is the required cam profile.

A method to reduce the side thrust present in a roller follower is to offset the follower, such that the centerline of the follower does not coincide with a radial centerline of the cam. A follower of this type is referred to as an offset follower and is shown in Figure 5.8. The points 0 to 4 on the follower are the follower displacements for a given type of motion. The total displacement of the follower is to occur in 120° of cam rotation.

A circle, called the *offset circle*, is drawn with its center at C and with a radius equal to the distance between the cam centerline and the follower centerline. The 120° of this circle, during which the follower motion is to occur, is divided into 4 equal parts of 30° each. Then, points 0′, 1′, 2′, 3′, and 4′ are located, beginning on the horizontal axis of the offset circle.

Lines tangent to the offset circle (they are also perpendicular to the radial lines of the circle) are next drawn through the primed points. When the measured distance 0′1 is laid off along the tangent line from 1′, point 1″ is located. The distance from 0′ to 2 is then laid off from 2′ to locate point 2″, and so on. The double-primed points, thus obtained, are used as the centers of circles having the same radius as the roller radius (r). A smooth curve tangent to these circles is the required cam profile. It is worth repeating once again that in the examples worked out in this chapter, only a few points are used to plot the motions and cam profiles, whereas in most actual cam designs, many more points are required to produce the required accuracy.

5.3 ANALYSIS OF CAM FOLLOWER MOTION

In this section, an analysis of cam follower motion will be made. Since the velocity, acceleration, and jerk are frequently the factors that determine the type of motion a follower is to have, they will be discussed in some detail.

Jerk is defined as equal to the time rate of change of acceleration. Since acceleration is directly proportional to force, jerk is associated with the time rate of change of force. Jerk is therefore important to cam designers because of its effect in terms of wear, noise, and stress.

The types of motion to be described can be used separately or in combination when specifying the various segments of an overall cam cycle. Combinations of motions are often used and provide a high degree of versatility in cam design.

5.3.1 Uniform Motion

A very simple but impractical follower motion is uniform motion, during which the follower is given a constant rate of rise or return; in other words, the follower has constant velocity. This, of course, means that the acceleration is zero.

Figure 5.9 shows the displacement diagram for a follower undergoing uniform motion. In this example, the full motion cycle of the follower consists of a dwell during 60° of cam rotation, followed by a uniform motion rise

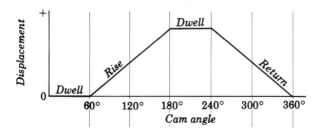

Figure 5.9 Displacement diagram for uniform motion. The follower dwells between 0 and 60°, rises with constant velocity between 60 and 180°, dwells from 180 to 240°, and returns with constant velocity from 240 to 360°. The abrupt motion changes inherent in this design result in high acceleration and jerk values, which adversely affect follower (and linkage) life and smoothness of response.

during the next 120°, then a second dwell for 60°, and a uniform motion return during the final 120° of the cam cycle. During a dwell, the follower is at rest. This is an extremely useful capability of cams. The displacement diagram shows that the follower rise and return follow a straight-line path. This type of motion is the simplest of the follower motions.

For uniform motion, the velocity goes from zero to its maximum (constant) velocity in zero time. To accomplish this, the acceleration would have to be infinite. While the physical conditions in a cam system prevent infinite accelerations, the values for the acceleration are high. The jerk will also theoretically approach infinity. Conditions of high acceleration and jerk result in a large amount of shock. Uniform motion is therefore seldom, if ever, used in cam design.

5.3.2 Modified Uniform Motion

It is possible to reduce the shock effects inherent in uniform motion cams by modifying the motion. The usual modification is to have the follower undergoing uniform acceleration at the start of the constant velocity interval, and uniform deceleration at the end of the constant velocity interval, so that the velocity curve is continuous.

Assume that the displacement diagram of Figure 5.9 is to be modified to correspond to the sequence indicated above. From the diagram, it is evident that the follower rise occurs between 60 and 180°; that is, during 120° of cam rotation. Figure 5.10 illustrates one example of modified uniform motion. Constant positive acceleration occurs from 60 to 90° and raises the follower through displacement $h/6$, where h is the total follower lift. Constant velocity occurs from 90 to 150° and raises the follower another $4h/6$. Constant

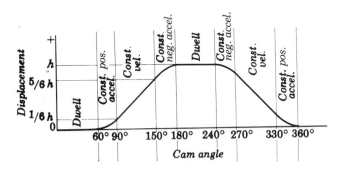

Figure 5.10 Diagram for displacement for modified uniform motion. The follower dwells from 0 to 60°, rises with constant positive acceleration from 60 to 90°, continues to rise with constant velocity from 90 to 150°, and completes the rise with constant negative acceleration from 150 to 180°. After a dwell from 180 to 240°, the follower returns with constant negative acceleration from 240 to 270°, continues the return with constant velocity from 270 to 330°, and completes the return with constant positive acceleration from 330 to 360°. This modification results in manageable acceleration.

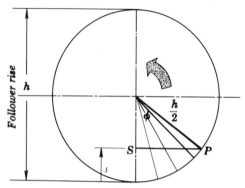

Figure 5.11 Simple harmonic motion. Point *S*, the projection of point *P*, moves up and down the vertical diameter with simple harmonic motion as *P* rotates about a circle of radius $h/2$.

negative acceleration occurs from 150 to 180°, during which interval the follower is raised the final $h/6$ as the velocity gradually decreases to zero. For this motion, the accelerations are no longer infinite. However, they may still be undesirably large. The jerk values are still theoretically infinite. This means that, while shock has been reduced somewhat, it still exists. As a result, this type of modified uniform motion is used only for slow-speed applications.

5.3.3 Simple Harmonic Motion

From Figure 5.11, the displacement equation for simple harmonic motion can be written in the form

$$s = \frac{h}{2} - \frac{h}{2} \cos \phi \tag{5.1a}$$

where $h/2$ is the radius of the generating circle and ϕ is the angle turned through by the generating radius. The total follower lift is h. It would be convenient to have the equation in terms of the cam angle rather than the generating circle angle. Let α be the angle the cam turns through while the generating circle is turning through 180°, or π radians. When the generating circle angle is some arbitrary angle ϕ, the cam angle is θ. The following equation, relating these angles, holds, where all angles are expressed in radians:

$$\frac{\phi}{\theta} = \frac{\pi}{\alpha}$$

Therefore,

$$\phi = \left(\frac{\pi}{\alpha}\right)\theta$$

The displacement equation (5.1a) can thus be written as

$$s = \frac{h}{2} - \frac{h}{2} \cos\left(\frac{\pi}{\alpha}\theta\right) \qquad 0 \le \theta \le \alpha \tag{5.1b}$$

where s is the instantaneous follower displacement, θ is the instantaneous cam angle, h is the total follower travel during the simple harmonic motion, and α is the total cam rotation angle during the motion. (h and α are constants.)

The velocity is then

$$v = \frac{ds}{dt} = -\frac{h}{2}\left(-\sin\frac{\pi}{\alpha}\theta\right)\left(\frac{\pi}{\alpha}\frac{d\theta}{dt}\right)$$

But $d\theta/dt = \omega$, where ω is the angular velocity of the cam, which is assumed here to be constant. Therefore, the velocity equation becomes

$$v = \left(\frac{h\pi}{2\alpha}\right)\omega \sin\left(\frac{\pi}{\alpha}\theta\right) \tag{5.2}$$

The acceleration equation is then obtained by differentiating the velocity equation:

$$a = \frac{dv}{dt} = \left(\frac{h\pi^2\omega^2}{2\alpha^2}\right)\cos\left(\frac{\pi}{\alpha}\theta\right) \tag{5.3}$$

Differentiating the acceleration equation gives the equation for jerk:

$$j = \frac{da}{dt} = -\left(\frac{h\pi^3\omega^3}{2\alpha^3}\right)\sin\left(\frac{\pi}{\alpha}\theta\right) \tag{5.4}$$

The corresponding motion curves for simple harmonic motion are drawn in Figure 5.12 for a dwell-rise-dwell-return example.

As can be seen from Figure 5.12, there are points in the motion for which the jerk is infinite (theoretically). For this reason, harmonic motion is not used for high-speed applications but, rather, is limited to medium-speed applications.

5.3.4 Cycloidal Motion

A *cycloidal curve* is the path traced by a point on a circle as the circle rolls on a straight line. Figure 5.13 shows such a cycloidal curve. As the circle of radius r rolls on the straight line, the point P traces out a cycloidal curve. Cycloidal motion is experienced by point S, the horizontal projection of point P. For a total follower lift h, the radius r of the generating circle is $r = h/(2\pi)$. It can then be shown that the equation for the displacement is

$$s = h\left(\frac{\theta}{\alpha}\right) - \left(\frac{h}{2\pi}\right)\sin\left(2\pi\frac{\theta}{\alpha}\right) \qquad 0 \le \theta \le \alpha \tag{5.5}$$

where α is the angle the cam turns through while the follower receives its total lift, and θ is the cam angle during which the displacement s occurs. Figure 5.14 illustrates the terms used in Eq. 5.5.

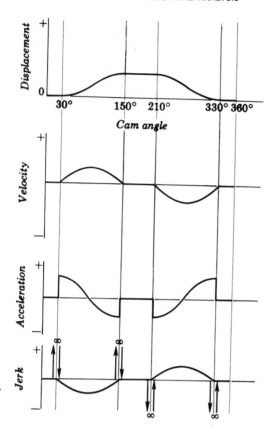

Figure 5.12 The follower displacement, velocity, acceleration, and jerk diagrams for simple
harmonic motion are shown. The follower dwells
from 0 to 30°, rises with simple harmonic
motion from 30 to 150°, dwells from 150 to
210°, returns with simple harmonic motion from
210 to 330°, and dwells from 330 to 360°. Note
the effect on acceleration and jerk.

Differentiating the displacement equation, Eq. 5.5, gives the expression for
velocity:

$$v = \frac{ds}{dt} = \frac{h\,d\theta}{\alpha\,dt} - \left(\frac{h}{2\pi}\right)\cos\left(2\pi\frac{\theta}{\alpha}\right)\left(\frac{2\pi\,d\theta}{\alpha\,dt}\right)$$

But $d\theta/dt = \omega$; therefore,

$$v = \frac{h\omega}{\alpha} - \left(\frac{h\omega}{\alpha}\right)\cos\left(2\pi\frac{\theta}{\alpha}\right) \tag{5.6}$$

The acceleration is obtained by differentiating the velocity equation:

$$a = \frac{dv}{dt} = h\left(\frac{2\pi\omega^2}{\alpha^2}\right)\sin\left(2\pi\frac{\theta}{\alpha}\right) \tag{5.7}$$

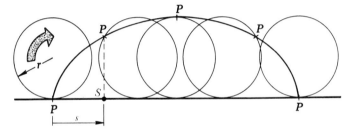

Figure 5.13 Cycloidal motion. As the circle shown rolls on the straight line, a point P on its circumference traces out a cycloidal curve. The curve shown results when the generating circle makes one complete revolution. Projected point S has cycloidal motion.

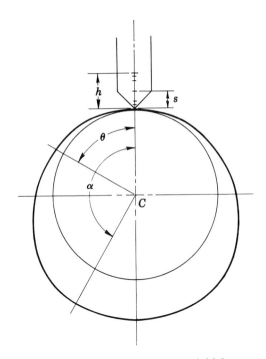

Figure 5.14 The cam imparts cycloidal motion to the follower. To impart the total displacement h, the cam must rotate through an angle α. Displacement s at any time is a function of the cam rotation θ.

Finally, the jerk is obtained by differentiating the acceleration equation:

$$j = \frac{da}{dt} = \left(\frac{h4\pi^2\omega^3}{\alpha^3}\right)\cos\left(2\pi\frac{\theta}{\alpha}\right) \tag{5.8}$$

The corresponding diagrams for displacement, velocity, acceleration, and jerk are shown in Figure 5.15. The advantage of the cycloidal curve as the

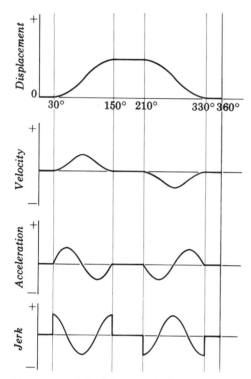

Figure 5.15 The follower displacement, velocity, acceleration, and jerk diagrams for cycloidal motion are shown. The follower dwells from 0 to 30°, rises with cycloidal motion from 30 to 150°, dwells from 150 to 210°, returns with cycloidal motion from 210 to 330°, and then dwells from 330 to 360°.

displacement motion of the follower can be seen by considering the acceleration diagram. The acceleration starts smoothly, with no vertical jumps in the curve. The result is very good operating characteristics. Wear, shock, stress, and noise are quite low. As a result, cycloidal cams are suited for high-speed applications.

5.3.5 Parabolic Motion

The displacement equation for parabolic motion is the equation of a parabola, which can be written

$$s = C\theta^2 \tag{5.9a}$$

where θ is the cam angle and C is a constant. If C is positive, Eq. 5.9a will give a constant positive acceleration. The equation is valid only between the

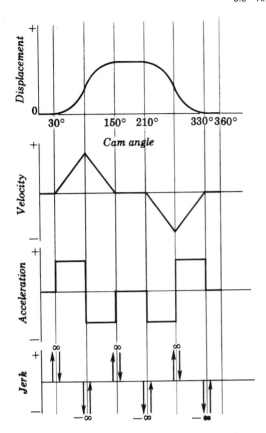

Figure 5.16 The follower displacement, velocity, acceleration, and jerk diagrams for parabolic motion are shown. The follower dwells from 0 to 30°, rises with constant positive acceleration from 30 to 90°, continues to rise with constant negative acceleration from 90 to 150°, dwells from 150 to 210°, returns with constant negative acceleration from 210 to 270°, completes the return with constant positive acceleration from 270 to 330°, and then dwells from 330 to 360°.

initial point and the inflection point on the displacement diagram. Once the inflection point is passed, the follower has negative acceleration, rather than positive acceleration. The inflection point, or point where the acceleration changes sign, occurs midway during the follower rise. See Figure 5.16. The following equations therefore apply only between the initial and halfway points.

Let α be the cam angle turned through while the follower is given total rise h. When the follower displacement is equal to $h/2$, the cam has turned through an angle $\alpha/2$. Therefore, from Eq. 5.9a,

$$\frac{h}{2} = C\left(\frac{\alpha}{2}\right)^2$$

$$C = \left(\frac{h}{2}\right)\left(\frac{2}{\alpha}\right)^2 = \frac{2h}{\alpha^2}$$

We have thus arrived at the value of the constant C during that interval of the parabolic curve which exhibits constant positive acceleration (from the initial point to the inflection point). The displacement equation can be rewritten, replacing the constant C by an expression in terms of the maximum displacement h and the corresponding cam angle α:

$$s = \left(\frac{2h}{\alpha^2}\right)\theta^2 \qquad 0 \le \theta \le \frac{\alpha}{2} \tag{5.9b}$$

The velocity equation for the interval is obtained by differentiating the displacement equation:

$$v = \frac{ds}{dt} = \left(\frac{2h}{\alpha^2}\right)\left(\frac{2\theta\,d\theta}{dt}\right)$$

But $d\theta/dt = \omega$; therefore,

$$v = \frac{4h\omega\theta}{\alpha^2} \tag{5.10a}$$

Since the maximum velocity occurs at $\theta = \alpha/2$,

$$v_{\max} = \left(\frac{4h\omega}{\alpha^2}\right)\left(\frac{\alpha}{2}\right) = \frac{2h\omega}{\alpha} \tag{5.10b}$$

Differentiating the velocity gives the acceleration:

$$a = \frac{dv}{dt} = \frac{4h\omega^2}{\alpha^2} \tag{5.11}$$

The derivative of the acceleration is the jerk:

$$j = \frac{da}{dt} = 0 \tag{5.12}$$

except at changes in acceleration, where the jerk is infinite.

The set of equations for that part of the curve between the inflection point and the maximum displacement can be written

$$s = C_0 + C_1\theta + C_2\theta^2 \tag{5.13a}$$

$$v = \frac{ds}{dt} = C_1\omega + 2C_2\theta\omega \tag{5.14a}$$

$$a = \frac{dv}{dt} = 2C_2\omega^2 \tag{5.15}$$

where the constants C_0, C_1, and C_2 are to be determined.

The following relationships between θ and the displacement and velocity exist:

1. $s = h$, when $\theta = \alpha$.
2. $v_{max} = 2h\omega/\alpha$, when $\theta = \alpha/2$.
3. $v = 0$, when $\theta = \alpha$.

By substituting these three relationships into Eqs. 5.13a and 5.14a, values for the constants C_0, C_1, and C_2 can be determined in terms of the (known) total displacement h and the corresponding (known) cam angle α. When $\theta = \alpha$, Eq. 5.13a becomes

$$h = C_0 + C_1\alpha + C_2\alpha^2 \tag{5.13b}$$

When $\theta = \alpha/2$, Eq. 5.14a becomes

$$\frac{2h\omega}{\alpha} = C_1\omega + 2C_2\left(\frac{\alpha\omega}{2}\right) \tag{5.14b}$$

When $\theta = \alpha$, Eq. 5.14a becomes instead

$$0 = C_1\omega + 2C_2\omega\alpha \tag{5.14c}$$

We can now solve the above equations to determine the values of C_0, C_1, and C_2. Using Eq. 5.14c, we obtain the following expression for C_1 in terms of C_2: $C_1 = -2C_2\alpha$. Substituting this expression for C_1 in Eq. 5.14b we obtain

$$\frac{2h\omega}{\alpha} = C_2\alpha\omega - 2C_2\alpha\omega = -C_2\alpha\omega$$

Therefore,

$$C_2 = -\frac{2h}{\alpha^2}$$

and

$$C_1 = \frac{4h}{\alpha}$$

Finally, substituting the expressions for C_1 and C_2 into Eq. 5.13b, we can determine the value of C_0:

$$h = C_0 + \left(\frac{4h}{\alpha}\right)\alpha + \left(-\frac{2h}{\alpha^2}\right)\alpha^2 = C_0 + 4h - 2h$$

$$C_0 = -h$$

Having arrived at expressions for C_0, C_1, and C_2 in terms of the maximum follower rise h and the corresponding cam angle α, we can substitute them into the displacement equation for the second half of the parabolic motion (between the inflection point and the maximum displacement), Eq. 5.13a, and differentiate to find velocity, acceleration, and jerk.

$$s = C_0 + C_1\theta + C_2\theta^2 = -h + \left(\frac{4h}{\alpha}\right)\theta + \left(-\frac{2h}{\alpha^2}\right)\theta^2$$

$$= h\left[-1 + \left(\frac{4}{\alpha}\right)\theta - \left(\frac{2}{\alpha^2}\right)\theta^2\right] \qquad \frac{\alpha}{2} \le \theta \le \alpha \qquad \text{(5.16)}$$

The velocity is then the derivative of the displacement:

$$v = \frac{ds}{dt} = 0 + \left(\frac{4h}{\alpha}\right)\frac{d\theta}{dt} - \left(\frac{2h}{\alpha^2}\right)(2\theta)\frac{d\theta}{dt}$$

$$= \left(\frac{4h\omega}{\alpha}\right)\left(1 - \frac{\theta}{\alpha}\right) \qquad \text{(5.17)}$$

The acceleration is found by taking the derivative of the expression for the velocity:

$$a = \frac{dv}{dt} = \left(\frac{4h\omega}{\alpha}\right)\left[0 - \left(\frac{1}{\alpha}\right)\frac{d\theta}{dt}\right] = \frac{-4h\omega^2}{\alpha^2} \qquad \text{(5.18)}$$

Finally, the jerk is given by

$$j = \frac{da}{dt} = 0 \qquad \text{(5.19)}$$

except at changes in acceleration, where the jerk is infinite.

As can be seen from Eq. 5.18, the acceleration for the second (latter) half of the rise is the same as that for the first half, except for the minus sign. The motion diagrams for parabolic motion are shown in Figure 5.16.

The advantage of the parabolic cam is that the maximum acceleration needed to produce a given follower displacement during a given cam angle is the smallest of all the curves discussed. However, as can be seen from the vertical jumps on the acceleration diagram, there are large, abrupt changes in acceleration at several points. This results in undesirable wear, stress, shock, and noise characteristics. The result is that parabolic cams are used only for low- or medium-speed applications.

5.3.6 Polynomial Motion

An important observation from the previous sections is that high values of acceleration or jerk, which tend to limit the design, often occur as the follower makes the transition from one phase of its motion to another phase (e.g., from a dwell to a rise, or from a constant positive-acceleration rise to a constant negative-acceleration rise). Discontinuities in velocity, as in the case of a transition from dwell to uniform motion (see Figure 5.9), lead to theoretically infinite accelerations, while discontinuities in acceleration, as occurs during the transition from dwell to either simple harmonic motion (see Fig. 5.12) or parabolic motion (see Figure 5.16), lead to infinite values of jerk. As was pointed out earlier, since acceleration and jerk relate to force and time rate of change of force, respectively, high values of these quantities will seriously restrict the speeds for successful cam operation.

Thus, smoother operation at higher speeds can be achieved by eliminating discontinuities in the follower motion time derivatives. An example of this for the dwell-rise-dwell-return motion, which we have been discussing, is cycloidal motion (see Figure 5.15). In that case, there are no discontinuities in either velocity or acceleration, and therefore, both acceleration and jerk remain finite. Hence, cycloidal motion is better suited for high-speed operation than are the other motions discussed.

The analysis of parabolic motion suggests a general class of follower motions with improved higher-order effects. Recall that in the case of parabolic motion examined, the constant coefficients in the quadratic follower displacement functions, Eqs. 5.9a and 5.13a, were adjusted to satisfy specific motion requirements related to displacement and velocity at the beginning and end of the travel. By a broadening of the displacement equation to include more terms, more motion requirements, including those dealing with acceleration and jerk, can be satisfied. The displacement equation for general polynomial motion can be written as

$$
\begin{aligned}
s &= C_0 + C_1(\theta - \theta_i) + C_2(\theta - \theta_i)^2 + C_3(\theta - \theta_i)^3 \\
&\quad + C_4(\theta - \theta_i)^4 + \cdots + C_N(\theta - \theta_i)^N \\
&= \sum_{k=0}^{N} C_k(\theta - \theta_i)^k
\end{aligned}
\tag{5.20}
$$

where s is the follower displacement, θ is the angular position of the cam, and θ_i is the initial cam angle at the beginning of the polynomial motion. In other words, the difference $(\theta - \theta_i)$ is the cam rotation occurring during displacement s. N is referred to as the degree of the polynomial, and $(N + 1)$ is the number of terms in the polynomial expression.

The velocity and acceleration are obtained by successive differentiations of Eq. 5.20:

$$
\begin{aligned}
v = \frac{ds}{dt} &= \omega\Big[C_1 + 2C_2(\theta - \theta_i) + 3C_3(\theta - \theta_i)^2 + 4C_4(\theta - \theta_i)^3 \\
&\quad + \cdots + NC_N(\theta - \theta_i)^{N-1} \Big] = \omega \sum_{k=1}^{N} kC_k(\theta - \theta_i)^{k-1}
\end{aligned}
\tag{5.21}
$$

$$
\begin{aligned}
a = \frac{dv}{dt} &= \omega^2\Big[2C_2 + 6C_3(\theta - \theta_i) + 12C_4(\theta - \theta_i)^2 + \cdots \\
&\quad + N(N-1)C_N(\theta - \theta_i)^{N-2} \Big] = \omega^2 \sum_{k=2}^{N} k(k-1)C_k(\theta - \theta_i)^{k-2}
\end{aligned}
\tag{5.22}
$$

where the angular velocity of the cam, $\omega = d\theta/dt$, is assumed to be constant. It can be seen that uniform motion and parabolic motion are special cases of Eq. 5.20 with $N = 1$ and $N = 2$, respectively. The following example illustrates how the coefficients $C_0, C_1, \ldots, C_N$ are selected to satisfy another specific set of motion requirements.

◆ **EXAMPLE PROBLEM 5.1** *Polynomial Motion Specification*
A disk cam is to give its follower a rise through travel h during a cam rotation α. This motion is preceded by a dwell and is also followed by a dwell. Determine the polynomial motion that will satisfy all displacement, velocity, and acceleration conditions at the beginning and end of the travel.

Solution. During dwell periods, the follower has zero velocity and acceleration. Thus, the following boundary conditions for the polynomial rise are identified:

    1. $s = 0$ when $\theta = \theta_i$.
    2. $v = 0$ when $\theta = \theta_i$.
    3. $a = 0$ when $\theta = \theta_i$.
    4. $s = h$ when $\theta = \theta_i + \alpha$.
    5. $v = 0$ when $\theta = \theta_i + \alpha$.
    6. $a = 0$ when $\theta = \theta_i + \alpha$.

In order to satisfy these six conditions, the polynomial must contain at least six terms ($N = 5$):

$$s = C_0 + C_1(\theta - \theta_i) + C_2(\theta - \theta_i)^2 + C_3(\theta - \theta_i)^3$$
$$+ C_4(\theta - \theta_i)^4 + C_5(\theta - \theta_i)^5$$

From Eqs. 5.21 and 5.22, the corresponding expressions for velocity and acceleration are

$$v = \omega\left[C_1 + 2C_2(\theta - \theta_i) + 3C_3(\theta - \theta_i)^2 + 4C_4(\theta - \theta_i)^3 + 5C_5(\theta - \theta_i)^4\right]$$

and

$$a = \omega^2\left[2C_2 + 6C_3(\theta - \theta_i) + 12C_4(\theta - \theta_i)^2 + 20C_5(\theta - \theta_i)^3\right]$$

Substituting the six required conditions leads to the following set of equations for the polynomial coefficients C_0, C_1, C_2, C_3, C_4, and C_5:

$$0 = C_0$$
$$0 = \omega C_1$$
$$0 = 2\omega^2 C_2$$
$$h = C_0 + C_1\alpha + C_2\alpha^2 + C_3\alpha^3 + C_4\alpha^4 + C_5\alpha^5$$
$$0 = \omega\left(C_1 + 2C_2\alpha + 3C_3\alpha^2 + 4C_4\alpha^3 + 5C_5\alpha^4\right)$$
$$0 = \omega^2\left(2C_2 + 6C_3\alpha + 12C_4\alpha^2 + 20C_5\alpha^3\right)$$

It is obvious from the first three equations that $C_0 = C_1 = C_2 = 0$. Solving the remaining three equations for C_3, C_4, and C_5 yields

$$C_3 = \frac{10h}{\alpha^3}$$

$$C_4 = -\frac{15h}{\alpha^4}$$

$$C_5 = \frac{6h}{\alpha^5}$$

Therefore,

$$s = 10h\left(\frac{\theta - \theta_i}{\alpha}\right)^3 - 15h\left(\frac{\theta - \theta_i}{\alpha}\right)^4 + 6h\left(\frac{\theta - \theta_i}{\alpha}\right)^5 \qquad \theta_i \leq \theta \leq \theta_i + \alpha$$

(5.23)

$$v = \omega\left[\left(\frac{30h}{\alpha}\right)\left(\frac{\theta - \theta_i}{\alpha}\right)^2 - \left(\frac{60h}{\alpha}\right)\left(\frac{\theta - \theta_i}{\alpha}\right)^3 + \left(\frac{30h}{\alpha}\right)\left(\frac{\theta - \theta_i}{\alpha}\right)^4\right] \quad (5.24)$$

$$a = \omega^2\left[\left(\frac{60h}{\alpha^2}\right)\left(\frac{\theta - \theta_i}{\alpha}\right) - \left(\frac{180h}{\alpha^2}\right)\left(\frac{\theta - \theta_i}{\alpha}\right)^2 + \left(\frac{120h}{\alpha^2}\right)\left(\frac{\theta - \theta_i}{\alpha}\right)^3\right]$$

(5.25)

The jerk, obtained by differentiating the acceleration, is

$$j = \omega^3\left[\frac{60h}{\alpha^3} - \left(\frac{360h}{\alpha^3}\right)\left(\frac{\theta - \theta_i}{\alpha}\right) + \left(\frac{360h}{\alpha^3}\right)\left(\frac{\theta - \theta_i}{\alpha}\right)^2\right] \qquad (5.26)$$

Because of the form of the displacement function, this particular follower motion is referred to as 3-4-5 polynomial motion. The corresponding diagrams for displacement, velocity, acceleration, and jerk are shown in Figure 5.17. Note that equations similar to Eqs. 5.23 to 5.26 can be derived for the return portion of the motion. If the return displacement is from $s = h$ at cam angle $\theta = \theta_j - \alpha'$ to $s = 0$ at cam angle $\theta = \theta_j$, with conditions of zero velocity and acceleration at the ends of the travel, then the governing displacement equation over the cam rotation interval of duration α' will be

$$s = 10h\left(\frac{\theta_j - \theta}{\alpha'}\right)^3 - 15h\left(\frac{\theta_j - \theta}{\alpha'}\right)^4 + 6h\left(\frac{\theta_j - \theta}{\alpha'}\right)^5 \qquad \theta_j - \alpha' \leq \theta \leq \theta_j$$

(5.27)

There are no vertical jumps in the displacement, velocity, and acceleration curves of Figure 5.17, resulting in good dynamic characteristics. The curves are similar in many respects to those for cycloidal motion in Figure 5.15, and the two motions are comparable. However, they do differ somewhat; for example, for the same rise h, cam speed ω, and cam rotation α, cycloidal motion will

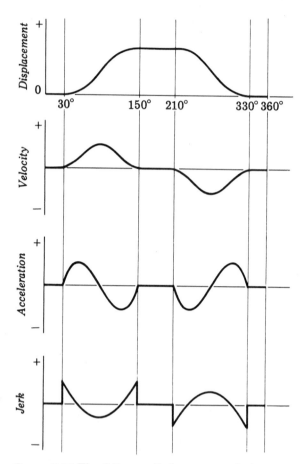

Figure 5.17 The follower displacement, velocity, acceleration, and jerk diagrams for 3-4-5 polynomial motion. The follower dwells from 0 to 30°, rises with polynomial motion from 30 to 150°, dwells from 150 to 210°, returns with polynomial motion from 210 to 330°, and then dwells from 330 to 360°.

have higher values of maximum acceleration but lower values of maximum jerk than 3-4-5 polynomial motion. ◆

Using the approach illustrated in the above example, polynomial motions can be derived for a wide range of different boundary conditions for displacements and derivatives, and complete follower motions can then be synthesized by combining properly matched polynomial motion segments. For example, a 4-5-6-7 polynomial can be generated that satisfies the conditions of Example Problem 5.1 plus two additional requirements of zero jerk at the beginning and end of the travel. However, manufacturing precision requirements generally increase as the number of boundary conditions increases.

5.4 ANALYTICAL CAM DESIGN

In many design applications, graphical methods of cam design have been replaced by mathematical or analytical design methods. This process has been facilitated by the digital computer, which provides some important advantages when compared to graphical design. First, the computer, once it has been programmed, can generate cam profiles and other design information in a much shorter period of time than required for a graphical cam layout on a drawing board. Since a number of trials are often necessary in arriving at an acceptable cam geometry, computer speed can lead to substantial savings of time in the design process.

A second advantage is the numerical precision of the computer, which far exceeds the precision that can be expected from graphical design. Furthermore, the precision of most computers is considerably greater than that of the machine tools used to produce cams. Thus, the computer-aided cam design process contributes a negligible amount of error to the final design. The reserve may be true in graphical design, where the machine tool error may be negligible compared to the errors inherent in the drafting and measurement techniques used to lay out the cam. The importance of accuracy in the design and production of high-speed cams, as it relates to dynamic performance effects, has been discussed in previous sections.

A third advantage of analytical cam design is that the computer used for designing a cam can also supply the necessary information for manufacturing that cam. Instructions can be generated for operation of numerically controlled machine tools used in cam production. The cam design process is one of many areas of design that have been heavily impacted by computer-aided design (CAD) and computer-aided manufacturing (CAM) technology. See Figure 5.18. There are several commercially available computer programs for designing cams.

5.4.1 Theory of Envelopes

The process of analytically generating a profile for a disk cam closely parallels the graphical approach that has been presented. As in the graphical approach, the desired positions of the follower are determined for an inversion of the cam-follower system in which the cam is held stationary. The cam that will produce the desired motion is then obtained by fitting a tangent curve to the follower positions. Unlike the graphical approach, however, the analytical approach can consider a virtually unlimited number of continuous follower positions as opposed to a finite number of discrete positions in a graphical layout.

The basis for the approach is the theory of envelopes from calculus. Consider the series of follower positions shown in Figure 5.19a. The follower is shown as a circular roller follower. However, any follower shape, including a flat face, can be considered. The set of all follower positions describes a family

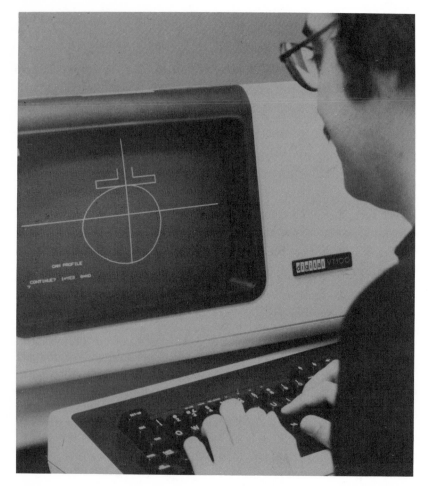

Figure 5.18 Cam design using interactive computer graphics. Computer-aided design and manufacturing (CAD/CAM) technology has extensive application in mechanism design.

of curves, circles in this case. The boundary of the family of follower curves is referred to as the envelope and is the cam profile. Notice that, for this example, the envelope consists of two curves, indicating that there are two possible cam profiles, an inner profile and an outer profile. The mathematical theory for determining the envelope is described in the following paragraphs.

A family of curves in the xy plane can be expressed mathematically as

$$F(x, y, \lambda) = 0 \tag{5.28}$$

where λ is called the parameter of the family. λ distinguishes the member curves from one another, and for a particular value of λ, Eq. 5.28 defines one member of the family of curves. For example, in Figure 5.19a, the parameter λ represents the location of the center of the constant-radius circular follower. In Figure 5.19b, two member curves, corresponding to arbitrary parameter

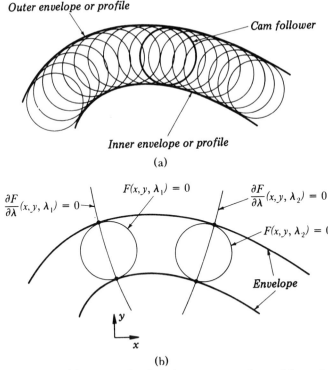

Figure 5.19 (a) The family of circles represents the positions of a roller follower as it moves relative to a cam. The boundary or envelope of this family of curves is therefore the cam profile. (b) The mathematical basis for describing the cam profile by means of the theory of envelopes is shown.

values λ_1 and λ_2, are shown. It can be seen that points lying on the envelope also lie on the curves, and, therefore, the x and y coordinates of the envelope must satisfy Eq. 5.28.

Consider the following equation involving the partial derivative of function F with respect to parameter λ:

$$\frac{\partial F}{\partial \lambda}(x, y, \lambda) = 0 \qquad (5.29)$$

Eq. 5.29 represents a second family of curves with parameter λ. It can be shown that each member curve of Eq. 5.29 intersects the corresponding member of Eq. 5.28 at the envelope (see Figure 5.19b). Therefore, the simultaneous solution of Eqs. 5.28 and 5.29 defines the envelope. This solution is accomplished either by eliminating the parameter λ or by expressing x and y in terms of λ.

◆ **EXAMPLE PROBLEM 5.2** *The Envelope of a Family of Circles*
Figure 5.20 contains a family of circles, each having a radius of 1.0 and a center lying on a 45° line in the xy plane. Determine the envelope for this family of curves.

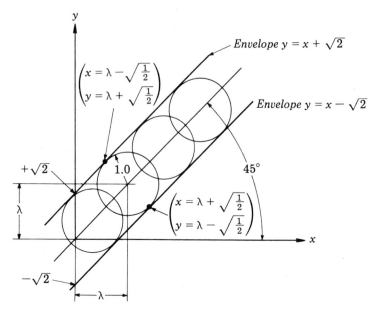

Figure 5.20 A family of circles, each having a radius of 1.0 and a center lying on the line $y = x$. As shown, the envelope of this family of curves consists of two straight lines.

Solution. The curves can be expressed mathematically as follows:

$$F(x, y, \lambda) = (x - \lambda)^2 + (y - \lambda)^2 - 1 = 0 \qquad \text{(5.28a)}$$

where a particular value of λ defines a circle of radius 1.0 centered at point $x = \lambda$, $y = \lambda$. Eq. 5.29 takes the form

$$\frac{\partial F}{\partial \lambda}(x, y, \lambda) = -2(x - \lambda) - 2(y - \lambda) = 0 \qquad \text{(5.29a)}$$

From Eq. 5.29a,

$$\lambda = \frac{x + y}{2}$$

Substituting into Eq. 5.28a,

$$\left[x - \left(\frac{x + y}{2} \right) \right]^2 + \left[y - \left(\frac{x + y}{2} \right) \right]^2 - 1 = 0$$

and solving for y, we have

$$y = x \pm \sqrt{2}$$

This equation defines the envelope: a pair of straight lines with 45° slopes and y intercepts of $+\sqrt{2}$ and $-\sqrt{2}$, as shown in Figure 5.20.

Alternatively, the envelope can be found in parametric form as a function of λ. From Eq. 5.29a,

$$y = 2\lambda - x$$

Substituting into Eq. 5.28a and solving for x, we have

$$x = \lambda \pm \sqrt{\frac{1}{2}}$$

and then,

$$y = \lambda \mp \sqrt{\frac{1}{2}}$$

Substituting a range of values for λ, these equations define the same envelope as given previously and shown in Figure 5.20. ◆

The following sections present the application of the theory of envelopes to common cam-follower types.

5.4.2 Disk Cam with Translating Flat-Faced Follower

A cam with a translating flat-faced follower is shown in Figure 5.21. This figure depicts an inversion of the cam-follower mechanism where the cam is fixed and the follower moves relative to it. In normal operation, the cam would rotate and the follower would translate in a guideway along the y axis. In either case, the relative motion between cam and follower is the same. The cam, having base circle radius r_b, is assumed to rotate in the clockwise direction under normal operation. Thus, for a cam rotation θ, the follower will rotate counterclockwise relative to the cam through angle θ while experiencing a translational displacement s, as shown in the figure. In this and the following sections, it is assumed that the follower displacement is a known function of the cam angle, as would be true in an actual design situation.

The equation of the family of straight lines (the follower face) generating the cam profile envelope is given by

$$y = mx + b \tag{5.30}$$

where m is the slope and b is the y intercept of the line. The origin of the xy coordinate system is positioned at the center of the base circle, which is also the pivot point of the cam. From inspection of Figure 5.21,

$$m = \tan \theta$$

Coordinates of point P, the intersection of the face of the follower and its axis, are given by

$$x = -(r_b + s)\sin \theta \tag{5.31}$$

$$y = (r_b + s)\cos \theta \tag{5.32}$$

where s, the displacement of the follower, is a prescribed function of cam angle θ. Point P is on the line described by Eq. 5.30. Substituting Eqs. 5.31 and 5.32 into Eq. 5.30, and solving for b, we have

$$b = \frac{(r_b + s)}{\cos \theta}$$

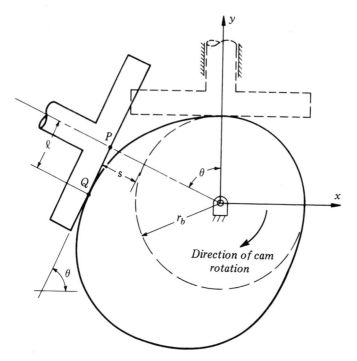

Figure 5.21 A disk cam with a translating flat-faced follower. The figure shows the motion of the follower relative to the cam. This motion consists of follower translation s during cam rotation θ. Point P is the point on the follower face coinciding with the follower centerline, and point Q is the instantaneous point of contact between the cam and follower.

and Eq. 5.30 becomes

$$y = \frac{x \sin \theta + (r_b + s)}{\cos \theta}$$

Rearranging, the family of straight lines (follower positions) generating the cam profile envelope is given by

$$F(x, y, \theta) = y \cos \theta - x \sin \theta - r_b - s = 0 \tag{5.33}$$

where θ is the parameter of the family; that is, each value of θ represents a different follower position and corresponding straight line.

Differentiating Eq. 5.33 yields

$$\frac{\partial F}{\partial \theta} = -y \sin \theta - x \cos \theta - \frac{ds}{d\theta} = 0 \tag{5.34}$$

where the quantity $ds/d\theta$ can be evaluated from the known displacement function. Solving Eqs. 5.33 and 5.34 simultaneously leads to the following expressions for the cam profile coordinates:

$$x = -(r_b + s)\sin \theta - \frac{ds}{d\theta} \cos \theta \tag{5.35}$$

$$y = (r_b + s)\cos \theta - \frac{ds}{d\theta} \sin \theta \tag{5.36}$$

Eqs. 5.35 and 5.36 give the coordinates of the cam-follower contact point (point Q in Fig. 5.21) for cam angle θ. The distance l between points P and Q is the perpendicular distance from the follower centerline to the contact point. From Eqs. 5.31, 5.32, 5.35, and 5.36,

$$l = \sqrt{(x_P - x_Q)^2 + (y_P - y_Q)^2} = \frac{ds}{d\theta} \tag{5.37}$$

The maximum value of l can be used in determining dimensions for the follower face. Eq. 5.37 can be rewritten as

$$l = \frac{ds}{dt} \frac{dt}{d\theta} = \frac{v}{\omega} \tag{5.38}$$

which is the translational follower velocity divided by the rotational cam velocity. Thus, for a constant-angular-velocity cam, the maximum value of l occurs when the follower velocity is at a maximum. It is noteworthy that l is independent of the base circle radius r_b.

◆ **EXAMPLE PROBLEM 5.3**   *Design of a Disk Cam with a Translating Flat-Faced Follower*

Design a disk cam to produce the following motion of a translating flat-faced follower: a rise through distance h with simple harmonic motion during 180° of rotation, followed by a return, also with simple harmonic motion, during the remaining 180° of cam rotation.

Solution.   From Eq. 5.1b, simple harmonic follower motion is given by

$$s = \frac{h}{2} - \left(\frac{h}{2}\right)\cos \theta$$

where $\alpha = \pi$ in this case. This equation holds for both the rise and return motions. The derivative of displacement s with respect to θ is

$$\frac{ds}{d\theta} = \left(\frac{h}{2}\right)\sin \theta$$

Substitution into Eqs. 5.35 and 5.36 yields the following expressions for the cam profile coordinates:

$$x = -\left(r_b + \frac{h}{2}\right)\sin \theta$$

$$y = \left(r_b + \frac{h}{2}\right)\cos \theta - \frac{h}{2}$$

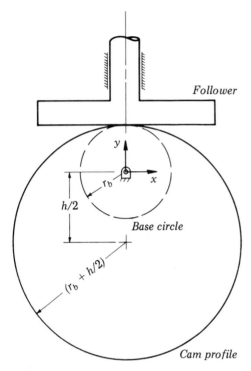

Figure 5.22 The cam and follower of Example Problem 5.3. The cam profile in this case is an offset circle, which will produce simple harmonic rise and return motion of the follower.

Alternatively, these equations can be rearranged to eliminate θ, leading to

$$x^2 + \left(y + \frac{h}{2}\right)^2 = \left(r_b + \frac{h}{2}\right)^2$$

It can be seen that, given a base circle radius r_b, the cam profile is a circle with its center at ($x = 0$, $y = -h/2$) and a radius equal to $r_b + h/2$. From Eq. 5.37,

$$l = \frac{ds}{d\theta} = \left(\frac{h}{2}\right)\sin\theta$$

which has a maximum absolute value of $h/2$ when θ equals 90 and 270°. Thus, the width of the follower face should be made greater than h. Figure 5.22 shows the offset circular cam that will produce the prescribed motion. ◆

◆ **EXAMPLE PROBLEM 5.4**   *Design of a Disk Cam with a Translating Flat-Faced Follower*

Design a disk cam to produce the following motion of a translating flat-faced follower: a dwell during 30° of cam rotation, a 2-in rise with parabolic motion during the next 150° of rotation, a second dwell during the next 60° of rotation, and a 2-in return with simple harmonic motion during the final 120° of cam rotation. The base circle radius is to be 3 in.

Solution. Figures 5.23a and 5.23b show the results of a typical computer program used to generate the required cam profile. Such a program would include analytical expressions for profile coordinates and follower motions. In this example, equations for simple harmonic motion (Section 5.3.3) and parabolic motion (Section 5.3.5) have been combined with Eqs. 5.35 and 5.36 to produce the results shown.

Figure 5.23a lists some of the computed profile coordinates and shows a computer-generated drawing of the cam and follower. Figure 5.23b contains

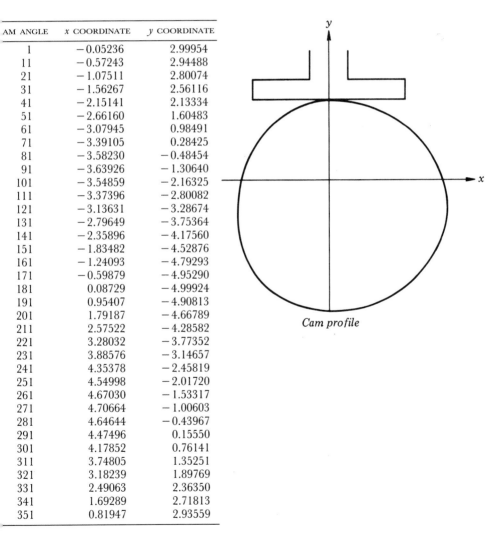

AM ANGLE	x COORDINATE	y COORDINATE
1	−0.05236	2.99954
11	−0.57243	2.94488
21	−1.07511	2.80074
31	−1.56267	2.56116
41	−2.15141	2.13334
51	−2.66160	1.60483
61	−3.07945	0.98491
71	−3.39105	0.28425
81	−3.58230	−0.48454
91	−3.63926	−1.30640
101	−3.54859	−2.16325
111	−3.37396	−2.80082
121	−3.13631	−3.28674
131	−2.79649	−3.75364
141	−2.35896	−4.17560
151	−1.83482	−4.52876
161	−1.24093	−4.79293
171	−0.59879	−4.95290
181	0.08729	−4.99924
191	0.95407	−4.90813
201	1.79187	−4.66789
211	2.57522	−4.28582
221	3.28032	−3.77352
231	3.88576	−3.14657
241	4.35378	−2.45819
251	4.54998	−2.01720
261	4.67030	−1.53317
271	4.70664	−1.00603
281	4.64644	−0.43967
291	4.47496	0.15550
301	4.17852	0.76141
311	3.74805	1.35251
321	3.18239	1.89769
331	2.49063	2.36350
341	1.69289	2.71813
351	0.81947	2.93559

Cam profile

Figure 5.23 (a) Cam profile coordinates and layout for Example Problem 5.4.

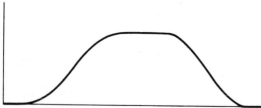

Follower displacement versus cam angle

Base radius: 3.0000
Type of motion and follower type: Tran flat
Number of segments: 4

Motion	Degrees	Displacement
DWLL	30	0.
PRAB	150	2.000
DWLL	60	0.
SMHM	120	−2.000

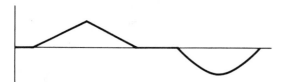

Follower velocity versus cam angle

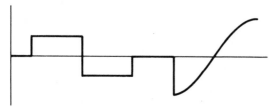

Follower acceleration versus cam angle

Figure 5.23 (b) Follower displacement, velocity, and acceleration for Example Problem 5.4.

computer graphics plots of the follower displacement, velocity, and acceleration corresponding to the prescribed motion for a complete cam rotation. ♦

5.4.3 Disk Cam with Translating, Offset Roller Follower

The configuration for this follower type is shown in Figure 5.24. Here the follower has a roller of radius r_f and eccentricity or offset e; radial follower motion is a special case of this configuration, where offset $e = 0$. As before, the base circle radius is r_b and the follower displacement is s which is a prescribed function of cam angle θ.

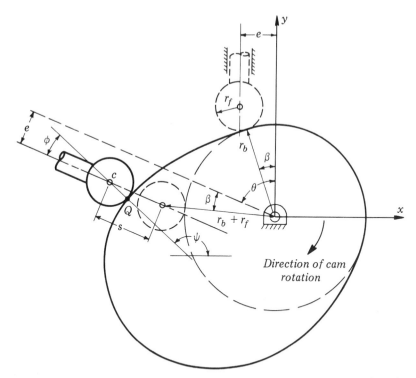

Figure 5.24 A disk cam with a translating, offset roller follower. Angle β is a function of the base circle radius r_b, the roller radius r_f, and the offset e. Angle ϕ is the pressure angle. The follower translates through distance s as the cam rotates through angle θ.

The equation for the family of circles described by the follower roller is

$$F(x, y, \theta) = (x - x_c)^2 + (y - y_c)^2 - r_f^2 = 0 \tag{5.39}$$

where x_c and y_c are the x and y coordinates, respectively, of the roller center c. For the arbitrary position shown in Figure 5.24 corresponding to cam angle θ,

$$x_c = -(r_b + r_f)\sin(\theta + \beta) - s \sin \theta \tag{5.40}$$

$$y_c = (r_b + r_f)\cos(\theta + \beta) + s \cos \theta \tag{5.41}$$

where

$$\beta = \arcsin\left(\frac{e}{r_b + r_f}\right) \tag{5.42}$$

Substituting Eqs. 5.40 and 5.41 into Eq. 5.39, we have

$$F(x, y, \theta) = \left[x + (r_b + r_f)\sin(\theta + \beta) + s \sin \theta \right]^2$$
$$+ \left[y - (r_b + r_f)\cos(\theta + \beta) - s \cos \theta \right]^2 - r_f^2 = 0 \qquad (5.43)$$

From Eq. 5.39, the partial derivative equation is

$$\frac{\partial F}{\partial \theta} = -2(x - x_c)\frac{dx_c}{d\theta} - 2(y - y_c)\frac{dy_c}{d\theta} = 0 \qquad (5.44)$$

where, from Eqs. 5.40 and 5.41,

$$\frac{dx_c}{d\theta} = -(r_b + r_f)\cos(\theta + \beta) - s \cos \theta - \frac{ds}{d\theta} \sin \theta \qquad (5.45)$$

$$\frac{dy_c}{d\theta} = -(r_b + r_f)\sin(\theta + \beta) - s \sin \theta + \frac{ds}{d\theta} \cos \theta \qquad (5.46)$$

Solving Eqs. 5.39 and 5.44 simultaneously gives the coordinates of the cam profile:

$$x = x_c \pm r_f \left(\frac{dy_c}{d\theta} \right) \left[\left(\frac{dx_c}{d\theta} \right)^2 + \left(\frac{dy_c}{d\theta} \right)^2 \right]^{-1/2} \qquad (5.47)$$

$$y = y_c \mp r_f \left(\frac{dx_c}{d\theta} \right) \left[\left(\frac{dx_c}{d\theta} \right)^2 + \left(\frac{dy_c}{d\theta} \right)^2 \right]^{-1/2} \qquad (5.48)$$

Note the plus and minus signs in Eqs. 5.47 and 5.48. This reflects the fact that there are two envelopes, an inner profile (shown in Figure 5.24) and an outer profile. Also observe that the plus sign in Eq. 5.47 goes with the minus sign in Eq. 5.48 and vice versa.

Pressure Angle

A mathematical expression can also be derived for the pressure angle ϕ. See Figure 5.24. Recall that the pressure angle is defined as the angle between the common normal at the cam-follower contact point and the line of travel of the follower. The common normal is the straight line passing through the contact point $Q(x, y)$, Eqs. 5.47 and 5.48, and the roller center $c(x_c, y_c)$, Eqs. 5.40 and 5.41. The angle ψ that this line makes with the x direction is given by

$$\psi = \arctan\left(\frac{y_c - y}{x_c - x} \right)$$

From Figure 5.24, angle ψ can also be expressed in terms of the pressure angle ϕ and the cam angle θ as

$$\psi = \frac{\pi}{2} + \theta - \phi$$

Therefore, equating the two expressions for ψ, the pressure angle is

$$\phi = \frac{\pi}{2} + \theta - \arctan\left(\frac{y_c - y}{x_c - x}\right) \tag{5.49a}$$

This equation is based on the inner-envelope cam profile coordinates. The pressure angle for the outer envelope is equal to that for the inner envelope. As discussed earlier, the pressure angle is an important design characteristic in cam-follower systems.

A useful relationship between the pressure angle ϕ and the cam dimensions can be derived from Eq. 5.49a. Rewriting that equation, we have

$$\frac{y_c - y}{x_c - x} = \tan\left[\frac{\pi}{2} + (\theta - \phi)\right] = -\cot(\theta - \phi) = \frac{-1}{\tan(\theta - \phi)}$$

from which

$$\tan(\theta - \phi) = -\left(\frac{x - x_c}{y - y_c}\right)$$

Substituting Eqs. 5.47 and 5.48 for x and y leads to

$$\tan(\theta - \phi) = -\left(\frac{\pm y_c'}{\mp x_c'}\right) = \frac{y_c'}{x_c'}$$

where the prime notation denotes differentiation with respect to angle θ (i.e., $x_c' = dx_c/d\theta$, etc.). The derivatives x_c' and y_c' are given by Eqs. 5.45 and 5.46, and upon substitution,

$$\tan(\theta - \phi) = \frac{\sin(\theta - \phi)}{\cos(\theta - \phi)} = \frac{(r_b + r_f)\sin(\theta + \beta) + s\sin\theta - s'\cos\theta}{(r_b + r_f)\cos(\theta + \beta) + s\cos\theta + s'\sin\theta}$$

Employing trigonometric identities, this equation reduces to the form

$$\frac{\sin\phi}{\cos\phi} = \tan\phi = \frac{s' - (r_b + r_f)\sin\beta}{s + (r_b + r_f)\cos\beta}$$

But from Eq. 5.42,

$$\sin\beta = \frac{e}{r_b + r_f}$$

and therefore,

$$\cos\beta = \sqrt{1 - \sin^2\beta} = \sqrt{1 - \left(\frac{e}{r_b + r_f}\right)^2}$$

Finally, making these substitutions, we have the desired relationship:

$$\tan\phi = \frac{s' - e}{s + \sqrt{(r_b + r_f)^2 - e^2}} \tag{5.49b}$$

Given a required follower displacement function s and a limit on how large the pressure angle ϕ can be, Eq. 5.49b can be used to size the cam base circle, the roller radius, and the follower offset. Note that the above equation also applies to the case of zero offset ($e = 0$).

The analytical cam synthesis equations derived in this section for a translating roller follower, as well as those equations derived for other follower types, are most effectively implemented in design practice through the use of a digital computer or a programmable calculator. The intention of the following examples is to illustrate the type of computation that would be performed at a relatively large number of positions in such a process.

◆ **EXAMPLE PROBLEM 5.5**   *Disk Cam with a Translating Roller Follower*
One segment of a prescribed cam-follower motion calls for a lift of 30 mm with cycloidal motion during cam rotation from $\theta = 0$ to $\theta = 90°$. The disk cam base circle radius is 40 mm. The follower type is a translating roller follower with a roller radius of 10 mm and an offset of 20 mm. Determine the profile coordinates and the pressure angle, corresponding to a cam angle of $\theta = 60°$.

Solution.   Applying Eq. 5.5 for cycloidal motion,

$$s = \left(\frac{h}{\alpha}\right)\theta - \left(\frac{h}{2\pi}\right)\sin\left(\frac{2\pi\theta}{\alpha}\right)$$

where $h = 30$ mm and $\alpha = \pi/2$ rad, leading to

$$s = \left(\frac{60}{\pi}\right)\theta - \left(\frac{15}{\pi}\right)\sin 4\theta$$

Differentiating with respect to θ,

$$s' = \frac{ds}{d\theta} = \frac{60}{\pi} - \left(\frac{60}{\pi}\right)\cos 4\theta$$

When $\theta = 60° = \pi/3$ rad,

$$s = \left(\frac{60}{\pi}\right)\left(\frac{\pi}{3}\right) - \left(\frac{15}{\pi}\right)\sin\left(\frac{4\pi}{3}\right) = 24.13 \text{ mm}$$

and

$$s' = \frac{60}{\pi} - \left(\frac{60}{\pi}\right)\cos\left(\frac{4\pi}{3}\right) = 28.65 \text{ mm/rad}$$

From Eq. 5.42,

$$\beta = \arcsin\left(\frac{20}{40 + 10}\right) = \arcsin(0.4) = 23.6°$$

Now the quantities x_c, y_c, x_c', and y_c' can be determined from Eqs. 5.40, 5.41,

5.45, and 5.46, respectively:

$$x_c = -(40 + 10)\sin(60° + 23.6°) - 24.13 \sin 60° = -70.59 \text{ mm}$$

$$y_c = (40 + 10)\cos(60° + 23.6°) + 24.13 \cos 60° = 17.64 \text{ mm}$$

$$x_c' = -(40 + 10)\cos(60° + 23.6°) - 24.13 \cos 60°$$

$$- 28.65 \sin 60° = -42.45 \text{ mm/rad}$$

$$y_c' = -(40 + 10)\sin(60° + 23.6°) - 24.13 \sin 60°$$

$$+ 28.65 \cos 60° = -56.26 \text{ mm/rad}$$

Substituting these values into Eqs. 5.47 and 5.48, we have the profile coordinates:

$$x = -70.59 \pm 10(-56.26)\left[(-42.45)^2 + (-56.26)^2\right]^{-1/2}$$

$$= -78.57, -62.61$$

$$y = 17.64 \mp 10(-42.45)\left[(-42.45)^2 + (-56.26)^2\right]^{-1/2} = 23.66, 11.62$$

From inspection of these results, it is clear that the coordinates of the contact point on the inner profile corresponding to this cam angle ($\theta = 60°$) are

$$x = -62.61 \text{ mm} \qquad y = 11.62 \text{ mm}$$

and the outer profile coordinates are

$$x = -78.57 \text{ mm} \qquad y = 23.66 \text{ mm}$$

The complete cam profile can be generated by repeating the above procedure over the total range of angle θ.

The pressure angle is obtained from either Eq. 5.49a or Eq. 5.49b. From the first of these equations,

$$\phi = 90° + 60° - \arctan\left[\frac{17.64 - 11.62}{-70.59 - (-62.61)}\right]$$

$$= 150° - \arctan\left(\frac{6.02}{-7.98}\right) = 150° - 143.0° = 7.0°$$

where the quadrant of the arctan function was determined from the signs of the numerator and denominator. As a check, using Eq. 5.49b,

$$\tan \phi = \frac{28.65 - 20}{24.13 + \sqrt{(40 + 10)^2 - (20)^2}} = 0.124$$

and

$$\phi = 7.0°$$

This pressure angle may be acceptable. However, the pressure angle varies with position, and the above value is probably not the maximum that occurs during the overall motion. ◆

♦ **EXAMPLE PROBLEM 5.6** *Translating Roller Follower with Zero Offset*
Repeat Example Problem 5.5 but with a follower offset of zero (i.e., a radial
follower).

Solution.   As before, at $\theta = 60°$, the values of the displacement and displacement derivative are $s = 24.13$ mm and $s' = 28.65$ mm/rad. But now, $\beta = 0$
because $e = 0$. Therefore, from Eqs. 5.40, 5.41, 5.45, and 5.46,

$$x_c = -50 \sin 60° - 24.13 \sin 60° = -64.20 \text{ mm}$$

$$y_c = 50 \cos 60° + 24.13 \cos 60° = 37.06 \text{ mm}$$

$$x'_c = -50 \cos 60° - 24.13 \cos 60° - 28.65 \sin 60° = -61.87 \text{ mm/rad}$$

$$y'_c = -50 \sin 60° - 24.13 \sin 60° + 28.65 \cos 60° = -49.88 \text{ mm/rad}$$

This leads to the following profile coordinate values from Eqs. 5.47 and 5.48:

$$x = -64.20 \pm 10(-49.88)\left[(-61.87)^2 + (-49.88)^2\right]^{-1/2}$$

$$= -70.48, -57.92$$

$$y = 37.06 \mp 10(-61.87)\left[(-61.87)^2 + (-49.88)^2\right]^{-1/2} = 44.84, 29.27$$

The inside profile coordinates are $x = -57.92$ mm and $y = 29.27$ mm, and the
outside profile coordinates are $x = -70.48$ mm and $y = 44.84$ mm.
  The pressure angle will be calculated from Eq. 5.49b:

$$\tan \phi = \frac{28.65 - 0}{24.13 + \sqrt{(40 + 10)^2 - (0)^2}} = 0.386$$

yielding

$$\phi = 21.1°$$

Whereas the pressure angle value for the offset case ($\phi = 7.0°$) is less than
this value, it should be realized that there may be other parts of the cam
mechanism cycle where the offset arrangement has higher pressure angle
values than the radial arrangement. However, the offset configuration may be
designed to utilize the pressure angle advantage during that portion of the
cycle when loads are large and, in turn, when side forces in the follower
guideway are most critical (especially during follower lift). Note that the
follower may be offset either to the left, as in Figure 5.24, or to the right, in
which case e would have a negative value. ♦

5.4.4 Disk Cam with Rotating Flat-Faced Follower

The oscillating or pivoted flat-faced follower is shown in Figure 5.25, where
the various parameters are defined. The family of follower positions are
straight lines.

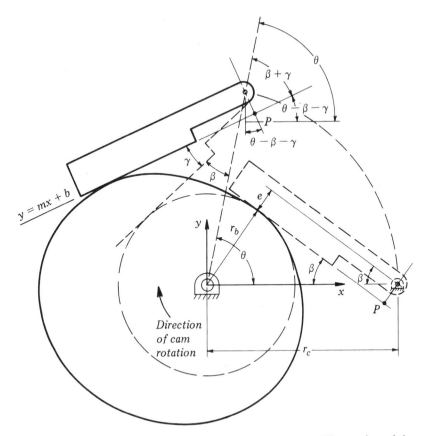

Figure 5.25 A disk cam with a rotating flat-faced follower. The motion of the follower relative to the cam consists of rotation γ during cam rotation θ. Angle β is a function of the base circle radius r_b, the distance r_c between the centers of rotation of the cam and follower, and the perpendicular distance e from the follower pivot to point P on the extension of the follower face.

The equation for the family of lines that will determine the envelope is the same as Eq. 5.30:

$$y = mx + b$$

From Figure 5.25, the slope m is given by

$$m = \tan(\theta - \beta - \gamma) \tag{5.50}$$

where γ is the angular displacement of the follower and β is the initial follower angle given by

$$\beta = \arcsin\left(\frac{r_b + e}{r_c}\right)$$

In a typical design situation, follower displacement γ would be a prescribed function of cam angle θ. The coordinates of point P on the extension of the follower face are

$$x = r_c \cos \theta + e \sin(\theta - \beta - \gamma) \tag{5.51}$$

$$y = r_c \sin \theta - e \cos(\theta - \beta - \gamma) \tag{5.52}$$

Substituting Eqs. 5.50, 5.51, and 5.52 into Eq. 5.30 and solving for b yields

$$b = r_c \sin \theta - e \cos A - (r_c \cos \theta + e \sin A)\tan A$$

where

$$A = \theta - \beta - \gamma$$

Then

$$F(x, y, \theta) = y + (r_c \cos \theta + e \sin A - x)\tan A - r_c \sin \theta + e \cos A$$

$$= 0 \tag{5.53}$$

The partial derivative expression is

$$\frac{\partial F}{\partial \theta} = \left(-r_c \sin \theta + e \cos A \frac{dA}{d\theta}\right)\tan A + (r_c \cos \theta + e \sin A - x)\sec^2 A \frac{dA}{d\theta}$$

$$- r_c \cos \theta - e \sin A \frac{dA}{d\theta} = 0 \tag{5.54}$$

where

$$\frac{dA}{d\theta} = 1 - \frac{d\gamma}{d\theta}$$

Solving Eqs. 5.53 and 5.54 simultaneously gives the following profile coordinates:

$$x = e \sin A + r_c \left[\cos \theta - \frac{\cos A \cos(\theta - A)}{dA/d\theta}\right] \tag{5.55}$$

$$y = -e \cos A + r_c \left[\sin \theta - \frac{\sin A \cos(\theta - A)}{dA/d\theta}\right] \tag{5.56}$$

5.4.5 Disk Cam with Rotating Roller Follower

The family of circles for this follower type is expressed by Eq. 5.39:

$$F(x, y, \theta) = (x - x_c)^2 + (y - y_c)^2 - r_f^2 = 0$$

The parameters are defined as shown in Figure 5.26: r_a is the length of the follower arm, r_f is the roller radius, r_c is the distance between the center of cam rotation and the follower pivot, and γ is the follower angular displacement, which is a prescribed function of cam angle θ. From the figure it can be seen that the coordinates of the center c of the roller are

$$x_c = r_c \cos \theta - r_a \cos(\theta - \beta - \gamma) \tag{5.57}$$

$$y_c = r_c \sin \theta - r_a \sin(\theta - \beta - \gamma) \tag{5.58}$$

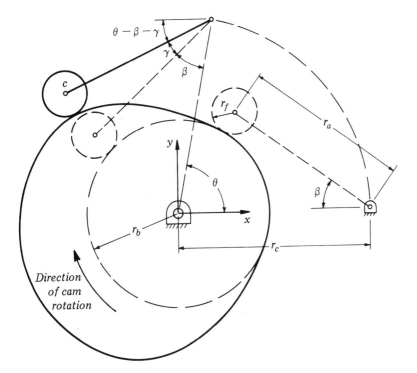

Figure 5.26 A disk cam with a rotating roller follower. The motion of the follower relative to the cam consists of rotation γ during cam rotation θ. Initial follower angle β is a function of the base circle radius r_b, the roller radius r_f, the center distance r_c between the cam and follower pivots, and the length r_a of the follower arm.

where the initial follower angle β is determined by the law of cosines:

$$\beta = \arccos\left[\frac{r_a^2 + r_c^2 - (r_b + r_f)^2}{2r_a r_c}\right]$$

The solution parallels that for the translating roller follower given earlier, and the cam profile coordinates are

$$x = x_c \pm r_f\left(\frac{dy_c}{d\theta}\right)\left[\left(\frac{dx_c}{d\theta}\right)^2 + \left(\frac{dy_c}{d\theta}\right)^2\right]^{-1/2}$$

$$y = y_c \mp r_f\left(\frac{dx_c}{d\theta}\right)\left[\left(\frac{dx_c}{d\theta}\right)^2 + \left(\frac{dy_c}{d\theta}\right)^2\right]^{-1/2}$$

where x_c and y_c are given by Eqs. 5.57 and 5.58, respectively, and

$$\frac{dx_c}{d\theta} = -r_c \sin\theta + r_a\left(1 - \frac{d\gamma}{d\theta}\right)\sin(\theta - \beta - \gamma) \tag{5.59}$$

$$\frac{dy_c}{d\theta} = r_c \cos\theta - r_a\left(1 - \frac{d\gamma}{d\theta}\right)\cos(\theta - \beta - \gamma) \tag{5.60}$$

As in the previous case of a roller follower, there are two envelopes, designated by the plus and minus signs in the coordinate equations, representing inner and outer cam profiles.

5.4.6 Cam Curvature

Another important factor affecting cam size and performance is cam curvature. If not limited by the pressure angle or some other consideration, the minimum size that a cam can have for a given application will be dictated by cam curvature. As one attempts to make the base circle radius, and therefore the cam, smaller, eventually a point is reached where both the graphical and analytical approaches will show the presence of cusps in the cam profile. Obviously, such a cam will not perform satisfactorily. However, there are other less obvious situations where the curvature can adversely affect performance, which can be identified more readily by using the analytical approach.

From calculus, the parametric expression for the radius of curvature ρ of a plane curve in the xy plane is

$$\rho = \frac{\left[(dx/d\theta)^2 + (dy/d\theta)^2\right]^{3/2}}{(dx/d\theta)(d^2y/d\theta^2) - (dy/d\theta)(d^2x/d\theta^2)} \tag{5.61}$$

The interpretation of the sign of Eq. 5.61 is as follows. In moving along the curve in the direction corresponding to increasing values of parameter θ, if the sign of ρ is positive, then the center of curvature is along the perpendicular to the curve on the left side, or, if the sign of ρ is negative, then the center of curvature of the curve is to the right. A straight-line portion of a curve has an infinite radius of curvature, and a cusp has a radius of curvature equal to zero. A change of sign for ρ indicates a transition from a convex portion of a curve to a concave portion, or vice versa (see Figure 5.27). Eq. 5.61 can be utilized in examining the cam curvature for any of the follower types that have been considered. The following sections show the application to the translating follower systems.

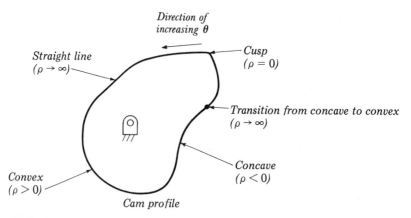

Figure 5.27 Various cases of radius of curvature ρ for a cam profile.

Translating Flat-Faced Follower

The radius of curvature of the cam profile in a translating flat-faced follower system is obtained by substituting Eqs. 5.35 and 5.36 into Eq. 5.61. The various derivatives are as follows. From Eq. 5.35,

$$\frac{dx}{d\theta} = -\left(r_b + s + \frac{d^2s}{d\theta^2}\right)\cos\theta$$

$$\frac{d^2x}{d\theta^2} = \left(r_b + s + \frac{d^2s}{d\theta^2}\right)\sin\theta - \left(\frac{ds}{d\theta} + \frac{d^3s}{d\theta^3}\right)\cos\theta$$

From Eq. 5.36,

$$\frac{dy}{d\theta} = -\left(r_b + s + \frac{d^2s}{d\theta^2}\right)\sin\theta$$

$$\frac{d^2y}{d\theta^2} = -\left(r_b + s + \frac{d^2s}{d\theta^2}\right)\cos\theta - \left(\frac{ds}{d\theta} + \frac{d^3s}{d\theta^3}\right)\sin\theta$$

Substitution into Eq. 5.61 yields

$$\rho = r_b + s + \frac{d^2s}{d\theta^2} \qquad (5.62)$$

It can be seen from Eq. 5.62 that the radius of curvature of the cam is dependent on the base circle radius and the prescribed follower motion. Figure 5.27 shows a cam profile depicting various cases of ρ as given by Eq. 5.62.

Obviously, cusps are to be avoided; in other words, the condition where $\rho = 0$ should not occur. Furthermore, the flat-faced follower will not operate properly on a concave portion of a cam profile. Therefore, from Eq. 5.62, the cam should be designed so that the following inequality is maintained at all points on the profile:

$$r_b + s + \frac{d^2s}{d\theta^2} > 0 \qquad (5.63)$$

Given the desired follower motion s, the inequality of Eq. 5.63 can be used to determine an acceptable value for the base circle radius. In this way, much of the trial-and-error process inherent in graphical cam layout can be avoided.

◆ **EXAMPLE PROBLEM 5.7** *Cam Curvature in a Flat-Faced Follower System*
Determine the minimum allowable base circle radius based on curvature for the cam design of Example Problem 5.4. Recall that the motion requirements for the translating flat-faced follower in that example were: a dwell during 30° of cam rotation, a two-inch rise with parabolic motion during the next 150° of rotation, a second dwell during the next 60° of rotation, and a two-inch return with simple harmonic motion during the remaining 120° of the cam rotation cycle.

Solution. We will determine the minimum value of the quantity $Q = s + s''$ (where the prime notation refers to differentiation with respect to θ) over the entire cam cycle, and then use the inequality of 5.63 to size the base circle. This will be accomplished by analyzing the four prescribed motion segments separately.

First dwell: Here, $s = s'' = 0$, and therefore, $Q = 0$ throughout this segment. A dwell period corresponds to a circular cam profile arc with a constant radius of curvature, which here is equal to the base circle radius.

Parabolic rise: In this range, we will utilize Eqs. 5.9b and 5.16 for parabolic motion. Note that the designation of a particular cam position as the zero angle position ($\theta = 0$) is arbitrary. For convenience in analyzing this motion segment, we will assume that $\theta = 0$ corresponds to the beginning of the rise. Therefore, Eqs. 5.9b and 5.16 apply directly, with the following values substituted: $h = 2$ in and $\alpha = 5\pi/6$ rad (150 degrees). Then, for the first half of the rise,

$$s = \left(\frac{2h}{\alpha^2}\right)\theta^2 = \left[\frac{2(2)}{(5\pi/6)^2}\right]\theta^2 = 0.584\theta^2$$

$$s' = 1.17\theta$$

$$s'' = 1.17$$

and

$$Q = 0.584\theta^2 + 1.17$$

Obviously, Q will be greater than zero during this range and we need not pursue this case further.

For the second half of the rise, using Eq. 5.16,

$$s = h\left[-1 + \left(\frac{4}{\alpha}\right)\theta - \left(\frac{2}{\alpha^2}\right)\theta^2\right] = -2 + 3.06\theta - 0.584\theta^2$$

$$s' = 3.06 - 1.17\theta$$

$$s'' = -1.17$$

and

$$Q = -3.17 + 3.06\theta - 0.584\theta^2$$

Examining the associated range of θ ($\theta = 1.31$ to $\theta = 2.62$), the minimum value of Q occurs at $\theta = 1.31$ and is

$$Q = -3.17 + 3.06(1.31) - 0.584(1.31)^2 = -0.164$$

Second dwell: As for the first dwell, quantity Q is constant during this range with a value of $Q = s + s'' = 2 + 0 = 2$.

Simple harmonic return: Once again, for convenience, we will assume that cam angle $\theta = 0$ refers to the beginning of the return motion. Since the motion goes from a value of $s = h$ at $\theta = 0$ to a value of $s = 0$ at $\theta = \alpha$, we must modify Eq. 5.1b for simple harmonic motion accordingly (recall that Eq. 5.1b was derived for motion from $s = 0$ to $s = h$). This is easily accomplished by subtracting that equation from the lift h:

$$s = h - \left[\frac{h}{2} - \left(\frac{h}{2}\right)\cos\left(\frac{\pi}{\alpha}\theta\right)\right] = \frac{h}{2} + \left(\frac{h}{2}\right)\cos\left(\frac{\pi}{\alpha}\theta\right)$$

Substituting values ($h = 2$ and $\alpha = 2\pi/3$ or $120°$) and differentiating, we have

$$s = 1 + \cos 1.5\theta$$
$$s' = -1.5 \sin 1.5\theta$$
$$s'' = -2.25 \cos 1.5\theta$$

and

$$Q = s + s'' = 1 - 1.25 \cos 1.5\theta$$

By inspection, the minimum value of Q for this segment occurs at $\theta = 0$ and is $Q = 1 - 1.25 = -0.25$.

Therefore, the minimum value of Q over the entire motion is (-0.25), and from the inequality of 5.63,

$$r_b + (-0.25) > 0$$

which indicates that the minimum limit for the base circle radius is $r_b = 0.25$ in. The radius selected in Example Problem 5.4 ($r_b = 3$ in) is well above this limit and should provide good curvature characteristics (see Figure 5.23a). However, a smaller cam could probably be used. ◆

Translating Roller Follower

Generally, the curvature of the pitch curve is analyzed when considering roller followers. The equation for the radius of curvature of the pitch curve is easier to obtain than that of the cam profile, and, as will be seen, the necessary design conditions can be expressed in terms of the pitch curve. The radius of curvature of the pitch curve of a translating roller follower can be determined from Eqs. 5.40, 5.41, and 5.61.

Consider the special case of a radial follower where the offset $e = 0$. Then, from Eq. 5.42, $\beta = 0$ and the pitch curve coordinates, Eqs. 5.40 and 5.41, are

$$x_c = -(r_b + r_f + s)\sin \theta$$
$$y_c = (r_b + r_f + s)\cos \theta$$

and the derivatives are as follows:

$$\frac{dx_c}{d\theta} = -(r_b + r_f + s)\cos \theta - \frac{ds}{d\theta} \sin \theta$$

$$\frac{dy_c}{d\theta} = -(r_b + r_f + s)\sin \theta + \frac{ds}{d\theta} \cos \theta$$

$$\frac{d^2x_c}{d\theta^2} = \left(r_b + r_f + s - \frac{d^2s}{d\theta^2}\right)\sin \theta - 2\frac{ds}{d\theta} \cos \theta$$

$$\frac{d^2y_c}{d\theta^2} = -\left(r_b + r_f + s - \frac{d^2s}{d\theta^2}\right)\cos \theta - 2\frac{ds}{d\theta} \sin \theta$$

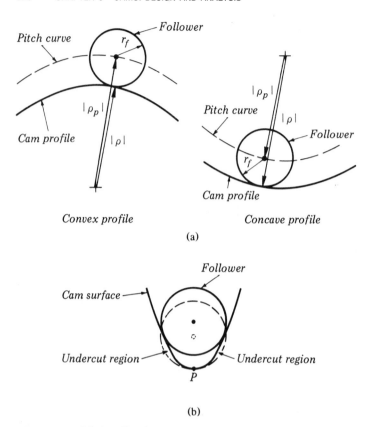

Convex profile   Concave profile

(a)

(b)

Figure 5.28 (a) A roller follower can operate on both convex and concave portions of cam profiles. (b) There are situations where unacceptable operation can occur, as shown here.

Substituting into Eq. 5.61, we have the radius of curvature for a radial roller follower as

$$\rho_p = \frac{\left[(r_b + r_f + s)^2 + (ds/d\theta)^2\right]^{3/2}}{(r_b + r_f + s)^2 + 2(ds/d\theta)^2 - (r_b + r_f + s)(d^2s/d\theta^2)} \tag{5.64}$$

Here, ρ_p is the radius of curvature of the pitch curve, where a positive value refers to a convex portion of the pitch curve and a negative value refers to a concave portion, based on the previous definition of parameter θ.

Roller followers may move along concave cam profiles as well as convex cam profiles. The two cases are depicted in Figure 5.28a. For the convex case, the absolute value of the radius of curvature ρ of the cam profile is

$$|\rho| = \rho_p - r_f$$

The cam will become pointed (i.e., there will be a cusp) when

$$|\rho| = \rho_p - r_f = 0$$

For concave portions of the profile,

$$|\rho| = -\rho_p + r_f$$

and the cam will become undercut when

$$|\rho| \leq r_f$$

Undercutting refers to the situation where the roller is too large for a concave section of cam profile, as characterized by two distinct points of contact between the cam and follower in these positions. See Figure 5.28b. As shown by the dashed follower position, the cam would have to be undercut in order to produce contact at cam profile points such as P. Thus, the conditions for acceptable cam curvature for a roller follower, in terms of the pitch curve radius, are

Convex cam profile: $\rho_p > r_f$ **(5.65)**

Concave cam profile: $\rho_p < 0$ **(5.66)**

and unacceptable profiles are defined by

$$0 \leq \rho_p \leq r_f \qquad\qquad\qquad (5.67)$$

These inequalities can be used for sizing the cam base circle and the roller.

◆ **EXAMPLE PROBLEM 5.8**   *Cam Curvature in a Roller Follower System*
Evaluate the curvature of the disk cam with a radial roller follower of Example Problem 5.6 for the position corresponding to a cam angle of $\theta = 60°$.

Solution.   Recall from Example Problems 5.5 and 5.6 that $r_b = 40$ mm, $r_f = 10$ mm, and the follower displacement function is

$$s = \left(\frac{60}{\pi}\right)\theta - \left(\frac{15}{\pi}\right)\sin 4\theta$$

Differentiating this function twice with respect to angle θ, we have

$$\frac{ds}{d\theta} = \frac{60}{\pi} - \left(\frac{60}{\pi}\right)\cos 4\theta$$

and

$$\frac{d^2s}{d\theta^2} = \left(\frac{240}{\pi}\right)\sin 4\theta$$

Evaluating these functions at $\theta = \pi/3$ rad,

$$s = 24.13, \qquad \frac{ds}{d\theta} = 28.65, \qquad \frac{d^2s}{d\theta^2} = -66.16$$

Summing terms,

$$(r_b + r_f + s) = (40 + 10 + 24.13) = 74.13$$

Substituting into Eq. 5.64, we see that the radius of curvature of the pitch curve at this point is

$$\rho_p = \frac{\left[(74.13)^2 + (28.65)^2\right]^{3/2}}{(74.13)^2 + 2(28.65)^2 - (74.13)(-66.16)} = 41.69 \text{ mm}$$

At this position, the cam profile is convex, and referring to inequality 5.65, the cam curvature is acceptable at this location. Of course, since the curvature will in general vary with position on the cam profile, it is necessary to examine the entire cam surface for adverse curvature effects, much as was done in Example Problem 5.7. However, this process can be expedited by the fact that the extreme curvature positions during standard follower motions can be determined in general, and then only these positions need be considered in specific cam design cases. This means that only a few points on the total cam profile will have to be considered. ◆

5.5 POSITIVE-MOTION CAMS

The cam types discussed up to this point depend on the force of gravity or a spring force to maintain contact between the cam and the follower during the return stroke. There are many applications in which it is necessary for the cam to exert positive control over the follower during the return as well as during the rise. In this section, some of the more common types of positive-motion cams are discussed.

5.5.1 Face Cam

One method of achieving positive motion is to cut a groove into the face of a cam. The roller follower then rides in this groove. During the rise, the inner surface of the groove (the side of the groove nearest the cam axis) causes the follower to move up, while on the return stroke, the outer surface of the

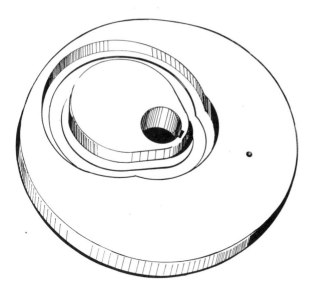

Figure 5.29 Face cam. This type of cam imparts positive motion to the follower at all times. The follower is alternately driven by the inner and outer sides of the groove.

groove forces the follower down. This type of cam, known as a face cam, is shown in Figure 5.29.

The graphical layout of a face cam is very similar to that for the roller follower, and is equivalent to the generation of both the inner and outer envelopes by the analytical method.

A clearance between the follower and the groove surfaces is obviously necessary. Because of this clearance, a reversal of force occurs when the follower moves from contact on one surface of the groove to contact on the other surface. For reduction of shock, the clearance should be kept as small as possible.

5.5.2 Constant-Breadth Cam

Another method for obtaining positive motion is the constant-breadth cam. This type of cam is shown in Figure 5.30. As seen in the illustration, the cam surface is always in contact with the two parallel surfaces of the follower. The follower "boxes in" the cam. The distance between the parallel surfaces, d, is equal to the base circle diameter plus the follower rise. Every point on the cam profile is a distance d from the point 180° from it. It is clear that the follower motion desired must occur between 0 and 180°. The cam profile between 180 and 360° must be such as to keep the distance d constant. In other words, the follower motion during the rise is the reverse of that during the return. This

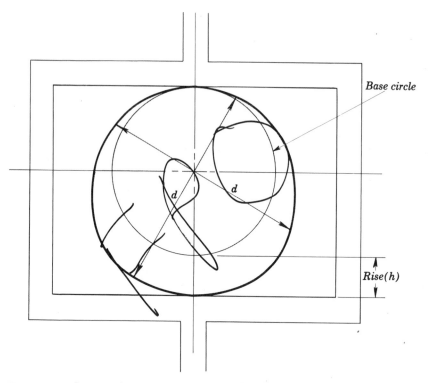

Figure 5.30 Constant-breadth cam. While this type of cam provides positive motion, the disadvantage is that the motion from 180 to 360° of cam rotation must be the reverse of the motion from 0 to 180°.

type of cam obviously cannot be used for applications where the rise and return motions are different. Design methods for this type of cam are identical to those for cams with flat-face followers.

5.5.3 Conjugate Cams

Conjugate cams are a variation of constant-breadth cams. The system consists of two roller followers, 180° apart, and two cams mounted on the same shaft. See Figure 5.31. The first cam is designed to give the first follower the required motion between 0 and 180° of camshaft rotation. The second cam has a profile that is obtained by drawing circles whose centers are a fixed distance d from the follower No. 1 circles. In the interval from 180 to 360°, the first cam profile is obtained by drawing circles whose centers are a fixed distance d from the follower No. 2 circles. In this interval, however, the *second* cam gives its follower the required motion.

The advantage of this type of positive-motion cam over the constant-breadth cam is that wear is reduced because of the rolling action, and the force reversal is kept to a minimum because the clearance can be made

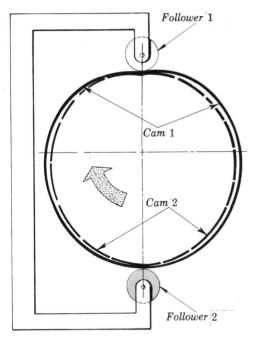

Figure 5.31 Conjugate cam. The conjugate cam system shown achieves continuous positive motion by making use of two cams and two followers. From 0 to 180°, cam 1 drives follower 1 to give the required motion. Between 180 and 360°, cam 2 gives follower 2 the required motion.

extremely small. The biggest disadvantage is probably the fact that this type of cam may be more expensive than other equally suitable cam designs.

5.6 CYLINDRICAL CAMS

A cylindrical or barrel cam (see Figure 5.1) is used to drive a translating roller follower, which moves parallel to the axis of the cam. A groove cut into the side of the cylinder provides the path for the follower. The follower may be cylindrical but is usually conical because the wear is reduced for this type of follower (see Figure 5.32). Cylindrical cams are positive-motion cams, except for the type known as end cams. It is possible to get a wide variety of motions from cylindrical cams.

Analytical methods for the design of three-dimensional cams, such as cylindrical cams, are available. The utility of these techniques is evident from the fact that graphical layout methods can become very complex.

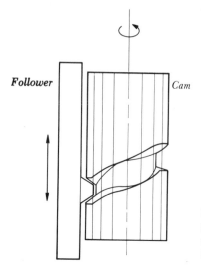

Figure 5.32 Cylindrical cam. The groove, when cut into the cylinder, will give the follower shown the required rise and fall with positive motion.

5.7 PRACTICAL CONSIDERATIONS IN CAM DESIGN

Practical considerations usually are the primary factors in the decision a designer must make with regard to the type of cam system to be used to solve a particular problem. Space limitation and the speed of operation are perhaps the most common factors that govern the decision. The force required to keep the follower and cam in contact is another important consideration. The system required may be a gravity type or a spring type. Another important factor the designer must consider is the accuracy of the machining operations used in manufacturing the cam. Machining errors may cause kinematic variations from the required operating conditions. For this reason, highly precise machinery is required.

With the need for high-speed cams, the effect of vibrations on cam performance has also become more important. A vibration investigation requires information concerning the elasticity of the system, and the determination of the elastic response of a cam system is an extremely difficult problem. It is very often desirable to construct a prototype of the cam system. As a result of tests on the prototype, modifications can be made to ensure a satisfactory solution to the given problem.

References

1. Chen, F. Y., *Mechanics and Design of Cam Mechanisms*, Pergamon, New York, 1982.
2. Rothbart, H. A., *Cams*, Wiley, New York, 1956.
3. Tesar, D., and G. K. Matthew, *The Dynamic Synthesis, Analysis, and Design of Modeled Cam Systems*, Lexington Books, Lexington, MA, 1976.

PROBLEMS

5.1 A follower rises 50 mm in 120° with constant velocity, dwells for 60°, returns in 120° with constant velocity, and dwells for 60°. Draw the follower displacement diagram.

5.2 A follower rises 50 mm in 120° with simple harmonic motion, dwells for 60°, returns in 120° with simple harmonic motion, and dwells for 60°. Draw the follower displacement diagram.

5.3 A follower rises 50 mm in 120° with cycloidal motion, dwells for 60°, returns in 120° with cycloidal motion, and dwells for 60°. Write a computer program to calculate and plot the displacement diagram.

5.4 A follower rises 50 mm in 120° with parabolic motion, dwells for 60°, returns in 120° with parabolic motion, and dwells for 60°. Write a computer program to calculate and plot the displacement diagram.

5.5 A follower rises 2 in in 120° with constant velocity, returns 1 in in 90° with constant velocity, dwells for 60°, and returns 1 in in 90° with constant velocity. Draw the displacement diagram.

5.6 A follower rises 1 in in 60° with simple harmonic motion, dwells for 30°, rises another 1 in in 60° with simple harmonic motion, dwells for 30°, then returns in 150° with simple harmonic motion and dwells for the remaining 30°. Draw the displacement diagram.

5.7 A follower rises 2 in in 210° and returns in 150°, both with cycloidal motion. Write a computer program to calculate and plot the displacement diagram.

5.8 A follower rises $\frac{1}{2}$ in in 30° with constant acceleration, rises 1 in in 30° with constant velocity, rises $\frac{1}{2}$ in in 30° with constant deceleration, dwells for 30°, returns 2 in in 180° with parabolic motion, and dwells for 60°. Write a computer program to calculate and plot the displacement diagram.

5.9 A follower rises 2 in in 90° with simple harmonic motion, dwells for 45°, returns in 180° with parabolic motion, and dwells for 45°. Write a computer program to calculate and plot the displacement diagram.

5.10 A follower rises 2 in in 150° with constant velocity, dwells for 30°, returns in 150° with cycloidal motion, and dwells for 30°. Write a computer program to calculate and plot the displacement diagram.

5.11 A follower rises 40 mm in 150° with simple harmonic motion, dwells for 90°, and returns 40 mm in 120° with cycloidal motion. Write a computer program to calculate and plot the displacement diagram.

5.12 A follower rises 40 mm in 120° with parabolic motion, returns 20 mm in 90° with simple harmonic motion, dwells for 30°, returns the final 20 mm in 90° with cycloidal motion, and dwells for 30°. Write a computer program to calculate and plot the displacement diagram.

For Problems 5.13 Through 5.16

Graphically lay out the profile of a disk cam for clockwise rotation of the cam. The base circle diameter of the cam is to be 100 mm. For those problems involving roller followers, the roller diameter is to be 25 mm.

5.13 Lay out the cam described in Problem 5.1. Use a translating, radial roller follower.

5.14 Lay out the cam described in Problem 5.2. Use a translating, radial roller follower. Determine the maximum pressure angle.

5.15 Lay out the cam described in Problem 5.2. Use a translating flat-faced follower.

5.16 Lay out the cam described in Problem 5.2. Use a translating roller follower with an offset of 25 mm to the right of the center of the camshaft. Determine the maximum pressure angle.

• **5.17** Write a computer program to generate the disk cam for the motion described in Problem 5.3. Use a translating, radial roller follower. The cam rotation is to be clockwise. The base circle diameter is 100 mm, and the roller diameter is 25 mm. Determine the maximum pressure angle.

5.18 Write a computer program to generate the disk cam for the motion described in Problem 5.4. Use a translating, radial roller follower. The cam rotation is to be clockwise. The base circle diameter is 100 mm, and the roller diameter is 25 mm. Determine the maximum pressure angle.

For Problems 5.19 Through 5.23

Write a computer program to generate the profile of a disk cam for clockwise rotation of the cam. The base circle diameter of the cam is to be 4 in. For those problems involving roller followers, the roller diameter is to be 1 in.

5.19 Lay out the cam described in Problem 5.5. Use a translating, radial roller follower.

5.20 Lay out the cam described in Problem 5.6. Use a translating roller follower with an offset of 1 in to the right of the center of the camshaft. Determine the maximum pressure angle.

5.21 Lay out the cam described in Problem 5.7. Use a translating flat-faced follower.

5.22 Lay out the cam described in Problem 5.8. Use a translating flat-faced follower with an offset of 1 in to the right of the center of the camshaft.

5.23 Lay out the cam described in Problem 5.9. Use a translating flat-faced follower.

5.24 Design a cam for an oscillating, pivoted flat-faced follower (like that shown in Figure P5.1) to provide the following motion sequence: the follower rotates clockwise for 15° with simple harmonic motion in 150° of cam rotation, dwells for 30°, returns with cycloidal motion in 150° of cam rotation, and dwells for 30°.

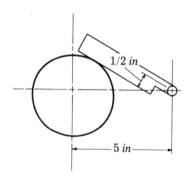

Base circle diam. = 4 in   **Figure P5.1**

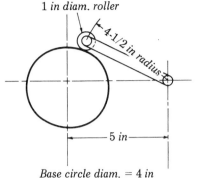

1 *in diam. roller*

4-1/2 *in radius*

5 *in*

Base circle diam. = 4 *in*   **Figure P5.2**

5.25 Design a cam for a pivoted roller follower (like that shown in Figure P5.2) to provide the following motion sequence: follower rotates clockwise 20° with simple harmonic motion in 150° of cam rotation, dwells for 90°, and returns 20° with parabolic motion in 120° of cam rotation.

For Problems 5.26 Through 5.31

A follower rises 4 *in in* 180° *of cam rotation, and then returns during the next* 180° *of cam rotation. The cam rotates at* 60 *rev/min. For the follower motions given below, determine the following: (a) mathematical expressions for the displacement, velocity, acceleration, and jerk of the follower; (b) the magnitudes and locations of maximum velocity and maximum acceleration of the follower; (c) the follower displacement, velocity, acceleration, and jerk when the cam angle is* 120°; *and (d) the follower displacement, velocity, acceleration, and jerk when the follower displacement is* 3 *in during the rise segment of the motion.*

5.26 Uniform motion.

5.27 Modified uniform motion (with constant velocity from 45 to 135° and from 225 to 315°).

5.28 Simple harmonic motion.

5.29 Cycloidal motion.

5.30 Parabolic motion.

• **5.31** 3-4-5 polynomial motion.

For Problems 5.32 Through 5.36

A follower rises 50 *mm in* 120° *of cam rotation, dwells for* 60°, *returns in* 120°, *and dwells for* 60°. *The cam rotational speed is* 60 *rev/min. For the follower motions given below, determine the following: (a) mathematical expressions for the displacement, velocity, acceleration, and jerk of the follower; and (b) the magnitudes and locations of maximum velocity and maximum acceleration of the follower.*

5.32 Uniform motion.

5.33 Simple harmonic motion.

5.34 Cycloidal motion.

5.35 Parabolic motion.

5.36 3-4-5 polynomial motion.

5.37 A 3-4-5 polynomial cam imparts the following motion to a follower: a rise of 22 mm in 80° of cam rotation, a dwell for 70° of cam rotation, and a return of 22 mm in 80° of cam rotation. Find the maximum follower velocity. The cam speed is 3200 rev/min.

5.38 A cam follower rises 2 in in 180° of cam rotation. The constant acceleration for the first part of the rise is three times as great as the constant deceleration for the second part of the lift period. If the cam is rotating at 300 rev/min, determine the value of the acceleration.

5.39 A disk cam rotates at 375 rev/min. The follower rises $\frac{3}{4}$ in with constant acceleration in 80° of cam rotation, then rises an additional $\frac{3}{4}$ in with constant deceleration in the next 80° of cam rotation. Find the acceleration and maximum velocity of the follower.

5.40 The follower of a disk cam rises $\frac{1}{2}$ in with constant acceleration, rises an additional $\frac{3}{4}$ in with constant velocity, and then rises an additional $\frac{1}{2}$ in with constant deceleration. The cam rotates at 200 rev/min, and the follower has a maximum velocity of 35 in/s. Calculate the acceleration and deceleration of the follower. How many degrees has the cam rotated while this motion is being performed?

5.41 A disk cam is to give its follower a rise through travel h during a cam rotation α. This motion is preceded by a dwell and is followed by a dwell. Derive the displacement, velocity, acceleration, and jerk expressions that will satisfy the boundary conditions for all four of these characteristics at the beginning and end of the travel.

For Problems 5.42 Through 5.46

Write a computer program to calculate the displacement, velocity, acceleration, and jerk of a follower for the motion types given. During the motion segment, the cam rotates from angle ANG1 to angle ANG2, and the follower moves from position S1 to position S2 = S1 + H. Note that travel H may be positive for a rise motion, negative for a return motion, or zero for a dwell. Also, the follower motion may be either translation or rotation. Check out the program for the case of a follower that is to have a translational displacement from 1 to 3 in during a cam rotation from 30 to 150°.

5.42 Uniform motion.

5.43 Simple harmonic motion.

5.44 Cycloidal motion.

5.45 Parabolic motion.

5.46 3-4-5 polynomial motion.

5.47 A family of circles has centers along a straight line passing through the origin of an xy coordinate system at a clockwise angle of 45° from the x axis. The radius of each circle is equal to one-half of the distance from the origin to the center of the circle. Determine the envelope for this family of curves:

(a) As a function relating x and y

(b) In parametric form, with x and y as functions of a parameter λ

5.48 A translating flat-face follower is to move through a distance h with cycloidal motion during 180° of clockwise cam rotation. Determine expressions for the x

and y coordinates of that portion of a cam profile that will produce this motion. For a travel of 50 mm and a base circle radius of 100 mm, calculate and plot the cam profile. Based on this portion of the cam design, what is the minimum allowable width of the follower face?

5.49 Repeat Problem 5.48 for uniform motion.

5.50 Repeat Problem 5.48 for parabolic motion.

5.51 Design a disk cam to produce the following motion of a translating, radial roller follower: a rise through a distance of 50 mm with simple harmonic motion during 180° of rotation, followed by a return, also with simple harmonic motion, during the remaining 180° of cam rotation. The base circle radius is 100 mm and the roller radius is 25 mm. Determine the pressure angle corresponding to cam angles of 30, 60, and 90°.

5.52 Determine an expression for the curvature of the cam profile segment synthesized in Problem 5.48. Find the minimum and maximum curvature values.

5.53 Determine an expression for the curvature of the cam profile segment synthesized in Problem 5.49. Find the minimum and maximum curvature values.

5.54 Determine an expression for the curvature of the cam profile segment synthesized in Problem 5.50. Find the minimum and maximum curvature values.

5.55 Determine an expression for the curvature of the cam profile segment synthesized in Problem 5.51. Find the minimum and maximum curvature values.

Spur Gears: Design and Analysis

6.1 BASIC CONSIDERATIONS

This chapter is devoted to a discussion of the power-transmitting machine element known as the spur gear. The terminology, kinematics, and force analysis of spur gears will be presented.

There are a variety of machine elements available for transmitting power from one shaft to another. If the distance between the shafts is relatively small, a pair of friction wheels (rolling cylinders) may be used. In Figure 6.1, which shows the cross-sectional view of such a pair, wheel 1 is the driver, while wheel 2 is the follower.

The force that cylinder 1 can transmit to cylinder 2 depends on the friction that can be developed between the two cylinders. Assuming that the frictional resistance between the two wheels is sufficiently large to prevent slipping of one cylinder relative to the other, the following kinematic relationship holds:

$$v_P = r_1\omega_1 = r_2\omega_2$$

or

$$\frac{\omega_1}{\omega_2} = \frac{r_2}{r_1}$$

(6.1)

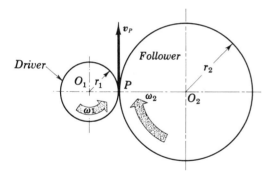

Figure 6.1 Two friction wheels (external cylinders). As shown, cylinder 1 rotates counterclockwise, driving cylinder 2 in the clockwise direction. If no slipping occurs, the velocity of point P on cylinder 1 is equal to the velocity of point P on cylinder 2.

where v_P is the instantaneous velocity of the point of contact, ω is angular velocity, and r is cylinder radius.

Eq. 6.1 indicates that the angular speed ratio of the cylinders is inversely proportional to the ratio of their radii. Another important fact to be observed from Figure 6.1 is that the rotations of the cylinders are in opposite directions (wheel 1 rotates counterclockwise, while wheel 2 rotates clockwise).

The cylinders shown in Figure 6.1 are external to each other. Figure 6.2 shows a similar situation, except that one of the cylinders is internal to the other. The only difference between the external and internal cylinder pairs is that the direction of rotation for both cylinders of the internal pair is the same. In Figure 6.2, cylinder 1 and cylinder 2 are both rotating counterclockwise.

The big disadvantage of friction wheels is the possibility that slipping may occur between the cylinders. Therefore, when exact angular velocity ratios are required or a constant-phase relationship must be maintained between the driver and follower, a machine element known as a gear is commonly used.

A gear may be thought of as a friction wheel with teeth cut around the circumference. Thus the transmission of motion from one shaft to another is

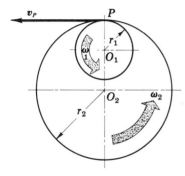

Figure 6.2 Two friction wheels (internal contact). Both the driver (1) and the follower (2) rotate in the same direction. As before, the velocity of point P for both wheels is equal if no slipping occurs.

independent of the frictional resistance between the wheels, and positive-motion transmission can be achieved.

6.2 GEAR TYPES

There are several types of gears in common use. Among the more important types are the following:

1. Spur gears (Figure 6.3a, b, c) and helical gears (Figure 6.4a and b), used when the driver and follower shafts are parallel to each other;

(a)   (b)

(c)

Figure 6.3 (a) An external spur gearset. Spur gears are easily identified by their straight teeth, which are parallel to the gear axis. (*Source:* Boston Gear Works.) (b) An internal spur gear. Internal gears permit a closer positioning of the gear shafts. (*Source:* Richmond Gear, Wallace Murray Corporation.) (c) A spur gear and rack set. This gearset permits precise translations and is frequently used as the drive train in manual controls. (*Source:* Browning Manufacturing Company.)

(a)　　　　　　　　　　　　　　　　　　　　　(b)

Figure 6.4 (a) A single helical gearset. Helical gears provide less shock and smoother, quieter operation than straight spur gears. (*Source:* Browning Manufacturing Company.) (b) Herringbone gear, or double helical gear. In some cases, the presence of end thrust inherent in helical gears is undesirable. Gears with opposing helices neutralize the end thrust of each helix. (*Source:* Horsburgh & Scott Company.)

(a)　　　　　　　　　　　　　　　　　(b)

Figure 6.5 (a) Straight bevel gears. Bevel gears are used in cases where the driver and driven shaft centerlines intersect. (*Source:* Richmond Gear, Wallace Murray Corporation.) (b) Spiral bevel gears. The spiral bevel gear is to the plain bevel gear as the helical gear is to the spur gear. However, the spiral bevel gear has teeth with a circular rather than a helical curvature. (*Source:* Richmond Gear, Wallace Murray Corporation.)

Figure 6.6 The worm and worm gear set. The worm gear is a special helical gear, used where large speed reductions are to be transmitted. (*Source:* Cleveland Gear Company, subsidiary of Vesper Corporation.)

Figure 6.7 Crossed helical gears are used with shafts that are nonparallel and nonintersecting. (*Source:* Browning Manufacturing Company.)

2. Bevel gears (Figure 6.5a, b), used when the shaft axes are intersecting;

3. Worm gears (Figure 6.6) and crossed helical gears (Figure 6.7), used when the shaft axes are nonintersecting and nonparallel.

As mentioned earlier, this chapter will deal with spur gears. The other gear types will be considered in Chapter 7.

6.3 SPUR GEAR TERMINOLOGY

A spur gear can be visualized as a right-circular cylinder that has teeth cut on its circumference parallel to the axis of the cylinder. Its design is the least complicated of gear designs. For this reason, the spur gear offers a convenient

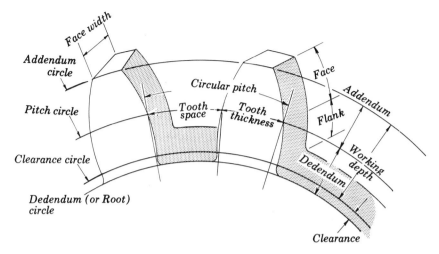

Figure 6.8 Spur gear nomenclature. This figure illustrates some of the more important terms and dimensions associated with spur gears.

starting point for the study of gears, since the terms introduced will also apply to more complex gears discussed in the next chapter. Spur gears, which are used to transmit power between parallel shafts, are the most common type of gear in use. When two gears are in mesh, it is customary to refer to the smaller as the *pinion* and the larger as the *gear*.

The most important parts of gears are identified below and illustrated in Figure 6.8.

> *Pitch circle*: the circle on a gear that corresponds to the contact surface of a friction wheel. Thus, for two gears in contact, the respective pitch circles can be imagined to roll on each other in the same manner as the circles of two friction wheels in contact. A gear may be thought of as similar to a friction cylinder with the face width of the gear equal to the length of the cylinder, and the diameter of the pitch circle of the gear equal to the diameter of the cylinder. The pitch circle location will be defined in more detail in the following section.
>
> *Addendum circle*: the circle drawn through the top of the gear tooth; its center is at the gear center.
>
> *Addendum*: the radial distance from the pitch circle to the addendum circle.
>
> *Root or dedendum circle*: the circle drawn through the bottom of the gear tooth; its center is at the gear center.
>
> *Dedendum*: the radial distance from the pitch circle to the root circle.
>
> *Clearance circle*: the largest circle centered at the gear center, which is not penetrated by the teeth of the mating gear.
>
> *Clearance*: the radial distance from the clearance circle to the root circle. Since the clearance is also equal to the distance between the root of the

tooth and the top of the tooth of the mating gear, it can also be defined as the difference between the dedendum of one gear and the addendum of the mating gear.

Whole depth: the radial distance between the addendum and dedendum circles.

Working depth: the radial distance between the addendum and clearance circles. The working depth is also equal to the sum of the addendums of the two meshing gears.

Circular pitch: the circular pitch *p* is the arc distance *measured along the pitch circle* from a point on one tooth to the corresponding point on the adjacent tooth of the gear. Therefore,

$$p = \frac{\pi d}{N} \tag{6.2}$$

where d is the diameter of the *pitch circle* in inches or millimeters and N is the number of teeth of the gear. It is therefore equal to the circumference of the pitch circle divided by the number of teeth.

Diametral pitch: the diametral pitch P is equal to the number of teeth of a gear divided by the diameter of the pitch circle in inches. Thus,

$$P = \frac{N}{d} \tag{6.3}$$

The circular and diametral pitches are extremely important in gear analysis. The pitch is an indication of the spacing and sizes of the gear teeth. Additionally, in order for two gears to mesh, they must have the same pitch. There are other dimensions of gears that must be equal in order for two gears to mesh, but these will be discussed later.

A simple relationship exists between the circular and diametral pitches:

$$Pp = \pi \tag{6.4}$$

That is, the product of the diametral pitch and the circular pitch is always equal to a constant, namely, π.

Figure 6.9 shows a gauge that illustrates the comparative sizes of gear teeth for standard values of diametral pitch.

In SI units, the module m is used to express gear tooth size rather than the diametral pitch P used in the U.S. customary system. The module is defined as

$$m = \frac{d}{N} \tag{6.5}$$

where d and m have units of millimeters. Clearly, the module is the reciprocal of the diametral pitch; keep in mind, however, that since the diametral pitch is a function of inches, and the module is a function of millimeters, a unit

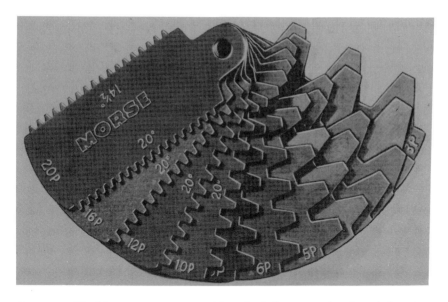

Figure 6.9 The Morse gear gauge provides a basis for comparing the relative sizes of gear teeth for standard values of diametral pitch. (*Source:* Morse Chain Division, Borg Warner.)

conversion is necessary in using one to replace the other. Obviously,

$$m = \frac{25.4}{P} \tag{6.6}$$

The circular pitch p, in millimeters, is

$$p = \frac{\pi d}{N} = \pi m \tag{6.7}$$

◆ **EXAMPLE PROBLEM 6.1**   *Spur Gear Properties*

A spur gear, having 32 teeth and a diametral pitch of 4, is rotating at 400 rev/min. Determine its circular pitch and its pitch line velocity.

Solution.   Since we know the diametral pitch, the circular pitch can be obtained directly from Eq. 6.4:

$$p = \frac{\pi}{P} = \frac{\pi}{4} = 0.7854 \text{ in}$$

To find pitch line velocity (equal to $r\omega$), we will first have to find the pitch diameter of the gear. From Eq. 6.3,

$$d = \frac{N}{P} = \frac{32}{4} = 8 \text{ in}$$

Converting angular velocity of the gear from revolutions per minute to radians per second, we have

$$\omega = 400 \text{ rev/min}\left(\frac{2\pi \text{ rad}}{1 \text{ rev}}\right)\left(\frac{1 \text{ min}}{60 \text{ s}}\right) = 41.9 \text{ rad/s}$$

Finally, the pitch line velocity is the product of the pitch circle radius and the angular velocity of the gear, Eq. 6.1:

$$v_p = r\omega = \frac{d\omega}{2} = \frac{8}{2} \text{ in } (41.9 \text{ rad/s}) = 167.6 \text{ in/s} \quad \blacklozenge$$

♦ **EXAMPLE PROBLEM 6.2** *Spur Gear Properties*
Repeat Example Problem 6.1 for a gear manufactured with a module of 1.5 mm rather than a diametral pitch of 4.

Solution. From Eq. 6.7,

$$p = \pi m = \pi(1.5) = 4.71 \text{ mm}$$

From Eq. 6.5,

$$d = mN = 1.5(32) = 48 \text{ mm}$$

Finally,

$$v_p = r\omega = \frac{48}{2} \text{ mm}(41.9 \text{ rad/s}) = 1005.6 \text{ mm/s} \quad \blacklozenge$$

Backlash

If tooth spaces were exactly equal to tooth thicknesses, it would be extremely difficult for the gears to mesh. Any inaccuracies in manufacturing would cause the gears to jam. It is also very often necessary to lubricate gears. For these reasons, space must be provided between the meshing teeth. This is accomplished by making the tooth thickness less than the tooth space. The difference between tooth space and tooth thickness is known as *backlash*.

Backlash, which is measured on the pitch circle, is then equal to the distance between the nondriving side of a tooth and the side of the corresponding tooth of the meshing gear. If one of a pair of meshing gears is held stationary, the amount of backlash is then proportional to the angle the other gear can be rotated through. Figure 6.10 shows the backlash between two gears.

The cutting tool used to manufacture gears can be set further into the gear blank, thus decreasing the tooth thickness and increasing the tooth space. This is the most common method of providing backlash for gears. Slight variations in backlash can also be obtained by changing the center distance of the gears.

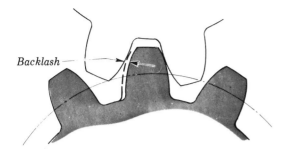

Figure 6.10 Backlash between two meshing teeth is described as proportional to the angle the driving gear can be rotated through before contacting the mating face of the driven gear. (*Source:* Precision Industrial Components Corporation, Wells-Benrus Corporation.)

It should be emphasized that, while some backlash is necessary, too much backlash can result in large shock loads. Excessive backlash will also result in inaccurate gear motion.

6.4 FUNDAMENTAL LAW OF GEARING

An important reason for the use of gears is to maintain a constant angular velocity ratio. The fundamental law of gearing states the condition that the gear tooth profiles must satisfy in order to maintain a constant angular velocity ratio. The law may be stated as follows: *The shape (profile) of the teeth of a gear must be such that the common normal at the point of contact between two teeth always passes through a fixed point on the line of centers of the gears.* The fixed point is called the *pitch point*. When the fundamental law is satisfied, the gears in mesh are said to produce *conjugate action*.

In Figure 6.11, O_1 and O_2 are the centers of the two gears in mesh, r_1 and r_2 are the radii of the pitch circles, P is the pitch point, and A is the point at which the gears are in contact.

Before proceeding with the discussion of conjugate action, we must define the velocity ratio. The velocity ratio is equal to the angular speed (ω) of the *follower* (driven gear) divided by the angular speed of the driving gear. However, the ratio can also be defined in terms of revolutions per minute, pitch radii, and the number of gear teeth.

In the following equations, the subscript 1 refers to the driver, and subscript 2 refers to the follower or driven gear:

$$r_v = \frac{\omega_2}{\omega_1} = \frac{n_2}{n_1} = \frac{r_1}{r_2} = \frac{N_1}{N_2} \tag{6.8}$$

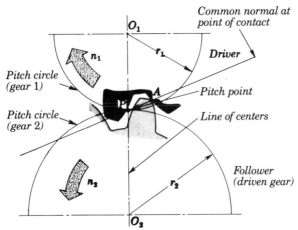

Figure 6.11 Two gears are shown in mesh, with the pitch point at P. The meshing gear teeth are shown in contact at point A. The circles centered at points O_1 and O_2 and passing through the pitch point are the pitch circles.

where r_v = velocity ratio

ω = angular velocity (rad/s)

n = angular velocity (rev/min)

r = pitch circle radius

N = number of teeth

The velocity ratio is less than one when the pinion is the driving gear (the usual case) and greater than one when the gear is the driver.

♦ **EXAMPLE PROBLEM 6.3** *Analysis of a Spur Gearset*

Two spur gears have a velocity ratio of $\frac{1}{4}$. The driven gear has a module of 6 mm, 96 teeth, and rotates at 500 rev/min. Determine the revolutions per minute of the driver, the number of teeth of the driver, and the pitch line velocity.

Solution. The angular velocity (in revolutions per minute) is obtained directly from the velocity ratio, Eq. 6.8:

$$r_v = \frac{n_2}{n_1}$$

Thus,

$$n_1 = \frac{n_2}{r_v} = \frac{500}{1/4} = 2000 \text{ rev/min}$$

The number of teeth on the driver also follows directly from the velocity ratio:

$$r_v = \frac{N_1}{N_2}$$

$$N_1 = r_v N_2 = \tfrac{1}{4}(96) = 24 \text{ teeth}$$

Pitch line velocity is given by Eq. 6.1, the product of pitch circle radius and angular velocity, just as in Example Problem 6.2:

$$r_2 = \frac{d_2}{2} = \frac{mN_2}{2} = \frac{6(96)}{2} = 288 \text{ mm}$$

$$\omega_2 = n_2\left(\frac{2\pi}{60}\right) = 500 \text{ rev/min}\left(\frac{2\pi \text{ rad}}{1 \text{ rev}}\right)\left(\frac{1 \text{ min}}{60 \text{ s}}\right) = 52.3 \text{ rad/s}$$

and

$$v_p = r_2\omega_2 = 288 \text{ mm}(52.3 \text{ rad/s}) = 15{,}062 \text{ mm/s}$$

Check: Since v_p is also $= r_1\omega_1$, and

$$r_v = \frac{r_1}{r_2} = \frac{\omega_2}{\omega_1}$$

therefore,

$$r_1 = \tfrac{1}{4}(288 \text{ mm}) = 72 \text{ mm}$$

$$\omega_1 = \frac{\omega_2}{r_v} = \frac{52.3}{1/4} = 209.2 \text{ rad/s}$$

and

$$v_p = r_1\omega_1 = 72 \text{ mm}(209.2 \text{ rad/s}) = 15{,}062 \text{ mm/s} \quad \blacklozenge$$

6.4.1 Conjugate Action and the Involute Curve

Figure 6.12 is a magnified view of the point of contact of two gears. Line *tt* is tangent to each of the two teeth at the point of contact *A*. Line *nn* is perpendicular to *tt* and is the common normal at the point of contact *A*. In

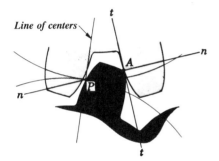

Figure 6.12 Two gears in contact at point *A*. The tangent drawn at the contact point *A* is *tt*, while the normal, *nn*, at point *A* passes through the pitch point *P*. In order for conjugate action to take place, the common normal at the point of contact must always pass through the same pitch point.

order for the fundamental law to be satisfied, *nn* must always pass through a fixed point *P* on the line of centers.

As the gears continue to rotate, other points on the teeth will come into contact. However, for each successive point of contact, the common normal at the contact point must continue to pass through the fixed point *P* in order for conjugate action to take place (to maintain a constant angular velocity ratio).

When gear profiles are cut in such a way as to produce conjugate action, the curves are known as conjugate curves. Most gears are cut using the involute curve to obtain conjugate action. However, for certain specialized applications, gears are sometimes cut using a cycloidal curve as the tooth profile.

The Base Circle of an Involute

Consider a cylinder with a string wrapped around it. An involute curve is the curve traced out by a point on the string as the string is unwrapped from the cylinder. In gear terminology, the cylinder around which the string is wrapped is known as the *base circle*. In order to better understand what the involute curve looks like, consider Figure 6.13a. The base circle, of radius r_b, has a string wrapped around it. Point *A* is the point on the end of the string, while point *B* is the corresponding point on the circle at which the string leaves the circle.

Initially, while the string is still completely wrapped around the base circle, points *A* and *B* coincide. This condition is represented by points A_1 and B_1. As the string is unwrapped, point *A* moves to position A_2, while B_2 is the point at which the string leaves the base circle. As the string is further unwrapped, the positions A_3, B_3, A_4, and B_4 are likewise determined. The curve then drawn through points A_1, A_2, A_3, and A_4 is the involute curve. A special case is the rack, for which the involute tooth profile is a straight line.

An important property possessed by the involute curve is that a normal drawn to it is tangent to the base circle. Referring again to Figure 6.13a, OB_1, OB_2, OB_3, and OB_4 are radii of the base circle, while A_2B_2, A_3B_3, and A_4B_4 are tangent to the base circle and perpendicular to the radii. The distance A_2B_2 is the radius of curvature of the involute at that instant, since point A_2 is rotating about point B_2 at that instant. Similar reasoning shows that A_3B_3 and A_4B_4 are also instantaneous radii of curvature. Therefore, it is clear that the radius of curvature of an involute curve is continuously varying. But, since at any given point on a curve, the radius of curvature is normal to the curve, lines A_2B_2, A_3B_3, and A_4B_4 are normals drawn to the involute curve and are also tangent to the base circle. Thus, the earlier statement, that a normal to the involute curve is tangent to the base circle, is correct.

From the involute geometry depicted in Figure 6.13a, a useful equation can be developed relating the tooth thicknesses at any two arbitrary radial positions on an involute gear tooth. Figure 6.13b shows a gear tooth that has thicknesses t_i and t_j at radial locations R_i and R_j, respectively. The tooth thickness is measured as an arc length along a circle that is centered at the

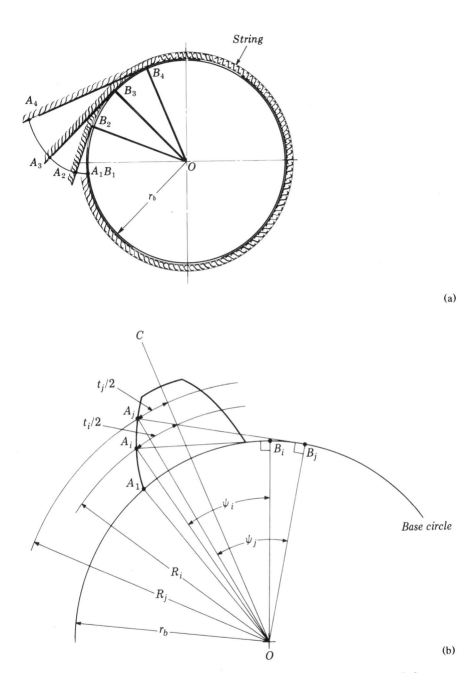

Figure 6.13 (a) Involute curve generation. As a string is unwrapped from a cylinder, the curve traced by a point on the string is an involute curve (A_1, A_2, A_3, A_4). The tangents to the base circle are the instantaneous radii of curvature of the involute. (b) A gear tooth with an involute profile generated by the process depicted in Figure 6.13a.

gear center. Thus, from the figure,

$$\frac{t_i}{2} = R_i |\angle A_i OC|$$

and

$$\frac{t_j}{2} = R_j |\angle A_j OC|$$

where $\angle A_i OC$ is the angle corresponding to $\text{arc}(t_i/2)$ and $\angle A_j OC$ is the angle corresponding to $\text{arc}(t_j/2)$. Combining these equations, we have

$$\frac{t_i}{2R_i} - \frac{t_j}{2R_j} = \angle A_i OC - \angle A_j OC = \angle A_i OA_j$$

But

$$\angle A_i OA_j = \angle A_1 OA_j - \angle A_1 OA_i$$

Therefore

$$\frac{t_i}{2R_i} - \frac{t_j}{2R_j} = \angle A_1 OA_j - \angle A_1 OA_i$$

Now we wish to determine alternative expressions for the angles $\angle A_1 OA_j$ and $\angle A_1 OA_i$. This can be accomplished by using the involute properties that were described above. In particular, we see from Figure 6.13b that

$$\angle A_1 OA_j = \angle A_1 OB_j - \angle A_j OB_j = \angle A_1 OB_j - \psi_j$$

where the angle ψ_j is referred to as the involute angle and is defined in the figure. Also,

$$\angle A_1 OB_j = \frac{\text{arc } A_1 B_j}{r_b}$$

Now recall that, based on the string analogy, the arc length $A_1 B_j$ is equal to the distance $A_j B_j$ (see Figure 6.13a). Therefore,

$$\angle A_1 OB_j = \frac{A_j B_j}{r_b}$$

But from Figure 6.13b,

$$\tan \psi_j = \frac{A_j B_j}{r_b}$$

which leads to

$$\angle A_1 OB_j = \tan \psi_j$$

Combining equations, we have

$$\angle A_1 OA_j = \tan \psi_j - \psi_j = \text{inv } \psi_j$$

wherein the involute function, inv ψ, is defined (note that the angle ψ must be expressed in radians). In a similar manner,

$$\angle A_1 OA_i = \tan \psi_i - \psi_i = \text{inv } \psi_i$$

Finally, combining equations, we obtain the desired relationship between tooth thicknesses:

$$\frac{t_i}{2R_i} - \frac{t_j}{2R_j} = \text{inv } \psi_j - \text{inv } \psi_i \tag{6.9a}$$

where, from Figure 6.13b,

$$\cos \psi_i = \frac{r_b}{R_i} \tag{6.9b}$$

and

$$\cos \psi_j = \frac{r_b}{R_j} \tag{6.9c}$$

Eqs. 6.9a to 6.9c can be used to determine the tooth thickness at any location in terms of that at another location, as, for example, at the pitch circle for which standard values are given later in this chapter.

It is now appropriate to consider the action that occurs when two gear teeth, cut with involute curve profiles, are in contact.

In Figure 6.14, gear 1 is the driver and is rotating clockwise, while gear 2, the follower, rotates counterclockwise. The distance c in the diagram is called the *center distance* and represents the spacing between the centers of the shafts upon which the gears are mounted. The following equation may be used to determine the center distance:

$$c = \frac{d_1 + d_2}{2} \tag{6.10}$$

where d_1 and d_2 are the diameters of the pitch circles. In practice, the center distance between two shafts and the speed ratio are usually specified. With this information, it is possible to determine the required pitch diameters. Since $d = N/P$ in English units and $d = mN$ in SI units, c can also be found by

$$c = \frac{1}{2P}(N_1 + N_2) \tag{6.11a}$$

where c is in inches, and

$$c = \frac{m(N_1 + N_2)}{2} \tag{6.11b}$$

where c is in millimeters.

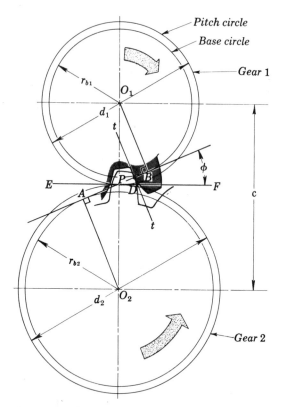

Figure 6.14 Two gears in contact. The center distance c is equal to one-half the sum of the pitch diameters. As the gears continue to rotate, the other points of contact must have their common normal passing through the pitch point P. For involute profiles, all contact points lie on line AB, called the *line of action*. Line AB is also called the *pressure line*, and ϕ is referred to as the *pressure angle*.

6.4.2 The Line of Action

When two curves are in contact at a point, they must have the same tangent and normal at the point of contact (Figure 6.12). In Figure 6.14, point D is the point of contact between the two involute curves to which the teeth of gears 1 and 2 have been cut. The common tangent to the tooth surfaces is tt, while line AB, which is perpendicular to tt at the point of contact, is the common normal.

According to the fundamental law of gearing, this common normal must pass through a fixed point on the line of centers $O_1 O_2$. But according to the properties of the involute curve discussed previously, the normal to each of the two involutes at the point of contact is tangent to the respective generating, or base, circle. It follows that the common normal must be simultaneously tangent to both of the base circles and therefore is a unique line. This means that as the gears continue to rotate and other points become contact points, these contact points will always lie on line *AB*, which will always be the common normal. Therefore, the intersection *P* of the common normal and the line of centers is a fixed point, and the gears have conjugate action.

The line *AB* is often called the *line of action* because the contact points of two gears in mesh must lie along it. The force that one gear tooth exerts on the tooth of the meshing gear acts along the common normal, which is also line *AB*. Therefore, *pressure line* is another name commonly given to line *AB*.

6.4.3 The Pressure Angle

The angle ϕ, between the pressure line *AB* and the common tangent to the pitch circles *EF* in Figure 6.14, is known as the *pressure angle*. The pressure line is located by rotating the common tangent to the pitch circles, *EF*, through the angle ϕ in a direction opposite to the direction of rotation of the driver. Referring again to Figure 6.14, since gear 1, the driver, is rotating clockwise, the pressure line *AB* is located by rotating the common tangent *EF* counterclockwise through the angle ϕ.

While gears may be manufactured with a wide range of pressure angles, most gears are made with standard angles of 20 or 25°. Gears with $14\frac{1}{2}^\circ$ pressure angles were once prevalent, but they are now almost obsolete. Although gears with pressure angles of $14\frac{1}{2}^\circ$ are still manufactured, they are mainly replacements for older gear trains still in use.

Although gears are designated by their pressure angle, it should be emphasized that the actual pressure angle between two gears in contact may differ from the designated value. Changes in center distance *c* will result in corresponding changes of the actual pressure angles. In other words, two nominally 20° gears actually may have a slightly larger pressure angle by increasing their center distance.

Example Problem 6.6 will illustrate the difference between designated and actual pressure angles. However, at this point, a diagram can illustrate the effect of increasing the center distance. In Figure 6.15a, gears 1 and 2 have their centers at O_1 and O_2. The pressure angle is 20°, and the base circle radii are r_{b_1} and r_{b_2}. The center distance is then increased so that the centers of the gears are at O_1' and O_2'. Figure 6.15b shows the new situation with the increased center distance and unchanged base circle radii. As in Figure 6.15a, the pressure line is drawn tangent to the base circles and through the pitch point. As can be seen from the illustration, the pressure angle is now increased to approximately 35°. The increase in the pressure angle has been made much larger than would normally be the case, for purposes of illustration.

In order to obtain a better understanding of gear tooth action, consider Figure 6.16. Two gears, 1 and 2, are shown in mesh. The pitch radii r_1 and r_2, as well as the base circle radii r_{b_1} and r_{b_2}, are shown. The base circles were determined by drawing circles tangent to the pressure line *AB*. Therefore, the

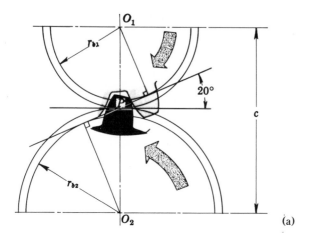

(a)

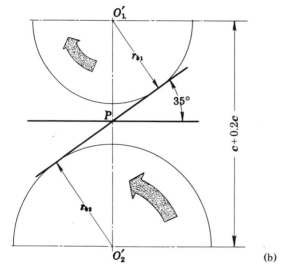

(b)

Figure 6.15 (a) Two gears in mesh showing center distance c, pressure angle of 20°, and base circle radii r_{b_1} and r_{b_2}. (b) The center distance of the gears in Figure 6.15a is shown increased by 20 percent. The base radii remain unchanged; however, the pressure angle is now increased to 35°.

radii r_{b_1} and r_{b_2} are each perpendicular to the pressure line at points D and C, respectively.

By considering the right triangles O_1PD and O_2PC, a simple relationship between the pitch circle radius and the base circle radius is seen to exist:

$$\cos \phi = \frac{r_{b_1}}{r_1} = \frac{r_{b_2}}{r_2} \text{ or } r_b = r \cos \phi \qquad (6.12)$$

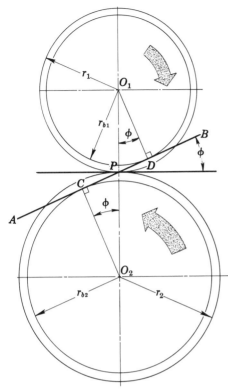

Figure 6.16 This illustration shows the relationship between the base circle radii, the pressure angle, and the pitch circle radii: $r_b = r \cos \phi$.

◆ **EXAMPLE PROBLEM 6.4** *Analysis of a Gear Pair*
Two meshing spur gears with diametral pitch of 4, pressure angle of 20°, and a velocity ratio of $\frac{1}{4}$ have their centers 15 in apart. Determine the number of teeth of the driver (pinion) and the base circle radius of the gear.

Solution. As usual, subscript 1 will refer to the pinion and subscript 2 will refer to the gear. From the velocity ratio, we know that

$$r_v = \frac{n_2}{n_1} = \frac{N_1}{N_2} = \frac{1}{4}$$

or

$$N_2 = 4N_1$$

We can set up a second equation in N_1 and N_2 (giving us two equations in two unknowns) and determine N_1 and N_2. Knowing the diametral pitch P and the distance between centers c, we can use Eq. 6.11a:

$$15 = \frac{1}{2 \times 4}(N_1 + N_2)$$

But $N_2 = 4N_1$; therefore,

$$120 = N_1 + 4N_1 = 5N_1$$

$$N_1 = 24 \text{ teeth}$$

To solve for the base circle radius of the gear, we simply solve for r_1 and then use the velocity ratio and pressure angle formulas to find r_{b_2}:

$$d_1 = \frac{N_1}{P} = \frac{24}{4} = 6 \text{ in}$$

Therefore, $r_1 = 3$ in, and

$$r_2 = \frac{r_1}{n_2/n_1} = \frac{3}{1/4} = 12 \text{ in}$$

And from Eq. 6.12, the base circle radius, r_{b_2}, is given as

$$r_{b_2} = r_2 \cos \phi = 12 \cos 20° = 11.3 \text{ in} \quad \blacklozenge$$

♦ **EXAMPLE PROBLEM 6.5** *Analysis of a Gear Pair*
Repeat Example Problem 6.4 for a module of 3 mm and a center distance of 180 mm.

Solution.

$$r_v = \frac{N_1}{N_2} = \frac{1}{4}$$

Therefore,

$$N_2 = 4N_1$$

From Eq. 6.11b,

$$180 = \frac{3(N_1 + 4N_1)}{2}$$

Therefore, $N_1 = 24$ teeth and $N_2 = 96$ teeth.
From Eq. 6.5,

$$d_2 = mN_2 = 3(96) = 288 \text{ mm}$$

$$r_2 = \frac{d_2}{2} = \frac{288}{2} = 144 \text{ mm}$$

From Eq. 6.12,

$$r_{b_2} = r_2 \cos \phi = 144 \cos 20° = 135.3 \text{ mm} \quad \blacklozenge$$

♦ **EXAMPLE PROBLEM 6.6** *Pressure Angle Determination*
Two 20° gears have a diametral pitch of 4. The pinion has 28 teeth, while the gear has 56 teeth. Determine the center distance for an actual pressure angle of 20°. What is the actual pressure angle if the center distance is increased by 0.2 in?

Solution. For the data given, the center distance is found by using Eq. 6.3 to obtain the pitch circle diameters, and then using these values in Eq. 6.10:

$$d_1 = \frac{N_1}{P} = \frac{28}{4} = 7 \text{ in}$$

$$d_2 = \frac{N_2}{P} = \frac{56}{4} = 14 \text{ in}$$

Thus,

$$c = \frac{1}{2}(d_1 + d_2) = \frac{1}{2}(7 + 14) = 10.5 \text{ in}$$

The base circle radius was determined when the gears were cut, and changing the center distance does not change the base circle radius. *However, increasing the center distance does increase the pitch radius, which in turn results in a larger pressure angle.* To find the actual pressure angle resulting from an increased center distance, we must first find the base circle radius and the new pitch radius. From Eq. 6.12,

$$r_{b_1} = r_1 \cos \phi = 3.5 \cos 20° = 3.29 \text{ in}$$

The new center distance c' is

$$10.5 + 0.2 = 10.7 \text{ in}$$

The new pitch radii, r_1' and r_2', although changed numerically, will maintain the same proportion held by the original pitch radii. Thus, from Eq. 6.8,

$$r_v = \frac{N_1}{N_2} = \frac{r_1}{r_2} = \frac{r_1'}{r_2'} = \frac{1}{2}$$

or

$$r_2' = 2r_1'$$

But

$$c' = r_1' + r_2' = 10.7$$

Therefore,

$$r_1' + 2r_1' = 10.7$$

$$r_1' = 3.57 \text{ in}$$

Since base circle radius does not change, we can now finally calculate the new pressure angle using Eq. 6.12:

$$r_{b_1} = r_1' \cos \phi$$

$$\cos \phi = \frac{r_{b_1}}{r_1'} = \frac{3.29}{3.57} = 0.922$$

or

$$\phi = 22.8° \quad \blacklozenge$$

6.4.4 Contact Length

As was stated earlier, all the points of contact between two gear teeth with involute profiles lie along the pressure line. The initial contact between two teeth will occur when the tip of the driven gear tooth is acted on by the flank of the driver tooth. The final point of contact will be at the flank of the driven gear tooth and the tip of the driver tooth. Another way of describing the interval of contact is to say that initial contact occurs where the addendum circle of the driven gear intersects the pressure line, and final contact occurs at the point where the addendum circle of the driver intersects the pressure line.

Figure 6.17 illustrates the important points of the previous discussion. Initial contact occurs at point E, which is the intersection of the addendum circle of the driven gear and the pressure line. Point F, the intersection of the

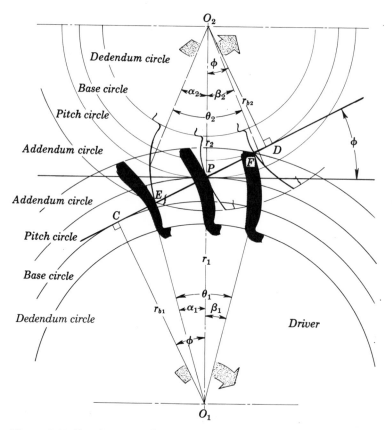

Figure 6.17 For the contacting gear teeth shown, initial contact occurs at point E and final contact occurs at point F. Line EF is the length of the line of action. Angles θ_1 and θ_2 are the angles of action, α_1 and α_2 are the angles of approach, and β_1 and β_2 are the angles of recess. The angle of action is equal to the sum of the angle of approach and the angle of recess.

addendum circle of the driver and the pressure line, is the final contact point of the interval during which the teeth contact.

The distance between points E and F is known as the length of the line of action or the contact length.

The angles θ_1 and θ_2 shown in Figure 6.17 are known as the angles of action. They are the angles turned through by the driver and follower gears, respectively, during the contact between a pair of gear teeth. The *arc of action* is the arc, measured on the pitch circle, turned through by a gear as a pair of meshing gear teeth go from initial to final contact. Note that θ_1 and θ_2 subtend the arcs of action.

Angles α_1 and α_2 are known as the angles of approach. The angle of approach is the angle turned through by a gear as measured from the initial contact point to contact at the pitch point.

The angle of recess is defined as the angle turned through by a gear while the contact between the teeth goes from the pitch point to the point of final contact. In Figure 6.17 β_1 and β_2 are the angles of recess. It should be clear from the definitions as well as from the illustration that the angle of action is equal to the sum of the angle of approach and the angle of recess.

6.4.5 Contact Ratio

The circular pitch, as defined in an earlier section, is equal to the distance, measured on the pitch circle, between corresponding points of adjacent teeth. Let γ be the angle determined by the circular pitch AB, as shown in Figure 6.18. The angle γ is known as the *pitch angle*. We now define the *contact ratio* as *the angle of action divided by the pitch angle*, or

$$\text{Contact ratio} = \frac{\theta}{\gamma} = \frac{\alpha + \beta}{\gamma} \qquad (6.13)$$

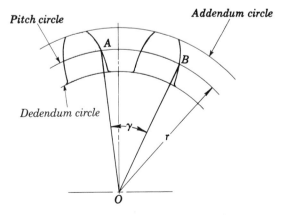

Figure 6.18 The pitch angle γ is simply the angular equivalent of the circular pitch.

If the contact ratio were equal to one, Eq. 6.13 would indicate that the angle of action is equal to the pitch angle. A contact ratio of one means that one pair of teeth are in contact at all times. If the contact ratio were less than one, there would be an interval during which *no* teeth would be in contact. Gears are usually designed with a contact ratio between 1.2 and 1.6. For a contact ratio of 1.2, one pair of teeth are always in contact, and two pairs of teeth are in contact 20 percent of the time. Therefore, the contract ratio is also commonly defined as the average number of teeth in contact.

The contact ratio can be defined in still another way. However, in order to understand this definition, we must first define the base pitch. The *base pitch* p_b *is equal to the distance between corresponding points of adjacent teeth measured on the base circle.* Therefore,

$$p_b = \frac{\pi d_b}{N} \tag{6.14}$$

where d_b is the diameter of the base circle and N is the number of teeth. Since $r_b = r \cos \phi$, it follows that $d_b = d \cos \phi$. Therefore,

$$p_b = \frac{\pi d \cos \phi}{N}$$

But since $P = N/d$,

$$p_b = \frac{\pi \cos \phi}{P}$$

Also, $p = \pi/P$; therefore,

$$p_b = p \cos \phi \tag{6.15}$$

In SI units,

$$p_b = m\pi \cos \phi$$

The contact ratio can now also be defined as the ratio of the length of action divided by the base pitch. Referring again to Figure 6.17, the line of action is equal to the distance between points E and F. This definition is valid since the length of action, EF, is the same distance the base circle rolls through as the involute is being generated. Therefore,

$$\text{Contact ratio} = \frac{EF}{p_b} \tag{6.16}$$

The length of action can be computed from the following formulas:

$$EF = EP + PF, \; EP = ED - PD, \; PF = CF - CP$$

To obtain EF, we must determine component lengths EP and PF. In right triangle O_2DE, length O_2E is equal to the sum of the pitch radius and the addendum of driven gear 2:

$$O_2E = r_2 + a_2$$

O_2D is the base radius of gear 2 and is given by

$$O_2D = r_{b_2} = r_2 \cos \phi$$

We can therefore determine length ED by using the Pythagorean theorem:

$$ED = \sqrt{(O_2E)^2 - (O_2D)^2}$$
$$= \sqrt{(r_2 + a_2)^2 - r_2^2 \cos^2 \phi}$$

Also,

$$PD = r_2 \sin \phi$$

Therefore,

$$EP = ED - PD = \sqrt{(r_2 + a_2)^2 - r_2^2 \cos^2 \phi} - r_2 \sin \phi$$

The distance PF is found in a similar manner. In triangle O_1CF, length O_1F equals the sum of the pitch radius and addendum of the driver:

$$O_1F = r_1 + a_1$$

O_1C, the base radius of the driver, is given as

$$O_1C = r_{b_1} = r_1 \cos \phi$$

As for ED, CF is found by using the Pythagorean theorem:

$$CF = \sqrt{(O_1F)^2 - (O_1C)^2}$$
$$= \sqrt{(r_1 + a_1)^2 - r_1^2 \cos^2 \phi}$$

Also,

$$CP = r_1 \sin \phi$$

Therefore,

$$PF = CF - CP = \sqrt{(r_1 + a_1)^2 - r_1^2 \cos^2 \phi} - r_1 \sin \phi$$

Finally, we obtain the length of the line of action solely in terms of the pitch radii, the addendums, and the common pressure angle:

$$EF = EP + PF$$
$$= \sqrt{(r_2 + a_2)^2 - r_2^2 \cos^2 \phi} - r_2 \sin \phi$$
$$+ \sqrt{(r_1 + a_1)^2 - r_1^2 \cos^2 \phi} - r_1 \sin \phi$$

Thus, the formula for the contact ratio, Eq. 6.16, becomes

$$\text{Contact ratio} = \frac{\sqrt{(r_2 + a_2)^2 - r_2^2 \cos^2 \phi} \; - r_2 \sin \phi}{p_b}$$

$$+ \; \frac{\sqrt{(r_1 + a_1)^2 - r_1^2 \cos^2 \phi} \; - r_1 \sin \phi}{p_b} \tag{6.17}$$

◆ **EXAMPLE PROBLEM 6.7**   *Contact Ratio Determination*

Two 25° full-depth spur gears have a velocity ratio of 1/3. The diametral pitch is 5 and the pinion has 20 teeth. Determine the contact ratio (number of teeth in contact) for the gears. (The formula for the addendum of a 25° full-depth gear is given in Table 6.1 in Section 6.6 as $a = 1/P$.)

Solution.   Before using Eq. 6.17 to determine the contact ratio, we will first have to find the following values: r_1, r_2, a_1, a_2, p_b, sin ϕ, and cos ϕ. Thus, from Eq. 6.3,

$$d_1 = \frac{N_1}{P} = \frac{20}{5} = 4 \text{ in or } r_1 = 2 \text{ in}$$

And from the velocity ratio equation,

$$r_v = \frac{n_2}{n_1} = \frac{r_1}{r_2} = \frac{1}{3} \text{ or } r_2 = 6 \text{ in}$$

The addendum is given by

$$a_1 = a_2 = \frac{1}{P} = \frac{1}{5} = 0.2 \text{ in}$$

And

$$\cos \phi = \cos 25° = 0.906$$

$$\sin \phi = \sin 25° = 0.423$$

The last of the unknowns is the base pitch. From Eq. 6.4, $Pp = \pi$,

$$p = \frac{\pi}{P} = \frac{\pi}{5} = 0.628 \text{ in}$$

and from Eq. 6.15,

$$p_b = p \cos \phi = 0.628(0.906) = 0.57 \text{ in}$$

Thus the contact ratio formula, Eq. 6.17, is solved with the values determined above:

$$\text{Contact ratio} = \frac{\sqrt{(r_2 + a_2)^2 - r_2^2 \cos^2 \phi} - r_2 \sin \phi}{p_b}$$

$$+ \frac{\sqrt{(r_1 + a_1)^2 - r_1^2 \cos^2 \phi} - r_1 \sin \phi}{p_b}$$

$$= \frac{\sqrt{(6 + 0.2)^2 - 6^2(0.906)^2} - 6(0.423)}{0.57}$$

$$+ \frac{\sqrt{(2 + 0.2)^2 - 2^2(0.906)^2} - 2(0.423)}{0.57}$$

$$= \frac{\sqrt{38.44 - 29.55} - 2.54 + \sqrt{4.84 - 3.28} - 0.85}{0.57}$$

$$= \frac{0.84}{0.57} = 1.47 \quad \blacklozenge$$

6.5 INTERNAL GEARS

There are many applications, such as in epicyclic gear trains (to be discussed in a later chapter), for which the use of an internal gear is highly desirable. An internal gear has its teeth cut on the inside of the rim rather than on the outside.

Figure 6.19 shows a typical internal gear in mesh with an external pinion. The important terms and dimensions associated with internal gears are shown. As can be seen from the illustration, the directions of rotation for an internal and external gear in mesh are the same, whereas two external gears in contact have opposite directions of rotation.

Since the internal gear has a concave tooth profile, while the external gear's tooth profile is convex, the surface contact between the gears is increased, thus decreasing the contact stress. The center distance between the gears is less, thus making a more compact arrangement than for external gearsets. Internal-external gearsets also have a greater number of teeth in contact, resulting in smoother and quieter operation.

6.6 STANDARD GEARS

It is economically desirable to standardize gears so that they can be interchanged. Gears with any number of teeth, but having the same pitch and pressure angle, are interchangeable if their tooth profiles have been cut to the same standard tooth system. Thus standard tooth systems have specified values for addendum, dedendum, clearance, and tooth thickness.

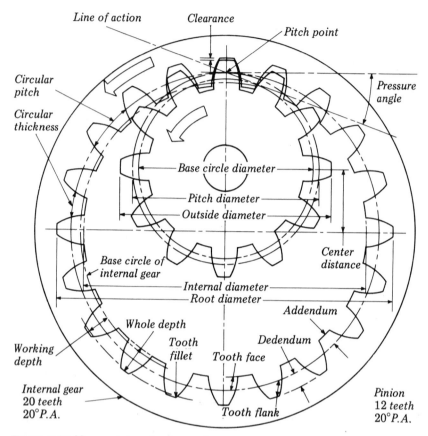

Figure 6.19 Nomenclature for internal gears.

Table 6.1 lists some of the most commonly used standard gear profile systems. The composite system has tooth profiles that are involute curves for a short distance and cycloidal for the remainder.

The standard gear profiles given in Table 6.1 can be expressed in SI units by replacing P by $1/m$. For example, for a 20° full-depth involute (coarse pitch), the addendum $= m$, the dedendum $= 1.25m$, and so on.

Commonly used values of diametral pitch are the following: coarse pitch, 1, $1\frac{1}{4}$, $1\frac{1}{2}$, $1\frac{3}{4}$, 2, $2\frac{1}{4}$, $2\frac{1}{2}$, $2\frac{3}{4}$, 3, $3\frac{1}{2}$, 4, 5, 6, 7, 8, 9, 10, 12, 14, 16, 18; fine pitch, 20, 22, 24, 26, 32, 40, 48, 64, 72, 80, 96, 120. Some of the standard values for metric module are as follows: 0.3, 0.5, 1, 1.25, 1.5, 2, 2.5, 3, 4, 5, 6, 8, 10, 12, 16, 20, 25, 32, 40, 50. Obviously, standard values of module do not correspond with standard values of diametral pitch (see Eq. 6.6). Therefore, the two systems are not interchangeable.

TABLE 6.1 STANDARD GEAR PROFILES

System	Addendum	Dedendum	Clearance	Whole depth	Tooth thickness
$14\frac{1}{2}°$ full-depth involute	$\dfrac{1}{P}$	$\dfrac{1.157}{P}$	$\dfrac{0.157}{P}$	$\dfrac{2.157}{P}$	$\dfrac{1.5708}{P}$
$14\frac{1}{2}°$ composite	$\dfrac{1}{P}$	$\dfrac{1.157}{P}$	$\dfrac{0.157}{P}$	$\dfrac{2.157}{P}$	$\dfrac{1.5708}{P}$
20° full-depth involute (coarse pitch)	$\dfrac{1}{P}$	$\dfrac{1.25}{P}$	$\dfrac{0.25}{P}$	$\dfrac{2.25}{P}$	$\dfrac{1.5708}{P}$
20° full-depth involute (fine pitch)	$\dfrac{1}{P}$	$\dfrac{1.2}{P} + 0.002$ in	$\dfrac{0.2}{P} + 0.002$ in	$\dfrac{2.2}{P} + 0.002$ in	$\dfrac{1.5708}{P}$
20° stub-tooth involute	$\dfrac{0.8}{P}$	$\dfrac{1}{P}$	$\dfrac{0.2}{P}$	$\dfrac{1.8}{P}$	$\dfrac{1.5708}{P}$
25° full-depth involute	$\dfrac{1}{P}$	$\dfrac{1.25}{P}$	$\dfrac{0.25}{P}$	$\dfrac{2.25}{P}$	$\dfrac{1.5708}{P}$

◆ **EXAMPLE PROBLEM 6.8** *Standard Gear Calculations*
A 20° full-depth spur gear has 35 teeth and a diametral pitch of 5. Determine the addendum circle diameter, the dedendum circle diameter, and the working depth.

Solution. The addendum and dedendum circle radii are found by adding the addendum to, and subtracting the dedendum from, the pitch circle radius, respectively. Thus, we must find the pitch radius. From Eq. 6.3, the pitch circle diameter is

$$d = \frac{N}{P} = \frac{35}{5} = 7 \text{ in}$$

From Table 6.1, we obtain the expression for the addendum of the gear profile system being considered:

$$\text{Addendum} = \frac{1}{P} = \frac{1}{5} = 0.2 \text{ in}$$

The addendum circle diameter is

$$d_a = d + 2(\text{addendum}) = 7 + 2(0.2) = 7.4 \text{ in}$$

Similarly, the dedendum is given by

$$\text{Dedendum} = \frac{1.25}{P} = \frac{1.25}{5} = 0.25 \text{ in}$$

and the dedendum circle diameter is

$$d_d = d - 2(\text{dedendum}) = 7 - 2(0.25) = 6.5 \text{ in}$$

Working depth = whole depth − clearance. The clearance, from Table 6.1, is

$$\text{Clearance} = \frac{0.25}{P} = \frac{0.25}{5} = 0.05 \text{ in}$$

Or, clearance = dedendum − addendum = 0.25 − 0.20 = 0.05 in.

Whole depth = addendum + dedendum = 0.20 + 0.25 = 0.45 in

Finally,

working depth = whole depth − clearance = 0.45 − 0.05 = 0.40 in

Or, working depth = 2 × addendum = 2(0.2) = 0.4 in. ◆

6.7 GEAR MANUFACTURE

Gears are usually manufactured by one of the following processes: milling, generating, or molding. In the milling method of manufacture, the milling cutter is shaped so as to conform to the shape of the space between the teeth. The cutter is then moved across the face of the gear blank, thus cutting out a space between teeth. The gear blank is then automatically rotated until the

Figure 6.20 The milling cutter shown, like all milling cutters, is completely accurate only for a specific gear of fixed pitch and fixed number of teeth. (*Source:* Horsburgh & Scott Company.)

Figure 6.21 Hobbing a gear. This method of gear production is more accurate than milling since a given hobbing cutter can cut with equal accuracy gears of various tooth numbers. (*Source:* Horsburgh & Scott Company.)

next space to be cut lines up with the cutter. This process is continued until all the spaces have been cut out, completely forming the gear. Figure 6.20 shows a typical milling cutter.

The disadvantage of the milling cutter is that a different cutter must be used not only for different pitches but also for different numbers of teeth. Gear manufacturers usually shape the milling cutters so that they are correct for the gear with the smallest number of teeth, in each of eight ranges of tooth numbers, for a given pitch. This means that when gears having a greater number of teeth are cut with this milling cutter, an error in the tooth profile results. The error increases toward the high end of each range of tooth numbers, but is usually acceptable for most applications.

Gears to be used for high-speed, high-load applications are not cut accurately enough by the milling process. The generating method to be described next should be used whenever high accuracy is required.

The generating process of gear cutting entails the use of either a hob or a shaper. A hob, the cutting tool used in the hobbing process, is shown in Figure 6.21. The cutting process is accomplished by moving the hob across the gear blank parallel to the gear axis as both the gear blank and the hob are rotated. Better accuracy results from this method because of the simultaneous rotation of both the cutter and the gear blank. The cutter acts in much the same manner to the gear being cut, as will the eventual meshing gear. The generating method has the further advantage of not requiring a different cutter for gears of like pitch but different numbers of teeth.

A second method of generating gears is by the shaping process. The cutting tool used in the shaping method is either a rack cutter or a pinion cutter. The rack cutter, which has teeth with straight sides, has its addendum made equal to the dedendum of the gear being cut. The angular orientation of the side of the rack tooth is equal to the pressure angle ϕ of the gear to be cut. Figure 6.22 shows a gear blank and rack cutter. The cutting process is

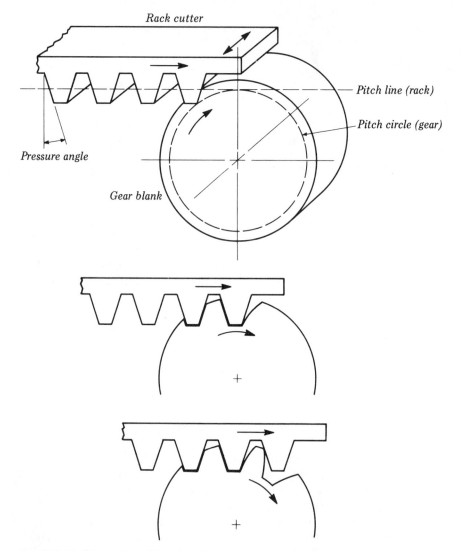

Figure 6.22 Generation of involute gear teeth with a rack cutter. The pitch line of the rack is tangent to the pitch circle of the gear being cut.

Figure 6.23 Generation of a spur gear with a pinion cutter. (*Source:* Fellows Gear Shaper Company.)

started when the gear blank has been moved into the cutter until the pitch circle of the gear blank is tangent to the pitch line of the rack cutter. The cutter is then given a reciprocating motion across the face of the gear blank, as the gear blank slowly rotates and the rack translates. The cutting of the space between two teeth is not accomplished in one pass but rather requires several passes of the cutter parallel to the gear axis. The cutting and rolling action is continued until the end of the rack is reached. At this point, the gear blank and the cutting rack are repositioned, and the rolling and cutting action continued until all the teeth on the gear have been cut.

The pinion cutter, as the name implies, is in the form of a gear rather than a rack. See Figure 6.23. The cutting operation for the pinion cutter is basically the same as that for the rack cutter; both are shaper operations in which the cutting tool passes back and forth across the rotating gear blank. The pinion cutter is the tool used to cut internal gears.

There are two principal advantages to using generating cutters rather than milling cutters. The first is that a much higher degree of accuracy can be obtained in the cutting process. Second, a single cutter can be used to cut gears with any number of teeth of the same pitch.

The third general method of manufacturing gears is molding. While the casting of gears was once a very common method, its present-day application is limited to gears used at very low speeds. Injection molding and die casting are used when a large number of gears are required. Injection molding is used when the material is a plastic, while die casting is often the process used for metals such as brass and aluminum.

If gears are to be used in applications involving high speeds and high loads, it is usually necessary to have a higher degree of accuracy than that obtained from the cutting process. One method of finishing is shaving. In this method, the gear is run with a hard mating gear or a shaving cutter. This results in the removal of small amounts of the surface of the gear.

Another popular method used for finishing gears is grinding. In this method, a form grinder or grinding wheel is used to obtain a high degree of accuracy. Of the two, the grinding method produces the more accurate finish. Other finishing methods include honing, lapping, and burnishing.

6.8 SLIDING ACTION OF GEAR TEETH

In the earlier discussion of friction wheels, the motion between the cylinders was pure rolling, except when the transmitted force was large enough to cause sliding. For gears, however, pure rolling motion occurs only when the contact point between gear teeth occurs at the pitch point (for spur gears, bevel gears, and helical gears on parallel shafts). Every other point of contact along the line of action results in sliding of one tooth on the other. Relative sliding velocity is of importance in the design of gears because sliding is one of the factors contributing to gear tooth wear.

Figure 6.24 shows two teeth in contact at the pitch point. The pitch line velocity, v_p, is given by

$$v_p = r_1\omega_1 = r_2\omega_2$$

Since the coincident contact points, P_1 and P_2, have the same velocity, it follows that the normal and tangential components of velocity will also be equal ($v_1^n = v_2^n, v_1^t = v_2^t$). The identical normal components represent the velocity of the gears along the line of action. As you will recall from Chapter 3, the difference in the tangential components of velocity of the two gears represents the *relative motion*, or the sliding. Since in this case the tangential components of velocity are equal, there is no relative motion and therefore no sliding velocity. The motion is pure rolling at this instant.

Figure 6.25 shows two teeth in contact at a point K on the line of action other than the pitch point. The velocities v_1 and v_2 are again given by $v_1 = R_1\omega_1$ and $v_2 = R_2\omega_2$, in the directions shown (note that R_1 and R_2 are not pitch radii). The components of velocities v_1 and v_2 acting along the line of action are equal and are shown in the figure as v^n. Note that if the components along the line of action were unequal, the gear teeth would be separating. The velocities v_1^t and v_2^t are the components of v_1 and v_2 in the tangential direction. As can be seen from the figure, v_1^t and v_2^t are unequal, and therefore one tooth must slide on the other.

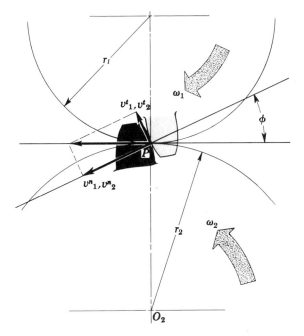

Figure 6.24 The contact between the gears at the instant shown is at the pitch point. The pitch line velocity is identical for P_1 and P_2. Since normal and tangential velocity components are also identical, there can be no relative, or sliding, motion. The motion is pure rolling at this point.

The sliding velocity becomes greater as the point of contact between the teeth occurs further from the pitch point. The sliding velocity will therefore be maximum at the points where the teeth first come into contact and where they leave contact. Since the motion is pure rolling when contact occurs at the pitch point, the pitch point is also the instant center. Since the sliding velocity is directly proportional to the distance from the instant center, it follows that the sliding velocity will be zero at the pitch point and maximum at either extreme point.

The sliding velocity can be determined by any of the procedures discussed in Chapter 3. Using instant centers, suppose, for a moment, that gear 1 is fixed and gear 2 rolls about it at an angular velocity $\omega_2 - \omega_1$ (the angular velocity of 2 relative to 1). The relative motion is the same; only absolute motion has changed. Rolling contact occurs at P, but at all other points there is sliding at a velocity proportional to the distance from P to the instantaneous point of contact. We may find sliding velocity magnitude by the expression.

$$|v_2^t - v_1^t| = |\omega_2 - \omega_1| PK \tag{6.18}$$

where K is any point of contact and $|\omega_2 - \omega_1|$ is the absolute value of the

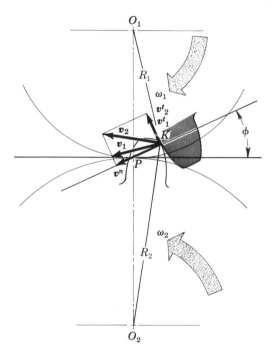

Figure 6.25 Unlike the contact point in the preceding illustration, the contact point K does not coincide with the pitch point P. In this case, the tangential velocities are unequal, and sliding of one gear on the other must occur.

difference between the two angular velocities. For two external gears, angular velocities ω_1 and ω_2 have different signs and $|\omega_2 - \omega_1|$ is the sum of the magnitudes of the two angular velocities. It is seen again that sliding velocity reaches an extreme value when contact occurs at the first or last point of contact.

6.9 INTERFERENCE

Involute gear teeth have involute profiles between the base circle and the addendum circle. If the dedendum circle lies inside the base circle, the portion of the tooth between the base circle and the dedendum circle will not be an involute and is usually made in a radial line. For this reason, if contact between two gears occurs below the base circle of one of the gears, interference is said to occur. The contact is then between two nonconjugate curves, and the fundamental law of gearing will be broken.

In other words, interference occurs whenever the addendum circle of a gear intersects the line of action beyond the *interference point*, which is the point where the line of action is tangent to the base circle of the other gear.

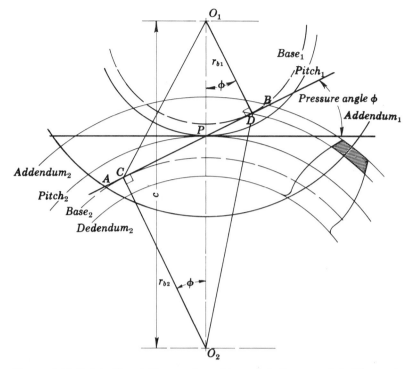

Figure 6.26 Points C and D are referred to as *interference points*. The part of the tooth that would have to be removed in order to prevent interference is shown shaded.

In order to better understand interference, consider Figure 6.26. Points A and B are the points of intersection of the addendum circles with the line of action. Points C and D are the points where the base circles are tangent to the line of action. Interference will occur for the gears shown since point B lies outside point D and point A lies outside point C. The interfering portion of the tooth of gear 2 is shown as the shaded area in Figure 6.26.

So that interference is prevented, the size of the addendum circle diameter must be limited. Referring again to Figure 6.26, the maximum radii that the addendum circles may have to avoid interference are O_1C and O_2D, respectively. From right triangle, O_1CD, we observe the following relationships:

$$O_1C = r_{a_1} \text{ (radius of addendum circle)} = \sqrt{(O_1D)^2 + (CD)^2}$$

$$O_1D = O_1P \cos \phi = r_1 \cos \phi = r_{b_1}$$

$$PD = O_1P \sin \phi$$

$$CP = O_2P \sin \phi$$

$$CD = CP + PD = (O_1P + O_2P)\sin \phi$$

Since the distance between gear centers is

$$c = O_1 P + O_2 P$$

the expression for CD becomes

$$CD = c \sin \phi$$

The expression for the radius of the addendum circle of gear 1 then becomes

$$O_1 C = \sqrt{r_1^2 \cos^2 \phi + c^2 \sin^2 \phi}$$

or

$$r_{a_1}(\max) = \sqrt{r_1^2 \cos^2 \phi + c^2 \sin^2 \phi} = \sqrt{r_{b_1}^2 + c^2 \sin^2 \phi} \qquad \text{(6.19a)}$$

Similarly, for the other gear,

$$r_{a_2}(\max) = \sqrt{r_2^2 \cos^2 \phi + c^2 \sin^2 \phi} = \sqrt{r_{b_2}^2 + c^2 \sin^2 \phi} \qquad \text{(6.19b)}$$

If the addendum circle radius exceeds the value as calculated from the above equation, interference will occur. If it is equal to or less than the calculated value, no interference will occur. However, clearance must still be provided for normal operation.

♦ **EXAMPLE PROBLEM 6.9** *Detection of Gear Interference*
Two standard 20° full-depth gears have a module of 8 mm. The larger gear has 30 teeth, while the pinion has 15 teeth. Will the gear interfere with the pinion?

Solution. Eq. 6.19b will indicate the maximum permissible addendum circle radius for the gear in this problem. We must therefore calculate the base circle radius for the gear, r_{b_2}, and the center distance c in order to use the formula. Thus, from Eq. 6.5,

$$d_2 = mN_2 = 8(30) = 240 \text{ mm}$$

$$d_1 = mN_1 = 8(15) = 120 \text{ mm}$$

Therefore,

$$c = \frac{1}{2}(d_1 + d_2) = \frac{1}{2}(120 + 240) = 180 \text{ mm}$$

The radius of the base circle of the gear is therefore

$$r_{b_2} = r_2 \cos \phi = \frac{240}{2} \cos 20° = 112.8 \text{ mm}$$

We can now use Eq. 6.19b to determine the maximum permissible addendum radius. By comparing the result to the actual value as determined from Table

6.1, we find whether interference exists in this case.

$$r_{a_2}(\max) = \sqrt{r_{b_2}^2 + c^2 \sin^2 \phi}$$

$$= \sqrt{(112.8)^2 + (180)^2 (0.342)^2} = 128.5 \text{ mm}$$

From Table 6.1, for the 20° full-depth gear,

$$a_2 = m = 8 \text{ mm}$$

and the addendum circle radius is the sum of the pitch radius and the addendum, or

$$r_{a_2} = r_2 + a_2 = 120 + 8 = 128 \text{ mm}$$

Comparing the maximum permissible addendum radius to the actual value, we find that there is no interference since the actual addendum circle radius is slightly less than the maximum allowable addendum circle radius. If r_{a_2} had been greater than $r_{a_2}(\max)$, interference would have occurred. ◆

Note that, for a pair of standard gears, if the addendum of the larger gear does not interfere with the flank of the smaller gear, then it follows that the addendum of the smaller gear will not interfere with the larger gear. That is, if r_{a_2} is less than $r_{a_2}(\max)$ from Eq. 6.19b, then r_{a_1} will be less than $r_{a_1}(\max)$ from Eq. 6.19a, where subscript 1 designates the smaller of the two gears. This can be illustrated by reconsidering the preceding example, where it was determined that

$$r_{a_2} = 128 \text{ mm and } r_{a_2}(\max) = 128.5 \text{ mm}$$

and therefore, no interference occurs, although the condition is close to interference. Examining the smaller gear, we have

$$r_{a_1} = r_1 + a_1 = 60 + 8 = 68 \text{ mm}$$

and

$$r_a(\max) = \sqrt{r_1^2 \cos^2 \phi + c^2 \sin^2 \phi}$$

$$= \sqrt{(60)^2 (0.940)^2 + (180)^2 (0.342)^2} = 83.5 \text{ mm}$$

As can be seen, the actual addendum circle radius of 68 mm for the pinion is well within the limit of 83.5 mm. Therefore, the governing equation for standard gears is Eq. 6.19b.

The following limiting condition based on Eq. 6.19b can be used to determine the number of gear teeth necessary to avoid interference:

$$(r_2 + a_2)^2 \le r_2^2 \cos^2 \phi + c^2 \sin^2 \phi$$

where, again, subscript 2 represents the larger of the two meshing gears.

Substituting the following expressions,

$$r_2 = \frac{N_2}{2P}, \, a_2 = \frac{k}{P}, \, c = \frac{1}{2P}(N_1 + N_2)$$

leads to

$$\left(\frac{N_2}{2P} + \frac{k}{P}\right)^2 \leq \left(\frac{N_2}{2P}\right)^2 \cos^2\phi + \left(\frac{N_1 + N_2}{2P}\right)^2 \sin^2\phi$$

where the value for k in the expression for the addendum is obtained from Table 6.1. Factoring P from the above inequality and simplifying yields

$$4kN_2 + 4k^2 \leq N_1^2 \sin^2\phi + 2N_1 N_2 \sin^2\phi \tag{6.20}$$

For any given pressure angle ϕ and addendum constant k, this inequality can be used to determine the necessary number of teeth on one of the gears in terms of the number of teeth on the other gear in order to avoid interference. For example, rearranging, we have

$$4k - 2N_1 \sin^2\phi \leq \frac{N_1^2 \sin^2\phi - 4k^2}{N_2}$$

If we consider the extreme case where N_2 approaches infinity, that is, the gear becomes a rack, then

$$4k - 2N_1 \sin^2\phi \leq 0$$

or

$$N_1 \geq \frac{2k}{\sin^2\phi}$$

for noninterference with a rack. The equality

$$N_1 = \frac{2k}{\sin^2\phi} \tag{6.21}$$

defines the minimum number of teeth that a pinion can have without interference occurring when the pinion meshes with a rack. The following values of N_1 for standard tooth forms are obtained from this equation, where the fractional value calculated is rounded to the next highest integer:

$14\frac{1}{2}^\circ$ full depth: $N_1 = 32$

20° full depth: $N_1 = 18$ (coarse pitch)

20° stub: $N_1 = 14$

25° full depth: $N_1 = 12$

These values of N_1 satisfy the general inequality condition of 6.20 not only for $N_2 \to \infty$ but also for any value of $N_2 \geq N_1$. However, the values from above are

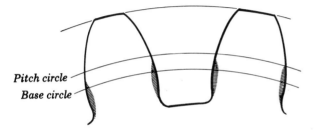

Figure 6.27 When gears are undercut, the shaded portions of the teeth are removed. Since a part of the profile that lies above the base circle is removed in the process, the result is a reduced length of contact.

no longer the minimum values required. That is, smaller numbers of teeth can be used if the pinion meshes with a gear other than a rack.

The results above for minimum tooth numbers illustrate a major disadvantage of the now obsolete $14\frac{1}{2}^\circ$ system; that is, the gears must be larger than for the other pressure angles in order to avoid interference problems.

Therefore, selecting a pinion with a number of teeth greater than or equal to the minimum value is one means for eliminating interference in the design of a gearset. Other ways of eliminating interference will be discussed in the following paragraphs.

In Section 6.7, the generating gear cutters (rack and pinion) were discussed. During the cutting process, the cutter and gear blank act in a manner similar to two meshing gears. It is therefore possible for interference to occur. However, since one of the elements is a cutting tool the portion of the gear that would be interfering is cut away. A gear that has had material removed in this manner is said to be *undercut*.

Figure 6.27 shows undercut gear teeth with the undercut portion shaded. As can be seen, undercutting also removes a portion of the tooth profile *above* the base circle. This is a definite disadvantage, since the length of contact is reduced. Reducing the length of contact, of course, decreases the contact ratio and results in rougher and noisier gear action, since fewer teeth are in contact. Since undercutting removes material from the base of the tooth, it also may seriously weaken the tooth.

Methods other than undercutting are also available to reduce the amount of interference. Interference can also be eliminated if the height of the tooth is reduced by cutting off a portion of its tip. Interference occurs when the tip of one gear is in contact below the base circle of the mating gear. Removing a portion of the tooth tip will therefore prevent contact below the base circle. This type of gear is called a stub-tooth gear.

Increasing the pressure angle will also decrease the problem of interference. A larger pressure angle decreases the diameter of the base circle and thus increases the involute portion of the tooth profile. The disadvantage of both of these procedures is that the active length of contact is reduced, which again leads to rougher, noisier gear operation.

An equation for determining the minimum pressure angle for which no interference will be present can be derived from Eqs. 6.19a and 6.19b.

$$r_a(\max) = \sqrt{r_b^2 + c^2 \sin^2 \phi}$$

$$r_a^2(\max) = r_b^2 + c^2 \sin^2 \phi$$

and since $r_b = r \cos \phi$,

$$r_a^2(\max) = r^2 \cos^2 \phi + c^2 \sin^2 \phi$$

But $\cos^2 \phi + \sin^2 \phi = 1$; therefore,

$$r_a^2(\max) = r^2(1 - \sin^2 \phi) + c^2 \sin^2 \phi = r^2 + (c^2 - r^2)\sin^2 \phi$$

Therefore,

$$\sin^2 \phi = \frac{r_a^2(\max) - r^2}{c^2 - r^2}$$

$$\sin \phi = \sqrt{\frac{r_a^2(\max) - r^2}{c^2 - r^2}} \qquad (6.22)$$

In the use of Eq. 6.22, the standard value for the addendum is to be used in determining $r_a(\max)$. Therefore,

$$\sin \phi = \sqrt{\frac{r_a^2 - r^2}{c^2 - r^2}} \qquad (6.23)$$

Of course, the value for the pressure angle ϕ must still provide an acceptable contact ratio.

The final method commonly used to eliminate interference is to cut the gears with unequal addendum and dedendum teeth. This is accomplished by increasing the addendum of the driver while decreasing its dedendum. The mating gear is then cut with a decreased addendum and increased dedendum. The result of this procedure is to increase the length of action for which involute action is obtainable. Gears of this type are usually called *long-and-short addendum teeth gears*.

The disadvantage of this procedure is an increase in the cost of the gear and the fact that gears cut in this manner are noninterchangeable. Gears of this type are also known as *nonstandard gears*.

Finally, it should be clear from the previous discussion that the method used to eliminate interference depends on the application for which the gear is to be used.

6.10 GEAR TOOTH FORCES

Aside from the kinematic considerations already discussed, the forces and torques acting on spur gears are of vital importance to the designer. As was discussed earlier, the normal force one gear exerts upon another always lies along the line of action (also known as the pressure line).

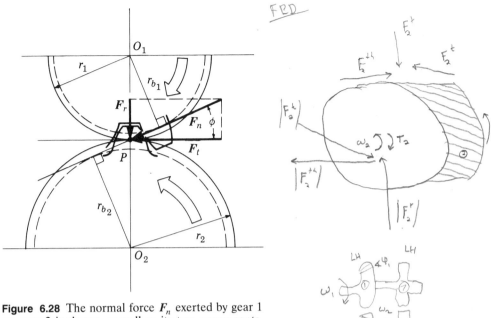

Figure 6.28 The normal force F_n exerted by gear 1 on gear 2 is shown, as well as its two components, the tangential force F_t and the radial force F_r. Gear 1 is the driver.

In Figure 6.28, with gear 1 the driver, the normal force, F_n, exerted by gear 1 on gear 2 is shown. Keep in mind that no matter where on the line of action the two teeth are in contact, the normal force will always pass through the pitch point P. At the pitch point, the normal force can be resolved into the two components F_t, the tangential force, and F_r, the radial force. Thus

$$F_t = F_n \cos \phi \tag{6.24}$$

and

$$F_r = F_n \sin \phi \tag{6.25}$$

or

$$F_r = F_t \tan \phi \tag{6.26}$$

where ϕ is the pressure angle, and the forces are expressed in pounds in the English system or in newtons in the SI system.

Clearly, the normal force exerted by gear 2 on gear 1 would act along the pressure line but in the direction opposite to that shown in Figure 6.28.

The radial force on external spur gears always acts in toward the center of the gear. On the other hand, the radial force acting on internal spur gears always acts out from the center of rotation. Obviously, the radial force tends to move the gear out of contact with its meshing gear. Hence, the radial force is sometimes known as the separating force.

The direction of the normal force, either up or down along the pressure line, is determined by the direction of rotation of the driver. Follow the direction of rotation of the driver along the pressure line to determine the direction of the normal force on the driven gear. Clearly, the normal force exerted by the driven gear on the driver acts in the opposite direction.

The torque that is produced about the center of a gear is given by

$$T = F_n\left(\frac{d_b}{2}\right) = F_n\left(\frac{d}{2}\right)\cos\phi = F_t\left(\frac{d}{2}\right) \tag{6.27}$$

where F_n = normal force

F_t = tangential force

d = pitch circle diameter

d_b = base circle diameter

T = torque

The power that can be transmitted from one gear to another is determined from

$$hp = \frac{Tn}{63,025} \tag{6.28}$$

or

$$hp = \frac{F_t v_p}{33,000} \tag{6.29}$$

where, in the English system,

T = torque in inch-pounds

n = revolutions per minute

F_t = tangential force in pounds

v_p = pitch line velocity in feet per minute

hp = power in units of horsepower

In the SI system, power is

$$kW = \frac{T\omega}{1000} \tag{6.30}$$

or

$$kW = \frac{F_t v_p}{1,000,000} \tag{6.31}$$

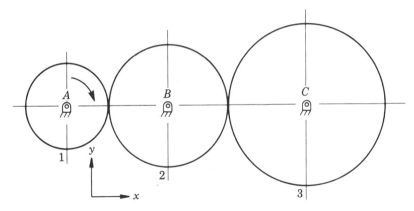

Figure 6.29 A gear train consisting of three spur gears.

where T = torque in newton-meters

   ω = rotational speed in radians per second

   F_t = tangential force in newtons

   v_p = pitch line velocity in millimeters per second

   kW = power in kilowatts

Finally, it should be noted that the efficiency of power transmission from one gear to another is not 100 percent. Generally, we can expect about 1 or 2 percent power loss at maximum transmitted power for spur gears. Unless stated otherwise, we will assume, when doing problems, that there is no power loss.

◆ **EXAMPLE PROBLEM 6.10**   *Force Analysis of Spur Gears*
For the three gears shown in Figure 6.29, gear 1, the driver, rotates at 1000 rev/min clockwise and delivers 30 kW. Gear 1 has a module of 10 mm, a pressure angle of 20°, and 35 teeth, while gear 2 has 45 teeth, and gear 3 has 60 teeth.

(a) What is the distance between A and C?
(b) What is the speed ratio between gears 1 and 3?
(c) If gear 2 were not in the train, what would be the speed ratio between gears 1 and 3? Based upon this answer, what purpose does the idler gear, gear 2, serve?
(d) Draw and completely label the free-body force diagram of each gear.

Solution.   (a) From Eq. 6.5,

$d_1 = mN_1 = 10(35) = 350$ mm

$d_2 = mN_2 = 10(45) = 450$ mm

$d_3 = mN_3 = 10(60) = 600$ mm

By Eq. 6.10,

$$c_{AB} = \frac{1}{2}(d_1 + d_2) = \frac{1}{2}(350 + 450) = 400 \text{ mm}$$

$$c_{BC} = \frac{1}{2}(d_2 + d_3) = \frac{1}{2}(450 + 600) = 525 \text{ mm}$$

Therefore,

$$c_{AC} = c_{AB} + c_{BC} = 400 + 525 = 925 \text{ mm}$$

Check: Upon considering the figure, it is clear that the distance between shafts A and C is

$$r_1 + r_2 + r_2 + r_3 = \frac{350}{2} + \frac{450}{2} + \frac{450}{2} + \frac{600}{2} = 925 \text{ mm}$$

(b) From Eq. 6.8,

$$r_{v_{1-3}} = \frac{N_1}{N_2} \times \frac{N_2}{N_3} = \frac{35}{45} \times \frac{45}{60} = 0.583$$

(c) Without gear 2,

$$r_v = \frac{N_1}{N_3} = \frac{35}{60} = 0.583$$

Obviously, gear 2 does not affect the speed ratio between gears 1 and 3. However, with gear 2 in the system, gear 3 rotates clockwise. Without gear 2, gear 3 would rotate counterclockwise. Therefore, the purpose in introducing an idler gear between the driver and driven gears is to permit the driver and driven gears to have the same direction of rotation.

(d) The angular velocity of gear 1 is

$$\omega_1 = 1000 \text{ rev/min}\left(\frac{2\pi \text{ rad}}{1 \text{ rev}}\right)\left(\frac{1 \text{ min}}{60 \text{ s}}\right) = 104.7 \text{ rad/s}$$

From Eq. 6.1,

$$v_p = r_1\omega_1 = \frac{350}{2} \text{ mm}(104.7 \text{ rad/s}) = 18{,}322 \text{ mm/s}$$

By Eq. 6.31,

$$F_{t_{1-2}} = \frac{kW(1{,}000{,}000)}{v_p} = \frac{30(1{,}000{,}000)}{18{,}322} = 1637 \text{ N}$$

From Eq. 6.27,

$$T_1 = F_{t_{1-2}}\left(\frac{d_1}{2}\right) = 1637 \text{ N}\left(\frac{350}{2} \text{ mm}\right)\left(\frac{1 \text{ m}}{1000 \text{ mm}}\right) = 286.5 \text{ N} \cdot \text{m}$$

By Eq. 6.26,

$$F_{r_{1-2}} = F_{t_{1-2}} \tan \phi = 1637 \tan 20° = 1637(0.364) = 595.9 \text{ N}$$

The pressure line is located by rotating the common tangent through the pressure angle in a direction opposite to the direction of rotation of the driver. The force F_{A_x}, which is equal in magnitude to force $F_{r_{1-2}}$, and the force F_{A_y}, which is equal in magnitude to force $F_{t_{1-2}}$, are the forces that shaft A exerts on gear 1. The torque T_A, which is equal in magnitude to torque T_1, is the torque the shaft exerts on gear 1. Figure 6.30 shows these forces and torque, with proper directions, on a free-body diagram of gear 1.

Using Newton's third law (action and reaction), and since the net torque on gear 2 must be zero, it follows that

$$F_{t_{2-3}} = F_{t_{1-2}} \quad \text{and} \quad F_{r_{2-3}} = F_{r_{1-2}}$$

(see Figure 6.30). And for gear 3,

$$F_{C_y} = F_{t_{2-3}} = 1637 \text{ N}$$

$$F_{C_x} = F_{r_{2-3}} = 595.9 \text{ N}$$

and

$$T_C = T_3 = F_{t_{2-3}}\left(\frac{d_3}{2}\right) = 1637 \text{ N}\left(\frac{600}{2} \text{ mm}\right)\left(\frac{1 \text{ m}}{1000 \text{ mm}}\right) = 491.1 \text{ N} \cdot \text{m}$$

Check: From Eq. 6.8,

$$\frac{\omega_3}{\omega_1} = \frac{N_1}{N_3}$$

Therefore,

$$\omega_3 = \omega_1\left(\frac{N_1}{N_3}\right) = 104.7 \text{ rad/s}\left(\frac{35}{60}\right) = 61.1 \text{ rad/s}$$

and

$$v_p = r_3\omega_3 = \frac{600}{2} \text{ mm}(61.1 \text{ rad/s}) = 18,330 \text{ mm/s}$$

Using Eq. 6.31, we have

$$F_{t_{2-3}} = \frac{kW(1,000,000)}{v_p} = \frac{30(1,000,000)}{18,330} = 1637 \text{ N}$$

The complete force diagrams of the individual gears are shown in Figure 6.30. Note that each gear satisfies force and moment equilibrium under the forces and torques shown in Figure 6.30 and determined above. Therefore, $F_{B_y} = F_{t_{1-2}} + F_{t_{2-3}} = 2F_{t_{1-2}}$, and $F_{B_x} = 0$, where F_{B_x} and F_{B_y} are the bearing forces on idler gear 2 at shaft B. Torque T_c represents the load on shaft C, which the gear train is driving against.   ◆

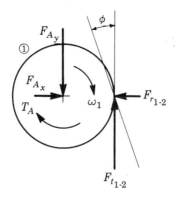

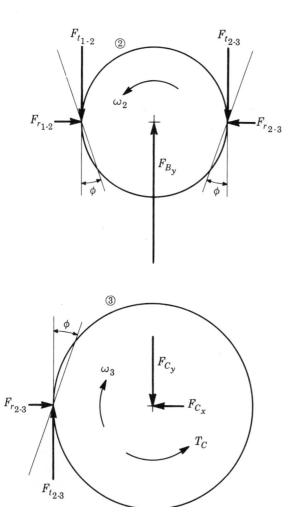

Figure 6.30 Free-body force diagrams of the gears in the gear train of Figure 6.29 and Example Problem 6.10.

References

1. "AGMA Standards," American Gear Manufacturers Association.
2. Colbourne, J. R., *The Geometry of Involute Gears*, Springer-Verlag, New York, 1987.
3. Dudley, D. W. (ed.), *Gear Handbook*, McGraw-Hill, New York, 1962.
4. Shigley, J. E., and C. R. Mischke, *Mechanical Engineering Design*, McGraw-Hill, New York, 1989.

PROBLEMS

6.1 For what reasons are gears preferred over friction drives?

6.2 State the fundamental law of gearing.

6.3 Define the following terms:
  (a) Pitch circle
  (b) Diametral pitch
  (c) Circular pitch
  (d) Pitch point
  (e) Addendum
  (f) Dedendum
  (g) Backlash
  (h) Base circle
  (i) Pressure angle
  (j) Angle of approach
  (k) Angle of recess
  (l) Contact ratio
  (m) Interference
  (n) Speed ratio
  (o) Module

• **6.4** What are the methods used to eliminate interference?

6.5 What is the pitch diameter of a 40-tooth spur gear having a circular pitch of 1.5708 in?

6.6 How many revolutions per minute is a spur gear turning at, if it has 28 teeth, a circular pitch of 0.7854 in, and a pitch line velocity of 12 ft/s?

6.7 How many revolutions per minute is a spur gear turning at if it has a module of 2 mm, 40 teeth, and a pitch line velocity of 2500 mm/s?

6.8 A spur gear having 35 teeth is rotating at 350 rev/min and is to drive another spur gear at 520 rev/min.
  (a) What is the value of the velocity ratio?
  (b) How many teeth must the second gear have?

• **6.9** An external 20°, full-depth spur gear has a diametral pitch of 3. The spur gear drives an internal gear with 75 teeth to produce a velocity ratio of $\frac{1}{3}$. Determine the center distance.

6.10 A standard 20°, full-depth spur gear has 24 teeth and a circular pitch of 0.7854 in. Determine **(a)** the working depth, **(b)** the base circle diameter, **(c)** the outside diameter, **(d)** the tooth thickness at the base circle, and **(e)** the tooth thickness at the outside diameter.

6.11 A spur gear is rotating at 300 rev/min and is in mesh with a second spur gear having 60 teeth and a diametral pitch of 4. The velocity ratio of the pair of meshing gears is $\frac{1}{3}$. What is the magnitude of the pitch line velocity?

6.12 Two meshing spur gears have a diametral pitch of 4, a velocity ratio of $\frac{1}{5}$, and a center distance of 15 in. How many teeth do the gears have?

● **6.13** Two meshing spur gears have a module of 1.25 mm, a center distance of 87.5 mm, and a velocity ratio of 0.4. How many teeth do the gears have? If the pinion speed is 1000 rev/min, what is the pitch line velocity?

6.14 A pair of spur gears have a circular pitch of 15.708 mm, a pitch line velocity of 4000 mm/s, and a center distance of 350 mm. The larger gear has 84 more teeth than the smaller gear. Determine the rotational speeds of the gears in revolutions per minute.

6.15 A pinion has 32 teeth, a diametral pitch of 4, and a 20° pressure angle. The driven gear is such that the velocity ratio is $\frac{1}{3}$.
   (a) What is the center distance?
   (b) What is the base circle radius of the driven gear?

6.16 A standard pinion has 30 teeth, a module of 6 mm, and a 20° pressure angle. The velocity ratio is 0.3.
   (a) What is the center distance?
   (b) What are the base circle radii?
   (c) What are the tooth thicknesses at the respective base circles?

6.17 Two meshing, 20° spur gears have a diametral pitch of 5. The pinion has 35 teeth, and the center distance is 15 in.
   (a) How many teeth does the driven gear have?
   (b) To what value should the center distance be increased in order for the actual pressure angle to become 24°?

6.18 Two meshing, 20° spur gears have a module of 5 mm. The pinion has 32 teeth, and the center distance is 200 mm. To what value should the center distance be increased in order for the actual pressure angle to become 23°?

● **6.19** Two meshing, standard full-depth spur gears have 20° pressure angles, addendums of $\frac{1}{3}$ in, and a velocity ratio of $\frac{1}{2}$. The pinion has 24 teeth. Calculate the contact ratio.

6.20 Two meshing, standard full-depth spur gears have 25° pressure angles. The pinion has 32 teeth and a base pitch of 0.712 in, while the gear has 48 teeth. How many teeth are in contact?

6.21 A pair of standard 20°, full-depth spur gears having a diametral pitch of 6 are to produce a speed ratio of $\frac{1}{3}$. What is the minimum allowable center distance if neither gear can have fewer than 20 teeth and the contact ratio cannot be less than 1.7?

● **6.22** Two meshing, standard 20° stub spur gears have a module of 5 mm and a center distance of 200 mm. The pinion has 32 teeth. Calculate the contact ratio.

6.23 Two meshing, standard 20°, full-depth spur gears have a module of 4 mm and a speed ratio of $\frac{1}{3}$. The gear has 60 teeth. Calculate the contact ratio.

6.24 Two meshing, standard 20°, full-depth spur gears have a module of 5 mm and a center distance of 200 mm. The pinion has 32 teeth. If the pinion speed is 1000 rev/min, calculate the extreme values of the relative sliding velocity of the gear teeth.

6.25 A pinion has 32 teeth, a diametral pitch of 4, and a 25° pressure angle. The velocity ratio is $\frac{1}{2}$, and the pinion speed is 200 rev/min. Calculate the extreme values of the relative sliding velocity of the gear teeth. The gears are standard.

● **6.26** Two meshing, standard 20°, full-depth gears have a diametral pitch of 3, the pinion has 12 teeth, and the velocity ratio is to be $\frac{1}{2}$. How much of the addenda must be removed in order to prevent interference?

6.27 The same data as for Problem 6.26. What value must the pressure angle be made in order to prevent interference?

6.28 A 25°, full-depth spur gear with 45 teeth has an addendum of $\frac{1}{3}$ in. What is the minimum number of teeth the pinion may have without interference occurring?

6.29 In Figure P6.1, two base circles are shown. Gear 1 drives.
(a) Show the line of action.
(b) Label interference points I_1, I_2 and pitch point P.
(c) Show maximum permissible addenda on gears 1 and 2 without interference.
(d) Find the contact ratio with maximum addenda.

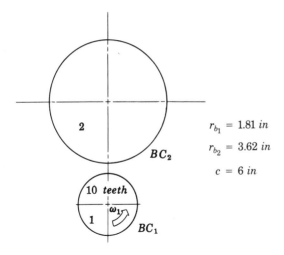

$r_{b_1} = 1.81\ in$

$r_{b_2} = 3.62\ in$

$c = 6\ in$

Figure P6.1

6.30 A spur gear with 32 teeth and a diametral pitch of 4 is meshed with a second gear having a pitch diameter of 16 in. Both gears are external, standard, full-depth, $14\frac{1}{2}°$ involute gears. Determine (a) the number of teeth on the second gear, (b) the standard addendum, and (c) the maximum addendum for gear 1 for which no interference will occur.

6.31 Two meshing, standard 20°, full-depth gears have a module of 3 mm. The pinion has 13 teeth and the velocity ratio is $\frac{1}{2}$. Determine whether the gears interfere. If so, how much addenda must be removed from each gear in order to prevent interference?

● **6.32** A standard 20°, full-depth spur gear has 40 teeth and a module of 8 mm. What is the minimum number of teeth that a meshing pinion may have without interference occurring?

6.33 A standard 20°, full-depth pinion has 15 teeth and a module of 2 mm. What is the maximum number of teeth that a meshing gear may have without interference occurring?

6.34 Two meshing, standard 20°, full-depth spur gears, with a module of 4 mm, have 14 teeth and 45 teeth. What must be the actual operating pressure angle, obtained by increasing the center distance, in order to avoid interference?

6.35 A shaft rotating at 2000 rev/min has a 20-tooth, 5-diametral-pitch pinion gear keyed to it. The pinion meshes with another spur gear whose center is 6 in from the centerline of the first shaft. Compounded with the gear is a double-threaded, left-hand worm that drives a 56-tooth worm wheel keyed to the shaft of an 8-in-diameter hoisting drum. Calculate the distance a load, attached to a cable wrapped around the drum, moves through in 1 min.

6.36 Find the maximum addendum of a $14\frac{1}{2}°$ pressure angle rack that is to mesh with a 6-in-diameter pinion. Draw the gears and label the rack addendum A.

6.37 In Problem 6.36, assume that the pinion has 15 full-depth teeth. (Note that addendum $= 1/P$ *for pinion only.*)
(a) Label the beginning and end of contact with the pinion driving clockwise.
(b) Determine the contact ratio.
(c) Is the contact ratio adequate?

6.38 Find the minimum pressure angle required for a rack to mesh with a 6-in-diameter pinion with 15 teeth, if both the rack and pinion are to have full-depth teeth. (Addendum $= 1/P$.)

6.39 Two spur gears, an external pinion and an internal gear, have a diametral pitch of 4. The center distance is 10 in and the speed ratio is $\frac{1}{5}$. How many teeth do the gears have? If the pinion speed is 250 rev/min, what is the pitch line velocity?

● **6.40** The gearset shown in Figure P6.2 consists of three 20°, full-depth spur gears with a diametral pitch of 4 and the following tooth numbers: $N_1 = 20$, $N_2 = 60$, and

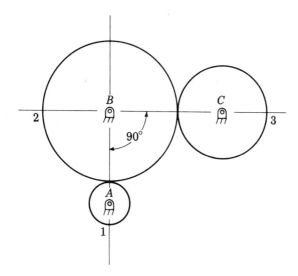

Figure P6.2

$N_3 = 40$. Gear 2 is an idler. The unit transmits 10 hp with the drive shaft to gear 1 rotating at 800 rev/min counterclockwise. Determine the following:

(a) The center distances

(b) The rotational velocity (magnitude and direction) of each gear

(c) The pitch line velocity

(d) The torque transmitted to each shaft

(e) All forces (magnitude and direction) on each gear

6.41 The gearset shown in Figure P6.3 consists of three 20°, full-depth spur gears with a module of 8 mm and the following tooth numbers: $N_1 = 20$, $N_2 = 40$, and $N_3 = 100$. Gear 2 is an idler. The unit transmits 20 kW at a rotational speed of 1000 rev/min of gear 1 in the clockwise direction. Determine the following:

(a) The center distances

(b) The rotational velocity (magnitude and direction) of each gear

(c) The torque transmitted to each shaft

(d) All forces (magnitude and direction) on each gear

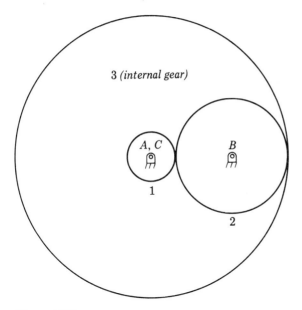

Figure P6.3

6.42 A double-reduction gearset has a configuration as shown in Figure P6.4, where gear 1 is rigidly attached to shaft A, gears 2 and 3 are rigidly fastened to shaft B, and gear 4 is fastened to shaft C. Gears 1 and 2 have 24 teeth and 60 teeth, respectively, and are 25°, full-depth spur gears of diametral pitch 5. Gears 3 and 4 have 20 teeth and 60 teeth, respectively, and are 20°, full-depth spur gears of diametral pitch 4. The system transmits 20 hp at a rotational speed of 1000 rev/min of gear 1 in the clockwise direction. Find the speed of each shaft and the torque transmitted by each shaft. Determine the forces acting on each shaft.

6.43 A double-reduction gearset has a configuration as shown in Figure P6.4, where gear 1 is rigidly attached to shaft A, gears 2 and 3 are rigidly fastened to shaft B, and gear 4 is fastened to shaft C. Gears 1 and 2 have 32 teeth and 80 teeth,

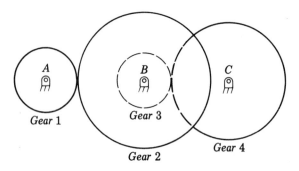

Figure P6.4

respectively, and are 20°, full-depth spur gears with a module of 4 mm. Gears 3 and 4 have 24 teeth and 50 teeth, respectively, and are 25°, full-depth spur gears with a module of 6 mm. The system transmits 30 kW at a rotational speed of 1200 rev/min of gear 1 in the counterclockwise direction. Find the speed of each shaft and the torque transmitted by each shaft. Determine the forces acting on each shaft.

Helical, Worm, and Bevel Gears: Design and Analysis

7.1 HELICAL GEARS ON PARALLEL SHAFTS

While the previous chapter was devoted exclusively to spur gears, they are by no means the only type of gears in common use. There are many applications in which the gears used cannot be mounted on parallel shafts, in which noise elimination is important, or in which large speed reductions are required. This chapter is devoted to a discussion of the more important types of gears other than spur gears. Figure 7.1a to c shows some of the gear types to be discussed.

A helical gear is an outgrowth of a type of gear known as a stepped gear. A stepped gear consists of a number of spur gears placed side by side. Each successive gear is rotated on its axis through a small angle relative to the adjacent gear. A helical gear may be thought of as the limiting case of a stepped gear, with the width of each individual spur gear extremely thin and the angular rotations extremely small. Thus the surface of a helical gear follows a helix, rather than a straight line parallel to the axis, as is the case for spur gears.

Helical gears are called right-hand or left-hand, depending on the direction in which the helix slopes away from the viewer. The line of sight is parallel to the axis of the gear. Thus Figure 7.2a illustrates a left-handed helical gear,

(a)

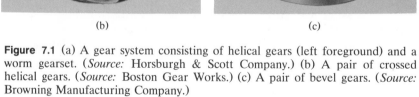

(b)   (c)

Figure 7.1 (a) A gear system consisting of helical gears (left foreground) and a worm gearset. (*Source:* Horsburgh & Scott Company.) (b) A pair of crossed helical gears. (*Source:* Boston Gear Works.) (c) A pair of bevel gears. (*Source:* Browning Manufacturing Company.)

(a)

(b)

Figure 7.2 (a) A left-hand helical gear. The teeth slope to the left (or counter-clockwise) when viewed along the axis. (*Source:* Horsburgh & Scott Company.) (b) A right-hand helical gear. The teeth slope to the right (or clockwise). (*Source:* Barber-Colman Company.)

while Figure 7.2b shows a right-handed helical gear alongside a gear blank from which such a gear would be cut.

In order for two helical gears on parallel shafts to mesh, they must be of different hands. For example, if the driver is right-handed, the driven gear, or follower, must be left-handed.

7.1.1 Helical Gear Tooth Contact

Spur gears have an initial line contact, with the result that the impact (shock) that occurs when two teeth come into contact is much larger than for helical gears. The initial contact between two helical gear teeth is a point. As the motion continues, the contact between the teeth becomes a line whose length gradually increases until the teeth are in full contact. To illustrate the contact that occurs between two meshing helical gears, consider Figure 7.3a and b. The action that occurs as the gears mesh can be approximated by moving gear 2 toward gear 1. The first contact for teeth AA' and BB' occurs at the point where A and B touch. As the gears continue to rotate, or as gear 2 continues to move toward gear 1, more and more points come into contact. Figure 7.3b, for example, shows the contact to a point C. Continued rotation brings more of the teeth into contact, until points A' and B' are touching. The tooth elements of each gear form an angle ψ with the axis of the gear, but the gears are of opposite hand.

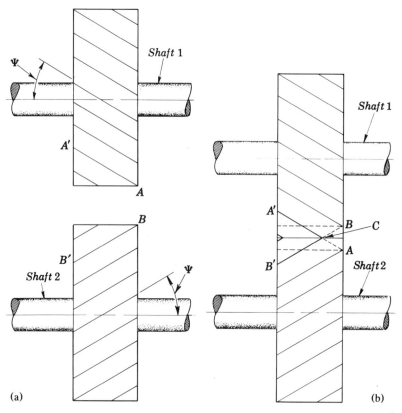

Figure 7.3 (a) The contact between two helical gears can be approximated by considering the contact as the two shafts are brought together. As shafts 1 and 2 approach each other, the initial contact between teeth AA' and BB' occurs when points A and B coincide. Thus initial contact is a point. (b) As shafts 1 and 2 are brought closer, additional points come into contact. In the drawing, the contact has progressed to point C. Finally, points A' and B' coincide and the teeth are fully in line contact. Of course, the shafts cannot be brought together as shown; this concept is merely a device to help visualize the progressing contact as both gears rotate in mesh.

The gradual engagement of helical gear teeth permits larger load transmission, smoother operation, and quieter transmission of power compared to spur gears of similar size. For these reasons, helical gears are often preferred over spur gears, even though they are usually more expensive and more difficult to manufacture.

7.1.2 Helical Gear Terminology and Geometry

The terminology used for helical gears is very similar to that used for spur gears. In fact, most of the relationships developed for spur gears are equally applicable to helical gears on parallel shafts. Several additional terms are necessary, however.

Helix Angle

Figure 7.4 shows the pitch surface of a helical gear, with some of the more important terms expressed as symbols. As has been stated earlier, the teeth on a helical gear are not parallel to the axis of the shaft on which they are mounted. The angle the teeth make with the axis is known as the helix angle, ψ. When two meshing helical gears are on parallel shafts, their helix angles must be equal.

Normal Pitch

The circular pitch p is defined, as for spur gears, as the distance between corresponding points on adjacent teeth, as measured on the pitch surface. However, the pitch of a helical gear can be measured in two different ways. Thus the *transverse* circular pitch p is measured along the pitch circle, just as for spur gears. The *normal* circular pitch p^n is measured normal to the helix of the gear, as shown in Figure 7.4.

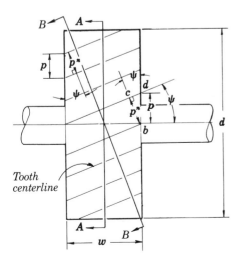

Figure 7.4 Helical gear terms: the diagram shows the helix angle ψ, the pitch diameter d, the transverse circular pitch p, the normal circular pitch p^n, and the gear width w.

The diametral pitch P, just as for spur gears, is given by the formula

$$P = \frac{N}{d} \tag{7.1}$$

where N is the number of teeth on the gear and d is the diameter of the pitch circle in inches.

As before, in SI units, the module m is used to express gear tooth size rather than the diametral pitch P used in the English system. As was the case for spur gears,

$$m = \frac{d}{N} \tag{7.2}$$

where d is expressed in millimeters and m is the module in the transverse plane.

The relationship between the normal diametral pitch and the normal circular pitch is identical to that between the diametral pitch and the circular pitch as given in the last chapter. Thus,

$$P = \frac{\pi}{p} \quad \text{and} \quad P^n = \frac{\pi}{p^n} \tag{7.3}$$

In the SI system,

$$p = \pi m \quad \text{and} \quad p^n = \pi m^n \tag{7.4}$$

where m^n is the module in the normal plane. The normal pitch of a helical gear is an important dimension because it becomes the circular pitch of the hob cutter used to manufacture the gear. When the cutting is done, instead, by a gear shaper, the transverse circular pitch of the gear becomes the circular pitch of the cutter. It can thus be seen why both pitches are of importance. In the following paragraphs, we will derive the relationships between the transverse and normal pitches.

Face Width

The face width of the gear is w. As can be seen from Figure 7.4, which shows a gear proportioned for the recommended minimum width,

$$\tan \psi = \frac{p}{w} \quad \text{or} \quad w = \frac{p}{\tan \psi} \tag{7.5}$$

While the above equation is the formula for the minimum gear width, the usual practice is to increase the width from 10 to 20 percent. It should be clear that when the face width exceeds the minimum width, the formula above no longer applies.

As seen from Figure 7.4, the relationship between the transverse and normal circular pitch is

$$p^n = p \cos \psi \tag{7.6}$$

Substituting the expression for the circular pitch employed in the last chapter ($p = \pi d/N$), we can also derive an expression for the pitch diameter in terms of the number of teeth N, the *normal* circular pitch p^n, and the helix angle ψ:

$$d = \frac{pN}{\pi} = \frac{p^n N}{\pi \cos \psi} \tag{7.7}$$

From Eqs. 7.3 and 7.6, we can express the *normal diametral pitch* in terms of the diametral pitch P and the helix angle:

$$P^n = \frac{\pi}{p^n} = \frac{\pi}{p \cos \psi} = \frac{P}{\cos \psi} \tag{7.8}$$

In SI units,

$$\cos \psi = \frac{m^n}{m} \tag{7.9}$$

Having derived the above equations from Figure 7.4, we may now determine most of the important dimensions of the helical gear.

◆ **EXAMPLE PROBLEM 7.1** *Properties of a Helical Gear*
A 30-tooth helical gear, with a 25° helix angle, has a module of 10 mm. Determine the pitch diameter, the normal module, and the normal and transverse circular pitches.

Solution. The pitch diameter is obtained directly from Eq. 7.2:

$$d = mN = 10(30) = 300 \text{ mm}$$

From Eq. 7.9,

$$m^n = m \cos \psi = 10 \cos 25° = 9.063 \text{ mm}$$

From Eq. 7.4,

$$p = \pi m = \pi(10) = 31.42 \text{ mm}$$

and

$$p^n = \pi m^n = \pi(9.063) = 28.47 \text{ mm}$$

Check: Using Eq. 7.6, we have

$$p^n = p \cos \psi = 31.42 \cos 25° = 28.47 \text{ mm} \quad ◆$$

◆ **EXAMPLE PROBLEM 7.2** *Properties of a Helical Gear*
A helical gear has 25 teeth, a helix angle of 25°, and a transverse circular pitch of $\pi/5$ in. Determine the pitch diameter, the diametral pitch, and the normal circular and diametral pitches.

Solution.   The pitch diameter is directly obtained from Eq. 7.7:

$$d = \frac{pN}{\pi} = \frac{(\pi/5)(25)}{\pi} = 5 \text{ in}$$

The diametral pitch P is given by Eq. 7.3:

$$P = \frac{\pi}{p} = \frac{\pi}{\pi/5} = 5$$

The normal diametral pitch and the normal circular pitch are given by Eqs. 7.8 and 7.3, respectively. Thus, from Eq. 7.8,

$$P^n = \frac{P}{\cos \psi} = \frac{5}{\cos 25°} = 5.52$$

and the normal circular pitch is given by

$$p^n = \frac{\pi}{P^n} = \frac{\pi}{5.52} = 0.569 \text{ in} \quad \blacklozenge$$

Pressure Angle

While spur gears were identified by means of one pressure angle, the geometry of helical gears requires the use of two pressure angles. Figure 7.5a shows the tooth profile for section *AA* of the gear in Figure 7.4, obtained by passing a plane perpendicular to the shaft axis. The pressure angle for this tooth profile is the transverse pressure angle ϕ. Figure 7.5b shows the tooth profile for section *BB* of Figure 7.4, obtained by passing a plane normal to the tooth line. The pressure angle for this tooth profile is the *normal pressure angle* ϕ^n. Note that the actual tooth profiles are not straight lines; the purpose of Figure 7.5a and b is simply to illustrate the relationship between pressure angles ϕ and ϕ^n.

In order to demonstrate that the angles ϕ and ϕ^n are pressure angles, consider Figure 7.5c, which depicts contact between two teeth on two racks. The pitch circles for gears become pitch *lines* for racks or stretched-out gears (line *pp* in Figure 7.5c). Line *FE* is perpendicular to the tooth profiles in contact and represents the projection of the line of action of the force between the teeth. The pressure angle was previously defined as the angle between the line of action between two teeth in contact and the tangent to the pitch circles at the pitch point. Therefore, the pressure angle is defined as the angle ϕ between lines *FE* and *pp*. Simple geometry can then be used to show that angle *bad* is also equal to the pressure angle.

A trigonometric relationship between the pressure angles and the helix angle of the gear can also be obtained. From Figure 7.5a and b,

$$\tan \phi = \frac{bd}{ab} \quad \text{and} \quad \tan \phi^n = \frac{bc}{ab}$$

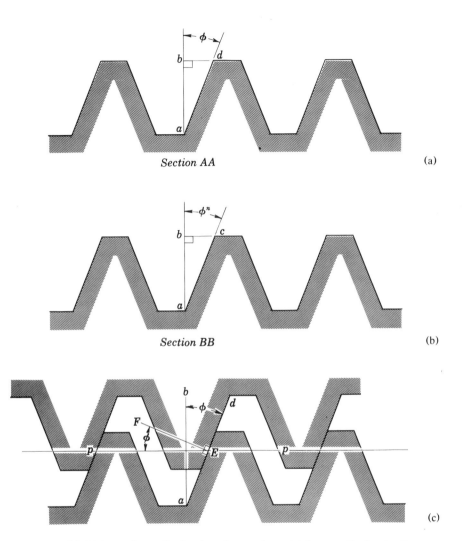

Section AA   (a)

Section BB   (b)

(c)

Figure 7.5 (a) This tooth profile is given by section *AA* (perpendicular to the gear axis) of Figure 7.4, showing the transverse pressure angle ϕ. (b) Section *BB* (normal to the helix) of the gear of Figure 7.4 illustrates the normal pressure angle ϕ^n. (c) Note that the pressure angles *bad* and *bac* of parts a and b are consistent with the definition of the pressure angle given in the last chapter. The pressure angle ϕ for the rack teeth shown is the angle between the projection of the line of action *FE* and the tangent *pp* to the pitch line. From the geometry of the diagram, ϕ and *bad* are equal; thus *bad* does indeed represent the pressure angle.

Therefore,

$$\frac{\tan \phi^n}{\tan \phi} = \frac{bc}{bd}$$

And, from Figure 7.4,

$$\cos \psi = \frac{bc}{bd}$$

Therefore,

$$\cos \psi = \frac{\tan \phi^n}{\tan \phi} \tag{7.10}$$

It is clear from Eq. 7.10 that the transverse pressure angle must always be larger than the normal pressure angle.

Center Distance

An equation for determining the center distance c between two meshing helical gears is extremely useful in gear design. The formulas for determining center distance will be seen to be functions of the normal pitch, the teeth numbers, and the common helix angle. These quantities are usually known for any given pair of gears. The center distance thus can be immediately calculated by the use of these formulas. For example, a designer may arbitrarily have chosen two gears based upon some required speed ratio. The formulas to be derived below will enable him to easily calculate the center distance to determine the space requirements (often a critical factor) for the particular gearset.

From the formulas for pitch diameter and center distance as derived in the last chapter, we obtain

$$d_1 = \frac{N_1 p}{\pi} \quad \text{and} \quad d_2 = \frac{N_2 p}{\pi}$$

and

$$c = \frac{d_1 + d_2}{2} = \frac{p}{2\pi}(N_1 + N_2) \tag{7.11}$$

But $p^n = p \cos \psi$; therefore,

$$c = \frac{p^n(N_1 + N_2)}{2\pi \cos \psi} \tag{7.12}$$

And since $p^n P^n = \pi$, another expression for the center distance is

$$c = \frac{N_1 + N_2}{2 P^n \cos \psi} \tag{7.13}$$

In the SI system,

$$d_1 = mN_1, \qquad d_2 = mN_2$$

$$c = \frac{d_1 + d_2}{2} = \frac{m(N_1 + N_2)}{2} = \frac{m^n(N_1 + N_2)}{2 \cos \psi} \qquad (7.14)$$

As an example, suppose the gears to be designed are to be mounted on shafts 10 in apart. The center distance is thus seen to be fixed. Since the normal pitch and helix angle are usually chosen first, the formulas may be used, instead of for determining center distance, to determine the required value for the sum of N_1 and N_2. This value, together with the speed ratio, will enable the designer to determine the appropriate values for N_1 and N_2.

◆ **EXAMPLE PROBLEM 7.3**   *Analysis of a Helical Gear Pair*
A pair of meshing helical gears have a normal pressure angle of $20°$, a diametral pitch of 5, and a normal circular pitch of 0.55 in. The driver has 18 teeth and the follower has 36 teeth. Determine the pressure angle ϕ and the center distance c.

Solution.   From Eq. 7.3 and the given data, the circular pitch is

$$p = \frac{\pi}{P} = \frac{\pi}{5} = 0.628 \text{ in}$$

From Eq. 7.6, the helix angle is given by

$$\cos \psi = \frac{p^n}{p} = \frac{0.55}{0.628} = 0.876, \qquad \psi = 28.8°$$

From Eq. 7.10, the pressure angle is

$$\tan \phi = \frac{\tan \phi^n}{\cos \psi} = \frac{0.364}{0.876} = 0.416$$

$$\phi = 22.6°$$

Finally, the center distance is found by using Eq. 7.12:

$$c = \frac{p^n(N_1 + N_2)}{2\pi \cos \psi} = \frac{0.55(18 + 36)}{2\pi(0.876)} = 5.4 \text{ in} \quad ◆$$

7.1.3 Tredgold's Approximation for Helical Gears

A better understanding of helical gear geometry can be obtained by considering the tooth profiles in the transverse and normal planes. In Figure 7.6a, if a vertical plane is passed perpendicular to the axis of the gear (section AA), the pitch circle shown in Figure 7.6b results. The radius of curvature is equal to the radius of the transverse pitch circle. The tooth profile in this plane would be the same as that for the tooth profile of a spur gear having a pitch radius r.

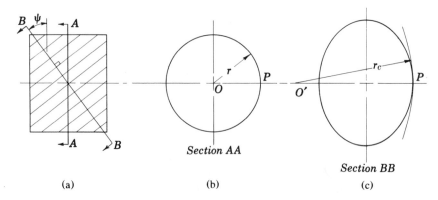

Section AA

Section BB

(a) (b) (c)

Figure 7.6 (a) A helical gear, which is to be sectioned perpendicular to its axis and perpendicular to the helix of the teeth. (b) Section *AA* results in the pitch circle of the gear. (c) Section *BB* results in an ellipse. The radius of curvature r_c is the pitch radius of a spur gear that would approximate the characteristics of the helical gear shown in part a.

When a plane (section *BB* of Figure 7.6a) is passed perpendicular to the helix at an angle ψ to the vertical plane, an ellipse, shown in Figure 7.6c, results from the intersection of this plane and the pitch surface. The radius of curvature of the ellipse at point *P* is given by the formula

$$r_c = \frac{r}{\cos^2 \psi}$$

where r is the pitch radius. The tooth profile in this plane is the same as that for the tooth of a spur gear having a pitch radius r_c. Thus the properties of a helical gear are similar to those of an equivalent spur gear having a pitch radius equal to r_c. For example, the beam strength of helical gear teeth is obtained from the equivalent spur gear. Using the properties of this equivalent spur gear to approximate the properties of a helical gear is known as Tredgold's approximation for helical gears.

The number of teeth on this equivalent spur gear is known as the equivalent, virtual, or formative number or teeth for the helical gear. The formative number of teeth, designated by the symbol N_f, is another important parameter to the designer and can be determined from the following equation:

$$N_f = P^n d_c$$

But $d_c = 2r_c = 2r/\cos^2 \psi$; thus,

$$N_f = P^n \frac{2r}{\cos^2 \psi} = \frac{P^n d}{\cos^2 \psi}$$

But $P^n = P/\cos \psi$; therefore,

$$N_f = \frac{Pd}{\cos^3 \psi}$$

or, since $Pd = N$, the expression becomes

$$N_f = \frac{N}{\cos^3 \psi} \tag{7.15}$$

where N_f is the formative number of teeth for the imaginary spur gear and N is the *actual* number of teeth for the helical gear. Note that N_f need not be an integer while N must be.

7.1.4 Velocity Ratio of Helical Gears

The velocity ratio for spur gears was defined as the ratio of the angular speed of the follower divided by the angular speed of the driver. Another form for the velocity ratio is the number of teeth of the driver divided by the number of teeth of the follower. A similar situation holds for helical gears on parallel shafts.

In Figure 7.7a, two helical gears are shown in mesh. Their shafts are parallel and thus they have opposite hands and equal helix angles. The pitch line velocities are v_1 and v_2. From Figure 7.7b, the normal components of the two velocities are given as v_1^n and v_2^n. The normal components of velocity of any two meshing gears must be equal. Thus,

$$v_1^n = v_2^n$$

From Figure 7.7b,

$$v_1^n = v_1 \cos \psi_1 \quad \text{and} \quad v_2^n = v_2 \cos \psi_2$$

Since the helix angles of the two gears are equal, $\cos \psi_1 = \cos \psi_2$; therefore,

$$v_1 = v_2$$

And since

$$\omega_1 = \frac{v_1}{d_1/2} \quad \text{and} \quad \omega_2 = \frac{v_2}{d_2/2}$$

we obtain the velocity ratio of two helical gears as the ratio of their angular velocities:

$$r_v = \frac{\omega_2}{\omega_1} \tag{7.16}$$

or

$$r_v = \frac{v_2/(d_2/2)}{v_1/(d_1/2)} = \frac{v_2 d_1}{v_1 d_2} = \frac{d_1}{d_2} \tag{7.17}$$

Since

$$d_1 = \frac{N_1}{P_1} = \frac{N_1}{P_1^n \cos \psi}$$

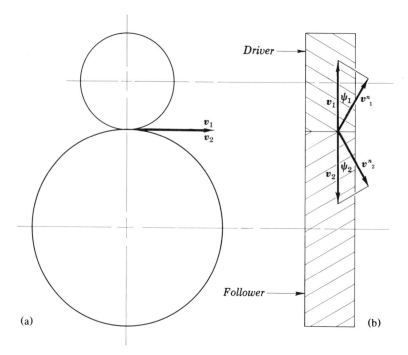

Figure 7.7 (a) The pitch line velocities of meshing helical gears. (b) The normal components of the pitch line velocities are equal for any pair of meshing gears. The pitch line velocities have been rotated through 90° from their actual lines of action for purposes of illustration. The velocities shown are for contact at the pitch point.

and

$$d_2 = \frac{N_2}{P_2} = \frac{N_2}{P_2^n \cos \psi}$$

$$r_v = \frac{d_1}{d_2} = \frac{N_1 P_2^n \cos \psi}{N_2 P_1^n \cos \psi} = \frac{N_1 P_2^n}{N_2 P_1^n}$$

And since $P_1^n = P_2^n$,

$$r_v = \frac{N_1}{N_2} \tag{7.18}$$

Thus, as with any pair of gears on fixed centers, the velocity ratio can be determined simply from the number of teeth of the meshing gears.

Helical gears are rarely used interchangeably, so that there is no need for standardized tooth systems. However, some standardization is desirable in terms of the cutting tools required. The helix angle for most helical gears varies between 15 to 30°, although some herringbone gears have helix angles as high as 45°. Since the thrust load varies directly with the tangent of the helix angle, the magnitude of the helix angle must be limited. In other words, if the helix angle becomes too large, the thrust load will be excessive. Helical gears are cut by the same general methods discussed in the previous chapter.

7.1.5 Helical Gear Forces

As in all direct contact mechanisms, the force that one helical gear exerts on its meshing gear acts normal (perpendicularly) to the contacting surfaces, if friction forces are neglected. This normal force can be resolved into tangential

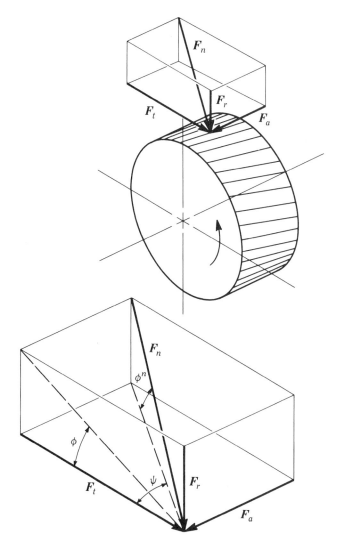

Figure 7.8 The normal force F_n acting on a helical gear consists of three components: tangential force F_t, radial force F_r, and axial or thrust force F_a.

and radial components as was the case with spur gear forces. However, for helical gears, there is a third component known as the axial or thrust load. This force acts parallel to the axis of the shaft the gear is mounted on. Figure 7.8 shows the forces acting on a helical gear. Let us assume the gear shown is the driver and is rotating in the counterclockwise direction. The forces shown in the figure are those exerted on the driver by the driven gear.

As is shown in Figure 7.8, the magnitudes of the components of the normal force, F_n, are

$$F_t = \text{tangential force} = F_n \cos \phi^n \cos \psi \tag{7.19}$$

$$F_r = \text{radial force} = F_n \sin \phi^n = F_t \tan \phi \tag{7.20}$$

and

$$F_a = \text{thrust or axial load} = F_n \cos \phi^n \sin \psi = F_t \tan \psi \tag{7.21}$$

The direction of the thrust load is determined by the right- or left-hand rule (depending on the hand of the driver) applied to the driver: *Fingers point in the direction of rotation; then the thumb points in the direction of the thrust load.* The *driven* thrust load is then *opposite* to the direction of the thrust load on the driver. Figure 7.9 shows the direction of the thrust load for helical gears on parallel shafts. Actually, the thrust force is applied at the pitch circle, leading to a combination of a thrust force and a thrust moment at the shaft center.

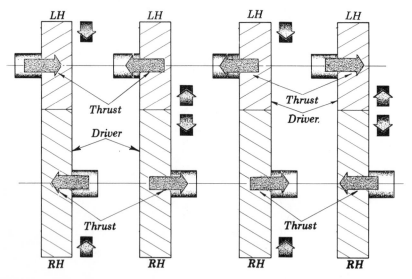

Figure 7.9 Direction of thrust load for helical gears mounted on parallel shafts.

The equations for torque and power are the same as those used for spur gears, namely,

$$T = F_t\left(\frac{d}{2}\right) \tag{7.22}$$

$$hp = \frac{F_t v_p}{33,000} \tag{7.23}$$

in the English system, where hp = horsepower, F_t = tangential force in pounds, and v_p = pitch velocity in feet per minute, and

$$kW = \frac{F_t v_p}{1,000,000} \tag{7.24}$$

in the SI system, where kW = power in kilowatts, F_t = tangential force in newtons, and v_p = pitch velocity in millimeters per second.

Eliminating Thrust with Herringbone Gears

A herringbone gear can be thought of as a helical gear with half of its face cut right-handed and the other half cut left-handed. A space is usually cut between the two halves to permit easier manufacture. Thus, the thrust loads generated by the left-hand and right-hand teeth cancel each other. Figure 7.10a shows a herringbone gear with no space provided between the right-hand and left-hand teeth, while Figure 7.10b shows a double-helical or herringbone gear with a rather wide space between the two halves.

7.2 CROSSED HELICAL GEARS

Helical gears can also be used when power is to be transmitted from one shaft to another nonparallel, nonintersecting shaft. These helical gears are referred to as *crossed helical gears*. Any helical gear can be used as a crossed helical gear; a helical gear becomes a crossed helical gear when it is meshed with another helical gear whose shaft is nonparallel and nonintersecting with the shaft of the first gear. A typical set of crossed helical gears is illustrated in Figure 7.11.

While helical gears on parallel shafts must be of opposite hand, two crossed helical gears usually have the *same* hand. And while helical gears on parallel shafts must have identical helix angles, the helix angles for crossed helical gears do *not* have to be equal.

Unlike helical gears on parallel shafts, crossed helical gears have point rather than line contact. However, after a wearing-in period, the point contact becomes line contact. While line contact does eventually occur, it is much poorer contact than that for the gear types previously discussed. Because of this poor contact, crossed helical gears are used only when small loads are to be transmitted.

(a)   (b)

Figure 7.10 (a) A herringbone (or double-helical) gear with no space between left-hand and right-hand teeth. (*Source:* Horsburgh & Scott Company.) (b) A herringbone gear with a large space between the two halves. (*Source:* Horsburgh & Scott Company.)

Figure 7.11 A pair of crossed helical gears, used when the shafts are not parallel. Usually, crossed helical gears have the same hand. (*Source:* Richmond Gear, Wallace Murray Corporation.)

An advantage of crossed helical gears is that the alignment of the gears does not have to be perfect in order to obtain smooth operation. For crossed helical gears, the normal pitch rather than the transverse pitch is usually referred to when specifying pitch. The reason for this is that while the normal pitches for meshing helical gears must be equal, the transverse pitches will be unequal if the helix angles are unequal.

7.2.1 Crossed Helical Gear Geometry

Some of the more important relationships involving crossed helical gears can be obtained by considering Figure 7.12. The two helical gears shown have different helix angles, ψ_1 and ψ_2. Both gears are right-handed, and Σ is the angle between the shafts. For crossed helical gears, the angle between the shafts always equals the sum or difference of the helix angles of the two gears. As can be seen from Figure 7.12,

$$\Sigma = \psi_1 + \psi_2 \qquad (7.25a)$$

If the gears were of opposite hand, the shaft angle would be instead

$$\Sigma = \psi_1 - \psi_2 \qquad (7.25b)$$

The pitch line velocities, v_1 and v_2, act as shown in the illustration. The normal components of v_1 and v_2 must be equal, are perpendicular to the axis t-t (the axis tangent to the teeth in contact), and are both labeled v^n. As seen from the illustration, a sliding velocity exists for crossed helical gears, even when they contact at the pitch point.

Center Distance

A procedure similar to that used to find the center distance for helical gears on parallel shafts, Eqs. 7.12 to 7.14, can be followed to obtain the formulas for the center distance between two crossed helical gears. Thus the pitch diameters are again given by

$$d_1 = \frac{N_1 p_1}{\pi} = \frac{N_1 p^n}{\pi \cos \psi_1}$$

$$d_2 = \frac{N_2 p_2}{\pi} = \frac{N_2 p^n}{\pi \cos \psi_2}$$

where the normal circular pitches of the two meshing gears must be equal. The center distance is given by

$$c = \frac{d_1 + d_2}{2}$$

Substituting the expressions for the pitch diameters into the center distance

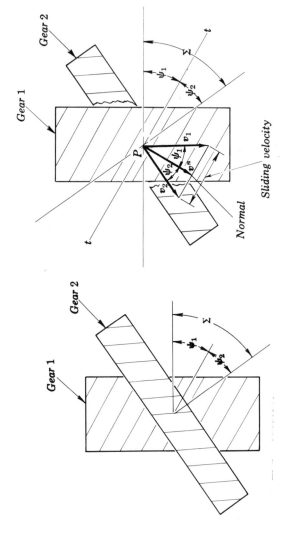

Figure 7.12 The two crossed helical gears are of the same hand but have different helix angles. The shaft angle Σ is equal to the sum of the helix angles. The normal components of the pitch line velocities are equal, but a sliding velocity exists.

formula, we obtain

$$c = \frac{p^n}{2\pi} \left(\frac{N_1}{\cos \psi_1} + \frac{N_2}{\cos \psi_2} \right)$$

(7.26)

Since $p^n P^n = \pi$, another form for the center distance is

$$c = \frac{1}{2P^n} \left(\frac{N_1}{\cos \psi_1} + \frac{N_2}{\cos \psi_2} \right)$$

(7.27)

while in the SI system

$$c = \frac{m^n}{2} \left(\frac{N_1}{\cos \psi_1} + \frac{N_2}{\cos \psi_2} \right)$$

(7.28)

7.2.2 Velocity Ratio of Crossed Helical Gears

The velocity ratio for crossed helical gears can be obtained by following the procedure used to obtain Eqs. 7.16 to 7.18. Thus $v_1^n = v_2^n$, and from Figure 7.12,

$$v_1^n = v_1 \cos \psi_1$$

$$v_2^n = v_2 \cos \psi_2$$

Therefore,

$$v_1 \cos \psi_1 = v_2 \cos \psi_2 \quad \text{or} \quad v_1 = v_2 \left(\frac{\cos \psi_2}{\cos \psi_1} \right)$$

We now have an expression for the pitch line velocities in terms of the separate helix angles of the crossed gears. We can use this relationship to obtain the velocity ratio of crossed helical gears in terms of their numbers of teeth. The angular velocities of the two gears are given by

$$\omega_1 = \frac{v_1}{d_1/2} \quad \text{and} \quad \omega_2 = \frac{v_2}{d_2/2}$$

And since the velocity ratio r_v is the ratio of the angular velocities, we obtain

$$r_v = \frac{\omega_2}{\omega_1} = \frac{v_2/(d_2/2)}{v_1/(d_1/2)} = \frac{v_2 d_1}{v_1 d_2}$$

or, since $v_1 = v_2 \cos \psi_2 / \cos \psi_1$,

$$r_v = \frac{d_1 \cos \psi_1}{d_2 \cos \psi_2}$$

(7.29)

Since $d_1 = N_1/P_1$ and $d_2 = N_2/P_2$,

$$r_v = \frac{N_1 P_2 \cos \psi_1}{N_2 P_1 \cos \psi_2}$$

But $P^n = P/\cos \psi$; therefore,

$$r_v = \frac{N_1 P_2^n}{N_2 P_1^n} = \frac{N_1 m_1^n}{N_2 m_2^n}$$

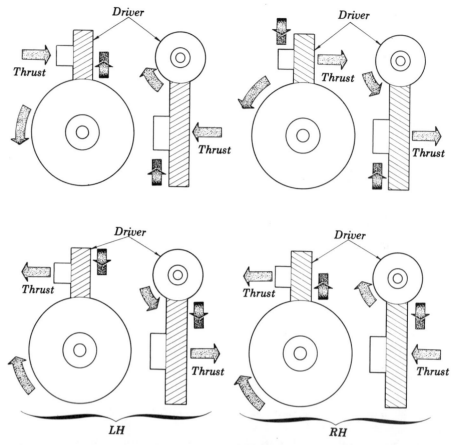

Figure 7.13 Direction of thrust loads for crossed helical gears. A diagram of this type is convenient for quickly determining the direction in which thrust loads act. (*Source:* Precision Industrial Components Corporation, Wells-Benrus Corporation.)

Since, as mentioned earlier, the normal diametral pitches and normal modules of crossed helical gears must be equal ($P_1^n = P_2^n$), we finally obtain

$$r_v = \frac{N_1}{N_2} \tag{7.30}$$

It can thus be seen that the velocity ratio defined *in terms of the number of teeth* is identical for both parallel helical gears and crossed helical gears. On the other hand, while parallel helical gears have a velocity ratio equal to the ratio of the pitch diameters, Eq. 7.17, crossed helical gears in general have their velocity ratio not equal to the pitch diameter ratio. See Eq. 7.29. Figure 7.13 shows the directions for the thrust loads for crossed helical gears. The force equations of Section 7.1.5 apply to this gear type as well.

◆ **EXAMPLE PROBLEM 7.4** *Analysis of Crossed Helical Gears*

Two crossed helical gears have a normal module of 15 mm. The driver has 20 teeth and a helix angle of 20°. The angle between the shafts of the driver and follower is 50°, and the velocity ratio is $\frac{1}{2}$. The driver and the follower are both right-handed. Determine the center distance.

Solution. The helix angle of the follower is found by using the known helix angle of the driver, the shaft angle, and Eq. 7.25a:

$$\psi_2 = \Sigma - \psi_1 = 50° - 20° = 30°$$

The number of teeth for the follower is found by using the number of teeth of the driver and the velocity ratio formula, Eq. 7.30:

$$N_2 = \frac{N_1}{r_v} = \frac{20}{1/2} = 40 \text{ teeth}$$

Finally, the center distance for a pair of crossed helical gears is found by using Eq. 7.28:

$$c = \frac{m^n}{2} \left(\frac{N_1}{\cos \psi_1} + \frac{N_2}{\cos \psi_2} \right) = \frac{15 \text{ mm}}{2} \left(\frac{20}{\cos 20°} + \frac{40}{\cos 30°} \right) = 506 \text{ mm} \quad ◆$$

7.3 WORM GEARS

If large speed reduction ratios are necessary between nonparallel shafts, crossed helical gears with a small driver and large follower are required. However, the magnitude of the load that can be transmitted by these gears is very limited. A better solution to the problem is the use of a worm and worm gear. However, it should be stated that worm gearsets can be considered a special case of crossed helical gears.

In Figure 7.14, a typical worm gearset is shown. As can be seen from the figure, the worm is very similar to a screw. In fact, the teeth on a worm are often spoken of as threads. The worm gear, sometimes called a worm wheel, is a helical gear.

If an ordinary cylindrical helical gear is used, the contact is a point. However, worm gears are usually cut with a concave rather than a straight width. See Figure 7.14. This results in the worm gear partially enclosing the worm, thus giving line contact. Such a set, which is called a single-enveloping worm gearset, can transmit much more power. If the worm is also manufactured with its length concave rather than straight, the worm teeth will partially enclose the gear teeth, as well as the gear teeth partially enclosing the worm teeth. Such a gearset, shown in Figure 7.15, is known as a double-enveloping worm gearset and will provide still more contact between gears, thus permitting even greater power transmission.

Figure 7.14 A cylindrical worm gear-set. This gearset is single-enveloping. (*Source:* Horsburgh & Scott Company.)

Figure 7.15 A double-enveloping worm gear-set. The worm body is concave so that the worm encloses, in addition to being enclosed by, the gear. (*Source:* Excello Corporation.)

Alignment is extremely important for proper operation of worm gearsets. For single-enveloping sets, the worm gear must be accurately mounted, while for double-enveloping sets, both the worm and worm gear must be accurately mounted. Worm gearsets are usually manufactured by using a hobbing tool to cut the worm gear teeth, and a milling cutter or a lathe to cut the worm.

7.3.1 Worm Gear Terminology and Geometry

A consideration of the geometry of worm gearsets would be informative at this point. The *axial pitch* of a worm is equal to the distance between corresponding points on adjacent threads measured along the axis of the worm. See Figure 7.16a. The axial pitch of the worm and the circular pitch of the gear are equal in magnitude if the shaft axes are 90° apart.

Lead

The lead of a worm is equal to the apparent axial distance that a thread advances in one revolution of the worm. For a single-threaded worm (a worm with one tooth), the lead is equal to the pitch. A double-threaded worm, a worm with two teeth, has a lead equal to twice the pitch. The worm of Figure 7.16a has a triple thread and thus a lead equal to three times the pitch. Therefore, the lead and the pitch of a worm are related by the following equation:

$$l = p_w N_w \tag{7.31}$$

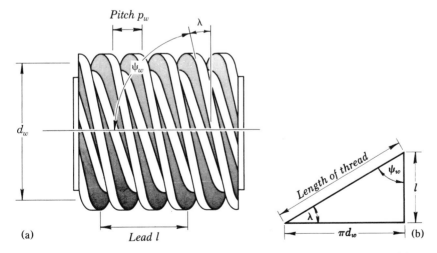

(a)   Pitch p_w λ ψ_w d_w Lead l

(b)   Length of thread ψ_w l λ πd_w

Figure 7.16 (a) The relationship between the pitch and lead for a worm with a triple thread. (b) One tooth of a worm is shown unwrapped to illustrate the relationship between the lead, the lead angle, the pitch diameter, and the helix angle. Note that the lead and helix angles are complementary.

where $l = $ lead

$\quad\quad p_w = $ axial pitch of the worm

$\quad\quad N_w = $ number of teeth of the worm

In the SI system

$$l = \pi m_w N_w \tag{7.32}$$

where $m_w = $ module of the worm. The triangle shown in Figure 7.16b represents the unwrapping of one tooth of the worm shown in Figure 7.16a. As seen from Figure 7.16b,

$$\tan \lambda = \frac{l}{\pi d_w} \tag{7.33}$$

Since the lead angle λ and the helix angle ψ_w of the worm are complementary to each other, Eq. 7.33 can be rewritten as

$$\cot \psi_w = \frac{l}{\pi d_w} \tag{7.34}$$

For shafts 90° apart (the usual case for worm gearsets), the lead angle of the worm and the helix angle of the gear are equal.

Once the worm pitch diameter has been determined, the gear pitch diameter can be found by using the equation for the center distance:

$$c = \frac{d_w + d_g}{2} \tag{7.35}$$

7.3.2 Velocity Ratio of Worm Gearsets

The velocity ratio for worm gearsets is derived as was the velocity ratio for crossed helical gears. In the following derivation, the subscript w refers to the worm while the subscript g refers to the worm gear. Thus

$$r_v = \frac{\omega \text{ (follower)}}{\omega \text{ (driver)}} = \frac{\omega_g}{\omega_w} = \frac{N_w}{N_g} \tag{7.36}$$

From Eq. 7.31, the number of teeth on the worm is given by

$$N_w = \frac{l}{p_w}$$

The number of teeth on the worm gear is given by

$$N_g = P_g d_g$$

But $P_g = \pi/p_g$; therefore,

$$N_g = \frac{\pi d_g}{p_g}$$

Substituting the above values for N_w and N_g into the expression for the velocity ratio, we obtain

$$r_v = \frac{N_w}{N_g} = \frac{l p_g}{p_w \pi d_g}$$

Since the axial pitch of the worm, p_w, and the circular pitch of the gear, p_g, are equal, the velocity ratio for the worm gearset becomes

$$r_v = \frac{l}{\pi d_g} \tag{7.37}$$

In most worm gearsets, the worm is the driver. The set is therefore a speed-reduction unit. It is possible for the gear to be the driver, thus making the set a speed-increasing unit. Whether a given set is reversible or not depends on how much frictional force exists between the worm and gear. Almost all worm gearsets are irreversible because of the frictional force developed. Gearsets that are irreversible are usually referred to as self-locking. Small lead angles (10° or less) usually result in irreversible gearsets.

There are some applications where a self-locking gearset is a distinct advantage. For example, in a hoisting machine application, a self-locking gearset would be an advantage because of the braking action it provides. The designer, however, must be certain that the braking capacity of a gearset is sufficient to perform satisfactorily as a self-locking unit and should include a secondary braking device to ensure safety.

♦ **EXAMPLE PROBLEM 7.5** *Analysis of a Worm Gearset*
A quadruple-threaded worm has an axial pitch of 1 in and a pitch diameter of 2 in. The worm drives a gear having 42 teeth. Determine the lead angle of the worm and the center distance between worm and gear.

Solution. Since we know that the worm has four threads ($N_w = 4$) and that the axial pitch is 1 in ($p_w = 1$), we can easily determine the lead of the worm from Eq. 7.31:

$$l = p_w N_w = 1(4) = 4 \text{ in}$$

Substituting this value into Eq. 7.33, we can determine the lead angle (λ) of the worm:

$$\tan \lambda = \frac{l}{\pi d_w} = \frac{4}{\pi(2)} = 0.637$$

$$\lambda = 32.5°$$

We can now determine the center distance. Since we must have the pitch diameters of the gear and worm in order to use the formula for the center distance, Eq. 7.35, we must first determine the pitch diameter of the gear. Eq. 7.37 gives us an expression containing this unknown. The value of the velocity ratio can be found by using Eq. 7.36:

$$r_v = \frac{N_w}{N_g} = \frac{4}{42} = 0.095$$

And from Eq. 7.37,

$$d_g = \frac{l}{\pi r_v} = \frac{4}{\pi (0.095)} = 13.4 \text{ in}$$

Finally, having found the pitch diameter of the gear, d_g, we can now use Eq. 7.35 to find the center distance:

$$c = \frac{d_w + d_g}{2} = \frac{2 + 13.4}{2} = 7.7 \text{ in} \quad \blacklozenge$$

7.3.3 Forces in Worm Gearsets

The forces acting on worms and worm gears are the same as those for helical gears, except that the shaft axes are 90° apart in almost all practical applications. The free-body diagrams of a worm and worm gear shown in Figure 7.17 depict the relationships of the forces for a normal force F_n exerted between the worm and gear.

First, considering the worm gear and using Eqs. 7.19 to 7.21, we see that the components of the resultant force F_n consist of a tangential force,

$$F_{tg} = F_n \cos \phi'' \cos \psi_g$$

a radial force,

$$F_{rg} = F_n \sin \phi'' = F_{tg} \tan \phi$$

and an axial or thrust force,

$$F_{ag} = F_n \cos \phi'' \sin \psi_g = F_{tg} \tan \psi_g$$

where ϕ'' and ϕ are the normal and transverse pressure angles, respectively, and ψ_g is the helix angle of the worm gear. Since the helix angle of the gear is equal to the lead angle of the worm (for 90° shafts), these forces can be expressed as

$$F_{tg} = F_n \cos \phi'' \cos \lambda$$

$$F_{rg} = F_n \sin \phi'' = F_{tg} \tan \phi$$

and

$$F_{ag} = F_n \cos \phi'' \sin \lambda = F_{tg} \tan \lambda$$

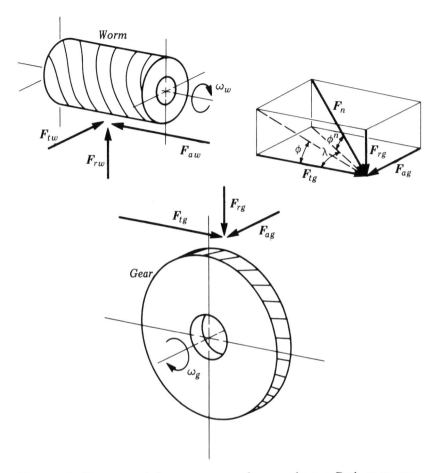

Figure 7.17 Forces exerted on a worm and engaged gear. Both gears are right-handed with the worm driving in the direction shown. The force diagram shows the component breakdown of the resultant normal force F_n on the gear. Note that $\lambda = \psi_g$.

Next, from an examination of the free body of the worm in Figure 7.17, it is clear that the magnitude of the axial or thrust force, F_{aw}, on the worm is equal to that of the tangential gear force F_{tg} (directions are opposite), the magnitude of the tangential force F_{tw} on the worm is equal to that of the axial gear force F_{ag}, and the magnitudes of the two radial force components, F_{rw} and F_{rg}, are equal.

The relationships of these forces to the shaft torques about the respective axes of rotation are

$$T_w = F_{tw} r_w$$

and

$$T_g = F_{tg} r_g$$

where r_w and r_g are the pitch radii of the worm and the gear. Note that, as pointed out above, the forces F_{tw} and F_{tg} are not equal, as is the case for a pair of spur gears or a pair of helical gears on parallel shafts. Instead, observing that $F_{tw} = F_{ag}$, we have

$$T_w = F_{ag} r_w = F_{tg} \tan \lambda r_w$$

leading to the following equation for torque ratio, utilizing Eqs. 7.33 and 7.37:

$$\frac{T_g}{T_w} = \frac{F_{tg} r_g}{F_{tg} \tan \lambda r_w} = \frac{1}{\tan \lambda} \left(\frac{r_g}{r_w} \right) = \frac{2\pi r_w}{l} \left(\frac{r_g}{r_w} \right) = \frac{\pi d_g}{l} = \frac{1}{r_v}$$

This relationship follows from the fact that, in the absence of friction and other losses, the power into a gearset $(T_w \omega_w)$ is equal to the power out $(T_g \omega_g)$. The torque ratio derived therefore assumes a power transmission efficiency of 100 percent. Actually, efficiencies of considerably less than 100 percent can occur in worm gear units.

7.4 BEVEL GEARS

When power is to be transmitted between two shafts that intersect, the type of gear usually used in this situation is a bevel gear. The pitch surfaces of two mating bevel gears are rolling cones, rather than the rolling cylinders that two mating spur gears have. Figure 7.18 shows a typical pair of meshing bevel gears. While the shafts that bevel gears are mounted on are usually 90° apart, there are applications for which the shaft angle is greater or less than 90°.

7.4.1 Bevel Gear Terminology and Geometry

Some of the more common terms used in bevel gearing are illustrated in Figure 7.19 and are described below. As seen in the illustration, the tooth size decreases along the face width as the apex of the pitch cone is approached. The pressure angle for most straight bevel gears is 20°. Due to manufacturing difficulties, bevel gear teeth are not involute but instead have a profile referred to as an octoid, based on a conjugate path of contact that is part of a figure-8 form.

   Pitch cone: the geometric shape of bevel gears based on equivalent rolling contact. The pitch cone is analogous to the pitch cylinders of spur gears.

   Apex of pitch cone: the intersection of the elements making up the pitch cone.

   Cone distance: the cone distance is the slant height of the pitch cone. In other words, it is the length of a pitch cone element.

   Face cone: the cone formed by the elements passing through the top of the teeth and the apex.

Figure 7.18 A pair of bevel gears. The pitch surfaces for the two gears are rolling cones, which can be clearly visualized. (*Source:* Browning Manufacturing Company.)

Root cone: the cone formed by the elements passing through the bottom of the teeth and the apex.

Face angle: the angle between an element of the face cone and the axis of the gear.

Pitch angle: the angle between an element of the pitch cone and the axis of the gear.

Root angle: the angle between an element of the root cone and the axis of the gear.

Face width: the width of a tooth.

Addendum: the distance from an element on the pitch cone to an element on the face cone, measured on the outside of the tooth.

Dedendum: the distance from an element on the pitch cone to an element on the root cone, measured on the outside of the tooth.

Addendum angle: the angle between an element on the pitch cone and an element on the face cone.

Dedendum angle: the angle between an element on the pitch cone and an element on the root cone.

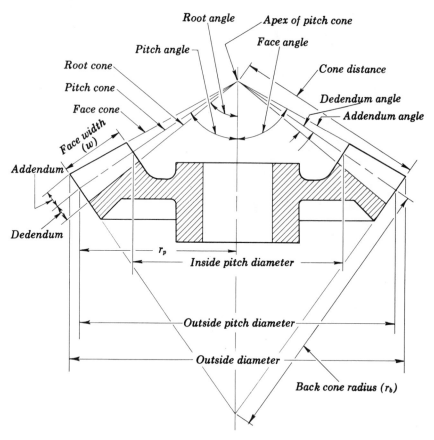

Figure 7.19 Terminology associated with bevel gears.

Inside pitch diameter: the pitch diameter measured on the inside of the tooth.

Outside pitch diameter: the pitch diameter measured on the outside of the tooth.

Back cone: the cone formed by elements perpendicular to the pitch cone elements at the outside of the teeth.

Back cone radius: the length of a back cone element, r_b.

Since most bevel gears are mounted on intersecting shafts, at least one is usually mounted *outboard*. That is, one gear is mounted on the cantilevered end of a shaft. Because of the outboard mounting, the deflection of the shaft where the gear is attached may be rather large. This would result in the teeth at the small end moving out of mesh. The load would thus be unequally distributed, with the larger ends of the teeth taking most of the load. For a reduction of this effect, the tooth face width is usually made no greater than $\frac{1}{3}$ of the cone distance.

Classifying Bevel Gears by Pitch Angle

Bevel gears are usually classified according to their pitch angle. A bevel gear having a pitch angle of 90° and a plane for its pitch surface is known as a *crown gear*. A crown gear is shown in Figure 7.20.

When the pitch angle of a bevel gear exceeds 90°, it is called an *internal bevel gear*. Internal bevel gears, like that shown in Figure 7.21, cannot have pitch angles very much greater than 90° because of the problems incurred in manufacturing such gears. In fact, these manufacturing difficulties are the main reason why internal bevel gears are rarely used.

Bevel gears with pitch angles less than 90° are the type most commonly used. Figure 7.19 illustrates such an external bevel gear.

When two meshing bevel gears have a shaft angle of 90° and have the same number of teeth, they are called *miter gears*. Miter gears have a speed ratio of one. Each of the two gears has a 45° pitch angle.

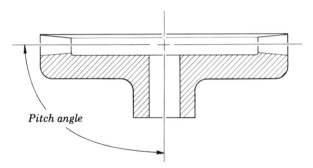

Pitch angle

Figure 7.20 A crown gear is a bevel gear with a pitch angle of 90°. The entire pitch surface lies in a single plane perpendicular to the gear axis.

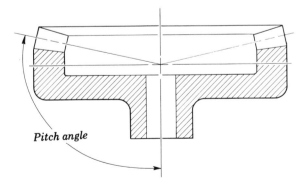

Pitch angle

Figure 7.21 An internal bevel gear is a bevel gear with a pitch angle greater than 90°.

7.4.2 Tredgold's Approximation for Bevel Gears

The profile of bevel gear teeth on the back cone is very similar to the profile of a spur gear having a pitch radius r_b. See Figure 7.22. Using the properties of this equivalent spur gear of pitch radius r_b to approximate the properties of a bevel gear is known as Tredgold's approximation for bevel gears.

It is clear that the imaginary spur gear has more teeth than the bevel gear that it approximates. The number of teeth on the equivalent spur gear is known as the formative or virtual number of teeth and is given by

$$N_f = d_b P = 2r_b P$$

where r_b is the back cone radius of the bevel gear, and P is the diametral pitch. From Figure 7.22, it can be seen that

$$\cos \Gamma = \frac{r_p}{r_b} \text{ or } r_b = \frac{r_p}{\cos \Gamma}$$

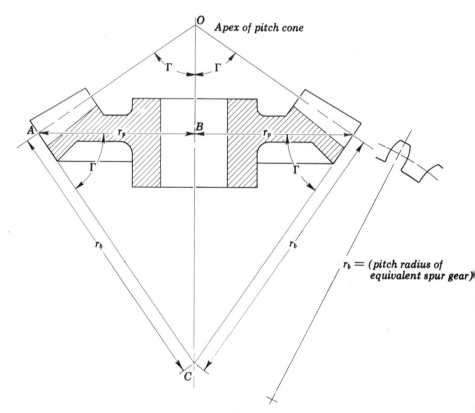

Figure 7.22 The geometry for determining the relationship (Tredgold's approximation) between the actual number of teeth for a bevel gear, the pitch angle, and the formative number of teeth. The back cone radius r_b becomes the pitch radius for the equivalent spur gear.

and N_f becomes

$$N_f = \frac{2r_p}{\cos \Gamma} P$$

But $2r_p P = N$; therefore,

$$N_f = \frac{N}{\cos \Gamma} \tag{7.38}$$

In other words, the virtual or formative number of teeth of the equivalent spur gear is equal to the actual number of teeth for the bevel gear divided by the cosine of the pitch angle of the bevel gear. As with helical gears, the formative number of teeth need not be an integer, whereas the actual number of teeth must be an integer.

7.4.3 Velocity Ratio of Bevel Gears

The velocity ratio for bevel gears is given by the same expressions used to determine the velocity ratio for spur gears, where the subscripts 1 and 2 refer to the driver and follower:

$$r_v = \frac{\omega_2}{\omega_1} = \frac{r_1}{r_2} = \frac{N_1}{N_2} \tag{7.39}$$

where r is the pitch circle radius and N is the number of teeth.

At this point it is desirable to derive some relationships between numbers of teeth and pitch angles for bevel gears. In Figure 7.23, which shows two external bevel gears in mesh, Σ is the shaft angle, γ and Γ are the pitch angles, and r_p and r_g are the pitch radii for the pinion and gear, respectively. From Figure 7.23,

$$\Sigma = \Gamma + \gamma$$

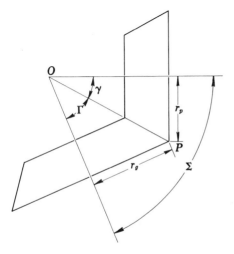

Figure 7.23 Two meshing external bevel gears, illustrating the relationships among the shaft angle, the pitch angles, and the pitch radii.

and

$$\sin \Gamma = \frac{r_g}{OP} \quad \text{and} \quad \sin \gamma = \frac{r_p}{OP}$$

$$OP = \frac{r_g}{\sin \Gamma} = \frac{r_p}{\sin \gamma}$$

$$\sin \Gamma = \frac{r_g}{r_p} \sin \gamma = \frac{r_g}{r_p} \sin(\Sigma - \Gamma)$$

$$= \frac{r_g}{r_p}(\sin \Sigma \cos \Gamma - \cos \Sigma \sin \Gamma)$$

Dividing this expression for $\sin \Gamma$ by $\cos \Gamma$, we obtain

$$\frac{\sin \Gamma}{\cos \Gamma} = \tan \Gamma = \frac{r_g}{r_p}(\sin \Sigma - \cos \Sigma \tan \Gamma)$$

$$\tan \Gamma = \frac{(r_g/r_p)\sin \Sigma}{1 + (r_g/r_p)\cos \Sigma} = \frac{\sin \Sigma}{(r_p/r_g) + \cos \Sigma}$$

At this point, we can utilize our original velocity ratio formula to arrive at the desired formula involving both the pitch angles and the numbers of teeth for both bevel gears. From Eq. 7.39,

$$\frac{r_p}{r_g} = \frac{N_p}{N_g}$$

Therefore,

$$\tan \Gamma = \frac{\sin \Sigma}{(N_p/N_g) + \cos \Sigma} \tag{7.40}$$

Similarly,

$$\tan \gamma = \frac{\sin \Sigma}{(N_g/N_p) + \cos \Sigma} \tag{7.41}$$

Finally, for shaft angle $\Sigma = 90°$ (which is usually the case), we obtain

$$\tan \Gamma = \frac{1}{(N_p/N_g) + 0} = \frac{N_g}{N_p} \tag{7.42}$$

and

$$\tan \gamma = \frac{1}{(N_g/N_p) + 0} = \frac{N_p}{N_g} \tag{7.43}$$

The formulas just derived are quite useful to the designer. The shaft angle, or angle at which one shaft intersects the other, and the required speed ratio are usually known to the designer. Since the speed ratio is also equal to the ratio of the number of teeth, it should be obvious that the formulas can thus be used to calculate the pitch angle required for each gear.

♦ **EXAMPLE PROBLEM 7.6** *Analysis of Bevel Gears*
A pair of straight-tooth bevel gears are mounted on shafts that intersect each other at an angle of 70°. The velocity ratio of the gears is $\frac{1}{2}$. Determine the pitch angles of the gears.

Solution. Knowing the velocity ratio, we can use Eqs. 7.40 and 7.41 to find the pitch angles for bevel gears with a shaft angle other than 90°. Thus

$$\tan \Gamma = \frac{\sin \Sigma}{(N_p/N_g) + \cos \Sigma}$$

and since $N_p/N_g = r_p/r_g = \frac{1}{2}$,

$$\tan \Gamma = \frac{\sin 70°}{1/2 + \cos 70°} = \frac{0.940}{0.842} = 1.12$$

$$\Gamma = 48.2°$$

Similarly,

$$\tan \gamma = \frac{\sin \Sigma}{(N_g/N_p) + \cos \Sigma} = \frac{0.940}{2.342} = 0.401$$

$$\gamma = 21.8° \quad ♦$$

7.4.4 Other Types of Bevel Gears

There are a number of other types of bevel gears in addition to straight-toothed gears. Spiral bevel gears, Figure 7.24, are used when high-speed, high-load applications occur. For these gears, the transmission of power is much smoother than for straight bevel gears, since there is gradual tooth contact in addition to more teeth being in contact at any instant.

While a wide variety of spiral angles are used, the 35° spiral angle is most common. The spiral angle is the angle of the spiral relative to the axis of the gear, as measured at the middle of the face width of the tooth. It is helpful, in understanding spiral bevel gears, to compare them to straight bevel gears in the same manner that helical gears are compared to spur gears.

For shafts that are nonintersecting, hypoid gears (Figure 7.25) are used. While hypoid gears are similar to spiral bevel gears, they are stronger and quieter in operation. Since the shafts on which they are mounted are nonintersecting, they may be placed between bearing supports. Gears used on automobile axles are usually hypoid gears because they allow the drive shaft to be placed closer to ground level, and they also permit large speed reduction ratios in a relatively compact space.

Figure 7.24 Spiral bevel gears feature a spiral rather than a helical tooth design. However, for simplification of the manufacture of spiral bevel gears, the tooth is usually made circular instead of spiral. (*Source:* Arrow Gear Company.)

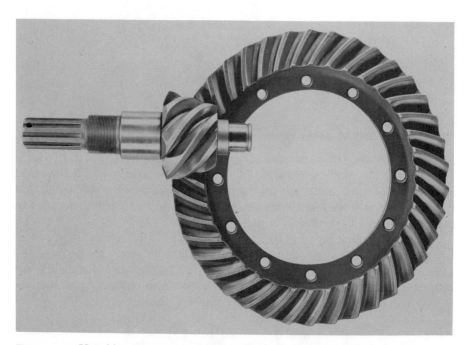

Figure 7.25 Hypoid gears are spiral bevel gears designed to operate on nonintersecting shafts. (*Source:* Richmond Gear, Wallace Murray Corporation.)

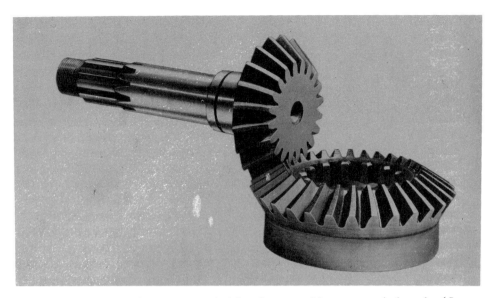

Figure 7.26 Zerol bevel gears are spiral bevel gears with a zero spiral angle. (*Source:* Richmond Gear, Wallace Murray Corporation.)

Zerol bevel gears, shown in Figure 7.26, are the same as spiral gears, except that the spiral angle is zero. Zerol gears are used when it is desirable to reduce the thrust loads that occur when spiral bevel gears are used.

While there are other types of bevel gears, those that have been discussed are the types most often used. It is suggested that the various gear manufacturers' catalogs be consulted when information about other specialized gears is desired.

7.4.5 Forces on Bevel Gears

Let us consider the forces exerted by a bevel gear on its mating gear (see Figure 7.27). We will draw the force diagram for one of the gears, assuming we have straight-tooth bevel gears mounted on shafts that are 90° apart. A reasonable assumption, and one that we will make, is that the resultant tooth load acts at the center of the tooth. In other words, the force acts at the mean pitch radius, r_m, given by

$$r_m = \frac{d_i + d_o}{4}$$

where d_i is the inside pitch diameter and d_o is the outside pitch diameter of the gear (see Figure 7.19).

In Figure 7.27, the force F_n is the normal force exerted by the driven gear on the driving pinion teeth for the direction of rotation shown. A force of

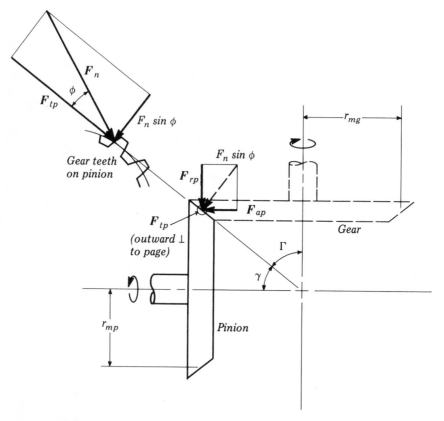

Figure 7.27 Tooth loads on straight-toothed bevel gears. The forces shown are those acting on the driving pinion, rotating in the direction shown. Equal and opposite forces act on the driven gear.

equal magnitude and opposite direction will act on the driven gear. If friction is neglected, these are the resultant gear mesh forces on the individual gears.

For a pressure angle ϕ, the component forces on the pinion, from Figure 7.27, are a tangential force

$$F_{tp} = F_n \cos \phi \tag{7.44}$$

a radial force,

$$F_{rp} = F_n \sin \phi \cos \gamma = F_{tp} \tan \phi \cos \gamma \tag{7.45}$$

and an axial force,

$$F_{ap} = F_n \sin \phi \sin \gamma = F_{tp} \tan \phi \sin \gamma \tag{7.46}$$

For the gear, the tangential, radial, and axial force components are as follows:

$$F_{tg} = F_n \cos \phi = F_{tp} \tag{7.47}$$

$$F_{rg} = F_n \sin \phi \cos \Gamma = F_{tg} \tan \phi \cos \Gamma \tag{7.48}$$

$$F_{ag} = F_n \sin \phi \sin \Gamma = F_{tg} \tan \phi \sin \Gamma \tag{7.49}$$

It should be obvious from Figure 7.27 that the radial force component for the pinion is equal in magnitude to the thrust or axial component for the gear, while the thrust component for the pinion is equal in magnitude to the radial component for the gear. Therefore, for shafts that are 90° apart,

$$F_{ap} = F_{rg} = F_n \sin \phi \sin \gamma = F_n \sin \phi \cos \Gamma$$

$$F_{rp} = F_{ag} = F_n \sin \phi \cos \gamma = F_n \sin \phi \sin \Gamma$$

This follows from the fact that

$$\gamma + \Gamma = 90°$$

and therefore,

$$\sin \gamma = \sin(90° - \Gamma) = \cos \Gamma$$

and

$$\cos \gamma = \cos(90° - \Gamma) = \sin \Gamma$$

Finally, the gear tooth forces produce the following torques on the pinion and gear:

$$T_p = F_{tp} r_{mp} \tag{7.50}$$

and

$$T_g = F_{tg} r_{mg} = F_{tp} r_{mg} \tag{7.51}$$

where T_p is the torque acting on the pinion and T_g is the torque on the gear. The force and torque expressions above can be used to determine the power transmitted by a bevel gearset.

References

1. "AGMA Standards," American Gear Manufacturers Association.
2. Dudley, D. W. (ed.), *Gear Handbook*, McGraw-Hill, New York, 1962.
3. Shigley, J. E., and C. R. Mischke, *Mechanical Engineering Design*, McGraw-Hill, New York, 1989.

PROBLEMS

7.1 Two helical gears, mounted on parallel shafts, are in mesh. They have a diametral pitch of 5, a 40° helix angle, and 20 and 30 teeth, respectively. Calculate the normal circular pitch and the virtual number of teeth.

7.2 Two meshing helical gears have a 25° transverse pressure angle and a 20° normal pressure angle. They have a diametral pitch of 10, with 15 and 45 teeth, respectively. Determine the center distance of the equivalent spur gears (spur gears having the same tooth properties as the helical gears).

7.3 Two parallel shafts are spaced 5 in apart. A pair of helical gears are to be selected to provide a velocity ratio of about $\frac{1}{2}$. The normal diametral pitch is to be 6, the normal pressure angle is to be 20°, and the gears are to have at least 20 teeth. Determine the number of teeth for the gears and the transverse pressure angle.

7.4 A helical pinion has a normal pressure angle of 20°, a transverse pressure angle of 25°, and rotates at 2000 rev/min. It is to drive a meshing helical gear so that the speed ratio is $\frac{1}{4}$. The centers of the shafts are 10 in apart. Determine the normal diametral pitch and the pitch diameters if the pinion has 20 teeth.

7.5 For the gears shown in Figure P7.1, label the rotations of each gear and also indicate the direction of the thrust load for each gear if shaft C is the input and rotates counterclockwise as observed from the left.

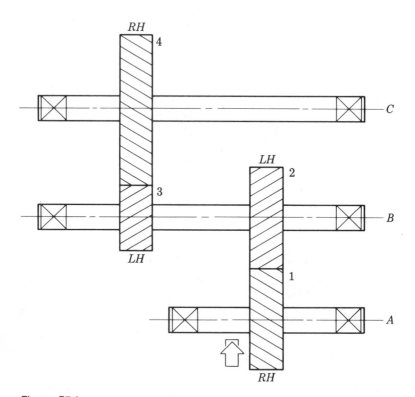

Figure P7.1

7.6 Two meshing helical gears on parallel shafts have a normal pressure angle of 20° and a transverse pressure angle of 23°. The normal circular pitch is 0.6 in. If the speed ratio is to be 0.4, determine the number of teeth for each gear. The center distance is 8 in.

7.7 Two meshing helical gears are mounted on parallel shafts which have rotational speeds of 1000 and 400 rev/min. The helix angle is 30° and the center distance is 252 mm. The gears have a module of 6 mm. Determine the normal circular pitch and the transverse circular pitch. Also determine the actual number of teeth and the formative number of teeth for each gear.

7.8 Two helical gears on parallel shafts have a normal circular pitch of 15 mm and a pitch line velocity of 4500 mm/s. If the rotational speed of the pinion is 800 rev/min and the number of pinion teeth is 20, what must be the helix angle?

7.9 Two helical gears on parallel shafts have 30 teeth and 60 teeth and a normal module of 5 mm. The normal and transverse pressure angles are 20° and 24°, respectively. Determine the center distance and the formative number of teeth of each gear.

7.10 Two helical gears on parallel shafts have a helix angle of 20°, a normal pressure angle of 25°, and a normal diametral pitch of 4. Numbers of teeth are 30 and 50. The pinion rotates at 800 rev/min and the gearset transmits 100 hp. Determine the tangential, radial, and axial gear tooth loads. Show these forces on a sketch of the gears. The pinion is left-handed and rotates counterclockwise.

7.11 Two helical gears on parallel shafts have a normal pressure angle of 20° and a normal module of 6 mm. The center distance is 200 mm and the tooth numbers are 20 and 40. The gearset transmits 50 kW at a pinion speed of 1200 rev/min. Determine the tangential, radial, and thrust loads on the gear teeth, and show these forces on a sketch of the gears. The pinion is right-handed and rotates clockwise.

• **7.12** The double-reduction helical gear train in Figure P7.1 employs helical gears with a helix angle of 30° and a normal pressure angle of 20°. The normal module is 8 mm and the numbers of teeth are $N_1 = 30$, $N_2 = 40$, $N_3 = 20$, and $N_4 = 50$. The pinion 1 has a rotational speed of 1000 rev/min and the power transmitted is 20 kW. Determine the torque carried by each shaft, the magnitude and direction of the thrust force on each shaft, and the resultant gear load on each gear.

7.13 Two helical gears of the same hand are used to connect two shafts that are 90° apart. The smaller gear has 24 teeth and a helix angle of 35°. Determine the center distance between the shafts if the speed ratio is $\frac{1}{2}$. The normal circular pitch is 0.7854 in.

7.14 Repeat Problem 7.13, but assume that the gears are of opposite hand and that the shaft angle is 10°.

7.15 Two left-handed helical gears having the same helix angle are used to connect two shafts 60° apart. The velocity ratio is to be 0.4 and the gears have a normal diametral pitch of 4. If the center distance is to be about 12 in, determine the numbers of teeth for each gear.

7.16 Two right-handed helical gears, with a normal module of 4 mm, connect two shafts that are 60° apart. The pinion has 32 teeth and the velocity ratio is $\frac{1}{2}$. The center distance is 220 mm. Determine the helix angles of the two gears.

7.17 Determine the pitch diameters of a worm gearset having a velocity ratio of 0.1 and a center distance of $2\frac{1}{2}$ in, if the worm has 3 teeth and a lead of 1 in.

● **7.18** A worm gearset has a velocity ratio of 0.05. The worm has 2 teeth, a lead of 3 in, and a pitch diameter of $1\frac{1}{2}$ in. Determine the helix angle and pitch diameter of the worm gear.

7.19 A worm gearset is to have a velocity ratio of 0.05. The worm has 3 teeth, a lead angle of 20°, and a pitch of 1.5 in. Determine the center distance.

7.20 A worm gearset has a velocity ratio of 0.04. Find the center distance if the worm has 3 teeth, a pitch diameter of 2.5 in, and an axial pitch of 0.5 in.

7.21 A worm gearset has a speed ratio of 0.05, a lead angle of 20°, and a center distance of 10 in. Determine the pitch diameters.

7.22 A single-threaded worm has an axial pitch of 20 mm and a pitch diameter of 50 mm. The worm rotates at 500 rev/min and drives a gear having 40 teeth and a 25° transverse pressure angle. The power transmitted is 0.5 kW. Determine **(a)** the lead angle of the worm, **(b)** the center distance between worm and gear, and **(c)** the tangential, radial, and thrust forces on the worm gear.

7.23 For the gear train of Figure 7.1a, assume that the small parallel helical gear is the input and rotates counterclockwise at 1000 rev/min. The power transmitted is 1.0 kW. The helical gears on parallel shafts have a 20° normal pressure angle, a 20° helix angle, a center distance of 120 mm, and teeth numbers of 24 and 48. The worm is single-threaded with a pitch of 20 mm, and the worm gear has 40 teeth and a 25° transverse pressure angle. The center distance of the worm gearset is 150 mm. Determine the torque and thrust load for each shaft, and sketch the intermediate shaft showing the gear force components acting on it.

7.24 Two bevel gears are to be used to connect two shafts that are 90° apart. The pinion has 18 teeth and a diametral pitch of 6. If the velocity ratio is to be 0.4, determine **(a)** the pitch angles, **(b)** the back cone radii, and **(c)** the virtual number of teeth for the gear.

7.25 Repeat Problem 7.24 for two shafts that are 60° apart.

7.26 Repeat Problem 7.24 for two shafts that are 110° apart.

● **7.27** Two 20° straight bevel gears have a diametral pitch of 4, and 24 and 48 teeth. The tooth face width is 2 in. The pinion rotates at 1000 rev/min and transmits 50 hp. The shafts are at 90°. Determine the components of the gear tooth force and show these on a sketch of the gears.

Drive Trains: Design and Analysis

8.1 INTRODUCTION

Most electric motors, internal-combustion engines, and turbines operate efficiently and produce maximum power at high rotating speeds—speeds much higher than the optimum speeds for operating machinery. For this reason, gear trains and other speed reducers are commonly used with industrial and domestic engines. In this chapter, various drive trains will be examined in terms of their operation, that is, the manner in which they effect a speed reduction.

Some gear trains permit us to change output speed even though the input speed remains constant. A pair of gears may be removed from the train and replaced by a pair having a different speed ratio. When the speed ratio change is required only occasionally, a design of this type is satisfactory. For rapid or frequent speed ratio changes, pairs of gears having different ratios are engaged by shifting the location of the gears themselves and by employing bands and clutches within the transmission. Other continuously variable (stepless) transmissions are available, employing belts, chains, and friction drives.

8.2 VELOCITY RATIOS FOR SPUR AND HELICAL GEAR TRAINS

8.2.1 Spur Gear Trains

We know from the preceding chapters that for any pair of gears with fixed centers, the angular velocity ratio is given by

$$\left| \frac{n_2}{n_1} \right| = \frac{N_1}{N_2} \tag{8.1}$$

where $|n_2/n_1|$ is the absolute value of the ratio of speeds of rotation, and N_1 and N_2 represent the number of teeth in each gear. Since the diametral pitch P (the number of teeth per inch of pitch diameter) must also be equal in the case of meshing gears on parallel shafts, pitch diameters d_1 and d_2 must be proportional, respectively, to N_1 and N_2:

$$P = \frac{N_1}{d_1} = \frac{N_2}{d_2} \quad \text{and} \quad \frac{N_1}{N_2} = \frac{d_1}{d_2}$$

Thus, for straight spur gears, the above equation may be written as

$$\frac{n_2}{n_1} = -\frac{N_1}{N_2} = -\frac{d_1}{d_2} \tag{8.2}$$

The minus sign refers to the fact that the rotaton direction changes for any pair of external gears. If one of the gears in the pair is a ring gear (an internal gear), then

$$\frac{n_2}{n_1} = \frac{N_1}{N_2} = \frac{d_1}{d_2} \tag{8.3}$$

and there is no change in the direction of rotation.

The result is the same if the gears conform to standard metric sizes, where the module m is defined as the pitch diameter (millimeters) divided by the number of teeth. The module is common to a pair of meshing gears from which

$$m = \frac{d_1}{N_1} = \frac{d_2}{N_2}$$

leading to Eqs. 8.2 and 8.3.

8.2.2 Helical Gear Trains

Helical gears are often used in place of spur gears to reduce vibration and noise levels. A pair of helical gears on *parallel shafts* will have the *same* helix angle, but one will be right-hand and the other left-hand. The hand is defined as it is for screw threads. As in the case of spur gears, the speed ratio for helical gears on parallel shafts is $n_2/n_1 = -N_1/N_2 = -d_1/d_2$, where N is again the number of teeth and d the pitch diameter.

In the case of *crossed* helical gears (gears on nonparallel shafts), the velocity ratio is still equal to the inverse ratio of the number of teeth: $|n_2/n_1| = N_1/N_2$. While helical gears on parallel shafts must have equal helix angles (of opposite hand), *crossed* helical gears usually have *unequal* helix angles (usually the same hand). This means that the velocity ratio for crossed helical gears is, in general, *not* equal to the ratio of the pitch diameters. In general, the velocity ratio for helical gears is given by

$$\left| \frac{n_2}{n_1} \right| = \frac{d_1 \cos \psi_1}{d_2 \cos \psi_2} \tag{8.4}$$

where d_1 and d_2 are the pitch diameters and ψ_1 and ψ_2 are the respective helix angles. See the section on crossed helical gears in Chapter 7.

8.2.3 Worm Drives

A worm drive is a special case of a pair of helical gears on crossed shafts. Most worm drives have a 90° shaft angle. In that case, using subscript 1 for the worm and subscript 2 for the worm gear, we have

$$\frac{n_2}{n_1} = \frac{d_1 \cos \psi_1}{d_2 \sin \psi_1} = \frac{d_1/d_2}{\tan \psi_1} = \left(\frac{d_1}{d_2} \right) \tan \lambda$$

where λ is the lead angle of the worm.

Helical gears may be designed to transmit power between a pair of shafts at any angle to one another as long as the shaft centerlines do not intersect. Bevel gears may be used to transmit power between shafts that have intersecting centerlines. Note that Eqs. 8.1 and 8.4 apply to any pair of helical gears (except in planetary trains), while Eqs. 8.2 and 8.3 require that the gears be on parallel shafts.

In the case of a worm drive, the direction of worm gear rotation may be determined by analogy to the screw and nut. If a right-hand worm turns clockwise, the worm gear teeth in contact with the worm move toward the observer. Rotation direction should be marked on each gear when solving gear train problems.

8.2.4 Idlers

An idler may be described as a gear placed between, and meshing with, the input and output gears. Its purpose is to reverse the direction of the output. Thus, an idler gear affects the sign of the angular velocity ratio. For the gear train of Figure 8.1,

$$\frac{n_2}{n_1} = -\frac{N_1}{N_2} \quad \text{and} \quad \frac{n_3}{n_2} = -\frac{N_2}{N_3}$$

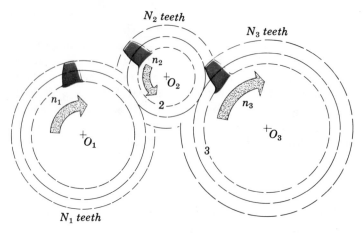

Figure 8.1 Gear train with idler. The idler affects the direction of rotation but not the numerical value of the speed ratio.

Multiplying the first equation by the second, we obtain

$$\left(\frac{n_3}{n_2}\right)\left(\frac{n_2}{n_1}\right) = \left(-\frac{N_2}{N_3}\right)\left(-\frac{N_1}{N_2}\right)$$

which reduces to

$$\frac{n_3}{n_1} = \frac{N_1}{N_3} \tag{8.5a}$$

The number of teeth in gear 2, the idler, does not affect the velocity ratio of the train. However, the idler does affect the direction of rotation of the output gear and, of course, takes up space. Since gear 2 meshes with both gears 1 and 3, all three gears must have the same diametral pitch P or the same module m. Thus, for spur gears, we may also write

$$\frac{n_3}{n_1} = \frac{d_1}{d_3} \tag{8.5b}$$

since $P = N_1/d_1 = N_3/d_3$ and $m = d_1/N_1 = d_3/N_3$.

Occasionally, several idlers are used to transmit power between shafts that are too far apart for the use of a single pair of gears. With an odd number of idlers, the driver and driven shafts rotate in the same direction. With an even number of idlers, one shaft turns clockwise and the other counterclockwise.

8.2.5 Reversing Direction

It is seen in the above work that the introduction of an idler in a spur gear train changes the velocity direction of the output. Both the driver and the driven gear rotate in the same direction with one idler in a simple train. When

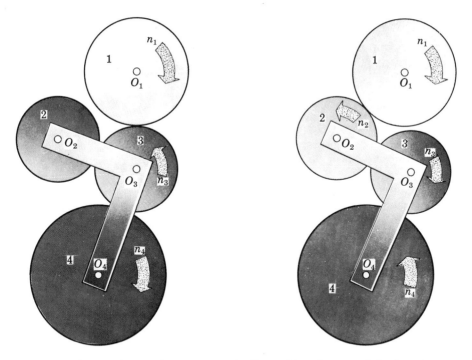

Figure 8.2 A reversing gearbox power train. In position shown at *left*, there is one idler in the train, and input and output shafts turn in the same direction. There are two idlers in the train when the arm is moved to the position shown at *right*, and the output shaft direction is reversed.

two idlers are inserted in the train, the driver and driven gear rotate in opposite directions. In Figure 8.2, gears 1 and 4 have fixed centers O_1 and O_4, while gears 2 and 3 rotate in bearings that are held in an arm. Reversing trains of this type have been used in lathes. The velocity ratio when the arm is fixed in the position shown in Figure 8.2, *left*, is given by

$$\frac{n_4}{n_1} = \frac{N_1}{N_4}$$

where n refers to rotation speed and N refers to number of teeth.

When the arm is rotated about O_4 to the position shown in Figure 8.2, *right*, the velocity ratio becomes

$$\frac{n_4}{n_1} = -\frac{N_1}{N_4}$$

This type of train is satisfactory for occasional direction changing but unsatisfactory for frequent direction changes because the gears do not always correctly engage when shifted. In some cases, the gears must be manipulated by hand before they will fully mesh.

8.2.6 Double Reductions

The minimum number of teeth that a spur gear may have is limited by considerations of contact ratio and interference. While there is no theoretical *maximum* number of teeth, practical considerations like cost and overall size may prevent the designer from specifying a gear with more than, say, one hundred teeth. A speed reduction of the order of one hundred to one can be accomplished in two to four steps with a transmission requiring as many pairs of spur gears and considerable space. A double or triple reduction is thus used in preference to a single pair of gears in cases where the required speed reduction is so great that the output gear of a single pair of spur gears would have to be unreasonably large.

The double reduction of Figure 8.3a is called a *reverted* gear train because the output shaft is in line with the input shaft. Examining this gear train, we see that gears 2 and 3 are keyed to the same shaft and have the same angular velocity:

$$n_3 = n_2 = -n_1\left(\frac{N_1}{N_2}\right)$$

and that

$$n_4 = -n_3\left(\frac{N_3}{N_4}\right)$$

from which the ratio of output speed to input speed becomes

$$\frac{n_4}{n_1} = \frac{N_1 N_3}{N_2 N_4} \tag{8.6a}$$

Since gears 1 and 2 must have the same diametral pitch, and since gears 3 and 4 must also have the same diametral pitch, Eq. 8.6a may be rewritten as

$$\frac{n_4}{n_1} = \frac{d_1 d_3}{d_2 d_4} \tag{8.6b}$$

Note that gear 1 drives gear 2, and that gear 3 drives gear 4. By adding more pairs of gears and examining the results, we find the general relationship to be

$$\left|\frac{n_{\text{output}}}{n_{\text{input}}}\right| = \frac{\text{product of driving gear teeth}}{\text{product of driven gear teeth}} \tag{8.7}$$

which applies to all gear trains in which the shaft centers are fixed in space. The torque-arm speed reducer shown in Figure 8.3b is a practical example of such a double reduction.

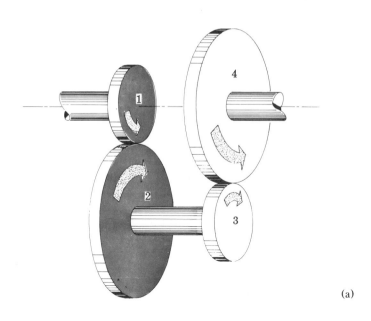

(a)

(b)

Figure 8.3 (a) A reverted gear train. The input and output shafts have the same centerline. The speed ratio equals the product of the tooth numbers of the driving gears divided by the product of the tooth numbers of the driven gears. (b) Torque-arm speed reducer. This double reduction is available with 1 : 15 and 1 : 25 output-to-input speed ratios. The helical involute teeth have an elliptoid form, slightly narrower at the ends for more even load distribution. (*Source:* Reliance Electric Company.)

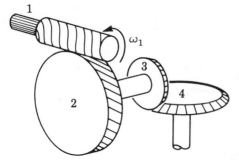

Figure 8.4 Speed reducer.

◆ **EXAMPLE PROBLEM 8.1**   *Speed Reducer*

Figure 8.4 shows a gear train that is to produce a 50-to-1 speed reduction. Given data are as follows:

	Gear 1, Worm	Gear 2 Worm Gear	Gear 3, Straight Bevel	Gear 4, Straight Bevel
Tooth numbers	2		20	40
rev/min	1000			20

Find the number of worm gear teeth. Find speeds and directions.

Solution.   Using Eq. 8.7, we have $|n_{\text{output}}/n_{\text{input}}| =$ (product of driving gear teeth)/(product of driven gear teeth), from which $20/1000 = 2 \times 20/(N_2 \times 40)$, yielding $N_2 = 50$ teeth. Using Eq. 8.1, $|n_2/n_1| = N_1/N_2$, or $n_2/1000 = 2/50$, from which $n_2 = n_3 = 40$ rev/min. Gears 2 and 3 turn counterclockwise and gear 4 turns counterclockwise as viewed from above in Figure 8.4.   ◆

8.3 SPEED RATIO CHANGE

Idlers may also be used to permit the changing of speed ratios. An arm holds idler gear 2 to drive gear 1 in Figure 8.5. Gear 1 is keyed or splined to the input shaft and turns with it. The arm A does not rotate with the input shaft and gear 1 but is connected to gear 1 by a sleeve. The arm is moved axially along the input shaft with gear 1 when a new speed ratio is required. The speed ratios available are

$$\frac{n_{\text{output}}}{n_{\text{input}}} = \frac{N_1}{N_3}, \frac{N_1}{N_4}, \frac{N_1}{N_5}, \cdots$$

A typical industrial lathe may have a "cone" of as many as 12 gears on the output shaft of a train similar to Figure 8.5. This train is partly responsible for the wide variation of feeds that are available. (Feed refers to the movement of the cutting tool along or into the workpiece.) With two other speed selectors, the total number of feeds available and therefore the number of

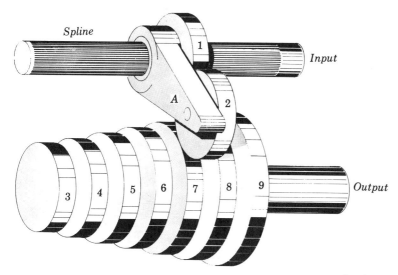

Figure 8.5 Speed changer employing an idler. Gear 1 turns the input shaft. Arm *A* keeps idler gear 2 in contact with gear 1 and one of the gears on the output shaft. Arm *A* is fixed in space except when changing gears.

different pitches of screw threads that can be cut would be 48. Since thread cutting requires a high degree of accuracy, a gear drive, which provides precise speed ratios, is ideal for this application.

The operator of a lathe or other machine tool may find it quite satisfactory to interrupt production in order to change speed ratio. Practical, efficient, and safe vehicle operation, however, requires a smooth, quick transition from one speed ratio to another without completely stopping the machine; different types of transmissions are therefore necessary.

8.3.1 Transmissions with Axial Shifting

Automotive transmissions include gear trains with axial shifting, fluid drive units, planetary gear trains, and combinations of the above. The transmission shown in Figure 8.6a is called a three-speed transmission even though it offers a reverse speed ratio in addition to three forward speed ratios and a neutral position. The axial distance between gears in the sketch has been exaggerated for clarity; a typical transmission (like the one shown in Figure 8.6b) would be more compact.

In Figure 8.6a, shafts *A*, *B*, and *D* and the shaft of idler gear 6 turn in bearings mounted in the transmission housing, but bearings and parts of the shafts have been omitted from the sketch. Gear 1 is an integral part of input shaft *A*. Output shaft *D* is not directly connected to the input, but power may be transmitted to it through clutch *C* or through gears on the countershaft *B*. In the position shown, no torque is transmitted between input and output

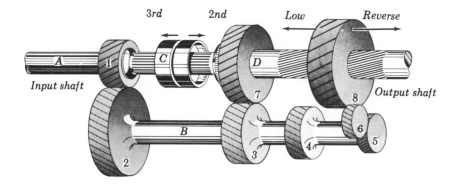

(a)

AUTOMATIC TRANSAXLE

PLANET
CARRIER
ASSEMBLY

PLANETARY
RING GEAR

SUN GEAR

SPLITTER GEAR

(b)

Figure 8.6 (a) A three-speed transmission. (b) Automatic transaxle. A three-speed automatic transaxle designed for certain front-wheel-drive automobiles. A "splitter" gearset within the torque converter improves efficiency. In third (or drive) gear, 93 percent of the torque is transmitted mechanically as in a manual transmission. Only 7 percent is transmitted through the torque converter where slippage reduces efficiency. (*Source:* Ford Motor Company.)

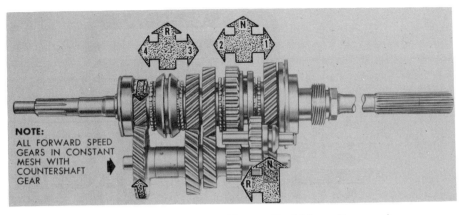

Figure 8.6 (c) A four-speed transmission. (*Source:* Ford Motor Company.)

shafts because gear 7 is not rigidly connected to output shaft D but turns freely on it. This position is called *neutral*.

Clutch C and gear 1 have cone-shaped internal and external mating faces, respectively. Clutch C and the end of the output shaft on which it rides are splined so that the two turn together, even though C may be moved axially on the shaft. When C is made to contact the conical face of gear 1, the two begin to move at the same speed. In addition to the conical friction faces, C and gear 1 have internal and external mating "teeth" that ensure a positive drive after the initial contact synchronizes the input and output shaft speeds. The synchronizer teeth are smaller than the gear teeth since all of the synchronizer teeth are engaged at once. The resulting direct-drive, or one-to-one speed ratio, is called *high* or *third gear* in the three-speed transmission.

The engine is not directly connected to the input shaft, but a disk clutch (not shown) is installed between the two. Synchronizing the input and output shafts is essentially independent of the engine speed when the disk clutch is disengaged. The word *clutch* above ordinarily refers to the disk clutch, while clutch C is called a *synchromesh clutch*. The synchromesh clutch is shifted by a fork that rides in an annular groove.

We see in Figure 8.6a that gears 1 and 2, gears 3 and 7, and gears 5 and 6 mesh at all times. Gears 2, 3, 4, and 5 are integral parts of countershaft B. Therefore, all gears except gear 8 turn at all times when the input shaft is in motion. One face of gear 7, like gear 1, mates with clutch C. Since output shaft D and the internal surface of gear 7 are smooth where they contact, clutch C must engage the clutch face of gear 7 when power is transmitted through that gear. The effect is a reverted gear train of gears 1, 2, 3, and 7, or symbolically, the path of power transmission from input to output is A–1–2–3–7–C–D. Output-to-input speed ratio is given by

$$\frac{n_D}{n_A} = \frac{N_1 N_3}{N_2 N_7}$$

where N_1 is the number of teeth in gear 1, and so on. Ratios N_1/N_2 and N_3/N_7 are both less than one and there is a speed reduction. This position is called *second gear*. The disk clutch is disengaged when shifting so that the speed of gear 7 can be synchronized with the output shaft speed.

Gear 4 is made smaller than gear 3 so that engaging gear 8 with gear 4 produces an even lower speed ratio called *low* or *first gear*, symbolically noted as A–1–2–4–8–D. The speed ratio is given by

$$\frac{n_D}{n_A} = \frac{N_1 N_4}{N_2 N_8}$$

Of course, the shifting mechanism must be designed so that clutch C is first disengaged from gear 7 or we would have simultaneously two different speed ratios. (Actually, the result would be a locked gear train or a broken transmission.) If there is no synchromesh in first gear, it is best to shift when the output and input shafts are stationary. This is the case when the vehicle is stationary and the disk clutch has been disengaged a few moments before shifting. Shifting of gear 8 along the splined portion of the output shaft is accomplished by a fork that rides in a grooved ring (not shown). The ends of the teeth of gears 4, 6, and 8 are rounded to facilitate engagement.

Gear 5 is made slightly smaller than gear 4 so that gear 8 cannot mesh with it but may be shifted to mesh with idler gear 6. For first, second, and third gears examined above, the ratios were positive; the output shaft turned in the same direction as the input shaft. But when engaged with gear 5, idler gear 6 causes an odd number of direction changes, and the output-to-input speed ratio is given by

$$\frac{n_D}{n_A} = -\frac{N_1 N_5}{N_2 N_8}$$

This arrangement is the *reverse gear*, symbolically given by A–1–2–5–6–8–D. When one is shifting to reverse, as when shifting to first gear, the disk clutch is disengaged, and both input and output shafts are stationary. Except for reverse and first gear ratios, gear 8 must turn freely, engaging neither gear 4 nor gear 6.

Helical gears are often selected for transmissions because of their greater strength and smoother, quieter operation. Meshing helical gears on parallel shafts are of opposite hand, where a right-hand helical gear resembles a right-hand screw. If gear 1 of Figure 8.6a is left-hand, then gear 2 must be right-hand. If gear 1 turns counterclockwise (as seen from the right), creating a thrust to the left on gear 1, there is a thrust to the right on the countershaft at gear 2. When in second gear, gear 3 will have a balancing thrust (to the left) if it is a right-hand helix. Gear 4 is also a right-hand helix so that countershaft thrust is balanced in first gear. Since gear 8 meshes with both gears 4 and 6, gear 6 must be right-hand, making gear 5 left-hand. Thus thrust is not balanced when in reverse. Finally, all speed ratios and paths of power transmission for the transmission of Figure 8.6a are summarized in Table 8.1. Automatic transmissions including transaxles employ several gear trains. Fig-

TABLE 8.1 POSITION, PATH, AND SPEED RATIOS FOR A THREE-SPEED
TRANSMISSION, FIGURE 8.6a

Gear	Position of synchro-mesh clutch, C	Position of gear 8	Path of transmitted power	Output-to-Input speed ratio n_D / n_A
Neutral	Center	Center	—	—
Third (high)	Left	Center	A–1–C–D	$+1$
Second	Right	Center	A–1–2–3–7–C–D	$+\dfrac{N_1 N_3}{N_2 N_7}$
First (low)	Center	Left	A–1–2–4–8–D	$+\dfrac{N_1 N_4}{N_2 N_8}$
Reverse	Center	Right	A–1–2–5–6–8–D	$-\dfrac{N_1 N_5}{N_2 N_8}$

ure 8.6b shows a transaxle with a torque converter, differential, and planetary
gear trains. Methods of analysis of planetary gear trains and differentials are
discussed in the following pages.

Automotive transmissions with four or more forward speed ratios are
available, some including synchromesh in all forward gears (see Figure 8.6c).
While there are many innovations among manufacturers, the basic principles
are often the same. Although the above discussion centers around speed
reduction, in a few instances it is desirable to increase speed. Kinematically,
the equations apply to speed increases as well as reductions. Friction losses,
however, make large increases in speed impossible.

8.3.2 Designing for a Particular Speed Ratio

For many applications, it is necessary to have a specific relationship between
output and input speed. When the required relationship is a ratio of small
whole numbers, say one-half or five-sevenths, we have a wide selection of
satisfactory pairs of gears to choose from. However, some speed ratios are
impossible to obtain exactly and others may be impractical to obtain exactly.
An example of a speed ratio that cannot be obtained with gears is the square
root of two, an irrational number. An example of a speed ratio difficult to
obtain exactly (from a practical standpoint) is 503/2003, the ratio of two prime
numbers.

In the first case, we cannot express the desired ratio as a fraction made up
of whole numbers and thus cannot select a corresponding set of gear tooth
numbers. The second example involves a pair of numbers, neither of which can
be factored. An exact solution involving a pair of gears with 503 and 2003 teeth
might be prohibitively expensive. Either problem, however, may be solved if a
small variation from the desired ratio is permitted.

In some instances, a table of factors may be used to advantage. Suppose, for example, we needed the exact speed ratio $n_o/n_i = 1501/1500$, where n_o and n_i represent the output and input speeds. For a gear train similar to Figure 8.3a, $n_4/n_1 = +N_1 N_3/(N_2 N_4)$, leading us to try $N_1 N_3 = 1501$ and $N_2 N_4 = 1500$. Fortunately, both numbers are factorable; $1501 = 19 \times 79$ and $1500 = 2^2 \times 3 \times 5^3$. Letting $N_1 = 19$, $N_3 = 79$, $N_2 = 2^2 \times 5 = 20$, and $N_4 = 3 \times 5^2 = 75$, we have the desired ratio exactly. When the desired ratio consists of a pair of large numbers that cannot be factored to give reasonable gear sizes, we may then be forced to try a more complicated and expensive solution or an approximate solution.

8.4 PLANETARY GEAR TRAINS

Gearsets of the type shown in Figure 8.7a through 1 are called *epicyclic* or *planetary* gear trains. In planetary trains, one or more gears are carried on a rotating planet carrier rather than on a shaft that rotates on a fixed axis. Several types of gear trains may be shifted manually to obtain greater or lesser values of speed reduction. The shifting process, however, is difficult to accomplish automatically with gears that rotate about fixed centers. On the other hand, planetary gear trains are readily adapted to automatic control. Some planetary gear trains are designed to change ratios simply by using electrically or hydraulically operated band brakes to keep one or more of the gears stationary. Other planetary trains operating with fixed gear ratios are selected for their compact design and high efficiency.

Lévai[7] identified twelve possible variations of planetary trains. These are shown in Figure 8.7a through 1 using simplified schematics. Other possible variations include planetary trains in which the planet carrier rotates freely, serving as neither input nor output of the mechanism.

The simple planetary gear train sketched in Figure 8.7a consists of a sun gear (S) in the center, a planet gear (P), a planet carrier or arm (C), and an internal, or ring, gear (R). The sun gear, ring gear, and planet carrier all rotate about the same axis. The planet gear is mounted on a shaft that turns in a bearing in the planet carrier; the planet gear meshes with both the sun gear and the ring gear. Since the planet gear does not rotate about a fixed center, some of the rules developed for gears rotating about fixed centers must be reexamined.

8.4.1 Formula Method (Train Value Formulation Method) for Solving Planetary Trains

We noted earlier that for gears in which the shaft centers are fixed in space, the train value is given by

$$\left| \frac{n_{\text{output}}}{n_{\text{input}}} \right| = \frac{\text{product of driving gear teeth}}{\text{product of driven gear teeth}}$$

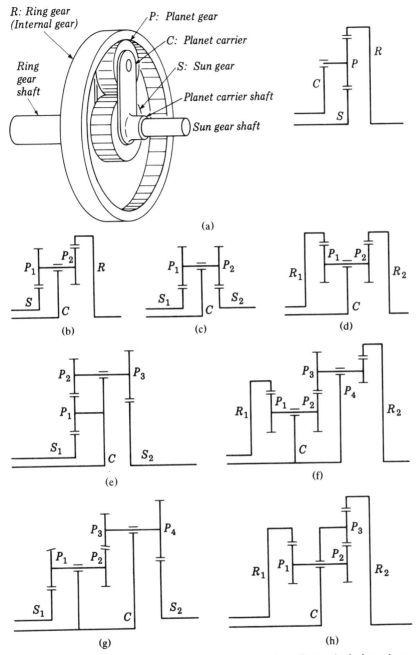

Figure 8.7 Planetary gear train types identified by Lévai. Part a includes a key to the skeleton diagrams.

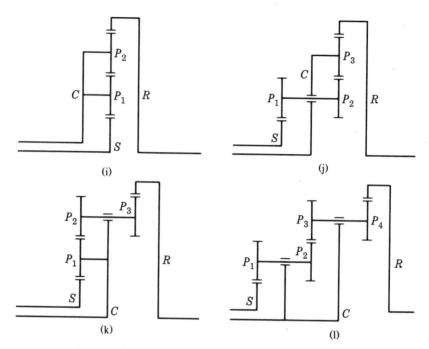

Figure 8.7 *Continued.*

This equation does not directly apply to planetary trains, but we will use the train value to aid in solving planetary train problems.

If the actual speeds of two of the gears in a planetary train are known, we arbitrarily designate one of them as the input gear and the other as the output gear. We then find the train value with the carrier fixed:

$$r^* = (\pm) \frac{\text{product of driving gear teeth}}{\text{product of driven gear teeth}} \tag{8.8}$$

based on our arbitrary designation of input and output gears. The asterisk is used to designate the condition that the planet carrier does not rotate. The sign of r^* is positive if the arbitrarily designated input and output gears would rotate in the same direction with the carrier fixed. Of course, we must correct for planet carrier rotation.

Let gears x and y be designated as input and output gears, respectively. With the concept of relative velocity, n_x^*, the speed of gear x relative to carrier C, is given by the difference in rotation speeds:

$$n_x^* = n_x - n_c$$

where n_x and n_c are, respectively, actual speeds of gear x and carrier C. Likewise, for gear y, the speed relative to the carrier is

$$n_y^* = n_y - n_c$$

from which the train value with the carrier fixed is

$$r^* = \frac{n_y^*}{n_x^*} = \frac{n_y - n_c}{n_x - n_c} \tag{8.9}$$

Eqs. 8.8 and 8.9 are the basis for the formula method of solving for planetary train speed ratios. To apply them, we first designate gears x and y. The designation is arbitrary except that we would designate gears whose speed we knew or wished to know. Planet gears would not ordinarily be so designated. Let the number of teeth in each gear be given. The fixed carrier train ratio r^* is determined as if x and y were input and output gears, respectively. Then, with Eq. 8.9, if two of the speeds n_x, n_y, and n_c are given, the third can be found.

Suppose, for example, that the speeds n_S and n_R are given for the sun gear and ring gear, respectively, in the simple planetary train shown in Figure 8.7a. Let the sun and ring gears be arbitrarily designated as input and output gears, respectively. Then,

$$n_x = n_S \qquad n_y = n_R$$

From Eq. 8.8, noting that the direction of rotation changes when two external gears mesh but does not change when an external gear meshes with an internal gear, we obtain

$$r^* = \left(\frac{-N_S}{N_P} \right) \left(\frac{+N_P}{N_R} \right) = -\frac{N_S}{N_R}$$

From Eq. 8.9,

$$r^* = \frac{n_R - n_C}{n_S - n_C}$$

Combining the two, the result for the train shown in Figure 8.7a is

$$\frac{n_R - n_C}{n_S - n_C} = -\frac{N_S}{N_R}$$

◆ **EXAMPLE PROBLEM 8.2**   *Planetary Train Analyzed by the Formula Method*
For the planetary train of Figure 8.7a, let tooth numbers be $N_S = 40$, $N_P = 20$, and $N_R = 80$. Find planet carrier speed if the sun gear rotates counterclockwise at 100 rev/min and the ring gear clockwise at 300 rev/min. Use the formula method.

Solution.   If we arbitrarily select the sun as the input gear and the ring as the output gear, the train value (Eq. 8.8) is

$$r^* = -\frac{N_S}{N_R} = -0.5$$

Using Eq. 8.9, where $n_x = -100$ and $n_y = 300$, we find

$$r^* = \frac{300 - n_C}{-100 - n_C}$$

If we equate the two expressions for r^*, the result is $n_c = 166.7$ rev/min clockwise. ◆

In some cases, the formula method must be applied in two steps. For the planetary speed reducer of Figure 8.9 appearing later in this chapter, for example, we may let $n_x = n_S$ and $n_y = n_{R_1}$ in Eq. 8.9 to find n_C if n_S and n_{R_1} are known. Then, using the value of n_C we have just found and letting $n_x = n_S$ and $n_y = n_{R_2}$ in Eq. 8.9, we may find n_{R_2}.

8.4.2 Input and Output Shafts

In most cases, gear trains are used to obtain a speed reduction. Other applications include a reversal in direction, a differential effect, a speed increase, or even a one-to-one, input-to-output relationship. In the skeleton diagrams of 12 planetary train types identified by Lévai, Figure 8.7a to l, carrier, sun gear, and ring gear shafts serve as potential input and output shafts. Referring to Figure 8.7a, for example, the ring gear may be fixed by a band brake. Then, if the sun gear shaft is used as the input and the carrier shaft as output, there will be a speed reduction. If the band brake is released, allowing the ring gear to rotate freely, and if a clutch engages the sun gear to the carrier, a one-to-one speed ratio results. If, instead, the sun gear in Figure 8.7a is fixed, the ring gear shaft can be used as the input and the planet carrier shaft as output. Considering the planetary trains shown, with different combinations of gear tooth numbers, the number of possible output-to-input speed ratios is virtually limitless.

8.4.3 Tabular Analysis (Superposition), an Alternative Method for Analysis of Planetary Trains

As observed above, the rotation of the planet carrier complicates the problem of determining gear speeds in a planetary train. However, by the simple device of calculating rotation *relative to the carrier* and combining it with the rotation of the entire train turning as a unit, we can find velocities in two steps. If the planet carrier is kept stationary so that the centers of all gears are fixed, gear speed ratios equal the inverse of the ratios of tooth numbers.

If the sun or ring gear is actually fixed, the constraint is (theoretically) temporarily relaxed and that gear is given one turn. The effect on the entire train is calculated. Of course, since the net rotation of the fixed sun or ring gear must be zero, a compensating rotation must take place to correct its position. That compensating motion is one rotation of the entire locked train *in the opposite direction*.

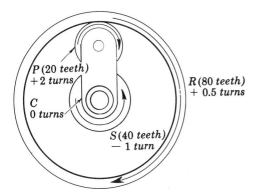

Figure 8.8 Planetary train. This figure corresponds to step 2 in Table 8.2.

As an example, consider the planetary train of Figure 8.8 where the sun, planet, and ring gears have, respectively, 40, 20, and 80 teeth, with the sun gear fixed. The ring gear is the driver and rotates at 300 rev/min clockwise. As a first step in the solution, let the entire train be locked together and given one clockwise rotation. The sun and ring gears and the planet carrier will each have turned through one clockwise rotation about their common center of rotation.

The above motion has violated the requirement that the sun gear be fixed. In the second step, the sun gear will be given one counterclockwise rotation, yielding a net sun gear motion of zero. While the sun gear is rotated counterclockwise, the carrier will be fixed so that all gears rotate on fixed centers. The results up to this point are tabulated by denoting clockwise rotations as positive ($+$) and counterclockwise rotations as negative ($-$). Thus we are able to construct Table 8.2.

TABLE 8.2 SUPERPOSITION METHOD FOR SOLVING PLANETARY TRAIN SPEED RATIOS, FIGURE 8.8

Gear	Sun	Planet	Ring	Planet carrier
No. of teeth	40	20	80	
Step 1: rotations with train locked	$+1$	$+1$	$+1$	$+1$
Step 2: rotations with planet carrier fixed	-1	$+2$	$+0.5$	0
Total rotations	0	$+3$	$+1.5$	$+1$
Speed (rev/min)	0	600 cw	300 cw	200 cw

As noted above, the planet carrier will be fixed as we complete the problem. One counterclockwise rotation of the sun gear results in

$$n_P = n_S\left(-\frac{N_S}{N_P}\right) = -1\left(-\frac{40}{20}\right) = +2$$

rotations of the planet and

$$n_R = n_P\left(\frac{N_P}{N_R}\right) = +2\left(+\frac{20}{80}\right) = +0.5$$

rotations of the ring, as shown in Figure 8.8. The planet carrier is given zero rotations. Including these values in Table 8.2 and adding each column, we obtain the total rotations. We see that the ring gear makes 1.5 turns for each turn of the planet carrier. Hence, if the ring gear is on the input shaft and the planet carrier on the output shaft, the ratio of output to input is given by $n_o/n_i = 1/1.5 = +2/3$. Actual rotation speeds bear the same relationship. Thus, we divide 300 rev/min, the given speed of the ring, by 1.5, the total rotations, to obtain a factor of 200. Multiplying each figure in the total rotations row by 200, we obtain the actual speeds in revolutions per minute. The reader will observe that, in the second step of the calculations, the planet acts as an idler and affects only the sign of the speed ratio. Thus, it is not necessary to compute rotations of the planet to obtain output to input speed ratios.

Using the data from this example in the formula method, the train value is $r^* = -0.5$. Selecting the sun and ring as the arbitrary input and output gears, respectively (although the ring is the actual input gear and the carrier is the actual output), we have

$$r^* = \frac{300 - n_C}{0 - n_C}$$

from which $n_C = 200$ rev/min (clockwise).

As a second example, let the ring gear in Figure 8.8 be held stationary while the planet carrier, sun, and planet gears are permitted to rotate. In order that the results be more general, the number of teeth in the sun, planet, and ring gears will be represented, respectively, by N_S, N_P, and N_R. In tabulating the solution in Table 8.3, the first step is identical with the previous example. In this case, however, the net rotation of the ring gear must be zero. To accomplish this, the ring gear is given one counterclockwise rotation in the second step of Table 8.3 while the planet carrier remains fixed. Noting that the planet gear acts as an idler in the second step, we obtain $+N_R/N_S$ rotations of the sun gear for -1 rotations of the ring gear. Tabulating these values and adding as before, we obtain the total rotations.

The last line of the Table 8.3 indicates that, with the ring gear fixed, the ratio of sun gear speed to planet carrier speed is

$$\frac{n_S}{n_C} = 1 + \frac{N_R}{N_S}$$

TABLE 8.3 SPEED RATIOS FOR A PLANETARY TRAIN WITH THE RING
FIXED, FIGURE 8.8

Gear	Sun	Planet	Ring	Planet carrier
No. of teeth	N_S	N_P	N_R	
Step 1: rotations with train locked	$+1$	$+1$	$+1$	$+1$
Step 2: rotations with planet carrier fixed	$+\dfrac{N_R}{N_S}$	$-\dfrac{N_R}{N_P}$	-1	0
Total rotations	$1+\dfrac{N_R}{N_S}$	$1-\dfrac{N_R}{N_P}$	0	$+1$

The speed relationship obtained above for this train by the formula method is

$$\frac{n_R - n_C}{n_S - n_C} = -\frac{N_S}{N_R}$$

With the ring gear fixed, we have

$$\frac{0 - n_C}{n_S - n_C} = -\frac{N_S}{N_R}$$

from which we obtain the same result:

$$\frac{n_S}{n_C} = 1 + \frac{N_R}{N_S}$$

Repeating the first example, using symbols to represent the actual numbers of teeth, the ratio of ring gear speed to planet carrier speed becomes

$$\frac{n_R}{n_C} = 1 + \frac{N_S}{N_R}$$

If the gear train operates instead with the planet carrier stationary, it no longer acts as a planetary train. Then, the ratio of sun gear speed to ring gear speed becomes $-N_R/N_S$ to 1 by inspection.

8.4.4 Compound Planetary Trains

A gear train that may be designed for extremely low ratios of output to input speed employs *two* planet gears and *two* ring gears. The reducer of Figure 8.9 is made up of a sun gear S, two ring gears, R_1 and R_2, and two planet gears, P_1 and P_2, which rotate at the same speed. The planets are held in the common planet carrier C, which is free to rotate. The input is the sun gear, and the ring gear R_2 is the output. Ring gear R_1 is fixed.

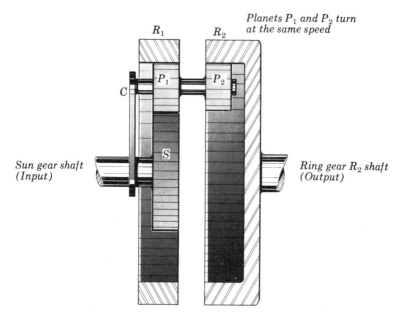

Figure 8.9 Planetary speed reducer. This planetary train employs two planets and two ring gears.

As in the previous examples, we begin by rotating the entire train one turn clockwise (step 1, Table 8.4). In the second step, the planet carrier is fixed, while ring gear R_1 is returned to its original position by one counterclockwise turn. The number of rotations of each gear is entered in Table 8.4, noting that both planets turn at the same speed. From the sum of the motions of steps 1 and 2, it is seen that the ratio of output to input speed is given by

$$\frac{n_{R_2}}{n_S} = \frac{1 - \left[N_{R_1} N_{P_2} / \left(N_{P_1} N_{R_2} \right) \right]}{1 + \left(N_{R_1} / N_S \right)} \tag{8.10}$$

TABLE 8.4   SPEED RATIOS FOR A COMPOUND PLANETARY SPEED
REDUCER WITH RING GEAR R_1 FIXED, FIGURE 8.9

Gear	S	P_1	P_2	R_1	R_2	C
No. of teeth	N_S	N_{P_1}	N_{P_2}	N_{R_1}	N_{R_2}	
Step 1: rotations with train locked	$+1$	$+1$	$+1$	$+1$	$+1$	$+1$
Step 2: rotations with planet carrier fixed	$+\dfrac{N_{R_1}}{N_S}$	$-\dfrac{N_{R_1}}{N_{P_1}}$	$-\dfrac{N_{R_1}}{N_{P_1}}$	-1	$-\dfrac{N_{R_1}}{N_{P_1}}\dfrac{N_{P_2}}{N_{R_2}}$	0
Total rotations	$1+\dfrac{N_{R_1}}{N_S}$			0	$1-\dfrac{N_{R_1}}{N_{P_1}}\dfrac{N_{P_2}}{N_{R_2}}$	$+1$

If the gears are chosen so that the value of the term $N_{R_1}N_{P_2}/(N_{P_1}N_{R_2})$ is very near to unity, the output speed will be very low, making a very high output *torque* available. Of course, if the term equals one (exactly), the reducer will be useless since the output shaft will not turn.

8.4.5 Balanced Planetary Trains

In the above examples, only one planet gear was shown meshing with the ring gear. Kinematically, this is sufficient, but better balancing of gear tooth loads and inertia forces will result if two or more planets mesh with each ring gear. See Figures 8.10, 8.11a, and 8.11b. The planetary gear train of Figure 8.8, for example, can be redesigned with four planets. The planet shafts are mounted in bearings in a planet carrier equivalent to the arrangement in Figure 8.10. Kinematically, the gear train of Figure 8.10 is identical to the train shown in Figure 8.8.

For a simple train of the type shown in Figures 8.8 and 8.10, the pitch diameter of the ring gear is obviously equal to the sum of the pitch diameter of the sun gear plus twice the planet gear pitch diameter. Since the metric module or the diametral pitch must be the same on all of the gears in order that they mesh, tooth numbers are related by the following equation:

$$N_R = N_S + 2N_P \qquad (8.11a)$$

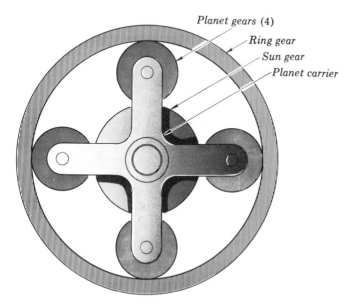

Planet gears (4)
Ring gear
Sun gear
Planet carrier

Figure 8.10 Planetary train with four planets. Forces are more readily balanced in a planetary train with three or four planets. Kinematically, there is no difference; speed ratio remains the same whether one or several planets are used.

(b)

(a)

Figure 8.11 (a) Simple planetary train. A train typical of the wheel planetaries used in drive axles of loaders and scrapers. Although this drive with three planets is kinematically equivalent to a drive with only one planet, this configuration balances the loading and has greater capacity. The planet carrier is not shown. (*Source:* Fairfield Manufacturing Company, Inc.) (b) Compound planetary train. This assembly with two sets of planet gears is used in a motorized wheel. The sun gear is driven by the shaft of an electric motor rated at 400 hp. The outside diameter of the ring gear is over 39 in, and the unit is designed for heavy-duty service. (*Source:* Fairfield Manufacturing Company, Inc.)

where subscripts R, S, and P refer to the ring, sun, and planet, respectively. When several equally spaced planet gears are to be used, as in Figure 8.10, the designer must ensure that it is possible to assemble the train. For example, if $N_S = 25$, $N_P = 20$, $N_R = 65$, Eq. 8.11 is satisfied. However, a layout will show that four planets cannot be equally spaced in the train. Equal spacing is possible only if

$$\frac{N_S + N_R}{\#_P} = \text{an integer} \tag{8.11b}$$

where $\#_P$ = the number of planets. The number of planets is also limited by the requirement that there be clearance between addendum circles of adjacent planets. Figure 8.11a illustrates a balanced train employing three planets.

8.4.6 Planetary Differential Drives

An alternate method of obtaining low ratios of output to input speed employs two sun gears and two planet gears as shown in Figure 8.12. The planet carrier C is keyed to the input shaft, which goes through the center of sun gear S_1. Sun gear S_1 is fixed, and sun gear S_2 is keyed to the output shaft. Planet gears P_1 and P_2 are both keyed to the same shaft, which turns freely in the planet carrier. Ring gears are not used in this speed reducer.

Speed ratios are again found by rotating the entire locked gear train and then correcting the position of sun gear S_1 while the planet carrier remains stationary. Thus we obtain Table 8.5. The result, as given in Table 8.5, is an

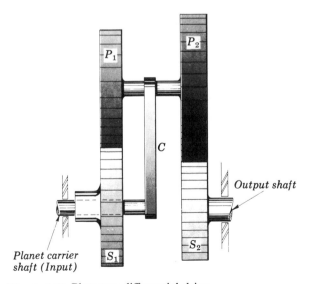

Figure 8.12 Planetary differential drive.

TABLE 8.5 SPEED RATIOS FOR A PLANETARY DIFFERENTIAL DRIVE
WITH SUN GEAR S_1 FIXED, FIGURE 8.12

Gear	S_1	P_1	P_2	S_2	C
No. of teeth	N_{S_1}	N_{P_1}	N_{P_2}	N_{S_2}	
Step 1: rotations with train locked	$+1$	$+1$	$+1$	$+1$	$+1$
Step 2: rotations with planet carrier fixed	-1	$+\dfrac{N_{S_1}}{N_{P_1}}$	$+\dfrac{N_{S_1}}{N_{P_1}}$	$-\dfrac{N_{S_1}N_{P_2}}{N_{P_1}N_{S_2}}$	0
Total rotations	0			$1-\dfrac{N_{S_1}N_{P_2}}{N_{P_1}N_{S_2}}$	1

output-to-input speed ratio:

$$\frac{n_o}{n_i} = \frac{n_{S_2}}{n_C} = 1 - \frac{N_{S_1}N_{P_2}}{N_{P_1}N_{S_2}} \tag{8.12}$$

This result may be checked by the formula method. Considering rotations relative to the carrier, we have

$$\frac{n_{S_2}^*}{n_{S_1}^*} = \frac{N_{S_1}N_{P_2}}{N_{P_1}N_{S_2}} = \frac{n_{S_2} - n_C}{n_{S_1} - n_C}$$

If sun gear S_1 is fixed, then

$$\frac{N_{S_1}N_{P_2}}{N_{P_1}N_{S_2}} = \frac{n_{S_2} - n_C}{0 - n_C}$$

from which

$$\frac{n_{S_2}}{n_C} = 1 - \frac{N_{S_1}N_{P_2}}{N_{P_1}N_{S_2}}$$

as above.

Gear sizes can be specified to produce a wide variety of output-to-input ratios. By selecting gears so that the fraction $N_{S_1}N_{P_2}/(N_{P_1}N_{S_2})$ is approximately one (but not exactly one), we obtain very great speed reductions. For example, let

$$N_{S_1} = 102 \qquad N_{P_2} = 49$$

$$N_{P_1} = 50 \qquad N_{S_2} = 100$$

Then, from Eq. 8.12,

$$\frac{n_o}{n_i} = 1 - \frac{4998}{5000} = + \frac{1}{2500}$$

In considering the above example, the reader may come to an extraordinary conclusion. According to the calculations, if the shaft of S_2 is given one turn, the planet carrier C will make 2500 revolutions. In an actual gear train possessing this speed ratio, however, friction would prevent gear S_2 from driving the train, since a small friction torque on the planet carrier shaft would be so greatly magnified. As is also the case with most worm drives, this particular gear train may only be used to reduce speed. Commercially available planetary trains of this type may be used to *step up* speed by using sun gear S_2 for the input and the planet carrier C for the output. Ordinarily, n_o/n_i is limited to about 10 or 20.

A planetary speed reducer of the type sketched in Figure 8.12 is very flexible; most output-to-input ratios may be approximated simply by selecting the appropriate gear sizes. Exact ratios, however, are not always obtainable. Suppose, for example, that a speed ratio $n_o/n_i = 1/2000$ is called for, where n_o is the speed of S_2, the output, and n_i is the speed of C, the input. Instead of using a trial-and-error method to select a suitable combination of gears, we will examine Eq. 8.12. If the term

$$\frac{N_{S_1} N_{P_2}}{N_{P_1} N_{S_2}} = \frac{1999}{2000}$$

then

$$\frac{n_o}{n_i} = \frac{1}{2000} \text{ (exactly)}$$

Let us, therefore, try to specify the gear sizes so that $N_{S_1} \times N_{P_2} = 1999$, and $N_{P_1} \times N_{S_2} = 2000$. Referring to a table of factors and primes such as that given by Selby and Weast,[8] we see that $2000 = 2^4 \times 5^3 = 2 \times 2 \times 2 \times 2 \times 5 \times 5 \times 5$, which suggests $N_{P_1} = 5 \times 5 = 25$ and $N_{S_2} = 2 \times 2 \times 2 \times 2 \times 5 = 80$. Unfortunately, 1999 has no factors except 1 and 1999; it is a prime number.

Since we cannot manufacture a pair of gears with 1 and 1999 teeth for this application, we refer again to the table of factors and primes. Noting that $2001 = 3 \times 23 \times 29$, we may specify $N_{S_1} = 3 \times 23 = 69$ and $N_{P_2} = 29$, so that $N_{S_1} \times N_{P_2} = 2001$. Then, from Eq. 8.12, $n_o/n_i = 1 - 2001/2000 = -1/2000$. The minus sign indicates that the output and input shafts rotate in opposite directions. If this is objectionable, an additional gear may be added to the train to change the output direction.

Alternatively, the desired speed ratio of $1/2000$ may be approximated by using the same factors, except that in this case

$$N_{P_1} = 29 \qquad N_{P_2} = 25$$

$$N_{S_2} = 69 \qquad N_{S_1} = 80$$

From Eq. 8.12, the result is

$$\frac{n_o}{n_i} = 1 - \frac{80 \times 25}{29 \times 69} = +\frac{1}{2001}$$

a value that in most cases would be close enough to be acceptable.

Manufacturing difficulties are sometimes encountered with reverted gear trains. For the gear train of Figure 8.12, the distance between centers of gears S_1 and P_1 is the same as for S_2 and P_2 since S_1 and S_2 turn about the same axis. Therefore, pitch diameters are related by the equation

$$d_{S_1} + d_{P_1} = d_{S_2} + d_{P_2}$$

(Recall that gears that mesh must have the same diametral pitch.) If the loading is such that a diametral pitch of 10 teeth per inch of diameter is satisfactory for gears S_2 and P_2 in the last example, then

$$d_{S_1} + d_{P_1} = d_{S_2} + d_{P_2} = \frac{25}{10} + \frac{69}{10} = 9.4 \text{ in}$$

The diametral pitch of gears S_1 and P_1 will be

$$P = \frac{N_{S_1} + N_{P_1}}{d_{S_1} + d_{P_1}} = \frac{80 + 29}{9.4} = 11\frac{56}{94}$$

teeth per inch of diameter. Since the above value is not a standard diametral pitch, special cutters must be manufactured, increasing the cost of this gear train. On the other hand, the use of only standard gears limits the number of different speed ratios that may be obtained. The planetary differential drive shown in Figure 8.13a and b offers a wide range of speed ratios due to a broad selection of available component gears.

Nonstandard gears may also result when metric sizes are used. Suppose, for example, we require an output speed of $\frac{1}{16}$ rev/min for an input speed of 100 rev/min. Let us select a transmission similar to that shown in Figure 8.12, using the carrier C as input, sun gear S_2 as output, with sun gear S_1 fixed. Referring to Table 8.5, we see that

$$1 - \frac{N_{S_1} N_{P_2}}{N_{P_1} N_{S_2}} = 0.000625$$

After several attempts, we find that one possible combination of spur gears is as follows:

$$N_{S_1} = 82 \qquad N_{S_2} = 80$$

$$N_{P_1} = 40 \qquad N_{P_2} = 39$$

Speeds are as follows:

$$n_{S_1} = 0 \qquad n_{P_1} = n_{P_2} = 305 \text{ rev/min (cw)}$$

$$n_{S_2} = 0.0625 \text{ rev/min cw} \qquad n_C = 100 \text{ rev/min cw}$$

(a)

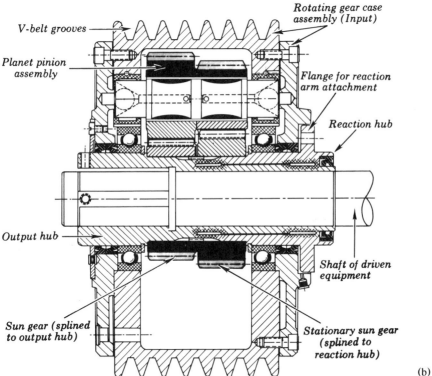

V-belt grooves

Rotating gear case
assembly (Input)

Planet pinion
assembly

Flange for reaction
arm attachment

Reaction hub

Output hub

Shaft of driven
equipment

Sun gear (splined
to output hub)

Stationary sun gear
(splined to
reaction hub)

(b)

Figure 8.13 (a) Planetary differential drive. Mounted within the driven pulley and directly on the shaft of the driven equipment, this speed reducer requires little additional space. A wide range of speed ratios and output torque capacities result from changes in tooth number. The available output-to-input speed ratios range from 1 : 1.79 to 1 : 482, with output direction opposite in some cases. (*Source:* Airborne Accessories Corporation.) (b) Section view of planetary differential drive. The driven pulley of a V-belt drive forms the rotating gear case of the compound planetary drive. The gear case carries the planet assemblies, each consisting of two helical planet gears splined to the same shaft. Power is transmitted to a helical sun gear on the output shaft through differential action between the planets and a fixed helical sun gear.

If we use a standard metric module $m_1 = 4$ for gears S_1 and P_1, the center distance between the shafts is

$$c = \frac{d_{S_1} + d_{P_1}}{2} = \frac{m_1(N_{S_1} + N_{P_1})}{2} = 244 \text{ mm}$$

Then, the metric module m_2 for gears P_2 and S_2 is given by

$$m_2 = \frac{2c}{N_{S_2} + N_{P_2}} = 4.1008$$

which is not a standard value.

8.4.7 Tandem Planetary Trains

For most planetary trains, there is no loss in generality when speed ratios are analyzed by giving the entire train one rotation and then superimposing a correcting rotation with the planet carrier fixed. When the gear train has more than one planet carrier, this procedure, if applied to the train as a whole, would arbitrarily restrict the planet carrier to the same speed—a result that

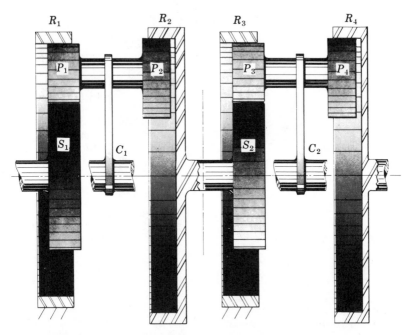

Figure 8.14 Train with two planet carriers. With two independent planet carriers in a train, the reader should assume that they do *not* rotate at the same speed. To solve for the overall speed ratio, the reader can solve for the speed ratio of each half of this train separately, thus correctly using the superposition method. The speed ratio for the entire train is then the *product* of the ratios of the two halves.

may contradict the actual problem constraints. A convenient method of solving tandem planetary trains is to divide the train at some convenient point and solve for the speed ratios for each half. The speed ratio for the entire train is then the product of the separate speed ratios.

Figure 8.14 illustrates a tandem planetary train that divides conveniently between R_2 and R_3. Following the usual procedure for the left side of the train, which includes S_1, P_1, P_2, R_1, R_2 and C_1, we obtain the equivalent of Eq. 8.10:

$$\frac{n_{R_2}}{n_{S_1}} = \frac{1 - \left[N_{P_2} N_{R_1} / (N_{P_1} N_{R_2}) \right]}{1 + (N_{R_1}/N_{S_1})}$$

Similarly, for the last half of the train, we obtain

$$\frac{n_{R_4}}{n_{S_2}} = \frac{1 - \left[N_{R_3} N_{P_4} / (N_{P_3} N_{R_4}) \right]}{1 + (N_{R_3}/N_{S_2})}$$

Since $n_{R_2} = n_{S_2}$, the output-to-input ratio for the entire train is the product of the above ratios:

$$\frac{n_o}{n_i} = \frac{n_{R_2}}{n_{S_1}} \cdot \frac{n_{R_4}}{n_{S_2}}$$

Using a table of factors and primes as in a previous example, we may demonstrate the flexibility and the extreme speed reductions available with this type of gear train. For example, letting

$$N_{R_1} = N_{R_3} = 87 \text{ teeth}$$

$$N_{P_1} = N_{P_3} = 20 \text{ teeth}$$

$$N_{S_1} = N_{S_2} = 47 \text{ teeth}$$

$$N_{P_2} = N_{P_4} = 23 \text{ teeth}$$

$$N_{R_2} = N_{R_4} = 100 \text{ teeth}$$

the speed ratio of each part of the gear train separately is

$$\frac{n_{R_4}}{n_{S_2}} = \frac{n_{R_2}}{n_{S_1}} = -\frac{47}{268,000}$$

The product of the two ratios is the output-to-input speed ratio:

$$\frac{n_o}{n_i} = +30.756 \times 10^{-9} = \frac{+1}{32,514,260}$$

The above set of values is only an illustration of the planetary train's potential for speed reduction. We might never need such extreme ratios in actual practice, but trains of this type are commonly used for speed reductions of one to several thousand.

Figure 8.15 (a) Free-floating planetary transmission with fixed ring gear removed to show details of planet spindles. (*Source:* The Free-Floating Planetary Transmission is a proprietary concept of the Curtiss-Wright Corp., Wood-Ridge, N.J., and is covered by a number of patents including No. 3,540,311 and others pending.)

8.4.8 Free-floating Planetary Transmission

A recent development in planetary transmissions (Figure 8.15a and b) consists of a sun gear S (the input) which drives five planet spindles. Each planet spindle has three planets: P_1, P_2, and P_3. Ring gear R_1 is the output gear, and ring gear R_2 is fixed. There is no planet carrier as such; the planets are constrained by the gear meshes and the spindles that roll on cylindrical rings. All planets rotate at the same speed. All forces and reactions are transmitted through the gear meshes and through rolling cylinders. Planet gears are so spaced that tangential gear tooth forces keep the planet spindles in equilibrium. Tooth separating and centifugal forces are balanced out by the cylindrical rings.

Using the superposition method, we may find relative speeds for the free-planet transmission. Answers are given in Table 8.6 in terms of tooth

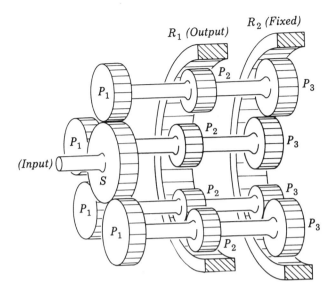

Figure 8.15 (b) Schematic of free-floating planetary transmission. (Schematic shows only four of five planetary spindles.)

numbers N_S, N_{P_1}, and so on, where C refers to motion of the planet spindle centers about the centerline of the transmission since there is no planet carrier.

8.4.9 Speed Changing in Planetary Transmissions

When smooth and rapid changing of speed ratios is required, planetary transmissions are often selected. Automotive requirements, for example, may call for a gear train with two or more positive (forward) speed ratios and one

TABLE 8.6 FREE PLANET TRANSMISSION

	Input		Output	Stationary	
Gear	S	$P_1 P_2 P_3$	R_1	R_2	C
Number of teeth	N_S	$N_{P_1} N_{P_2} N_{P_3}$	N_{R_1}	N_{R_2}	—
1: Train locked together	$+1$	$+1$	$+1$	$+1$	$+1$
2: Correct stationary gear with spindle centers fixed	$+\dfrac{N_{R_2} N_{P_1}}{N_{P_3} N_S}$	$-\dfrac{N_{R_2}}{N_{P_3}}$	$-\dfrac{N_{R_2} N_{P_2}}{N_{P_3} N_{R_1}}$	-1	0
Σ (true relative speed)	$1 + \dfrac{N_{R_2} N_{P_1}}{N_{P_3} N_S}$	$1 - \dfrac{N_{R_2}}{N_{P_3}}$	$1 - \dfrac{N_{R_2} N_{P_2}}{N_{P_3} N_{R_1}}$	0	$+1$

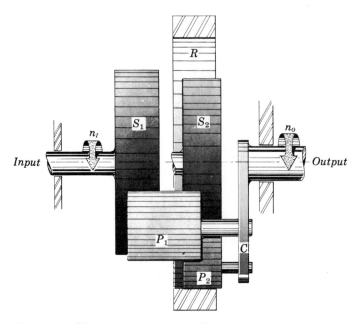

Figure 8.16 Planetary transmission. Planets P_1 and P_2 are carried by the *common* planet carrier. P_1 meshes only with S_1 and P_2. When both S_2 and R are free to turn, there is no output; the transmission is in neutral. S_2 may be locked to the input shaft by engaging a clutch, giving a one-to-one ratio of output to input. See Tables 8.7 and 8.8 for other ratios. (*Note:* Gear positions in an actual transmission do not correspond with this schematic.)

negative (reverse) speed ratio. The speed changes can be accomplished by clutches and brakes in a planetary transmission.

Figure 8.16 illustrates, schematically, such a transmission. With the gears as shown, there is no output since both sun gear S_2 and the ring gear R are free to turn. Gear P_1 meshes only with S_1 and P_2. Thus the carrier C could remain stationary even though the input shaft was turning.

If sun gear S_2 is locked to the input shaft by engaging a clutch (not shown), a direct drive is obtained. Since the two sun gears rotate at the same speed (and in the same direction), the planet gears are, in effect, locked to the sun gears. The planets are also locked to each other since the sun gears tend to rotate the meshed planets in the same direction (preventing any relative motion). Any motion of planets P_1 and P_2 relative to the sun gears would be incompatible with the fact that both planets are held in the same planet carrier. Thus, when S_2 is locked to the input shaft, all gears lock together and turn as a unit. The ratio of output to input speed becomes $n_o/n_i = 1$.

In order to fix either the ring gear or sun gear S_2 to the frame, band brakes may be used. The brakes would engage the outer surface of the ring gear or a drum attached to sun gear S_2. Both bands and drums are omitted

TABLE 8.7 SPEED RATIOS FOR THE PLANETARY TRANSMISSION
OF FIGURE 8.16 WITH SUN GEAR S_2 FIXED

Gear	S_1	P_1	P_2	S_2	R	C
No. of teeth	N_{S_1}	N_{P_1}	N_{P_2}	N_{S_2}	N_R	
Step 1: rotations with train locked	$+1$	$+1$	$+1$	$+1$	$+1$	$+1$
Step 2: rotations with planet carrier fixed	$+\dfrac{N_{S_2}}{N_{S_1}}$	$-$	$-$	-1	$-$	0
Total rotations	$1+\dfrac{N_{S_2}}{N_{S_1}}$	$-$	$-$	0	$-$	$+1$

from Figure 8.16 so that we may examine the kinematics of the problem without additional complications. The effect of locking sun gear S_2 to the frame is given by Table 8.7. In the first step, the entire train is given one clockwise rotation, just as in each of the previous examples. In the second step, the carrier is fixed while the sun gear S_2 is given one counterclockwise rotation so that it has zero net motion.

By adding steps 1 and 2 in Table 8.7, we obtain an output-to-input speed reduction of

$$\frac{n_o}{n_i} = \frac{1}{1 + \left(N_{S_2}/N_{S_1} \right)} \tag{8.13}$$

when gear S_2 is fixed.

When the ring gear is free to rotate (Figure 8.16), it serves no purpose in the transmission. When the ring is fixed and gear S_2 is free to turn, however, a negative or reverse ratio is obtained. The results are given in Table 8.8, which is obtained by proceeding as in Table 8.7 except that, in this case, the ring gear R is given the zero net rotation.

TABLE 8.8 SPEED RATIOS FOR THE PLANETARY TRANSMISSION
OF FIGURE 8.16 WITH THE RING GEAR R FIXED

Gear	S_1	P_1	P_2	S_2	R	C
No. of teeth	N_{S_1}	N_{P_1}	N_{P_2}	N_{S_2}	N_R	
Step 1: rotations with train locked	$+1$	$+1$	$+1$	$+1$	$+1$	$+1$
Step 2: rotations with planet carrier fixed	$-\dfrac{N_R}{N_{S_1}}$	$-$	$-$	$-$	-1	0
Total rotations	$1-\dfrac{N_R}{N_{S_1}}$	$-$	$-$	$-$	0	1

In this case, sun gear S_2 turns freely and serves no useful purpose. The output-to-input speed ratio is

$$\frac{n_o}{n_i} = \frac{1}{1 - (N_R/N_{S_1})} \tag{8.14}$$

with the ring fixed. Since N_R must be greater than N_{S_1}, this ratio is always negative, indicating that the output direction of rotation is opposite the input.

8.4.10 Planetary Trains with More Than One Input

In some planetary gear train applications, *none* of the gears are fixed. As an example, it might be necessary to find the angular velocity n_C of the planet carrier of Figure 8.8 being given n_S and n_R, the velocities of the sun and ring, respectively. A problem of this type was solved above in an example illustrating the formula method. As an alternative, the data already calculated in Tables 8.2 and 8.3 enable us to solve this problem by superposition.

Let the sun have 40 teeth and the ring 80 teeth. Given speeds will be $n_S = +900$ rev/min and $n_R = +1500$ rev/min (both clockwise). Table 8.2 gives gear speeds when the sun is fixed. Since these speeds are only relative, the total rotations line of Table 8.2 may be multiplied by 1000 to give the correct ring speed. We have, then, the following speeds (in revolutions per minute), adjusted to give the correct speed for n_R:

$$n_S = 0$$

$$n_R = 1.5 \times 1000 = 1500$$

$$n_C = 1 \times 1000 = 1000$$

Next, the values of N_S and N_R are substituted into the last line of Table 8.3 and it is multiplied by 300 to give the correct sun gear speed when the ring is fixed. The result is

$$n_S = 3 \times 300 = 900$$

$$n_R = 0$$

$$n_C = 1 \times 300 = 300$$

Adding the values for the case where the sun gear is fixed to the values for the case where the ring is fixed, we obtain

$$n_S = 0 + 900 = +900$$

$$n_R = 1500 + 0 = +1500$$

$$n_C = 1000 + 300 = +1300$$

TABLE 8.9   SPEED RATIOS FOR THE PLANETARY TRAIN OF FIGURE 8.8
WITH NO FIXED GEARS

Gear	Sun	Planet	Ring	Carrier
No. of teeth	40	20	80	
Step 1: rotations with train locked	$+v$	$+v$	$+v$	$+v$
Step 2: rotations with carrier fixed	$+w$	$-\left(\dfrac{N_S}{N_P}\right)w$	$-\left(\dfrac{N_S}{N_R}\right)w$	0
Total rotations	$v+w$	$v-\left(\dfrac{N_S}{N_P}\right)w$	$v-\left(\dfrac{N_S}{N_R}\right)w$	v

The following method avoids making use of previous results. We begin the problem by rotating the entire train $+v$ revolutions (where the value of v is to be found later) instead of $+1$ revolutions as done previously. Thus we have step 1 of Table 8.9. In step 2, the sun gear S is given $+w$ rotations with the carrier fixed (w is also unknown at this time). Noting the gear ratio, we obtain $(-N_S/N_R) \times w$ rotations of the ring gear in step 2. The last line of Table 8.9 is the sum of steps 1 and 2, from which

$$n_S = v + w$$

$$n_R = v - \left(\frac{N_S}{N_R}\right)w$$

$$n_C = v$$

Eliminating v and w from the above equations, we obtain

$$n_R = n_C - \left(\frac{N_S}{N_R}\right)(n_S - n_C) \tag{8.15}$$

from which

$$n_C = \frac{n_R + n_S(N_S/N_R)}{1 + (N_S/N_R)} \tag{8.16}$$

and

$$n_S = n_C + \left(\frac{N_R}{N_S}\right)(n_C - n_R) \tag{8.17}$$

These three equations describe the motion of the planetary train shown in

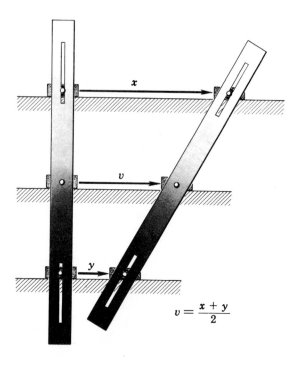

$$v = \frac{x+y}{2}$$

(a)

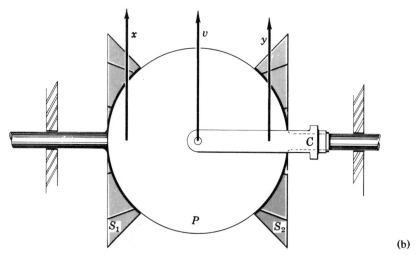

(b)

Figure 8.17 (a) A mechanical-averaging linkage. (b) A bevel gear differential acting as a mechanical averaging linkage.

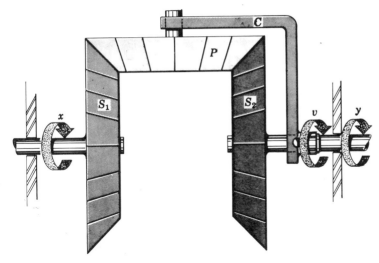

Figure 8.17 (c) When x and y are given as *angular* velocities of S_1 and S_2, respectively, v is the angular velocity of the planet carrier and $v = (x + y)/2$. Table 8.10 illustrates how this result may be obtained by using a variation of the superposition method.

Figures 8.7a and 8.8, *for any input or output given in terms of angular velocity.* When the data of the above problems are substituted into the equation for n_C, we again obtain $n_C = 130$ rev/min clockwise.

8.4.11 Differentials

While engineering problem solving often suggests the use of electronic calcula-tors and computers, there are many specialized applications for mechanical computing devices. Automatic control of a certain process may depend on the addition of two input functions. Mechanically computed products and sums of signals from sensing devices have been used for guidance of aircraft. We can see by inspection of Figure 8.17a that a simple mechanical linkage can average two functions. If the paths of the upper and lower sliders are equidistant from the center slider,

$$v = \frac{x + y}{2}$$

where x and y are the velocities of the upper and lower sliders. Of course, the range of this linkage is very limited. If the displacements of the upper and lower sliders differ appreciably, the linkage will not operate.

The same principle applies to the bevel gear differential. The straight link in the averager of Figure 8.17a is replaced by a planetary bevel gear P in

TABLE 8.10   SPEED RATIO FOR THE BEVEL GEAR
DIFFERENTIAL OF FIGURE 8.17C WITH TWO INPUTS

Gear	S_1	S_2	C
Step 1: rotations with train locked	v	v	v
Step 2: rotations with planet carrier fixed	$+w$	$-w$	0
Total rotations	$x = v + w$	$y = v - w$	v

Figure 8.17b. Input velocities to the planet are x and y, the pitch line velocities of gears S_1 and S_2, respectively, in Figure 8.17b. If $N_{S_2} = N_{S_1}$, it is seen that the velocity of the planet center is

$$v = \frac{x + y}{2}$$

where v lies in the plane of x and y.

Alternatively, the tabular method may be employed with a modification to incorporate two *angular* inputs, Figure 8.17c. In the first step (Table 8.10), the entire locked train is given $+v$ rotations. In the second step, the planet carrier is fixed while gear S_1 is given $+w$ rotations. If S_1 and S_2 have the same number of teeth, then S_2 makes $-w$ rotations. The results are given in Table 8.10, where x and y represent the actual total rotations of S_1 and S_2, respectively.

On adding the values of x and y, we see that $x + y = 2v$, or

$$v = \frac{x + y}{2} \quad \text{or} \quad n_C = \frac{n_{S_1} + n_{S_2}}{2} \tag{8.18}$$

In this mechanism x, y, and v may represent angular displacement, angular velocity, or angular acceleration. Figure 8.18a illustrates a miniature bevel gear differential. The range of this mechanism is unlimited; there are no physical stops such as are present in the linkage sketched in Figure 8.17a.

If the sum of the two inputs, x and y, is required, it is only necessary to double the output (the rotation of the planet carrier shaft). This may be accomplished simply by a scale change or by using gears to effect a two-to-one speed increase. The value of the function $v = Ax + By$, where A and B are constants, is obtained by additional pairs of gears driving the input shafts. See Figure 8.18b. Since rotation direction changes with each meshing pair of external spur gears, idlers may be used to obtain the correct sign.

When an automobile is making a turn, the wheels on the outside of the turn travel farther than the wheels on the inside. A differential allows the rear wheels to rotate at different velocities so that the tires are not required to drag along the road during a turn. A typical automotive differential is similar to the sketch in Figure 8.17c, except that *several* bevel planet gears are carried on the same planet carrier, which is driven by a pair of hypoid gears (Figure 8.19).

(a)

(b)

Figure 8.18 (a) The large bevel gears that act as sun gears are mounted on bearings. Each of the sun gears may be directly attached to a spur gear through which input motion is transmitted. The planet bevel gear drives the junction block, which is rigidly connected to the shaft through the sun gears. (*Source:* Precision Industrial Components Corporation, Wells-Benrus Corporation.) (b) When this mechanism is used to *add* or *subtract*, the shafts at the right and left foreground act as inputs (to the pair of gears driving the bevel sun gears) and the planetary bevel gear cross arm drives the output shaft. When used as a *phase shifter*, the shafts at the right and left foreground become the input and output, respectively. Rotation of the junction block changes the relative position (phase) between input and output. (*Source:* Precision Industrial Components Corporation, Wells-Benrus Corporation.)

Kinematically, there is no difference in the differential itself. The input to the planet carrier (C of Figure 8.17c) is transmitted through the planet P to the rear axles represented by the shafts of S_1 and S_2. The sum of the axle speeds is equal to twice the planet carrier speed: $x + y = 2v$. On a straight, dry road $x = y = v$. When turning, x does not equal y; however, the average of x and y is v. If there is no provision for positive traction, the torque delivered to both axles is equal. Then, with one wheel on ice and the other on dry concrete, the tire on the ice may turn freely at a speed of $2v$ while the other tire does not turn at all. This is the price we pay to avoid excessive tire wear. This obvious disadvantage can be remedied by devices that deliver most of the torque to the wheel which is *not* slipping. See Figure 8.19.

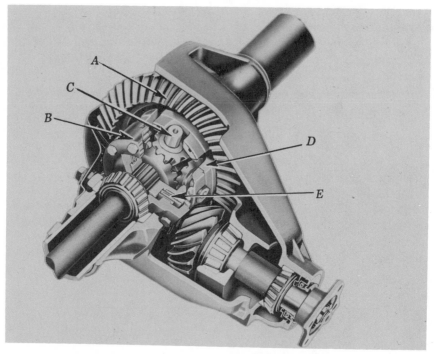

Figure 8.19 Limited-slip automotive differential. Engine power is transmitted to the large hypoid gear (A), which drives the differential case (planet carrier B). The driving force moves cross pins (C) up cam surfaces (D), engaging disk clutches (C). This type of differential *applies the greatest amount of torque to the wheel with the most traction* to prevent the other wheel from spinning on ice or snow. (*Source:* Dana Corporation.)

Using the formula method to describe the motion of a bevel gear differential, we have

$$\frac{n_{S_2}^*}{n_{S_1}^*} = \frac{n_{S_2} - n_C}{n_{S_1} - n_C} = -\frac{N_{S_1}}{N_{S_2}}$$

for $N_{S_1} = N_{S_2}$, the carrier speed is given by

$$n_C = \frac{n_{S_1} + n_{S_2}}{2}$$

as in Eq. 8.18.

A mechanism kinematically equivalent to the bevel gear differential may be manufactured by using spur gears. See Figure 8.20. While one planet gear is sufficient to obtain this motion in a bevel gear differential, the spur gear differential of Figure 8.20 requires a train of two planets held in the planet carrier. The number of teeth in sun gear S_1 in Figure 8.20 will equal the number of teeth in sun gear S_2. Then, all of the equations that apply to the

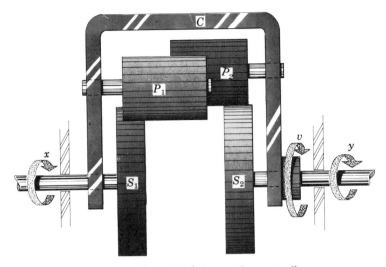

Figure 8.20 Spur gear differential (Figure 1.8 repeated).

bevel gear differential of Figure 8.17a apply also to the spur gear differential of Figure 8.20.

Figure 8.21 illustrates the use of a different type of spur gear differential in a final drive assembly. This configuration is equivalent to Figure 8.7i. The differential drive pinion is splined directly to an automatic transmission. In this particular application, a front-wheel drive, a pair of spiral bevel gears were chosen to avoid offsetting the pinion from the center of the planetary train. (Hypoid gears, which are used to drive conventional differentials in order to locate the drive shaft below the axles, would be undesirable here.)

The spiral bevel gears produce a speed reduction of 3.21, and when the car is traveling in a straight path, the entire planetary train turns as a unit. A right turn causes the planet carrier (driving the left axle) to overspeed, and a left turn causes the sun gear (driving the right axle) to overspeed. During either turn, there is relative motion between the ring, planets, and sun, providing the differential action. The planet gears must be in pairs, one contacting the ring and the other the sun, for proper rotation direction. The number of teeth in the gears are as follows: sun, 36; inner planets, 16; outer planets, 16; internal ring, 72; large spiral bevel, 45; and the spiral bevel pinion, 14. So that the axles are flexible, Rzeppa-type, constant-velocity universal joints are used in the drive assembly.

8.5 OTHER DRIVE TRAIN ELEMENTS

8.5.1 Chain Drives

Chain drives, like gear drives, result in definite output-to-input speed ratios and have high power transmission capacity—over 1000 hp for some designs

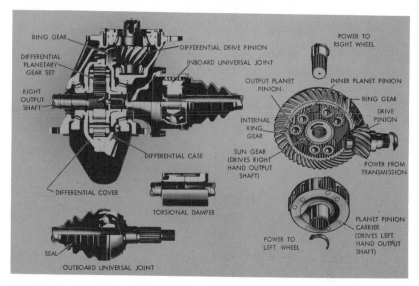

Figure 8.21 An automotive final drive assembly. The differential is made up entirely of spur gears. When the car is traveling in a straight path, the entire train turns as a unit. A right turn causes the planet carrier (driving the left axle) to overspeed; a left turn causes the sun gear (driving the right axle) to overspeed. During either turn, there is relative motion between the internal ring, planets, and sun, providing the differential action. (*Source:* General Motors Corporation.)

operating at high speed. Unlike gear trains, however, distance between shafts is not a critical factor; shaft distance may even fluctuate during operation of a chain drive without adverse effect. Chain length is changed by simply adding or removing a link.

There is a slight speed fluctuation in chain drives due to chordal action. This effect is most significant when sprockets with a small number of teeth are used. Chain drives tend to be noisy due to impact of the links with sprocket teeth. Inverted-tooth chain (sometimes called silent chain) is usually quieter than roller chain of the same capacity. Chain drives should be guarded to prevent injury.

Roller chain consists of side plates and pin and bushing joints that mesh with toothed sprockets. Roller chain sprockets have modified involute teeth. Unwrapping the chain from the sprocket tooth resembles involute generation, although the involute form is modified by the nonzero roller diameter. Most high-speed chain drives use roller-type chain or an inverted-tooth type made up of pin-connected, inward-facing teeth. See Figure 8.22. For maximum chain life, both sprockets are vertical and in the same plane. Although there is little restriction on center distances, the chain should wrap around at least 120° on both sprockets. Chain tension adjustment, if required, is made by moving one of the shafts or by using an idler sprocket.

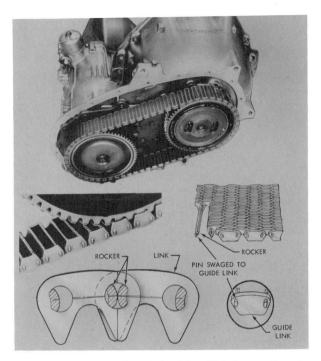

Figure 8.22 Automotive chain drive. This inverted-tooth chain transmits power from torque converter (*left*) to a planetary transmission (*right*). It was selected in preference to a gear drive, which would have required an idler. The links are designed for smooth engagement with the sprockets. Although timing chains are common in automobiles, development of a drive chain posed several problems, including excessive noise. The manufacturer arrived at the above design after considerable analytical and experimental study, including the use of high-speed motion pictures for analysis. (*Source:* General Motors Corporation.)

For instrument drives and other very light-load, low-speed applications, bead chains (Figure 8.23a) may be used. Sprockets need not be in the same plane if idlers are provided where necessary to maintain chain-to-sprocket contact. The complex drive shown in Figure 8.23b employs both bead chain and roller chain drives.

Sprocket speeds are inversely proportional to the number of teeth in each sprocket (or the number of pockets or indentations in bead chain sprockets). Both sprockets rotate in the same direction. Speed ratio is given by

$$\frac{n_2}{n_1} = +\frac{N_1}{N_2}$$

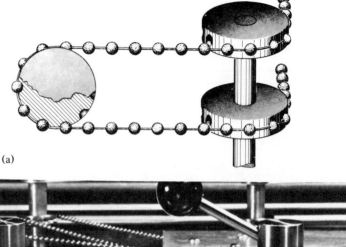

(a)

(b)

Figure 8.23 (a) Bead chain drive. Shown is a close-tolerance bead chain on pocketed sprockets for a positive, low-speed drive. The beads swivel on pins so that sprockets need not be in the same plane. (b) Complex plane instrument dial drive. Both bead chain and roller chain are used in this drive, assuring precise speed ratios with maximum flexibility. (*Source:* Voland Corporation.)

where n_1 and n_2 are, respectively, driver and driven sprocket speeds, and N_1 and N_2 are, respectively, the numbers of teeth in driver and driven sprockets. Idler sprockets and center distance do not affect speed. When bead chain is used, the chain may be crossed between sprockets to reverse rotation direction.

Variable-Speed Chain Drives

Speed ratio changing is generally *not* practical with high-speed or high-load roller chain drives except by using two or more chains and sprocket pairs, with clutches to engage one set at a time. For low load and low speed, a movable idler sprocket may be used to shift the chain from one sprocket to another without interrupting the driving. One application is a bicycle **derailleur** type transmission having two driving sprockets with different numbers of teeth and

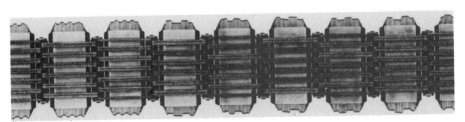

Figure 8.24 Variable-speed chain drive. This drive provides essentially stepless speed ratio variation with a positive drive. Effective sheave (driving surface) diameter is changed by increasing or decreasing the spacing between the sides of the grooved sheaves. The chain has floating laminations that conform to the grooves, and thus output-to-input speed ratio depends on effective sheave diameters only. Speed ratio is independent of the number of grooves in the sheaves. (*Source:* FMC Corporation.)

a "cone" of five driven sprockets at the rear wheel with graduated tooth numbers. Thus, there are ten sprocket combinations available, yielding ten different speed ratios. Drives of this type are also available with additional sprockets, producing 12, 15, 18, or 21 different speed ratios. For most of the ratios, the sprockets and chain are not exactly in line (i.e., the sprockets in use are not in the same plane), but in this application, loads are light enough so that chain wear due to misalignment is slight.

A variable-speed chain drive that provides essentially stepless speed ratio variation is shown in Figure 8.24. Effective sheave diameter is changed by the spacing between the sides of the grooved sheaves. The chain has floating laminations that conform to the grooves. Thus the output-to-input speed ratio is related to effective sheave diameter by Eq. 8.3: $n_2/n_1 = d_1/d_2$.

8.5.2 Belt Drives

While the low cost of belt drives is the most common reason for their selection over other means of power transmission, belts have other positive features. Belts absorb shock that might otherwise be transmitted from the driven shaft back to the driver, and prevent vibrations in the driver from being transmitted to the driven shaft, thus reducing noise and damage from these sources.

In addition, a belt may act as a clutch, disengaging driver and driven shafts when loose and engaging them when tight. Belt tightening may be accomplished by moving either driver or driven shaft or, if both shafts are fixed, a movable idler pulley may be used to tighten the belt. This method is used to engage some power lawn mower drives to the gasoline engine and to drive punch presses and shears, in which case a belt is momentarily engaged to a large flywheel for the working stroke of the machine. An idler is the less desirable of the two methods from the standpoint of belt wear; if used, it should be as large as practical and, preferably, ride on the inside of the slack span of the belt.

V Belt Drives

V belts, the most common type, are made of rubber reinforced with load-carrying cords. They operate in V-grooved pulleys or sheaves. The belt sidewalls wedge into the pulley; the belt does not ride in the bottom of the groove. The effect of this wedging action is to improve driving traction by more than tripling the friction force over what it would be if the pulleys were flat. Output-to-input speed ratio is given by Eq. 8.3:

$$\frac{n_2}{n_1} = \frac{d_1}{d_2}$$

where d_1 and d_2 refer, respectively, to the driver and driven pulley pitch diameters, the diameters of the neutral surface of the belt riding in the pulley. (The neutral surface of the belt is the surface that is neither compressed nor extended due to bending around the pulleys.)

Since belt tension varies as an element of the belt turns about a pulley, there is some slippage; therefore, speed ratios are not precise except with specially designed positive-drive belts and pulleys. If the belt does not cross itself, driver and driven pulleys turn in the same direction. For best results, the pulleys should be as large as practical and the belt should operate in a single vertical plane. Heavy loads may be carried by a multigrooved V pulley with a single-backed multiple-V belt.

Variable-Speed Belt Drives

For occasional speed ratio changing on light machinery, a "cone" of graduated diameter V pulleys is matched with a similar "cone" (large pulley on one opposite the small pulley on the other). The belt is manually transferred from one pair of pulleys to the other when the machine is stopped.

Figure 8.25 Pneumatically controlled belt drive. The pneumatic control regulator at left responds to pressure signals, continuously varying speed for process control. Speeds may be varied or adjusted to remain constant, or a predetermined cycle of speed variation may be programmed. The linkage simultaneously moves both halves of the variable-pitch pulley (*foreground*) in order to maintain belt alignment as speeds are changed. Units are available with horsepowers to 30 and speed ranges to 10:1. (*Source:* Lewellen Manufacturing Corporation.)

A pair of belt-drive variable pulleys, similar in appearance and operation to the variable-speed chain drive pulleys shown in Figure 8.24, may be used to change speed ratios on moving machinery without interrupting power. The sides of one pulley are moved apart to decrease its pitch diameter, while the sides of the opposite pulley are moved together to increase its pitch diameter. Output-to-input speed ratio again equals the inverse ratio of pitch diameters and may be continuously varied. Each variable pulley may change diameter by as much as a factor of three or four.

The variable pulleys may be varied manually, pneumatically, or hydraulically. Speed ratio change may be entirely automatic in response to a speed governor on the output shaft, output torque, engine vacuum, or a combination of the above. See Figure 8.25.

Some variable-speed drives use a fixed-pitch-diameter pulley opposite a *spring-loaded pulley* like that shown in Figure 8.26. The sides of the spring-loaded pulley are separated simply by increasing the distance between the driver and driven shafts. A cam may be incorporated into the variable pulley to prevent separation of the sides when there is an increase in the driven load. The cam is visible on the left side of the pulley shown in Figure 8.26.

Countershaft Pulleys

If the distance between driver and driven shafts is fixed, a variable double pulley may be used on a countershaft, as shown in Figure 8.27a. By moving the

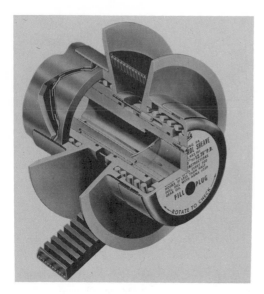

Figure 8.26 Variable-pitch pulley. This smooth-faced pulley is part of a variable-speed *belt* drive. When the distance between pulley centers is increased, the spring-loaded sides of the variable pulley separate to decrease its effective diameter. A cam at the left side of the pulley increases pressure on the pulley sides with increases in torque, compensating for the tendency of the pulley to separate at increased load. (*Source:* T. B. Wood's Sons Company.)

countershaft toward either the driver or driven shaft in Figure 8.27a, the pitch diameters of the intervening double pulley change simultaneously. One increases while the other decreases, giving a wide range of speed ratios.

When one or more countershafts are used, the output-to-input speed ratio again depends on the pitch diameters of the pulleys and is given by

$$\left| \frac{n_o}{n_i} \right| = \frac{\text{product of driving pulley pitch diameters}}{\text{product of driven pulley pitch diameters}} \qquad (8.19)$$

In this case, output-to-input speed ratio is given by

$$\frac{n_4}{n_1} = \frac{d_1 d_3}{d_2 d_4}$$

where d is pitch diameter and the subscripts 1 to 4 refer to driver pulley, countershaft driven pulley, countershaft driver pulley, and driven pulley, in that order. The two variable diameters are d_2 and d_3. Figure 8.27b is a section view of a spring-loaded, cam-aligned, countershaft double pulley.

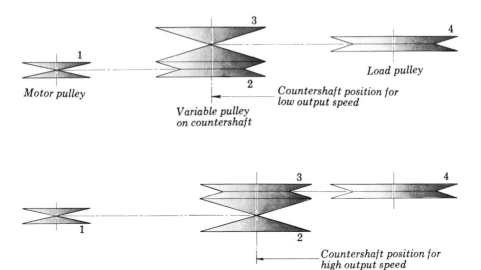

Motor pulley

Variable pulley
on countershaft

Load pulley

Countershaft position for
low output speed

Countershaft position for
high output speed

(a)

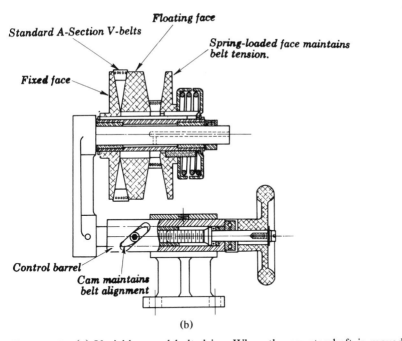

Standard A-Section V-belts

Floating face

Spring-loaded face maintains
belt tension.

Fixed face

Control barrel

Cam maintains
belt alignment

(b)

Figure 8.27 (a) Variable-speed belt drive. When the countershaft is moved toward the load pulley, the spring-loaded faces of the countershaft double-pulley change the effective pitch diameters d_2 and d_3. Pitch diameter d_3 increases, while d_2 decreases. (*Source:* Speed Selector, Inc.) (b) A section view of a spring-loaded, cam-aligned, countershaft double pulley for a compound variable-speed belt drive. (*Source:* Speed Selector, Inc.)

Flat Belts

When a single steam engine was used to drive dozens of separate pieces of factory machinery, power was transmitted by large flat belts turning pulleys on overhead countershafts. Belts were shifted, removed, and replaced by skilled operators who worked while the shafts were turning. While far less prominent now, flat belts have many modern, specialized applications. Light fabric belts and flat rubber-covered fabric belts are used at speeds up to 10,000 ft/min and higher—speeds that are well above the operating range of ordinary V belt drives. When used on balancing machine drives, lightweight flat belts do not significantly affect the rotating mass, and, in many high-speed applications, light flat belts are least likely to excite and transmit vibration.

Output-to-input speed ratio for flat belts is again given by $n_2/n_1 = d_1/d_2$, where pitch diameter d is measured to the center of the belt and is equal to the pulley diameter plus one thickness of belt. As before, subscripts 1 and 2 refer to driver and driven pulleys, respectively. Either the driver or driven pulley of a flat belt drive is usually crowned to prevent the belt from riding off.

Positive-Drive Belts

A special type of flat belt called a *timing belt* is used as a positive drive, ensuring exact speed ratios. Teeth on the inside surface of the belt mesh with a grooved pulley, and there is no slippage. See Figure 8.28. Output-to-input speed ratio is given by Eq. 8.1, $n_2/n_1 = N_1/N_2$, where N_1 and N_2 are, respectively, the *number of grooves* in the input and output pulleys. Timing belts may operate at high speeds and require only light tensioning since the drive does not depend on friction.

A patented drive system for instruments and other light-duty applications uses a reinforced plastic belt with side lugs (Figure 8.29), which meshes with a

Figure 8.28 Timing belt drive. Teeth on the inside surface of the belt mesh with grooves in the pulleys. Speed ratio depends on the number of grooves on the input and output pulleys. (*Source:* T. B. Wood's Sons Company.)

Figure 8.29 Positive-drive belt used with geared pulleys. The reinforced plastic belt was originally developed as a silent drive for sound and recording systems. The geared pulleys may be directly meshed with spur gears, eliminating intermediate components. An idler is usually used as shown to take up the slack. The flexibility of this type of belt allows for considerable shaft misalignment, and, with idlers or guides, the belt may even be used between pulleys on perpendicular intersecting shafts. (*Source:* Precision Industrial Components Corporation, Wells-Benrus Corporation.)

gearlike pulley. The pulley has a central groove to accommodate the belt but may be meshed with a spur gear. Output-to-input speed ratio is equal to the inverse ratio of the numbers of pulley teeth as with timing belts.

8.5.3 Friction Disk Drives

For certain low-torque applications, disks may be used to transmit power from one shaft to another. See Figure 8.30. When a pair of disks on parallel shafts contact at their outer edges, output-to-input speed ratio is equal to the inverse ratio of the disk diameters. For any number of disks, output-to-input speed ratio is given by

$$\left| \frac{n_o}{n_i} \right| = \frac{\text{product of driving disk diameters}}{\text{product of driven disk diameters}}$$

To maintain adequate friction force between the disks, alternate disks, or possibly all of the disks, might have rubber-covered driving surfaces.

Speed change in discrete steps may be accomplished by using a stepped set of disks on the driven shaft and changing the location of the driving disk to

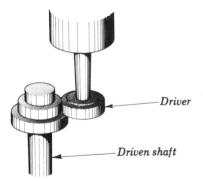

— Driver

— Driven shaft

Figure 8.30 Friction drive. For this type of drive, the *speed ratio equals the inverse ratio of disk diameters*. The drive is shifted to a new position to change ratios, and no clutch is required.

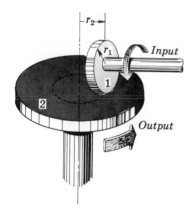

Figure 8.31 Wheel-disk drive. The output-to-input speed ratio is $n_2/n_1 = r_1/r_2$, where r_2 is measured from disk centerline to point of contact. The input and output shafts are kept perpendicular and in the same plane.

mesh with the driven disk, giving the desired speed. Alternatively, disk speed changers and reversers may employ an idler as is done with simple geared speed changers and reversing gear trains. Possible slippage between disks makes this type of drive unsuitable for most instrument applications but permits a simple and inexpensive drive system for low-torque applications, eliminating the need for a clutch. Speed ratio may be changed by shifting disks from one location to another without stopping the drive since there is no problem of meshing teeth.

Variable-Speed Wheel-Disk Drives

A wheel-and-disk type of transmission provides stepless speed changing and reversing. The wheel (1) of Figure 8.31 traces a path of radius r_2 on the disk (2), and output-to-input speed ratio is given by

$$\frac{n_2}{n_1} = \frac{r_1}{r_2}$$

where r_1 is the radius of the driver and r_2 is the distance from the axis of the driven disk to the point of contact of the wheel and disk. When the wheel contacts the disk to the left of the disk center, the output direction is reversed.

An alternative transmission employs two disks and an idler wheel. Only the idler wheel is shifted while the input and output shafts rotate in place, as in Figure 8.32. Since the velocity ratio is unaffected by idler diameter, speed ratio is again given by $n_2/n_1 = r_1/r_2$, where r_1 and r_2 vary simultaneously when the idler is shifted.

If a drive of this type were used on a phonograph, for example, the idler could be slid along between the input and output disks, producing turntable speeds of 16, $33\frac{1}{3}$, 45, and 78 rev/min with a single input speed. The drive shown is not reversible, but by simply moving the input and output shafts closer together, or by increasing the output disk diameter, one could design a drive with same-direction, zero, or reverse output. When the idler is shifted to

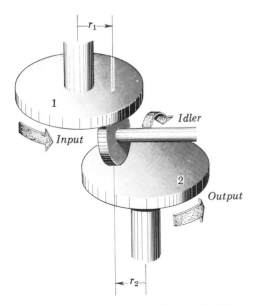

Figure 8.32 Friction disk drive with idler wheel. As in the preceding illustration, the output-to-input speed is given by $n_2/n_1 = r_1/r_2$, where both radii are measured from the respective disk centerlines to the points of contact. Input and output shafts are parallel, and the axes of all three shafts lie in the same vertical plane. The idler shaft is perpendicular to the other two and is moved axially to change speed.

the center of the input disk, it does not turn and output speed is zero. When shifted farther to the left, the output is reversed.

Variable-Speed Cone-Roller Disk Drives

A disk drive that is similar in principle uses one or more *cone rollers*, shifted by an adjusting screw when the unit is running, Figure 8.33a. Cam surfaces on the split output shaft respond to high output torque by increasing the normal force between the roller and disks to ensure adequate driving traction. When the load decreases, the normal force is decreased so that wear is not excessive. A disk drive with four cone rollers is diagramed in Figure 8.33b.

Variable-Speed Ball-Disk Drives

Friction drives require relatively high normal forces where driver and driven parts contact. Optimum conditions for shifting, however, include low normal forces for drives of the type shown in Figures 8.31 and 8.32, neither of which would be satisfactory if speed changing were frequent or continuous.

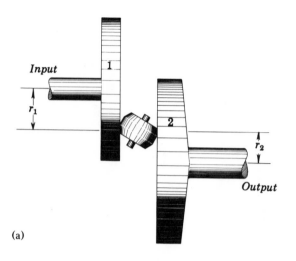

(a)

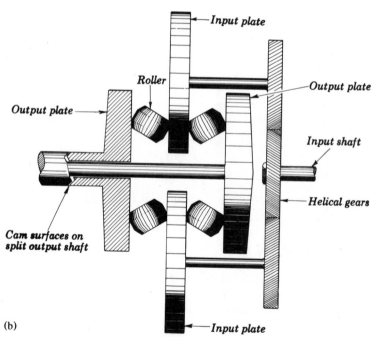

(b)

Figure 8.33 (a) Friction disk drive with cone roller. The cone roller is shifted by means of an adjusting screw to change the radius of the rolling paths on disks 1 and 2. As with the other disk drives, output-to-input speed ratio is given by $n_2/n_1 = r_1/r_2$. (b) Disk drive with four cone rollers. By the use of four adjustable rollers instead of one, the maximum power capacity is increased.

Input

Output

Figure 8.34 Ball-disk drive. The output is zero when the ball cage is at the disk center. Direction reverses when the ball cage is moved past (to the left of) the disk center.

The ball-disk drive of Figure 8.34, however, does not suffer extensive wear when continuously varied because the two balls in the cage *roll* on one another and on the input disk and output shaft when shifted. (Note that a single ball could not roll on both surfaces as radius r_1 was changed.) As for the previously discussed drives, the equation

$$\frac{n_2}{n_1} = \frac{r_1}{r_2}$$

can also be used to find the output-to-input speed ratio for the ball-disk drive. When the ball cage is shifted axially to the center of the input disk, there is no output motion but there is still no significant wear on the drive members. Shifting farther to the left results in an output in the opposite direction from that shown in the figure.

Wheel-disk and ball-disk drives may be used as multipliers. From the familiar speed ratio equation, we have

$$\theta_2 = \left(\frac{r_1}{r_2}\right)\theta_1$$

where θ_1 and θ_2 are input and output angular displacement, respectively (measured in degrees, radians, or revolutions with both θ_1 and θ_2 in the same units). To multiply two numbers by using the ball-disk drive, we adjust r_1, the ball cage position, so that one of the numbers is represented by r_1/r_2, where r_2 is the radius of the output shaft. The driver is given a rotation θ_1 representing the other number and the product is given by θ_2, the rotation of the output shaft.

When r_1 is varied as a function of θ_1, then θ_2 is the integral of r_1/r_2 with respect to θ_1. Thus, the ball-disk drive may be used as an integrator. See Figure 8.35. For example, let the ball cage position be automatically adjusted so that r_1/r_2 is proportional to the flow of liquid through a meter. Disk 1 is

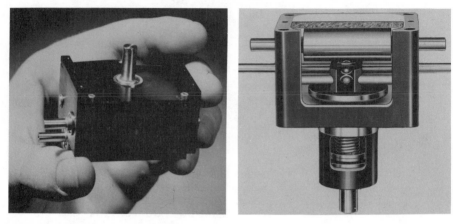

Figure 8.35 Two views of a ball-disk integrator are shown. At *right* is a cutaway view. (*Source:* The Singer Company, Librascope Division.)

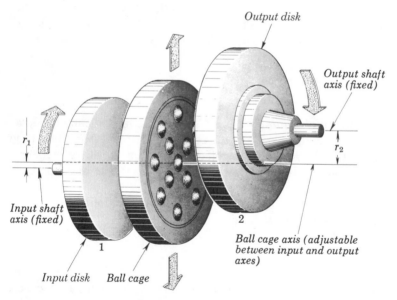

Output disk

Output shaft axis (fixed)

r_1

r_2

Input shaft axis (fixed)

1

2

Input disk   *Ball cage*

Ball cage axis (adjustable between input and output axes)

Figure 8.36 The Graham *ball-disk* drive. This variation of the ball-disk drive employs a number of balls held in a rotating cage. The balls contact both the input and output disks. Output-to-input ratio is again given by $n_2/n_1 = r_1/r_2$ and may vary from 0 to 1.5. Output speed is increased as the ball cage centerline is moved toward the output disk axis. (*Source:* Graham Company.)

turned at a constant rate and the rotation of shaft 2 is proportional to the total amount of liquid that passed through the meter during the measuring period. Measurements could, of course, be scaled to keep both input and output values within a reasonable range.

Consider the friction disk drive of Figure 8.32 with two offset disks but with the idler wheel replaced instead by a single ball restrained in a cage. The variable-speed, ball-disk drive sketched in Figure 8.36 operates on this principle except that, instead of a single ball, a cage of several steel balls rotates about its own center between the steel input and output disks. Each ball contacts both input and output disks at all times, permitting greater tractive forces between the two offset disks.

As far as input and output speeds are concerned, the rotating ball cage is kinematically equivalent to a single ball at the cage axis. When the ball cage axis coincides with the input axis, the output speed is zero. As the ball cage is moved toward the output shaft axis, output speed increases until, at the limit for this design, the output speed is 1.5 times the input speed. Speed ratios are changed when the unit is running. When recommended torque is not exceeded, there is pure rolling without sliding between the balls and disks except during speed changes. In general, the ratio of output to input speed is given by $n_2/n_1 = r_1/r_2$, where r_1 and r_2 are defined in Figure 8.26. The angular velocity of the cage is half the sum of the input and output speeds:

$$n_{cage} = \frac{n_1}{2}\left(\frac{r_1}{r_2} + 1\right)$$

Variable-Speed Axle-Mounted Ball-Disk Drives

Another compact metal-to-metal friction drive employs axle-mounted balls that transmit power between beveled disks, as shown in Figure 8.37. The ball axles are tilted uniformly by a slotted iris plate (acting as a set of cams) when it is desired to change speed ratios (see Figure 8.38a). In analyzing this drive kinematically, we need consider only one ball and observe effective radii of driving and driven paths. When the ball axle is parallel to the input and output shafts (Figure 8.38a) the distances r_1 and r_2 (from the ball axle to the points of contact of the ball with the beveled input and output disks) are equal. As the ball axle tilts either up or down (Figure 8.38b and c), the radii r_1 and r_2 change simultaneously. Using the relationship for output-to-input speed ratio,

$$\frac{n_o}{n_i} = \frac{\text{product of effective driving member radii}}{\text{product of effective driven member radii}}$$

we obtain

$$\frac{n_o}{n_i} = \frac{r_i r_2}{r_1 r_o} = \frac{r_2}{r_1}$$

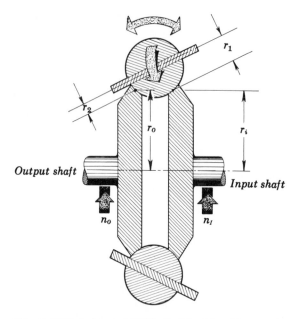

Figure 8.37 Axle-mounted ball-disk drive. Since output and input disk radii are equal ($r_o = r_i$), output-to-input speed is given by $n_o/n_i = r_2/r_1$. A change in the angle of the ball axle simultaneously changes r_1 and r_2, the distances from ball axle to the point of contact of the ball surface with the beveled faces of the input and output disks, respectively. (*Source:* Cleveland Gear Company, subsidiary of Vesper Corporation.)

after noting that the effective radii of the input and output disks are equal. Reversing is not possible with this mechanism alone, but speed ratio may be varied from $\frac{1}{3}$ to 3 on commercially available units (while the unit is running).

Variable-Speed Roller-Torus Drives

A variation of the axle-mounted ball-disk drive is the roller-torus drive shown in Figure 8.39a. This drive employs two tiltable rollers within a split torus. The roller axes are shifted equally by gearing, which includes a worm drive to provide fine speed adjustment. See Figure 8.39b. The rollers act as idlers between the input and output tori. The output-and-input speed ratio is given by

$$\frac{n_2}{n_1} = -\frac{r_1}{r_2}$$

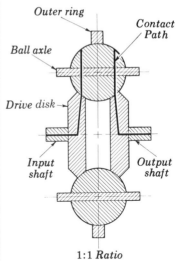

(a)

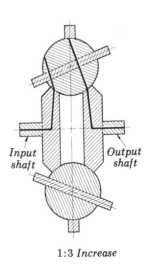

(b)

Figure 8.38 (a) With the ball axles in the position shown, input speed equals output speed. (*Source:* Cleveland Gear Company, subsidiary of Vesper Corporation.) (b) The precision speed adjustment device (worm drive) at the top of the transmission is manually adjusted. The iris plate tilts the ball axles, simultaneously changing r_1 and r_2 as shown in the insert. The effect is a 1:3 input-to-output ratio. (*Source:* Cleveland Gear Company, subsidiary of Vesper Corporation.)

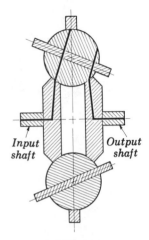

Input shaft   Output shaft

3:1 *Decrease*

Figure 8.38 (c) With the axles tilted to the position shown, the input-to-output ratio becomes 3:1. For automatic process control or remote speed adjustment, the handwheel drive may be replaced by an electric or pneumatic actuator. The ring that encircles the balls balances radial forces but runs freely and does not affect the speed ratio. (*Source:* Cleveland Gear Company, subsidiary of Vesper Corporation.)

where r_1 and r_2 are measured from the drive shaft centerline to points of roller contact on the input torus and output torus, respectively. The minus sign indicates that output rotation direction is opposite input. The roller-torus drive by itself cannot have a zero output and cannot be shifted to change the output direction.

Friction drives and gearing are often combined to provide greater flexibility. Consider a simple planetary gear train with no fixed gears. From the section on planetary trains with more than one input, planet carrier speed is given by Eq. 8.16:

$$n_C = \frac{n_R + n_S(N_S/N_R)}{1 + (N_S/N_R)}$$

where n represents speed in revolutions per minute (or any other unit provided we are consistent), N represents numbers of teeth, and subscripts C, R, and S refer to the carrier, ring, and sun, respectively. Let the output torus of a roller-torus drive be directly connected to the ring gear of the planetary train, and let the input torus shaft pass through the output torus to drive the sun gear. Then, the planet carrier may be used as output of the combined transmission so that a zero speed and a reverse are possible. Recall that, for all roller positions, the output torus rotates in a direction opposite the input

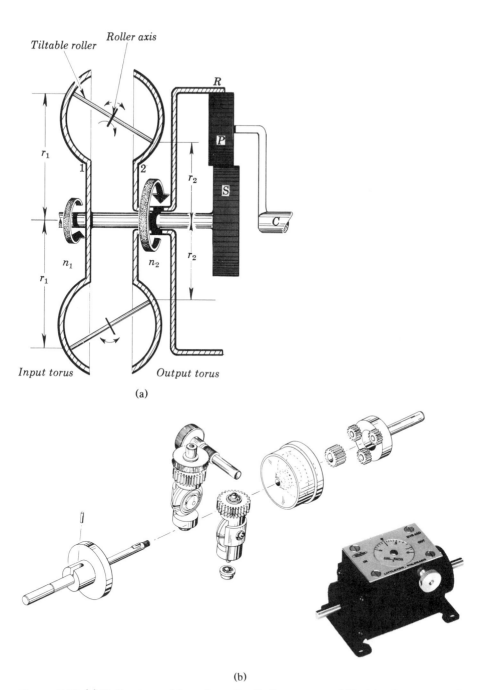

Figure 8.39 (a) Roller-torus drive schematic. Roller axes are shifted to change r_1 and r_2. The input torus drives the sun gear S; the output torus drives the ring R. (*Source:* Metron Instruments, Inc.) (b) Roller-torus drive. The roller-torus drive is shown combined with a planetary gear transmission in a low-power application. (*Source:* Metron Instruments, Inc.)

torus. Referring to the roller-torus equation, $n_2/n_1 = -r_1/r_2$, and to the above equation for n_C, we have the ratio of output to input speed for the combined transmission:

$$\frac{n_C}{n_1} = \frac{-(r_1/r_2) + (N_S/N_R)}{1 + (N_S/N_R)}$$

By rotating the rollers until r_1/r_2 is less than the fraction N_S/N_R, we obtain output and input in the same direction (with a considerable speed reduction). The roller adjustment at which $r_1/r_2 = N_S/N_R$ results in a zero output speed. Most of the range of the transmission is obtained with r_1/r_2 greater than N_S/N_R, resulting in an output direction opposite the input.

8.5.4 Planetary Traction Drives

Analysis of planetary traction drives differs little from the analysis of planetary gear trains. The same equations are applicable except that ratios of gear teeth are replaced by ratios of rolling path radii.

Planetary Cone Transmission

One variable-speed planetary friction drive employs *cone* planets that make contact with a variable-position reaction ring (Figure 8.40a). The drive motor may be directly connected to the planet carrier. A set of planet pinions (integral with the planet cones) drives a ring gear, which serves as the output. At most input speeds, inertia forces hold the cones against the friction ring, but if the input speed is low, auxiliary springs are used to ensure contact. For a change of speeds, the reaction ring is shifted axially (Figure 8.40b) to contact the planet cones at a different location (changing the effective planet radius, r_{P_1}). Speeds may be changed when the unit is operating if the cone planets are spring-loaded. Speed change is stepless.

Superposition will be used to solve for relative speeds. In the first step of Table 8.11, the entire transmission is given one positive rotation. In step 2, the reaction ring rotation is corrected by giving it one negative rotation with the carrier fixed. The effect is calculated for the cone planet and planet pinion (which turn as a unit) and for the ring gear. Adding steps 1 and 2 of Table 8.11, we have the required rotation of the reaction ring (zero) and one rotation of the planet carrier. The output-to-input speed ratio is given by the total rotations of the ring gear divided by the carrier rotations:

$$\frac{n_{R_2}}{n_C} = 1 - \frac{r_{R_1} N_{P_2}}{r_{P_1} N_{R_2}} \tag{8.20}$$

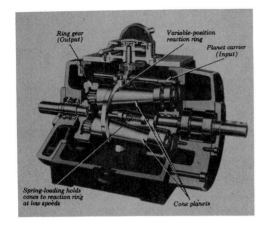

(a)

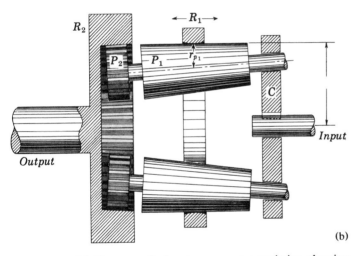

(b)

Figure 8.40 (a) Cutaway of planetary cone transmission showing speed control mechanism. The mechanism at the top of the transmission permits fine adjustment of reaction ring position and a wide range of speed ratios through zero and reverse speeds. Most models of this transmission depend on centrifugal force to hold the cone rollers against the ring. This model, however, is designed with a spring-loading assembly in the center to ensure traction betweeen the cones and ring, even at low speeds. (*Source:* Graham Company.) (b) Schematic of the planetary cone transmission. The element of the planet cone P_1 that contacts ring R_1 is parallel to the input axis. The ring R_1 is moved left or right to increase or decrease the effective radius of the planet cone r_{p_1} at the point of contact. Table 8.11 gives the output-to-input speed ratio.

TABLE 8.11   SPEED RATIOS FOR A PLANETARY CONE TRANSMISSION
WITH THE REACTION RING R_1 FIXED, FIGURE 8.40b

Component	C	R_1	P_1	P_2	R_2
Step 1: rotations with train locked	+1	+1	+1	+1	+1
Step 2: rotations with carrier fixed [R_1 given (-1) rotations]	0	-1	$-\dfrac{r_{R_1}}{r_{P_1}}$	$-\dfrac{r_{R_1}}{r_{P_1}}$	$-\left(\dfrac{r_{R_1}}{r_{P_1}}\right)\left(\dfrac{N_{P_2}}{N_{R_2}}\right)$
Total rotations (R_1 fixed)	+1	0			$1-\dfrac{r_{R_1}N_{P_2}}{r_{P_1}N_{R_2}}$

In one model, when the reaction ring (also called the control ring) is adjusted to approximately the middle of its range, the ratio of reaction ring radius to cone radius is equal to the ratio of the ring gear teeth to the pinion teeth (N_{R_2}/N_{P_2}). Then, the output speed is zero from Eq. 20. As we move the reaction ring to the left, r_{P_1} increases and output speed is in the same direction as input until, at its maximum value, $n_{R_2}/n_C = \frac{1}{5}$. Moving the control ring to the right of the zero output position gives us an output rotation opposite the input until we reach a limiting value of $n_{R_2}/n_C = -\frac{1}{5}$. A similar transmission with a range of output to input speeds between $-\frac{1}{100}$ and $+\frac{1}{3}$ is also available.

8.5.5 Impulse Drives and Flexible-Spline Drives

A gear-and-linkage impulse drive driven by an eccentric is shown in Figure 8.41a. This particular drive is no longer commercially available. The sketch of the drive (Figure 8.41b) is incomplete, showing only one of the three pinion-and-linkage assemblies of the unit shown in Figure 8.41a, but we will first examine the drive as shown in part before considering the entire mechanism.

The eccentric drives a roller follower pinned to link 1. The follower in turn drives links 2 and 3. Link 3 turns a pinion counterclockwise through a one-way clutch, and the pinion drives the output gear. It is seen that the cam and follower will cause link 3 to oscillate through the same angle whether the input cam turns clockwise or counterclockwise. Thus, the orientation of the one-way clutch determines the directions the pinion and output gears will turn. In this case, the output is clockwise whether the input is clockwise or counterclockwise.

When links 2 and 3 are nearly perpendicular, the oscillation of link 3 is greatest and output speed is maximum. The pivot point of link 1 may be rotated about the drive axis, making he angle between links 2 and 3 quite small so that link 3 oscillates through a small angle and output speed is reduced.

(a)

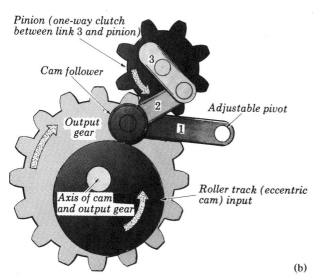

(b)

Figure 8.41 (a) Cutaway view of the gear-and-linkage impulse drive. (*Source:* Morse Chain Division, Borg Warner.) (b) Schematic of the gear-and-linkage impulse drive (showing only one of the three pinion-and-linkage assemblies). The eccentric roller track drives the cam follower, which turns the pinion through a linkage. A one-way clutch permits link 3 to drive the pinion counterclockwise only. The location of the pivot of link 1 may be adjusted to change the amount of pinion rotation per cam revolution.

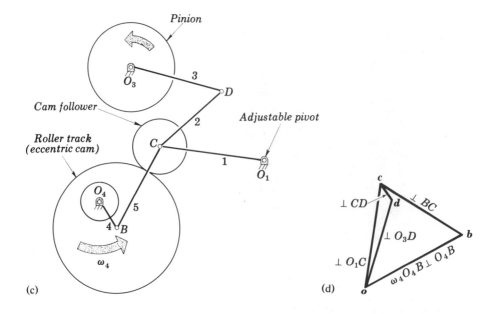

Figure 8.41 (c) Equivalent-linkage drawing for the pinion-and-linkage assembly, showing the principle of operation. (d) The velocity polygon for the pinion-and-linkage assembly.

The output shaft would be stationary for about half of each input shaft revolution if the drive had only one pinion-and-linkage assembly. The actual drive (Figure 8.41a) with three such assemblies turns continuously, each linkage driving one-third of the time. For a given linkage, during two-thirds of each input cycle, link 3 is either rotating opposite the pinion direction or it is rotating slower than the pinion. (In the latter case, the pinion is being driven by one of the other linkages through the output gear.)

Instantaneous output speed is found by sketching the cam and follower (or its equivalent mechanism) and the linkage that is driving at a given instant. The procedure is illustrated in Figure 8.41c. Since the roller track, which acts as a cam, and the cam follower are circular, the distance from the roller track center B to cam follower center C is fixed and will be considered rigid link 5. The axis of rotation of the roller track is labeled O_4, and O_4B is designated as link 4. For a specified position of adjustable pivot O_1, we form a four-bar linkage with links 4, 5, and 1, and the frame. Given ω_4, the angular velocity of the roller track, we may draw velocity polygon obc by using the methods of Chapter 3. See Figure 8.41d. We then note that links 1, 2, and 3 and the frame also form a four-bar linkage, permitting us to complete the polygon with velocity point d. The linkage shown is obviously driving the transmission at this instant (through a one-way clutch whereby link 3 drives the pinion counterclockwise). Thus pinion angular velocity is given by $\omega_3 = od/O_3D$.

Angular velocity of the output gear is related to the angular velocity of link 3 by the negative ratio of pinion teeth to output gear teeth. This process is repeated at intervals for the entire time a given linkage drives to obtain a smooth plot of output speed. The same roller track drives the three identical linkages and thus the output motion is repeated for every 120° of cam rotation.

If it is not obvious which of two linkages is driving the output gear at a particular instant, the problem is solved for both linkages; the linkage giving the highest speed is the actual driver. It is seen that the output is pulsating, making a drive of this type practical only for relatively low speeds.

Flexible-Spline Drives

The *flexible-spline drive* of Figure 8.42a operates on a unique principle, employing a wave generator to deflect a flexible external spline. The flexible spline meshes with a slightly larger, rigid, stationary member having an internal circular spline. The major components are shown in exploded view in Figure 8.42a. The wave generator, usually an ellipselike cam, is ordinarily the input to the drive. The flexible spline serves as the output, rotating in a direction opposite the input, as shown below. Both splines have the same circular pitch, but the stationary spline has, typically, two or four more teeth than the flexible spline.

The wave generator causes the flexible spline to "walk" around the inside of the stationary spline. Since the splines mesh tooth for tooth, when the wave generator turns one rotation clockwise, the flexible spline will advance slightly counterclockwise. The difference in pitch circumference between the splines results in a tangential motion with a magnitude of two or four circular pitches. In general, then, there are

$$\frac{N_f - N_s}{N_f}$$

rotations of the flexible spline (output) for each wave generator rotation (input), where N_f is the number of the flexible spline teeth and N_s is the number of stationary internal spline teeth (see Figure 8.42b).

The speed ratio can also be obtained by using the superposition method used previously for planetary drives. In Table 8.12, the locked train is given one rotation for step 1. In step 2, the rigid, internal spline is given one negative rotation with the wave generator fixed. The two steps are then added as shown in the table.

The wave generator of the flexible-spline drive has two or more equally spaced lobes or rollers; the two splines mate where the lobes or rollers press the flexible-spline unit against the rigid splines. The difference in tooth numbers, $N_s - N_f$, is a whole-number multiple of the number of lobes. An output-to-input speed ratio of $-1/80$, for example, is obtained by using a two-roller or two-lobe wave generator input and a flexible spline of 160 teeth

(a)

Rigid circular spline
(part of stationary housing)

Flexible spline
(output member)
Ellipse-like wave generator
(input member)

Output shaft

(b)

Figure 8.42 (a) Components of the flexible-spline drive are a housing with an integral 162-tooth circular spline (fixed); a plastic, 160-tooth flexible spline (the output member); a two-lobe, ball-bearing wave generator (mounted on the output shaft). Other flexible spline drives are available for heavy-duty and high-precision applications. A combination of two rollers or an ellipselike cam is used as the wave generator for drives of this type. (*Source:* Harmonic Drive Division, USM Corporation.) (b) The flexible-spline drive analyzed. The wave generator makes the flexible spline "walk" around the inside of the stationary circular spline. Output-to-input speed ratio (given by Table 8.12) is $(N_f - N_s)/N_f$, a small negative fraction.

TABLE 8.12 SPEED RATIOS FOR THE FLEXIBLE-SPLINE DRIVE, FIGURE 8.42a

Component	Flexible spline	Circular spline	Wave generator
Step 1: rotations with train locked	$+1$	$+1$	$+1$
Step 2: rotations with wave generator fixed	$-\dfrac{N_s}{N_f}$	-1	0
Total rotations	$1-\dfrac{N_s}{N_f}$	0	$+1$

meshing with a 162-tooth, fixed internal spline. If output and input direction are to be the same, the wave generator is again used as the input member, but the output is taken from the rigid circular spline with the flexible spline fixed. Using the data from the above example, this condition would yield an output-to-input ratio of

$$\frac{N_s - N_f}{N_s} = +\frac{1}{81}$$

Speed can be reduced to very low values with correspondingly great increases in output torque with compact, one- or two-stage, flexible-spline drives. For this reason, they have been used in aircraft and space applications including vertical-takeoff aircraft, short-takeoff-and-landing aircraft, and a meteor-detection satellite. In the latter application, a two-stage, 3-by-$4\frac{1}{2}$-in flexible spline drive provides a 1-to-3840 speed reduction for a secondary drive to extend 15-by-48-ft detector "wings" (while orbiting).

8.5.6 Selection of Drive Train Components

A given set of design requirements might be satisfied by more than one of the gear train configurations, packaged speed reducers, or drive train components considered above. A few of the most important design objectives will be discussed as an aid to optimum selection.

Shaft Geometry Considerations

One of the first requirements might involve *arrangement* of the input and output shafts. If the shaft centerlines intersect as in an outboard motor drive, bevel gears might be used. For nonintersecting perpendicular shafts, an automobile rear end, for example, hypoid gears are used. Power may also be transmitted between nonintersecting perpendicular shafts by a worm and worm gear if considerable speed reduction is desirable. If shaft centerlines do not intersect and are neither parallel nor perpendicular, crossed helical gears, bead chain, and some types of belt drives are used. When the design requires parallel or colinear input and output axes, packaged speed reducers, belt-and-chain drives, and spur and parallel helical gear trains may be considered.

Speed Ratio Considerations

If speed must be considerably reduced, worm and worm gear drives and some compound planetary drives are used. Rapid *speed ratio changing* (in discrete steps) suggests the use of clutches engaging and disengaging spur gears or helical gears on parallel shafts. An alternate method employs a planetary train with band brakes and clutches to lock certain elements.

When speed must be continuously variable, that is, if stepless variation is required, variable-speed belt drives may be employed. For low-power applications (phonographs and instrument-type applications), disk and wheel-disk drives are used. Some disk, roller-disk, and ball-disk drives are variable through zero output speed to negative speed ratios.

Control Considerations

The method of *speed control* is an important factor in selection of a transmission. Spur and helical gear trains are ordinarily shifted manually through mechanical linkages, but speed changing may be actuated or assisted pneumatically, hydraulically, or electrically. The bands and clutches in most planetary trains are automatically actuated by speed-sensitive and load-sensitive controls.

Load (Power) Considerations

Bead chain and plastic belts with lugs are limited to very low torque applications (chart drives, etc.). *High-torque or -power requirements* preclude the selection of these drives and most traction drives for many industrial applications. Drives available for high horsepower include chains, belts, and gears.

The most severe power transmission and torque requirements are met by various forms of gearing. Using high-speed double helical gears, a single mesh has been used for loads as high as 22,000 to 30,000 kW (30,000 to 40,000 hp). Two or more input pinions are used (as in ship propulsion units) for higher-horsepower drives.

Precision-Drive Considerations

A requirement for *precision* in the transfer of motion is usually met by gear or power screw drives. Most high-precision instrument drives employ spur gear trains, but for high-velocity applications, helical gears are used. When considerable speed reduction is required, a single, precision worm and worm gear pair may be used.

Sources of position error in gear drives include error in tooth form and location; gear and shaft eccentricity, misalignment, and deflection under load; thermal expansion and contraction; and looseness and deflection in couplings, bearings, and other train components. When the output of a train must accurately reflect the input motion, the designer may use one or more of the following methods to control error:

1. Select precision components.
2. Use as few pairs of gears as possible to reduce integrated errors.
3. Use larger diameter or shorter shafting and rigid supports to limit deflection under load.
4. Replace alternate gears in the train with split antibacklash gears. The split gear is spring-loaded so that the halves rotate slightly with respect to one another and eliminate play.
5. Attach one gear to a floating shaft that moves slightly to eliminate play between meshing teeth.

In multistage speed reducers, angular error from each stage is reduced by the succeeding stages. If the final (output) stage has a high reduction ratio, precision components may be required in that stage only.

Noise, Shock, and Vibration Considerations

Noise, shock, and vibration effects are usually reduced by nonmetallic drives. Belt drives absorb vibration and shock rather than transmitting these effects between the driver and driven shaft. For low loads, nylon or fiber gears are used to reduce noise.

Noise and vibration result as the load is transferred from one pair of gear teeth to another. The load transfer in helical gears on parallel shafts is smoother and more pairs of teeth are in contact at a given time than for similar spur gears. For this reason, helical gears are often chosen for high-power, high-speed drives.

Efficiency Considerations

Efficiency is an important factor in continuously operated drive trains. Over a period of years, losses of a few percent of transmitted power can be expensive. Low efficiency may exact additional penalties since friction losses represent (1) *heat*, which must be dissipated, and (2) additional fuel *weight* to be carried (a problem in surface vehicle design as well as aircraft and space vehicles).

Spur gears, helical gears on parallel shafts, and bevel gears have efficiencies of 98 to 99 percent at rated power (for each mesh). If lubrication is poor or if the drive operates at lower-than-rated power, the efficiency will be lower. Efficiency is considerably reduced in pairs of gears with high sliding velocity; with high-reduction worm drives, losses may exceed 50 percent of transmitted power. Belt drive efficiencies are typically about 95 percent for single V belts. Traction drives have efficiencies from 65 to 90 percent (lower at very low output speeds).

Miscellaneous Considerations

Other major selection criteria include *size, weight, reliability, and cost*. On a cost-per-transmitted-horsepower basis, belt drives usually have the advantage. If speed is to be reduced by a large factor, a worm and worm wheel drive may be selected on the basis of its compactness and low cost, as compared with more complicated drive trains. Reliability is an important positive characteristic of gear trains, particularly spur gears and helical gears on parallel shafts. Planetary gear trains offer a compact package, taking up little space in an axial direction, even for high-power drives.

8.6 VELOCITY AND ACCELERATION IN A PLANETARY TRAIN

The velocities and acceleration of a point on a planet gear may be determined by vector methods. We begin by describing the gear train in terms of a fixed and a moving coordinate system, using the equations of Section 4.1.4.

Consider, for example, the simple planetary gear train in Figure 8.43a. A fixed XYZ coordinate system with unit vectors, I, J, and K, respectively, is oriented with its origin at the center of the gear train. Moving coordinate system xyz with unit vectors i, j, and k, respectively, rotates with the planet carrier and has its center at the planet center. Let the planet carrier (4) rotate counterclockwise with constant angular velocity $\omega_4 = \omega_4 k$ (given). Note, in Figure 8.43b, that while I and i and J and j differ, in this problem, K and k are identical. Ring gear (3) is to be fixed. It was shown that the speed ratios are

$$\frac{\omega_1}{\omega_4} = 1 + \frac{N_3}{N_1} \quad \text{and} \quad \frac{\omega_2}{\omega_4} = 1 - \frac{N_3}{N_2}$$

for this train.

Since the sun gear and planet carrier rotate about fixed centers, velocities and accelerations of points on these members are given, respectively, by

$$\dot{R} = \omega \times R \quad \text{and} \quad \ddot{R} = \omega \times (\omega \times R)$$

when the sun gear and planet carrier have no angular acceleration.

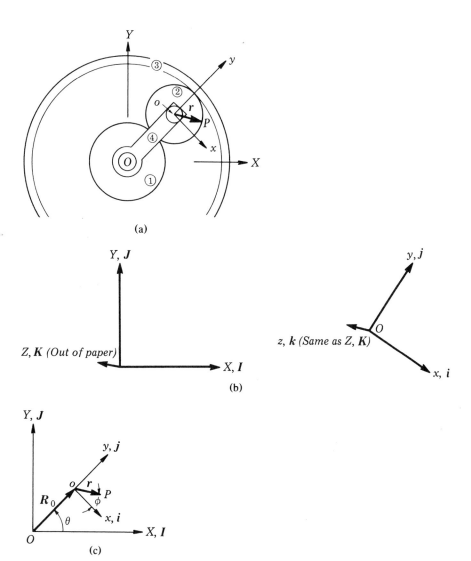

Figure 8.43 (a) Planetary gear train. (b) *Left*: The fixed coordinate system and its corresponding unit vectors. *Right*: The moving coordinate system and its unit vectors, for the position shown. The moving system rotates with the planet carrier (4) at ω_4 rad/s and zero angular acceleration. (c) The position of the moving coordinate system relative to the fixed system. The absolute position of point P is given by $R_0 + r$.

The motion of a point on the planet is more complicated; we will apply Eq. 4.7 with the following substitutions:

$$\dot{R}_0 = \omega_4 k \times R_0 j = -\omega_4 R_0 i$$

There are no angular acceleration terms corresponding to $\dot{\omega}$ since the angular velocity of the gears is constant due to the (given) constant angular velocity of the carrier. Position vector r defines the position of point P on the planet. The relative velocity of P is given by

$$\dot{r}_r = \Omega \times r$$

where Ω, the angular velocity of the planet relative to the carrier, is given by

$$\Omega = \omega_2 - \omega_4$$

The velocity of P relative to xyz is therefore given by

$$\dot{r}_r = (\omega_2 - \omega_4) \times r$$

The relative velocity of P with respect to point o is given by

$$\dot{r} = \dot{r}_r + \omega_4 \times r$$

$$= (\omega_2 - \omega_4) \times r + \omega_4 \times r = \omega_2 \times r$$

Continuing with the unit vector substitutions, the relative velocity of P with respect to point o can be expressed in terms of the angular velocity (ω_4) of the rotating system and the components of r. From Figure 8.43c, we obtain

$$r = r(i \cos \phi + j \sin \phi)$$

Substituting these vector expressions into the speed ratios for the sun and planet, we obtain

$$\omega_2 = \left(1 - \frac{N_3}{N_2}\right)\omega_4$$

$$\omega_2 \times r = \left(1 - \frac{N_3}{N_2}\right)(\omega_4 \times r)$$

$$= \left(1 - \frac{N_3}{N_2}\right)\omega_4 k \times r(i \cos \phi + j \sin \phi)$$

Since, by the right-hand rule, $k \times i = j$ and $k \times j = -i$,

$$\omega_2 \times r = \left(1 - \frac{N_3}{N_2}\right)(\omega_4 r)(j \cos \phi - i \sin \phi)$$

Combining terms, then, the velocity of point P on the planet with respect to the fixed coordinates of the train (XYZ) is finally given by

$$\dot{R} = \dot{R}_0 + \dot{r}$$

$$= -\omega_4 R_0 i + \left(1 - \frac{N_3}{N_2}\right)(\omega_4 r)(j \cos \phi - i \sin \phi)$$

$$= \omega_4 \left[\left(\frac{N_3}{N_2} - 1\right)(r \sin \phi) - R_0\right]i - \omega_4 \left[\left(\frac{N_3}{N_2} - 1\right)(r \cos \phi)\right]j \qquad (8.21)$$

The acceleration of point P is determined as follows:

$$\ddot{R} = \ddot{R}_0 + \dot{\omega} \times r + \omega \times (\omega \times r) + \ddot{r}_r + 2\omega \times \dot{r}_r$$

(Eq. 4.15 repeated). Recalling that the given angular velocity of the carrier is constant, the $\dot{\omega}$ nd $\dot{\Omega}$ terms in the acceleration equation are zero. We can further make the following substitutions:

$$\ddot{R}_0 = \omega \times (\omega \times R_0) = \omega_4 k \times (\omega_4 k \times R_0 j)$$

$$= \omega_4 k \times (-\omega_4 R_0 i) = \omega_4^2 R_0(-j) = -\omega_4^2 R_0(j)$$

(using the right-hand rule where the term in parentheses is evaluated first). The next term, $\dot{\omega} \times r$, is zero (ω is constant). The next term, $\omega \times (\omega \times r)$, becomes

$$\omega \times (\omega \times r) = \omega_4 k \times [\omega_4 k \times r(i \cos \phi + j \sin \phi)]$$

$$= \omega_4 k \times [\omega_4 rj \cos \phi - \omega_4 ri \sin \phi]$$

$$= \omega_4^2 r(-i)\cos \phi - \omega_4^2 rj \sin \phi$$

$$= -\omega_4^2 r(i \cos \phi + j \sin \phi)$$

The next term to be considered is the derivative of the relative velocity, $\ddot{r}_r$. Recall that $\dot{r}_r = \Omega \times r$. The derivative of this expression is then given by

$$\ddot{r}_r = \Omega \times (\Omega \times r)$$

where relative angular velocity

$$\Omega = \omega_2 - \omega_4 = -\left(\frac{N_3}{N_2}\right)\omega_4 k$$

Thus

$$\ddot{r}_r = \left[-\left(\frac{N_3}{N_2}\right)\omega_4 k \right] \times \left[-\left(\frac{N_3}{N_2}\right)\omega_4 k \times r \right]$$

$$= -\left[\left(\frac{N_3}{N_2}\right)\omega_4\right]^2 r(i\cos\phi + j\sin\phi)$$

Finally, the last term of Eq. 4.15 is given by

$$2\omega \times \dot{r}_r = 2\omega_4 k \times \left[-\left(\frac{N_3}{N_2}\right)\omega_4 k \times r(i\cos\phi + j\sin\phi) \right]$$

$$= 2\left(\frac{N_3}{N_2}\right)(\omega_4^2 r)(i\cos\phi + j\sin\phi)$$

Combining all of the above terms, we obtain the total acceleration of a point on the planet with respect to the fixed coordinates:

$$\ddot{R} = \omega_4^2\left\{ -R_0 j + \left[2\left(\frac{N_3}{N_2}\right) - \frac{N_3^2}{N_2^2} - 1 \right](r)(i\cos\phi + j\sin\phi) \right\}$$

which may be written as

$$\ddot{R} = \omega_4^2(-R_0) + \omega_4^2\left[\frac{N_3^2}{N_2^2} - 2\left(\frac{N_3}{N_2}\right) + 1 \right](-r)$$

We see that the term $N_3^2/N_2^2 - 2(N_3/N_2) + 1 = (1 - N_3/N_2)^2$, which is equal to $(\omega_2/\omega_4)^2$ from the tabular analysis. Using this, we find the following simple expression for the acceleration of P:

$$\ddot{R} = \omega_4^2(-R_0) + \omega_2^2(-r) \tag{8.22}$$

◆ **EXAMPLE PROBLEM 8.3** *Velocity and Acceleration of a Point on a Planet Gear*

In Figure 8.43a let, the planet carrier speed be $\omega_4 = 100$ rad/s counterclockwise (constant). The ring gear is fixed. Numbers of teeth are $N_1 = 30$, $N_2 = 25$, and $N_3 = 80$. The distance Oo on the planet carrier is $R_0 = 5.5$ in, and the planet pitch radius is $r = 2.5$ in. Find the velocity and acceleration of a point P on the planet *at the instant when P is the pitch point of the planet and ring* (i.e., O, o, and P lie on one line).

Solution. The vector directions of R_0 and r are the same when we consider the instant when P is the pitch point of the planet and ring. For the velocity of the moving coordinate system (the axis of the planet gear) relative to the fixed axis, we obtain

$$\dot{R}_0 = \omega_4 \times R_0$$

For the point P on the planet relative to the moving axis (moving system xyz), we obtain

$$\dot{r}_r = \Omega \times r = (\omega_2 - \omega_4) \times r = \omega_2 \times r - \omega_4 \times r$$

Using $\omega_2/\omega_4 = 1 - N_3/N_2$, we obtain

$$\dot{r}_r = \left(1 - \frac{N_3}{N_2}\right)\omega_4 \times r - \omega_4 \times r$$

$$= \omega_4 \times r\left(1 - \frac{N_3}{N_2} - 1\right) = \omega_4 \times r\left(-\frac{N_3}{N_2}\right)$$

Using Eq. 4.7, the velocity of point P relative to the fixed axis XYZ becomes

$$\dot{R} = \dot{R}_0 + \dot{r}_r + \omega_4 \times r$$

$$= \omega_4 \times R_0 - \left(\frac{N_3}{N_2}\right)\omega_4 \times r + \omega_4 \times r$$

$$= \omega_4 \times \left[R_0 + r + \left(-\frac{N_3}{N_2}\right)r\right]$$

But $R_0 + r$ is the pitch radius of the ring gear, which is precisely $(N_3/N_2)r$. Since R_0 and r have the same direction in this problem, the term in brackets is zero, and we have $\dot{R} = 0$. We obtain the same result from Eq. 8.21 with $\phi = 90°$.

Acceleration of P is given by Eq. 8.22. Noting that

$$\omega_2 = \left(1 - \frac{N_3}{N_2}\right)\omega_4 = 220 \text{ rad/s}$$

we obtain

$$\ddot{R} = (100)^2(5.5) + (220)^2(2.5)$$

the first term in the $-R_0$ direction, the second in the $-r$ direction. The result is

$$\ddot{R} = 176,000 \text{ in/s}^2$$

toward O, the center of the planetary train. ◆

8.7 FORCES, TORQUES, AND TRANSMITTED POWER IN GEAR TRAINS

The torque T on a gear is given by the tangential force F^t times the pitch radius r of the gear. If a gear makes contact with two or more other gears (e.g., as in a planetary train), each contact contributes to the torque. The net torque on an idler gear due to gear tooth contacts is zero.

Radial force on a gear is given by

$$F^r = F^t \tan \phi \qquad (8.23)$$

where ϕ is the pressure angle. The radial force is directed toward the center of a gear except for a ring gear, in which case the radial force is directed outward. Axial forces as well as radial and tangential forces occur in helical gears (including worms and worm gears), bevel gears, and hypoid gears.

One of the reasons for selection of balanced planetary trains over other drive trains is the elimination of shaft bending loads due to unbalanced radial and tangential forces on the sun and ring gears. Another advantage is the reduction in tooth loads for a given torque capacity due to the presence of two or more planet gears.

Noting that $1 \text{ W} = 1 \text{ N} \cdot \text{m/s}$, transmitted power is given by

$$P \text{ (kW)} = T \text{ (N} \cdot \text{mm)} \, \omega \text{ (rad/s)} \times 10^{-6} \qquad (8.24)$$

The equivalent expression in English units is

$$P \text{ (hp)} = \frac{T \text{ (in} \cdot \text{lb)} \, n \text{ (rev/min)}}{63,025} \qquad (8.25)$$

◆ **EXAMPLE PROBLEM 8.4** *Forces and Torques in a Planetary Train*

Forces, torques, and transmitted power in a planetary train are considered. Figure 8.44a shows a balanced planetary train (the carrier is not shown). Spur gears of 2 module ($d/N_t = 2$ mm) and $\phi = 20°$ are used. The sun gear is the input gear and the carrier is the output. The ring gear is fixed.

$$N_S = 36 \quad N_P = 32 \quad \omega_S = 400 \text{ rad/s} \quad P = 14.4 \text{ kW}$$

Determine angular velocities. Tabulate the results. Determine forces and torques. Draw free-body diagrams. Do external torques balance? Does power in equal power out? Do forces balance?

Solution. For this planetary train,

$$d_R = d_S + 2d_P$$

from which

$$N_R = N_S + 2N_P = 36 + 2 \times 32 = 100 \text{ teeth}$$

For a module $m = 2$ mm, the pitch radius of each gear is given by

$$r_P = \frac{1}{2} m N_t = \frac{1}{2} \times 2 \times N_t$$

The radius of the carrier is

$$r_C = r_S + r_P = 36 + 32 = 68 \text{ mm}$$

Rotation speeds are obtained by the superposition method, as shown in Table 8.13.

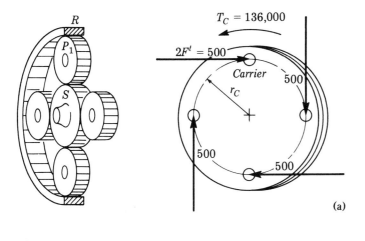

(a)

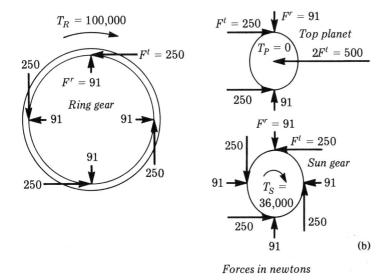

Forces in newtons
Torques in newton-millimeters

Figure 8.44 (a) A balanced planetary train. (b) Forces and torques in a planetary train.

Free-body diagrams are the key to the remainder of the problem. The torque on the sun gear shaft is

$$T_S = \frac{10^6 P}{\omega} = \frac{10^6 \times 14.4 \text{ kW}}{400 \text{ rad/s}} = 36,000 \text{ N} \cdot \text{mm}$$

This torque is balanced by tangential forces at each of the four planets:

$$T_S = 4F^t r_S$$

TABLE 8.13 SPEEDS, TORQUES, AND TRANSMITTED POWER
IN A PLANETARY TRAIN

	Equation / procedure	Gear S (input)	P	R (fixed)	C (output)
N_t	—	36	32	100	—
r_{pitch} (mm)	$r_p = \frac{1}{2}mN_t$	36	32	100	$r_C = 68$
1	Lock train	$+1$	$+1$	$+1$	$+1$
2	Correct R, C fixed	$+\dfrac{100}{36}$	$-\dfrac{100}{32}$	-1	0
Sum	$1+2$	$+3.7777$	-2.125	0	$+1$
ω (rad/s)	$\dfrac{400}{3.777} \times$ sum	400 cw	225 ccw	0	105.88 cw
Shaft T (N · mm)	From free body	36,000 cw	0	100,000 cw	136,000 ccw
Shaft P (kW)	$10^{-6}T\omega$	14.4	0	0	14.4

from which the tangential force is

$$F^t = \frac{T_S}{4r_S} = \frac{36,000}{4 \times 36} = 250 \text{ N}$$

and radial force is

$$F^r = F^t \tan \phi = 250 \tan 20° = 91 \text{ N}$$

as shown. The 250-N tangential force and 91-N radial force are each applied at four locations on the sun gear, two locations on each of the four planet gears, and four locations on the ring gear, as shown on the free-body diagrams in Figure 8.44b. The direction of the tangential forces on the sun and the planet, where they make contact, are opposite. The planet gear in this train acts as an idler. Thus planet torque $T_P = 0$, enabling us to obtain the tangential force magnitude on the planet where it contacts the ring gear. The tangential force on the ring is equal and opposite.

A reaction torque is required on the ring gear to prevent it from turning. It is given by

$$T_R = 4F^t r_R = 4 \times 250 \times 100 = 100,000 \text{ N} \cdot \text{mm cw}$$

The reaction torque may be provided by bolting the ring in place if the transmission is designed for a fixed speed ratio. If speed changing is required, the reaction torque may be provided by a band brake about the outside of the ring.

We note that forces are balanced in the horizontal and vertical directions on the sun and ring gears (and on the planet carrier, as will be seen later).

Thus, these elements of the gear train do not cause bending moments on their respective shafts. For the planet shown (in the top position at this instant), the two tangential forces must be balanced by a horizontal force of $2F' = 2 \times 250 = 500$ N to the left. Thus each planet causes an equal and opposite force on the planet carrier. Torque equilibrium on the planet carrier requires

$$T_C = 4 \times 2F'r_C = 4 \times 2 \times 250 \times 68 = 136{,}000 \text{ N} \cdot \text{mm ccw}$$

on the free-body diagram of the carrier. Of course, the carrier produces a 136,000-N · mm clockwise torque on the output shaft, which turns clockwise.

External torques on the transmission as a whole are in balance since

$$T_S + T_C + T_R = 36{,}000 - 136{,}000 + 100{,}000 = 0$$

Note that friction losses have been neglected. Output power is given by

$$P_C = 10^{-6}T_C\omega_C = 10^{-6} \times 136{,}000 \text{ N} \cdot \text{mm} \times 105.88 \text{ rad/s} = 14.4 \text{ kW} \quad \blacklozenge$$

8.8 SPREADSHEETS APPLIED TO THE DESIGN OF GEAR TRAINS

The design of a gear train may require many trials in order to meet a given set of specifications. The use of a computer permits a designer to consider several different combinations when attempting to optimize a design. A number of computer programs are available for gear train design. Electronic spreadsheets are also ideally suited to this task.

Spreadsheet techniques were introduced briefly in Chapters 3 and 4. Detailed information is given in the reference and tutorial manuals that accompany the software (e.g., Stephenson, in the Chapter 3 References).

The superposition method of planetary gear train analysis fits easily into the spreadsheet format. Calculations should be checked by a second method whenever practical. The formula method for analyzing planetary trains can be coded into the spreadsheet as a check.

♦ **EXAMPLE PROBLEM 8.5**   *Design of a Two-Speed Transmission*
Design a transmission with input-to-output speed ratios of approximately 1 : 3.5 and 1 : 1.

Solution.   There are many possible solutions. Since we have just considered planetary transmissions, let us attempt to use a simple planetary transmission of the type shown in Figure 8.44. We will try sun gears with 24 to 26 teeth and planet gears with 17 to 19 teeth, using either three or four planet gears (eighteen possible combinations). The number of teeth in the ring gear is given by

$$N_R = N_S + 2N_P$$

(Note that the solution will be outlined using symbols for convenience, whereas spreadsheet formulas are coded in terms of cell addresses.)

We want to consider only balanced planetary trains, in order to avoid lateral loads on the sun and carrier shafts. The train can be balanced with equally spaced planet gears if

$(N_S + N_R)/$ (the number of planets) $=$ an integer

An IF statement and modulo (remainder) function may be used. The test for a balanced train is based on the following statement:

$$\text{IF}\{\text{MODULO}[(N_S + N_R)/(\text{number of planets})] < 10^{-6}, \text{"yes", "no"}\}$$

which returns "yes" if the remainder is zero (within a small roundoff error). If this test returns "no," the train cannot be assembled with equally spaced planets, and we reject that combination of gears.

Let the sun gear be integral with the input shaft, and let the planet carrier shaft be the output. The 1:1 speed ratio requirement can be met by incorporating a clutch in the system. If the ring gear is free to rotate, and the clutch engages the sun gear to the planet carrier, then the input and output speeds are equal.

A speed reduction is possible if the clutch is disengaged and the ring gear is held fixed. One way to stop rotation of the ring gear is to tighten a band brake on the outer surface of the gear.

Combining two steps in the superposition method, for one rotation of the carrier and zero rotations of the ring gear, we have the following results:

$$n_P/n_C = 1 - N_R/N_P$$

and

$$n_S/n_C = 1 + N_R/N_S$$

Speed ratios are calculated for various combinations of gears.

If the speed ratios are divided by n_S/n_C, we obtain the ratios n/n_S. The result is also checked by the formula method, from which

$$n_C/n_S = N_S/[N_S + N_R]$$

A spreadsheet program was used to obtain the results shown in Table 8.14. It can be seen that seven of the eighteen combinations tested form balanced trains. The speed ratio n_C/n_S is precisely 1:3.5 with 24, 18, and 60 teeth in the sun, planet, and ring gears, respectively. This combination will be selected as our tentative design. Note that either three or four planet gears may be used. ◆

If we wish to calculate shaft torques and gear tooth loading, the spreadsheet may be expanded for that purpose. If it is decided to evaluate a change in any design parameter, the spreadsheet program automatically recalculates all values related to the change. It is again recommended that checks be incorporated to detect programming errors.

TABLE 8.14 PLANETARY GEAR TRAIN WITH STATIONARY RING
Superposition Method (Check by formula method.)

	Sun	Planet	Ring	Carrier	Planets 3 Balanced	4
Teeth	24	17	58		No	No
Speed $n/n(c)$	3.416667	−2.41176	0	1		
Speed $n/n(s)$	1	−.705882	0	.2926829		
Speed check	1		0	.2926829		
Teeth	24	18	60		Yes	Yes
Speed $n/n(c)$	3.5	−2.33333	0	1		
Speed $n/n(s)$	1	−.666667	0	.2857143		
Speed check	1		0	.2857143		
Teeth	24	19	62		No	No
Speed $n/n(c)$	3.583333	−2.26316	0	1		
Speed $n/n(s)$	1	−.631579	0	.2790698		
Speed check	1		0	.2790698		
Teeth	25	17	59		Yes	Yes
Speed $n/n(c)$	3.36	−2.47059	0	1		
Speed $n/n(s)$	1	−.735294	0	.297619		
Speed check	1		0	.297619		
Teeth	25	18	61		No	No
Speed $n/n(c)$	3.44	−2.38889	0	1		
Speed $n/n(s)$	1	−.694444	0	.2906977		
Speed check	1		0	.2906977		
Teeth	25	19	63		No	Yes
Speed $n/n(c)$	3.52	−2.31579	0	1		
Speed $n/n(s)$	1	−.657895	0	.2840909		
Speed check	1		0	.2840909		
Teeth	26	17	60		No	No
Speed $n/n(c)$	3.307692	−2.52941	0	1		
Speed $n/n(s)$	1	−.764706	0	.3023256		
Speed check	1		0	.3023256		
Teeth	26	18	62		No	Yes
Speed $n/n(c)$	3.384615	−2.44444	0	1		
Speed $n/n(s)$	1	−.722222	0	.2954545		
Speed check	1		0	.2954545		
Teeth	26	19	64		Yes	No
Speed $n/n(c)$	3.461538	−2.36842	0	1		
Speed $n/n(s)$	1	−.684211	0	.2888889		
Speed check	1		0	.2888889		

TABLE 8.15 PLANETARY GEAR TRAIN WITH STATIONARY RING
Module: 5   Power: 12 kW   Pressure angle: 20°
Superposition method. Check by formula method
Shaft torque in N · mm

	Sun	Planet	Ring	Carrier	Planets 3	4
					Balanced?	
Teeth	24	21	66		Yes	No
Radius (mm)	60	52.5	165	112.5		
Relative speed	3.75	−2.14286	0	1		
Speed (rad/s)	120	−68.5714	0	32		
Speed check	120		0	32		
Shaft torque	100000	0	275000	−375000		
Tangential force (N)					555.5556	416.6667
Radial force (N)					202.2057	151.6543
Shaft power (kW)	12	0	0	−12		
Torque balance (N · mm)					0 OK	

♦ **EXAMPLE PROBLEM 8.6**   *Power, Torque and Tooth Loading in a Tentative Transmission Design*

Design a planetary transmission to transmit 12 kW of mechanical power with an input speed of 120 rad/s and a speed reduction of 1 : 3.75.

Solution. A simple planetary train will be selected for this application. Let the sun gear be the input and the carrier the output, and let the ring gear be held stationary. After investigating a number of combinations, it was found that 24, 21, and 66 teeth in the sun, planet, and ring gears, respectively, produce the required ratio. The results are shown in Table 8.15. Three (but not four) equally spaced planet gears may be used. Design decisions include selection of a 20° pressure angle and a module of $m = 5$ mm. Gear pitch radii are given by

$$r = mN/2$$

The radius of the planet carrier measured at the planet centers is

$$r_C = r_S + r_P$$

Speed is calculated relative to the carrier speed, using the superposition method. We see that $n_S/n_C = 3.75$ as required. Multiplying this row by 120/3.75, we obtain the actual speed of each gear when $\omega_S = 120$ rad/s. The result is checked by the formula method. Torque (N · mm) on the input shaft is given by

$$T_S = 10^6 P/\omega_S$$

where P = power (kW). As the design calls for three planet gears, tangential force is given by

$$F^t = T_S/[3r_S]$$

The radial force is

$$F^r = F^t \tan \phi$$

where $\phi = 20°$, the pressure angle.

Torque on the ring gear is given by

$$T_R = 3F^t r_R$$

and torque on the carrier

$$T_C = 2 \times 3F^t r_C$$

Free-body sketches should be used to check directions. Note that the torques in a free-body diagram of the assembly balance. Of course, there are many other aspects to the design of a transmission, including considerations of strength and wear. ◆

References

1. Adams, A. E., *Plastics Gearing: Selection and Application*, Marcel Dekker, New York, 1986.
2. Chironis, N. P. (ed.), *Gear Design and Application*, McGraw-Hill, New York, 1967.
3. Coy, J. J., D. Townsend, and E. Zaretsky, *Gearing*, NASA Scientific and Technical Branch, Springfield, VA, 1985.
4. Drago, R. J., *Fundamentals of Gear Design*, Butterworths, Boston, 1988.
5. Dudley, D. W., *Handbook of Practical Gear Design*, McGraw-Hill, New York, 1984.
6. Dyson, A., *A General Theory of the Kinematics and Geometry of Gears in Three Dimensions*, Clarendon Press, Oxford, 1969.
7. Lévai, Zoltan, *Theory of Epicyclic Gears and Epicyclic Change-Speed Gearing*, Budapest: Technical University of Building, Civil and Transport Engineering, 1966.
8. Selby, S. M., and R. C. Weast (eds.), Table of Factors and Primes in *Handbook of Chemistry and Physics*, 47th ed., Cleveland: Chemical Rubber, 1967, pp. A167–A176.
9. White, G., "Epicyclic Gears Applied to Early Steam Engines," *Mechanism and Machine Theory*, Penton, Cambridge, vol. 23, no. 1, pp. 25–38, 1988.

PROBLEMS

• **8.1** In Figure 8.3a (reverted gear train), input shaft 1 rotates at 2000 rev/min. $N_1 = 20$ and $N_4 = 40$. Find the highest and lowest output speeds obtainable if the tooth numbers for gears 2 and 3 are to be no fewer than 20 and no more than 60. (Note that shafts 1 and 4 lie on the same centerline.) All gears have the same module.

• **8.2** In Figure P8.1, find the speed and direction of rotation of gear 4. Gear 1 rotates at 1000 rev/min as shown.

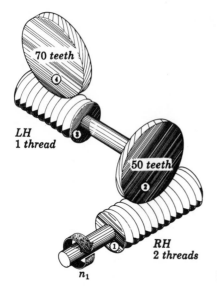

n_1

Figure P8.1

8.3 Specify the gears for a speed changer similar to that shown in Figure 8.5 (speed changer employing idler), having available output-to-input speed ratios of 0.8, 0.75, 0.6, 0.5, and 0.4. The smallest gear cannot have fewer than 20 teeth.

8.4 Sketch a transmission similar to that shown in Figure 8.6a (three-speed transmission). A synchromesh clutch is to be used in all gears. For the transmission above, specify tooth numbers to produce (approximately) the following output-to-input speed ratios: $n_o/n_i = +1, +0.5, +0.25, -0.25$. No gear may have fewer than 18 teeth.

• **8.5** Specify the gears required to obtain an output-to-input speed ratio of exactly 1131/2000 with a gear train similar to that shown in Figure 8.3a (reverted gear train). Use gears of not fewer than 20 nor more than 50 teeth.

8.6 In Figure P8.2, tooth numbers are $N_P = 30$, $N_{S_1} = 50$, $N_{S_2} = 49$, and $N_{S_3} = 51$. Gear S_1 is fixed, and the carrier speed is 100 rev/min clockwise. Find the rotation

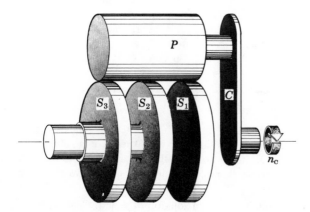

Figure P8.2

speed and the direction of S_2 and S_3, which are mounted on separate shafts. Use the tabular method. (*Note:* If the three sun gears have the same diameter, their pitches will be slightly different and the drive will not, theoretically, be precise.)

• **8.7** In Figure P8.3, assume that the sun gear is fixed. Find the speed of each gear by the tabular method if the planet carrier speed, n_c, is 200 rev/min clockwise. Answer in terms of tooth numbers: N_S, and so on.

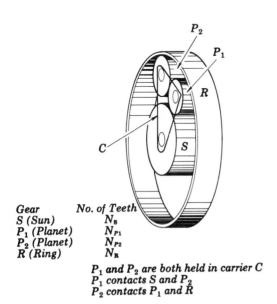

Gear	No. of Teeth
S (Sun)	N_S
P_1 (Planet)	N_{P1}
P_2 (Planet)	N_{P2}
R (Ring)	N_R

P_1 and P_2 are both held in carrier C
P_1 contacts S and P_2
P_2 contacts P_1 and R

Figure P8.3

8.8 Repeat Problem 8.7 for the case where the ring gear is fixed.

8.9 In Figure 8.9 (planetary speed reducer), ring gear R_1 is fixed. The sun gear rotates at 100 rev/min clockwise.
   (a) Find the output speed if tooth numbers are $N_{R_1} = 95$, $N_{P_2} = 21$, $N_{P_1} = 20$, $N_{R_2} = 100$, and $N_S = 55$.
   (b) Consider the same problem, but let R_2 serve as the input gear rotating at 100 rev/min, and let S serve as the output gear. Will a gear train actually operate in this manner?

• **8.10** Using a speed reducer similar to Figure 8.10 (planetary train with four planets), and with the ring gear fixed, obtain an output speed of approximately 1000 rev/min with input speed 3500 rev/min.
   (a) Specify the tooth numbers, letting the smallest gear have 20 teeth.
   (b) Determine the exact output speed for the gears you selected.
   (c) If the planets are to be equally spaced, how many will the design call for?

8.11 In Figure P8.4, carrier C rotates at 1000 rev/min clockwise and sun gear S_1 is fixed. Tooth numbers are $N_{S_1} = 51$, $N_{P_1} = 20$, $N_{P_2} = 19$, and $N_{S_2} = 50$. Find the speed and direction of rotation for each gear, using the tabular method.

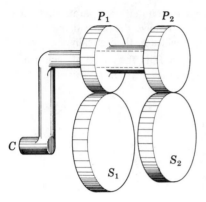

Figure P8.4

8.12 Repeat Problem 8.11 for the case where sun gear S_2 is fixed.

8.13 In Figure P8.5, tooth numbers are $N_{S_1} = N_{S_3} = 101$, $N_{S_2} = N_{S_4} = 100$, $N_{P_1} = N_{P_3} = 100$, and $N_{P_2} = N_{P_4} = 99$. Sun gears S_1 and S_3 are fixed. If the input shaft turns at 60 rev/min, how long will it take for one revolution of the output shaft?

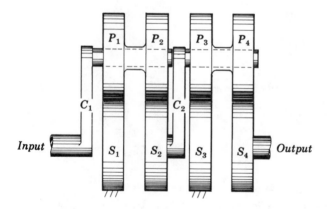

Figure P8.5

8.14 In Figure P8.3, $n_S = 400$ rev/min clockwise, and $n_R = 300$ rev/min clockwise. Let $N_S = 40$ teeth and $N_R = 100$ teeth. Find the carrier speed.

8.15 In Figure P8.4, let speeds n_C and n_{S_1} be given. Find n_{S_2} in terms of speeds n_C and n_{S_1} and tooth numbers N_{S_1}, N_{P_1}, N_{P_2}, and N_{S_2}.

Problems 8.16 Through 8.49 Refer to the Simplified Schematics of Figure 8.7

Find carrier speed n_C in terms of input speed n_i and tooth numbers N_S, N_P, and so on. Use the formula method to solve Problems 8.16 through 8.33.

8.16 Use Figure 8.7b, input gear S, fixed gear R.

8.17 Use Figure 8.7c, input gear S_1, fixed gear S_2.

8.18 Use Figure 8.7d, input gear R_1, fixed gear R_2.

8.19 Use Figure 8.7f, input gear R_1, fixed gear R_2.

8.20 Use Figure 8.7g, input gear S_1, fixed gear S_2.

8.21 Use Figure 8.7h, input gear R_1, fixed gear R_2.

8.22 Use Figure 8.7i, input gear S, fixed gear R.

8.23 Use Figure 8.7j, input gear S, fixed gear R.

8.24 Use Figure 8.7k, input gear S, fixed gear R.

8.25 Use Figure 8.7l, input gear S, fixed gear R.

8.26 Use Figure 8.7b, input gear R, fixed gear S.

8.27 Use Figure 8.7e, input gear S_1, fixed gear S_2.

8.28 Use Figure 8.7e, input gear S_2, fixed gear S_1.

8.29 Use Figure 8.7h, input gear R_2, fixed gear R_1.

8.30 Use Figure 8.7i, input gear R, fixed gear S.

8.31 Use Figure 8.7j, input gear R, fixed gear S.

8.32 Use Figure 8.7k, input gear R, fixed gear S.

● **8.33** Use Figure 8.7l, input gear R, fixed gear S.

Use the Tabular (Superposition) Method to Solve Problems 8.34 Through 8.49.

8.34 Use Figure 8.7b. The input gear is S and the fixed gear R.

8.35 Use Figure 8.7c. The input gear is S_1 and the fixed gear S_2.

8.36 Use Figure 8.7d. The input gear is R_1 and the fixed gear R_2.

8.37 Use Figure 8.7f. The input gear is R_1 and the fixed gear R_2.

8.38 Use Figure 8.7g. The input gear is S_1 and the fixed gear S_2.

8.39 Use Figure 8.7h. The input gear is R_1 and the fixed gear R_2.

8.40 Use Figure 8.7i. The input gear is S and the fixed gear R.

8.41 Use Figure 8.7j. The input gear is S and the fixed gear R.

8.42 Use Figure 8.7k. The input gear is S and the fixed gear R.

8.43 Use Figure 8.7l. The input gear is S and the fixed gear R.

8.44 Use Figure 8.7b. The input gear is R and the fixed gear S.

8.45 Use Figure 8.7h. The input gear is R_2 and the fixed gear R_1.

8.46 Use Figure 8.7i. The input gear is R and the fixed gear S.

8.47 Use Figure 8.7j. The input gear is R and the fixed gear S.

8.48 Use Figure 8.7k. The input gear is R and the fixed gear S.

● **8.49** Use Figure 8.7l. The input gear is R and the fixed gear S.

● **8.50** Design a gear train that produces one rotation of the output shaft in approximately 16 minutes for an input speed of 100 rev/min. Use a planetary train similar to Figure 8.7c with the center distance between the planet shaft and the input and output shafts = 244 mm. Find the speeds of all gears and the modules (which may not be standard for this problem). More than one trial solution may be required.

8.51 **(a)** Design a gear train similar to Figure P8.6 so that $\omega_C = 100$ rad/s ccw for $\omega_S = 300$ rad/s cw. Let the sun gear diameter be 80 mm and the module 4 mm. Give the possible range of planet sizes. Find the speed of each gear after selecting planets.

(b) Check the speed ratio using the formula method.

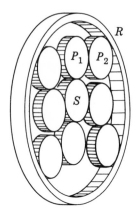

Figure P8.6

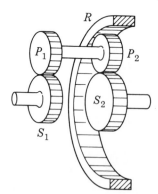

Figure P8.7

8.52 In the planetary train shown in Figure P8.7, the input gear is sun S_1 and the output gear is sun S_2. Ring gear R is fixed. Planets P_1 and P_2 rotate at the same speed. Find output-to-input speed ratio n_{S_2}/n_{S_1} by the superposition method (in terms of tooth numbers N_{S_1}, etc.).

8.53 In the planetary train of Figure P8.7, let $d_{S_1} = 100$, $d_{P_1} = 100$, $d_{P_2} = 95$, $d_{S_2} = 105$, $d_R = 295$ mm, and the module $m = 5$ mm. Find n_{S_2}/n_{S_1} by the formula method if the ring gear is stationary. *Suggestion*: Find $n_R^*/n_{S_1}^*$ to obtain n_C/n_{S_1}. Then find $n_{S_2}^*/n_{S_1}^*$.

8.54 Repeat Problem 8.53, using the superposition method.

8.55 Suppose a gear train was designed similar to the transmission shown in Figure 8.15, but with the $d_S = 60$, $d_{P_1} = 60$, $d_{P_2} = 40$, $d_{P_3} = 44$, $d_{R_1} = 160$, $d_{R_2} = 164$ mm, and $m = 2$ mm for all gears. Find speed ratio ω_{R_1}/ω_S if ring gear R_2 is fixed. Use the formula method (in two steps).

8.56 Repeat Problem 8.55 by the tabular method.

8.57 In the planetary train of Figure 8.10, $m = 2$ mm, $d_S = 36$, $d_P = 32$, and $d_R = 100$ mm. Find the speed of all gears if $\omega_S = 400$ rad/s cw and $\omega_R = 0$. Use the tabular method.

8.58 Repeat Problem 8.57 by the formula method.

8.59 In Figure 8.17c (bevel gear differential), let the planet P have 20 teeth and both sun gears 30 teeth. Gear S_1 drives the left rear axle and S_2 the right rear axle of a vehicle making a right turn at 20 mi/h. The 26-in-diameter tires are 56 in apart (center to center). The right wheel rolls in a 30-ft-radius path. Find the speed of the carrier and each sun gear, and the speed of the planet with respect to the carrier. Use the tabular method.

8.60 Repeat Problem 8.59, using the formula method.

8.61 In Figure P8.8, tooth numbers are $N_{S_1} = 35$, $N_P = 22$, and $N_{S_2} = 25$.

(a) If gear S_1 makes 40 rotations clockwise and S_2 makes 15 rotations counterclockwise, find the angular displacement of the carrier by the tabular method. How many rotations does the planet make about its own axis?

(b) Represent the motion of S_1 and S_2 by x and y, respectively. Write an expression for carrier motion in terms of x and y.

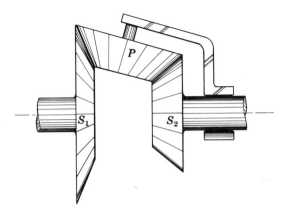

Figure P8.8

8.62 Repeat Problem 8.61 by the formula method.

8.63 In Figure 8.27a (variable-speed belt drive), pulley pitch diameters are $d_1 = 4$ in and $d_4 = 10$ in. Both countershaft pulleys are to have a minimum pitch diameter of 5 in. Determine the range of d_2 and d_3 so that output speed may be varied from 400 to 1400 rev/min with a motor speed of 1800 rev/min.

8.64 Repeat Problem 8.63 with $d_1 = 80$ mm, $d_4 = 200$ mm, and the minimum values of d_2 and $d_3 = 100$ mm.

8.65 In Figure 8.32 (disk drive with idler wheel), the input shaft rotates at 90 rev/min counterclockwise. Input and output shaft centerlines are 2 in apart. Find idler positions for output speeds of 16, $33\frac{1}{3}$, 45, and 78 rev/min counterclockwise.

8.66 Repeat Problem 8.65 if input and output shaft centerlines are 38 mm apart.

8.67 Design a variable-speed drive similar to Figure 8.31 (wheel-disk drive) to produce output speeds from 2000 rev/min through 0 to 2000 rev/min reverse with constant input speed of 1000 rev/min. *Hint*: Disk 2 may be used as input and disk 1 as output.

8.68 In Figure 8.40b (planetary cone transmission), let the ring gear have 45 teeth and the planets 15 teeth. If the reaction ring has an inside diameter of 5 in, what would the required cone diameters be for output speeds ranging from 450 rev/min clockwise to 180 rev/min counterclockwise, with 1800 rev/min input clockwise. Give cone diameters at points of contact for extreme positions of the reaction ring. (*Note:* The data for this problem do not represent actual dimensions of any commercially available transmission.)

8.69 Repeat Problem 8.68 for a 200-mm planet ring inside diameter.

8.70 A reverted gear train is to be designed similar to that of Figure 8.3a with input and output shafts colinear. The distance from the input shaft (gear 1) to the countershaft (gears 2 and 3) is 112.5 mm. Module $m = 5$ mm. Find gear diameters for minimum and maximum values of speed ratio if no gear is to have less than 20 teeth.

8.71 How many different speed ratios are possible in Problem 8.70? Calculate them.

8.72 Calculate the ratio of crankshaft speed to rotor speed for the rotating combustion engine (Wankel engine) described in Chapter 1. Note that the sun gear is fixed, the crankshaft is equivalent to the carrier, and the internal gear integral with the rotor acts as a planet. $N_P/N_S = 1.5$. Use the tabular method.

8.73 Repeat Problem 8.72 by the formula method.

8.74 A planetary gear train similar to Figure 8.44a, but with only two planets, transmits 20 hp. Input is to the sun gear and the carrier drives the load. The ring gear is fixed. The sun gear has 25 teeth and the planets 20. The diametral pitch is 5 and the pressure angle 20°. The sun gear rotates at 630 rev/min cw. Find speeds and draw free-body diagrams of each gear and the carrier.

8.75 Refer to Figure 8.12. Let $n_C = 1960$ rev/min cw. $N_{S_1} = 25$, $N_{P_2} = 39$, $N_{S_2} = 49$. Find the required number of teeth in P_1 so that $n_{S_2} = 10$ rev/min cw. The module may be nonstandard.

• 8.76 Refer to Figure P8.9. The sun gear is fixed and has a radius of 20 mm. The planets each have 20 teeth, 25° pressure angle and a 10-mm radius. A clockwise torque of 12,000 $N \cdot$ mm is applied to the ring, the input gear, and it rotates at 200 rev/min clockwise. Determine speeds. Show free-body diagrams of each gear and determine transmitted power. The planet carrier is the transmission output.

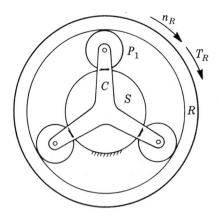

Figure P8.9

8.77 Design a two-speed transmission with output-to-input speed ratios of approximately 1:4.2 and 1:1. There are many possible designs that will satisfy this requirement. *Suggestions:* Use a computer to evaluate possible solutions. Consider a simple planetary train with 18 to 20 sun gear teeth and 19 to 21 planet gear teeth. Use three or four planet gears.

8.78 Design a two-speed transmission with speed ratios of approximately $1:3.8$ and $1:1$. There are many possible designs that will satisfy this requirement. *Suggestions*: Use a computer to evaluate possible solutions. Consider a simple planetary train with 21 to 23 sun gear teeth and 20 to 22 planet gear teeth. Use three or four planet gears.

8.79 Design a two-speed transmission with output-to-input speed ratios of approximately $3:10$ and $1:1$. There are many possible designs that will satisfy this requirement. *Suggestions*: Use a computer to evaluate possible solutions. Consider a simple planetary train with 18 to 26 sun gear teeth and 18 planet gear teeth. Use three or four planet gears.

8.80 A transmission is to be designed to transmit 2.8 kW of mechanical power. Speed is to be reduced from 220 rad/s to about 58 rad/s. A $1:1$ speed ratio should be available as well. Evaluate the kinematic and dynamic aspects of a simple planetary transmission with 20 sun gear teeth, 18 planet gear teeth, a 20° pressure angle, and a module of 1.5. Find:
  (a) The number of ring gear teeth
  (b) The actual output speed
  (c) The number of planets that will result in a balanced train
  (d) Pitch radii of the gears
  (e) Shaft torques
  (f) Tangential force at each gear mesh
  (g) Radial force at each gear mesh
  (h) Check torque balance.

8.81 A transmission is to be designed to transmit 12 kW of mechanical power. Speed is to be reduced from 400 rad/s to about 107 rad/s. A $1:1$ speed ratio should be available as well. Evaluate the kinematic and dynamic aspects of a simple planetary transmission with 24 sun gear teeth, 21 planet gear teeth, a 20° pressure angle, and a module of 5. Find:
  (a) The number of ring gear teeth
  (b) The actual output speed
  (c) The number of planets that will result in a balanced train
  (d) Pitch radii of the gears
  (e) Shaft torques
  (f) Tangential force at each gear mesh
  (g) Radial force at each gear mesh
  (h) Check torque balance.

8.82 A transmission is to be designed to transmit 3 kW of mechanical power. Speed is to be reduced from 400 rad/s to about 105 rad/s. A $1:1$ speed ratio should be available as well. Evaluate the kinematic and dynamic aspects of a simple planetary transmission with 22 sun gear teet, 20 planet gear teeth, a 20° pressure angle, and a module of 2.5. Find:
  (a) The number of ring gear teeth
  (b) The actual output speed
  (c) The number of planets that will result in a balanced train
  (d) Pitch radii of the gears
  (e) Shaft torques
  (f) Tangential force at each gear mesh

(g) Radial force at each gear mesh

(h) Check torque balance.

8.83 A transmission is to be designed to transmit 1.75 kW of mechanical power. Speed is to be reduced from 400 rad/s to about 100 rad/s. A 1 : 1 speed ratio should be available as well. Evaluate the kinematic and dynamic aspects of a simple planetary transmission with 20 sun gear teeth, 20 planet gear teeth, a 20° pressure angle, and a module of 1.5. Find:

(a) The number of ring gear teeth

(b) The actual output speed

(c) The number of planets that will result in a balanced train

(d) Pitch radii of the gears

(e) Shaft torques

(f) Tangential force at each gear mesh

(g) Radial force at each gear mesh

(h) Check torque balance.

PROJECTS

8.1 Design a light-duty transmission with six forward speed ratios and six reverse speed ratios. Output-to-input speed ratios are to range from 1.5 : 1 to 0.1 : 1 and −0.1 : 1 to −1 : 1.

8.2 Design a light-duty transmission with continuously variable speed ratios. Output-to-input speed ratios are to range from 2 : 1 to −0.75 : 1.

Static-Force Analysis

9.1 INTRODUCTION

A machine is a device that performs work and, as such, transmits energy by means of mechanical force from a power source to a driven load. It is necessary in the design of machine mechanisms to know the manner in which forces are transmitted from the input to the output, so that the components of the machine can be properly sized to withstand the stresses that are developed. If the members are not designed to be strong enough, then failure will occur during machine operation; if, on the other hand, the machine is overdesigned to have much more strength than required, then it may not be competitive with others in terms of cost, weight, size, power requirements, or other criteria.

Various assumptions must be made during the course of a force analysis reflecting the specific characteristics of the particular system being investigated. These assumptions should be verified as the design proceeds. A major assumption concerns dynamic or inertia forces. All machines have mass, and if parts of a machine are accelerating, there will be inertia forces associated with this motion. If the magnitudes of these inertia forces are small relative to externally applied loads, then they can be neglected in the force analysis. Such an analysis is referred to as static-force analysis and is the topic of this chapter.

For example, during normal operation of a front-end loader, such as that shown later in the chapter in Figure 9.13a, the bucket load and static weight loads may far exceed any dynamic loads due to accelerating masses, and a static-force analysis may be justified. An analysis that includes inertia effects is called a dynamic-force analysis and will be discussed in the next chapter. An example of an application where a dynamic-force analysis would be required is in the design of an automatic sewing machine, where, due to high operating speeds, the inertia forces may be greater than the external loads on the machine.

Another assumption deals with the rigidity of the machine components. No material is truly rigid, and all materials will experience significant deformation if the forces, either external or inertial in nature, are great enough. It will be assumed in this chapter and the next that deformations are so small as to be negligible and, therefore, the members will be treated as though they are rigid. The subject of mechanical vibrations, which is beyond the scope of this book, considers the flexibility of machine components and the resulting effects on machine behavior.

A third major assumption that is often made is that friction effects are negligible. Friction is inherent in all devices, and its degree is dependent upon many factors, including types of bearings, lubrication, loads, environmental conditions, and so on. Friction will be neglected in the first few sections of this chapter, with an introduction to the subject presented in Section 9.5.

In addition to assumptions of the types discussed above, other assumptions may be necessary, and some of these will be addressed at various points throughout the chapter.

The first part of this chapter is a review of general force analysis principles and will also establish some of the convention and terminology to be used in succeeding sections. The remainder of the chapter will then present both graphical and analytical methods for static-force analysis of machines.

9.2 BASIC PRINCIPLES OF FORCE ANALYSIS

9.2.1 Force and Torque

A force is a vector quantity; it has a magnitude, a direction or line of action, and a sense. Figure 9.1 shows a force vector F, which can be expressed in terms of Cartesian coordinates as follows:

$$F = F_x i + F_y j + F_z k \tag{9.1}$$

where F_x, F_y, and F_z are the components of the force in the x, y, and z directions, respectively, with these directions represented by unit vectors i, j, and k. The resultant force F of two forces F_1 and F_2 is the vector sum of those forces. This is expressed graphically in Figure 9.2 and mathematically as

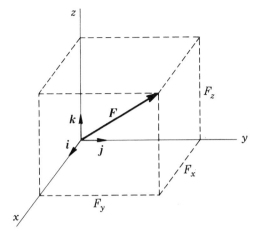

Figure 9.1 A force vector F in a Cartesian coordinate system.

follows:

$$F = F_1 + F_2 = (F_{1x} + F_{2x})i + (F_{1y} + F_{2y})j + (F_{1z} + F_{2z})k \qquad (9.2)$$

where F_{1x} is the x component of force F_1, and so on.

A torque or moment T is defined as the moment of a force about a point, and it also is a vector quantity. Using the vector cross product notation, we have

$$T = R \times F \qquad (9.3)$$

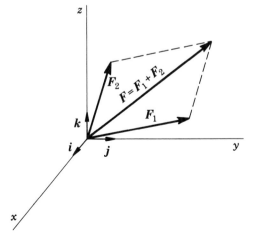

Figure 9.2 The resultant force F of two forces F_1 and F_2.

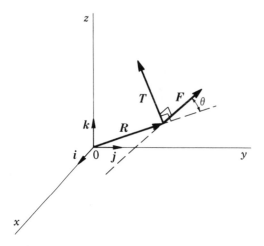

Figure 9.3 Torque T is the moment of force F about point 0. Vector R locates the line of action of the force relative to point 0.

where R is a position vector directed from the point about which the moment is taken to any point on the line of action of force F. See Figure 9.3. The magnitude of T is

$$T = RF|\sin \theta|$$

where θ is the angle between vectors R and F, and R and F are the magnitudes of the vectors. The direction of T is perpendicular to the plane containing R and F and the sense is given by the right-hand rule. Alternatively, in determinant form,

$$T = \begin{vmatrix} i & j & k \\ R_x & R_y & R_z \\ F_x & F_y & F_z \end{vmatrix}$$

$$= (R_y F_z - R_z F_y)i + (R_z F_x - R_x F_z)j + (R_x F_y - R_y F_x)k \qquad (9.4)$$

An infinite number of combinations of a force vector F and a moment arm vector R exist that will produce the same moment T; that is, different values of vectors R and F can lead to the same cross product as given by Eq. 9.3. The resultant of two or more moments is the vector sum of the moments.

Figure 9.4 shows two forces, F_1 and F_2, which have equal magnitudes but different lines of action. Furthermore, the two forces have parallel directions and are of opposite sense. Such a pair of forces is called a couple. The resultant force of a couple is zero. However, the resultant moment about an arbitrary point is not zero.

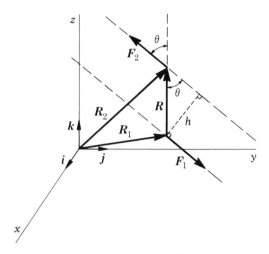

Figure 9.4 Forces F_1 and F_2 form a couple, which has zero resultant force but a nonzero resultant moment.

For example, if moments about the origin in Figure 9.4 are summed, the resultant moment T is

$$T = R_1 \times F_1 + R_2 \times F_2$$

But $F_1 = -F_2$, and, therefore,

$$T = R_1 \times (-F_2) + R_2 \times F_2 = (R_2 - R_1) \times F_2 = R \times F_2 \qquad (9.5)$$

where $R = R_2 - R_1$ is a vector from any point on the line of action of F_1 to any point on the line of action of F_2. The direction of the torque is perpendicular to the plane of the couple and the magnitude is given by

$$T = RF_2|\sin \theta| = hF_2 \qquad (9.6)$$

where $h = R|\sin \theta|$ is the perpendicular distance between the lines of action. It can be seen that the resultant moment of a couple, Eq. 9.5, is independent of the point about which moments are taken. Conversely, the moment of a couple about a particular point is independent of the position of the couple relative to the point. For these reasons, a couple is sometimes referred to as a pure moment or pure torque. As will be seen, the concept of a couple is very useful in force analysis applications.

9.2.2 Free-Body Diagrams

Free-body diagrams are extremely important and useful in force analysis. A free-body diagram is a sketch or drawing of part or all of a system, isolated in order to determine the nature of forces acting on that body. Sometimes a free-body diagram may take the form of a mental picture; however, actual

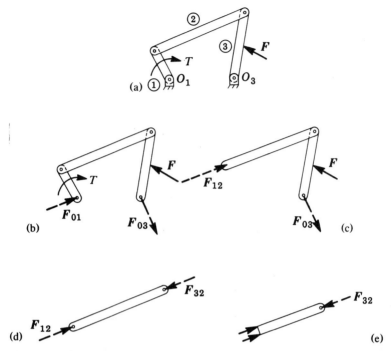

Figure 9.5 (a) A four-bar linkage. (b) Free-body diagram of the three moving links. (c) Free-body diagram of two connected links. (d) Free-body diagram of a single link. (e) Free-body diagram of part of a link.

sketches are strongly recommended, especially for complex mechanical systems.

Generally, the first step (and one of the most important) in a successful force analysis is the identification of the free bodies to be used. Figures 9.5b through 9.5e show examples of various free bodies that might be considered in the analysis of the four-bar linkage shown in Figure 9.5a. In Figure 9.5b, the free body consists of the three moving members isolated from the frame; here, the forces acting on the free body include a driving force or torque, external loads, and the forces transmitted from the frame at bearings O_1 and O_3. The force convention is defined as follows: F_{ij} represents the force exerted by member i on member j. This convention will be used throughout the text. Figure 9.5c is a free-body diagram of two links, which are acted upon by the forces transmitted from adjoining links as well as other applied loads. Probably the most commonly used form of a free-body diagram is that of a single link. See Figure 9.5d. Most force analyses can be accomplished by examining each of the individual members that make up the system. Such an approach leads to the determination of all of the bearing forces between members as well as the required input force or torque for a given output load or set of loads. For investigating internal forces or stresses in members, free bodies consisting of portions of members, as in Figure 9.5e, are useful.

9.2.3 Static Equilibrium

For a free body in static equilibrium, the vector sum of all forces acting on the body must be zero and the vector sum of all moments about any arbitrary point must also be zero. These conditions can be expressed mathematically as follows:

$$\sum F = 0 \qquad (9.7a)$$

$$\sum T = 0 \qquad (9.7b)$$

Since each of these vector equations represents three scalar equations, there are a total of six independent scalar conditions that must be satisfied for the general case of equilibrium under three-dimensional loading.

There are many situations where the loading is essentially planar, in which case, forces can be described by two-dimensional vectors. If the xy plane designates the plane of loading, then the applicable form of Eqs. 9.7a and 9.7b is

$$\sum F_x = 0 \qquad (9.8a)$$

$$\sum F_y = 0 \qquad (9.8b)$$

$$\sum T_z = 0 \qquad (9.8c)$$

Eqs. 9.8a to 9.8c are three scalar equations that state that, for the case of two-dimensional xy loading, the summations of forces in the x and y directions must individually equal zero and the summation of moments about any arbitrary point in the plane must also equal zero. The remainder of this chapter deals with two-dimensional force analysis. A common example of three-dimensional forces is gear forces, which were discussed in Chapter 7.

9.2.4 Superposition

The principle of superposition of forces is an extremely useful concept, particularly in graphical force analysis. Basically, the principle states that, for linear systems, the net effect of multiple loads on a system is equal to the superposition (i.e., vector summation) of the effects of the individual loads considered one at a time. Physically, linearity refers to a direct proportionality between input force and output force. Its mathematical characteristics will be discussed in the section on analytical force analysis. Generally, in the absence of Coulomb or dry friction, most mechanisms are linear for force analysis purposes, despite the fact that many of these mechanisms exhibit very nonlinear motions. Examples and further discussion in later sections will demonstrate the application of this principle.

9.3 GRAPHICAL FORCE ANALYSIS

Graphical force analysis employs scaled free-body diagrams and vector graphics in the determination of unknown machine forces. The graphical approach is best suited for planar force systems. Since forces are normally not constant during machine motion, analyses may be required for a number of mechanism positions; however, in many cases, critical maximum-force positions can be identified and graphical analyses performed for these positions only. An important advantage of the graphical approach is that it provides useful insight as to the nature of the forces in the physical system.

This approach suffers from disadvantages related to accuracy and time. As is true of any graphical procedure, the results are susceptible to drawing and measurement errors. Further, a great amount of graphics time and effort can be expended in the iterative design of a machine mechanism for which fairly thorough knowledge of force-time relationships is required. In recent years, the physical insight of the graphics approach and the speed and accuracy inherent in the computer-based analytical approach have been brought together through computer graphics systems, which have proved to be very effective engineering design tools.

There are a few special types of member loadings that are repeatedly encountered in the force analysis of mechanisms. These include a member subjected to two forces, a member subjected to three forces, and a member subjected to two forces and a couple. Other loading cases with more forces can usually be reduced to one of these situations by combining known individual forces into equivalent resultant forces. These special cases will be considered in the following paragraphs, before proceeding to the graphical analysis of complete mechanisms.

9.3.1 Analysis of a Two-Force Member

A *member subjected to two forces is in equilibrium if and only if the two forces (1) have the same magnitude, (2) act along the same line, and (3) are opposite in sense.* Figure 9.6a shows a free-body diagram of a member acted upon by forces F_1 and F_2, where the points of application of these forces are points A

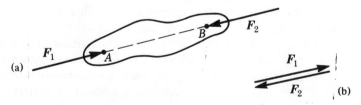

Figure 9.6 (a) A two-force member. The resultant force and the resultant moment both equal zero. (b) Force summation for a two-force member.

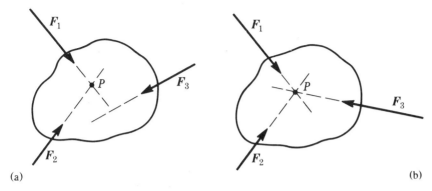

(a)   (b)

Figure 9.7 (a) The three forces on the member do not intersect at a common point and there is a nonzero resultant moment. (b) The three forces intersect at the same point P, called the *concurrency point*, and the net moment is zero.

and B. For equilibrium, the directions of F_1 and F_2 must be along line AB and F_1 must equal $(-F_2)$. A graphical vector addition of forces F_1 and F_2 is shown in Figure 9.6b, and, obviously, the resultant net force on the member is zero when $F_1 = -F_2$. The resultant moment about any point will also be zero, as can be seen from inspection or by application of Eq. 9.5.

Thus, if the load application points for a two-force member are known, the line of action of the forces is defined, and if the magnitude and sense of one of the forces are known, then the other force can immediately be determined. Such a member will be in either tension or compression.

9.3.2 Analysis of a Three-Force Member

A *member subjected to three forces is in equilibrium if and only if* (*1*) *the resultant of the three forces is zero, and* (*2*) *the lines of action of the forces all intersect at the same point*. The first condition guarantees equilibrium of forces, while the second condition guarantees equilibrium of moments. The second condition can be understood by considering the case when it is not satisfied. See Figure 9.7a. If moments are summed about point P, the intersection of forces F_1 and F_2, then the moments of these forces will be zero, but F_3 will produce a nonzero moment, resulting in a nonzero net moment on the member. On the other hand, if the line of action of force F_3 also passes through point P (Figure 9.7b), the net moment will be zero. This common point of intersection of the three forces is called the point of concurrency.

A typical situation encountered is that when one of the forces, F_1, is known completely, magnitude and direction, a second force, F_2, has known direction but unknown magnitude, and force F_3 has unknown magnitude and direction. The graphical solution of this case is depicted in Figure 9.8a through c. First, the free-body diagram is drawn to a convenient scale and the points of

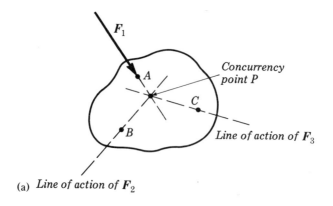

(a) *Line of action of F_2*

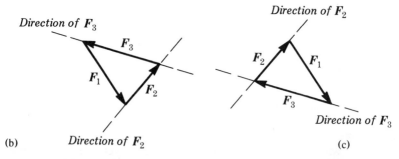

(b)

(c)

Figure 9.8 (a) Graphical force analysis of a three-force member. (b) Force polygon for the three-force member. (c) An equivalent force polygon for the three-force member.

application of the three forces are identified. These are points A, B, and C. Next, the known force F_1 is drawn on the diagram with the proper direction and a suitable magnitude scale. The direction of force F_2 is then drawn, and the intersection of this line with an extension of the line of action of force F_1 is the concurrency point P. For equilibrium, the line of action of force F_3 must pass through points C and P and is therefore as shown in Figure 9.8a.

The force equilibrium condition states that

$$F_1 + F_2 + F_3 = 0$$

Since the directions of all three forces are now known and the magnitude of F_1 was given, this equation can be solved for the remaining two magnitudes. A graphical solution follows from the fact that the three forces must form a closed vector loop, called a force polygon. The procedure is shown in Figure 9.8b. Vector F_1 is redrawn. From the head of this vector, a line is drawn in the direction of force F_2, and from the tail, a line is drawn parallel to F_3. The intersection of these lines closes the vector loop and determines the magnitudes of forces F_2 and F_3. Note that the same solution is obtained if, instead, a line parallel to F_3 is drawn from the head of F_1, and a line parallel to F_2 is

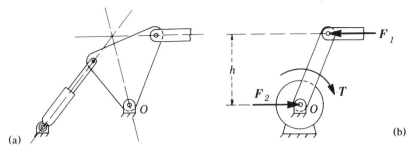

Figure 9.9 (a) The crank is driven by a piston and cylinder and is a three-force member. (b) The crank is driven by an electric motor and is subjected to two forces and a pure torque.

drawn from the tail of F_1. See Figure 9.8c. This is so because vector addition is commutative, and, therefore, both force polygons are equivalent to the vector equation above.

It is important to remember that, by the definition of vector addition, the force polygon corresponding to the general force equation

$$\sum F = 0$$

will have adjacent vectors connected head to tail. This principle is used in identifying the sense of forces F_2 and F_3 in Figure 9.8b and c. Also, if the lines of action of F_1 and F_2 are parallel, then the point of concurrency is at infinity, and the third force F_3 must be parallel to the other two. In this case, the force polygon collapses to a straight line.

9.3.3 Analysis of a Member with Two Forces and a Couple

In performing force analyses, it is imperative that we know the nature of forces that drive the system or that act as loads on the system. Only by knowing where these forces act and how they act can we proceed with a complete force analysis of the system.

This point can be illustrated by considering two ways by which the input crank of a four-bar linkage can be driven. See Figure 9.9a and b. In Figure 9.9a, the bell crank is driven by a hydraulic cylinder attached at the point shown. In this case, the crank is a three-force member and the analysis proceeds according to the preceding section. If, on the other hand, the crank is driven by a shaft with a direct connection to an electric motor, as shown in Figure 9.9b, then the applied shaft torque takes the form of a pure torque, and the crank is subjected to two forces plus an input couple. Both of these drive systems can be designed to produce the same torque about pivot O, but the forces experienced by the crank will differ in the two cases.

For equilibrium of a member subjected to two forces F_1 and F_2 plus an applied couple, the forces F_1 and F_2 must form a couple that is equal and opposite to the applied couple. Hence, if force F_1 is known in magnitude and direction, then force F_2 will be equal in magnitude, parallel in direction, and

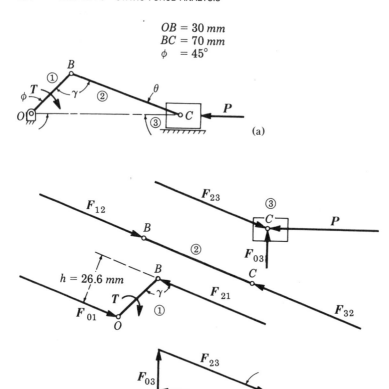

$$OB = 30 \ mm$$
$$BC = 70 \ mm$$
$$\phi \ = 45°$$

(a)

(b)

Figure 9.10 (a) Graphical force analysis of a slider-crank mechanism, which is acted on by piston force P and crank torque T. (b) Static force balances for the three moving links, each considered as a free body.

opposite in sense, and the moment of the applied couple must be equal and opposite to the moment of couple F_1, F_2. This is illustrated in Figure 9.9b, in which the magnitude of couple T is equal to the product $hF_1 = hF_2$, where F_1 and F_2 are the magnitudes of forces F_1 and F_2, respectively.

9.3.4 Graphical Force Analysis of the Slider-Crank Mechanism

The slider-crank mechanism finds extensive application in reciprocating compressors, piston engines, presses, toggle devices, and other machines where force characteristics are important. The force analysis of this mechanism employs most of the principles described in previous sections, as demonstrated by the following example.

◆ **EXAMPLE PROBLEM 9.1** *Static-Force Analysis of a Slider-Crank Mechanism*

Consider the slider-crank linkage shown in Figure 9.10a, representing a compressor, which is operating at so low a speed that inertia effects are negligible. It is also assumed that gravity forces are small compared with other forces and that all forces lie in the same plane. The dimensions are $OB = 30$ mm and $BC = 70$ mm. We wish to find the required crankshaft torque T and the bearing forces for a total gas pressure force $P = 40$ N at the instant when the crank angle $\phi = 45°$.

Solution. The graphical analysis is shown in Figure 9.10b. First, consider connecting rod 2. In the absence of gravity and inertia forces, this link is acted on by two forces only, at pins B and C. These pins are assumed to be frictionless and, therefore, transmit no torque. Thus, link 2 is a two-force member loaded at each end as shown. The forces F_{12} and F_{32} lie along the link, producing zero net moment, and must be equal and opposite for equilibrium of the link. At this point, the magnitude and sense of these forces are unknown.

Next, examine piston 3, which is a three-force member. The pressure force P is completely known and is assumed to act through the center of the piston (i.e., the pressure distribution on the piston face is assumed to be symmetric). From Newton's third law, which states that for every action there is an equal and opposite reaction, it follows that $F_{23} = -F_{32}$, and the direction of F_{23} is therefore known. In the absence of friction, the force of the cylinder on the piston, F_{03}, is perpendicular to the cylinder wall, and it also must pass through the concurrency point, which is the piston pin C. Now, knowing the force directions, we can construct the force polygon for member 3 (Figure 9.10b). Scaling from this diagram, the contact force between the cylinder and piston is $F_{03} = 12.7$ N, acting upward, and the magnitude of the bearing force at C is $F_{23} = F_{32} = 42.0$ N. This is also the bearing force at crankpin B, because $F_{12} = -F_{32}$. Further, the force directions for the connecting rod shown in the figure are correct, and the link is in compression.

Finally, crank 1 is subjected to two forces and a pure torque T (the shaft torque is assumed to be caused by a couple). The force at B is $F_{21} = -F_{12}$ and is now known. For force equilibrium, $F_{01} = -F_{21}$, as shown on the free-body diagram of link 1. However, these forces are not colinear, and for equilibrium, the moment of this couple must be balanced by torque T. Thus, the required torque is clockwise and has magnitude

$$T = F_{21}h = (42.0 \text{ N})(26.6 \text{ mm}) = 1120 \text{ N} \cdot \text{mm} = 1.12 \text{ N} \cdot \text{m}$$

It should be emphasized that this is the torque required for static equilibrium in the position shown in Figure 9.10a. If torque information is needed for a complete compression cycle, then the analysis must be repeated at other crank positions throughout the cycle. In general, the torque will vary with position.

♦

9.3.5 Graphical Force Analysis of the Four-Bar Linkage

The force analysis of the four-bar linkage proceeds in much the same manner as that of the slider-crank mechanism. However, in the following example, we consider the case of external forces on both the coupler and follower links and utilize the principle of superposition.

◆ **EXAMPLE PROBLEM 9.2**   *Static-Force Analysis of a Four-Bar Linkage*
The link lengths for the four-bar linkage of Figure 9.11a are given in the figure. In the position shown, coupler link 2 is subjected to force F_2 of magnitude 47 N, and follower link 3 is subjected to force F_3 of magnitude 30 N. Determine the shaft torque T_1 on input link 1 and the bearing loads for static equilibrium.

$O_1O_3 = 70\ mm$
$O_1B = 30\ mm$
$BC = 100\ mm$
$O_3C = 50\ mm$

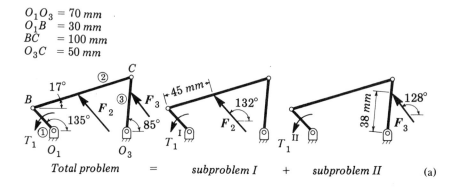

Total problem $=$ *subproblem I* $+$ *subproblem II*   (a)

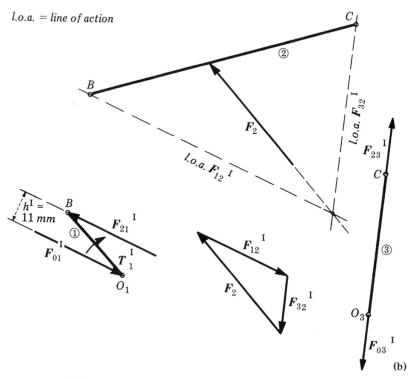

Figure 9.11 (a) Graphical force analysis of a four-bar linkage, utilizing the principle of superposition. (b) The solution of subproblem I.

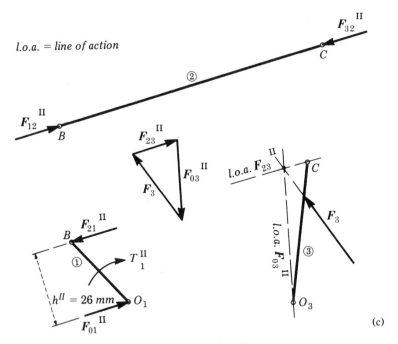

Figure 9.11 (c) The solution of subproblem II.

Solution. As shown in Figure 9.11a, the solution of the stated problem can be obtained by superposition of the solutions of subproblems I and II. In subproblem I, force F_3 is neglected, and in subproblem II, force F_2 is neglected. This process facilitates the solution by dividing a more difficult problem into two simpler ones.

The analysis of subproblem I is shown in Figure 9.11b, with quantities designated by superscript I. Here, member 3 is a two-force member because force F_3 is neglected. The direction of forces F_{23}^I and F_{03}^I are as shown, and the forces are equal and opposite (note that the magnitude and sense of these forces are as yet unknown). This information allows the analysis of member 2, which is a three-force member with completely known force F_2, known direction for F_{32}^I, and, using the concurrency point, known direction for F_{12}^I. Scaling from the force polygon, the following force magnitudes are determined (the force directions are shown in Figure 9.11b):

$$F_{32}^I = F_{23}^I = F_{03}^I = 21 \text{ N}$$

$$F_{12}^I = F_{21}^I = 36 \text{ N}$$

Link 1 is subjected to two forces and couple T_1^I, and for equilibrium,

$$F_{01}^I = -F_{21}^I$$

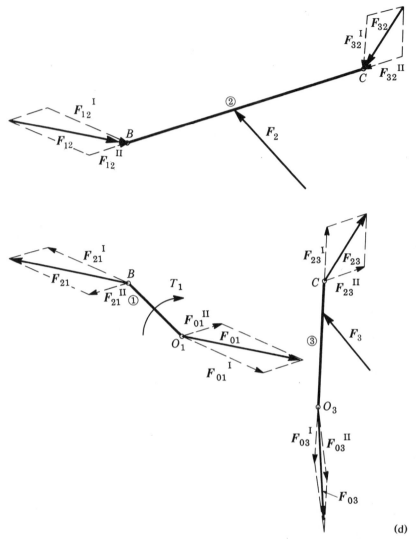

Figure 9.11 (d) The solutions combine to give the total solution.

and

$$T_1^I = F_{21}^I h^I = (36 \text{ N})(11 \text{ mm}) = 396 \text{ N} \cdot \text{mm cw}$$

The analysis of subproblem II is very similar and is shown in Figure 9.11c, where superscript II is used. In this case, link 2 is a two-force member and link 3 is a three-force member, and the following results are obtained:

$$F_{03}^{II} = 29 \text{ N}$$

$$F_{23}^{II} = F_{21}^{II} = F_{01}^{II} = 17 \text{ N}$$

and

$$T_1^{II} = F_{21}^{II} h^{II} = (17 \text{ N})(26 \text{ mm}) = 442 \text{ N} \cdot \text{mm cw}$$

The superposition of the results of Figure 9.11b and c is shown in Figure 9.11d. The results must be added vectorially, as shown. By scaling from the free-body diagrams, the overall bearing force magnitudes are

$$F_{01} = 50 \text{ N} \quad F_{23} = 31 \text{ N}$$

$$F_{12} = 50 \text{ N} \quad F_{03} = 49 \text{ N}$$

and the net crankshaft torque is

$$T_1 = T_1^I + T_1^{II} = 838 \text{ N} \cdot \text{mm cw}$$

The directions of the bearing forces are as shown in the figure. These resultant quantities represent the actual forces experienced by the mechanism.

It can be seen from the analysis that the effect of the superposition principle, in this example, was to create subproblems containing two-force members, from which the separate analyses could begin. In an attempt of a graphical analysis of the original problem without superposition, there is not enough intuitive force information to analyze three-force members 2 and 3, because none of the bearing force directions can be determined by inspection.

♦

9.3.6 Graphical Force Analysis of Complex Linkages

In this section, an example is presented involving the static-force analysis of a mechanical system that is somewhat more complex than the previous cases considered. This example will demonstrate that, although each force analysis problem has its own special characteristics, the solution procedure for a broad range of mechanisms is essentially unchanged, relying on the basic force analysis groundwork that has been developed.

♦ **EXAMPLE PROBLEM 9.3** *Force Analysis of a Door Mechanism*
Figure 9.12a shows the plan view of a fourfold industrial door. The door, which opens at the center, has four panels, two of which fold to the left side and two fold to the right side. The door is shown in the closed position with the open position of the panels inserted for reference as dashed lines. Figure 9.12a also shows the electrically powered operating system mounted above the doors and consisting of two symmetric linkages driven by the same motor.

Figure 9.12b is a schematic drawing of the right half of the system, drawn to scale for an intermediate position between the open and closed door positions. Including the door frame and the two door panels as links, the mechanism is an eight-bar linkage. Member 0 is the frame and members 1 and 2 are the door panels hinged together at point B. Slider 3 is pinned to panel 2 and moves along a fixed track as the doors open and close. Power is transmitted to the door by means of drive arm 7, connecting rod 6, and links 4 and 5.

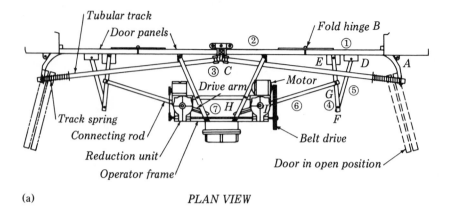

(a) PLAN VIEW

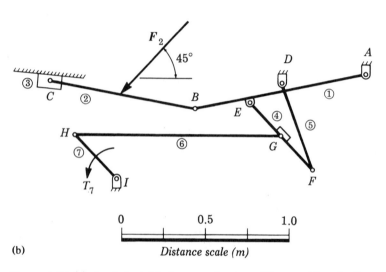

(b) Distance scale (m)

Figure 9.12 (a) An industrial door mechanism. (*Source:* Electric Power Door Company.) (b) Schematic linkage diagram.

Member 4 is connected to door panel 1 at point E, and member 5 is connected to the door frame at point D.

For the position of Figure 9.12b, determine the required shaft torque on drive arm 7 for static equilibrium against applied load F_2, which has a magnitude of 1000 N and acts on door panel 2 as shown in the figure.

Solution. It is assumed that inertia forces and friction effects are negligible. A planar force analysis will be performed considering those forces that act in planes parallel to that shown in Figure 9.12b. Gravity loads, which act perpendicular to these planes, are not included in the analysis.

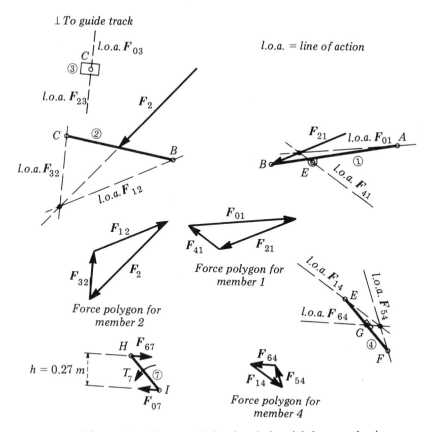

⊥ *To guide track*

l.o.a. = *line of action*

Figure 9.12 (c) Graphical force analysis of an industrial door mechanism.

The graphical analysis, presented in Figure 9.12c, starts with slider 3, which is a two-force member. Since friction is neglected, these forces must act perpendicular to the guide track, thus establishing the directions of forces F_{03} and F_{23}. Door panel 2 is a three-force member with known force F_2 and known direction of force F_{32}. From this information, the concurrency point can be found and the force polygon constructed, yielding the following force magnitudes:

$$F_{32} = 420 \text{ N} \quad F_{12} = 730 \text{ N}$$

Member 1 is also a three-force member, acted on by force F_{21}, which is now completely known, and the forces F_{01} and F_{41}, both of which have unknown direction and magnitude. In order for the analysis of this member to be completed, the direction of either F_{01} or F_{41} must be determined.

The direction of F_{41} can be found by considering link 4, which is another three-force member, acted upon by force F_{14} from link 1 at point E, force F_{54} from link 5 at point F, and force F_{64} from link 6 at point G. Since links 5 and

6 are two-force members, the lines of action of forces F_{54} and F_{64} are along the respective links, and the intersection of these lines is the concurrency point for member 4 (see Figure 9.12c). This leads to the line of action for F_{14} and, in turn, the direction of F_{41}.

The force polygon can now be constructed for member 1, as shown in Figure 9.12c, yielding the following force magnitudes:

$$F_{01} = 960 \text{ N} \quad F_{41} = 350 \text{ N}$$

Next, the polygon is constructed for member 4 (see Figure 9.12c), from which

$$F_{54} = 210 \text{ N} \quad F_{64} = 220 \text{ N}$$

Finally, member 7 is acted upon by two forces, known force F_{67} and equal and opposite force F_{07} (see Figure 9.12c), and shaft torque T_7, which must be equal and opposite to the moment of the couple F_{67}, F_{07}. Therefore, the torque is counterclockwise and, scaling the moment arm from Figure 9.12b, the magnitude is

$$T_7 = hF_{67} = (0.27 \text{ m})(220 \text{ N}) = 59.4 \text{ N} \cdot \text{m ccw}$$

If the other half of the door system is loaded symmetrically, a total torque double that above would be required. From this and knowledge of the speed reduction unit, the necessary motor torque can be found.

Similar analyses can be performed throughout the range of motion of the mechanism in order to size components for proper operation under various loading conditions, such as wind loads, which would be represented by external loads on both door panels. ◆

9.4 ANALYTICAL STATICS

Analytical methods for investigating static and dynamic forces in machines employ mathematical models that are solved either (1) for unknown forces and torques associated with known mechanism motion, or (2) for unknown motion of a given mechanism resulting from known driving forces or torques. This text deals almost exclusively with the former analysis category; however, there is a brief discussion of the latter category in Chapter 10. There are two approaches to formulating mathematical models; one approach is based on force and moment equilibrium, and the second is based on energy principles. Methods utilizing force and moment equilibrium equations parallel very closely the graphical method that has been presented. Both rely heavily on free-body diagrams, but the graphical force polygons are replaced in the analytical approach by equivalent vector equations. Energy methods utilize the principle of conservation of energy, one of the best-known examples being the method of virtual work.

The mathematical basis of the analytical approach lends itself well to computer implementation. Solutions can be obtained quickly and accurately for many positions of a mechanism, and the computer is particularly useful in design situations where many mechanism variations are to be considered. This facilitates design optimization, wherein those values of design parameters are determined such that selected performance criteria are optimized.

The designer may choose to write his or her own computer program for analysis or apply one of a number of general-purpose programs that are available. Examples of large programs that have been developed for various kinematic analysis, static-force analysis, and dynamic-force analysis tasks associated with planar and spatial machinery are Automatic Dynamic Analysis of Mechanical Systems (ADAMS)* and Dynamic Analysis and Design System (DADS).[†] These computer codes are very general and, therefore, are applicable to a broad range of mechanical systems. In addition, there are several commercially available programs developed for personal computers.

The following sections will introduce some of the basic theory involved in analytical methods for static-force analysis.

9.4.1 Static-Equilibrium Equations

The mathematical conditions for static equilibrium of a body were stated in Eqs. 9.7a and 9.7b, which are repeated here:

$$\sum F = 0$$

$$\sum T = 0$$

The detailed mathematical expression of these equations can take many forms depending on the vector representation used and, for example, may involve Cartesian vectors or complex number vectors, fixed or moving coordinate systems, and so on. Employing Cartesian vectors referenced to a fixed x, y, z coordinate frame, the component forms of Eqs. 9.7a and 9.7b become

$$\sum F_x = 0 \tag{9.9a}$$

$$\sum F_y = 0 \tag{9.9b}$$

$$\sum F_z = 0 \tag{9.9c}$$

$$\sum T_x = 0 \tag{9.9d}$$

$$\sum T_y = 0 \tag{9.9e}$$

$$\sum T_z = 0 \tag{9.9f}$$

which state that the net forces on a body in the x, y, and z directions must be

---

*Marketed by Mechanical Dynamics Inc., Ann Arbor, Michigan.
[†]Marketed by Computer Aided Design Software Inc., Oakdale, Iowa.

Figure 9.13 (a) A front-end loader. (*Source:* Sperry New Holland Company.)

zero and the net moments on the body about any three axes parallel to the x, y, and z directions must be zero. For two-dimensional force problems in the xy plane, Eqs. 9.9a to f reduce to the three conditions of Eqs. 9.8a to c presented earlier in this chapter:

$$\sum F_x = 0$$

$$\sum F_y = 0$$

$$\sum T_z = 0$$

For determinate force systems, Eqs. 9.9a to f will yield solutions to spatial force problems and Eqs. 9.8a to c will yield solutions to planar force problems.

♦ **EXAMPLE PROBLEM 9.4** *Force Analysis of a Front-End Loader*
Figure 9.13a is a photograph of a front-end loader showing the linkage arrangement for the boom mechanism. The boom is actuated by two hydraulic cylinders, one on each side of the machine, and the bucket is pivoted relative

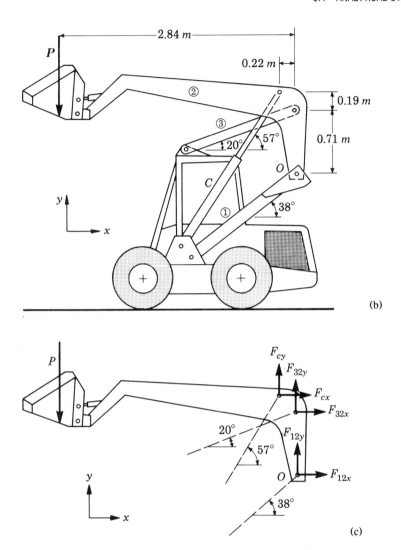

Figure 9.13 (b) Dimensions of the front-end loader. (c) Free-body diagram of the bucket and boom.

to the boom by a third hydraulic cylinder. Neglecting member weights and friction effects, determine the cylinder force required for static equilibrium of the boom in the position shown under a total bucket load of 4000 N.

Solution. In a thorough design analysis of the loader, the member weights would also be considered; they are neglected here in order to simplify the example. Also, it is assumed that the bucket load is evenly distributed between the two sides of the loader. Therefore, we will consider just one side under a vertical bucket load P having a magnitude of 2000 N.

Figure 9.13b is a drawing of the boom mechanism in the analysis position showing the force P and various dimensions and angular orientations. The xy coordinate system has been selected with x horizontal and y vertical.

A free body consisting of the bucket and boom is shown in Figure 9.13c, and, as indicated, four forces act on the free body. These forces, which are identified in x and y component form, are the vertical bucket load P, the force F_{12} from member 1, the force F_{32} from member 3, and the cylinder force F_c. Applying Eqs. 9.8a and 9.8b, we have

$$F_{cx} + F_{12x} + F_{32x} = 0 \qquad\qquad\qquad \text{(9.10a)}$$

$$F_{cy} + F_{12y} + F_{32y} - P = 0 \qquad\qquad\qquad \text{(9.10b)}$$

and summing moments about point O, Eq. 9.8c becomes

$$2.84P - 0.71F_{32x} - 0.90F_{cx} - 0.22F_{cy} = 0 \qquad\qquad\qquad \text{(9.10c)}$$

Eqs. 9.10a to c are a system of three equations in six unknowns: F_{cx}, F_{cy}, F_{12x}, F_{12y}, F_{32x}, and F_{32y}. However, links 1 and 3 and the hydraulic cylinder are two-force members, and, therefore, the directions of the forces that they exert on the boom will be along the links. Thus, the six unknowns can be expressed in terms of three unknowns as follows:

$$F_{cx} = F_c \cos(57°) \qquad F_{cy} = F_c \sin(57°)$$

$$F_{12x} = F_{12} \cos(38°) \qquad F_{12y} = F_{12} \sin(38°)$$

$$F_{32x} = F_{32} \cos(20°) \qquad F_{32y} = F_{32} \sin(20°)$$

where the angular orientations of the links are for the position under consideration. Substituting into Eqs. 9.10a to c, we have

$$(0.545) F_c + (0.788) F_{12} + (0.940) F_{32} = 0$$

$$(0.839) F_c + (0.616) F_{12} + (0.342) F_{32} - 2000 = 0$$

$$5680 - (0.667) F_{32} - (0.675) F_c = 0$$

Solving these equations for the three unknowns, we obtain

$$F_c = 6595 \text{ N} \qquad F_{12} = -6758 \text{ N} \qquad F_{32} = 1842 \text{ N}$$

or, in component form,

$$F_{cx} = 3594 \text{ N} \qquad F_{cy} = 5533 \text{ N}$$

$$F_{12x} = -5325 \text{ N} \qquad F_{12y} = -4163 \text{ N}$$

$$F_{32x} = 1731 \text{ N} \qquad F_{32y} = 630 \text{ N}$$

The signs indicate whether the force components are in the positive or negative coordinate directions. Thus, the actual directions of the components of forces F_c and F_{32} are as shown in Figure 9.13c, whereas the components of

F_{12} act in the negative coordinate directions. This means that member 3 and the cylinder experience compressive forces; whereas member 1 is in tension for the position analyzed. ◆

9.4.2 Analytical Solution for the Slider-Crank Mechanism

Because of its extensive use, the slider-crank mechanism deserves special attention. In the next chapter, a detailed dynamic-force analysis of this mechanism will be presented that will account for inertia effects, which are usually significant in machines such as engines and compressors. In this chapter, a graphical static-force analysis has already been presented to determine the relationship between piston force and crank torque. In this section, an equivalent analytical model will be developed.

An in-line slider-crank mechanism is shown in Figure 9.14a with crank length r, connecting rod length ℓ, and piston force P. A mathematical expression is sought relating force P to the crankshaft torque T required for

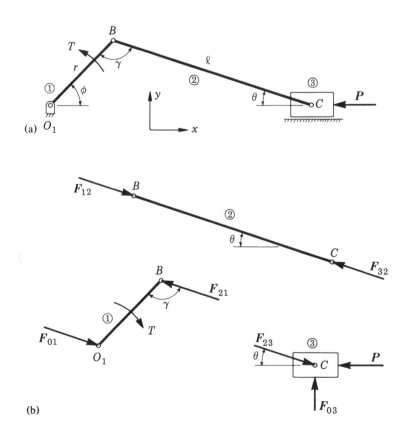

Figure 9.14 (a) An in-line slider-crank linkage. (b) Free-body diagrams of the moving members, employed in an analytical solution for torque T as a function of piston force P.

equilibrium. This expression will be a function of the mechanism position as given by crank angle ϕ.

Free bodies of the moving links are shown in Figure 9.14b. Connecting rod 2 is a two-force member, and, therefore, the force it exerts on the piston 3, F_{23}, will act at the angle θ of the connecting rod. Summing x direction forces on the piston, we have

$$F_{23} \cos \theta = P$$

or

$$F_{23} = \frac{P}{\cos \theta} \tag{9.11}$$

Next, considering the crank, it follows that

$$F_{21} = \frac{P}{\cos \theta} \tag{9.12}$$

and summing moments about point O_1, we have

$$T = -F_{21} r \cos(\gamma - 90°) = -F_{21} r \sin \gamma \tag{9.13}$$

A negative sign is used to denote a clockwise torque, which will occur for the force convention shown (this convention assumes that a positive piston force P acts to the left). Substituting $\gamma = 180° - (\phi + \theta)$ and Eq. 9.12 into Eq. 9.13, we have

$$T = -\frac{Pr}{\cos \theta} \sin[180° - (\phi + \theta)] = -Pr(\sin \phi + \cos \phi \tan \theta)$$

Finally, we wish to express angle θ as a function of crank angle ϕ. From the linkage geometry,

$$\sin \theta = \frac{r}{\ell} \sin \phi$$

and

$$\cos \theta = \sqrt{1 - \sin^2 \theta} = \sqrt{1 - \left(\frac{r}{\ell} \sin \phi\right)^2}$$

Upon substitution,

$$T = -Pr \sin \phi \left(1 + \frac{r \cos \phi}{\sqrt{\ell^2 - r^2 \sin^2 \phi}}\right) \tag{9.14}$$

As can be seen, even if force P is constant, the torque T will vary as the linkage orientation changes. Of course, in engines and compressors, the force P, representing the compression chamber pressure force, will also vary with position. This can be accounted for in Eq. 9.14 by expressing P as a function of ϕ. The torque will be zero when $\phi = 0$ or 180°, corresponding to the top and bottom dead-center positions of the mechanism. This means that the mechanism can lock in these positions under force P, unless acted upon by other forces such as inertia effects or the torque from other cylinders.

♦ **EXAMPLE PROBLEM 9.5** *Analysis of a Slider-Crank Mechanism*
Calculate the torque required for static equilibrium of an in-line slider-crank mechanism in the position when crank angle $\phi = 45°$. The dimensions are $r = 30$ mm and $\ell = 70$ mm, and the piston force is $P = 40$ N.

Solution. Substituting into Eq. 9.14, we have

$$T = -(40)(30) \sin(45°)\left[1 + \frac{(30)\cos(45°)}{\sqrt{(70)^2 - (30)^2 \sin^2(45°)}}\right]$$

$$= -1119 \text{ N} \cdot \text{mm}$$

This result agrees with that determined by graphical solution in Example Problem 9.1. ♦

9.4.3 Analytical Solution for the Four-Bar Linkage

In this section, a generalized analysis of the four-bar linkage will be presented. The equations to be derived are applicable to a wide variety of static-force situations, and in the next chapter, it will be shown that they also apply to dynamic-force analysis. In addition, it should be noted that the same procedure, with essentially the same equations, can be used to analyze more complex mechanism types.

A four-bar linkage is shown in Figure 9.15a, with the link lengths designated by ℓ_i and angular positions represented by the angles ϕ_i, $i = 1, 2, 3$. The distances r_i, $i = 1, 2, 3$, locate the points of intersection of the lines of action of applied forces F_1, F_2, and F_3 with the respective links. In addition to these loads, it is assumed that each link is acted upon by an externally applied couple C_i and input link 1 is driven by shaft torque T. The sign convention to be used for torques is that counterclockwise torques are positive and clockwise torques are negative. Employing this convention, torques, which in the strict sense are vectors perpendicular to the xy plane, will be treated as scalars. The forces F_i and couples C_i are assumed to be known quantities. Most typical forms of external loading can be represented by some combination of these forces and couples. Torque T will be treated as an unknown input required for equilibrium of the mechanism in the given position under specified loads.

Free-body diagrams of the three moving links are drawn in Figure 9.15b, which shows the bearing forces as well as the loads defined above. The location and orientation of the xy coordinate system are arbitrary. Angles ϕ_i are measured counterclockwise from the positive x direction. All forces are expressed in terms of x and y components.

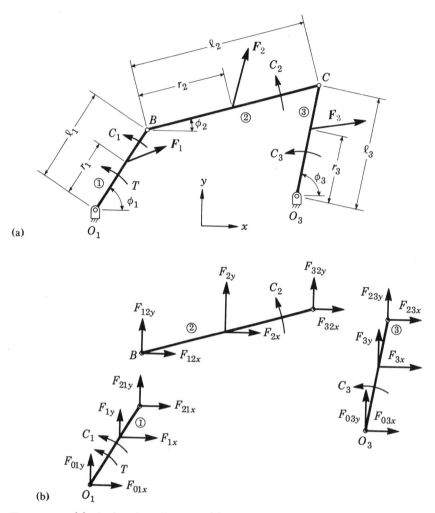

Figure 9.15 (a) A four-bar linkage. (b) Free-body diagrams of the moving members, employed in an analytical solution for forces and torques.

Since the mechanism is planar, a maximum of three independent equilibrium equations can be written for each link considered as a free body. Beginning with link 3, and summing forces in the x and y directions and moments about pivot O_3, we have

$$F_{03x} + F_{23x} + F_{3x} = 0 \quad \text{(9.15a)}$$

$$F_{03y} + F_{23y} + F_{3y} = 0 \quad \text{(9.15b)}$$

$$F_{23y}\ell_3 \cos \phi_3 - F_{23x}\ell_3 \sin \phi_3 + F_{3y}r_3 \cos \phi_3 - F_{3x}r_3 \sin \phi_3 + C_3 = 0 \quad \text{(9.15c)}$$

Eqs. 9.15a to c contain four unknowns ($F_{03x}, F_{03y}, F_{23x}, F_{23y}$) and therefore cannot be solved completely. Examining link 2, and writing a similar set of equations, where moments are summed about point B, we have

$$F_{12x} + F_{32x} + F_{2x} = 0$$

$$F_{12y} + F_{32y} + F_{2y} = 0$$

$$F_{32y}\ell_2 \cos \phi_2 - F_{32x}\ell_2 \sin \phi_2 + F_{2y}r_2 \cos \phi_2 - F_{2x}r_2 \sin \phi_2 + C_2 = 0$$

These three equations appear to introduce four new unknowns: F_{12x}, F_{12y}, F_{32x}, and F_{32y}. However, from Newton's third law,

$$F_{32x} = -F_{23x}$$

$$F_{32y} = -F_{23y}$$

Substituting these relationships into the equilibrium equations for link 2,

$$F_{12x} - F_{23x} + F_{2x} = 0 \quad \text{(9.16a)}$$

$$F_{12y} - F_{23y} + F_{2y} = 0 \quad \text{(9.16b)}$$

$$- F_{23y}\ell_2 \cos \phi_2 + F_{23x}\ell_2 \sin \phi_2 + F_{2y}r_2 \cos \phi_2 - F_{2x}r_2 \sin \phi_2 + C_2 = 0 \quad \text{(9.16c)}$$

Now, Eqs. 9.15a to 9.16c are a system of six equations in six unknowns: F_{12x}, F_{12y}, F_{23x}, F_{23y}, F_{03x}, and F_{03y}. The solution of these equations is simplified by observing that Eqs. 9.15c and 9.16c contain only two of the unknowns, F_{23x} and F_{23y}. Rearranging those equations, we have

$$a_{11}F_{23x} + a_{12}F_{23y} = b_1 \quad \text{(9.17a)}$$

$$a_{21}F_{23x} + a_{22}F_{23y} = b_2 \quad \text{(9.17b)}$$

where

$$a_{11} = -\ell_3 \sin \phi_3$$

$$a_{12} = \ell_3 \cos \phi_3$$

$$a_{21} = \ell_2 \sin \phi_2$$

$$a_{22} = -\ell_2 \cos \phi_2$$

$$b_1 = F_{3x}r_3 \sin \phi_3 - F_{3y}r_3 \cos \phi_3 - C_3$$

$$b_2 = F_{2x}r_2 \sin \phi_2 - F_{2y}r_2 \cos \phi_2 - C_2$$

Solving, we obtain

$$F_{23x} = \frac{a_{22}b_1 - a_{12}b_2}{a_{11}a_{22} - a_{12}a_{21}} \quad \text{(9.18a)}$$

$$F_{23y} = \frac{a_{11}b_2 - a_{21}b_1}{a_{11}a_{22} - a_{12}a_{21}} \quad \text{(9.18b)}$$

Returning to Eqs. 9.15a, 9.15b, 9.16a, and 9.16b, we can determine the other four unknown bearing force components as follows:

$$F_{03x} = -F_{23x} - F_{3x} \tag{9.19a}$$

$$F_{03y} = -F_{23y} - F_{3y} \tag{9.19b}$$

$$F_{12x} = F_{23x} - F_{2x} \tag{9.20a}$$

$$F_{12y} = F_{23y} - F_{2y} \tag{9.20b}$$

Negative values for any of the quantities indicate that their directions are actually in the negative coordinate directions opposite to the directions shown in Figure 9.15b.

Proceeding to member 1, the equilibrium equations are slightly different due to the presence of torque T (moments are summed about pivot O_1):

$$F_{01x} + F_{21x} + F_{1x} = 0$$

$$F_{01y} + F_{21y} + F_{1y} = 0$$

$$T + F_{21y}\ell_1 \cos\phi_1 - F_{21x}\ell_1\sin\phi_1 + F_{1y}r_1\cos\phi_1 - F_{1x}r_1\sin\phi_1 + C_1 = 0$$

Substituting $F_{21x} = -F_{12x}$ and $F_{21y} = -F_{12y}$ and rearranging, we solve these equations for F_{01x}, F_{01y}, and T:

$$F_{01x} = F_{12x} - F_{1x} \tag{9.21a}$$

$$F_{01y} = F_{12y} - F_{1y} \tag{9.21b}$$

$$T = F_{12y}\ell_1 \cos\phi_1 - F_{12x}\ell_1 \sin\phi_1 - F_{1y}r_1 \cos\phi_1 + F_{1x}r_1 \sin\phi_1 - C_1 \tag{9.22}$$

This completes the analysis of the four-bar linkage, having determined the x and y components of the four bearing forces, Eqs. 9.18a through 9.21b, and the required input torque, Eq. 9.22.

♦ **EXAMPLE PROBLEM 9.6**   *Analysis of a Four-Bar Linkage*
Solve Example Problem 9.2 by the analytical method.

Solution. The four-bar linkage of Figure 9.11 is redrawn in Figure 9.16 showing the various dimensions and forces F_2 and F_3 in component form. The following information is determined from the figure:

$$\ell_1 = 30 \text{ mm} \quad \ell_2 = 100 \text{ mm} \quad \ell_3 = 50 \text{ mm}$$

$$\phi_1 = 135° \qquad \phi_2 = 17° \qquad \phi_3 = 85°$$

$$r_1 = 0 \qquad r_2 = 45 \text{ mm} \qquad r_3 = 38 \text{ mm}$$

$$F_{2x} = 47\cos(132°) = -31.4 \text{ N} \quad F_{2y} = 47\sin(132°) = 34.9 \text{ N}$$

$$F_{3x} = 30\cos(128°) = -18.5 \text{ N} \quad F_{3y} = 30\sin(128°) = 23.6 \text{ N}$$

$$F_{1x} = F_{1y} = C_1 = C_2 = C_3 = 0$$

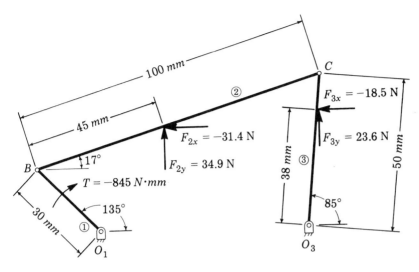

Figure 9.16 The four-bar linkage of Example Problem 9.6.

From these data, the coefficients and right-hand terms in Eqs. 9.17a and b are calculated to be as follows:

$a_{11} = -49.8 \quad a_{21} = 29.2 \quad b_1 = -778$

$a_{12} = 4.36 \quad\quad a_{22} = -95.6 \quad b_2 = -1910$

Then, from Eqs. 9.18a through 9.21b,

$F_{23x} = 17.8 \text{ N} \quad F_{23y} = 25.5 \text{ N}$

$F_{03x} = 0.64 \text{ N} \quad F_{03y} = -49.1 \text{ N}$

$F_{12x} = 49.2 \text{ N} \quad F_{12y} = -9.42 \text{ N}$

$F_{01x} = 49.2 \text{ N} \quad F_{01y} = -9.42 \text{ N}$

and the magnitudes of these forces are

$F_{23} = 31.1 \text{ N} \quad F_{12} = 50.2 \text{ N}$

$F_{03} = 49.1 \text{ N} \quad F_{01} = 50.2 \text{ N}$

Finally, the torque is calculated by Eq. 9.22:

$T = -845 \text{ N} \cdot \text{mm}$

Recall that a negative torque is clockwise. ◆

Each of the nine equations (i.e., three equilibrium equations for each of three members) that have been derived consists of a sum of multiples of the nine unknowns: F_{23x}, F_{23y}, F_{03x}, F_{03y}, F_{12x}, F_{12y}, F_{01x}, F_{01y}, and T. The coefficients of the unknowns in these equations do not depend on the applied forces, and there are no nonlinear terms in the unknowns or applied loads. Such a set of equations is said to be linear, and the principle of superposition applies. Recall that this principle states that the solution under combined loads is equal to the sum of the solutions for the individual loads that combine to produce the total load.

This characteristic of a linear set of equations can be demonstrated by examining Eqs. 9.17a and b. The coefficients a_{11}, a_{12}, a_{21}, and a_{22} are functions of the linkage configuration only and are not functions of the loads on the linkage. Right-hand term b_1 is a linear function of the loading on member 3, and term b_2 is a linear function of the loading on member 2. Consider the case, which we will refer to as subproblem I, where the forces on member 3 are zero; that is, $F_3 = C_3 = 0$ and, in turn, $b_1 = 0$. Then, Eqs. 9.17a and b become

$$a_{11}F^I_{23x} + a_{12}F^I_{23y} = 0 \tag{9.23a}$$

$$a_{21}F^I_{23x} + a_{22}F^I_{23y} = b_2 \tag{9.23b}$$

where F^I_{23x} and F^I_{23y} are the bearing forces resulting from this loading. Next, consider the case, subproblem II, where the forces on member 2, and therefore b_2, are zero:

$$a_{11}F^{II}_{23x} + a_{12}F^{II}_{23y} = b_1 \tag{9.24a}$$

$$a_{21}F^{II}_{23x} + a_{22}F^{II}_{23y} = 0 \tag{9.24b}$$

Here, F^{II}_{23x} and F^{II}_{23y} are the bearing forces resulting from the loading on member 3 only. Combining the two sets of equations, by adding Eq. 9.23a to Eq. 9.24a, and Eq. 9.23b to Eq. 9.24b,

$$a_{11}\left(F^I_{23x} + F^{II}_{23x}\right) + a_{12}\left(F^I_{23y} + F^{II}_{23y}\right) = b_1$$

$$a_{21}\left(F^I_{23x} + F^{II}_{23x}\right) + a_{22}\left(F^I_{23y} + F^{II}_{23y}\right) = b_2$$

By comparison with Eqs. 9.17a and 9.17b,

$$F_{23x} = F^I_{23x} + F^{II}_{23x}$$

$$F_{23y} = F^I_{23y} + F^{II}_{23y}$$

Thus, superposition of the solutions to subproblems I and II leads to the total solution for force F_{23}. It can be shown that the other unknowns, including torque T, can also be found by superposition.

9.4.4 The Method of Virtual Work

The method of virtual work is a method of force analysis that utilizes energy principles. This approach is considerably different from the force and moment equilibrium approach and, as such, offers advantages relative to that approach in certain types of analyses. For example, an entire mechanism can be examined as a whole without the need for dividing it up into a number of free bodies. This leads directly to a relationship between input and output forces and/or torques without the intermediate solution for bearing forces throughout the mechanism. The method is applicable to both static-force and dynamic-force analyses; the following paragraphs will discuss its application to static-force analysis.

As the name indicates, the method of virtual work derives from the concept of work. Work is defined as a force (or torque) acting through a displacement. In mathematical terms, work is the vector dot product of force and displacement, that is:

$$W = F \cdot S \qquad\qquad\qquad (9.25a)$$

where W is the work, F is the force, and S is the vector displacement at the point of application of the force. For rotational motion,

$$W = T\psi \qquad\qquad\qquad (9.25b)$$

where T is the torque and ψ is the angular displacement in a plane perpendicular to the torque. Since the result of the dot product operation is a scalar, work is a scalar quantity and it has units of energy.

The mathematical definition of work in Eq. 9.25a provides some useful insight into the nature of work. In particular, no net work is performed in the following cases:

1. When there is no displacement at the point of application of the force
2. When force F is perpendicular to displacement S
3. When equal but opposite forces act at the same point

Consider these statements as they pertain to the slider-crank mechanism shown in Figure 9.17. The mechanism experiences external piston force P and external crankshaft torque T, as well as bearing forces. We wish to determine the work performed as the mechanism travels through a small displacement from crank position ϕ, during which the angular displacement of the crank is $\delta\phi$ and the corresponding displacement of the piston is δx. See Figure 9.17. If we examine all forces acting on the mechanism, we see that the bearing force on the mechanism at crank pivot O_1 produces no work because there is no displacement of point O_1. Neglecting friction between the piston and cylinder, the force of the cylinder on the piston is perpendicular to the piston displacement, and hence no work is performed by this force. No work is performed by internal forces, including bearing forces at the crankpin and wrist pin, because equal and opposite forces act at all internal points. Thus the only work performed is that by torque T and force P:

$$W = T\delta\phi + P\delta x \qquad\qquad\qquad (9.26)$$

We assume here that torque T and force P are constant during the displacement of the mechanism. For small displacements, this is a reasonable approximation. If the force varies significantly during a displacement, then integration

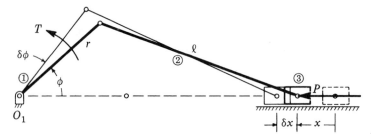

Figure 9.17 A slider-crank mechanism to be analyzed by the method of virtual work. The dashed-line configuration is the reference position for piston displacement x. Quantities $\delta\phi$ and δx define a virtual displacement of the mechanism from the analysis position corresponding to crank angle ϕ.

must be used to find the work. Note also that work is positive if the force acts in the same direction as the displacement and is negative if the force acts in a direction opposite to the displacement. Both T and P acting as shown in Figure 9.17 would produce positive work during the displacement shown.

In a true statics problem, there is no displacement. We therefore introduce a quantity called a virtual displacement, which is defined as an imaginary infinitesimal displacement of the system that is consistent with the constraints on the system. For example, the constraints on the slider-crank mechanism are that all members including the frame are rigid and all joints maintain contact. Thus Figure 9.17 depicts a virtual displacement of the mechanism, where $\delta\phi$ and δx are related by the kinematics of the rigid-membered linkage. Virtual work is defined as the work performed during a virtual displacement.

We now can state the principle of virtual work as it applies to equilibrium of mechanisms: *The work performed during a virtual displacement from equilibrium is equal to zero.* The interested student should consult a text on engineering mechanics for a complete derivation of this principle.

Let us now apply the principle for the determination of the torque T required in the slider-crank mechanism of Figure 9.17 for static equilibrium against applied force P for the mechanism position given by crank angle ϕ. For virtual displacements $\delta\phi$ and δx, it follows from Eq. 9.26 that

$$T\delta\phi + P\delta x = 0 \tag{9.27}$$

for equilibrium. Before solving this equation, we introduce one more characteristic of virtual displacements; that is, they are assumed to take place during the same time interval dt. Dividing Eq. 9.27 by dt,

$$T\frac{\delta\phi}{dt} + P\frac{\delta x}{dt} = 0$$

or

$$T\dot{\phi} + P\dot{x} = 0 \tag{9.28}$$

where $\dot{\phi}$ and $\dot{x}$ are the instantaneous velocities of the crank and piston. Solving for T, we have

$$T = -\frac{\dot{x}}{\dot{\phi}}P \tag{9.29}$$

Recall from Chapter 3 that

$$\dot{x} = r\dot{\phi}\sin\phi\left[1 + \frac{r\cos\phi}{\sqrt{\ell^2 - (r\sin\phi)^2}}\right]$$

and therefore,

$$T = -Pr\sin\phi\left[1 + \frac{r\cos\phi}{\sqrt{\ell^2 - (r\sin\phi)^2}}\right] \tag{9.30}$$

The negative sign indicates that, for equilibrium, torque T must produce a negative amount of work in Eq. 9.27 equal to the positive amount produced by force P, and, therefore, the torque must be clockwise in Figure 9.17.

It can be seen from this example that for a mechanism in equilibrium under the action of two forces, an input driving force and an output load, the ratio of the magnitude of the input force to the magnitude of the output force equals the inverse ratio of the corresponding velocity magnitudes. Of course, the method can also be used to analyze mechanisms with multiple loads, in which case the summation of all of the virtual work performed by the individual forces and torques must equal zero for equilibrium. Gravity loadings can be treated like any other force. Furthermore, inertia forces can be included so that the method can be employed in dynamic-force analysis.

9.5 FRICTION IN MECHANISMS

Whenever two connected mechanism members have relative motion, friction occurs at the joint that connects them. This friction produces heat and wear, which may eventually lead to bearing failure. It can also adversely affect the motion response of the mechanism; for example, it will slow the response of fast-action devices such as mechanical circuit breakers for electrical transmission lines, and it may alter the synchronization of automated systems such as multiple-input manipulators. In addition, the presence of friction can substantially increase the energy requirements of a machine.

The nature and amount of friction depend on the type of bearing employed. High-speed, high-load machinery is often designed with low-friction bearings: either rolling-contact bearings, utilizing balls or rollers to eliminate relative sliding, or thick-film bearings, in which the moving parts are separated by a layer of lubricant film. However, there exist many situations where direct physical contact between relatively sliding members occurs. This type of

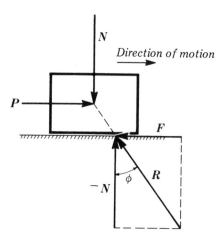

Figure 9.18 Dry friction in a translating bearing. Friction force *F* acting on the sliding block opposes the motion of the block relative to the contacting surface.

friction is referred to as dry or Coulomb friction, which will be examined in this section.

Dry friction can occur for various reasons. For example, equipment may be poorly lubricated, or there may be loss of lubricant due to leakage. In other cases, the environment places severe restrictions on the use of lubrication. For example, lubricants are prohibited in outer space applications and in certain food-processing areas; special self-lubricating bearing materials with low coefficients of friction have found successful application in these areas. Purely economic considerations sometimes rule against the use of low-friction bearings, resulting in machines or products in which dry-friction effects can be significant. Even in hydrodynamic thick-film bearings, there will be direct surface contact during machine start-up, before the lubricant layer develops, resulting in Coulomb friction.

Two common types of joints in which sliding friction can be present are prismatic or slider connections and revolute or journal bearings. Figure 9.18 shows a sliding block moving relative to a flat surface, which may be moving or fixed. In general, the applied force on the slider consists of components parallel and perpendicular to the direction of motion. These are shown in Figure 9.18 as components *P* and *N*, respectively. The Coulomb friction force *F* is related to the normal force *N* between contacting surfaces, and its magnitude is given by

$$F = \mu N \tag{9.31}$$

where μ is defined as the coefficient of sliding friction, which is a characteristic of the contacting materials and the operating conditions. Although precise values for coefficients of friction are extremely difficult, if not impossible, to obtain, handbooks contain extensive lists of approximate values that are acceptable for most design purposes. The direction of the friction force is always such that the relative motion is opposed. In Figure 9.18, the block is assumed to be sliding to the right. The friction force *F* acting on the block is therefore directed to the left.

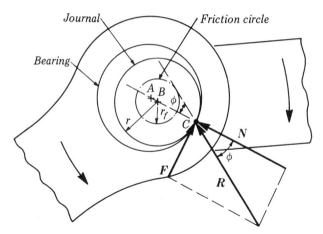

Figure 9.19 Dry friction in a journal bearing. The resultant force R acts tangent to the friction circle and includes a component F, which opposes relative motion, and a component N, which acts normal to the contacting surfaces.

The resultant force R that the surface exerts on the block has components of magnitude N and $F = \mu N$, and, therefore, the angle ϕ of this resultant from the normal direction is given by

$$\tan \phi = \frac{F}{N} = \frac{\mu N}{N} = \mu \qquad\qquad (9.32)$$

Angle ϕ is referred to as the friction angle, and it is evident that the direction of R is known once the coefficient of friction has been determined. As a limiting case, when there is no friction ($\mu = 0$), $\phi = 0$ and the resultant force is normal to the surfaces.

Friction in a journal bearing, Figure 9.19, is essentially the same as that for a sliding block. Shown in the figure is a journal or pin of radius r attached to one member, which rotates in a bearing or sleeve in another member. The center of the bearing is at point A and the center of the journal is at point B; normally, the clearance in the joint is much less than that depicted in the figure, since large clearances can lead to serious impact problems. The two members instantaneously contact at point C, where there is a resultant force R of the bearing on the journal. This force consists of a compressive normal component N and a friction component F, which opposes the sliding motion of the journal relative to the bearing.

As before, the two force components are related by the equation

$$F = \mu N$$

where μ is the coefficient of friction, and the friction angle ϕ is given by

$$\tan \phi = \mu$$

Thus, as long as the friction coefficient is constant, the resultant force will always act at a fixed angle ϕ with respect to the common normal to the two surfaces. It follows that the line of action of resultant R will always be tangent to a circle with center at point B and radius

$$r_f = r \sin \phi \tag{9.33}$$

See Figure 9.19. This circle is called the friction circle, and, with the following trigonometric identity, its radius can be expressed as

$$r_f = r \sin \phi = \frac{r \tan \phi}{\sqrt{1 + \tan^2 \phi}} = \frac{r\mu}{\sqrt{1 + \mu^2}} \tag{9.34a}$$

For small values of μ, which is often the case, the radius can be approximated as follows:

$$r_f = r \sin \phi \cong r \tan \phi = r\mu \tag{9.34b}$$

In addition to being tangent to the friction circle, the resultant force R will be directed so as to produce a moment about journal center B which opposes the relative motion. As will be demonstrated in the following sections, the friction circle is a useful concept for force analysis.

9.5.1 Graphical Solution for a Slider-Crank Mechanism Including Friction

The material of the preceding section will be illustrated here by investigation of the slider-crank linkage shown in Figure 9.20a. The force analysis will consider the presence of friction at all four connections, the three turning connections and one sliding connection. A load P is applied to the piston and the various bearing forces and the required input torque for static equilibrium are to be determined. Crank 1 is rotating in the clockwise direction.

Figure 9.20b shows the free-body diagrams of the members. The friction circles are constructed for the three rotating joints by means of Eq. 9.34a or Eq. 9.34b. The forces at these joints must be tangent to the friction circles. Furthermore, the forces must act in such a way as to oppose relative motion at the joints. By examination, the relative link motions are as shown in Figure 9.20a for clockwise crank rotation.

First, consider the connecting rod 2. This link is a two-force member and obviously is in compression for the loading and position being analyzed. Therefore, the forces F_{12} and F_{32} must be colinear, with force F_{12} producing a clockwise moment about pin B opposing the counterclockwise rotation of link 2 relative to link 1, and force F_{32} producing a clockwise moment about pin C opposing the counterclockwise rotation of link 2 relative to piston 3. The line of action of the compressive forces, which is shown on the free body of link 2, will satisfy these conditions. Notice that there are four possible straight lines that can be drawn tangent to the two friction circles but that the one shown is the only one that will satisfy all of the conditions stated. Generally, some intuitive guessing or trial and error is necessary in properly locating lines of action.

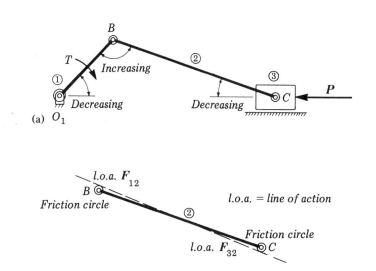

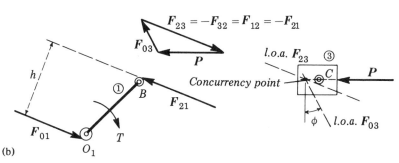

Figure 9.20 (a) Graphical force analysis of a slider-crank mechanism including friction. The crank rotates clockwise. (b) Free-body diagrams and force polygon.

Having found the line of action for member 2, we can now proceed with the analysis of the piston. Force P is known completely and the line of action of force F_{23} is now known. Force F_{03}, exerted by the frame on the piston, must pass through the concurrency point, given by the intersection of P and F_{23}, and must act at the friction angle ϕ with respect to the normal to the surfaces. This angle is measured as shown, accounting for a friction force to the left opposing the sliding of the block to the right. From this information, the force polygon is constructed, yielding forces F_{03} and F_{23}, and, in turn, forces F_{32}, F_{12}, and F_{21}.

Force F_{21} has been drawn on the free body of crank 1. Note that this force will produce a counterclockwise moment about the joint that opposes the clockwise rotation of link 1 relative to link 2. Force F_{01} must be equal in magnitude and opposite in direction. It is drawn tangent to the friction circle at pin O_1 and is properly placed to oppose the clockwise rotation of the crank relative to the frame. Finally, the moment created by this couple must be opposed by the driving torque $T = hF_{21}$, the direction of which is clockwise by inspection of the free-body diagram.

The graphical force analysis of other mechanism types with friction would proceed in a similar fashion. However, the more complex the mechanism is, the greater is the probability that trial-and-error solution will be required in the determination of forces.

The performance of a machine with friction is evaluated relative to ideal friction-free operation by means of a quantity called instantaneous efficiency. Instantaneous efficiency is defined as the ratio of the required input torque or force in the absence of friction to that in the presence of friction. For the slider-crank mechanism under consideration,

$$e = \frac{T_o}{T} \qquad\qquad (9.35)$$

where e is the instantaneous efficiency, T_o is the required input torque when there is no friction present, and T is the required input torque with friction included. For the case analyzed above, T is greater than T_o because the input driving torque must overcome mechanism friction as well as piston force P.

As the name implies, instantaneous efficiency may vary with the position of the mechanism, as would be the case for the slider-crank mechanism of Figure 9.20a. Also, for this mechanism, if one considers the case of counter-clockwise rotation of crank 1, with the same applied piston force P, the instantaneous efficiency will apparently be greater than one. This can be interpreted to mean that, in this case, mechanism friction assists the input torque in offsetting the piston force P, and, therefore, T is less than T_o. Looking at it another way, the piston force may more logically be thought of as the input in this situation, and for a given torque load T, the efficiency

$$e = \frac{P_o}{P}$$

will be less than one.

9.5.2 Analytical Solution for a Slider-Crank Mechanism Including Friction

There are some disadvantages inherent in the analysis procedure of the previous section. First, friction circles may be much smaller than those utilized for illustrative purposes in the slider-crank example. This can lead to graphical inaccuracies. Second, since guesswork on force placements is necessary in some analyses, the amount of time-consuming graphical construction may be great. An alternative is the following analytical approach, which can be computerized for faster and more accurate solution.

Figure 9.21a shows a machine member i that is connected to a member j (not shown) by means of a journal bearing, which has radius r and assumed negligible clearance. The friction circle is shown, and the bearing force F_{ji} of member j on member i acts along a direction tangent to the friction circle and is positioned so as to oppose the rotation of member i relative to member j.

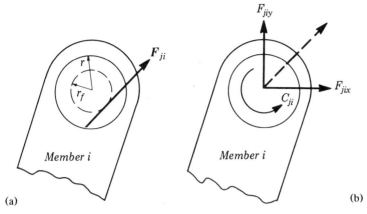

Figure 9.21 (a) The force transmitted at a journal bearing with friction. The diagram shows the actual line of action of the resultant bearing force. (b) An equivalent combination of a force and a torque.

In Figure 9.21b, the bearing force has been replaced by an equivalent combination of a force and a couple. These two force systems will be equivalent if they produce the same net force and the same net moment about any arbitrary point. The force, represented by perpendicular x and y components, has the same magnitude and direction as the original force but has a line of action through the center of the joint. The couple C_{ji} accounts for the offset r_f of the original force and can be expressed as

$$C_{ji} = r_f |F_{ji}| \operatorname{sign}(\omega_j - \omega_i) \cong \mu r \left(F_{jix}^2 + F_{jiy}^2 \right)^{1/2} \operatorname{sign}(\omega_j - \omega_i) \qquad \textbf{(9.36)}$$

The sign function is defined as follows:

$$\operatorname{sign}(\omega_j - \omega_i) = \begin{cases} -1 & \text{if } (\omega_j - \omega_i) < 0 \\ 0 & \text{if } (\omega_j - \omega_i) = 0 \\ +1 & \text{if } (\omega_j - \omega_i) > 0 \end{cases} \qquad \textbf{(9.37)}$$

where ω_i and ω_j are the angular velocities of members i and j, respectively. The difference in the angular velocities, $\omega_j - \omega_i$, is an indication of the motion of link i relative to link j. Therefore, the force system in Figure 9.21b has a net force equal to F_{ji} and a net moment about the bearing center that is equal in both magnitude and direction to that in Figure 9.21a, and the two systems are equivalent.

The representation of bearing friction in Figure 9.21b eliminates any guesswork in the analysis and is well suited for computer implementation. For example, let us return to the slider-crank mechanism of the previous section. The mechanism and free bodies are redrawn in Figure 9.22a and b with the new force representation. Summing forces on the connecting rod 2 and

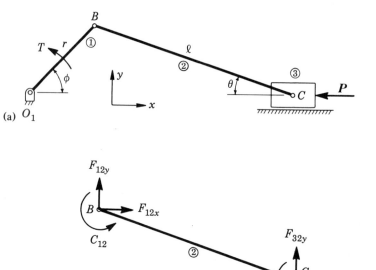

(a) O_1

(b)

Figure 9.22 (a) A slider-crank mechanism. (b) Free-body diagrams to be used for an analytical solution including friction effects.

moments about point B, we have

$$F_{12x} + F_{32x} = 0 \qquad (9.38a)$$

$$F_{12y} + F_{32y} = 0 \qquad (9.38b)$$

$$F_{32y}\ell \cos\theta + F_{32x}\ell \sin\theta + \mu_{32}r_{32}\left(F_{32x}^2 + F_{32y}^2\right)^{1/2}\text{sign}(\omega_3 - \omega_2)$$

$$+ \mu_{12}r_{12}\left(F_{12x}^2 + F_{12y}^2\right)^{1/2}\text{sign}(\omega_1 - \omega_2) = 0 \qquad (9.38c)$$

where ℓ is the connecting rod length, and μ_{ij} and r_{ij} are the coefficient of friction and bearing radius, respectively, for the joint connecting members i and j. Note that, in this case, the angular velocity ω_3 of member 3, the piston, is zero.

A couple of observations should be made. First, the equilibrium conditions have produced three equations in four unknowns. Therefore, more information, which can be obtained from examination of the free bodies of other members, is necessary before solution can take place. For the piston in Figure 9.22b,

$$F_{03x} + F_{23x} = \mu_{03}|F_{03y}|\text{sign}(-\dot{s}) - F_{32x} = P \tag{9.39a}$$

and

$$F_{03y} + F_{23y} = F_{03y} - F_{32y} = 0 \tag{9.39b}$$

where $\dot{s}$ is the piston velocity (assumed to be positive to the right), and $\text{sign}(-\dot{s})$ is defined analogous to Eq. 9.37. Thus, when the piston is moving to the right, $\dot{s}$ will be positive, and friction force F_{03x} will be negative or to the left. Eqs. 9.39a and b are two new equations introducing only one new unknown, F_{03y}, and the set of five equations can now be solved simultaneously.

The second observation to be made is that the equations are nonlinear in the unknowns due to the presence of the square root terms, which, by the way, disappear if the friction is zero. This nonlinearity has serious ramifications in that the principle of superposition no longer holds, and also the solution procedure is much more difficult. In fact, in general, numerical techniques or approximation methods must be employed in solving the equations.

The solution is facilitated in the present case by the fact that the connecting rod is a two-force member. Solving Eqs. 9.38a and b for F_{12x} and F_{12y} and then substituting these expressions into Eq. 9.38c and factoring yields

$$F_{32y}\ell \cos \theta + F_{32x}\ell \sin \theta + k\sqrt{F_{32x}^2 + F_{32y}^2} = 0 \tag{9.40}$$

where

$$k = \mu_{12}r_{12}\,\text{sign}(\omega_1 - \omega_2) + \mu_{32}r_{32}\,\text{sign}(\omega_3 - \omega_2)$$

Next, we express the force components F_{32x} and F_{32y} in terms of polar coordinates as

$$F_{32x} = F_{32} \cos \psi \tag{9.41a}$$

$$F_{32y} = F_{32} \sin \psi \tag{9.41b}$$

where F_{32} is the magnitude of the force transmitted from member 3 to member 2 and angle ψ is its argument with respect to the positive x axis (see Figure 9.23). The value of angle ψ will depend on whether the connecting rod is in tension or compression and on the relative motion at joints B and C. As mentioned earlier, there are four possible orientations of the force line of action tangent to each of the two friction circles.

Substituting Eqs. 9.41a and b into Eq. 9.40, we have

$$F_{32}\ell \sin \psi \cos \theta + F_{32}\ell \cos \psi \sin \theta + kF_{32} = 0$$

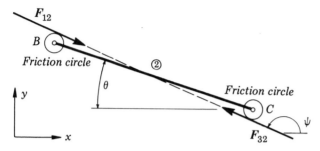

Figure 9.23 Free-body diagram of connecting rod 2 showing the orientation of the forces at bearings B and C.

or

$$\ell \sin(\psi + \theta) + k = 0$$

Solving for angle ψ, we have

$$\psi = \arcsin\left(-\frac{k}{\ell}\right) - \theta \tag{9.42}$$

There are two solutions for ψ from the above equation, one corresponding to compression of the connecting rod and the other corresponding to tension; note that the relative joint motions are accounted for by the sign functions in the expression for k. For example, when there is no friction, the solutions are

$$\psi = -\theta$$

and

$$\psi = \pi - \theta$$

From inspection of Figures 9.22a and 9.23, the first of these solutions represents tension of the connecting rod, as would occur when force P on the piston is directed to the right. The second solution represents the situation depicted in Figure 9.22a with piston force P acting to the left and the connecting rod therefore in compression.

Eq. 9.42 gives the argument of force F_{32}; however, the magnitude of this force is as yet unknown. Combining Eqs. 9.39a and b,

$$\mu_{03}|F_{32y}|\text{sign}(\dot{s}) + F_{32x} + P = 0$$

where $\text{sign}(\dot{s}) = -\text{sign}(-\dot{s})$. Substituting Eqs. 9.41a and b into this equation,

$$\mu_{03}|F_{32} \sin \psi|\text{sign}(\dot{s}) + F_{32} \cos \psi + P = 0$$

or

$$F_{32} = \frac{-P}{\cos \psi + \mu_{03}|\sin \psi|\text{sign}(\dot{s})} \tag{9.43}$$

Force F_{32} is now completely determined as given by Eqs. 9.42 and 9.43. Eq. 9.43 also serves as a check for the proper solution for angle ψ from Eq. 9.42. The correct value of ψ is that which will yield a positive value for the right side of Eq. 9.43, which represents the absolute value of force F_{32}.

The various x and y force components can now be expressed as follows:

$$F_{32x} = -F_{23x} = -F_{12x} = F_{21x} = F_{32} \cos \psi \qquad (9.44a)$$

$$F_{32y} = -F_{23y} = -F_{12y} = F_{21y} = F_{03y} = F_{32} \sin \psi \qquad (9.44b)$$

Once F_{21x} and F_{21y} have been determined, the analysis can be completed by solution of the following equations for crank 1 (see Figure 9.22b):

$$F_{01x} + F_{21x} = 0 \qquad (9.45a)$$

$$F_{01y} + F_{21y} = 0 \qquad (9.45b)$$

$$T + C_{01} + C_{21} + F_{21y} r \cos \phi - F_{21x} r \sin \phi = 0 \qquad (9.45c)$$

where r is the crank length, and

$$C_{01} = \mu_{01} r_{01} \left(F_{01x}^2 + F_{01y}^2 \right)^{1/2} \text{sign}(-\omega_1)$$

and

$$C_{21} = \mu_{21} r_{21} \left(F_{21x}^2 + F_{21y}^2 \right)^{1/2} \text{sign}(\omega_2 - \omega_1)$$

Note that the angular velocity of the frame 0 is zero. The instantaneous efficiency can then be determined as before.

◆ **EXAMPLE PROBLEM 9.7** *Analysis of a Slider-Crank Mechanism with Friction*

Determine the instantaneous efficiency for the slider-crank mechanism of Example Problems 9.1 and 9.5. The mechanism has a crank length r of 30 mm, a connecting rod length ℓ of 70 mm, and the analysis is to be performed, as before, for a crank angle $\phi = 45°$. Recall that, in the absence of friction, the crankshaft torque for a piston force $P = 40$ N was found to be $T = -1119$ N · mm, the negative sign indicating that the torque is clockwise. Assume that the coefficient of friction for all bearings is 0.1. The three journal bearings all have radii of 10 mm, and the crank is rotating in the clockwise direction.

Solution. From the specified information, we have the following:

$$r_{01} = r_{12} = r_{23} = 10 \text{ mm}$$

$$\mu_{03} = \mu_{01} = \mu_{12} = \mu_{23} = 0.1$$

The sign functions for a complex mechanism would require a kinematic velocity analysis to determine relative joint velocities. In this case, the functions can be determined by inspection (for example, see Figure 9.20a):

$$\text{sign}(\dot{s}) = +1$$

$$\text{sign}(\omega_3 - \omega_2) = \text{sign}(-\omega_2) = -1$$

$$\text{sign}(\omega_1 - \omega_2) = -\text{sign}(\omega_2 - \omega_1) = -1$$

$$\text{sign}(\omega_0 - \omega_1) = \text{sign}(-\omega_1) = +1$$

Angle θ (see Figure 9.22a) can be calculated from relationships developed earlier in the chapter:

$$\sin \theta = \frac{r}{\ell}\sin \phi = \frac{30}{70}\sin 45° = 0.303$$

$$\theta = 17.6°$$

Angle ψ, Eq. 9.42, can now be determined as follows:

$$k = 0.1(10)(-1) + 0.1(10)(-1) = -2$$

$$\psi = \arcsin\left[-\frac{(-2)}{70}\right] - 17.6° = -15.9°, 160.7°$$

Since the connecting rod is in compression, the angle ψ must be in the second quadrant, indicating that $\psi = 160.7°$ is the correct value. Substituting into Eq. 9.43,

$$F_{32} = \frac{-40}{\cos(160.7°) + 0.1|\sin(160.7°)|(+1)} = 43.9 \text{ N}$$

The positive value obtained for F_{32} confirms the choice of 160.7° for ψ. The force components, then, are as follows:

$$F_{32x} = -F_{23x} = -F_{12x} = F_{21x} = 43.9 \cos 160.7° = -41.4 \text{ N}$$

$$F_{32y} = -F_{23y} = -F_{12y} = F_{21y} = F_{03y} = 43.9 \sin 160.7° = 14.5 \text{ N}$$

Noting that the magnitudes of forces F_{32}, F_{12}, and F_{01} are all equal, the input crank torque can now be computed from Eq. 9.45c:

$$T = -0.1(10)(43.9)(+1) - 0.1(10)(43.9)(+1)$$

$$- (14.5)(30) \cos 45° + (-41.4)(30) \sin 45°$$

$$= -1273 \text{ N} \cdot \text{mm}$$

and the instantaneous efficiency is

$$e = \frac{-1119}{-1273} = 0.88 \quad (88 \text{ percent}) \quad \blacklozenge$$

9.6 FORCES IN GEAR AND CAM MECHANISMS

Up to this point, this chapter has dealt exclusively with forces in linkage mechanisms. However, the general force analysis principles that have been presented are applicable to mechanisms of all types, including gears and cams.

Due to the special nature of gears, associated forces are described in Chapters 6, 7, and 8, where terminology specific to gears is defined. In particular, Section 6.10 deals with spur gear loads, and Sections 7.1.5, 7.3.3, and 7.4.5 cover forces in helical, worm, and bevel gear systems, respectively. The latter three gear types are subject to three-dimensional loading, while forces on spur gears are essentially planar. The use of free-body diagrams of individual gears usually makes the force analysis treatment of these machine components relatively straightforward.

The analysis methods and principles described earlier in this chapter can also be applied to cam mechanisms, as the following example will illustrate.

◆ **EXAMPLE PROBLEM 9.8**   *Analysis of a Cam Mechanism*
The cam in Figure 9.24a drives the four-bar linkage against an applied load F_3, having a magnitude of 100 N. Determine the torque required on the camshaft for static equilibrium. Neglect friction.

Solution.   Figure 9.24b shows free-body diagrams of the individual moving members. The solution will be carried out graphically. Before the force polygon for member 3 can be constructed, we need to determine the direction of the force at pin C. This is accomplished by considering member 2, because the directions of two of its three forces are already known: the force F_{42} at point D is directed along link 4, which is a two-force member, and, in the absence of friction, the contact force F_{12} of the cam acts along the common normal at point B. The intersection of these two lines of action is concurrency point P for member 2, and line PC is the direction of force F_{32} and, in turn, force F_{23}.

Now the force polygon for member 3 can be determined, as shown in Figure 9.24b, by first establishing the location of concurrency point Q. If we measure from this polygon the magnitude of force F_{32}, the polygon for member 2 can then be constructed, and it is also shown in Figure 9.24b. Finally, considering the cam 1, which is acted on by two forces and a torque, we now know force F_{21}, and force F_{01} must be equal and opposite. Torque T_1 must balance this couple, and, measuring the perpendicular distance between these forces, we find that the magnitude is

$$T_1 = (70 \text{ N})(26 \text{ mm}) = 1820 \text{ N} \cdot \text{mm}$$

and, by inspection, the direction is counterclockwise.   ◆

As is true of other mechanism types, friction effects may be significant in gears and cams, depending on factors such as loads, speeds, lubrication, and

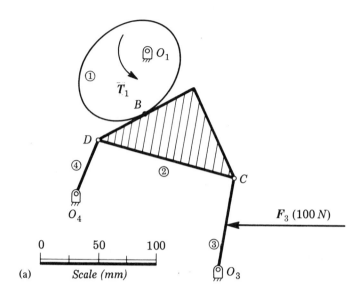

(a) Scale (mm)

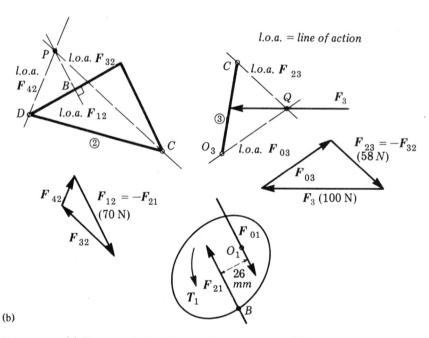

(b)

Figure 9.24 (a) Force analysis of a cam linkage system. (b) Free-body diagrams and force polygons.

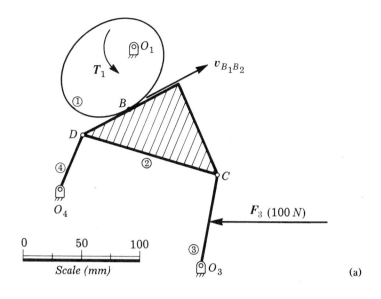

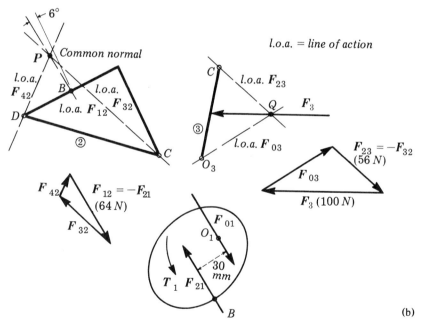

Figure 9.25 (a) Force analysis of a cam linkage system including friction at the cam surface. Friction at all other joints is neglected. (b) Free-body diagrams and force polygons.

operating conditions. Therefore, the assumption of no friction must be carefully imposed. Efficiencies of less than 100 percent are tabulated in gear design handbooks for gear drives to account for friction losses, which can be substantial in units such as worm gear drives, for example. In the following example, the mechanism of Example Problem 9.8 is reconsidered with the inclusion of friction at the cam surface.

♦ **EXAMPLE PROBLEM 9.9**   *Analysis of a Cam Mechanism with Friction*
The cam in Figure 9.25a rotates in the counterclockwise direction. Determine the camshaft torque T_1 required for static equilibrium against applied load F_3, which has a magnitude of 100 N. The coefficient of friction between cam 1 and coupler link 2 is 0.1, while friction in the linkage bearings and pin O_1 is negligible.

Solution.   The solution procedure is similar to that of Example Problem 9.8. The only difference is that the net force between members 1 and 2 does not act along the common normal in this case. Instead, it acts at an angle to the normal equal to the friction angle defined by Eq. 9.32:

$$\tan \phi = \mu = 0.1 \quad \text{or} \quad \phi = 6°$$

This angle is incorporated into the free-body diagrams of Figure 9.25b and must reflect the proper direction of the friction force. By inspection of Figure 9.25a, the sliding velocity $v_{B_1 B_2}$ will have the direction shown. This will therefore be the direction of the friction force acting on member 2, with the friction force on member 1 equal and opposite. Figure 9.25b shows the complete static-force analysis, from which torque T_1 is found to have a magnitude of

$$T_1 = (64 \text{ N})(30 \text{ mm}) = 1920 \text{ N} \cdot \text{mm}$$

and a counterclockwise direction. As expected, the presence of friction brings about increased torque requirements as compared to Example Problem 9.8.

♦

References

1. Beer, F. P., and E. R. Johnston, Jr., *Vector Mechanics for Engineers: Statics and Dynamics*, McGraw-Hill, New York, 1984.
2. Meriam, J. L., *Engineering Mechanics: Statics and Dynamics*, Wiley, New York, 1978.
3. Shames, I. H., *Engineering Mechanics: Statics and Dynamics*, Prentice-Hall, Englewood Cliffs, NJ, 1980.

PROBLEMS

For Problems 9.1 Through 9.17

Perform a graphical static-force analysis of the given mechanism. Construct the complete force polygon for determining bearing forces and the required input force or torque. Mechanism dimensions are given in the accompanying figures.

● **9.1** The applied piston load **P** on the in-line slider-crank mechanism of Figure P9.1 remains constant as angle ϕ varies and has a magnitude of 500 N. Determine the required input torque T_1 for static equilibrium at the following crank positions:
(a) $\phi = 45°$
(b) $\phi = 135°$
(c) $\phi = 270°$

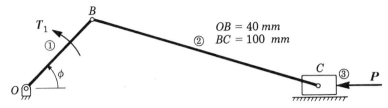

OB = 40 mm
BC = 100 mm

Figure P9.1

9.2 The applied piston load **P** on the offset slider-crank mechanism of Figure P9.2 remains constant as angle ϕ varies and has a magnitude of 100 lb. Determine the required input torque T_1 for static equilibrium at the following crank positions:
(a) $\phi = 45°$
(b) $\phi = 135°$
(c) $\phi = 270°$
(d) $\phi = 315°$

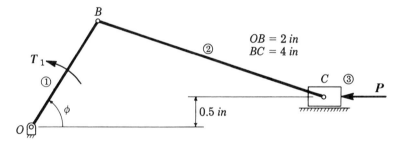

OB = 2 in
BC = 4 in

0.5 in

Figure P9.2

● **9.3** Determine the required input torque T_1 for static equilibrium of the mechanism shown in Figure P9.3. Force F_2 has a magnitude of 200 N.

OB = 30 mm
BC = 90 mm

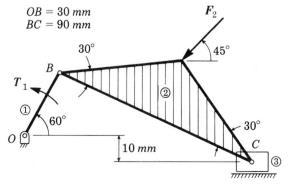

10 mm

Figure P9.3

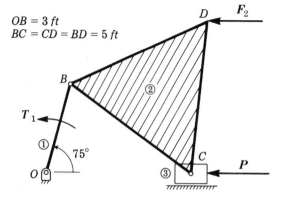

$OB = 3\ ft$
$BC = CD = BD = 5\ ft$

Figure P9.4

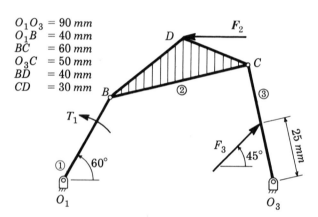

$O_1O_3 = 90\ mm$
$O_1B\ \ = 40\ mm$
$BC\ \ \ = 60\ mm$
$O_3C\ \ = 50\ mm$
$BD\ \ \ = 40\ mm$
$CD\ \ \ = 30\ mm$

Figure P9.5

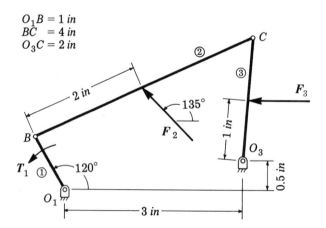

$O_1B = 1\ in$
$BC\ \ = 4\ in$
$O_3C = 2\ in$

Figure P9.6

9.4 Determine the required input torque T_1 for static equilibrium of the mechanism shown in Figure P9.4. Force F_2 has a magnitude of 100 lb, and the piston force P is 200 lb. Both forces are horizontal.

9.5 Determine the required input torque T_1 for static equilibrium of the mechanism shown in Figure P9.5. Forces F_2 and F_3 have magnitudes of 50 N and 75 N, respectively. Force F_2 acts in the horizontal direction.

9.6 Determine the required input torque T_1 for static equilibrium of the mechanism shown in Figure P9.6. Forces F_2 and F_3 have magnitudes of 20 lb and 10 lb, respectively. Force F_3 acts in the horizontal direction.

9.7 Determine the required input torque T_1 for static equilibrium of the mechanism shown in Figure P9.7. Torques T_2 and T_3 are pure torques, having magnitudes of 10 N · m and 7 N · m, respectively.

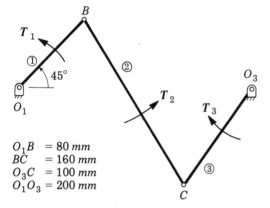

$O_1B = 80\ mm$
$BC = 160\ mm$
$O_3C = 100\ mm$
$O_1O_3 = 200\ mm$

Figure P9.7

9.8 Determine the required cylinder gage pressure for static equilibrium of the mechanism shown in Figure P9.8. Torque T_3 has a magnitude of 1000 in · lb. The diameter of the piston is 1.5 in.

$OA = 9\ in$
$O_1A = 6\ in$
$O_1B = 9\ in$
$BC = 15\ in$
$O_3C = 12\ in$

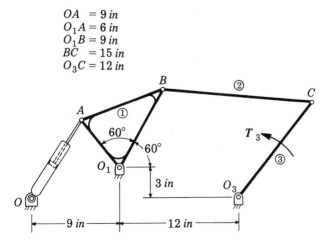

Figure P9.8

9.9 Figure P9.9 shows a four-bar linkage that has link masses of 7 kg, 15 kg, and 12 kg for members 1, 2, and 3, respectively. The centers of mass are at the link centers, and gravity acts in the negative y direction. Determine the required force in the horizontal spring attached to member 1 at point A for static equilibrium.

$$O_1A = 0.2 \text{ m}$$
$$O_1B = 0.3 \text{ m}$$
$$BC = 0.7 \text{ m}$$
$$O_3C = 0.6 \text{ m}$$

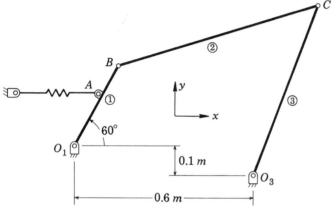

Figure P9.9

9.10 Determine the required input torque T_1 for static equilibrium of the mechanism shown in Figure P9.10. The force F_5 on the slider has a magnitude of 1000 lb.

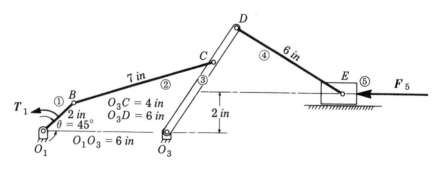

Figure P9.10

9.11 Determine the required input torque T_2 for static equilibrium of the quick-return mechanism shown in Figure P9.11. The force F_5 on the slider has a magnitude of 800 lb. Angle θ equals 105°.

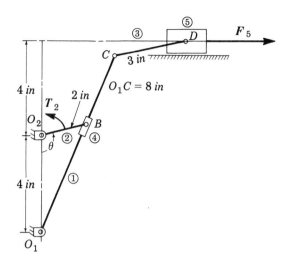

Figure P9.11

9.12 Determine the required input torque T_1 for static equilibrium of the mechanism shown in Figure P9.12. Torque T_5 has a magnitude of 600 in · lb.

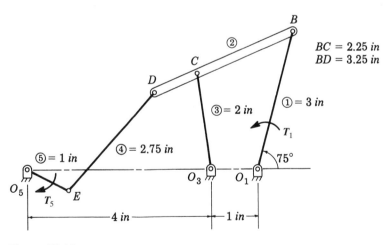

Figure P9.12

9.13 Figure P9.13 is a schematic diagram of the variable-stroke pump discussed in Section 1.11.9. Determine the torque T_1 required for static equilibrium when the piston load P has a magnitude of 500 N.

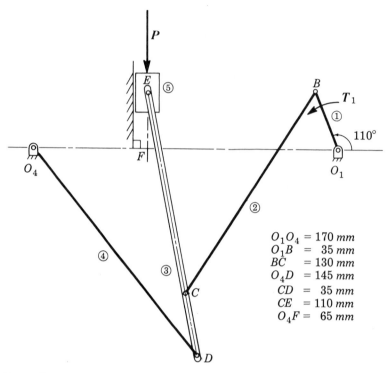

$$O_1O_4 = 170 \ mm$$
$$O_1B = 35 \ mm$$
$$BC = 130 \ mm$$
$$O_4D = 145 \ mm$$
$$CD = 35 \ mm$$
$$CE = 110 \ mm$$
$$O_4F = 65 \ mm$$

Figure P9.13

9.14 For the toggle mechanism shown in Figure P9.14, determine the force developed at the slider 5 for an input torque T_1 of 100 in · lb clockwise, assuming inertia

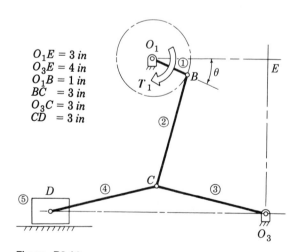

$$O_1E = 3 \ in$$
$$O_3E = 4 \ in$$
$$O_1B = 1 \ in$$
$$BC = 3 \ in$$
$$O_3C = 3 \ in$$
$$CD = 3 \ in$$

Figure P9.14

forces and friction forces are negligible. Perform the analysis for the following crank positions:

(a) $\theta = 0$
(b) $\theta = 30°$
(c) $\theta = 60°$

9.15 Determine the torque T_1 required for static equilibrium of the two-cylinder engine depicted in Figure P9.15. The piston forces F_4 and F_5 have magnitudes of

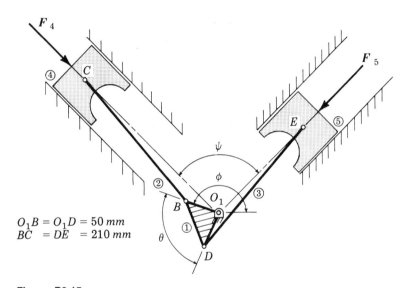

$O_1B = O_1D = 50\ mm$
$BC = DE = 210\ mm$

Figure P9.15

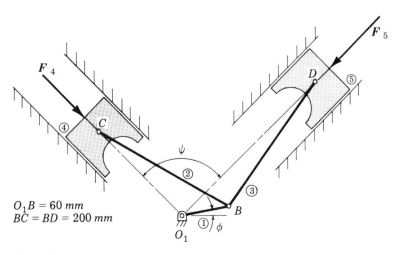

$O_1B = 60\ mm$
$BC = BD = 200\ mm$

Figure P9.16

6000 N and 2000 N, respectively. Consider the case where the cylinder V angle ψ is 90°, the crank spacing θ is 90°, and the crank angle ϕ is 160°.

9.16 Determine the torque T_1 required for static equilibrium of the two-cylinder engine depicted in Figure P9.16. The piston forces F_4 and F_5 have magnitudes of 3000 N and 7000 N, respectively. Consider the case where the cylinder V angle ψ is 90° and the crank angle ϕ is 20°.

9.17 Determine the handle force P necessary to produce a jaw force Q of 100 lb on the lever wrench shown in Figure P9.17.

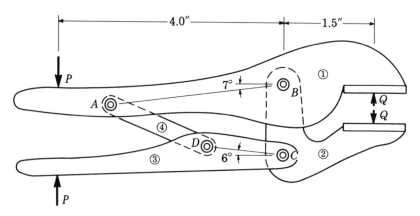

Figure P9.17 $AB = 3.0$ in, $BC = 1.2$ in, $CD = 1.3$ in. Force P, force Q, and line BC are vertical. Angles are measured from the horizontal.

For Problems 9.18 through 9.34

Obtain an analytical static-force solution for the given mechanism, using the force and moment equilibrium approach. Determine all bearing forces and the required input force or torque at the specified position(s). Mechanism dimensions are given in the accompanying figures.

9.18 Analyze the mechanism of Problem 9.1.

9.19 Analyze the mechanism of Problem 9.2. Derive a general function relating input torque T_1 to the piston force P and crank angle ϕ for an offset slider-crank mechanism with crank length r, connecting rod length ℓ, and offset h. Calculate the input torque and bearing forces at the following crank positions of the mechanism of Figure P9.2:
(a) $\phi = 45°$
(b) $\phi = 135°$
(c) $\phi = 270°$
(d) $\phi = 315°$

9.20 Analyze the mechanism of Problem 9.3.

9.21 Analyze the mechanism of Problem 9.4.

9.22 Analyze the mechanism of Problem 9.5.

9.23 Analyze the mechanism of Problem 9.6.

9.24 Analyze the mechanism of Problem 9.7.

9.25 Analyze the mechanism of Problem 9.8.

9.26 Analyze the mechanism of Problem 9.9.

9.27 Analyze the mechanism of Problem 9.10.

9.28 Analyze the mechanism of Problem 9.11.

9.29 Analyze the mechanism of Problem 9.12.

9.30 Analyze the mechanism of Problem 9.13.

9.31 Analyze the mechanism of Problem 9.14.

9.32 Analyze the mechanism of Problem 9.15.

9.33 Analyze the mechanism of Problem 9.16.

9.34 Analyze the mechanism of Problem 9.17.

9.35 Analyze the mechanism of Problem 9.1 by the method of virtual work.

9.36 Analyze the mechanism of Problem 9.2 by the method of virtual work.

9.37 Analyze the mechanism of Problem 9.3 by the method of virtual work.

9.38 Analyze the mechanism of Problem 9.4 by the method of virtual work.

9.39 Analyze the mechanism of Problem 9.5 by the method of virtual work.

9.40 Analyze the mechanism of Problem 9.6 by the method of virtual work.

9.41 Analyze the mechanism of Problem 9.7 by the method of virtual work.

9.42 Analyze the mechanism of Problem 9.8 by the method of virtual work.

9.43 Analyze the mechanism of Problem 9.10 by the method of virtual work.

9.44 Analyze the mechanism of Problem 9.11 by the method of virtual work.

9.45 Analyze the mechanism of Problem 9.15 by the method of virtual work.

9.46 Analyze the mechanism of Problem 9.16 by the method of virtual work.

9.47 Perform a graphical force analysis of the mechanism of Problem 9.1 including friction effects. The coefficients of friction are 0.1 at bearing O, 0.15 at pins B and C, and 0.2 at the slider. The bearing radii are 20 mm at O and 15 mm at B and C. The crank rotates clockwise. Determine the following for crank angle $\phi = 45°$:
(a) the torque T_1 required for static equilibrium
(b) the instantaneous efficiency

9.48 Do Problem 9.47 by the analytical method.

9.49 Perform a graphical force analysis of the mechanism of Problem 9.2 including friction effects. The coefficients of friction are 0.15 at bearing O, 0.25 at bearing B, 0.1 at bearing C, and 0.2 at the slider. The bearing radii all equal 1.0 in. The crank rotates clockwise. Determine the following for crank angle $\phi = 45°$:
(a) the torque T_1 required for static equilibrium
(b) the instantaneous efficiency

9.50 Do Problem 9.49 by the analytical method.

9.51 Perform a graphical force analysis of the mechanism of Problem 9.3 including friction effects. The friction circle radii are 4 mm at O, 5 mm at B, and 3 mm at C. The crank rotates clockwise. Determine the torque T_1 required for static equilibrium. Neglect friction between the slider and the frame.

9.52 Perform a graphical force analysis including friction effects of the mechanism of Problem 9.5 with force $F_2 = 0$. The coefficients of friction are 0.2 at bearings O_1 and O_2 and 0.15 at bearings B and C. The bearing radii are all equal to 15 mm. The crank rotates counterclockwise. Determine the torque T_1 required for static equilibrium.

9.53 Perform a graphical force analysis including friction of the mechanism of Problem 9.6 with force $F_2 = 0$. The coefficients of friction all equal 0.2, and the bearing radii are 0.3 in at O_1 and 0.5 in at B, C, and O_3. The crank rotates clockwise. Determine the torque T_1 required for static equilibrium.

9.54 Figure P9.18 shows a geared five-bar linkage, where the larger gear is fixed to the frame and the smaller gear is attached to link 2. The gears are spur gears with a pressure angle of 20° and pitch diameters of 6 in and 4 in. A force F_4 of 200 lb

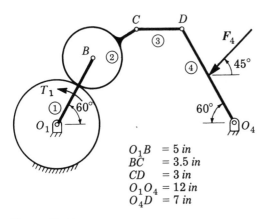

$$\begin{array}{ll} O_1B & = 5\ in \\ BC & = 3.5\ in \\ CD & = 3\ in \\ O_1O_4 & = 12\ in \\ O_4D & = 7\ in \end{array}$$

Figure P9.18

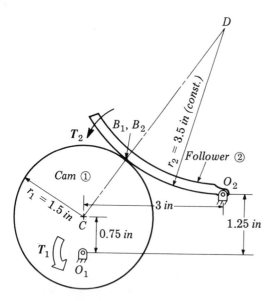

Figure P9.19 At the instant shown, point D, the center of curvature of the cam follower, lies directly above O_2 and point C lies directly above O_1.

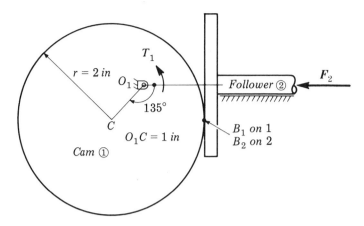

Figure P9.20

magnitude is applied as shown at the midpoint of member 4. Determine the required torque T_1 for static equilibrium.

9.55 Determine the camshaft torque T_1 required for static equilibrium of the cam and follower mechanism of Figure P9.19. Torque T_2 has a magnitude of 50 in · lb and a counterclockwise direction. Neglect friction.

9.56 Repeat Problem 9.55 including friction at the cam-follower interface. The coefficient of friction is 0.15, and the cam has clockwise rotation.

9.57 Determine the camshaft torque T_1 required for static equilibrium of the cam and follower mechanism of Figure P9.20. Force F_2 on the follower has a magnitude of 20 lb. Neglect friction.

9.58 Repeat Problem 9.57 including friction at both the cam-follower interface and the follower-guideway interface. The coefficient of friction is 0.15, and the cam has counterclockwise rotation.

PROJECTS

See Projects 1.1 to 1.6 and suggestions in Chapter 1.

Describe and plot representative forces and torques. Make use of computer software wherever practical. Check the results by a graphical method for at least one linkage position. Evaluate the linkage in terms of the performance requirements.

Dynamic-Force Analysis

10.1 INTRODUCTION

This chapter deals with dynamic forces in machines. Dynamic forces are associated with accelerating mass, and since virtually all machines contain accelerating parts, dynamic forces are always present.

Perhaps the simplest example of a dynamic force is the case of a mass mounted on a rod that is rotating about a fixed pivot with constant angular velocity. See Figure 10.1a. We know that in order to maintain a circular path of motion of the mass, the rod, and in turn the pivot, will experience a force acting radially outward. This force is commonly referred to as a centrifugal force and is a dynamic force because it results from the centripetal acceleration of the mounted mass. This type of dynamic force is very prevalent in machines. For example, a rotating shaft or rotor will experience centrifugal force if its center of mass does not lie exactly on the bearing centerline, as shown in Figure 10.1b. The mass will then rotate about the bearing centerline producing a centrifugal force as in Figure 10.1a. Discrepancies between center of mass and center of rotation locations can occur in actual rotors as the result of a number of factors, such as manufacturing inaccuracies, material inhomogeneity, and bowing or bending of the rotor.

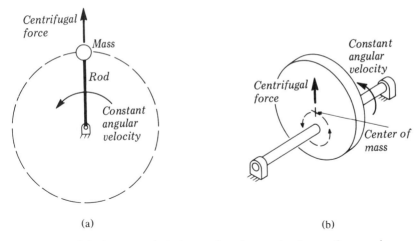

Figure 10.1 (a) A mass that is moving in a circular path experiences centripetal acceleration, and there is a dynamic force, referred to as centrifugal force, associated with this acceleration. This force acts radially outward and will be transmitted to the support, producing a bearing reaction. (b) A rotating shaft with an eccentric mass.

In the previous chapter it was pointed out that, in certain applications, dynamic forces are small compared to other forces acting on the system, and in those cases, dynamic forces may be neglected in the analysis. However, there are many situations where the reverse is true and dynamic forces are dominant or at least comparable in magnitude to external forces. And the group of problems in this class continues to grow with the trend toward machines with higher and higher operating speeds, as, for example, in industrial manufacturing processes where machine speed has a direct bearing on productivity. As new machines are designed to operate at higher speeds, dynamic-force effects become increasingly more important. For example, rotors have been designed and built that run at rotational speeds in excess of 100,000 rev/min. Even the slightest eccentricity of the center of mass from the axis of rotation will lead to significant dynamic forces. If dynamic forces are not accounted for in the design process, serious consequences may result, in the form of vibrations, noise, wear, or machine failure.

10.2 NEWTON'S SECOND LAW OF MOTION

The basis for investigating dynamic forces in machines is Newton's second law of motion, which states that *a particle acted on by forces whose resultant is not zero will move in such a way that the time rate of change of its momentum will at any instant be proportional to the resultant force.* This law is expressed mathematically, for the special case of invariant mass, as

$$F = ma \tag{10.1}$$

where F is the resultant force on the particle, m is the mass of the particle, and a is the acceleration of the particle. The law, as stated above for a particle, can be extended to a rigid body in the following form:

$$F_e = ma_G \tag{10.2a}$$

$$T_{eG} = \dot{H} \tag{10.2b}$$

Eq. 10.2a states that the resultant external force, F_e, on a rigid body is equal to the product of the mass of the body, m, and the acceleration of the center of mass, a_G, where G designates the location of the center of mass. Eq. 10.2b, which is the analogous statement for rotational motion, states that the resultant external moment on the body about its center of mass, T_{eG}, is equal to the time rate of change of the body's angular momentum H. This statement is often expressed in the following component form, known as Euler's equations of motion for a rigid body:

$$T_{eG_x} = I_x \alpha_x + \left(I_z - I_y \right) \omega_y \omega_z \tag{10.2c}$$

$$T_{eG_y} = I_y \alpha_y + \left(I_x - I_z \right) \omega_z \omega_x \tag{10.2d}$$

$$T_{eG_z} = I_z \alpha_z + \left(I_y - I_x \right) \omega_x \omega_y \tag{10.2e}$$

In this formulation, the xyz axes are the principal axes of the body, with their origin at the center of mass G of the body. The T_{eG}, α, and ω terms are the components of the external moment T_{eG}, the angular acceleration α, and the angular velocity ω, respectively, with respect to these axes. The I terms are the mass moments of inertia of the body about these axes.

The mass moment of inertia of a particle about a particular axis is defined as the product of the mass of the particle and the square of the distance of the particle from the axis. For a continuous body made up of an infinite number of mass particles, the moment of inertia I is the integral of all of the individual moments of inertia, as follows:

$$I = \int r^2 \, dm$$

where dm is a particle mass and r is the distance of that particle from the axis. This mathematical definition can be used to compute mass moments of inertia of machine members, and tables are available in mechanics texts and handbooks that contain expressions for moments of inertia determined this way for standard shapes. However, experimental methods of determination are often used for complex shapes, such as connecting rods and cams. Note that the moment of inertia of a body depends on the particular reference axis as well as on the shape of the body and the manner in which its mass is distributed. In the SI system, mass moment of inertia is usually expressed in units of kilogram-meter squared (kg · m^2). Commonly used English units are slug-foot squared (slug · ft^2) or pound force-second squared-inch (lb · s^2 · in).

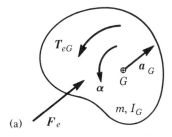
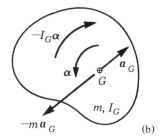

Figure 10.2 (a) A rigid body accelerating under the influence of external force F_e and external torque T_{eG}. (b) The inertia force and the inertia torque corresponding to planar motion of a rigid body.

For the special case of planar motion, Eq. 10.2b can be stated as

$$T_{eG} = I_G \alpha \tag{10.2f}$$

where vectors T_{eG} and α are perpendicular to the plane of motion. The subscript G has been used in Eq. 10.2f, indicating that the moment of inertia is with respect to an axis passing through the center of mass G and perpendicular to the plane of rotation of the body. There is a useful relationship between the moment of inertia I of a body about any axis and the moment of inertia I_G about a parallel axis through the center of mass:

$$I = I_G + mR^2$$

where m is the mass of the body and R is the perpendicular distance between the two axes. Eqs. 10.2a and 10.2f are depicted in Figure 10.2a.

An important observation from Eqs. 10.2a and 10.2b should be made at this point, based on the following two ways in which the equations can be interpreted:

1. The equations can be solved for the force and torque required to produce a known motion of a body.
2. The equations can be solved for the motion of a body resulting from the application of known forces and torques.

Strictly speaking, the second approach is the more accurate analysis procedure in machinery dynamics, because we are more apt to have knowledge of the nature of applied loads and input forces (e.g., torque-speed characteristics for electric motors) than knowledge of the exact motion response of complex nonlinear mechanisms. However, the solution by this approach is more difficult than by the first approach.

It is common practice in many design situations to utilize the first approach and to assume that the mechanism motion is equal to the ideal required motion for the specific application. The solution is then carried out for the resulting forces, and this information is used to size the mechanism members and bearings and to select an input power source. This is an iterative process because the forces are dependent on the member sizes. Also, it is generally impossible to select design components so that the forces required for the ideal assumed motion are exactly produced. Therefore, it may be necessary to evaluate the mechanism design obtained this way, either by construction and testing of a physical prototype or by examination of a computer model based on the second approach described above; these procedures can be used to "fine tune" the design through adjustment of final design parameters.

Most of the examples and problems of this chapter will be based on the approach of assumed mechanism motion. In the following section, it will be shown that dynamic-force analysis can be expressed in a form very similar to static-force analysis so that solution techniques from the preceding chapter can be applied.

10.3 D'ALEMBERT'S PRINCIPLE AND INERTIA FORCES

An important principle, known as d'Alembert's principle, can be derived from Newton's second law. In words, d'Alembert's principle states that *the reverse-effective forces and torques and the external forces and torques on a body together give statical equilibrium*. The equivalent mathematical expression can be obtained by transposing the right-hand terms of Eqs. 10.2a and 10.2b:

$$F_e + (-ma_G) = 0 \tag{10.3a}$$

$$T_{eG} + (-\dot{H}) = 0 \tag{10.3b}$$

The terms in parentheses in Eqs. 10.3a and 10.3b are called the reverse-effective force and the reverse-effective torque, respectively. These quantities are also referred to as inertia force and inertia torque. Thus we define the inertia force F_i as

$$F_i = -ma_G \tag{10.4a}$$

which reflects the fact that a body resists any change in its velocity by an inertia force proportional to the mass of the body and its acceleration. The inertia force acts through the center of mass G of the body. The inertia torque or inertia couple C_i is given by

$$C_i = -\dot{H} \tag{10.4b}$$

or, for the case of planar motion (Eq. 10.2f),

$$C_i = -I_G\alpha \tag{10.4c}$$

As indicated, the inertia torque is a pure torque or couple. Figure 10.2b shows these quantities for the case of planar motion, where, from Eqs. 10.4a and

10.4c, their directions are opposite to that of the accelerations. Substitution of Eqs. 10.4a and 10.4b into Eqs. 10.3a and 10.3b leads to equations that are similar to those used for static-force analysis:

$$\sum F = \sum F_e + F_i = 0 \tag{10.5a}$$

$$\sum T_G = \sum T_{eG} + C_i = 0 \tag{10.5b}$$

where $\sum F_e$ refers here to the summation of external forces and therefore is the resultant external force, and $\sum T_{eG}$ is the summation of external moments, or resultant external moment, about the center of mass G. Thus the dynamic analysis problem is reduced in form to a static force and moment balance where inertia effects are treated in the same manner as external forces and torques. In particular, for the case of assumed mechanism motion, the inertia forces and couples can be determined completely and thereafter treated as known mechanism loads. Any of the graphical or analytical solution techniques described in Chapter 9 can then be employed.

Furthermore, d'Alembert's principle facilitates moment summation about any arbitrary point P in the body, if we remember that the moment due to inertia force F_i must be included in the summation. Hence,

$$\sum T_P = \sum T_{eP} + C_i + R_{PG} \times F_i = 0 \tag{10.5c}$$

where $\sum T_P$ is the summation of moments, including inertia moments, about point P, $\sum T_{eP}$ is the summation of external moments about P, C_i is the inertia couple defined by Eq. 10.4b or Eq. 10.4c, F_i is the inertia force defined by Eq. 10.4a, and R_{PG} is a vector from point P to point G. It is clear that Eq. 10.5b is the special case of Eq. 10.5c, where point P is taken as the center of mass G (i.e., $R_{PG} = 0$).

For a body in plane motion in the xy plane with all external forces in that plane, Eqs. 10.5a and 10.5b become

$$\sum F_x = \sum F_{ex} + F_{ix} = \sum F_{ex} + (-ma_{Gx}) = 0 \tag{10.6a}$$

$$\sum F_y = \sum F_{ey} + F_{iy} = \sum F_{ey} + (-ma_{Gy}) = 0 \tag{10.6b}$$

$$\sum T_G = \sum T_{eG} + C_i = \sum T_{eG} + (-I_G\alpha) = 0 \tag{10.6c}$$

where a_{Gx} and a_{Gy} are the x and y components of a_G. These are three scalar equations, where the sign convention for torques and angular accelerations is based on a right-hand xyz coordinate system; that is, counterclockwise is positive and clockwise is negative. The general moment summation about arbitrary point P, Eq. 10.5c, becomes

$$\sum T_P = \sum T_{eP} + C_i + R_{PGx}F_{iy} - R_{PGy}F_{ix}$$

$$= \sum T_{eP} + (-I_G\alpha) + R_{PGx}(-ma_{Gy}) - R_{PGy}(-ma_{Gx}) = 0$$

$$\tag{10.6d}$$

where R_{PGx} and R_{PGy} are the x and y components of position vector $\boldsymbol{R}_{PG}$. This expression for dynamic moment equilibrium will be useful in the analyses to be presented in the following sections of this chapter.

Equivalent Offset Inertia Force

For purposes of graphical plane force analysis, it is convenient to define what is known as the equivalent offset inertia force. This is a single force that accounts for both translational inertia and rotational inertia corresponding to the plane motion of a rigid body. Its derivation will follow, with reference to Figures 10.3a through d.

Figure 10.3a shows a rigid body with planar motion represented by center of mass acceleration $\boldsymbol{a}_G$ and angular acceleration α. The inertia force and inertia torque associated with this motion are also shown. The inertia torque $(-I_G\alpha)$ can be expressed as a couple consisting of forces $\boldsymbol{Q}$ and $(-\boldsymbol{Q})$ separated by perpendicular distance h, as shown in Figure 10.3b. The necessary conditions for the couple to be equivalent to the inertia torque are that the sense and magnitude be the same. Therefore, in this case, the sense of the couple must be clockwise and the magnitudes of $\boldsymbol{Q}$ and h must satisfy the

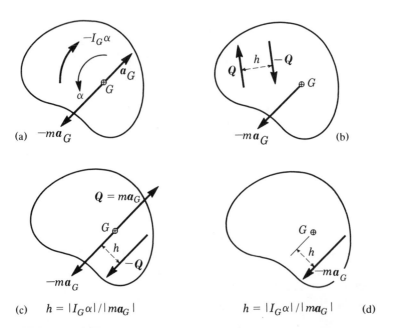

Figure 10.3 (a) Derivation of the equivalent offset inertia force associated with planar motion of a rigid body. (b) Replacement of the inertia torque by a couple. (c) The strategic choice of a couple. (d) The single force is equivalent to the combination of a force and a torque in (a).

relationship

$$|Qh| = |I_G\alpha|$$

Otherwise, the couple is arbitrary and an infinite number of possibilities will work. Furthermore, the couple can be placed anywhere in the plane.

Figure 10.3c shows a special case of the couple, where force vector Q is equal to ma_G and acts through the center of mass. Force $(-Q)$ must then be placed as shown to produce a clockwise sense and at a distance

$$h = \frac{|I_G\alpha|}{|Q|} = \frac{|I_G\alpha|}{|ma_G|} \qquad (10.7)$$

Force Q will cancel with the inertia force $F_i = -ma_G$, leaving the single equivalent offset force shown in Figure 10.3d, which has the following characteristics:

1. The magnitude of the force is $|ma_G|$.
2. The direction of the force is opposite to that of acceleration a_G.
3. The perpendicular offset distance from the center of mass to the line of action of the force is given by Eq. 10.7.
4. The force is offset from the center of mass so as to produce a moment about the center of mass that is opposite in sense to acceleration α.

The usefulness of this approach for graphical force analysis will be demonstrated in the following section. It should be emphasized, however, that this approach is usually unnecessary in analytical solutions, where Eqs. 10.6a to 10.6d, including the original inertia force and inertia torque, can be applied directly.

10.4 DYNAMIC ANALYSIS OF THE FOUR-BAR LINKAGE

The analysis of a four-bar linkage will effectively illustrate most of the ideas that have been presented. Furthermore, the extension to other mechanism types should become clear from the analysis of this mechanism.

♦ **EXAMPLE PROBLEM 10.1**   *Dynamic-Force Analysis of a Four-Bar Linkage*
The four-bar linkage shown in Figure 10.4a has the dimensions shown in the figure, where G refers to center of mass, and the mechanism has the following mass properties:

$$m_1 = 0.1 \text{ kg} \quad I_{G1} = 20 \text{ kg} \cdot \text{mm}^2$$

$$m_2 = 0.2 \text{ kg} \quad I_{G2} = 400 \text{ kg} \cdot \text{mm}^2$$

$$m_3 = 0.3 \text{ kg} \quad I_{G3} = 100 \text{ kg} \cdot \text{mm}^2$$

Determine the instantaneous value of drive torque T required to produce an

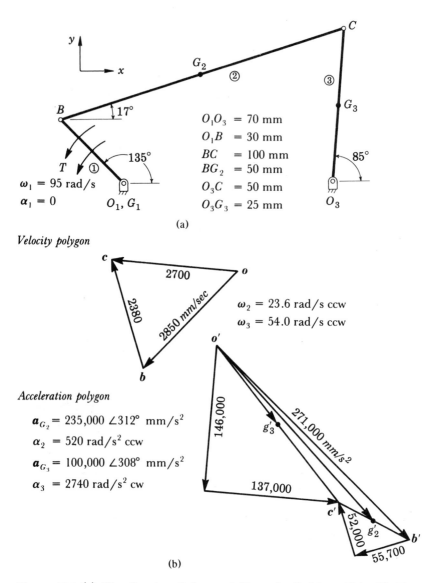

(a)

Velocity polygon

ω_2 = 23.6 rad/s ccw
ω_3 = 54.0 rad/s ccw

Acceleration polygon

a_{G_2} = 235,000 ∠312° mm/s²
α_2 = 520 rad/s² ccw
a_{G_3} = 100,000 ∠308° mm/s²
α_3 = 2740 rad/s² cw

O_1O_3 = 70 mm
O_1B = 30 mm
BC = 100 mm
BG_2 = 50 mm
O_3C = 50 mm
O_3G_3 = 25 mm

(b)

Figure 10.4 (a) The four-bar linkage of Example Problem 10.1. (b) The velocity and acceleration analyses necessary for determination of inertia forces and inertia torques.

assumed motion given by input angular velocity ω_1 = 95 rad/s counterclockwise and input angular acceleration α_1 = 0 for the position shown in the figure. Neglect gravity and friction effects.

Solution. This problem falls in the first analysis category described in Section 10.2; that is, given the mechanism motion, determine the resulting bearing forces and the necessary input torque. Therefore, the first step in the solution

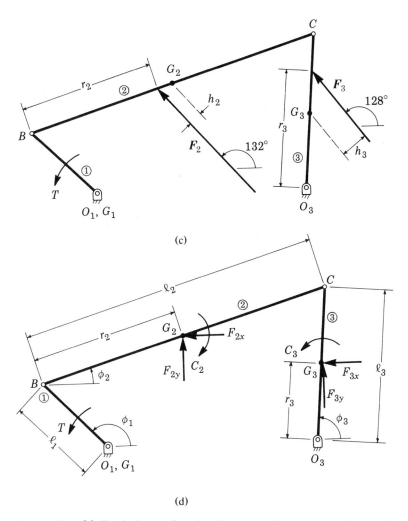

Figure 10.4 (c) Equivalent offset inertia forces for members 2 and 3. (d) Combinations of inertia forces and inertia torques for members 2 and 3.

process is to determine the inertia forces and inertia torques. Thereafter, the problem can be treated as though it were a static-force analysis problem.

Kinematic analysis of the mechanism can be accomplished by using any of the methods presented in Chapters 2, 3, and 4. Figure 10.4b shows a graphical analysis employing velocity and acceleration polygons. From the analysis, the following accelerations are determined:

$a_{G1} = 0$ (stationary center of mass) $\alpha_1 = 0$ (given)

$a_{G2} = 235,000 \angle 312°$ mm/s^2 $\alpha_2 = 520$ rad/s^2 ccw

$a_{G3} = 100,000 \angle 308°$ mm/s^2 $\alpha_3 = 2740$ rad/s^2 cw where the angles of the acceleration vectors are measured counterclockwise from the positive x direction shown in Figure 10.4a. From Eqs. 10.4a and 10.4c, the inertia forces and inertia torques are

$$F_{i1} = 0$$

$$F_{i2} = -m_2 a_{G2} = 47,000 \angle 132° \text{ kg} \cdot \text{mm/s}^2 = 47 \angle 132° \text{ N}$$

$$F_{i3} = -m_3 a_{G3} = 30,000 \angle 128° \text{ kg} \cdot \text{mm/s}^2 = 30 \angle 128° \text{ N}$$

$$C_{i1} = 0$$

$$C_{i2} = -I_{G2} \alpha_2 = 208,000 \text{ kg} \cdot \text{mm}^2/\text{s}^2 \text{ cw} = 208 \text{ N} \cdot \text{mm cw}$$

$$C_{i3} = -I_{G3} \alpha_3 = 274,000 \text{ kg} \cdot \text{mm}^2/\text{s}^2 \text{ ccw} = 274 \text{ N} \cdot \text{mm ccw}$$

The inertia forces have lines of action through the respective centers of mass, and the inertia torques are pure torques.

Graphical Solution.   In order to simplify the graphical force analysis, we will account for the inertia torques by introducing equivalent offset inertia forces. These forces are shown in Figure 10.4c, and their placement is determined according to the previous section. For link 2, the offset force F_2 is equal and parallel to inertia force F_{i2}. Therefore,

$$F_2 = 47 \angle 132° \text{ N}$$

It is offset from the center of mass G_2 by a perpendicular amount equal to

$$h_2 = \frac{|I_{G2} \alpha_2|}{|m_2 a_{G2}|} = \frac{208}{47} = 4.43 \text{ mm}$$

and this offset is measured to the left as shown to produce the required clockwise direction for the inertia moment about point G_2. In a similar manner, the equivalent offset inertia force for link 3 is

$$F_3 = 30 \angle 128° \text{ N}$$

at an offset distance

$$h_3 = \frac{|I_{G3} \alpha_3|}{|m_3 a_{G3}|} = \frac{274}{30} = 9.13 \text{ mm}$$

where this offset is measured to the right from G_3 to produce the necessary counterclockwise inertia moment about G_3. From the values of h_2 and h_3 and the angular relationships, the force positions r_2 and r_3 in Figure 10.4c are computed to be

$$r_2 = BG_2 - \frac{h_2}{\cos(132° - 17° - 90°)} = 45.1 \text{ mm}$$

$$r_3 = O_3 G_3 + \frac{h_3}{\cos(90° + 85° - 128°)} = 38.4 \text{ mm}$$

Now, we wish to perform a graphical force analysis for known forces F_2 and F_3. This has been done in Example Problem 9.2, and the reader is referred to that analysis. The required input torque was found to be

$$T = 838 \text{ N} \cdot \text{mm cw}$$

and the bearing forces were also determined in that example. Note that free-body diagrams of the three moving members are shown in Figure 9.11d.

Analytical Solution.   Having determined the equivalent offset inertia forces F_2 and F_3, the analytical solution could proceed according to Example Problem 9.6, which examined the same problem. However, it is not necessary to convert to the offset force, and here we will carry out the analytical solution in terms of the original inertia forces and inertia couples.

Figure 10.4d shows the linkage with the inertia torques and the inertia forces in xy coordinate form. Consistent with Figure 9.15a, we define the following quantities:

$$\ell_1 = 30 \text{ mm} \qquad \ell_2 = 100 \text{ mm} \qquad \ell_3 = 50 \text{ mm}$$

$$\phi_1 = 135° \qquad \phi_2 = 17° \qquad \phi_3 = 85°$$

$$r_1 = 0 \qquad r_2 = 50 \text{ mm} \qquad r_3 = 25 \text{ mm}$$

$$F_{2x} = 47 \cos(132°) = -31.4 \text{ N} \qquad F_{2y} = 47 \sin(132°) = 34.9 \text{ N}$$

$$F_{3x} = 30 \cos(128°) = -18.5 \text{ N} \qquad F_{3y} = 30 \sin(128°) = 23.6 \text{ N}$$

$$C_2 = -208 \text{ N} \cdot \text{mm} \qquad C_3 = 274 \text{ N} \cdot \text{mm}$$

$$F_{1x} = F_{1y} = C_1 = 0$$

These data lead to approximately the same values as determined in Example Problem 9.6 for the coefficients and right-hand terms in Eqs. 9.17a and 9.17b, where the differences are due to roundoff:

$$a_{11} = -49.8 \qquad a_{21} = 29.2 \qquad b_1 = -786$$

$$a_{12} = 4.36 \qquad a_{22} = -95.6 \qquad b_2 = -1920$$

Then, from Eqs. 9.18a through 9.22,

$$F_{23} = 31.3 \text{ N} \qquad F_{12} = 50.3 \text{ N}$$

$$F_{03} = 49.2 \text{ N} \qquad F_{01} = 50.3 \text{ N}$$

and

$$T = -851 \text{ N} \cdot \text{mm}$$

Thus it can be seen that the general analytical solution of the four-bar linkage presented in Chapter 9 for static-force analysis is equally well suited for dynamic-force analysis.

Before leaving this example, a couple of general comments should be made. First, the torque determined is the instantaneous value required for the prescribed motion, and the value will vary with position. Furthermore, for the position considered, the torque is opposite in direction to the angular velocity of the crank. This can be explained by the fact that the inertia of the mechanism in this position is tending to accelerate the crank in the counterclockwise direction, and, therefore, the required torque must be clockwise to maintain a constant angular speed. If a constant speed is to be maintained throughout the mechanism cycle, there will be other positions of the mechanism for which the required torque will be counterclockwise. The second comment is that it may be impossible to find a mechanism actuator, such as an electric motor, that will supply the required torque versus position behavior. This problem can be alleviated, however, in the case of a "constant" rotational speed mechanism through the use of a device called a flywheel, which is mounted on the input shaft and produces a relatively large mass moment of inertia for crank 1. The flywheel can absorb mechanism torque and energy variations with minimal speed fluctuation and thus maintain an essentially constant input speed. In such a case, the assumed-motion approach to dynamic-force analysis is appropriate. ◆

◆ **EXAMPLE PROBLEM 10.2**   *Dynamic-Force Analysis of an Air Pump*
In Chapter 4, an acceleration analysis was carried out for a curved-wing air pump, utilizing an equivalent four-bar linkage (see Example Problem 4.13 and Figure 4.15). The equivalent linkage considered in that example and the resulting acceleration polygon are reproduced here as Figure 10.5a and b. This linkage represents the actual pump shaft (link 1), one of the four curved wings (link 2), and the housing (link 3), which constrains the motion of the tip of the wing (point C). Point D is the center of mass of the wing. The angular velocity of the shaft is constant at 400 rev/min counterclockwise. The acceleration of the center of mass of the wing at the instant shown was found to have a magnitude of 5400 in/s^2 and a direction as shown in the acceleration polygon (Figure 10.5b). The instantaneous angular acceleration of the wing was determined to be 265 rad/s^2 counterclockwise.

Perform a dynamic-force analysis of the mechanism to determine the restraining force exerted by the housing on the wing at point C for the instant shown in the figure, and the necessary drive torque on link 1 associated with the inertia force and torque of the wing. The weight of the wing is 1 lb and the moment of inertia about its center of mass is 0.004 lb $\cdot$ s^2 $\cdot$ in. Furthermore, assume that the coefficient of sliding friction between the wing and housing is 0.2. Neglect friction at pins O_1 and B. As mentioned in Chapter 4, these dimensions may not correspond to an actual pump and are used here to illustrate general principles of force analysis only.

Solution.   The solution will be carried out graphically. First, we must determine the inertia forces and inertia torques. For link 1, we assume that the center of mass is coincident with the stationary point of rotation O_1, since the

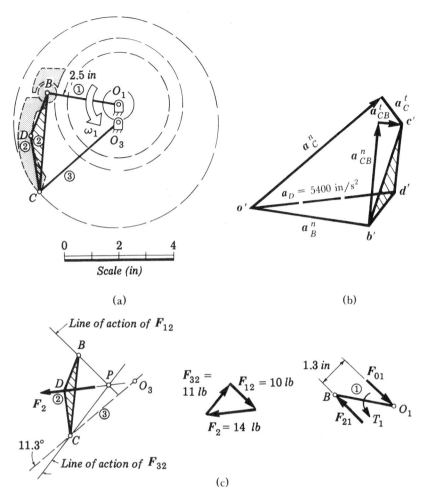

Figure 10.5 (a) Dynamic-force analysis of a curved-wing air pump. The diagram shows the equivalent four-bar linkage. (b) Acceleration polygon from Chapter 4. (c) The graphical force analysis.

member is symmetric (see Figure 4.15), and therefore the associated inertia force is zero. Further, the inertia torque for member 1 is also zero, because the angular acceleration is zero. Since a moving member 3 does not actually exist, there are obviously no corresponding inertia terms. Member 2 has an inertia force given by

$$F_{i2} = -m_2 a_D$$

which has a magnitude of

$$F_{i2} = m_2 a_D = \left(\frac{w_2}{g}\right) a_D = \left(\frac{1\ \text{lb}}{386.4\ \text{in/s}^2}\right)(5400\ \text{in/s}^2) = 14.0\ \text{lb}$$

and a direction opposite to that of acceleration a_D with a line of action through point D. The inertia torque of member 2 has a magnitude equal to

$$|C_2| = |I_{G2}\alpha_2| = (0.004 \text{ lb} \cdot \text{s}^2 \cdot \text{in})(265 \text{ rad/s}^2) = 1.06 \text{ lb} \cdot \text{in}$$

and a clockwise direction opposite to that of the angular acceleration.

Since the analysis will be graphical, we will make use of an equivalent offset inertia force F_2 for member 2. This force has a magnitude of 14.0 lb, a direction opposite to a_D, and a line of action that is displaced by the following perpendicular distance h_2 from the center of mass:

$$h_2 = \frac{I_{G2}\alpha_2}{m_2 a_D} = \frac{1.06}{14.0} = 0.076 \text{ in}$$

From inspection of Figure 10.5a, this offset should be downward from point D in order to produce a clockwise inertia torque. Note that the offset is negligible in this example. Figure 10.5c shows a free-body diagram of member 2 with the offset inertia force placed on the diagram to a suitable scale. The member is a three-force member with known force F_2 and unknown forces at points B and C. We can, however, determine the direction of the force exerted by the housing at point C. In the absence of friction, it would act normal to the contacting surfaces, that is, along imaginary link 3 toward point O_3. With friction present, it acts at an angle to the normal, where that angle is given by Eq. 9.32:

$$\tan \phi = 0.2 \quad \text{or} \quad \phi = 11.3°$$

The direction in which this angle is measured is based on the relative motion of the contacting surfaces and should produce a friction force that opposes that motion. In Figure 10.5a, point C on the wing moves counterclockwise relative to the housing, and therefore the angle is laid off as shown in Figure 10.5c to produce a clockwise-acting friction force of the housing on the wing. This establishes the line of action of force F_{32} in Figure 10.5c.

Now, knowing the lines of action of forces F_2 and F_{32}, we can find the location of the point of concurrency P at the intersection of these lines and thereby obtain the line of action of the third force F_{12}, applied at point B. Next, the force polygon for member 2 is constructed, as shown in Figure 10.5c. Finally, member 1 is subjected to two forces and a torque (see the free-body diagram of Figure 10.5c). Force F_{21} is equal and opposite to force F_{12}, and force F_{01} is, in turn, equal and opposite to force F_{21}.

We can now address the specific problem requirements.

From the force polygon of Figure 10.5c, the force F_{32} of the housing on the wing has a magnitude of approximately 11 lb and the direction shown. The components of this force are a normal force of magnitude $11 \cos(11.3°) = 10.8$ lb, and a friction force of magnitude $11 \sin(11.3°) = 2.2$ lb.

The required shaft torque T_1 is equal in magnitude to the moment produced by the force couple (F_{21}, F_{01}):

$$T_1 = (10 \text{ lb})(1.3 \text{ in}) = 13 \text{ lb} \cdot \text{in}$$

and by inspection the torque must be counterclockwise. It should be observed that a complete force analysis of the pump would also consider the other wings and forces produced by air pressure. ♦

10.5 DYNAMIC ANALYSIS OF THE SLIDER-CRANK MECHANISM

Dynamic forces are a very important consideration in the design of slider-crank mechanisms for use in machines such as internal-combustion engines and reciprocating compressors. Dynamic-force analysis of this mechanism can be carried out in exactly the same manner as for the four-bar linkage in the previous section. Following such a process, a kinematic analysis is first performed from which expressions are developed for the inertia force and inertia torque for each of the moving members. These quantities may then be converted to equivalent offset inertia forces for graphical analysis or they may be retained in the form of forces and torques for analytical solution, utilizing, in either case, the methods presented in Chapter 9. In fact, the analysis of the slider-crank mechanism is somewhat easier than that of the four-bar linkage because there is no rotational motion and, in turn, no inertia torque for the piston or slider, which has translating motion only. The following paragraphs will describe an analytical approach in detail.

Figure 10.6a is a schematic diagram of a slider-crank mechanism, showing crank 1, connecting rod 2, and piston 3, all of which are assumed to be rigid. The center of mass locations are designated by letter G, and the members have masses m_i and moments of inertia I_{Gi}, $i = 1, 2, 3$. The following analysis will consider the relationships of the inertia forces and torques to the bearing reactions and the drive torque on the crank, at an arbitrary mechanism position given by crank angle ϕ. Friction will be neglected.

Figure 10.6b shows free-body diagrams of the three moving members of the linkage. Applying the dynamic equilibrium conditions, Eqs. 10.6a to 10.6d, to each member yields the following set of equations. For the piston (moment equation not included):

$$F_{23x} + (-m_3 a_{G3}) = 0 \tag{10.8a}$$

$$F_{03y} + F_{23y} = 0 \tag{10.8b}$$

For the connecting rod (moments about point B):

$$F_{12x} + F_{32x} + (-m_2 a_{G2x}) = 0 \tag{10.8c}$$

$$F_{12y} + F_{32y} + (-m_2 a_{G2y}) = 0 \tag{10.8d}$$

$$F_{32x} \ell \sin \theta + F_{32y} \ell \cos \theta + (-m_2 a_{G2x}) \ell_G \sin \theta$$

$$+ (-m_2 a_{G2y}) \ell_G \cos \theta + (-I_{G2} \alpha_2) = 0 \tag{10.8e}$$

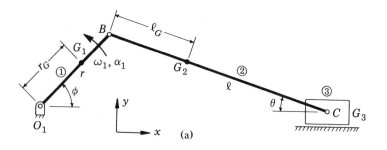

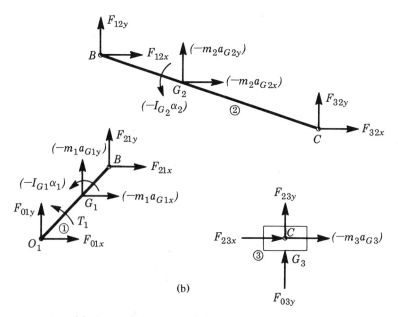

Figure 10.6 (a) Dynamic-force analysis of a slider-crank mechanism. (b) Free-body diagrams of the moving members.

For the crank (moments about point O_1):

$$F_{01x} + F_{21x} + (-m_1 a_{G1x}) = 0 \qquad \text{(10.8f)}$$

$$F_{01y} + F_{21y} + (-m_1 a_{G1y}) = 0 \qquad \text{(10.8g)}$$

$$T_1 - F_{21x} r \sin \phi + F_{21y} r \cos \phi - (-m_1 a_{G1x}) r_G \sin \phi$$

$$+ (-m_1 a_{G1y}) r_G \cos \phi + (-I_{G1} \alpha_1) = 0 \qquad \text{(10.8h)}$$

where T_1 is the input torque on the crank. This set of equations embodies both of the dynamic-force analysis approaches described in Section 10.2. However, its form is best suited for the case of known mechanism motion, as illustrated by the following example.

◆ **EXAMPLE PROBLEM 10.3** *Dynamic-Force Analysis of a Slider-Crank Mechanism*

Perform a dynamic-force analysis of the in-line slider-crank mechanism for which a kinematic acceleration analysis was carried out in Example Problem 4.8. That mechanism has dimensions $r = 2$ in and $l = 3.76$ in, and a known motion based on a crank angular velocity of $\omega_1 = +10$ rad/s and an angular acceleration of $\alpha_1 = +40$ rad/s^2 when the crank angle is $\phi = 70°$ (therefore, $\theta = 30°$). We assume for the purpose of this example that the center of mass locations are $r_G = 1$ in and $l_G = 1.88$ in, and that the mass properties are $m_1 = 1$ lbm, $m_2 = 2$ lbm, $m_3 = 3$ lbm, $I_{G1} = 0.01$ lb · s^2 · in, and $I_{G2} = 0.02$ lb · s^2 · in.

Solution. From the previous acceleration analysis in Chapter 4, the following quantities can be obtained:

$$a_{G1x} = -71.6 \text{ in/s}^2 \qquad a_{G2y} = -80.0 \text{ in/s}^2$$

$$a_{G1y} = -79.5 \text{ in/s}^2 \qquad \alpha_2 = 46 \text{ rad/s}^2$$

$$\alpha_1 = 40 \text{ rad/s}^2 \qquad a_{G3} = -70 \text{ in/s}^2$$

$$a_{G2x} = -106 \text{ in/s}^2$$

The inertia force components and inertia torques are calculated as follows:

$$-m_1 a_{G1x} = -\left(\frac{1 \text{ lbm}}{386.4 \text{ lbm} \cdot \text{in/s}^2 \cdot \text{lb}}\right)(-71.6 \text{ in/s}^2) = 0.185 \text{ lb}$$

$$-m_1 a_{G1y} = -\left(\frac{1}{386.4}\right)(-79.5) = 0.206 \text{ lb}$$

$$-I_{G1}\alpha_1 = -(0.01 \text{ lb} \cdot \text{s}^2 \cdot \text{in})(40 \text{ rad/s}^2) = -0.4 \text{ lb} \cdot \text{in}$$

$$-m_2 a_{G2x} = -\left(\frac{2}{386.4}\right)(-106) = 0.549 \text{ lb}$$

$$-m_2 a_{G2y} = -\left(\frac{2}{386.4}\right)(-80.0) = 0.414 \text{ lb}$$

$$-I_{G2}\alpha_2 = -(0.02)(46) = -0.92 \text{ lb} \cdot \text{in}$$

$$-m_3 a_{G3} = -\left(\frac{3}{386.4}\right)(-70) = 0.543 \text{ lb}$$

Substituting into Eqs. 10.8a and 10.8e,

$$F_{23x} = -0.543 \text{ lb} = -F_{32x}$$

$$F_{32y} = -F_{23y}$$

$$= \frac{-0.543(3.76)\sin 30° - 0.549(1.88)\sin 30° - 0.414(1.88)\cos 30° + 0.92}{3.76\cos(30°)}$$

$$= -0.396 \text{ lb}$$

and, from Eq. 10.8b,

$$F_{03y} = -0.396 \text{ lb}$$

Using Eqs. 10.8c and 10.8d, we have

$$F_{12x} = -F_{21x} = -0.543 - 0.549 = -1.09 \text{ lb}$$

$$F_{12y} = -F_{21y} = 0.396 - 0.414 = -0.018 \text{ lb}$$

From Eqs. 10.8f and 10.8g,

$$F_{01x} = -1.09 - 0.185 = -1.28 \text{ lb}$$

$$F_{01y} = -0.018 - 0.206 = -0.224 \text{ lb}$$

Finally, substituting into Eq. 10.8h, we find the required input torque T_1 for the prescribed motion to be

$$T_1 = (1.09)(2)\sin(70°) - (0.018)(2)\cos(70°)$$
$$+ (0.185)(1)\sin(70°) - (0.206)(1)\cos(70°) + 0.4$$
$$= 2.54 \text{ lb} \cdot \text{in}$$

This indicates that an instantaneous counterclockwise torque of this amount is required at the mechanism position considered. ◆

Instead of the totally general analysis procedure described above, it is possible in certain cases to make use of some simplifying approximations particular to the slider-crank linkage, which are often used in the design of this mechanism for applications such as engines and compressors. These approximations deal with the mass distribution of the connecting rod and the acceleration of the piston. These approximations are the topics of the following sections and will also be useful in a later section of this chapter on balancing reciprocating masses. In addition, we will analyze this mechanism from the two standpoints discussed earlier, namely, force analysis under assumed motion and motion analysis under assumed forces.

10.5.1 Equivalent Inertia

The slider-crank mechanism of Figure 10.6a is reproduced in Figure 10.7a. The crank and piston perform rotational motion and rectilinear motion, respectively, which are relatively simple motions to analyze. The connecting rod exhibits a more complex motion, except for the special points B, the crankpin, and C, the wrist pin. These points coincide with points on the crank and piston, and, therefore, point B has a circular path and point C follows a straight-line path.

Because of the simplified kinematics for points B and C, we will examine the possibility of representing the mass distribution of the original connecting rod by an equivalent body consisting of two point masses, m_B and m_C at

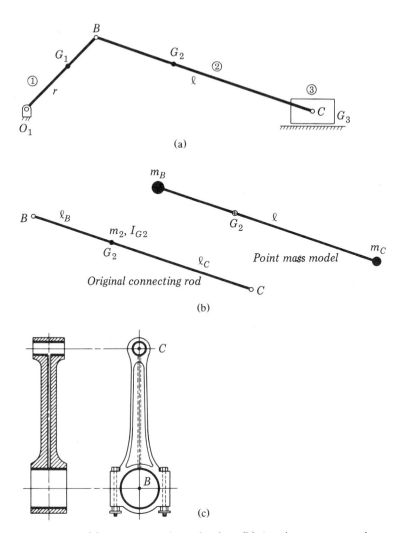

Figure 10.7 (a) A slider-crank mechanism. (b) A point-mass approximation of the continuous-mass connecting rod. (c) An engine connecting
rod.

points B and C, connected by a massless rigid rod. See Figure 10.7b. The two
members are equivalent for dynamic analysis purposes if the following three
conditions are satisfied:

1. The centers of mass must be at the same location, point G_2.
2. The total masses must be equal.
3. The moments of inertia with respect to the center of mass must be
   equal.

With reference to Figure 10.7b, these conditions can be expressed mathematically as follows:

1. $m_B \ell_B = m_C \ell_C$.
2. $m_B + m_C = m_2$.
3. $m_B \ell_B^2 + m_C \ell_C^2 = I_{G2}$.

Since there are only two adjustable quantities (m_B and m_C) in these equations, all three conditions will not be satisfied, in general. Therefore the point-mass representation is an approximation of the original connecting rod. Using the first two equations to set values for masses m_B and m_C, we have

$$m_B = \left(\frac{\ell_C}{\ell_B + \ell_C} \right) m_2 = \left(\frac{\ell_C}{\ell} \right) m_2 \tag{10.9a}$$

$$m_C = \left(\frac{\ell_B}{\ell_B + \ell_C} \right) m_2 = \left(\frac{\ell_B}{\ell} \right) m_2 \tag{10.9b}$$

The accuracy of the approximation employing these point masses will depend on how closely the third condition above is satisfied. There are many actual geometries for which this approximation is quite satisfactory. For example, Figure 10.7c shows an engine connecting rod that can be modeled this way.

10.5.2 Approximate Dynamic Analysis Equations

In this section, we make use of the point-mass approximation of the previous section to derive equations for the dynamic analysis of the slider-crank mechanism. Figure 10.8a shows the mechanism and includes pertinent information. Note that the center of mass location for the crank is positioned at fixed pivot O_1 for this analysis. As a result, the associated inertia force will be zero. The more general analysis of Eqs. 10.8a to 10.8h can be applied when the center of mass is located elsewhere.

Figure 10.8b contains free-body diagrams of the three moving links, including inertia forces and torques. The piston does not experience an inertia torque because its angular acceleration is zero. The two inertia forces shown for the equivalent connecting rod represent the combined inertia force and inertia torque of the actual connecting rod. Bearing forces are shown in the figure in terms of x and y components. Expressing the inertia forces in component form, we have

$$-m_B a_B = -m_B a_{Bx} i - m_B a_{By} j \tag{10.10}$$

$$-m_C a_C = -m_C a_{Cx} i \tag{10.11a}$$

$$-m_3 a_C = -m_3 a_{Cx} i \tag{10.11b}$$

where the accelerations a_B and a_C will, in general, be functions of the crank angular position, velocity, and acceleration, and the linkage dimensions. Sum-

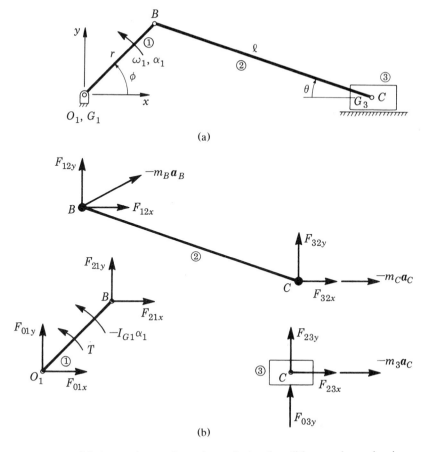

Figure 10.8 (a) Approximate dynamic analysis of a slider-crank mechanism. (b) Free-body diagrams of the moving members.

ming forces on the piston in the x and y directions, we have

$$F_{23x} - m_3 a_{Cx} = 0$$

$$F_{03y} + F_{23y} = 0$$

From the first of these equations,

$$F_{23x} = -F_{32x} = m_3 a_{Cx} \tag{10.12}$$

Writing a moment equation for link 2 about point B, we obtain

$$(F_{32x} - m_C a_{Cx})\ell \sin\theta + F_{32y}\ell \cos\theta = 0$$

from which

$$F_{32y} = -F_{23y} = F_{03y} = (m_C + m_3)a_{Cx}\tan\theta \tag{10.13}$$

Summing forces on the connecting rod, we have

$$F_{12x} = -F_{21x} = F_{01x} = m_B a_{Bx} + (m_C + m_3) a_{Cx} \qquad (10.14)$$

$$F_{12y} = -F_{21y} = F_{01y} = m_B a_{By} - (m_C + m_3) a_{Cx} \tan \theta \qquad (10.15)$$

Finally, summing moments on the crank about pivot O_1, we have

$$T = I_{G1} \alpha_1 + F_{21x} r \sin \phi - F_{21y} r \cos \phi \qquad (10.16)$$

Notice that, in the determination of bearing forces, it is necessary to distinguish, for example, between the mass of the connecting rod at point C and the mass of the piston at point C. Similarly, mass m_B is on the connecting rod side of crankpin B.

The analysis described above is similar to that of Eqs. 10.8a to 10.8h but includes the point-mass approximation of the connecting rod. The following sections will illustrate how this model, in conjunction with an approximation dealing with the mechanism kinematics, can be applied to a variety of mechanism operating conditions.

10.5.3 Dynamic-Force Analysis for an Assumed Mechanism Motion

Suppose we wish to determine the bearing forces and required crankshaft torque during a complete mechanism cycle for the case of assumed constant angular velocity ω_1 of the crank. We will substitute analytical acceleration expressions for this motion into the equations derived in the previous section. The result will be equations for the bearing forces and input torque as functions of the mechanism parameters (i.e., dimensions, masses, etc.) and the crank angle ϕ for constant crank speed ($\alpha_1 = 0$).

From Figure 10.8a, the acceleration a_B is determined by differentiating the following position vector:

$$R_B = r \cos \phi i + r \sin \phi j \qquad (10.17)$$

The velocity and acceleration so obtained, with ω_1 constant, are

$$v_B = -r\omega_1 \sin \phi i + r\omega_1 \cos \phi j \qquad (10.18)$$

$$a_B = -r\omega_1^2 \cos \phi i - r\omega_1^2 \sin \phi j \qquad (10.19)$$

and, therefore, the associated inertia force is

$$-m_B a_B = m_B r \omega_1^2 (\cos \phi i + \sin \phi j) \qquad (10.20)$$

This is a centrifugal force directed radially outward along the crank. Examination of Eqs. 10.14 to 10.16 shows that the torque T is independent of this inertia force.

From Figure 10.8a, the position vector for point C is

$$R_C = (r \cos \phi + \ell \cos \theta) i \qquad (10.21)$$

Angle θ can be expressed in terms of ϕ from the relationship

$$\ell \sin \theta = r \sin \phi \tag{10.22}$$

which leads to

$$\cos \theta = \sqrt{1 - \sin^2 \theta} = \sqrt{1 - \left(\frac{r}{\ell} \sin \phi\right)^2} \tag{10.23}$$

This expression can be substituted into the position vector and the result differentiated twice to obtain the exact acceleration. Instead, we introduce an approximate form that will reduce the mathematical complexity and is quite satisfactory for crank length-to-connecting rod length ratios commonly found in engines and compressors.

The following binomial series will be utilized:

$$\sqrt{1 - s} = 1 - \frac{1}{2}s - \frac{1}{8}s^2 - \frac{1}{16}s^3 - \cdots \tag{10.24}$$

This is an infinite series that holds for magnitudes of $s < 1$. Since r must be less than ℓ if the crank is to rotate completely, and since $\sin \phi$ cannot exceed a value of one, this series can be used in Eq. 10.23 with

$$s = \left(\frac{r}{\ell} \sin \phi\right)^2 \tag{10.25}$$

Furthermore, the square root may be approximated by a truncated partial sum of the infinite series. Using the first two terms of the series, we have

$$\cos \theta \cong 1 - \frac{1}{2}\left(\frac{r}{\ell}\right)^2 \sin^2 \phi \tag{10.26}$$

Because s is proportional to $(r/\ell)^2$, this approximation will be more accurate for smaller values of r/ℓ. For example, for $r/\ell \le 1/4$, the magnitude of the third term in Eq. 10.24 will be less than 0.0005, which is small enough to be neglected in most analyses.

Substituting the trigonometric identity

$$\sin^2 \phi = \frac{1}{2} - \frac{1}{2} \cos 2\phi$$

Equation 10.26 becomes

$$\cos \theta \cong 1 - \frac{1}{4}\left(\frac{r}{\ell}\right)^2 + \frac{1}{4}\left(\frac{r}{\ell}\right)^2 \cos 2\phi \tag{10.27}$$

and

$$\boldsymbol{R}_C = \left(\ell - \frac{r^2}{4\ell} + r \cos \phi + \frac{r^2}{4\ell} \cos 2\phi\right)\boldsymbol{i} \tag{10.28}$$

Differentiating twice with ω_1 constant, we have

$$v_C = \left(-r\omega_1 \sin \phi - \frac{\omega_1 r^2}{2\ell} \sin 2\phi\right)i \tag{10.29}$$

and

$$a_C = \left(-r\omega_1^2 \cos \phi - \frac{r^2\omega_1^2}{\ell} \cos 2\phi\right)i \tag{10.30}$$

The corresponding inertia forces are

$$-m_C a_C = m_C r\omega_1^2\left(\cos \phi + \frac{r}{\ell} \cos 2\phi\right)i \tag{10.31}$$

and

$$-m_3 a_C = m_3 r\omega_1^2\left(\cos \phi + \frac{r}{\ell} \cos 2\phi\right)i \tag{10.32}$$

Substitution of the acceleration expressions into the previous force analysis equations, Eqs. 10.12 to 10.16, leads to the following relationships for the bearing forces and crankshaft torque:

$$F_{32x} = m_3 r\omega_1^2\left(\cos \phi + \frac{r}{\ell} \cos 2\phi\right) \tag{10.33}$$

$$F_{32y} = F_{03y} = \frac{-(m_C + m_3)r^2\omega_1^2 \sin \phi[\cos \phi + (r/\ell)\cos 2\phi]}{\sqrt{\ell^2 - r^2 \sin^2 \phi}} \tag{10.34}$$

$$F_{12x} = F_{01x} = -m_B r\omega_1^2 \cos \phi - (m_C + m_3)r\omega_1^2\left(\cos \phi + \frac{r}{\ell} \cos 2\phi\right) \tag{10.35}$$

$$F_{12y} = F_{01y} = -m_B r\omega_1^2 \sin \phi$$
$$+ \frac{(m_C + m_3)r^2\omega_1^2 \sin \phi[\cos \phi + (r/\ell)\cos 2\phi]}{\sqrt{\ell^2 - r^2 \sin^2 \phi}} \tag{10.36}$$

and

$$T = (m_C + m_3)r^2\omega_1^2 \sin \phi\left(\cos \phi + \frac{r}{\ell} \cos 2\phi\right)\left(1 + \frac{r \cos\phi}{\sqrt{\ell^2 - r^2 \sin^2 \phi}}\right) \tag{10.37}$$

These are explicit relationships between forces, mechanism motion, and design parameters that are useful to the design engineer who must properly size a mechanism for a given set of performance requirements.

In the absence of friction, the superposition principle can be applied to determine the combined effects of cylinder pressure forces and mechanism inertia. The force analysis for an external piston force has been presented in Chapter 9.

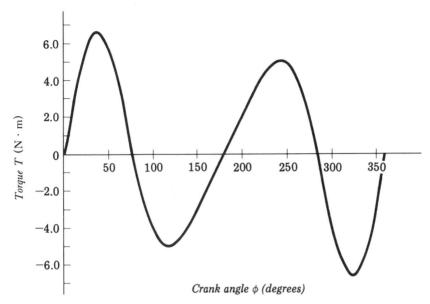

Figure 10.9 Input torque versus crank angle for the slider-crank mechanism of Example Problem 10.4. The input crank is assumed to have a constant velocity of 1000 rev/min.

◆ **EXAMPLE PROBLEM 10.4**   *Dynamic Analysis of a Slider-Crank Mechanism Under Assumed Input Motion*

An in-line slider-crank mechanism has a crank length $r = 0.04$ m, a connecting rod length $\ell = 0.16$ m, and a reciprocating mass $(m_C + m_3) = 0.6$ kg. Determine the torque variation required over a complete cycle in order to maintain a constant crank speed of 1000 rev/min.

Solution.   The angular velocity ω_1 in radians per second is

$$\omega_1 = \frac{1000 \text{ rev/min}}{60 \text{ s/min}} \cdot 2\pi \text{ rad/rev} = 104.72 \text{ rad/s}$$

Substituting this value and the given data into Eq. 10.37, we find that the crank torque, as a function of crank angle ϕ, is as follows:

$$T = 10.53 \sin \phi (\cos \phi + 0.25 \cos 2\phi) \left[1 + \frac{0.25 \cos \phi}{\sqrt{1 - (0.25 \sin \phi)^2}} \right] N \cdot m$$

The torque is plotted as a function of ϕ in Figure 10.9.   ◆

10.5.4 Dynamic-Motion Analysis for an Assumed Input Torque

In this section, the dynamic analysis equations derived previously will be employed in dynamic-motion analysis, sometimes referred to as time-response analysis. Here the mechanism motion is not assumed and is treated as an

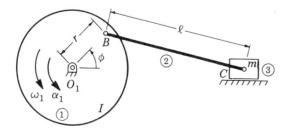

Figure 10.10 A slider-crank mechanism for which a dynamic-motion analysis is to be performed.

unknown. Instead, an assumption will be made about the input torque T representative of the behavior of a typical driving device. Once again, although a totally general analysis can be performed, the approximations, which were described above and which are typically valid, will be imposed to simplify the mathematical development somewhat.

Consider the slider-crank mechanism shown in Figure 10.10. The inertia effects of the rotating part of the connecting rod mass (mass m_B) can be nullified by counterweighting the crank; assuming that has been done in this case, we set $m_B = 0$. The net mass moment of inertia of the crank will then be referred to as I. Further, let $m = m_C + m_3$. The mechanism is to be driven by an electric motor, governed by the following torque-speed equation, which is a function of the motor characteristics:

$$T = A - B\omega_1 \tag{10.38}$$

where A and B are motor constants. This relationship, which is typical of some DC motors, indicates that the input torque varies with the crank speed; the torque increases as ω_1 decreases and decreases as ω_1 increases. We wish to determine the motion of the mechanism when driven by such a motor.

Substituting Eqs. 10.14, 10.15, and 10.38 into Eq. 10.16, with $m_B = 0$,

$$A - B\omega_1 = I\alpha_1 - mra_{Cx}\sin\phi - mra_{Cx}\cos\phi\tan\theta$$

Rearranging,

$$I\alpha_1 + B\omega_1 - mra_{Cx}(\sin\phi + \cos\phi\tan\theta) - A = 0 \tag{10.39}$$

Acceleration a_{Cx} is obtained by differentiating Eq. 10.29, where now $\alpha_1 \neq 0$:

$$a_{cx} = -r\alpha_1\sin\phi - r\omega_1^2\cos\phi - \left(\frac{\alpha_1 r^2}{2\ell}\right)\sin 2\phi - \left(\frac{\omega_1^2 r^2}{\ell}\right)\cos 2\phi$$

$$= -r\alpha_1\left[\sin\phi + \left(\frac{r}{2\ell}\right)\sin 2\phi\right] - r\omega_1^2\left[\cos\phi + \left(\frac{r}{\ell}\right)\cos 2\phi\right] \tag{10.40}$$

Substitution into Eq. 10.39 yields

$$K_1 \frac{d^2\phi}{dt^2} + B\frac{d\phi}{dt} + K_2\left(\frac{d\phi}{dt}\right)^2 - A = 0 \qquad (10.41)$$

where $\dfrac{d\phi}{dt} = \omega_1 \qquad \dfrac{d^2\phi}{dt^2} = \alpha_1$

$$K_1 = I + mr^2\left[\sin\phi + \left(\frac{r}{2\ell}\right)\sin 2\phi\right](\sin\phi + \cos\phi\tan\theta)$$

$$K_2 = mr^2\left[\cos\phi + \left(\frac{r}{\ell}\right)\cos 2\phi\right](\sin\phi + \cos\phi\tan\theta)$$

and

$$\tan\theta = \frac{r\sin\phi}{\sqrt{\ell^2 - r^2\sin^2\phi}}$$

This is a differential equation relating crank angle ϕ and time t. The equation is nonlinear, and an exact closed-form mathematical solution cannot be obtained. Therefore, numerical methods must be employed to solve the equation. A variety of computer methods, which are beyond the scope of this book, are available for solving differential equations. With a computer, an engineer can design, by means of an iterative process, the motor-mechanism system for a particular application by numerically solving the above equation for different combinations of the motor and linkage parameters until acceptable performance is achieved.

Once the motion $\phi(t)$ and its derivatives $\omega_1(t)$ and $\alpha_1(t)$ have been calculated, the bearing forces can be determined from the equations derived previously, Eqs. 10.12 to 10.15. Although quantitative results are difficult to obtain, some qualitative information follows from inspection of Eq. 10.41. For example, solving for the second derivative of ϕ, we have

$$\frac{d^2\phi}{dt^2} = \frac{A - B(d\phi/dt) - K_2(d\phi/dt)^2}{K_1} \qquad (10.42)$$

As the moment of inertia I increases in size, denominator K_1 will grow large and the magnitude of the acceleration will decrease. Therefore, for some suitably large value of I (as produced by a flywheel), α_1 will become negligibly small and the mechanism can be assumed to be operating at constant angular speed ω_1. On the other hand, the piston inertia is represented by the second term in denominator K_1, which varies between limiting values of zero and mr^2. Therefore, for small moment of inertia I, the acceleration will also vary and will have large values at certain crank positions, leading to considerable mechanism speed fluctuation.

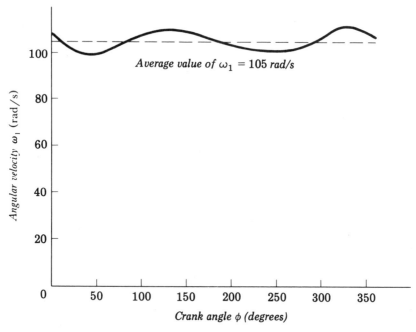

Figure 10.11 Crank angular velocity versus crank angle for the slider-crank mechanism of Example Problem 10.5. The average velocity is 105 rad/s (approximately 1000 rev/min).

♦ **EXAMPLE PROBLEM 10.5** *Dynamic Analysis of a Slider-Crank Mechanism under Assumed Input Torque*

An in-line slider-crank mechanism has a crank length $r = 0.04$ m, a connecting rod length $\ell = 0.16$ m, a reciprocating mass $m = 0.6$ kg, and a crank moment of inertia $I = 0.001$ kg $\cdot$ m^2. The mechanism is driven by a DC motor having the following constants: $A = 105$ $N \cdot m$ and $B = 1.0$ $N \cdot m \cdot s$. Determine how the crank velocity varies over a complete steady-state cycle of the mechanism.

Solution. Notice that this is the same mechanism as was studied in Example Problem 10.4; however, a more realistic input actuator is now considered. Figure 10.11 contains a plot of the crank angular velocity versus angle ϕ for a complete 360° rotation of the crank. This response was obtained by numerical solution of Eq. 10.41, starting from initial conditions of rest; that is, $\phi = d\phi/dt = 0$ when time $t = 0$. The data plotted in Figure 10.11 are for the eighth cycle of rotation, by which time steady-state motion had been established. ♦

10.6 BALANCING OF MACHINERY

Up to this point, little mention has been made of the dynamic forces that are transmitted to the frame of a machine. And yet, some of the more serious problems encountered in high-speed machinery are the direct result of these forces. As can be seen from the analyses presented previously, the forces exerted on the frame by moving machine members will, in general, be time varying, and they will therefore impart vibratory motion to the frame. This vibration, and the accompanying noise, can produce human discomfort, it can alter the desired machine performance, and it can adversely affect the structural integrity of the machine foundation. Furthermore, these problems are intensified both by increased operating speeds, which, as was pointed out earlier, lead to greater inertia forces, and by conditions of resonance or natural-frequency vibration.

A net unbalanced force acting on the frame of a machine or mechanism (i.e., the resultant of the forces transmitted at all of the connections between the machine and frame) is referred to as the shaking force. Likewise, a resultant unbalanced moment acting on the frame is called a shaking moment. Since the shaking force and shaking moment are unbalanced effects, they will cause the frame to vibrate, with the magnitude of the vibration dependent on the amount of unbalance. Clearly, then, an important design objective is the minimization of machine unbalance.

The process of designing or modifying machinery in order to reduce unbalance to an acceptable level, and possibly to eliminate it entirely, is called balancing. Since unbalance is caused in the first place by the inertia forces associated with moving machine mass, the most common approach to balancing is the redistribution of mass, accomplished by the addition of mass to or the removal of mass from various machine members. However, other techniques, such as those involving springs and dampers or balancing mechanisms, are also used.

The method of balancing employed depends to a considerable extent on the type of unbalance present in the machine. The two basic types are rotating unbalance and reciprocating unbalance, which may occur separately or in combination. These two types of unbalance are considered in detail in the following sections.

10.7 BALANCING OF RIGID ROTORS

In the introduction to this chapter, the example of a dynamic force that was presented was that of a centrifugal force associated with a mass attached to a rotating rod (see Figure 10.1a). The same kind of force occurs in an eccentric rotor (see Figure 10.1b). In both cases, the bearing mounts on the machine frame experience a net unbalanced shaking force. This type of unbalance due to rotating mass is referred to as rotating unbalance, and since virtually all machines contain rotating parts, this form of unbalance is very common. It occurs, for example, in such diverse applications as turbine rotors, engine crankshafts, washing machine drums, and window fans. Fortunately, rotating unbalance is relatively easy to deal with.

Before presenting general balancing methods, let us consider the above example in more detail. Figure 10.12 shows a rotor consisting of a disk of mass m attached to a rigid shaft of assumed negligible mass. We will also assume

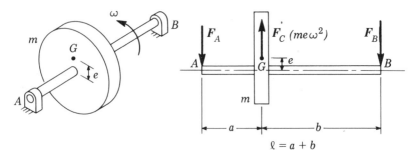

Figure 10.12 Static unbalance caused by an eccentric mass on a rotating shaft.

that the rotor has constant angular velocity ω. The center of mass G of the disk does not coincide with the bearing centerline AB, with the amount of this eccentricity represented by e. From d'Alembert's principle, the rotor experiences the centrifugal force F_C shown, which has a magnitude equal to $me\omega^2$. Summing forces and moments on the rotor, we find that the bearing reactions on the shaft are

$$F_A = \left(\frac{b}{\ell}\right)me\omega^2$$

and

$$F_B = \left(\frac{a}{\ell}\right)me\omega^2$$

which act in the directions shown in the figure. The net force on the frame will therefore be a force of magnitude F_s given by

$$F_s = \left(\frac{a}{\ell}\right)me\omega^2 + \left(\frac{b}{\ell}\right)me\omega^2 = \left(\frac{a+b}{\ell}\right)me\omega^2 = me\omega^2$$

and with a direction that rotates with speed ω. Thus, the centrifugal force is transmitted directly to the frame. Notice that if the rotational speed is doubled, this shaking force is quadrupled.

The unbalance described above is fairly easy to detect. For example, if the shaft was mounted horizontally on knife-edge bearings, due to gravity the rotor would always seek the static position with point G below the bearing centerline. Any rotating unbalance that can be detected in a static test, such as that described above, is referred to as a static unbalance. This is somewhat of a misnomer in that we are primarily concerned with the dynamic effects caused by such an unbalance. Not only can a static unbalance be detected through a static test, but it can also be corrected through a static procedure. The bubble balance sometimes used for automobile wheels is an example; this device statically determines the location and amount of unbalance so that corrective counterweights can be attached in such a way that the combined center of

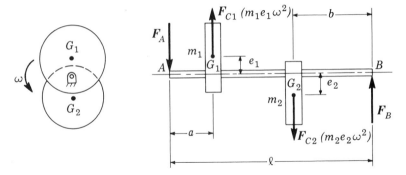

Figure 10.13 Dynamic unbalance due to eccentric masses at multiple axial locations on a rotating shaft.

mass is coincident with the bearing centerline (i.e., $e = 0$), and the resulting shaking force is zero.

Contrasted with the case of Figure 10.12 is the rotor of Figure 10.13, with two disks having masses m_1 and m_2 and eccentricities e_1 and e_2. Suppose that the centers of mass G_1 and G_2 of the disks are 180° apart, as shown in the figure. Again assuming constant angular velocity ω and negligible shaft mass, the bearing reactions are determined to be

$$F_A = \left(\frac{\omega^2}{\ell}\right)[m_1 e_1 (\ell - a) - m_2 e_2 b]$$

and

$$F_B = \left(\frac{\omega^2}{\ell}\right)[m_2 e_2 (\ell - b) - m_1 e_1 a]$$

The directions of these forces are based on the signs of the above expressions; if positive, then the force is oriented as shown in Figure 10.13; if negative, then the direction of the force is reversed. These forces rotate about the bearing centerline with angular speed ω. The shaking force magnitude is

$$F_s = |F_A - F_B| = \omega^2 |m_1 e_1 - m_2 e_2|$$

and if $m_1 e_1 = m_2 e_2$, the shaking force is zero. However, in this situation, even though the resultant force on the frame is zero, the individual bearing forces are nonzero, now having equal magnitudes given by

$$F_A = F_B = \left(\frac{m_1 e_1 \omega^2}{\ell}\right)[\ell - (a + b)]$$

but opposite directions, as shown in Figure 10.13. Thus, there is still a resultant couple, or shaking couple, acting on the frame and tending to produce rotational vibration of the frame. In general, shaking couples occur when unbalanced masses are located at multiple axial positions on a rotor.

The unbalance shown in Figure 10.13 with $m_1 e_1 = m_2 e_2$ could not be detected in a static test; for example, the rotor would take random orientations in the gravity test described earlier. Such an unbalance can only be detected by means of a dynamic test in which the rotor is spinning. Any unbalance that can be detected through such a dynamic test is referred to as a dynamic unbalance. (By this definition, the static unbalance defined previously is also one form, but not the only form, of dynamic unbalance.) A static balancing procedure will not completely correct this type of unbalance, and a statically balanced rotor may perform very poorly under actual operating conditions. In such a situation, dynamic balancing procedures are required, wherein the rotor is driven at an arbitrary speed and bearing forces are measured. From this information, magnitudes and locations of corrective counterweights are determined.

In general, static unbalance is characterized by a net shaking force, and dynamic unbalance is characterized by a combination of a net shaking force and a net shaking couple. Dynamic unbalance is more apt to be significant in cases of rotors having mass distributed over relatively large axial distances. For example, static balancing may be satisfactory for machine components such as automobile wheels or household window fans, which have short axial lengths, whereas dynamic balancing must be performed on equipment such as automotive crankshafts and multistage turbine rotors that have large axial lengths.

From the preceding discussion, general balancing procedures for sizing and positioning corrective masses on rotors are based on the following criteria:

1. For static balance, the shaking force must be zero.
2. For dynamic balance, the shaking force and the shaking moment must both be zero.

These procedures are illustrated in the following sections.

10.7.1 Static Balancing

Consider the rigid rotor shown in Figure 10.14a which is assumed to be rotating with constant angular velocity ω. Unbalanced masses are depicted as point masses m at radial distances r. These may represent a variety of actual rotating masses, including turbine or propeller blades, eccentric disks, crankthrows, and so on. In this case, there are three masses, but there could be any number. It is assumed here that all of these masses lie in a single transverse plane at the same axial location along the shaft, or close to the same plane. It will be shown that this arrangement can be balanced by a single counterbalance lying in this same plane and represented by dashed lines in Figure 10.14a.

Each of the original masses in Figure 10.14a produces a centrifugal force acting radially outward from the axis of rotation with a magnitude equal to $m_n r_n \omega^2$, $n = 1, 2, 3$. The vector sum of these forces will be transmitted through the support bearings to the frame, resulting in a shaking force F_s given by

$$F_s = m_1 \omega^2 r_1 + m_2 \omega^2 r_2 + m_3 \omega^2 r_3$$

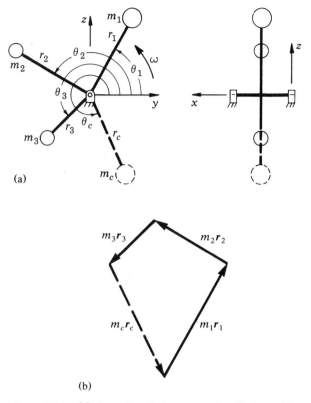

(a)

(b)

Figure 10.14 (a) A static unbalance can be eliminated by
the addition of a single counterweight m_c at the proper
radial distance r_c and angle θ_c. (b) Graphical determination of the counterweight size and location.

If this vector sum is zero then the rotor is balanced. In general, this will not be
the case, and we therefore introduce the counterweight of mass m_c shown in
Figure 10.14a at radial distance r_c. The magnitude and location of this
counterweight are determined from the condition that the resultant inertia
force must now be zero; that is,

$$m_1\omega^2 r_1 + m_2\omega^2 r_2 + m_3\omega^2 r_3 + m_c\omega^2 r_c = 0 \qquad (10.43)$$

The quantity ω^2 can be factored from Eq. 10.43, yielding the following
relationship for static balance:

$$m_1 r_1 + m_2 r_2 + m_3 r_3 + m_c r_c = 0 \qquad (10.44a)$$

This equation indicates that the combined center of mass must lie on the axis of rotation. In general, for N initial masses, the balancing condition is

$$\sum_{n=1}^{N} m_n r_n + m_c r_c = 0 \tag{10.44b}$$

Since all the vectors in Eq. 10.44a lie in a plane that is parallel to the yz plane in Figure 10.14a, this equation is a two-dimensional vector equation, and it therefore can be satisfied by the two parameters, that is, magnitude and direction, associated with the vector for the single counterweight m_c. The solution can be performed either graphically or mathematically. Figure 10.14b shows a graphical solution where vectors $m_1 r_1$, $m_2 r_2$, and $m_3 r_3$ are drawn in sequence to a suitable scale. The vector that closes this loop, and therefore satisfies Eq. 10.44a, is $m_c r_c$. The direction of this vector identifies the angular orientation of the counterweight relative to the other masses, and the magnitude of the vector is the required correction amount $m_c r_c$. Note that, because only the proper value of the product is required, either m_c or r_c can be selected arbitrarily. For example, if mass m_c is chosen, then the solution above determines vector location r_c of this counterweight.

Eq. 10.44a can also be solved mathematically by dividing it into y and z components, as follows:

$$m_1 r_1 \cos \theta_1 + m_2 r_2 \cos \theta_2 + m_3 r_3 \cos \theta_3 + m_c r_c \cos \theta_c = 0$$

$$m_1 r_1 \sin \theta_1 + m_2 r_2 \sin \theta_2 + m_3 r_3 \sin \theta_3 + m_c r_c \sin \theta_c = 0$$

where θ represents instantaneous angular orientation with respect to the y axis; see Figure 10.14a. Solving for $m_c r_c$ and θ_c, we have

$$m_c r_c = \left[\left(m_1 r_1 \cos \theta_1 + m_2 r_2 \cos \theta_2 + m_3 r_3 \cos \theta_3 \right)^2 \right.$$

$$\left. + \left(m_1 r_1 \sin \theta_1 + m_2 r_2 \sin \theta_2 + m_3 r_3 \sin \theta_3 \right)^2 \right]^{1/2} \tag{10.45}$$

$$\theta_c = \arctan\left(\frac{-m_1 r_1 \sin \theta_1 - m_2 r_2 \sin \theta_2 - m_3 r_3 \sin \theta_3}{-m_1 r_1 \cos \theta_1 - m_2 r_2 \cos \theta_2 - m_3 r_3 \cos \theta_3} \right) \tag{10.46}$$

Note that the signs of the numerator and denominator of the arctan function in Eq. 10.46 will identify the proper quadrant for angle θ_c. Rotors are often balanced through the removal of mass, as by drilling holes, rather than by adding counterweights. This is accomplished in the above procedure by specifying negative correction mass. Therefore the rotor in Figure 10.14a is also balanced by removing mass of amount $-m_c$ at position $-r_c$.

♦ **EXAMPLE PROBLEM 10.6**   *Static Balancing of a Rotor*

The rotor of Figure 10.14a has the following properties:

$m_1 = 3$ kg $r_1 = 80$ mm $\theta_1 = 60°$

$m_2 = 2$ kg $r_2 = 80$ mm $\theta_2 = 150°$

$m_3 = 2$ kg $r_3 = 60$ mm $\theta_3 = 225°$

Determine the counterweight amount and location required for static balance.

Solution.   Substituting the values given into Eq. 10.45, we have

$$m_c r_c = \left[(240 \cos 60° + 160 \cos 150° + 120 \cos 225°)^2 \right.$$

$$\left. + (240 \sin 60° + 160 \sin 150° + 120 \sin 225°)^2 \right]^{1/2}$$

$$= \left[(103.4)^2 + (203.0)^2 \right]^{1/2} = 227.8 \text{ kg} \cdot \text{mm}$$

This product will result, for example, from a counterweight mass of 2.85 kg at a radial distance of 80 mm. The angular position of the counterweight is calculated from Eq. 10.46:

$$\theta_c = \arctan\left(\frac{-203.0}{+103.4} \right) = 297.0°$$

where the angle is determined to be in the fourth quadrant from the signs of the numerator and denominator in the argument of the arctan function. The graphical solution of this example has been carried out to scale in Figure 10.14b.   ◆

10.7.2 Dynamic Balancing

Figure 10.15a shows a rotor with eccentric masses located at multiple axial locations. As a result, the rotor experiences general dynamic unbalance. As in the preceding section, the case of three initial masses will be presented here; in addition, the results will be generalized to any number of masses. The angular velocity ω is assumed to be constant.

We will examine the possibility of completely balancing this rotor through the addition of two counter masses, m_{c1} and m_{c2}, placed in transverse planes at arbitrarily selected axial locations, P and Q. For static balance, the sum of all the inertia forces must be zero; this condition yields the following equations which are similar to Eqs. 10.44a and 10.44b:

$$m_1 r_1 + m_2 r_2 + m_3 r_3 + m_{c1} r_{c1} + m_{c2} r_{c2} = 0 \tag{10.47a}$$

and for the general case of N original masses,

$$\sum_{n=1}^{N} m_n r_n + m_{c1} r_{c1} + m_{c2} r_{c2} = 0 \tag{10.47b}$$

However, a shaking couple will still exist if the inertia forces produce a net couple. Therefore, the additional condition for dynamic balance is that the sum of the moments of the inertia forces about any arbitrary point must be zero. For convenience in determining the required counterbalances, we will take moments about point P, the axial location of counterweight 1, thereby eliminating this unknown counterweight from the moment equation. The axial

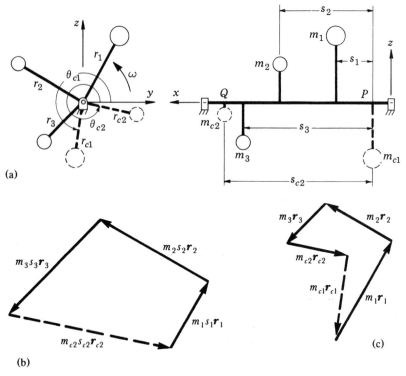

(a)

(b)

(c)

Figure 10.15 (a) In general, dynamic balancing requires the use of two counterweights. Here, counterbalances are placed in arbitrarily selected planes at axial positions P and Q. (b) Graphical determination of counterweight 2. (c) Graphical determination of counterweight 1.

distances of all other masses relative to point P are designated by symbol s in Figure 10.15a.

Recalling the definition for the moment of a force, Eq. 9.3, we find that the sum of the inertia force moments about point P is

$$\left(s_1 i \times m_1 \omega^2 r_1\right) + \left(s_2 i \times m_2 \omega^2 r_2\right) + \left(s_3 i \times m_3 \omega^2 r_3\right)$$

$$+ \left(s_{c2} i \times m_{c2} \omega^2 r_{c2}\right) = 0$$

Factoring terms, this equation can be rearranged as follows:

$$i \times \left(m_1 s_1 \omega^2 r_1 + m_2 s_2 \omega^2 r_2 + m_3 s_3 \omega^2 r_3 + m_{c2} s_{c2} \omega^2 r_{c2}\right) = 0 \qquad (10.48)$$

where vectors r_1, r_2, r_3, and r_{c2} do not have i components. The only way Eq. 10.48 can be satisfied is if the second factor in the cross product is zero. Factoring out ω^2, which appears in each term, leads to the following condition for dynamic balance:

$$m_1 s_1 r_1 + m_2 s_2 r_2 + m_3 s_3 r_3 + m_{c2} s_{c2} r_{c2} = 0 \qquad \textbf{(10.49a)}$$

which, when extended to the general case, yields

$$\sum_{n=1}^{N} m_n s_n r_n + m_{c2} s_{c2} r_{c2} = 0 \qquad \textbf{(10.49b)}$$

Thus, Eq. 10.47a or 10.47b and Eq. 10.49a or 10.49b are the requirements for complete rotor balancing and must be solved simultaneously for the necessary counterbalances. These are two-dimensional vector equations in two unknown vectors, $m_{c1} r_{c1}$ and $m_{c2} r_{c2}$.

The graphical solution is shown in Figure 10.15b and c for the case of three initial masses. Eq. 10.49a is solved first (see Figure 10.15b) by drawing the first three vectors, which are completely known, to an appropriate scale. The vector that closes this polygon is $m_{c2} s_{c2} r_{c2}$. The direction of this vector specifies the required angular orientation of the counterweight in the transverse plane at point Q, and the magnitude of the vector divided by the known distance s_{c2} is the required correction $m_{c2} r_{c2}$. The counterweight in the plane at point P can now be determined from Eq. 10.47a, since vector $m_{c2} r_{c2}$ is now known. See Figure 10.15c. The vector that closes this polygon is $m_{c1} r_{c1}$, which identifies the direction and magnitude of this counterbalance.

The mathematical solution parallels the graphical approach. First, Eq. 10.49a is divided into component form:

$$m_1 s_1 r_1 \cos \theta_1 + m_2 s_2 r_2 \cos \theta_2 + m_3 s_3 r_3 \cos \theta_3 + m_{c2} s_{c2} r_{c2} \cos \theta_{c2} = 0$$

$$m_1 s_1 r_1 \sin \theta_1 + m_2 s_2 r_2 \sin \theta_2 + m_3 s_3 r_3 \sin \theta_3 + m_{c2} s_{c2} r_{c2} \sin \theta_{c2} = 0$$

Solving for $m_{c2} r_{c2}$ and θ_{c2}, we have

$$m_{c2} r_{c2} = \frac{1}{s_{c2}} \Big[\big(m_1 s_1 r_1 \cos \theta_1 + m_2 s_2 r_2 \cos \theta_2 + m_3 s_3 r_3 \cos \theta_3 \big)^2$$

$$+ \big(m_1 s_1 r_1 \sin \theta_1 + m_2 s_2 r_2 \sin \theta_2 + m_3 s_3 r_3 \sin \theta_3 \big)^2 \Big]^{1/2}$$

$$\textbf{(10.50)}$$

$$\theta_{c2} = \arctan \left(\frac{-m_1 s_1 r_1 \sin \theta_1 - m_2 s_2 r_2 \sin \theta_2 - m_3 s_3 r_3 \sin \theta_3}{-m_1 s_1 r_1 \cos \theta_1 - m_2 s_2 r_2 \cos \theta_2 - m_3 s_3 r_3 \cos \theta_3} \right)$$

$$\textbf{(10.51)}$$

Next, Eq. 10.47a is solved for $m_{c1}r_{c1}$ and θ_{c1}, where computed values for $m_{c2}r_{c2}$ and θ_{c2} from the above expressions are substituted:

$$
m_{c1}r_{c1} = \left[\left(m_1 r_1 \cos \theta_1 + m_2 r_2 \cos \theta_2 + m_3 r_3 \cos \theta_3 + m_{c2}r_{c2} \cos \theta_{c2} \right)^2 \right.
$$
$$
\left. + \left(m_1 r_1 \sin \theta_1 + m_2 r_2 \sin \theta_2 + m_3 r_3 \sin \theta_3 + m_{c2}r_{c2} \sin \theta_{c2} \right)^2 \right]^{1/2}
$$

$$\tag{10.52}$$

$$
\theta_{c1} = \arctan\left(\frac{-m_1 r_1 \sin \theta_1 - m_2 r_2 \sin \theta_2 - m_3 r_3 \sin \theta_3 - m_{c2}r_{c2} \sin \theta_{c2}}{-m_1 r_1 \cos \theta_1 - m_2 r_2 \cos \theta_2 - m_3 r_3 \cos \theta_3 - m_{c2}r_{c2} \cos \theta_{c2}} \right)
$$

$$\tag{10.53}$$

The signs of the numerator and denominator in the arctan functions of Eqs. 10.51 and 1053 identify the correct quadrants of angles θ_{c2} and θ_{c1}.

◆ **EXAMPLE PROBLEM 10.7**   *Dynamic Balancing of a Rotor*
The rotor of Figure 10.15a has the following properties:

$$m_1 = 3 \text{ kg} \quad r_1 = 80 \text{ mm} \quad \theta_1 = 60°$$

$$m_2 = 2 \text{ kg} \quad r_2 = 80 \text{ mm} \quad \theta_2 = 150°$$

$$m_3 = 2 \text{ kg} \quad r_3 = 60 \text{ mm} \quad \theta_3 = 225°$$

The total axial length is 1000 mm between bearings. Counterweights are to be placed in planes that are 100 mm from each bearing. The axial distances in Figure 10.15a are then

$$s_1 = 200 \text{ mm} \quad s_2 = 500 \text{ mm} \quad s_3 = 700 \text{ mm} \quad s_{c2} = 800 \text{ mm}$$

Determine the counterweight amounts and locations in planes P and Q required for complete balance.

Solution.   From Eq. 10.50,

$$
m_{c2}r_{c2} = \frac{1}{800}\left[(104{,}679)^2 + (22{,}172)^2 \right]^{1/2} = 133.8 \text{ kg} \cdot \text{mm}
$$

One combination that will produce this product is $m_{c2} = 2.23$ kg and $r_{c2} = 60$ mm. The angle is computed from Eq. 10.51:

$$
\theta_{c2} = \arctan\left(\frac{-22{,}172}{104{,}679} \right) = 348.0°
$$

Now, $m_{c1}r_{c1}$ and θ_{c1} can be computed from Eqs. 10.52 and 10.53:

$$
m_{c1}r_{c1} = \left[(27.4)^2 + (175.3)^2 \right]^{1/2} = 177.4 \text{ kg} \cdot \text{mm}
$$

(e.g., 2.96 kg at 60 mm) and

$$\theta_{c1} = \arctan\left(\frac{-175.3}{-27.4}\right) = 261.1°$$

The negative numerator and denominator indicate that this angle is in the third quadrant. The graphical solution of this example is shown to scale in Figure 10.15b and c.   ◆

A couple of observations follow from the previous discussion. First, the question arises as to whether a rotor can be completely balanced by means of a single counterbalance. We saw earlier that a single counterweight will suffice for the case of static unbalance alone. In general, however, a rotor cannot be completely balanced by one counterweight, because there are four scalar conditions, Eqs. 10.47b and 10.49b, to be satisfied but only three design parameters, these being the axial location of the counterweight, the angular orientation θ_c, and the correction magnitude $m_c r_c$. One special case where a single counterbalance will work is the situation where all the initial masses lie in a single plane containing the shaft axis (i.e., all the angular orientations θ are either equal or differ by 180°) and an initial static unbalance exists. In this case, Eqs. 10.47b and 10.49b will reduce to two scalar equations for the unknown counterweight.

We saw that, because two counterweights represent a total of six design parameters, two values can be selected arbitrarily, those having been the axial locations P and Q in Figure 10.15a. Two different parameters could have been selected. However, an advantage of choosing the axial locations is that the counterweights can be placed near bearing supports in order to minimize the bending moments and resulting shaft deflection that they will produce. This leads to a second observation. Eqs. 10.44a, 10.44b, 10.47a, 10.47b, 10.49a, and 10.49b are independent of shaft speed ω. This means that the rotor will be balanced at any speed for which the initial assumptions, particularly that dealing with rotor rigidity, are valid. For a range of speeds, depending on the rotor material and size, deflections will be negligible and rigid-rotor balancing is satisfactory. However, as speeds are increased, eventually shaft flexibility will become significant. There is a wide range of such applications where balancing techniques similar to those described above, but which also account for elastic deflection, must be used. Since deflections will vary with speed, the balancing arrangement will then be speed-dependent and therefore must be designed for the specific operating speed.

10.8 BALANCING OF RECIPROCATING MACHINES

Another common type of machine unbalance is reciprocating unbalance, which is caused by the inertia forces associated with translating mass. Its effects are very evident in machines such as piston engines and compressors. Balancing of reciprocating machines is more difficult than balancing of rotors, and in many cases complete balancing cannot be achieved by practical means.

Figure 10.16a shows a slider-crank mechanism with crank length r, connecting rod length ℓ, and reciprocating mass m. Recall that mass m consists of the piston mass plus part of the connecting rod mass based on a lumped mass approximation (see Section 10.5.1). It will be assumed throughout the following material that the angular velocity ω of the crank is constant. It will also be assumed that the rotating unbalance associated with this mechanism is balanced by a counterweight mounted on the crank. This is commonly done and reduces the inertia forces produced by the crank and the rotating part of the connecting rod mass to such an extent that they may be neglected.

Figure 10.16b shows a free-body diagram of the frame. Based on the above assumptions and the analysis of Section 10.5.3, the forces transmitted to the mechanism supports are obtained from Eqs. 10.34, 10.35, and 10.36, with $m_B = 0$, $m_C + m_3 = m$, and $\omega_1 = \omega$:

$$F_{10x} = mr\omega^2\left[\cos \phi + \left(\frac{r}{\ell}\right)\cos 2\phi\right] \tag{10.54}$$

$$F_{10y} = \frac{-mr^2\omega^2 \sin \phi\left[\cos \phi + \left(\frac{r}{\ell}\right)\cos 2\phi\right]}{\left(\ell^2 - r^2 \sin^2 \phi\right)^{1/2}} \tag{10.55}$$

$$F_{30y} = \frac{mr^2\omega^2 \sin \phi\left[\cos \phi + \left(\frac{r}{\ell}\right)\cos 2\phi\right]}{\left(\ell^2 - r^2 \sin^2 \phi\right)^{1/2}} \tag{10.56}$$

The net effects of these forces are a shaking force F_{10x} and a shaking couple consisting of equal and opposite forces F_{10y} and F_{30y}. The shaking force produces translational vibration of the frame, whereas the shaking couple results in rotational vibration about an axis parallel to the crankshaft.

The largest, and probably the most critical, of the three forces defined above is force F_{10x}. We will therefore restrict our attention to balancing of this shaking force, which, upon substituting $\phi = \omega t$, is rewritten as follows:

$$F_s = mr\omega^2 \cos \omega t + mr\omega^2\left(\frac{r}{\ell}\right)\cos 2\omega t \tag{10.57}$$

This force has variable magnitude and sense, but its line of action is always along the cylinder centerline (i.e., line O_1C in Figure 10.16a); thus, the translational vibration induced will be in this direction. The first term in Eq. 10.57, which is the larger of the two terms, is called the primary part of the shaking force and has a frequency ω equal to the rotational frequency of the crank. The second term is referred to as the secondary part of the shaking force and has a frequency 2ω, which is twice that of the crank. In the following sections, we will explore ways of counteracting this shaking force.

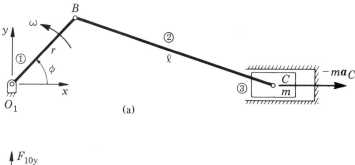

(a)

(b)

Figure 10.16 (a) Slider-crank mechanism. The inertia force of a reciprocating mass creates a shaking force and a shaking couple on the machine frame. (b) Free-body diagram of the machine frame.

10.8.1 Single-Cylinder Machines

One approach used to partially force balance single-cylinder engines and compressors is the addition of a rotating counterbalance to the crank. This counterweight is in addition to that described in the preceding section, which is used to counteract the rotating unbalance due to the crank mass and the rotating part of the connecting rod mass. Figure 10.17a shows the mechanism of Figure 10.16a with a counterweight of mass m_c mounted on the crank at a radial distance r_c from main bearing O_1 and at an angular position equal to $\phi + 180°$. This mass will create a constant-magnitude centrifugal force at O_1 that rotates with speed ω. The total shaking force will then be the vector sum of this force and the force of Eq. 10.57, as shown in Figure 10.17b. In terms of x and y unit vectors,

$$F_s = \left[mr\omega^2 \cos \phi + mr\omega^2 \left(\frac{r}{\ell}\right) \cos 2\phi - m_c r_c \omega^2 \cos \phi \right] i$$

$$- m_c r_c \omega^2 \sin \phi \, j \tag{10.58}$$

Clearly, this counterweight cannot eliminate the shaking force entirely, because it introduces a nonzero y component, and the x component, though reduced, will not be identically equal to zero. However, by properly sizing the correction $m_c r_c$, the maximum magnitude of the shaking force can be reduced considerably.

Correction amounts typically used range from $m_c r_c = mr/2$ to $m_c r_c = 2mr/3$. For example, consider the case of $m_c r_c = 0.6mr$ for a mechanism with a ratio of crank length to connecting rod length given by $r/\ell = 0.25$. The

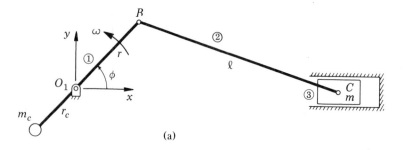

(a)

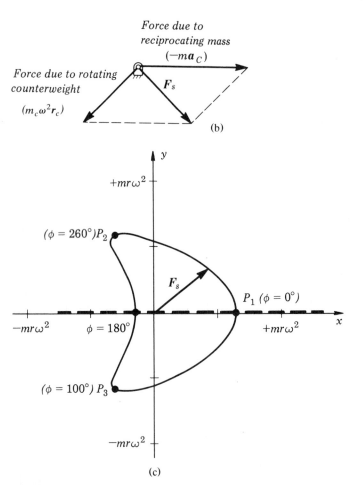

Force due to
reciprocating mass
$(-m\mathbf{a}_C)$

Force due to rotating
counterweight

$(m_c\omega^2 r_c)$

F_s

(b)

$+mr\omega^2$

$(\phi = 260°)P_2$

F_s

$P_1 (\phi = 0°)$

$-mr\omega^2$ $\phi = 180°$ $+mr\omega^2$

$(\phi = 100°) P_3$

$-mr\omega^2$

(c)

Figure 10.17 (a) Partial shaking-force balancing of a slider-crank mechanism by means of a rotating counterweight attached to the crank. (b) The net shaking force F_s. (c) Polar plot of the shaking force.

following expression for the shaking force results:

$$F_s = (0.4mr\omega^2 \cos \phi + 0.25mr\omega^2 \cos 2\phi)i - 0.6mr\omega^2 \sin \phi j$$

The magnitude of this force as a function of crank angle ϕ is

$$|F_s| = mr\omega^2\sqrt{(0.4 \cos \phi + 0.25 \cos 2\phi)^2 + (0.6 \sin \phi)^2}$$

Figure 10.17c shows a polar plot of this force, where each point on the curve defines the magnitude and direction of the shaking force for a corresponding value of ϕ. The maximum shaking-force magnitude is $|F_s|_{max} = 0.66mr\omega^2$ and occurs when ϕ equals 100 and 260°. Superimposed on Figure 10.17c is the initial shaking-force variation (dashed line) without the counterweight. The maximum shaking force is $|F_s|_{max} = 1.25mr\omega^2$ at $\phi = 0$. Thus a 47 percent reduction in magnitude has been achieved through the addition of a rotating counterweight. The optimum counterweight size would be that which produces equal shaking force magnitudes at points P_1, P_2, and P_3 on the polar force plot of Figure 10.17c. Examination of the figure shows that the correction used in this example is close to optimum and, therefore, little improvement beyond the 47 percent reduction could be obtained.

10.8.2 Multicylinder Machines

Many applications of the slider-crank mechanism in engines, pumps, and compressors involve the use of multiple mechanisms, which are synchronized to provide smoother flow of fluid or transmission of power than can be accomplished in a single-cylinder device. These multicylinder systems facilitate one of the more effective means of reducing the consequences of shaking forces. By a proper arrangement of the individual mechanisms, the shaking forces will partially, and perhaps totally, cancel one another. In the following paragraphs, we will first develop general shaking-force-balancing relationships for multicylinder machines and then examine some specific configurations.

Figure 10.18 depicts a general arrangement where the total number of cylinders is N (only three cylinders are shown in the figure). It is assumed that all the slider-crank mechanisms have the same crank length r, connecting rod length ℓ, and reciprocating mass m, and that the crank angular velocity ω is constant. The cylinder orientations are defined by angles θ_n, $n = 1, 2, \ldots, N$, which are fixed angular positions with respect to the y axis. The angular crank throw spacings with respect to crank 1 are represented by angles ψ_n, $n = 2, 3, \ldots, N$, which do not vary with time (i.e., each crank is rigidly attached to the same crankshaft).

Each slider-crank mechanism will generate a shaking force with a line of action along that particular cylinder's centerline (i.e., at angle θ_n with respect to the y axis). From Eq. 10.57, the expression for the individual shaking forces is

$$F_{sn} = mr\omega^2 \cos \phi_n + mr\omega^2 \left(\frac{r}{\ell}\right)\cos 2\phi_n \quad n = 1, 2, \ldots, N$$

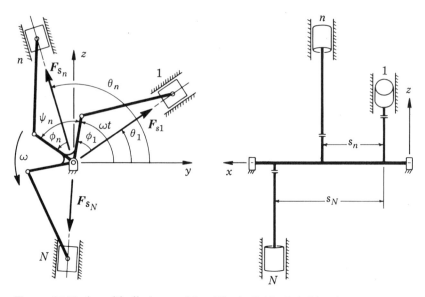

Figure 10.18 A multicylinder machine. The individual shaking forces combine vectorially to produce the net shaking force. Because of the axial distribution of the cylinders, the individual forces also produce a shaking moment.

Substituting the angle relationships from Figure 10.18, we can rewrite this equation as

$$F_{sn} = mr\omega^2 \cos(\omega t + \psi_n - \theta_n) + mr\omega^2 \left(\frac{r}{\ell}\right) \cos[2(\omega t + \psi_n - \theta_n)]$$

$$n = 1, 2, \ldots, N \quad \textbf{(10.59)}$$

where $\psi_1 = 0$ from the previous definition of angle ψ_n. The resultant shaking force will be the vector sum of all of the individual shaking forces:

$$\mathbf{F}_s = \sum_{n=1}^{N} (F_{sn} \cos \theta_n \mathbf{j} + F_{sn} \sin \theta_n \mathbf{k}) \qquad \textbf{(10.60)}$$

In order for the arrangement to be completely force-balanced, the y and z components of Eq. 10.60 must be identically zero; that is

$$\sum_{n=1}^{N} F_{sn} \cos \theta_n = 0 \quad \text{for all } t \qquad \textbf{(10.61a)}$$

and

$$\sum_{n=1}^{N} F_{sn} \sin \theta_n = 0 \quad \text{for all } t \qquad \textbf{(10.61b)}$$

Substituting Eq. 10.59, we see that the conditions become

$$mr\omega^2 \sum_{n=1}^{N} \left[\cos(\omega t + \psi_n - \theta_n)\cos\theta_n \right.$$

$$\left. + \left(\frac{r}{\ell}\right)\cos(2\omega t + 2\psi_n - 2\theta_n)\cos\theta_n \right] = 0$$

and

$$mr\omega^2 \sum_{n=1}^{N} \left[\cos(\omega t + \psi_n - \theta_n)\sin\theta_n \right.$$

$$\left. + \left(\frac{r}{\ell}\right)\cos(2\omega t + 2\psi_n - 2\theta_n)\sin\theta_n \right] = 0$$

Canceling $mr\omega^2$, which is nonzero, and factoring further, we have

$$\cos\omega t \sum_{n=1}^{N} \cos(\psi_n - \theta_n)\cos\theta_n - \sin\omega t \sum_{n=1}^{N} \sin(\psi_n - \theta_n)\cos\theta_n$$

$$+ \left(\frac{r}{\ell}\right)\cos 2\omega t \sum_{n=1}^{N} \cos(2\psi_n - 2\theta_n)\cos\theta_n$$

$$- \left(\frac{r}{\ell}\right)\sin 2\omega t \sum_{n=1}^{N} \sin(2\psi_n - 2\theta_n)\cos\theta_n = 0 \tag{10.62a}$$

and

$$\cos\omega t \sum_{n=1}^{N} \cos(\psi_n - \theta_n)\sin\theta_n - \sin\omega t \sum_{n=1}^{N} \sin(\psi_n - \theta_n)\sin\theta_n$$

$$+ \left(\frac{r}{\ell}\right)\cos 2\omega t \sum_{n=1}^{N} \cos(2\psi_n - 2\theta_n)\sin\theta_n$$

$$- \left(\frac{r}{\ell}\right)\sin 2\omega t \sum_{n=1}^{N} \sin(2\psi_n - 2\theta_n)\sin\theta_n = 0 \tag{10.62b}$$

The only way that these expressions can be identically zero is if the individual coefficients of the time-dependent sine and cosine functions are all zero. This yields the following eight necessary conditions for complete shaking force balance:

$$\sum_{n=1}^{N} \cos(\psi_n - \theta_n)\cos\theta_n = 0 \tag{10.63a}$$

$$\sum_{n=1}^{N} \sin(\psi_n - \theta_n)\cos\theta_n = 0 \tag{10.63b}$$

$$\sum_{n=1}^{N} \cos(\psi_n - \theta_n)\sin\theta_n = 0 \tag{10.63c}$$

$$\sum_{n=1}^{N} \sin(\psi_n - \theta_n)\sin\theta_n = 0 \tag{10.63d}$$

$$\sum_{n=1}^{N} \cos(2\psi_n - 2\theta_n)\cos\theta_n = 0 \tag{10.63e}$$

$$\sum_{n=1}^{N} \sin(2\psi_n - 2\theta_n)\cos\theta_n = 0 \tag{10.63f}$$

$$\sum_{n=1}^{N} \cos(2\psi_n - 2\theta_n)\sin\theta_n = 0 \tag{10.63g}$$

$$\sum_{n=1}^{N} \sin(2\psi_n - 2\theta_n)\sin\theta_n = 0 \tag{10.63h}$$

The first four of these conditions account for the primary parts of the shaking forces, and if these are all satisfied, then the primary shaking forces are balanced. The last four conditions represent the secondary parts, and if satisfied, then the secondary shaking forces are balanced. Note that the eight conditions are in terms of the cylinder orientations θ_n and the angular crank spacing ψ_n, and it follows that some arrangements of these parameters may produce force balancing while other arrangements will not. Further, some arrangements may result in only primary force balancing or only secondary force balancing. Of these two possibilities, primary balancing is preferred because this represents cancellation of the larger parts of the shaking forces.

In most multicylinder machines, the slider-crank mechanisms must be spaced axially along the crankshaft in order to avoid interference during operation. This axial spacing is represented in Figure 10.18 by distances s_n, $n = 1, 2, \ldots, N$, measured from that cylinder designated as number 1 (therefore, $s_1 = 0$). Since the individual shaking forces will not, in general, lie in a single transverse plane, they will produce a net shaking moment, as well as a net shaking force, which will tend to cause an end-over-end rotational vibration of the crankshaft.

A set of conditions for shaking moment balance can be established by imposing the requirement that the sum of shaking force moments about any arbitrary axial location must be zero. Taking moments about the axial location of cylinder 1,

$$\sum_{n=1}^{N} s_n \mathbf{i} \times (F_{sn} \cos \theta_n \mathbf{j} + F_{sn} \sin \theta_n \mathbf{k}) = 0 \qquad (10.64)$$

or, upon factoring,

$$\mathbf{i} \times \sum_{n=1}^{N} (s_n F_{sn} \cos \theta_n \mathbf{j} + s_n F_{sn} \sin \theta_n \mathbf{k}) = 0$$

In order for this equation to be satisfied, the individual $\mathbf{j}$ and $\mathbf{k}$ components of the second factor in the cross product must be identically zero; that is,

$$\sum_{n=1}^{N} s_n F_{sn} \cos \theta_n = 0 \quad \text{for all } t \qquad (10.65a)$$

and

$$\sum_{n=1}^{N} s_n F_{sn} \sin \theta_n = 0 \quad \text{for all } t \qquad (10.65b)$$

These equations are similar to Eqs. 10.61a and 10.61b and lead to the following similar set of conditions for shaking moment balance:

$$\sum_{n=1}^{N} s_n \cos(\psi_n - \theta_n)\cos \theta_n = 0 \qquad (10.66a)$$

$$\sum_{n=1}^{N} s_n \sin(\psi_n - \theta_n)\cos \theta_n = 0 \qquad (10.66b)$$

$$\sum_{n=1}^{N} s_n \cos(\psi_n - \theta_n)\sin \theta_n = 0 \qquad (10.66c)$$

$$\sum_{n=1}^{N} s_n \sin(\psi_n - \theta_n)\sin \theta_n = 0 \qquad (10.66d)$$

$$\sum_{n=1}^{N} s_n \cos(2\psi_n - 2\theta_n)\cos \theta_n = 0 \qquad (10.66e)$$

$$\sum_{n=1}^{N} s_n \sin(2\psi_n - 2\theta_n)\cos \theta_n = 0 \qquad (10.66f)$$

$$\sum_{n=1}^{N} s_n \cos(2\psi_n - 2\theta_n)\sin \theta_n = 0 \qquad (10.66g)$$

$$\sum_{n=1}^{N} s_n \sin(2\psi_n - 2\theta_n)\sin \theta_n = 0 \qquad (10.66h)$$

The first four conditions guarantee primary shaking moment balance, while the last four conditions yield secondary shaking moment balance. These equations account for the axial configuration of the cylinders as well as for their angular orientation and the angular crank spacing. Eqs. 10.63a to 10.63h and 10.66a to 10.66h can be used to investigate the balancing of any piston engine or compressor.

In-Line Engines

Consider an engine for which all the cylinders lie in a single plane and on one side of the crank axis. Suppose that these locations are given by $\theta_1 = \theta_2 = \cdots = \theta_n = \cdots = \theta_N = \pi/2$. Further, assume that the cylinders are equally spaced axially with a spacing s; then, $s_n = (n-1)s$, where the cylinders are numbered consecutively from one end of the crankshaft to the other end. Substituting this information, we see that Eqs. 10.63a to 10.63h and 10.66a to 10.66h reduce to the following conditions:

$$\sum_{n=1}^{N} \sin \psi_n = 0 \qquad \sum_{n=1}^{N} (n-1)\sin \psi_n = 0$$

$$\sum_{n=1}^{N} \cos \psi_n = 0 \qquad \sum_{n=1}^{N} (n-1)\cos \psi_n = 0$$

$$\sum_{n=1}^{N} \cos 2\psi_n = 0 \qquad \sum_{n=1}^{N} (n-1)\cos 2\psi_n = 0$$

$$\sum_{n=1}^{N} \sin 2\psi_n = 0 \qquad \sum_{n=1}^{N} (n-1)\sin 2\psi_n = 0$$

Figure 10.19 shows a two-cylinder, in-line arrangement with 180° cranks; that is, $N = 2$, $\psi_1 = 0$, and $\psi_2 = \pi$. Substituting into the above equations,

$$\left.\begin{array}{l} \displaystyle\sum_{n=1}^{2} \sin \psi_n = \sin(0) + \sin \pi = 0 \\[1.5em] \displaystyle\sum_{n=1}^{2} \cos \psi_n = \cos(0) + \cos \pi = 0 \end{array}\right\} \text{primary force}$$

$$\left.\begin{array}{l} \displaystyle\sum_{n=1}^{2} \cos 2\psi_n = \cos(0) + \cos 2\pi = 2 \\[1.5em] \displaystyle\sum_{n=1}^{2} \sin 2\psi_n = \sin(0) + \sin 2\pi = 0 \end{array}\right\} \text{secondary force}$$

$$\left.\begin{array}{l} \displaystyle\sum_{n=1}^{2} (n-1)\sin \psi_n = (0)\sin(0) + (1)\sin \pi = 0 \\[1.5em] \displaystyle\sum_{n=1}^{2} (n-1)\cos \psi_n = (0)\cos(0) + (1)\cos \pi = -1 \end{array}\right\} \text{primary moment}$$

$$\left.\begin{array}{l} \sum_{n=1}^{2} (n-1)\cos 2\psi_n = (0)\cos(0) + (1)\cos 2\pi = 1 \\[2mm] \sum_{n=1}^{2} (n-1)\sin 2\psi_n = (0)\sin(0) + (1)\sin 2\pi = 0 \end{array}\right\} \text{secondary moment}$$

Thus, the primary parts of the shaking forces are always equal and opposite and, therefore, cancel, but because they are offset axially, they form a nonzero couple. This is shown in Figure 10.19. On the other hand, the secondary parts of the shaking forces are always equal with the same sense, and they therefore combine to produce a net force and also cause a net moment. From Eq. 10.60, the net shaking force is

$$\begin{aligned} F_s &= \sum_{n=1}^{2} F_{sn}\mathbf{k} \\ &= \left\{ mr\omega^2 \cos\left(\omega t - \frac{\pi}{2}\right) + mr\omega^2\left(\frac{r}{\ell}\right)\cos\left[2\left(\omega t - \frac{\pi}{2}\right)\right] \right.\\ &\quad \left. + mr\omega^2 \cos\left(\omega t + \frac{\pi}{2}\right) + mr\omega^2\left(\frac{r}{\ell}\right)\cos\left[2\left(\omega t + \frac{\pi}{2}\right)\right] \right\}\mathbf{k} \\ &= -2mr\omega^2\left(\frac{r}{\ell}\right)\cos 2\omega t\,\mathbf{k} \end{aligned}$$

with a maximum magnitude of $2mr\omega^2(r/\ell)$. Although this shaking force is nonzero, it nevertheless represents a significant improvement compared to a

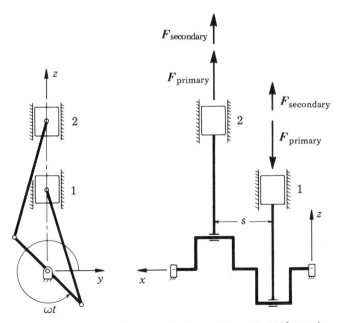

Figure 10.19 An in-line two-cylinder engine with 180° cranks.

single-cylinder engine when considering typical (r/ℓ) ratios. However, as noted, a shaking couple has been introduced.

Opposed Engines

In an opposed engine, all the cylinders lie in the same plane with half on each side of the crank axis. Selecting $\theta_1 = \cdots = \theta_{N/2} = \pi/2$ and $\theta_{N/2+1} = \cdots = \theta_N = 3\pi/2$, half of Eqs. 10.63a to 10.63h and 10.66a to 10.66h are automati-

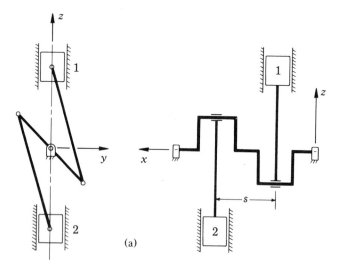

(a)

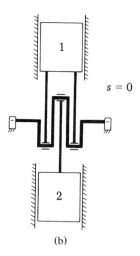

(b)

Figure 10.20 (a) An opposed two-cylinder engine with 180° cranks. (b) Double connecting rods for cylinder 1.

cally satisfied; these are Eqs. 10.63a, 10.63b, 10.63e, 10.63f, 10.66a, 10.66b, 10.66e, and 10.66f. This is due to the fact that there will be no y direction forces or z direction moments in the general force and moment equations. As an example, consider the two-cylinder opposed engine of Figure 10.20a with 180° cranks, where $N = 2$, $\theta_1 = \pi/2$, $\theta_2 = 3\pi/2$, $\psi_1 = 0$, $\psi_2 = \pi$, $s_1 = 0$, and $s_2 = s$. Substituting into Eqs. 10.63c, 10.63d, 10.63g, 10.63h, 10.66c, 10.66d, 10.66g, and 10.66h, we have

$$\left.\begin{aligned}
&\sum_{n=1}^{2} \cos(\psi_n - \theta_n)\sin\theta_n = \cos\left(-\frac{\pi}{2}\right)\sin\frac{\pi}{2} \\
&\quad + \cos\left(-\frac{\pi}{2}\right)\sin\frac{3\pi}{2} = 0 \\
&\sum_{n=1}^{2} \sin(\psi_n - \theta_n)\sin\theta_n = \sin\left(-\frac{\pi}{2}\right)\sin\frac{\pi}{2} \\
&\quad + \sin\left(-\frac{\pi}{2}\right)\sin\frac{3\pi}{2} = 0
\end{aligned}\right\} \begin{array}{l}\text{primary} \\ \text{force}\end{array}$$

$$\left.\begin{aligned}
&\sum_{n=1}^{2} \cos(2\psi_n - 2\theta_n)\sin\theta_n = \cos(-\pi)\sin\frac{\pi}{2} \\
&\quad + \cos(-\pi)\sin\frac{3\pi}{2} = 0 \\
&\sum_{n=1}^{2} \sin(2\psi_n - 2\theta_n)\sin\theta_n = \sin(-\pi)\sin\frac{\pi}{2} \\
&\quad + \sin(-\pi)\sin\frac{3\pi}{2} = 0
\end{aligned}\right\} \begin{array}{l}\text{secondary} \\ \text{force}\end{array}$$

$$\left.\begin{aligned}
&\sum_{n=1}^{2} s_n \cos(\psi_n - \theta_n)\sin\theta_n = (0) + s\cos\left(-\frac{\pi}{2}\right)\sin\frac{3\pi}{2} = 0 \\
&\sum_{n=1}^{2} s_n \sin(\psi_n - \theta_n)\sin\theta_n = (0) + s\sin\left(-\frac{\pi}{2}\right)\sin\frac{3\pi}{2} = s
\end{aligned}\right\} \begin{array}{l}\text{primary} \\ \text{moment}\end{array}$$

$$\left.\begin{aligned}
&\sum_{n=1}^{2} s_n \cos(2\psi_n - 2\theta_n)\sin\theta_n = (0) + s\cos(-\pi)\sin\frac{3\pi}{2} = s \\
&\sum_{n=1}^{2} s_n \sin(2\psi_n - 2\theta_n)\sin\theta_n = (0) + s\sin(-\pi)\sin\frac{3\pi}{2} = 0
\end{aligned}\right\} \begin{array}{l}\text{secondary} \\ \text{moment}\end{array}$$

The net shaking force is zero because both parts of the individual shaking forces cancel. This is an improvement over the two-cylinder, in-line engine of Figure 10.19. But there will be a significant shaking couple (both primary and secondary) due to the staggering of the crank throws. Clearly, the smaller the spacing s is, the better will be the design from the point of view of balancing. One method of reducing s to zero and thereby eliminating the shaking couple is to use double connecting rods for one of the cylinders, as shown in Figure 10.20b.

V Engines

Due to its compact form, this engine type is very common in automotive and other applications. Consider, for example, the V–8 engine of Figure 10.21a consisting of two banks of four cylinders with an angle of 90° between banks. The four-throw crankshaft has 90° cranks, with an axial spacing s between

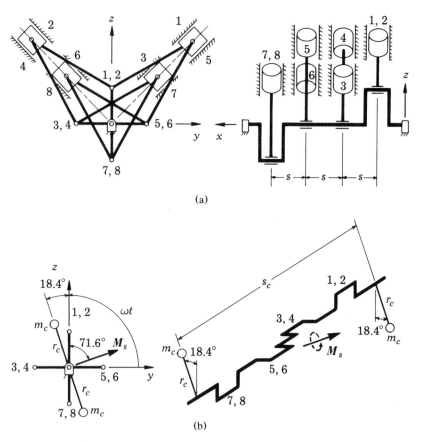

Figure 10.21 (a) A V-8 engine with 90° cranks. This arrangement can be completely balanced with the addition of rotating counterweights on the crankshaft. (b) Location of the counterweights.

cranks. The following quantities are determined from the figure:

$$\theta_1 = \theta_3 = \theta_5 = \theta_7 = \frac{\pi}{4}$$

$$\theta_2 = \theta_4 = \theta_6 = \theta_8 = \frac{3\pi}{4}$$

$$\psi_1 = \psi_2 = 0 \quad \psi_5 = \psi_6 = \frac{3\pi}{2}$$

$$\psi_3 = \psi_4 = \frac{\pi}{2} \quad \psi_7 = \psi_8 = \pi$$

$$s_1 = s_2 = 0 \qquad s_5 = s_6 = 2s$$

$$s_3 = s_4 = s \qquad s_7 = s_8 = 3s$$

The force-balance conditions as evaluated from Eqs. 10.63a to 10.63h are

$$\sum_{n=1}^{8} \cos(\psi_n - \theta_n)\cos\theta_n = \frac{1}{2} + \frac{1}{2} + \frac{1}{2} - \frac{1}{2} - \frac{1}{2} + \frac{1}{2} - \frac{1}{2} - \frac{1}{2} = 0$$

$$\sum_{n=1}^{8} \sin(\psi_n - \theta_n)\cos\theta_n = -\frac{1}{2} + \frac{1}{2} + \frac{1}{2} + \frac{1}{2} - \frac{1}{2} - \frac{1}{2} + \frac{1}{2} - \frac{1}{2} = 0$$

$$\sum_{n=1}^{8} \cos(\psi_n - \theta_n)\sin\theta_n = \frac{1}{2} - \frac{1}{2} + \frac{1}{2} + \frac{1}{2} - \frac{1}{2} - \frac{1}{2} - \frac{1}{2} + \frac{1}{2} = 0$$

$$\sum_{n=1}^{8} \sin(\psi_n - \theta_n)\sin\theta_n = -\frac{1}{2} - \frac{1}{2} + \frac{1}{2} - \frac{1}{2} - \frac{1}{2} + \frac{1}{2} + \frac{1}{2} + \frac{1}{2} = 0$$

$$\sum_{n=1}^{8} \cos(2\psi_n - 2\theta_n)\cos\theta_n = 0 + 0 + 0 + 0 + 0 + 0 + 0 + 0 = 0$$

$$\sum_{n=1}^{8} \sin(2\psi_n - 2\theta_n)\cos\theta_n$$

$$= -\frac{1}{\sqrt{2}} - \frac{1}{\sqrt{2}} + \frac{1}{\sqrt{2}} + \frac{1}{\sqrt{2}} + \frac{1}{\sqrt{2}} + \frac{1}{\sqrt{2}} - \frac{1}{\sqrt{2}} - \frac{1}{\sqrt{2}} = 0$$

$$\sum_{n=1}^{8} \cos(2\psi_n - 2\theta_n)\sin\theta_n = 0 + 0 + 0 + 0 + 0 + 0 + 0 + 0 = 0$$

$$\sum_{n=1}^{8} \sin(2\psi_n - 2\theta_n)\sin\theta_n$$

$$= -\frac{1}{\sqrt{2}} + \frac{1}{\sqrt{2}} + \frac{1}{\sqrt{2}} - \frac{1}{\sqrt{2}} + \frac{1}{\sqrt{2}} - \frac{1}{\sqrt{2}} - \frac{1}{\sqrt{2}} + \frac{1}{\sqrt{2}} = 0$$

Thus the engine is completely force-balanced. In fact, this configuration is force-balanced for any angle between cylinder banks, because each bank of four cylinders is force-balanced independently.

Examining the shaking moment conditions leads to the following results:

$$\sum_{n=1}^{8} s_n \cos(\psi_n - \theta_n)\cos\theta_n$$

$$= 0 + 0 + \frac{s}{2} - \frac{s}{2} - s + s - \frac{3s}{2} - \frac{3s}{2} = -3s$$

$$\sum_{n=1}^{8} s_n \sin(\psi_n - \theta_n)\cos\theta_n = 0 + 0 + \frac{s}{2} + \frac{s}{2} - s - s + \frac{3s}{2} - \frac{3s}{2} = -s$$

$$\sum_{n=1}^{8} s_n \cos(\psi_n - \theta_n)\sin\theta_n = 0 + 0 + \frac{s}{2} + \frac{s}{2} - s - s - \frac{3s}{2} + \frac{3s}{2} = -s$$

$$\sum_{n=1}^{8} s_n \sin(\psi_n - \theta_n)\sin\theta_n = 0 + 0 + \frac{s}{2} - \frac{s}{2} - s + s + \frac{3s}{2} + \frac{3s}{2} = 3s$$

$$\sum_{n=1}^{8} s_n \cos(2\psi_n - 2\theta_n)\cos\theta_n = 0 + 0 + 0 + 0 + 0 + 0 + 0 + 0 = 0$$

$$\sum_{n=1}^{8} s_n \sin(2\psi_n - 2\theta_n)\cos\theta_n$$

$$= 0 + 0 + \frac{s}{\sqrt{2}} + \frac{s}{\sqrt{2}} + \frac{2s}{\sqrt{2}} + \frac{2s}{\sqrt{2}} - \frac{3s}{\sqrt{2}} - \frac{3s}{\sqrt{2}} = 0$$

$$\sum_{n=1}^{8} s_n \cos(2\psi_n - 2\theta_n)\sin\theta_n = 0 + 0 + 0 + 0 + 0 + 0 + 0 + 0 = 0$$

$$\sum_{n=1}^{8} s_n \sin(2\psi_n - 2\theta_n)\sin\theta_n$$

$$= 0 + 0 + \frac{s}{\sqrt{2}} - \frac{s}{\sqrt{2}} + \frac{2s}{\sqrt{2}} - \frac{2s}{\sqrt{2}} - \frac{3s}{\sqrt{2}} + \frac{3s}{\sqrt{2}} = 0$$

There is a primary shaking couple but no secondary shaking couple; hence the engine arrangement, by itself, does not yield complete force and moment balance. However, the shaking couple has a special nature that facilitates total balancing by means of a relatively straightforward modification. To understand this, consider the shaking moment expression of Eq. 10.64, where M_s refers to the shaking moment:

$$M_s = \sum_{n=1}^{8} s_n i \times (F_{sn} \cos \theta_n j + F_{sn} \sin \theta_n k)$$

$$= \sum_{n=1}^{8} (-F_{sn} s_n \sin \theta_n j + F_{sn} s_n \cos \theta_n k)$$

$$= mr\omega^2 \sum_{n=1}^{8} [-s_n \cos(\omega t + \psi_n - \theta_n)\sin \theta_n j$$

$$+ s_n \cos(\omega t + \psi_n - \theta_n)\cos \theta_n k]$$

where the secondary parts of the shaking forces have been disregarded since they will cancel. Rearranging this expression and substituting the results obtained earlier, we have

$$M_s = mrs\omega^2[(3 \sin \omega t + \cos \omega t)j + (\sin \omega t - 3 \cos \omega t)k]$$

$$= mrs\omega^2\sqrt{10}\,[\cos(\omega t - 71.6°)j + \sin(\omega t - 71.6°)k]$$

The magnitude of this moment is $mrs\omega^2\sqrt{10}$, which is constant for all values of time t, and the direction of the moment is perpendicular to the crank axis and rotates with speed ω, where at any instant the angle of the moment vector with respect to the y direction is $(\omega t - 71.6°)$. This is exactly the same as a rotating, unbalanced dynamic couple discussed in Section 10.7. Thus, the net effect of this engine arrangement is what appears to be a rotating dynamic unbalance. Therefore, the shaking couple can be balanced by a set of rotating counterweights that produce an equal but opposite rotating couple. The magnitude of this couple is given by $m_c r_c s_c \omega^2 = mrs\omega^2\sqrt{10}$, and the locations are as depicted in Figure 10.21b, where m_c is the mass, r_c is the radial position, and s_c is the axial spacing of the counterweights. Because this engine can be completely balanced in this fashion, it exhibits smooth-running performance.

References

1. Beer, F. P., and E. R. Johnston, Jr., *Vector Mechanics for Engineers: Statics and Dynamics*, McGraw-Hill, New York, 1984.
2. Meriam, J. L., *Engineering Mechanics: Statics and Dynamics*, Wiley, New York, 1978.
3. Shames, I. H., *Engineering Mechanics: Statics and Dynamics*, Prentice-Hall, Englewood Cliffs, NJ, 1980.

PROBLEMS

10.1 Determine the magnitude and location of the equivalent offset inertia force for the connecting rod, link 2, of the slider-crank mechanism of Figure P10.1 for the position shown. The crank has a constant angular velocity of 100 rad/s counter-

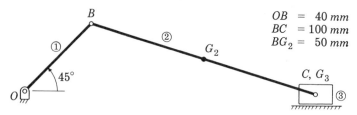

Figure P10.1

clockwise. The mass of the connecting rod is 0.2 kg and the moment of inertia about the center of mass G_2 is 300 kg · mm².

● **10.2** For the mechanism of Problem 10.1, determine the crank torque T_1 required for dynamic equilibrium at the position shown. Neglect external loads and the inertia of crank 1 and slider 3.

  10.3 For the mechanism of Problems 10.1 and 10.2, determine the crank torque T_1 required for dynamic equilibrium if slider 3 has a mass of 0.3 kg. Neglect external loads and the inertia of crank 1.

  10.4 Determine the magnitude and location of the equivalent offset inertia force for the coupler link 2 of the slider-crank mechanism at the position shown in Figure P10.2. The crank has a constant angular velocity of 60 rev/min clockwise. The weight of the coupler is 1000 lb, and the moment of inertia about the center of mass G_2 is 70 lb · s² · ft.

$OB = 3\ ft$
$BC = CD = BD = 5\ ft$

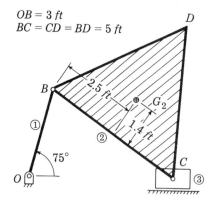

Figure P10.2

  10.5 The four-bar linkage of Figure P10.3 has a constant crank angular velocity $\omega_1 = 60$ rad/s clockwise. The coupler link 2 has a mass of 0.3 kg and a moment of inertia about the center of mass G_2 of 1000 kg · mm². The follower link 3 has a mass of 0.2 kg and a moment of inertia about its center of mass G_3 of 150 kg · mm². For the position shown, determine the following:
  (a) The equivalent offset inertia force for the coupler
  (b) The equivalent offset inertia force for the follower
  (c) The crank torque T_1 required for dynamic equilibrium

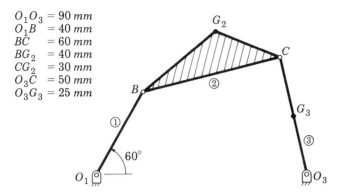

$O_1O_3 = 90\ mm$
$O_1B = 40\ mm$
$BC = 60\ mm$
$BG_2 = 40\ mm$
$CG_2 = 30\ mm$
$O_3C = 50\ mm$
$O_3G_3 = 25\ mm$

Figure P10.3

10.6 For the mechanism shown in Figure P10.4, member 2 has a weight of 2 lb and a moment of inertia of 0.04 lb · s² · in about its center of mass G_2. Sliding block 1 has a constant velocity of 10 ft/s upward.

(a) Determine the instantaneous force F_1 required to produce this motion, assuming that sliding block 3 is massless.

(b) Determine the required instantaneous force F_1 if sliding block 3 has a weight of 1 lb.

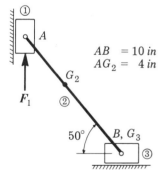

$AB = 10\ in$
$AG_2 = 4\ in$

Figure P10.4

10.7 The slider-crank mechanism of Figure P10.5 has a constant crank angular velocity of 50 rad/s counterclockwise. The acceleration polygon is shown in the figure. The connecting rod weighs 2 lb, with a mass moment of inertia about its center of mass G_2 of 0.009 lb · s² · in. The piston has a weight of 1.5 lb. Determine all bearing forces and the required input torque T_1 for the position shown.

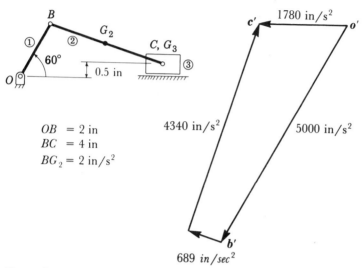

$OB = 2$ in
$BC = 4$ in
$BG_2 = 2$ in/s^2

Figure P10.5

10.8 The crank of the slider-crank mechanism of Figure P10.6 has an instantaneous angular velocity of 10 rad/s clockwise and an angular acceleration of 200 rad/s^2 clockwise. Acceleration information is given in the figure. The connecting rod has a mass of 15 kg and a mass moment of inertia of 7500 kg · mm^2 about its mass center G_2. The slider has a mass of 8 kg. The crank has a moment of inertia of 4000 kg · mm^2 about its stationary center of mass G_1. Determine all bearing forces and the input torque T_1 for the position shown.

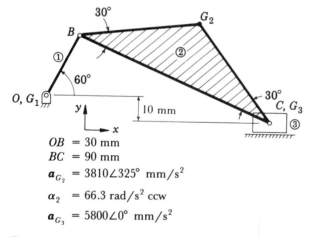

$OB = 30$ mm
$BC = 90$ mm
$a_{G_2} = 3810\angle 325°$ mm/s^2
$\alpha_2 = 66.3$ rad/s^2 ccw
$a_{G_3} = 5800\angle 0°$ mm/s^2

Figure P10.6

10.9 The four-bar linkage of Figure P10.7 has a constant input angular velocity $\omega_1 = 200$ rad/s clockwise. This results in the following accelerations: $a_{G2} = 34,500\angle 298°$ in/s^2, $\alpha_2 = 2670$ rad/s^2 ccw, $a_{G3} = 14,600\angle 294°$ in/s^2, and $\alpha_3 = 6940$ rad/s^2 cw. Coupler link 2 weighs 1.2 lb and has a mass moment of inertia about center of mass G_2 equal to 0.03 lb · s^2 · in. Follower link 3 weighs 1.0 lb and has a mass moment of inertia about its mass center G_3 equal to 0.002

lb · s² · in. Determine all bearing forces and instantaneous input torque T_1 by **(a)** graphical solution and   **(b)** analytical solution.

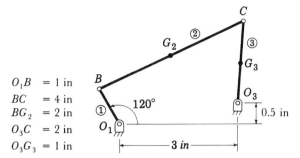

O_1B = 1 in
BC = 4 in
BG_2 = 2 in
O_3C = 2 in
O_3G_3 = 1 in

Figure P10.7

10.10 The four-bar linkage of Figure P10.8 has a constant input angular velocity $\omega_1 = 60$ rad/s counterclockwise. The acceleration polygon is shown in the figure. The masses of coupler link 2 and follower link 3 are 1.0 kg and 0.6 kg, respectively, and the moments of inertia about the centers of mass are 700 kg · mm² and 500 kg · mm², respectively. Determine all bearing forces and instantaneous torque T_1 by   **(a)** graphical solution and   **(b)** analytical solution.

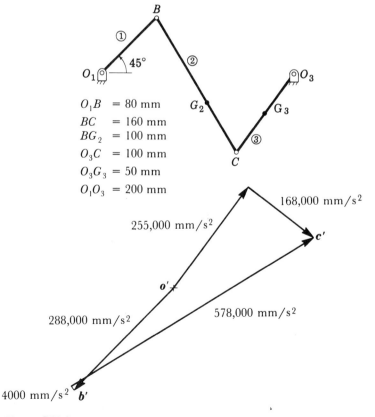

O_1B = 80 mm
BC = 160 mm
BG_2 = 100 mm
O_3C = 100 mm
O_3G_3 = 50 mm
O_1O_3 = 200 mm

Figure P10.8

10.11 For the mechanism of Problem 10.10, determine the bearing forces and input torque if the constant input angular velocity of 60 rad/s is clockwise rather than counterclockwise. How do the forces change if the input speed is doubled?

10.12 The four-bar linkage of Figure P10.9 has the following weights and moments of inertia: $w_1 = 3$ lb, $w_2 = 5$ lb, $w_3 = 4$ lb, $I_{G_1} = 0.1$ lb $\cdot$ s$^2 \cdot$ in, $I_{G_2} = 0.3$ lb $\cdot$ s$^2 \cdot$ in, $I_{G_3} = 0.3$ lb $\cdot$ s$^2 \cdot$ in. Input link 1 has instantaneous angular velocity and acceleration of 10 rad/s clockwise and 100 rad/s^2 counterclockwise, respectively, leading to the following accelerations: $a_{G_1} = 636\angle 195°$ in/s^2, $a_{G_2} = 1240\angle 189°$ in/s^2, $a_{G_3} = 610\angle 182°$ in/s^2, $\alpha_2 = 19$ rad/s^2 ccw, and $\alpha_3 = 77$ rad/s^2 ccw. Determine the bearing forces and instantaneous input torque T_1 by **(a)** graphical solution and **(b)** analytical solution.

O_1B = 9 in
O_1G_1 = 4.5 in
BC = 15 in
BG_2 = 7.5 in
O_3C = 12 in
O_3G_3 = 6 in

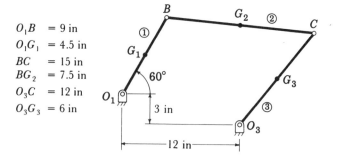

Figure P10.9

10.13 An in-line slider-crank mechanism has a crank length of 0.1 m and a connecting rod length of 0.5 m. The piston has a mass of 3 kg. The connecting rod has a mass of 2 kg with a center of mass located at a distance of 0.15 m from the crankpin end of the rod. Utilizing a lumped-mass approximation for the connecting rod, determine an expression for the torque required to maintain a constant crank speed of 200 rev/min. Evaluate the torque at the following crank angles ϕ: 0, 30, 60, 90, 120, 150, and 180°.

10.14 For the mechanism of Problem 10.13, determine the magnitudes and directions of all bearing forces for crank angle ϕ = 0, 45, 90, 135, and 180°. Assume that the center of mass of the crank is stationary.

10.15 An in-line slider-crank mechanism has a crank length of 2 ft and a connecting rod length of 7 ft. The piston has a weight of 100 lb. The connecting rod has a weight of 75 lb with a center of gravity located at a distance of 2 ft from the crankpin end of the rod. Utilizing a lumped-mass approximation for the connecting rod, determine an expression for the torque required to maintain a constant crank speed of 60 rev/min. Evaluate the torque at the following crank angles ϕ: 0, 30, 60, 90, 120, 150, 180°.

10.16 For the mechanism of Problem 10.15, determine the magnitudes and directions of all bearing forces for crank angle ϕ = 0, 45, 90, 135, and 180°. The crank is balanced.

10.17 Derive an expression for input torque T_1 similar to Eq. 10.37 for the offset slider-crank mechanism of Figure P10.10 with constant crank angular velocity ω_1. For the parameter values of Problem 10.13 and an offset e of 0.1 m, evaluate the torque at the following crank angles ϕ: 0, 30, 60, 90, 120, 150, and 180°.

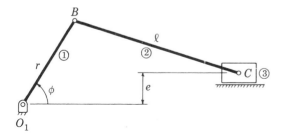

Figure P10.10

10.18 Determine the torque T_1 required to maintain a constant crank speed of 1000 rev/min ccw for the two-cylinder engine depicted in Figure P10.11. The individual pistons and connecting rods have masses of 1.0 kg and 0.8 kg, respectively. Consider the case where the cylinder V angle ψ is 90°, the crank spacing θ is 90°, and the crank angle ϕ is 160°. $BG_2 = DG_3 = 105$ mm.

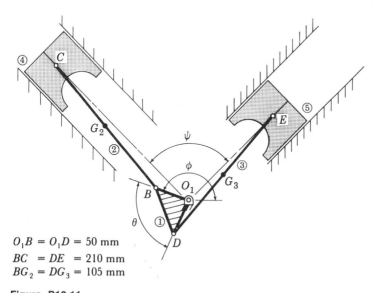

$O_1B = O_1D = 50$ mm
$BC = DE = 210$ mm
$BG_2 = DG_3 = 105$ mm

Figure P10.11

10.19 Determine the torque T_1 required to maintain a constant crank speed of 1000 rev/min ccw for the two-cylinder engine depicted in Figure P10.12. The individual pistons and connecting rods have masses of 1.0 kg and 0.8 kg, respectively.

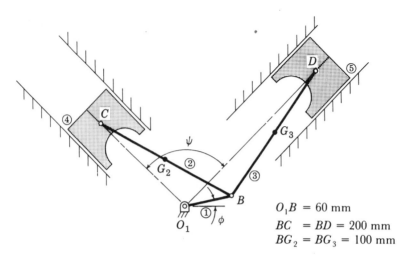

$$O_1B = 60 \text{ mm}$$
$$BC \; = BD = 200 \text{ mm}$$
$$BG_2 = BG_3 = 100 \text{ mm}$$

Figure P10.12

Consider the case where the cylinder V angle ψ is 90° and the crank angle ϕ is 20°. $BG_2 = BG_3 = 100$ mm.

10.20 Figure P10.13 is a schematic of a three-bladed propeller. Determine the location and correction amount of the counterweight that will balance the rotor. Perform the solution by using **(a)** the graphical method and **(b)** the analytical method.

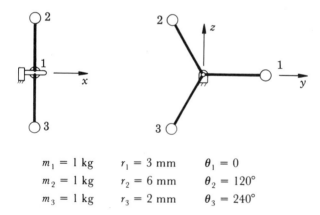

$m_1 = 1 \text{ kg}$	$r_1 = 3 \text{ mm}$	$\theta_1 = 0$
$m_2 = 1 \text{ kg}$	$r_2 = 6 \text{ mm}$	$\theta_2 = 120°$
$m_3 = 1 \text{ kg}$	$r_3 = 2 \text{ mm}$	$\theta_3 = 240°$

Figure P10.13

10.21 Determine the corrections needed in planes P and Q to balance the rotor shown in Figure P10.14. Carry out the solution by **(a)** the graphical method and **(b)** the analytical method.

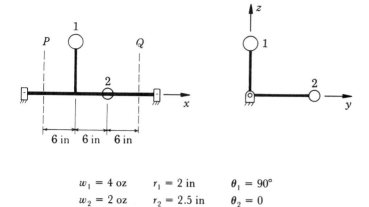

$$w_1 = 4 \text{ oz} \qquad r_1 = 2 \text{ in} \qquad \theta_1 = 90°$$
$$w_2 = 2 \text{ oz} \qquad r_2 = 2.5 \text{ in} \qquad \theta_2 = 0$$

Figure P10.14

10.22 For the rotor shown in Figure P10.15, unbalanced masses 1 and 2 have weights of 6 lb and 4 lb, respectively. The system is to be balanced by adding mass in the L plane at a radius of 3.0 in and removing mass in the R plane at a radius of 3.5 in. Determine the magnitudes and locations of the required corrections by **(a)** graphical solution and   **(b)** analytical solution.

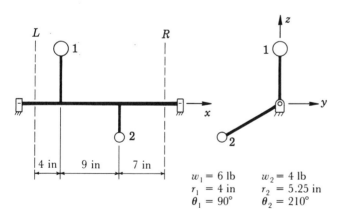

$$w_1 = 6 \text{ lb} \qquad w_2 = 4 \text{ lb}$$
$$r_1 = 4 \text{ in} \qquad r_2 = 5.25 \text{ in}$$
$$\theta_1 = 90° \qquad \theta_2 = 210°$$

Figure P10.15

10.23 The rotor of Figure P10.16 has unbalanced weights $w_1 = 2$ oz, $w_2 = 3$ oz, and $w_3 = 2$ oz, at radial positions $r_1 = 3$ in, $r_2 = 3$ in, and $r_3 = 2$ in. Determine the necessary counterweight correction amounts (in · oz) and locations in balance planes at locations P and Q for complete static and dynamic balance of the rotor by **(a)** graphical solution and   **(b)** analytical solution.

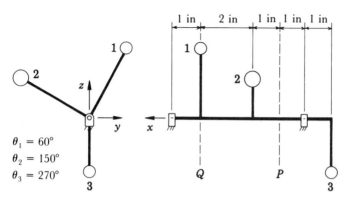

$\theta_1 = 60°$
$\theta_2 = 150°$
$\theta_3 = 270°$

Figure P10.16

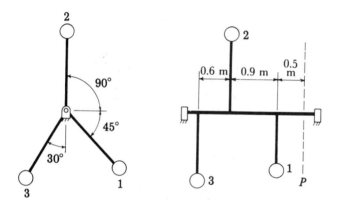

Figure P10.17

10.24 The rotor of Figure P10.17 has the following unbalanced amounts: $m_1r_1 = 1500$ kg · mm, $m_2r_2 = 2000$ kg · mm, and $m_3r_3 = 1500$ kg · mm. Balance the rotor by determining the angular orientation and correction amount for a counterweight in plane P and the axial location and angular orientation of a second counterweight having a correction amount of 1000 kg · mm by   **(a)** graphical solution and   **(b)** analytical solution.

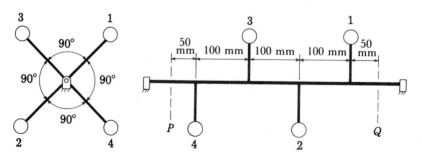

Figure P10.18

10.25 Figure P10.18 depicts a four-throw crankshaft that has the following properties: $m_1 = m_2 = m_3 = m_4 = 10$ kg and $r_1 = r_2 = r_3 = r_4 = 40$ mm. Determine the balancing arrangement in correction planes P and Q.

10.26 Figure P10.19 depicts a four-throw crankshaft that has the following properties: $m_1 = m_2 = m_3 = m_4 = 10$ kg and $r_1 = r_2 = r_3 = r_4 = 40$ mm. Determine the balancing arrangement in correction planes P and Q.

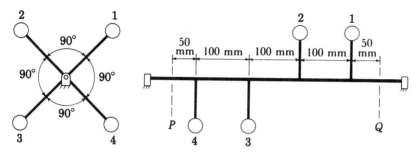

Figure P10.19

10.27 Repeat Problem 10.14 with the inclusion of a rotating counterweight on the crankshaft at an angle of 180° relative to the crank. The counterweight has a mass of 3.5 kg and a center of mass location at a distance of 0.1 m from the crankshaft axis.

10.28 Repeat Problem 10.16 with the addition of a rotating counterweight on the crankshaft at an angle of 180° relative to the crank. The counterweight has a weight of 100 lb and a center of mass location at a distance of 2.5 ft from the crankshaft axis.

10.29 Figure P10.20 shows an in-line, two-cylinder engine arrangement in which the cranks are spaced at 90°. Determine expressions for the net shaking force F_s and

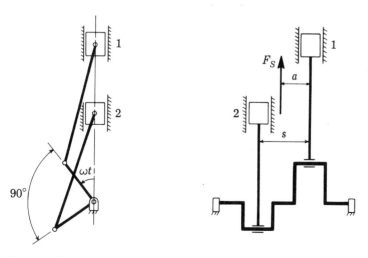

Figure P10.20

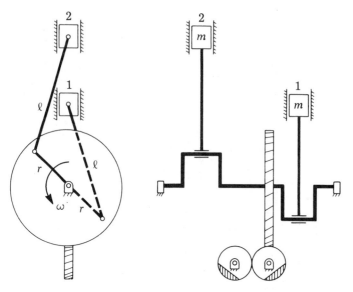

Figure P10.21

Figure P10.22

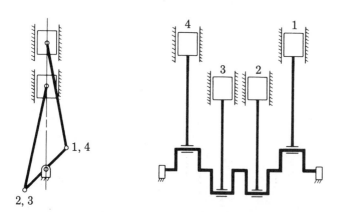

Figure P10.23

its axial position a as functions of angle ωt. For each cylinder, $mr\omega^2 = 2000$ lb and $r/\ell = 0.25$, where m is the reciprocating mass, r is the crank length, and ℓ is the connecting rod length. Axial distance $s = 4$ in. Which of the following are balanced: primary shaking force, secondary shaking force, primary shaking moment, secondary shaking moment?

10.30 Figure P10.21 shows an in-line, two-cylinder engine with 180° cranks. Also shown is a gear and rotating counterweight arrangement that is driven from the crankshaft. What gear ratio, correction amount, and counterweight orientation must be used to balance the net shaking force?

10.31 Examine the shaking force balance conditions as they apply to the two-cylinder opposed engine of Figure P10.22, which has a single crank and zero axial distance between cylinders.

10.32 Examine the balance conditions, both shaking force and shaking moment, as they apply to the four-cylinder, in-line engine of Figure P10.23.

10.33 Examine the balance conditions, both shaking force and shaking moment, as they apply to the four-cylinder, in-line engine of Figure P10.24.

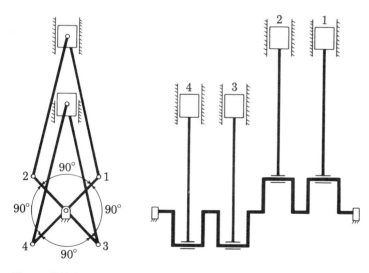

Figure P10.24

10.34 Examine the balance conditions, both shaking force and shaking moment, as they apply to the six-cylinder, in-line engine of Figure P10.25.

10.35 Examine the shaking force balance of the two-cylinder, V engine of Figure P10.11. The two cylinders lie in the same axial plane. Angle ψ is 90° and angle θ is 90°.

10.36 Examine the shaking force balance of the two-cylinder, V engine of Figure P10.12. The two cylinders lie in the same axial plane. Angle ψ is 90°.

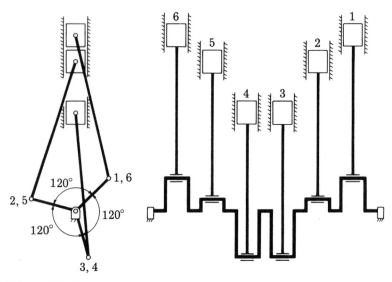

Figure P10.25

10.37 The four-cylinder radial engine depicted in Figure P10.26 is an excellent engine from the point of view of dynamic balance. Show that the engine can be balanced by means of a single rotating counterweight mounted on the crankshaft, and determine the location and magnitude of such a counterweight. The crank length is r, the connecting rod lengths all equal ℓ, and the reciprocating masses all equal m. The rotating masses are balanced, and all four cylinders lie in a single transverse plane.

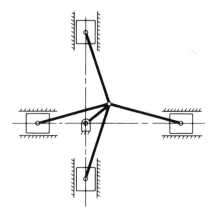

Figure P10.26

PROJECTS

See Projects 1.1 to 1.6 and suggestions in Chapter 1.

Review the results of the study of acceleration and angular acceleration character-istics of linkages in the design. Plot representative dynamic forces and torques in the linkages. Make use of computer software wherever practical. Check the results by a graphical method for at least one linkage position. Evaluate the linkage in terms of the performance requirements.

Synthesis

If the dimensions of a mechanism are given, and we attempt to determine its motion characteristics, the process is called **analysis**. **Synthesis** is the inverse process; given a set of performance requirements, we attempt to proportion a mechanism to meet those specifications. Ideally, all mechanisms would be designed by some mathematical synthesis process. However, because of the complexity of real-world machine requirements, ingenuity, judgment, and analysis are still a major part of the design process, while the application of formal synthesis is limited.

Type synthesis, number synthesis, and various forms of dimensional synthesis are described in the paragraphs that follow. Design methods involving less formal synthesis procedures are illustrated throughout the text.

In some situations, the performance requirements and our design decisions reduce a synthesis problem to a simple analysis task. In other cases, the synthesis approach yields many acceptable solutions. Where possible, a figure of merit is defined (e.g., minimum cost or weight) and the design is optimized. Then the proposed design is analyzed to confirm that the performance requirements are met.

11.1 TYPE SYNTHESIS

The process of deciding whether to use gears or cams or linkages or other machine elements to transmit motion is called type synthesis. For example, when a precise, constant speed ratio is required, as when driving a camshaft, a gear or chain drive might be selected. A four-bar linkage or slider-crank linkage might be selected to change continuous rotation to oscillation. Other considerations in type synthesis include static and inertial loading, wear, reliability, and cost. Type synthesis is largely dependent on experience and creativity of the designer. It may be followed by number synthesis, dimensional synthesis, material selection, and processing. As part of the adequacy assessment, analysis is an essential ingredient. Expert-system computer software packages, which are described in recent literature, may also be of interest to the designer.

11.2 NUMBER SYNTHESIS

Number synthesis is the determination of linkage configurations (the number of links and joints) that satisfy given criteria. It allows us to consider a variety of linkages for a particular application. Suppose, for example, we wish to identify planar, one-degree-of-freedom linkages with revolute joints. Grübler's criterion (see Chapter 1) relates the number of links and pairs as follows:

$$2n'_J - 3n_L + 4 = 0 \tag{11.1}$$

from which

$$n_L = (2n'_J + 4)/3 \tag{11.2}$$

where n_L = the number of links and n'_J = the number of revolute joints (or other lower pairs). Since the number of joints and the number of links must be integers, the following combinations satisfy the criterion:

n'_J	n_L	
1	2	
4	4	
7	6	
10	8	
13	10	etc.

The first combination above represents two links joined by a revolute (pin) joint. The second is a four-bar linkage. The Watt linkage and the Stephenson linkage (Figure 1.9) are each formed by six links and seven pin joints.

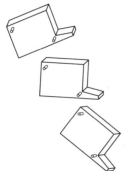

Figure 11.1 Design for specified positions.

11.3 TWO-POSITION SYNTHESIS

Consider the situation in which a link must assume certain prescribed positions, as, for example, in Figure 11.1. Such requirements are common in the design of materials handling equipment. If a link must assume three positions in a plane, the link may be guided as the coupler of a four-bar linkage. If only two planar positions are prescribed, the link may be rotated about a single pivot point, or a four-bar linkage may be used. This synthesis problem may be solved analytically or graphically.

Graphical Solution

Figure 11.2 shows a machine member in planar motion with only two positions prescribed. Two points B and C are arbitrarily selected within the member, where initial and final positions are identified by subscripts 1 and 2, respec-

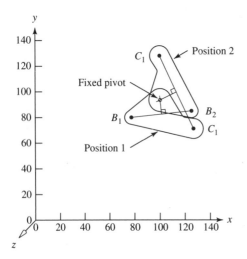

Figure 11.2 Two-position problem.

tively. If point B moves in a circular path between the initial and final positions of the member, the center of that circular path lies on the perpendicular bisector of line $B_1 B_2$. Likewise, if C moves in a circular path, the center lies on the perpendicular bisector of $C_1 C_2$. The link will assume the two prescribed positions if a fixed pivot (revolute joint) is located at the intersection of the perpendicular bisectors as shown in the figure. If translation in the z direction is required as well, a helical pair (a screw or helical spline) could be used instead of the revolute joint.

Computer Solution

An analytical solution for the planar two-position synthesis problem is outlined in the flowchart of Figure 11.3. Length L is the distance between points B and C, and angle β is the orientation of line BC. If the angle is input in degrees, it may be changed to radians by multiplying by $PI/180$, where $PI = 4 \times$ arctangent (1). Point C is located by the equations

$$C_x(I) = B_x(I) + L \cos \beta$$
$$C_y(I) = B_y(I) + L \sin \beta \qquad (11.3)$$

where the subscripts identify the x and y coordinates. It is suggested that input values and all relevant calculated values be printed out as an aid to debugging, and for graphical checking of results.

The midpoints of lines $B(1)B(2)$ and $C(1)C(2)$ are identified, respectively, by D_x, D_y and E_x, E_y, where

$$D_x = [B_x(1) + B_x(2)]/2 \qquad \text{etc.} \qquad (11.4)$$

The slopes of the perpendicular bisectors are given by M_B and M_C, where

$$M_B = [B_x(1) - B_x(2)]/[B_y(2) - B_y(1)] \qquad \text{etc.} \qquad (11.5)$$

We may now write the equations of the perpendicular bisectors in the form

$$y = M_B x + K_B$$
$$y = M_C x + K_C \qquad (11.6)$$

where the constants are given by

$$K_B = D_y - M_B D_x$$
$$K_C = E_y - M_C E_x \qquad (11.7)$$

Solving these equations simultaneously, the location of the fixed pivot point x, y is given by

$$x = [K_C - K_B]/[M_B - M_C]$$
$$y = M_B[K_C - K_B]/[M_B - M_C] + K_B \qquad (11.8)$$

If one of the selected points moves in the x direction, the perpendicular bisector will have an infinite slope. This special case can be handled by a

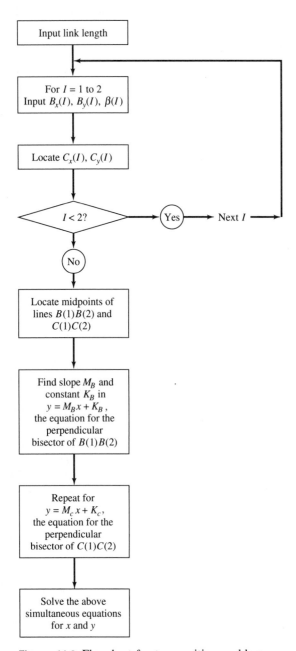

Figure 11.3 Flowchart for two-position problem.

branch in the program. A final step in the computer program may include drawing line BC in the specified positions and plotting the location of the pivot point.

◆ **EXAMPLE PROBLEM 11.1** *Computer-Aided Two-Position Synthesis*
A link identified by points B and C, which are 50 mm apart, is to assume the following positions:

Coordinates	B_x	B_y	Angle β
Position 1	75	80	$-10°$
Position 2	125	85	$125°$

Find the location of a fixed pivot that will permit the above motion.

Solution. A program based on the flowchart and equations described in the above paragraph is used to obtain the following results.

Coordinate	x	y
Point $C(1)$	124.24	71.32
Point $C(2)$	96.32	125.96
Midpoint of $B(1)B(2)$	100	82.5
Midpoint of $C(1)C(2)$	110.28	98.64
Fixed pivot location	98.96	92.86

The results agree with the graphical solution shown in Figure 11.2. ◆

Inaccessible Pivot Point

If the intersection of the displacement perpendicular bisectors is inaccessible, or other design considerations make its use as a pivot point impractical, a four-bar linkage may be used. Fixed revolute joints may be located anywhere along the displacement perpendicular bisectors. However, transmission angles should be considered when locating these pivots.

◆ **EXAMPLE PROBLEM 11.2** *Two-Position Synthesis Using a Four-Bar Linkage*
A link containing points B and C, which are 75 mm apart, is to assume the following positions:

Coordinates	B_x	B_y	β
Position 1	140	180	$0°$
Position 2	225	185	$15°$

Design a linkage that will permit the above motion.

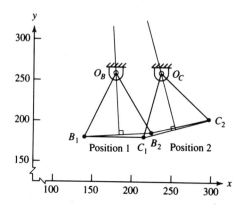

Figure 11.4 Two-position problem using a four-bar linkage.

Solution. A program based on the flowchart in Figure 11.3 and Eqs. 11.3 through 11.8 gives the following results:

Coordinates	x	y
Midpoint of $B(1)B(2)$	182.5	182.5
Midpoint of $C(1)C(2)$	256.2	192.2
Intersection of perpendicular bisectors	163.5	505.3

	Slope	Orientation
Perpendicular bisector of $B(1)B(2)$	-17	$93.37°$
Perpendicular bisector of $C(1)C(2)$	-3.377	$106.49°$

Use of the intersection of the perpendicular bisectors as a pivot point would probably result in an unacceptable design in this case. One possible design is four-bar linkage $O_B BCO_C$ shown in Figure 11.4, where fixed revolute joints O_B and O_C are arbitrarily located along the perpendicular bisectors. Note that the graphical solution shown in the figure verifies the computer solution. ◆

11.4 THREE-POSITION SYNTHESIS USING A FOUR-BAR LINKAGE

If three positions of a link are specified, it will not generally be possible to use a single pivot point. Figure 11.5 shows a link containing points B and C in three positions in a plane, identified by subscripts 1, 2, and 3. A four-bar linkage is used to produce the required motion. If we select points B and C as locations of revolute joints, the three specified locations of B determine a circle, and we locate a fixed revolute joint at O_B, the center of that circle. Likewise, fixed revolute joint O_C is located at the center of the circle determined by the specified locations of C.

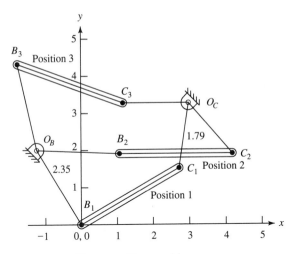

Figure 11.5 Three-position problem.

♦ **EXAMPLE PROBLEM 11.3** *Computer-Aided Three-Position Synthesis*
A link containing points B and C, which are 3.1 in apart, must assume the following positions:

Coordinates	B_x	B_y	Angle β
Position 1	0	0	$28°$
2	1.1	1.9	$0°$
3	-1.7	4.3	$-20°$

Design a linkage that will satisfy this requirement.

Solution. Figure 11.5 shows a graphical solution. Fixed pivot O_B is located at the intersection of the perpendicular bisectors of chords $B(1)B(2)$ and $B(2)B(3)$. Fixed pivot O_C is similarly located with respect to positions of point C. A computer-aided solution is outlined in the flowchart given in Figure 11.6. The equations represented in the flowchart are similar to those used in two-position synthesis. The coordinates of fixed pivot O_B are $-1.250, 1.992$ and the coordinates of O_C are $2.352, 1.786$. Crank lengths are $O_BB = 2.352$ and $O_CC = 1.786$. ♦

11.5 VELOCITY AND ACCELERATION SYNTHESIS

A four-bar linkage may be proportioned to produce specified values of angular velocity and acceleration for each link, using a procedure developed by Rosenauer.[8] Unfortunately, applications of this method of synthesis are limited since, in general, the specified values can be produced for only an instant (i.e., for only one linkage position).

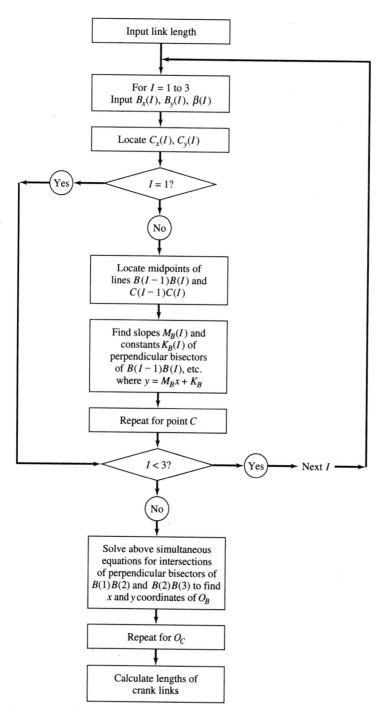

Figure 11.6 Flowchart for three-position problem.

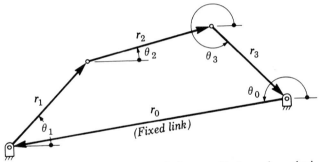

Figure 11.7 Mechanism synthesis for specified angular velocities and angular accelerations.

In this method, the links are described by complex numbers. Referring to Figure 11.7, with link 1 represented by vector r_1, and so on, the closed polygon formed by the four-bar linkage may be represented by the vector equation

$$r_1 + r_2 + r_3 + r_0 = 0 \tag{11.9}$$

Putting the fixed link vector on the right side of the equation and using the complex exponential form $r = re^{j\theta}$, we have

$$r_1 e^{j\theta_1} + r_2 e^{j\theta_2} + r_3 e^{j\theta_3} = -r_0 e^{j\theta_0} \tag{11.10}$$

Differentiating with respect to time, we obtain

$$j\omega_1 r_1 e^{j\theta_1} + j\omega_2 r_2 e^{j\theta_2} + j\omega_3 r_3 e^{j\theta_3} = 0 \tag{11.11}$$

where angular velocity $\omega = d\theta/dt$ and $\omega_0 = 0$. Differentiating with respect to time again, we obtain

$$\left(j\alpha_1 - \omega_1^2\right)r_1 e^{j\theta_1} + \left(j\alpha_2 - \omega_2^2\right)r_2 e^{j\theta_2} + \left(j\alpha_3 - \omega_3^2\right)r_3 e^{j\theta_3} = 0 \tag{11.12}$$

where angular acceleration $\alpha = d\omega/dt$.

After dividing all terms of Eq. 11.11 by j, Eqs. 11.10 to 11.12 are written in vector form:

$$r_1 + r_2 + r_3 = -r_0$$

$$\omega_1 r_1 + \omega_2 r_2 + \omega_3 r_3 = 0$$

$$\left(j\alpha_1 - \omega_1^2\right)r_1 + \left(j\alpha_2 - \omega_2^2\right)r_2 + \left(j\alpha_3 - \omega_3^2\right)r_3 = 0$$

which may be expressed by the matrix equation

$$\begin{bmatrix} 1 & 1 & 1 \\ \omega_1 & \omega_2 & \omega_3 \\ j\alpha_1 - \omega_1^2 & j\alpha_2 - \omega_2^2 & j\alpha_3 - \omega_3^2 \end{bmatrix} \begin{bmatrix} r_1 \\ r_2 \\ r_3 \end{bmatrix} = \begin{bmatrix} -r_0 \\ 0 \\ 0 \end{bmatrix} \tag{11.13}$$

The solution to this equation is given in determinant form by

$$
r_1 = \frac{-r_0}{D} \begin{vmatrix} 1 & 1 & 1 \\ 0 & \omega_2 & \omega_3 \\ 0 & j\alpha_2 - \omega_2^2 & j\alpha_3 - \omega_3^2 \end{vmatrix}
$$

and so on, where

$$
D = \begin{vmatrix} 1 & 1 & 1 \\ \omega_1 & \omega_2 & \omega_3 \\ j\alpha_1 - \omega_1^2 & j\alpha_2 - \omega_2^2 & j\alpha_3 - \omega_3^2 \end{vmatrix}
$$

The term $-r_0/D$ appears in the equations for r_2 and r_3, as well. For convenience, we may set $-r_0/D$ equal to unity in each r equation and still satisfy the specified conditions (given values of ω and α). Each link will change in length by the same proportion, and each angle will change by the same value. The resulting linkage is represented by vectors R_0, R_1, R_2, and R_3 (where $R_1 = r_1/(-r_0/D)$, etc.), which are given by the following determinants:

$$
R_1 = \begin{vmatrix} 1 & 1 & 1 \\ 0 & \omega_2 & \omega_3 \\ 0 & j\alpha_2 - \omega_2^2 & j\alpha_3 - \omega_3^2 \end{vmatrix} \tag{11.14}
$$

$$
R_2 = \begin{vmatrix} 1 & 1 & 1 \\ \omega_1 & 0 & \omega_3 \\ j\alpha_1 - \omega_1^2 & 0 & j\alpha_3 - \omega_3^2 \end{vmatrix} \tag{11.15}
$$

$$
R_3 = \begin{vmatrix} 1 & 1 & 1 \\ \omega_1 & \omega_2 & 0 \\ j\alpha_1 - \omega_1^2 & j\alpha_2 - \omega_2^2 & 0 \end{vmatrix} \tag{11.16}
$$

$$
R_0 = -R_1 - R_2 - R_3 \tag{11.17}
$$

◆ **EXAMPLE PROBLEM 11.4**   *Velocity and Acceleration Synthesis*
Synthesis of a linkage for specified angular velocities and accelerations is considered. The values are as follows:

Link:	0 (fixed)	1	2	3
ω (rad/s):	0	2	0	1
α (rad/s^2):	0	0	1	1

Specify a linkage that (instantaneously) satisfies the above values.

Solution.

$$R_1 = \begin{vmatrix} 1 & 1 & 1 \\ 0 & 0 & 1 \\ 0 & j & j-1 \end{vmatrix} = -j = e^{-j\pi/2}$$

$$R_2 = \begin{vmatrix} 1 & 1 & 1 \\ 2 & 0 & 1 \\ -4 & 0 & j-1 \end{vmatrix} = -2 - j2 = 2.828e^{-j2.356}$$

$$R_3 = \begin{vmatrix} 1 & 1 & 1 \\ 2 & 0 & 0 \\ -4 & j & 0 \end{vmatrix} = j2 = 2e^{j\pi/2}$$

$$R_0 = j + 2 + j2 - j2 = 2 + j = 2.236e^{j0.464}$$

The resulting linkage is sketched in Figure 11.8. This mechanism will satisfy the required conditions only when the links pass through the relative position shown.

An Alternate Solution Using the Matrix Inverse. Let the matrix equation 11.13 be represented by

$$AX = B \tag{11.18}$$

where

$$X = \begin{bmatrix} r_1 \\ r_2 \\ r_3 \end{bmatrix}, \qquad \text{etc.}$$

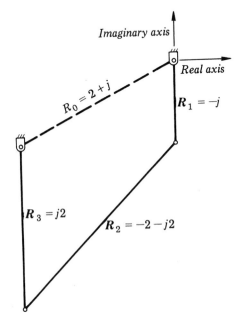

Imaginary axis

Real axis

$R_0 = 2 + j$

$R_1 = -j$

$R_3 = j2$

$R_2 = -2 - j2$

Figure 11.8 Synthesis solution to Example Problem.

Multiplying both sides of the equation by A^{-1}, the inverse of matrix A, the result is

$$X = A^{-1}B \qquad (11.19)$$

This form of solution is most convenient for computer calculations. A program for multiplication and inversion of complex matrices may be written by the user or purchased as part of a commercial software package. MathCAD$^{(TM)}$, called an "electronic scratchpad," is a software package with this capability (Anderson[1]).   ♦

♦ **EXAMPLE PROBLEM 11.5**   *Computer Solution Using the Matrix Inverse*
Use the matrix inverse method to solve the above example problem.

Solution.   It can be seen from the above results that one link vector may be chosen arbitrarily. For convenience, we let $r_0 = 1$. Using a software package to solve Eq. 11.19, we obtain X, from which $r_1 = -0.2 - j0.4$, $r_2 = -1.2 - j0.4$, and $r_3 = 0.4 + j0.8$. The proportions of the linkage are the same as in the previous solution, but the size and orientation are different. If each link vector is multiplied by $2 + j$, then the results are identical to the above example.   ♦

11.6 DESIGN OF A FUNCTION GENERATOR

Four-bar linkages may be used as function generators. Their low friction and higher load capacity make them preferable to cams for certain applications. An important disadvantage of the four-bar-linkage function generator, however, is its inability to represent an arbitrary function exactly except at a few points, called *precision points*. The range of input and output motion is further limited by the limiting positions of the linkage itself and sometimes by problems of mechanical advantage or transmission angle.

Consider the problem of designing a four-bar linkage so that output angle ϕ is a specified function of input angle θ, where links and angles are identified as in Figure 11.9. Each of the four links of the mechanism is considered a vector as shown in part b of the figure. The coupler link vector is equated to the vector sum of the other three link vectors, and a dot or scalar product is formed from the vectors on both sides of the equation. Thus, we have the *vector* relationship

$$-r_2 = r_1 + r_0 + r_3 \qquad (11.20)$$

and

$$r_2 \cdot r_2 = r_2^2 = (r_1 + r_0 + r_3) \cdot (r_1 + r_0 + r_3) \qquad (11.21)$$

where we have formed the dot product of each side of Eq. 11.20 with itself. Freudenstein[3] developed this method to obtain linkage dimensions of a

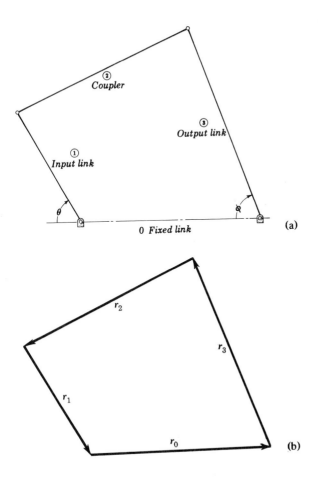

Figure 11.9 (a) Design of a function generator.
(b) Vectors representing the linkage.

function generator by using three, four, or five precision points. Freudenstein's three-point approximation is obtained as follows:

1. Select three input angles θ_1, θ_2, and θ_3 and the corresponding output angles ϕ_1, ϕ_2, and ϕ_3. Compute each output angle so that the required relationship between θ and ϕ is satisfactory at each of the three precision points for the prescribed function $\phi = f(\theta)$.

   Figure 11.10 shows the input link 1 and output link 3 at the three precision points. The desired relationship between θ and ϕ is achieved exactly at (θ_1, ϕ_1), (θ_2, ϕ_2), and (θ_3, ϕ_3) and satisfied approximately at other values of θ and ϕ.

2. In order to simplify the expressions for the lengths of links 1 and 3, compute the following angle relationships, which will be used in Eq.

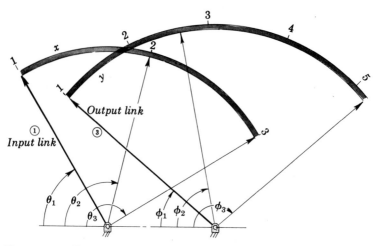

Figure 11.10 Proposed function generator. The input and output links of this four-bar linkage are to generate the function y (output) $= x^{1.5}$ for values of x (input) between 1 and 3. The coupler (link 2) is not shown. The sketch is not to scale since link dimensions are yet to be determined.

11.22 below:

$$A = \cos \theta_1 - \cos \theta_2$$

$$B = \cos \theta_1 - \cos \theta_3$$

$$C = \cos \phi_1 - \cos \phi_2$$

$$D = \cos \phi_1 - \cos \phi_3$$

$$E = \cos(\theta_1 - \phi_1) - \cos(\theta_2 - \phi_2)$$

$$F = \cos(\theta_1 - \phi_1) - \cos(\theta_3 - \phi_3)$$

3. If the fixed link is arbitrarily given a length r_0, the lengths of the input and output cranks are, respectively,

$$r_1 = \frac{BC - AD}{AF - BE} r_0 \quad \text{and} \quad r_3 = \frac{BC - AD}{CF - DE} r_0 \qquad \text{(11.22)}$$

If r_1 is found to be negative, link 1 is drawn in a direction opposite that shown in Figure 11.10. Similarly, if r_3 is negative, its direction is reversed.

♦ **EXAMPLE PROBLEM 11.6** *Design of a Function Generator*

Design a four-bar linkage to generate the function $y = x^{1.5}$ for values of x between 1 and 3.

Solution.

Step 1.   Angular displacement θ of the input crank will be proportional to x and angular displacement ϕ of the output crank proportional to y. At our disposal, we have the initial values of θ and ϕ, which will be taken as $\theta_0 = 60°$ and $\phi_0 = 40°$. We are also free to select the ranges of θ and ϕ, which will both be $90°$ for this example. See Figure 11.10. Measured from the initial values, the change in θ is proportional to the change in x, or, when $x_0 = 1$

$$\theta = \theta_0 + \frac{\text{range of } \theta}{\text{range of } x}(x - x_0) = 15 + 45x \qquad (11.23)$$

Similarly,

$$\phi = \phi_0 + \frac{\text{range of } \phi}{\text{range of } y}(y - y_0) \qquad (11.24)$$

As x varies from 1 to 3, y varies from $(1)^{1.5} = 1$ to $(3)^{1.5} = 5.196$, from which we obtain output crank angle:

$$\phi = 18.55 + 21.45y \quad \text{(both } \theta \text{ and } \phi \text{ measured in degrees)} \qquad (11.25)$$

Step 2.   For convenience, the precision points will be selected as

$$x_1 = 1, \quad x_2 = 2, \quad \text{and} \quad x_3 = 3$$

The above equations give the corresponding values of y, θ and ϕ:

Point	x	y	θ	ϕ
1	1.000	1.000	60°	40°
2	2.000	2.828	105	79.22
3	3.000	5.196	150	130

Step 3.   Using these values in Eq. 11.22, and arbitrarily letting $r_0 = 2$ in, we calculate the lengths of the input and output cranks:

$$r_1 = 10.38 \text{ in} \quad \text{and} \quad r_3 = 10.06 \text{ in}$$

The coupler length $r_2 = 2.57$ in may be found analytically or simply by drawing the input and output cranks in positions θ_1 and ϕ_1, respectively. The positions for the other precision points are also sketched to check for limiting positions, as in Figure 11.11. It can be seen that the limiting positions do not fall within the range of operation, making the design satisfactory from that standpoint. If the linkage proportions are not acceptable, the designer would try a new set of initial values or new ranges for θ and ϕ.

Step 4.   In order to transform crank angle ϕ to output y, we rewrite Eq. 11.25 to obtain

$$y = \frac{\phi - 18.55}{21.45}$$

where ϕ is measured in degrees. The linkage generates the exact function

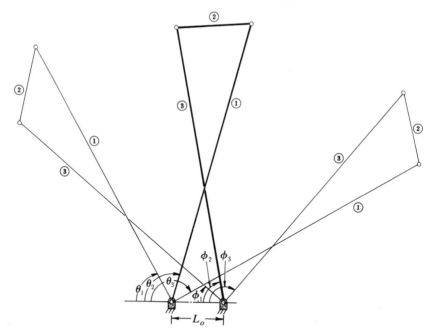

Figure 11.11 A linkage designed to approximate the relationship $y = x^{1.5}$.

$y = x^{1.5}$ for the precision points $x = 1$, 2, and 3 (corresponding to $\theta = 60$, 105, and 150°). Error in the generated function for any other value of x depends on the initial values and ranges of θ and ϕ as well as on the particular value of x.

A computer solution allows the designer of a function generator to try various design changes when link ratios or other parameters are unsatisfactory. A computer solution also makes it practical to evaluate the function generator as it moves through its range of positions. The solution to this example problem was made easier by using a spreadsheet that incorporated the position analysis equations of a four-bar linkage. Table 11.1 and Figure 11.12a show the value of y (as generated) compared with the ideal value ($y = x^{1.5}$). The greatest error is about 1.5%. The error is, of course, zero at the precision points corresponding to $x = 1$, 2, and 3. Figure 11.12b shows output angle ϕ and transmission angle plotted against input angle θ. The transmission angle ranges from about 120 to 50° as the input crank moves within its 60 to 150° operating range. Transmission angles in the range from about 40 or 45° to 135 or 140° are generally considered acceptable. Thus the function generator is satisfactory from the standpoint of transmission angle. ◆

Selection of Precision Points

The output of a function generator will differ from the desired function due to **mechanical error** and **structural error**. Mechanical error results from tolerances on link length, bearing clearance, and other characteristics of actual

TABLE 11.1   EVALUATION OF A FUNCTION GENERATOR

Theta, deg	Phi, deg	x	y	$x^{1.5}$	Error, %
60	40.000	1.000	1.000	1.000	0.000
65	43.932	1.111	1.183	1.171	1.034
70	47.947	1.222	1.371	1.351	1.429
75	52.056	1.333	1.562	1.540	1.461
80	56.268	1.444	1.758	1.736	1.296
85	60.595	1.556	1.960	1.940	1.035
90	65.044	1.667	2.168	2.152	0.743
95	69.625	1.778	2.381	2.370	0.459
100	74.347	1.889	2.601	2.596	0.206
105	79.217	2.000	2.828	2.828	0.000
110	84.240	2.111	3.063	3.067	− 0.154
115	89.423	2.222	3.304	3.313	− 0.254
120	94.767	2.333	3.553	3.564	− 0.303
125	100.273	2.444	3.810	3.822	− 0.305
130	105.938	2.556	4.074	4.085	− 0.270
135	111.756	2.667	4.346	4.355	− 0.209
140	117.717	2.778	4.623	4.630	− 0.133
145	123.805	2.889	4.907	4.910	− 0.058
150	130.000	3.000	5.196	5.196	0.000

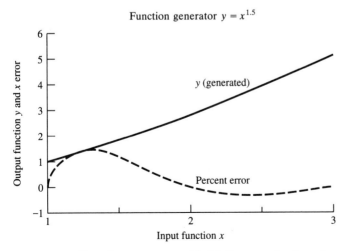

Function generator $y = x^{1.5}$

y (generated)

Percent error

Output function y and x error

Input function x

Figure 11.12 Function generator $x^{1.5}$. (a) Generated value of y and % error.

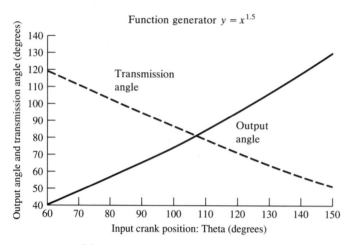

Figure 11.12 (b) Output angle and transmission angle.

mechanism components. Structural error results from the inability of a four-bar linkage to generate an arbitrary function precisely except at the precision points (although certain functions can be generated precisely over a continuous range of values).

In the previous example, the beginning, midpoint, and end of the range were selected as precision points. Other selections may be made in an attempt to reduce structural error (see Freudenstein[3,4]). One method of selection, called Chebychev spacing, is described by Hinkle.[5] For three precision points, the values would be as follows:

$$x_1 = x_{mid} - \frac{1}{2}x_{range} \cos 30°$$
$$x_2 = x_{mid} \qquad\qquad\qquad\qquad (11.26)$$
$$x_3 = x_{mid} + \frac{1}{2}x_{range} \cos 30°$$

where x_{mid} is the mean value of x in the range of the function generator.

◆ EXAMPLE PROBLEM 11.7   *Design of a Function Generator Using Chebychev Spacing*

Design a four-bar linkage to generate the function $y = e^x - x$ for values of x between $x_0 = 0$ and $x_f = 1$ (corresponding to $y_0 = 1$ and $y_f = 1.7183$).

Solution.   For $x_{range} = 1$ and $x_{mid} = 0.5$, we obtain

$$x_1 = 0.5 - \frac{1}{2} \cos 30° = 0.0670$$
$$y_1 = e^{x_1} - x_1 = 1.0023$$
$$x_2 = 0.5$$

$$y_2 = 1.1487$$

$$x_3 = 0.5 + \frac{1}{2} \cos 30° = 0.9330$$

$$y_3 = 1.6091$$

We will select the range of θ from $\theta_0 = 65°$ to $\theta_f = 125°$ and the (approximate) range of ϕ from $\phi_0 = 40°$ to $\phi_f = 80°$. For changes in θ and ϕ proportional to changes in x and y, respectively, we have

$$\theta = \theta_0 + \frac{\text{range of } \theta}{\text{range of } x}(x - x_0)$$

$$= 65° + \frac{125° - 65°}{1 - 0}(x - 0)$$

$$\theta \text{ (degrees)} = 60x + 65°$$

and

$$\phi = \phi_0 + \frac{\text{range of } \phi}{\text{range of } y}(y - y_0)$$

$$= 40° + \frac{80° - 40°}{1.7183 - 1}(y - 1)$$

$$\phi \text{ (degrees)} = 51.086y - 11.086°$$

The values of x, y, θ, and ϕ are as follows:

Position	x	$y = e^x - x$	θ	ϕ
0	0	1 (ideal)	65°	40° (ideal)
1 (precision point)	0.0670	1.0023	69.02°	40.13°
2 (precision point)	0.5	1.1487	95°	48.28°
3 (precision point)	0.9330	1.6091	120.98°	73.92°
f	1	1.7183 (ideal)	125°	80° (ideal)

Equations 11.22 are used to compute link lengths (where any convenient value of r_0 may be selected). The results are $r_1/r_0 = 1.65$ and $r_3/r_0 = 1.78$, from which $r_2/r_0 = 0.45$.

We note that the sum of the lengths of the longest and shortest links is less than the sum of the other link lengths and that the coupler is shortest. Thus, based on the Grashof criterion, the proposed linkage is a double rocker. Figure 11.13 shows the linkage in positions corresponding to the three precision points. It can be seen from the sketch that the proposed linkage is unsatisfactory. A computer solution is summarized in Table 11.2. The limiting positions of the linkage do not permit the input link to assume an angle of 65°. The computer indicates that the linkage will not close by signaling an error at this point. Output link position and transmission angle are plotted against input link position in Figure 11.14. The figure shows the unacceptable values of transmission angle.

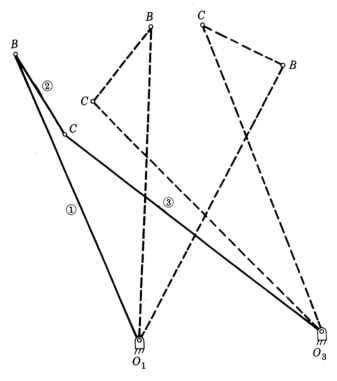

Figure 11.13 A linkage designed to approximate the relationship $y = e^x$.

TABLE 11.2   THE EFFECT OF LIMITING POSITIONS ON THE PERFORMANCE OF A FUNCTION GENERATOR

Theta, deg	Phi, deg	x	y	$e^x - x$	Error, %
65	ERR	0.000	ERR	1.000	ERR
70	39.811	0.083	0.997	1.004	−0.69
75	39.924	0.167	0.999	1.015	−1.58
80	41.198	0.250	1.022	1.034	−1.21
85	43.085	0.333	1.055	1.062	−0.65
90	45.455	0.417	1.098	1.100	−0.21
95	48.282	0.500	1.149	1.149	−0.00
100	51.595	0.583	1.208	1.209	−0.04
105	55.465	0.667	1.278	1.281	−0.26
110	60.018	0.750	1.359	1.367	−0.55
115	65.480	0.833	1.458	1.468	−0.69
120	72.327	0.917	1.581	1.584	−0.24
125	82.365	1.000	1.761	1.718	2.47

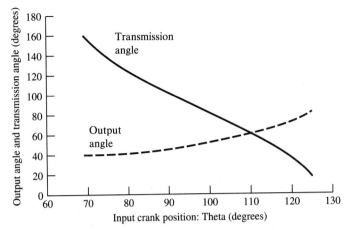

Figure 11.14 Function generator $y = e^x - x$. Output angle and transmission angle.

When an attempt to design a function generator fails, we have a number of alternatives. If there are to be three precision points, then the beginning and end values of θ and ϕ are at our disposal. We can also use a different scheme for selection of precision points. In this example, the nature of the function

$$y = e^x - x \qquad \text{for} \quad 0 \le x \le 1$$

is also a problem in that

$$dy/dx = 0 \qquad \text{at} \quad x = 0$$

This condition would (ideally) require link 3 to remain stationary when link 1 begins to move. Thus, we might consider excluding $x = 0$ from the range of operation of the function generator. An alternative is the use of a different type of configuration, a cam, for example, to generate the function. ◆

11.7 COUPLER CURVES

We are severely limited in flexibility of design if we consider only the motion of the two cranks in the four-bar linkage. *Coupler curves*, curves generated by points on the coupler (connecting rod) of a four-bar linkage, provide a variety of paths to choose from. For some applications, a point on the coupler itself is utilized directly as in a film drive mechanism or for a mixing device. In other cases, a point on the coupler is used to drive an added linkage. By using a catalog of coupler curves, we may select linkage dimensions to perform a specific function. An exhaustive catalog of coupler curves for the crank rocker mechanism (over 7000 curves made up of about half a million plotted locations and velocities) is given by Hrones and Nelson.[6] A work of this type might be used to solve a problem such as the following.

◆ **EXAMPLE PROBLEM 11.8** *An Application of Coupler Curves*
Design a linkage to meet the following requirements. One of the links is to rotate through an angle of 20° at a rate of one cycle per second with a dwell of 1/6 second between oscillations. The drive link is to rotate at a constant angular velocity.

Solution. Problems of this type are common in the design of machinery. The solution almost always involves a cam, but we will attempt to use a four-bar linkage for purposes of illustrating linkage design procedures. Examining a catalog of coupler curves, we see several possibilities, one of which is sketched in Figure 11.15a. During about 1/6 of each cycle, the coupler curve described

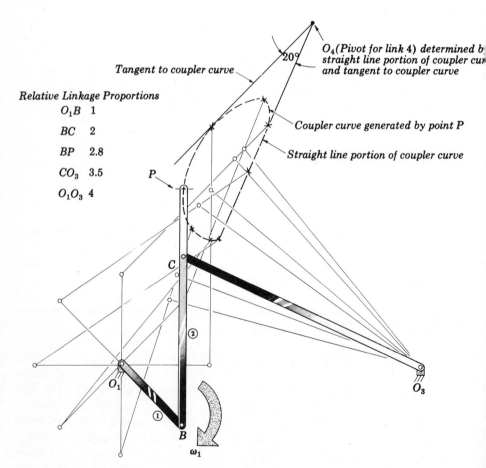

Figure 11.15 (a) Coupler curve of a four-bar linkage.

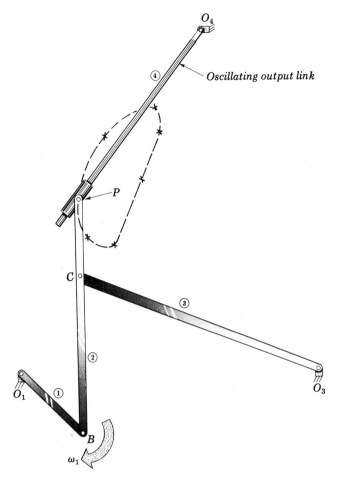

Figure 11.15 (b) Link 4 and a slider are added to the four-bar linkage. The output link oscillates with a dwell period.

by point P in that figure approximates a straight line. By combining the four-bar linkage with a sliding contact linkage (see Figure 11.15b), we may take advantage of the straight-line portion of the coupler curve to provide the required dwell.

Link 4, the output link, is located by using the coupler curve. The straight-line portion of the coupler curve defines one limiting position of link 4. The other limiting position is defined by a tangent to the coupler curve that intersects the first (limiting position) line at an angle of 20°. Link 4 is pivoted at that intersection, and its length must be sufficient to allow the slider to assume all positions on the coupler curve. Since a complete cycle of motion corresponds to one rotation of link 1, that link will be driven at 60 rev/min.

When link 1 turns with constant angular velocity, the time required for output link 4 to rotate from one position to another is proportional to the angle between corresponding positions of link 1. On this basis, the approximate motion of link 4 is as follows: dwell, 17 percent, clockwise rotation, 30 percent, counterclockwise rotation, 53 percent (expressed in terms of the time required for one complete cycle, i.e., one clockwise rotation of link 1).   ◆

Pattern Matching Synthesis

It is no longer necessary to use a catalog of coupler curves such as the one referred to above. The required coupler curves can now be generated by a personal computer with graphics software. The synthesis process can be further automated by pattern matching applied to coupler curves, using artificial intelligence programming techniques. We are thus able to use the computer to determine the parameters of a linkage that will produce a required function.

Pattern matching techniques may be applied to synthesis of both planar and spatial linkages, using visual or computer-aided pattern matching. Sodhi et al.[9] present a set of computer-generated curves for use in designing four-revolute spherical function generators. Their method involves matching a plot of the desired function with curves derived from computer solutions of the displacement equation.

References

1. Anderson, Richard B., *The Student Edition of MathCAD*$^{(TM)}$, Version 2.0, Addison-Wesley, Reading, MA, 1989.
2. Angeles, J., *Spatial Kinematic Chains: Analysis-Synthesis-Optimization*, Springer, New York, 1982.
3. Freudenstein, F., "Approximate Synthesis of Four-Bar Linkages," *Transactions of the American Society of Mechanical Engineers*, vol. 77, 1955, pp. 853–861.
4. Freudenstein, F., "Structural Error Analysis in Plane Kinematic Synthesis," *Transactions of the ASME, Ser. B, Journal of Engineering for Industry*, vol. 81, 1959, pp. 15–22.
5. Hinkle, R., *Kinematics of Machines*, 2nd ed., Prentice-Hall, Englewood Cliffs, NJ, 1960.
6. Hrones, J. A. and G. L. Nelson, *Analysis of the Four-Bar Linkage: Its Application to the Synthesis of Mechanisms*, Wiley, New York, 1951.
7. Kishore, A., and M. Keefe, "Synthesis of an Elastic Mechanism," *Mechanism and Machine Theory*, Penton, Cambridge, vol. 23, no. 4, 1988, pp. 305–312.
8. Rosenauer, N., "Complex Variable Method for Synthesis of Four-Bar Linkages," *Australian Journal of Applied Science*, vol. 5, no. 4, 1954.
9. Sodhi, R. S., A. J. Wilhelm, and T. E. Shoup, "Design of a Four-Revolute Spherical Function Generator with Transmission Effectiveness by Curve Matching," *Mechanism and Machine Theory*, Pergamon Press, vol. 20, no. 6, 1985, pp. 577–585.

PROBLEMS

• **11.1** A link containing points B and C, which are 150 mm apart, is to assume the following positions:

Coordinates	B_x	B_y	β (degrees)
Position 1	0	0	35
Position 2	250	0	75

Find the location of a fixed pivot that will permit the above motion. Use graphical methods.

11.2 A link containing points B and C, which are 6 in apart, is to assume the following positions:

Coordinates	B_x	B_y	β (degrees)
Position 1	5	4	10
Position 2	4	7.5	−40

Find the location of a fixed pivot that will permit the above motion. Use graphical methods.

11.3 A link containing points B and C, which are 4 in apart, is to assume the following positions:

Coordinates	B_x	B_y	β (degrees)
Position 1	2	3	30
Position 2	7	8	60

Design a four-bar linkage that will permit the above motion. Use graphical methods.

11.4 Write a computer program for two-position synthesis. Include enough output data for design of a four-bar linkage or single pivot linkage.

• **11.5** A link containing points B and C, which are 150 mm apart, is to assume the following positions:

Coordinates	B_x	B_y	β (degrees)
Position 1	0	0	35
Position 2	250	0.001	75

Find the location of a fixed pivot that will permit the above motion. Solve analytically, using a computer if available.

11.6 A link containing points B and C, which are 3.5 in apart, is to assume the following positions:

Coordinates	B_x	B_y	β (degrees)
Position 1	2	3	12
Position 2	5.5	3.2	190

Find the location of a fixed pivot that will permit the above motion. Solve analytically, using a computer if available.

11.7 A link containing points B and C, which are 4 in apart, is to assume the following positions:

Coordinates	B_x	B_y	β (degrees)
Position 1	2	3	30
Position 2	7	8	60

Design a four-bar linkage that will permit the above motion. Solve analytically, using a computer if available.

11.8 A link containing points B and C, which are 4.2 in apart, must assume the three following positions:

Coordinates	B_x	B_y	β (degrees)
Position 1	0	0	18
Position 2	1.3	1	8
Position 3	2.6	0.6	11

Design a linkage that will satisfy these requirements. Use graphical methods.

11.9 A link containing points B and C, which are 3 in apart, must assume the three following positions:

Coordinates	B_x	B_y	β (degrees)
Position 1	2	2	-30
Position 2	5.5	1.5	5
Position 3	7.5	0.5	40

Design a linkage that will satisfy these requirements. Use graphical methods.

11.10 Write a computer program for three-position synthesis.

11.11 A link containing points B and C, which are 60 mm apart, must assume the three following positions:

Coordinates	B_x	B_y	β (degrees)
Position 1	0	0	20
Position 2	20	40	5
Position 3	-1.5	4	-15

Design a linkage that will satisfy these requirements. Solve analytically. Use a computer if available.

11.12 A link containing points B and C, which are 1.5 in apart, must assume the three following positions:

Coordinates	B_x	B_y	β (degrees)
Position 1	1	1	15
Position 2	2	3	0
Position 3	0	5	-15

Design a linkage that will satisfy these requirements. Solve analytically. Use a computer if available.

11.13 A link containing points B and C, which are 100 mm apart, must assume the three following positions:

Coordinates	B_x	B_y	β (degrees)
Position 1	0	0	-10
Position 2	50	-40	5
Position 3	20	-50	30

Design a linkage that will satisfy these requirements. Solve analytically. Use a computer if available.

Problems 11.14 *through* 11.16 *are to be solved using determinants.*

11.14 Specify a linkage that (instantaneously) satisfies the following conditions:

Link:	0	1	2	3
ω (rad/s):	0	2	1	1
α (rad/s^2):	0	1	0	-1

Use Rosenauer's method.

11.15 (a) Design a four-bar linkage to satisfy the following velocity and acceleration requirements. Use Rosenauer's method of synthesis.

Link:	0	1	2	3
ω (rad/s):	0	-1	2	0
α (rad/s^2):	0	1	0	2

(b) Sketch the solution. Draw velocity and acceleration polygons. Compare with the given data.

• **11.16 (a)** Design a four-bar linkage to satisfy the following velocity and acceleration requirements. Use Rosenauer's method of synthesis.

Link:	0	1	2	3
ω (rad/s):	0	-1	2	0
α (rad/s^2):	0	0	1	2

(b) Sketch the solution. Draw velocity and acceleration polygons. Compare with the given data.

11.17 Letting $r_0 = 100$, design a linkage to satisfy the following requirements:

$$\omega_0 = 0 \qquad \omega_1 = 150 \qquad \omega_2 = -50 \qquad \omega_3 = -90$$
$$\alpha_0 = 0 \qquad \alpha_1 = 2500 \qquad \alpha_2 = -400 \qquad \alpha_3 = -1600$$

Suggestion: Solve by the matrix inverse method, using a computer.

11.18 The fixed link of a four-bar linkage is to be 1 in long, oriented in the positive real direction (i.e., $r_0 = 1$). Velocities and accelerations are as given in Problem 11.14.

(a) Design a linkage to satisfy these conditions, using the matrix inverse method.

(b) If you have already solved Problem 11.14, compare the results by multiplying the link vectors in this solution by the value of r_0 found previously.

11.19 The fixed link of a four-bar linkage is to be 1 in long, oriented in the positive real direction (i.e., $r_0 = 1$). Velocities and accelerations are as given in Problem 11.15.

(a) Design a linkage to satisfy these conditions, using the matrix inverse method.

(b) If you have already solved Problem 11.15, compare the results by multiplying the link vectors in this solution by the value of r_0 found previously.

11.20 The fixed link of a four-bar linkage is to be 1 in long, oriented in the positive real direction (i.e., $r_0 = 1$). Velocities and accelerations are as given in Problem 11.16.

(a) Design a linkage to satisfy these conditions, using the matrix inverse method.

(b) If you have already solved Problem 11.16, compare the results by multiplying the link vectors in this solution by the value of r_0 found previously.

11.21 Design a four-bar linkage to generate the function $y = e^x - 0.5x^{1.1}$ where $x_0 = 0.5$, $x_{range} = 1.0$, $\theta_0 = 70°$, $\phi_0 = 40°$, $\theta_{range} = 60°$, and $\phi_{range} = 60°$. A computer solution is suggested.

(a) Determine relative link lengths.

(b) Tabulate input angle, output angle, x, y (generated), y (ideal), and percent error.

(c) Plot output angle and transmission angle vs. input angle.

● **11.22** Design a four-bar linkage to generate the function $y = e^x - 0.2x^2$ where $x_0 = 0.5$, $x_{range} = 1.0$, $\theta_0 = 75°$, $\phi_0 = 40°$, $\theta_{range} = 60°$, and $\phi_{range} = 40°$. A computer solution is suggested.

(a) Determine relative link lengths.

(b) Tabulate input angle, output angle, x, y (generated), y (ideal), and percent error.

(c) Plot output angle and transmission angle vs. input angle.

PROJECT

Plastic tires are used on utility carts and garden equipment. Consider the problem of forming a tire from a continuous tube. It is necessary to measure and cut the tube, form it into a circle, and join the cut ends. Determine which of these processes can be automated. It may be possible to apply number and type synthesis, three-position synthesis, or motion synthesis using coupler curves. Do not limit the study to methods discussed in this text.

Introduction to Robotic Manipulators

12.1 INTRODUCTION

Robots are an important class of electromechanical systems, which have applications in a diverse range of fields, including manufacturing, construction, space exploration, care for the handicapped, and many others. A robot can be defined as a programmable machine, that is, a machine capable of executing a wide variety of tasks through the specification of sets of computer software commands. These systems are useful in flexible manufacturing, because they can adapt to changes in production operations such as the introduction of new product lines. They are also useful in environments that are hazardous to humans such as undersea exploration or handling of toxic materials. The above definition of a robot suggests an almost unlimited number of potential variations. Some examples of existing, commercially available robots are shown in Figure 12.1.

There are three major components of a robot system; a manipulator mechanism, a set of actuators, and a controller. The manipulator mechanism is a multi-degree-of-freedom mechanical device, capable of performing a broad range of planar or spatial motions. Most industrial robots have from three to six degrees of freedom. Each independent degree of freedom requires separate actuation, usually applied at the joints of the robot. Electric motors are often used, but hydraulic and pneumatic actuation are also common. Some robots use a combination of actuator types. The controller provides the means

(a)

(b)

(c)

(d)

Figure 12.1 Examples of industrial robots. (*Sources:* (a) Adept Technology, Inc.; (b) AEG Westinghouse, Industrial Automation Corporation; (c), (d) GMFanuc Corporation.)

for programming and executing specific motion tasks. There usually are different ways of programming a robot, from using a teach mode, in which the robot is physically placed in desired positions with the corresponding coordinates stored in controller memory, to the direct use of a computer command language in developing motion programs. Many robots have sensing and feedback capabilities for implementing control algorithms, which enable the robot to maintain a desired motion profile within some specified degree of accuracy.

The objective of this chapter is to introduce concepts and analysis methods pertaining to robotic manipulators. Background will be provided for carrying out the kinematic analysis of manipulator mechanisms. These methods, however, are generally applicable to planar and spatial mechanisms of various types. There are many excellent references on the subject of robotics; some of these are listed at the end of the chapter.[1,2,4,6]

12.2 ROBOT TERMINOLOGY AND DEFINITIONS

A robotic manipulator consists of a multi-degree-of-freedom kinematic chain. Either open-loop chains (e.g., see Figure 12.1a, b, and d) or closed-loop chains (e.g., Figure 12.1c) can be used. Revolute and prismatic pairs are commonly employed as joints, because these joint types are relatively easy to drive, for example, through the use of electric motors or pneumatic cylinders. At the free end of the manipulator is mounted an end effector, which is a device for performing the specific work function of the robot. For example, the end effector may be a tool, such as a welding torch or a painting nozzle, it may be a gripper, or it may possibly be another manipulator. Depending on the type of application and controller, the robot is programmed for either discrete positions (point-to-point motion) or continuous path motion of the end effector.

Robots have multiple degrees of freedom in order to produce general motion capabilities for carrying out different programmed tasks. For example, the planar manipulator in Figure 12.2a has three degrees of freedom, which is the minimum number required to produce general planar motion of a body (member 3). By means of the three rotational inputs shown, member 3, which represents the end effector in this case, can be placed at any position and angular orientation within the workspace or work envelope of the robot. This workspace is the total area or volume within which the end effector can move and is determined from the member dimensions and limits of joint motion of the robot. Similarly, the spatial robot of Figure 12.2b has six degrees of freedom, which is the minimum number required for general three-dimensional motion of a body.

The methods described in Chapter 1 (see Eqs. 1.1, 1.2, and 1.3) can be used for the determination of the number of degrees of freedom of a manipulator. Repeating Eq. 1.2 here,

$$DF_{spatial} = 6(n_L - n_J - 1) + \sum_{i=1}^{n_J} f_i \qquad (12.1)$$

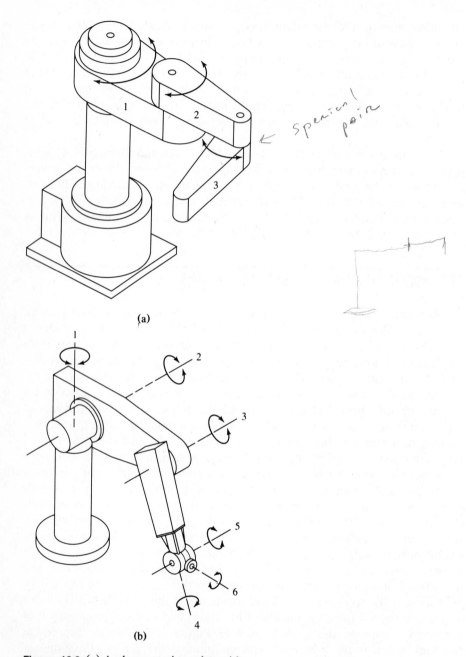

spherical pair

Figure 12.2 (a) A planar motion robot with three degrees of freedom. (b) A spatial motion robot with six degrees of freedom. (*Source:* AEG Westinghouse, Industrial Automation Corporation.)

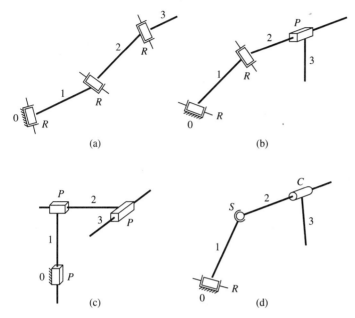

Figure 12.3 Examples of manipulators. (a) RRR. (b) RRP.
(c) PPP. (d) RSC.

where $DF_{spatial}$ is the total number of degrees of freedom of a spatial linkage, n_L is the number of links (including the fixed link), n_J is the number of joints, and f_i is the number of degrees of freedom for joint i. Consider the application of this equation to the open-chain manipulators shown in Figure 12.3. Each of these manipulators has four members and three joints, illustrating the following general condition: In an open-loop manipulator, the number of joints is one less than the number of links, and the number of degrees of freedom of the mechanism is therefore the sum of the joint degrees of freedom; that is,

$$DF_{spatial} = \sum_{i=1}^{n_J} f_i \qquad (12.2)$$

If the manipulator also contains only single-degree-of-freedom joints for which $f_i = 1$ (e.g., revolute and prismatic joints), then the number of degrees of freedom of the manipulator is equal to the number of joints. This is true for the three manipulators shown in Figure 12.3a, b, and c, and, therefore, DF = 3 in each of these cases. Applying Eq. 12.2 to the mechanism of Figure 12.3d,

$$DF = 1 + 3 + 2 = 6$$

Thus six independent actuations are required for constrained motion of this particular mechanism.

An important part of the analysis of robotic manipulators is the development of relationships between joint coordinates and end effector coordinates.

This leads to two classes of kinematic analysis, forward kinematics and inverse kinematics. In forward kinematics, the end effector coordinates are determined from knowledge of the joint coordinates. Specifically, the position and orientation of the end effector are computed from the angular positions (revolute joints) and translational positions (prismatic joints) of the joints. This, however, is not the normal procedure in programming and operating robots, where we would prefer to specify prescribed end effector coordinates, and then compute the corresponding joint coordinates which would be required for actuation. This process of going from end effector coordinate space to joint coordinate space is referred to as inverse kinematics. For most typical manipulators, the inverse kinematics problem is more difficult than the forward kinematics problem. Both forms of analysis will be considered in the following sections.

12.3 MATRIX METHOD OF ANALYSIS

Matrix methods are very useful for the analysis of robotic manipulators, and spatial mechanisms in general. They are also applicable to the special case of planar mechanisms. In this approach, multiple reference frames are utilized in describing the kinematic and dynamic behavior of a mechanism. The material to be presented assumes some background in linear algebra, for which a number of references are available (e.g., 5).

Consider the situation depicted in Figure 12.4, in which the position of a point in space is referred to two different frames of reference, frame i and frame j. Each of these frames is a Cartesian reference frame having a set of x, y, and z coordinate axes. The position of the point as described in the i frame can be expressed by the vector

$$^i\{P\} = \begin{Bmatrix} P_{x_i} \\ P_{y_i} \\ P_{z_i} \end{Bmatrix} \tag{12.3}$$

where P_{x_i}, P_{y_i}, and P_{z_i} are the x, y, and z coordinates of the position vector as measured in the i frame. Similarly, the position of the point referred to the j frame is

$$^j\{P\} = \begin{Bmatrix} P_{x_j} \\ P_{y_j} \\ P_{z_j} \end{Bmatrix} \tag{12.4}$$

where P_{x_j}, P_{y_j}, and P_{z_j} are the coordinates of the point in the j frame. There is a relationship between vectors $^i\{P\}$ and $^j\{P\}$, having the form

$$^i\{P\} = {}^i_j[T]\,{}^j\{P\} \tag{12.5}$$

where the matrix $^i_j[T]$, which has dimensions of 3×3 in this case, relates the

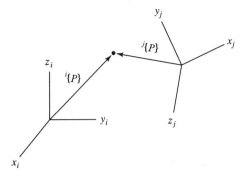

Figure 12.4 The vector location of a point in space with respect to two different reference frames.

relative positions and orientations of the two reference frames. This matrix is called a transformation matrix, because it produces a coordinate mapping transformation from the j frame to the i frame. The notation convention defined above for pre-subscripts and pre-superscripts will be used throughout the chapter. In particular, the pre-superscript on a vector defines the reference frame in which the vector is based. The subscript and superscript on the transformation matrix define the frame from which and the frame to which a vector is mapped, respectively.

Alternatively, the transformation can take place from the i frame to the j frame, in which case

$$^{j}\{P\} =\,_{i}^{j}[T]\,^{i}\{P\} \tag{12.6}$$

Comparing Eqs. 12.5 and 12.6, it is observed that matrix $_{i}^{j}[T]$ is the inverse of $_{j}^{i}[T]$; that is,

$$_{i}^{j}[T] =\,_{j}^{i}[T]^{-1} \tag{12.7}$$

or

$$_{j}^{i}[T]\,_{i}^{j}[T] = [I] \tag{12.8}$$

where $[I]$ is the identity matrix.

Based on the above concepts, multiple reference frames can be strategically located throughout a mechanism. For example, frame j in Figure 12.4 could be rigidly attached to the end effector of a manipulator, while frame i might be attached to the fixed base of the robot. Indeed, it will prove useful to have a reference frame embedded in each member of the mechanism, in which case Eq. 12.5 can be applied recursively in going successively from link to link throughout the mechanism. This use of multiple frames on adjacent links simplifies the determination of the specific forms of the transformation matrix, which is the subject of the next section. These matrices will be functions of the link dimensions and the relative joint displacements.

12.4 TRANSFORMATION MATRICES

Referring to Figure 12.4 and Eq. 12.5, the transformation matrix $^i_j[T]$ depends on the three degrees of translational displacement that define the position of the origin of frame j relative to the origin of frame i, and the three degrees of rotational displacement that define the angular orientation of frame j relative to that of frame i. This represents the most general case of coordinate mapping. However, we will consider some special cases before examining this general case.

12.4.1 Pure Rotation about the Origin

Figure 12.5 depicts the situation where the j frame is located relative to the i frame by a rotation through angle α about the x_i axis. From the geometry of the figure, the following equations are written relating the coordinates of any arbitrary point based on the two reference frames:

$$P_{x_i} = P_{x_j}$$

$$P_{y_i} = P_{y_j} \cos \alpha - P_{z_j} \sin \alpha$$

$$P_{z_i} = P_{y_j} \sin \alpha + P_{z_j} \cos \alpha$$

or, in matrix form,

$$\begin{Bmatrix} P_{x_i} \\ P_{y_i} \\ P_{z_i} \end{Bmatrix} = \begin{bmatrix} 1 & 0 & 0 \\ 0 & \cos \alpha & -\sin \alpha \\ 0 & \sin \alpha & \cos \alpha \end{bmatrix} \begin{Bmatrix} P_{x_j} \\ P_{y_j} \\ P_{z_j} \end{Bmatrix} \tag{12.9}$$

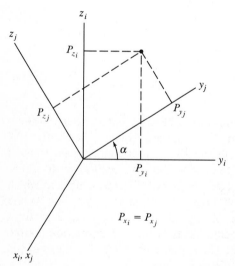

Figure 12.5 Rotation through angle α about the x_i axis.

Expressing this equation in condensed form,

$$^i\{P\} = {}^i_j[R]^j\{P\} \tag{12.10}$$

where the notation ${}^i_j[R]$ is used to represent a rotational transformation matrix. Specializing this notation further, we have

$$^i\{P\} = [R(x_i, \alpha)]^j\{P\} \tag{12.11}$$

where the rotational transformation $[R(x_i, \alpha)]$ defines a rotation of amount α about the x_i axis, and is given by

$$[R(x_i, \alpha)] = \begin{bmatrix} 1 & 0 & 0 \\ 0 & \cos\alpha & -\sin\alpha \\ 0 & \sin\alpha & \cos\alpha \end{bmatrix} \tag{12.12}$$

Similar derivations lead to the determination of transformation matrices for rotation about the y_i axis through angle β, $[R(y_i, \beta)]$, and rotation about the z_i axis through angle γ, $[R(z_i, \gamma)]$, yielding the following:

$$[R(y_i, \beta)] = \begin{bmatrix} \cos\beta & 0 & \sin\beta \\ 0 & 1 & 0 \\ -\sin\beta & 0 & \cos\beta \end{bmatrix} \tag{12.13}$$

$$[R(z_i, \gamma)] = \begin{bmatrix} \cos\gamma & -\sin\gamma & 0 \\ \sin\gamma & \cos\gamma & 0 \\ 0 & 0 & 1 \end{bmatrix} \tag{12.14}$$

Each of the three rotation matrices defined in Eqs. 12.12 to 12.14 has the useful property that its inverse is equal to its transpose. For example,

$$[R(x_i, \alpha)]^{-1} = [R(x_i, \alpha)]^T = \begin{bmatrix} 1 & 0 & 0 \\ 0 & \cos\alpha & \sin\alpha \\ 0 & -\sin\alpha & \cos\alpha \end{bmatrix} \tag{12.15}$$

Also, a general rotational transformation can be described in terms of a sequence of rotations about the x, y, and z axes. For example, suppose that the orientation of frame j is obtained from the orientation of frame i as a combination of a rotation through angle α about the x_i axis, followed by a rotation β about the y_i axis, followed by a rotation γ about the z_i axis, where the origins of frames i and j are coincident. Then, the transformation of the coordinates of any point from the j frame to the i frame is obtained from

$$^i\{P\} = {}^i_j[R]^j\{P\} \tag{12.16}$$

where

$${}^i_j[R] = [R(z_i, \gamma)][R(y_i, \beta)][R(x_i, \alpha)]$$

$$= \begin{bmatrix} \cos\beta\cos\gamma & (\sin\alpha\sin\beta\cos\gamma - \cos\alpha\sin\gamma) & (\cos\alpha\sin\beta\cos\gamma + \sin\alpha\sin\gamma) \\ \cos\beta\sin\gamma & (\sin\alpha\sin\beta\sin\gamma + \cos\alpha\cos\gamma) & (\cos\alpha\sin\beta\sin\gamma - \sin\alpha\cos\gamma) \\ -\sin\beta & \sin\alpha\cos\beta & \cos\alpha\cos\beta \end{bmatrix}$$

$$\tag{12.17}$$

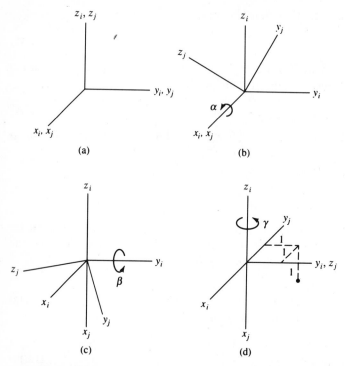

Figure 12.6 Rotation about the x_i axis, (a)–(b), followed by rotation about the y_i axis, (b)–(c), followed by rotation about the z_i axis, (c)–(d).

The order of the operations in Eq. 12.17 is important, since finite rotational displacements about multiple axes are not commutative. For example, in general,

$$[R(y_i, \beta)][R(x_i, \alpha)] \neq [R(x_i, \alpha)][R(y_i, \beta)]$$

It is helpful in forming the matrix $_j^i[R]$, and in forming transformation matrices in general, to think in terms of the steps that are required to move frame j from initial coincidence with frame i to its actual position and orientation. This is illustrated for the case of Eq. 12.17 in Figure 12.6. In Figure 12.6a, the two frames are coincident, and, therefore,

$$_j^i[R] = [I]$$

Next, an intermediate frame orientation is defined in Figure 12.6b, resulting from rotating the initial frame about the x_i axis. This yields

$$_j^i[R] = [R(x_i, \alpha)][I]$$

Now, the intermediate frame of Figure 12.6b is rotated about the y_i axis (see Figure 12.6c), leading to

$$_j^i[R] = [R(y_i, \beta)][R(x_i, \alpha)][I]$$

Finally, this second intermediate frame is rotated about the z_i axis into the final, actual position of frame j (Figure 12.6d):

$$_j^i[R] = [R(z_i, \gamma)][R(y_i, \beta)][R(x_i, \alpha)][I]$$

which is the same as the transformation matrix of Eq. 12.17. This process will be very useful in the following sections for analyzing manipulators. Note that the right-hand rule applies in assigning directions of rotation; that is, a right-hand rotation about a positive axis direction is positive and a left-hand rotation is negative.

♦ **EXAMPLE PROBLEM 12.1**   *An x–y–z Rotation Sequence*
Two reference frames, i and j, have the same origin. The orientation of frame j relative to frame i is given by a rotation of $60°$ about the x_i axis, followed by a rotation of $90°$ about the y_i axis, followed by a rotation of $150°$ about the z_i axis. A point has the following coordinates in frame j:

$$^j\{P\} = \begin{Bmatrix} 1 \\ 1 \\ 1 \end{Bmatrix}$$

Determine the coordinates of the point in frame i.

Solution.   This example is depicted in Figure 12.6. First, the transformation matrix is calculated as follows, using Eq. 12.17:

$$_j^i[R] = \begin{bmatrix} \cos 150° & -\sin 150° & 0 \\ \sin 150° & \cos 150° & 0 \\ 0 & 0 & 1 \end{bmatrix} \begin{bmatrix} \cos 90° & 0 & \sin 90° \\ 0 & 1 & 0 \\ -\sin 90° & 0 & \cos 90° \end{bmatrix}$$

$$\times \begin{bmatrix} 1 & 0 & 0 \\ 0 & \cos 60° & -\sin 60° \\ 0 & \sin 60° & \cos 60° \end{bmatrix}$$

$$= \begin{bmatrix} -\sqrt{3}/2 & -1/2 & 0 \\ 1/2 & -\sqrt{3}/2 & 0 \\ 0 & 0 & 1 \end{bmatrix} \begin{bmatrix} 0 & 0 & 1 \\ 0 & 1 & 0 \\ -1 & 0 & 0 \end{bmatrix} \begin{bmatrix} 1 & 0 & 0 \\ 0 & 1/2 & -\sqrt{3}/2 \\ 0 & \sqrt{3}/2 & 1/2 \end{bmatrix}$$

$$= \begin{bmatrix} -\sqrt{3}/2 & -1/2 & 0 \\ 1/2 & -\sqrt{3}/2 & 0 \\ 0 & 0 & 1 \end{bmatrix} \begin{bmatrix} 0 & \sqrt{3}/2 & 1/2 \\ 0 & 1/2 & -\sqrt{3}/2 \\ -1 & 0 & 0 \end{bmatrix}$$

$$= \begin{bmatrix} 0 & -1 & 0 \\ 0 & 0 & 1 \\ -1 & 0 & 0 \end{bmatrix}$$

Then, from Eq. 12.16,

$$^i\{P\} = \begin{bmatrix} 0 & -1 & 0 \\ 0 & 0 & 1 \\ -1 & 0 & 0 \end{bmatrix} \begin{Bmatrix} 1 \\ 1 \\ 1 \end{Bmatrix} = \begin{Bmatrix} -1 \\ 1 \\ -1 \end{Bmatrix}$$

This result is evident from Figure 12.6d. Note that the transformation matrix can be formed from different sets of rotations from the i frame. For example, we can convert frame j from the orientation in Figure 12.6a to that in Figure 12.6d by a combination of a rotation of 90° about the z_i axis, followed by a rotation of $-90°$ about the x_i axis, yielding

$$^i_j[R] = \begin{bmatrix} 1 & 0 & 0 \\ 0 & \cos(-90°) & -\sin(-90°) \\ 0 & \sin(-90°) & \cos(-90°) \end{bmatrix} \begin{bmatrix} \cos 90° & -\sin 90° & 0 \\ \sin 90° & \cos 90° & 0 \\ 0 & 0 & 1 \end{bmatrix}$$

$$= \begin{bmatrix} 0 & -1 & 0 \\ 0 & 0 & 1 \\ -1 & 0 & 0 \end{bmatrix}$$

which is the same transformation matrix as above.   ◆

Another approach to defining the rotational transformation matrix $^i_j[R]$ is to describe the orientation of frame j relative to frame i by means of a single rotation about a particular noncoordinate axis, rather than by means of a sequence of rotations about the coordinate axes. Figure 12.7 illustrates this case, where frame j is obtained from frame i as the result of a rotation through angle ϕ about an axis defined by unit vector $^i\{u\}$, given by

$$^i\{u\} = \begin{Bmatrix} u_{x_i} \\ u_{y_i} \\ u_{z_i} \end{Bmatrix} \qquad (12.18)$$

where

$$u_{x_i}^2 + u_{y_i}^2 + u_{z_i}^2 = 1 \qquad (12.19)$$

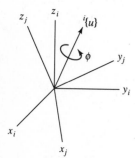

Figure 12.7 A general rotation about axis $^i\{u\}$ through angle ϕ.

The derivation of the rotational transformation matrix based on these quantities can be found in various sources on robotics.[4,6] Only the final result is presented here as the following matrix:

$$\substack{i\\j}[R] = \begin{bmatrix} (u_{x_i}^2 \text{ ver } \phi + \cos \phi) & (u_{x_i}u_{y_i} \text{ ver } \phi - u_{z_i} \sin \phi) & (u_{x_i}u_{z_i} \text{ ver } \phi + u_{y_i} \sin \phi) \\ (u_{x_i}u_{y_i} \text{ ver } \phi + u_{z_i} \sin \phi) & (u_{y_i}^2 \text{ ver } \phi + \cos \phi) & (u_{y_i}u_{z_i} \text{ ver } \phi - u_{x_i} \sin \phi) \\ (u_{x_i}u_{z_i} \text{ ver } \phi - u_{y_i} \sin \phi) & (u_{y_i}u_{z_i} \text{ ver } \phi + u_{x_i} \sin \phi) & (u_{z_i}^2 \text{ ver } \phi + \cos \phi) \end{bmatrix}$$

$$(12.20)$$

where

$$\text{ver } \phi = \text{versine } \phi = 1 - \cos \phi \qquad (12.21)$$

Eq. 12.20 is the most general form of the rotational transformation matrix, and, in fact, Eqs. 12.12, 12.13, and 12.14 are special cases of this matrix, wherein the axis of rotation $^i\{u\}$ is the x_i axis for the case of Eq. 12.12, the y_i axis for Eq. 12.13, and the z_i axis for Eq. 12.14.

◆ **EXAMPLE PROBLEM 12.2** *Rotation About a Noncoordinate Axis*
Frame j has an orientation relative to frame i given by a rotation of 120° about an axis defined by the unit vector

$$^i\{u\} = \left\{ \begin{array}{c} -1/\sqrt{3} \\ 1/\sqrt{3} \\ 1/\sqrt{3} \end{array} \right\}$$

Determine the transformation matrix $\substack{i\\j}[R]$.

Solution. Noting that ver 120° = 1 − cos 120° = 3/2 and sin 120° = $\sqrt{3}$/2, and substituting into Eq. 12.20, the result is

$$\substack{i\\j}[R] = \begin{bmatrix} 0 & -1 & 0 \\ 0 & 0 & 1 \\ -1 & 0 & 0 \end{bmatrix}$$

This is the same matrix as in Example 12.1, indicating that the orientation of frame j will be the same as shown in Figure 12.6d for the rotation specified in this example. ◆

The above examples illustrate the fact that, regardless of the path taken or the sequence of operations in going from frame i to a particular final orientation of frame j, the transformation matrix itself is unique. Therefore, given a sequence of x, y, and z rotations and the resulting $[R]$ matrix from Eq. 12.17, we can solve for the equivalent single axis rotation (that is, axis $^i\{u\}$ and angle ϕ) from Eq. 12.20. In the subsequent sections, the following general representation of $\substack{i\\j}[R]$ will be used, which can be obtained in any of the ways described above:

$$
{}_{j}^{i}[R] = \begin{bmatrix} R_{11} & R_{12} & R_{13} \\ R_{21} & R_{22} & R_{23} \\ R_{31} & R_{32} & R_{33} \end{bmatrix} \tag{12.22}
$$

which is a 3×3 matrix.

12.4.2 Pure Translation and Homogeneous Coordinates

Figure 12.8 shows the case where frame j is located relative to frame i by a three-dimensional translation, given by vector ${}^{i}\{Q\}$, which locates the origin of frame j with respect to that of frame i; that is,

$$
{}^{i}\{Q\} = \begin{Bmatrix} Q_{x_i} \\ Q_{y_i} \\ Q_{z_i} \end{Bmatrix} \tag{12.23}
$$

The coordinate axes of the two frames are parallel. Therefore, as shown in Figure 12.8, the coordinates of any point in space, expressed in the two reference frames, are related as follows:

$$
P_{x_i} = P_{x_j} + Q_{x_i}
$$

$$
P_{y_i} = P_{y_j} + Q_{y_i}
$$

$$
P_{z_i} = P_{z_j} + Q_{z_i}
$$

or

$$
{}^{i}\{P\} = {}^{j}\{P\} + {}^{i}\{Q\} \tag{12.24}
$$

Eq. 12.24 accurately and completely describes the transformation. However, we would like to continue the use of transformation matrices in the form of Eq. 12.5; that is,

$$
{}^{i}\{P\} = {}_{j}^{i}[T]{}^{j}\{P\}
$$

where matrix ${}_{j}^{i}[T]$ would somehow be a function of translation ${}^{i}\{Q\}$ in this case.

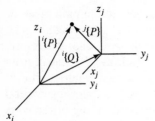

Figure 12.8 Frame j is located relative to frame i by means of translation vector ${}^{i}\{Q\}$.

We can accomplish this by revising the definition of vectors from the 3×1 dimensional form, as in Eqs. 12.3 and 12.4, to the following form with dimensions of four rows and one column:

$$
{}^{i}\{P\} = \begin{Bmatrix} P_{x_i} \\ P_{y_i} \\ P_{z_i} \\ 1 \end{Bmatrix}
\tag{12.25}
$$

and

$$
{}^{j}\{P\} = \begin{Bmatrix} P_{x_j} \\ P_{y_j} \\ P_{z_j} \\ 1 \end{Bmatrix}
\tag{12.26}
$$

where the fourth entry in each case is always unity. This modified vector definition is referred to as homogeneous coordinates, and from this point on, vectors will be assumed to be in homogeneous coordinate form, unless specified otherwise.

Now, Eq. 12.24 can be written equivalently as

$$
{}^{i}\{P\} = {}^{i}_{j}[T]\,{}^{j}\{P\}
\tag{12.27}
$$

where ${}^{i}\{P\}$ and ${}^{j}\{P\}$ are given by Eqs. 12.25 and 12.26, respectively, and matrix ${}^{i}_{j}[T]$, which now has dimensions of 4×4, is defined as

$$
{}^{i}_{j}[T] = \begin{bmatrix} 1 & 0 & 0 & Q_{x_i} \\ 0 & 1 & 0 & Q_{y_i} \\ 0 & 0 & 1 & Q_{z_i} \\ 0 & 0 & 0 & 1 \end{bmatrix}
\tag{12.28}
$$

As desired, the transformation matrix is a function of the relative position of the two reference frames, and is not a function of the coordinates of specific points. In this way, the transformation can be applied to any point for the purpose of converting from coordinates in frame j to coordinates in frame i.

The inverse of matrix ${}^{i}_{j}[T]$ of Eq. 12.28 is

$$
{}^{i}_{j}[T]^{-1} = \begin{bmatrix} 1 & 0 & 0 & -Q_{x_i} \\ 0 & 1 & 0 & -Q_{y_i} \\ 0 & 0 & 1 & -Q_{z_i} \\ 0 & 0 & 0 & 1 \end{bmatrix}
\tag{12.29}
$$

which, by definition, is the transformation matrix ${}^{j}_{i}[T]$ which maps the i frame into the j frame; that is,

$$^j\{P\} = {}^j_i[T]\,{}^i\{P\}$$
(12.30)

This is easily checked by carrying out the multiplication process in Eq. 12.30, which leads to the following relations:

$$P_{x_j} = P_{x_i} - Q_{x_i}$$
(12.31a)

$$P_{y_j} = P_{y_i} - Q_{y_i}$$
(12.31b)

$$P_{z_j} = P_{z_i} - Q_{z_i}$$
(12.31c)

$$1 = 1$$
(12.31d)

which are in agreement with Eqs. 12.24 and Figure 12.8. Note that the last equation resulting from matrix equations such as Eq. 12.30 is always an identity, because of the way in which homogeneous coordinates have been defined.

The rotational transformations of the previous section can be easily modified to incorporate the homogeneous coordinate form. In particular, the 4×4 transformation matrix for producing general rotation is now

$$
{}^i_j[T] =
\begin{bmatrix}
R_{11} & R_{12} & R_{13} & 0 \\
R_{21} & R_{22} & R_{23} & 0 \\
R_{31} & R_{32} & R_{33} & 0 \\
0 & 0 & 0 & 1
\end{bmatrix}
$$
(12.32)

where the 3×3 $[R]$ matrix is determined as in Section 12.4.1.

12.4.3 Combined Rotation and Translation

We can now consider the general case where frame j has a combination of a different rotational orientation and a different translational position from that of frame i (see Figure 12.9). Once again, it will be helpful in deriving the transformation matrix to break down the transformation into steps required to

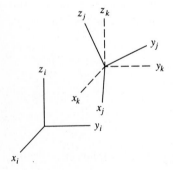

Figure 12.9 The location of frame j relative to frame i consists of a combination of rotation and translation.

move frame j from initial coincidence with frame i to its actual location. For example, suppose that the translational transformation is performed first. This is a translation $^i\{Q\}$ to the intermediate dashed location, frame k in Figure 12.9. This places the origin of the coordinate frame at its final position; however, the coordinate axes of this intermediate frame are still parallel to frame i. Mathematically, for any point, we would have

$$^i\{P\} = \begin{bmatrix} 1 & 0 & 0 & Q_{x_i} \\ 0 & 1 & 0 & Q_{y_i} \\ 0 & 0 & 1 & Q_{z_i} \\ 0 & 0 & 0 & 1 \end{bmatrix} {}^k\{P\} \tag{12.33}$$

Now consider the rotation of frame k about its origin to the orientation of frame j. Applying Eq. 12.32 to frames k and j,

$$^k\{P\} = \begin{bmatrix} R_{11} & R_{12} & R_{13} & 0 \\ R_{21} & R_{22} & R_{23} & 0 \\ R_{31} & R_{32} & R_{33} & 0 \\ 0 & 0 & 0 & 1 \end{bmatrix} {}^j\{P\} \tag{12.34}$$

where the 3×3 rotation matrix is $^k_j[R]$, which is also $^i_j[R]$ since the i and k frames are parallel. Substituting Eq. 12.34 into Eq. 12.33 for $^k\{P\}$,

$$^i\{P\} = \begin{bmatrix} 1 & 0 & 0 & Q_{x_i} \\ 0 & 1 & 0 & Q_{y_i} \\ 0 & 0 & 1 & Q_{z_i} \\ 0 & 0 & 0 & 1 \end{bmatrix} \begin{bmatrix} R_{11} & R_{12} & R_{13} & 0 \\ R_{21} & R_{22} & R_{23} & 0 \\ R_{31} & R_{32} & R_{33} & 0 \\ 0 & 0 & 0 & 1 \end{bmatrix} {}^j\{P\} \tag{12.35}$$

Carrying out the matrix multiplication, the final result is

$$^i\{P\} = {}^i_j[T] \, {}^j\{P\} \tag{12.36}$$

where the general transformation matrix for combined rotation and translation is

$$^i_j[T] = \begin{bmatrix} R_{11} & R_{12} & R_{13} & Q_{x_i} \\ R_{21} & R_{22} & R_{23} & Q_{y_i} \\ R_{31} & R_{32} & R_{33} & Q_{z_i} \\ 0 & 0 & 0 & 1 \end{bmatrix} \tag{12.37}$$

or, in partitioned form,

$$^i_j[T] = \left[\begin{array}{c|c} {}^i_j[R] & {}^i\{Q\} \\ \hline 0 & 1 \end{array} \right] \tag{12.38}$$

where $^i_j[R]$ is the 3×3 rotation matrix and $^i\{Q\}$ is the 3×1 translation vector.

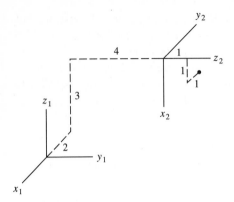

Figure 12.10 The situation for Example Problem 12.3.

♦ **EXAMPLE PROBLEM 12.3**   *Combined Rotation and Translation*

Two reference frames, 1 and 2, are shown in Figure 12.10. The orientation of frame 2 relative to frame 1 is given by a rotation of 60° about the x_1 axis, followed by a rotation of 90° about the y_1 axis, followed by a rotation of 150° about the z_1 axis. The position of the origin of frame 2 relative to the origin of frame 1 is

$$^1\{Q\} = \begin{Bmatrix} -2 \\ 4 \\ 3 \end{Bmatrix}$$

A point has the following coordinates in frame 2:

$$^2\{P\} = \begin{Bmatrix} 1 \\ 1 \\ 1 \end{Bmatrix}$$

Determine the coordinates of the point in frame 1.

Solution.   The rotational transformation between the two frames is the same as that considered in Example Problems 12.1 and 12.2. Therefore, referring to either of those examples,

$$^1_2[R] = \begin{bmatrix} 0 & -1 & 0 \\ 0 & 0 & 1 \\ -1 & 0 & 0 \end{bmatrix}$$

Substituting $^1_2[R]$, $^1\{Q\}$, and $^2\{P\}$ into Eqs. 12.36 and 12.37,

$$^1\{P\} = {}^1_2[T]^2\{P\} = \begin{bmatrix} 0 & -1 & 0 & -2 \\ 0 & 0 & 1 & 4 \\ -1 & 0 & 0 & 3 \\ 0 & 0 & 0 & 1 \end{bmatrix} \begin{Bmatrix} 1 \\ 1 \\ 1 \\ 1 \end{Bmatrix} = \begin{Bmatrix} -3 \\ 5 \\ 2 \\ 1 \end{Bmatrix}$$

This result can be verified through examination of Figure 12.10.   ♦

12.5 FORWARD KINEMATICS

Before applying the methods of the previous sections to robot manipulators, we first need to define a consistent set of conventions and notation, which will be applicable to a broad range of robots. In particular, we define key link parameters and joint variables, and establish a convention for systematically locating coordinate reference frames on each link. In this way, a single transformation matrix can be derived for transforming information between adjacent links, and this matrix can then be used repeatedly for analyzing entire mechanisms.

12.5.1 Definitions

Figure 12.11a shows a general member, link i, of a manipulator. It will be assumed that all members are binary members (i.e., having two joints), as shown, and that the joints are either revolute or prismatic. These assumptions are not overly restrictive but, instead, are consistent with the construction of most manipulators in common use. The joint axes are labeled m and n in the figure. For purposes of rigid-body kinematics, there are two key parameters, the link length and the link twist, which totally define the link geometry. These are constant quantities that describe the relative locations of axes m and n. In particular, the link length l_i is defined as the perpendicular distance between the axes, and the link twist τ_i is the angle between the axes, defined as follows. Axis n is projected along the common perpendicular until it intersects axis m (see Figure 12.11a). Angle τ_i is the angle measured from axis m to this projected line, with its sense given by the right-hand rule with respect to the common perpendicular.

Figure 12.11b shows the coordinate frame attached to member i. Its origin is at the intersection of the common perpendicular and axis m. The x axis of

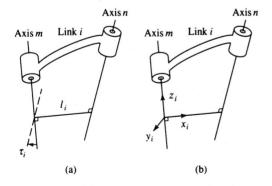

(a)   (b)

Figure 12.11 (a) The link parameters (l_i, τ_i) for link i of a manipulator. (b) The definition of the coordinate system for link i.

frame i is directed along the common perpendicular toward axis n. The z axis is along axis m. Note that there are two choices for the direction of this axis (the upward direction has been selected in Figure 12.11b), and once the choice has been made, all subsequent results will be consistent with it. Finally, the direction of the y axis is obtained from axes x_i and z_i and the right-hand rule, forming a right-hand coordinate system. In order to apply the analysis equations, which we will derive, each link reference frame must be defined in this manner.

There are instances where special definitions of reference frames are required. Specifically, if the manipulator is formed from an open kinematic chain, then the coordinate systems for the first link (the fixed base) and the last link (the end effector) cannot be defined as above, because each of these members has only one joint. For the base, it is common practice to define its frame such that it is coincident with the frame for the adjacent moving link at some predefined "home" position of the robot. In this way, the link length and link angle of the base are both zero. The frame for the last link in the chain is conveniently located such that its z axis is along the axis of the joint that connects the link to the preceding link and its x axis is along a key dimension of the link. There are occasions when other choices of reference frames for these links are also useful.

Figure 12.12a shows the joint variables associated with joint n between links i and j. The joint offset s_n is the distance along the joint axis from the common perpendicular for link i to that for link j, and is positive when in the direction of positive z_j. The joint angle θ_n is the angle around axis n from the x_i axis to the x_j axis, with its sign given by the right-hand rule with respect to the z_j axis. In the case of a revolute joint, angle θ_n is truly variable with time, while offset s_n would have a constant value fixed by the dimensions of the robot. On the other hand, for a prismatic joint, s_n is variable and θ_n is fixed. For cylindric and helical joints, both joint quantities are variable, independently of one another in the former case and dependently in the latter.

12.5.2 Link-to-Link Transformation Matrix

Consider Figure 12.12 and the derivation of transformation matrix ${}_j^i[T]$, which relates coordinates measured in the frame attached to link j and those measured in the frame attached to link i. In developing this matrix, we temporarily utilize an intermediate frame, frame k in Figure 12.12b. This frame has its origin at the intersection of axis n and the common perpendicular for link i, its x axis along axis x_i, and its z axis along axis z_j. Now consider the relation between frames j and k. Frame j is obtained from frame k by a rotation of angle θ_n about axis z_k and a translation of distance s_n along axis z_k (see Figure 12.12a). Using Eq. 12.14 to form ${}_j^k[R]$, and substituting into Eq. 12.37, the transformation from frame j to frame k is

$$^k\{P\} = {}_j^k[T]^j\{P\} \tag{12.39}$$

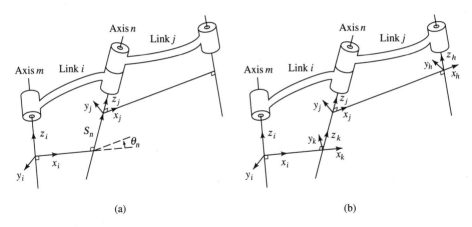

Figure 12.12 (a) The joint variables (s_n, θ_n) associated with joint n of a manipulator. (b) Intermediate frame k for the derivation of the transformation matrix from frame j to frame i.

where

$$
{}^k_j[T] = \begin{bmatrix} \cos\theta_n & -\sin\theta_n & 0 & 0 \\ \sin\theta_n & \cos\theta_n & 0 & 0 \\ 0 & 0 & 1 & s_n \\ 0 & 0 & 0 & 1 \end{bmatrix}
\tag{12.40}
$$

Next, consider the relation between frames k and i. Frame k is obtained by a rotation τ_i about axis x_i (Eq. 12.12) and a translation l_i along x_i (see Figure 12.11), leading to

$$
{}^i\{P\} = {}^i_k[T]{}^k\{P\}
\tag{12.41}
$$

where

$$
{}^i_k[T] = \begin{bmatrix} 1 & 0 & 0 & l_i \\ 0 & \cos\tau_i & -\sin\tau_i & 0 \\ 0 & \sin\tau_i & \cos\tau_i & 0 \\ 0 & 0 & 0 & 1 \end{bmatrix}
\tag{12.42}
$$

Substituting Eq. 12.39 into Eq. 12.41, the desired result is

$$
{}^i\{P\} = {}^i_j[T]{}^j\{P\}
\tag{12.43}
$$

where the final transformation matrix is

$$
{}^i_j[T] = {}^i_k[T]{}^k_j[T] = \begin{bmatrix} \cos\theta_n & -\sin\theta_n & 0 & l_i \\ \cos\tau_i\sin\theta_n & \cos\tau_i\cos\theta_n & -\sin\tau_i & -s_n\sin\tau_i \\ \sin\tau_i\sin\theta_n & \sin\tau_i\cos\theta_n & \cos\tau_i & s_n\cos\tau_i \\ 0 & 0 & 0 & 1 \end{bmatrix}
\tag{12.44}
$$

This transformation matrix will be applied to the analysis of some typical robotic manipulators in the following sections.

It can be shown that Eq. 12.44 is consistent with the well-known Denavit-Hartenberg representation[3], when one considers the different coordinate systems used. In their approach, a transformation matrix was derived relating frame k as the primary reference frame for link i, and a similarly defined frame, frame h in Figure 12.12b, for link j. Then the transformation matrix relating frames h and k is

$$_h^k[T] = {}_j^k[T]{}_h^j[T]$$

where ${}_j^k[T]$ is given in Eq. 12.40, and ${}_h^j[T]$ is defined for link j as ${}_k^i[T]$ was defined for link i; that is,

$$
{}_h^j[T] =
\begin{bmatrix}
1 & 0 & 0 & l_j \\
0 & \cos \tau_j & -\sin \tau_j & 0 \\
0 & \sin \tau_j & \cos \tau_j & 0 \\
0 & 0 & 0 & 1
\end{bmatrix}
$$

Substituting and multiplying,

$$
{}_h^k[T] =
\begin{bmatrix}
\cos \theta_n & -\sin \theta_n \cos \tau_j & \sin \theta_n \sin \tau_j & l_j \cos \theta_n \\
\sin \theta_n & \cos \theta_n \cos \tau_j & -\cos \theta_n \sin \tau_j & l_j \sin \theta_n \\
0 & \sin \tau_j & \cos \tau_j & s_n \\
0 & 0 & 0 & 1
\end{bmatrix}
$$

This is the Denavit-Hartenberg transformation matrix.

12.5.3 2R Planar Manipulator

The manipulator of Figure 12.13 will serve to illustrate the analysis approach for a case that is easily verified using traditional analysis methods. The manipulator has three members, including the fixed base (member 0), and two revolute joints whose axes are parallel. The mechanism therefore experiences planar motion and, from Eq. 12.2, has two degrees of freedom. These degrees of freedom are represented by independent rotations at joints A and B. The

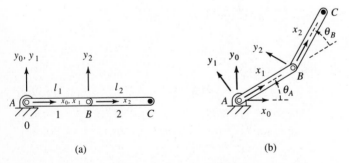

Figure 12.13 A two-degree-of-freedom planar manipulator.

objective here is to describe the location of point C on member 2 relative to the base of the manipulator.

The home position of the manipulator is shown in Figure 12.13a, and the link coordinate frames will be defined with reference to this figure. The frame for link 1 follows directly from the convention presented earlier. The x_1 axis is along the common normal between the joint axes at A and B, both of which are perpendicular to the page. The z_1 axis is along the joint A axis, and is chosen to point out of the page. The location of axis y_1 then follows as shown. The reference frame on the base, frame 0, is coincident with frame 1 in the home position and remains stationary as the robot moves. Frame 2 is defined with the z_2 axis along the axis of joint B, pointing out of the page. Axis x_2 is along link 2, and axis y_2 follows for a right-hand system.

Based on the frame locations and Figure 12.13b, the following values of the link parameters and joint variables are obtained:

$$l_0 = \tau_0 = \tau_1 = s_A = s_B = 0$$
$$l_1 = l_1$$
$$\theta_A = \theta_A$$
$$\theta_B = \theta_B$$

Recall that the joint angles θ_A and θ_B are the relative rotations between the two links at each joint, as shown in Figure 12.13b. Also, the coordinates of point C in frame 2 are given by

$$^2\{P\}_C = \begin{Bmatrix} l_2 \\ 0 \\ 0 \\ 1 \end{Bmatrix} \tag{12.45}$$

where this is a constant vector since both frame 2 and point C are attached to link 2. Substituting into Eq. 12.44, the transformation matrices are

$$^0_1[T] = \begin{bmatrix} \cos\theta_A & -\sin\theta_A & 0 & 0 \\ \sin\theta_A & \cos\theta_A & 0 & 0 \\ 0 & 0 & 1 & 0 \\ 0 & 0 & 0 & 1 \end{bmatrix} \tag{12.46}$$

and

$$^1_2[T] = \begin{bmatrix} \cos\theta_B & -\sin\theta_B & 0 & l_1 \\ \sin\theta_B & \cos\theta_B & 0 & 0 \\ 0 & 0 & 1 & 0 \\ 0 & 0 & 0 & 1 \end{bmatrix} \tag{12.47}$$

Applying these transformations to point C,

$$^1\{P\}_C = {}^1_2[T]^2\{P\}_C = \begin{Bmatrix} l_1 + l_2\cos\theta_B \\ l_2\sin\theta_B \\ 0 \\ 1 \end{Bmatrix} \tag{12.48}$$

and

$$
{}^{0}\{P\}_C = {}^{0}_{1}[T]{}^{1}\{P\}_C = \left\{ \begin{array}{c} l_1 \cos \theta_A + l_2 \cos \theta_A \cos \theta_B - l_2 \sin \theta_A \sin \theta_B \\ l_1 \sin \theta_A + l_2 \sin \theta_A \cos \theta_B + l_2 \cos \theta_A \sin \theta_B \\ 0 \\ 1 \end{array} \right\}
$$

$$
= \left\{ \begin{array}{c} l_1 \cos \theta_A + l_2 \cos(\theta_A + \theta_B) \\ l_1 \sin \theta_A + l_2 \sin(\theta_A + \theta_B) \\ 0 \\ 1 \end{array} \right\} \tag{12.49}
$$

It can easily be observed from Figure 12.13b that the first two terms in Eq. 12.49 are the x and y coordinates, respectively, of point C as referred to frame 0, which is fixed to the base of the robot. Obviously, these coordinates vary with the joint rotations, θ_A and θ_B, as the robot moves.

12.5.4 RP Manipulator

Figure 12.14 shows another simple manipulator consisting of three members and two joints, in this case a revolute joint and a prismatic joint. The figure shows the locations of the three reference frames attached to the three members, following the rules of Section 12.5.1. Note that in this case the two joint axes (A and B) intersect; therefore, length $l_1 = 0$. Also, for convenience, the origin of the fixed reference frame 0 has not been selected to be coincident

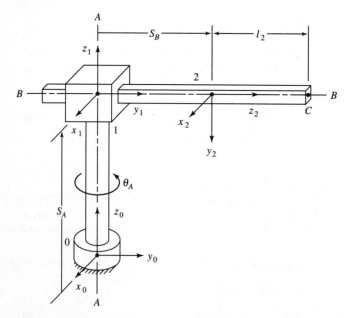

Figure 12.14 A manipulator with a revolute joint and a prismatic joint.

with the origin of frame 1 but instead is placed at a constant offset s_A along axis A. Then, from the figure, the following values are obtained:

$$l_0 = \tau_0 = \theta_B = 0$$

$$\tau_1 = -90°$$

$$\theta_A = \theta_A$$

$$s_B = s_B$$

From Eq. 12.44, the resulting transformation matrices are

$$
{}^0_1[T] =
\begin{bmatrix}
\cos\theta_A & -\sin\theta_A & 0 & 0 \\
\sin\theta_A & \cos\theta_A & 0 & 0 \\
0 & 0 & 1 & s_A \\
0 & 0 & 0 & 1
\end{bmatrix}
\tag{12.50}
$$

and

$$
{}^1_2[T] =
\begin{bmatrix}
1 & 0 & 0 & 0 \\
0 & 0 & 1 & s_B \\
0 & -1 & 0 & 0 \\
0 & 0 & 0 & 1
\end{bmatrix}
\tag{12.51}
$$

In frame 2, point C attached to the sliding link has the following definition:

$$
{}^2\{P\}_C =
\begin{Bmatrix}
0 \\
0 \\
l_2 \\
1
\end{Bmatrix}
\tag{12.52}
$$

Then, combining Eqs. 12.50 to 12.52, the coordinates of point C, as defined in reference frame 0, are determined as follows:

$$
{}^0\{P\}_C = {}^0_1[T]\,{}^1_2[T]\,{}^2\{P\}_C
$$

$$
=
\begin{bmatrix}
\cos\theta_A & -\sin\theta_A & 0 & 0 \\
\sin\theta_A & \cos\theta_A & 0 & 0 \\
0 & 0 & 1 & s_A \\
0 & 0 & 0 & 1
\end{bmatrix}
\begin{bmatrix}
1 & 0 & 0 & 0 \\
0 & 0 & 1 & s_B \\
0 & -1 & 0 & 0 \\
0 & 0 & 0 & 1
\end{bmatrix}
\begin{Bmatrix}
0 \\
0 \\
l_2 \\
1
\end{Bmatrix}
$$

$$
=
\begin{bmatrix}
\cos\theta_A & -\sin\theta_A & 0 & 0 \\
\sin\theta_A & \cos\theta_A & 0 & 0 \\
0 & 0 & 1 & s_A \\
0 & 0 & 0 & 1
\end{bmatrix}
\begin{Bmatrix}
0 \\
l_2 + s_B \\
0 \\
1
\end{Bmatrix}
$$

$$
=
\begin{Bmatrix}
-(s_B + l_2)\sin\theta_A \\
(s_B + l_2)\cos\theta_A \\
s_A \\
1
\end{Bmatrix}
\tag{12.53}
$$

Once again, this result can be verified from observation of Figure 12.14. In this mechanism, the time-dependent joint variables are rotation θ_A and translation s_B.

12.5.5 RRR Spatial Manipulator

Figure 12.15 depicts a robotic manipulator that has spatial motion with three degrees of freedom. There are three revolute axes, identified as axes A, B, and C, with corresponding joint rotations θ_A, θ_B, and θ_C, which would be produced by actuators such as electric motors, not shown in the figure. An end effector or tool would be mounted at location D on member 3. The robot is shown in its home position, and the reference frame for each of the four members is shown. As in the previous section, it will be convenient to locate the reference frame for the base, member 0, as shown, rather than following the guideline of making it coincident with frame 1 at the home position. Instead, the two frames are parallel with frame 1 offset a constant distance s_A along the z_0 axis.

From Figure 12.15, the following constant values of parameters are identified:

$$l_0 = l_1 = 0$$

$$l_2 = l_2$$

$$\tau_0 = \tau_2 = 0$$

$$\tau_1 = 90°$$

$$s_B = s_C = 0$$

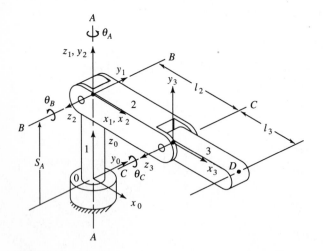

Figure 12.15 A three-degree-of-freedom spatial manipulator.

Substituting into Eq. 12.44, the transformation matrices are

$$
{}^0_1[T] =
\begin{bmatrix}
\cos\theta_A & -\sin\theta_A & 0 & 0 \\
\sin\theta_A & \cos\theta_A & 0 & 0 \\
0 & 0 & 1 & s_A \\
0 & 0 & 0 & 1
\end{bmatrix}
\tag{12.54}
$$

$$
{}^1_2[T] =
\begin{bmatrix}
\cos\theta_B & -\sin\theta_B & 0 & 0 \\
0 & 0 & -1 & 0 \\
\sin\theta_B & \cos\theta_B & 0 & 0 \\
0 & 0 & 0 & 1
\end{bmatrix}
\tag{12.55}
$$

and

$$
{}^2_3[T] =
\begin{bmatrix}
\cos\theta_C & -\sin\theta_C & 0 & l_2 \\
\sin\theta_C & \cos\theta_C & 0 & 0 \\
0 & 0 & 1 & 0 \\
0 & 0 & 0 & 1
\end{bmatrix}
\tag{12.56}
$$

Combining these transformation matrices, utilizing trigonometric identities,

$$
{}^0_3[T] =
\begin{bmatrix}
\cos\theta_A & -\sin\theta_A & 0 & 0 \\
\sin\theta_A & \cos\theta_A & 0 & 0 \\
0 & 0 & 1 & s_A \\
0 & 0 & 0 & 1
\end{bmatrix}
\begin{bmatrix}
\cos\theta_B & -\sin\theta_B & 0 & 0 \\
0 & 0 & -1 & 0 \\
\sin\theta_B & \cos\theta_B & 0 & 0 \\
0 & 0 & 0 & 1
\end{bmatrix}
$$

$$
\times
\begin{bmatrix}
\cos\theta_C & -\sin\theta_C & 0 & l_2 \\
\sin\theta_C & \cos\theta_C & 0 & 0 \\
0 & 0 & 1 & 0 \\
0 & 0 & 0 & 1
\end{bmatrix}
$$

$$
=
\begin{bmatrix}
\cos\theta_A & -\sin\theta_A & 0 & 0 \\
\sin\theta_A & \cos\theta_A & 0 & 0 \\
0 & 0 & 1 & s_A \\
0 & 0 & 0 & 1
\end{bmatrix}
$$

$$
\times
\begin{bmatrix}
\cos(\theta_B + \theta_C) & -\sin(\theta_B + \theta_C) & 0 & l_2\cos\theta_B \\
0 & 0 & -1 & 0 \\
\sin(\theta_B + \theta_C) & \cos(\theta_B + \theta_C) & 0 & l_2\sin\theta_B \\
0 & 0 & 0 & 1
\end{bmatrix}
$$

We now apply this transformation to point D, which has the following fixed coordinates in reference frame 3:

$$^3\{P\}_D = \begin{Bmatrix} l_3 \\ 0 \\ 0 \\ 1 \end{Bmatrix} \tag{12.58}$$

The resulting coordinates of point D, as expressed in terms of fixed frame 0, are

$$^0\{P\}_D = {}^0_3[T]^3\{P\}_D = \begin{Bmatrix} l_2 \cos\theta_A \cos\theta_B + l_3 \cos\theta_A \cos(\theta_B + \theta_C) \\ l_2 \sin\theta_A \cos\theta_B + l_3 \sin\theta_A \cos(\theta_B + \theta_C) \\ s_A + l_2 \sin\theta_B + l_3 \sin(\theta_B + \theta_C) \\ 1 \end{Bmatrix} \tag{12.59}$$

The quantities l_2, l_3, and s_A are fixed robot dimensions, and the joint angles θ_A, θ_B, and θ_C are time variant.

◆ **EXAMPLE PROBLEM 12.4**   *Analysis of an RRR Robot*
A robot of the type shown in Figure 12.15 has the following dimensions: $s_A = l_2 = l_3 = 10$ in. Determine the location of tip point D for the following joint rotations from the home position (shown in Figure 12.15): $\theta_A = 90°$, $\theta_B = 30°$, and $\theta_C = 60°$.

Solution.   Substituting into Eq. 12.59, the coordinates of point D are

$$^0\{P\}_D = \begin{Bmatrix} 10\cos 90° \cos 30° + 10\cos 90° \cos(30° + 60°) \\ 10\sin 90° \cos 30° + 10\sin 90° \cos(30° + 60°) \\ 10 + 10\sin 30° + 10\sin(30° + 60°) \\ 1 \end{Bmatrix} = \begin{Bmatrix} 0 \\ 8.66 \\ 25.0 \\ 1 \end{Bmatrix}$$

This position of the robot is illustrated in Figure 12.16, which confirms the above results.   ◆

The matrix approach that has been presented here for position analysis can be extended to velocity and acceleration analysis. It also facilitates force analysis and full dynamic analysis.

12.6 INVERSE KINEMATICS

Inverse kinematics refers to the process of determining joint coordinates from known values of the end effector coordinates. As mentioned earlier, this type of computation is necessary in programming robots, because actuation occurs at the joints, and therefore, the joint displacements corresponding to a prescribed end effector displacement must be known. However, as will be demonstrated in the following sections, inverse kinematic analysis is often

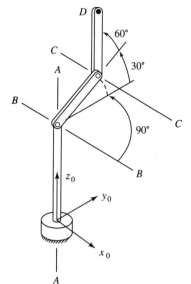

Figure 12.16 The manipulator of Example Problem 12.4.

more difficult than forward kinematic analysis. This is so because the mathematical relationships are nonlinear in the joint coordinates. Consequently, there is not a general solution procedure that governs all robot cases. Indeed, in some instances, it may be necessary to resort to numerical methods for the solution of the nonlinear equations.

Two inverse kinematics approaches will be illustrated. The first approach involves the use of forward kinematics equations, which are solved for joint variables as the unknowns. In the second approach, a more direct derivation of expressions for joint coordinates will be pursued. The success of the latter approach is heavily dependent upon the specific manipulator geometry.

12.6.1 2R Planar Manipulator

Consider the two-degree-of-freedom manipulator that was analyzed in Section 12.5.3 and is illustrated in Figure 12.13. Expressions for the robot tip coordinates as functions of the joint coordinates were derived, yielding Eq. 12.49, which is repeated here in the following form:

$$x = l_1 \cos \theta_A + l_2 \cos(\theta_A + \theta_B) \tag{12.60}$$

$$y = l_1 \sin \theta_A + l_2 \sin(\theta_A + \theta_B) \tag{12.61}$$

where x and y refer to the tip or end effector coordinates (i.e., point C in Figure 12.13) in terms of the fixed reference frame (i.e., frame 0). Subscripts and superscripts have been dropped in order to simplify the form of the equations.

Following the first approach outlined above, we will attempt to solve Eqs. 12.60 and 12.61 for θ_A and θ_B as functions of x and y. Squaring both sides of

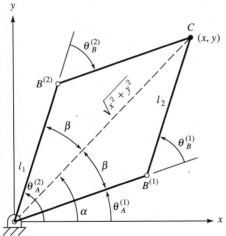

Figure 12.17 Multiple manipulator configurations obtained from inverse kinematics.

each equation and then adding the two equations,

$$x^2 + y^2 = l_1^2 + l_2^2 + 2l_1l_2[\cos\theta_A\cos(\theta_A + \theta_B) + \sin\theta_A\sin(\theta_A + \theta_B)]$$

$$= l_1^2 + l_2^2 + 2l_1l_2\cos\theta_B$$

Rearranging,

$$\cos\theta_B = \frac{x^2 + y^2 - l_1^2 - l_2^2}{2l_1l_2} \tag{12.62}$$

or

$$\theta_B = \cos^{-1}\left(\frac{x^2 + y^2 - l_1^2 - l_2^2}{2l_1l_2}\right) \tag{12.63}$$

There are two values of θ_B between $-180°$ and $+180°$ which satisfy this equation, indicating that there are multiple configurations that the robot can have and still maintain the same tip coordinates x, y. In fact, there are two such configurations of the robot in this example, which are illustrated in Figure 12.17. This characteristic of multiple solutions is common to most inverse kinematic analyses. The particular orientation that the robot would actually take would be dictated by the previous positions of the robot and the way in which the joint motions are programmed.

Having found θ_B, we can return to Eqs. 12.60 and 12.61 in order to determine θ_A. Rewriting those equations in a modified form,

$$(l_1 + l_2\cos\theta_B)\cos\theta_A - l_2\sin\theta_B\sin\theta_A = x \tag{12.64}$$

and

$$l_2\sin\theta_B\cos\theta_A + (l_1 + l_2\cos\theta_B)\sin\theta_A = y \tag{12.65}$$

where $\cos \theta_B$ is given by Eq. 12.62 and

$$\sin \theta_B = \pm \sqrt{1 - \cos^2 \theta_B} \tag{12.66}$$

The $\pm$ signs in Eq. 12.66 correspond to the two solutions for θ_B as noted above. Solving Eqs. 12.64 and 12.65 for $\cos \theta_A$ and $\sin \theta_A$,

$$\cos \theta_A = \frac{x(l_1 + l_2 \cos \theta_B) + y l_2 \sin \theta_B}{l_1^2 + l_2^2 + 2 l_1 l_2 \cos \theta_B} \tag{12.67}$$

$$\sin \theta_A = \frac{y(l_1 + l_2 \cos \theta_B) - x l_2 \sin \theta_B}{l_1^2 + l_2^2 + 2 l_1 l_2 \cos \theta_B} \tag{12.68}$$

and, hence,

$$\tan \theta_A = \frac{\sin \theta_A}{\cos \theta_A} = \frac{y(l_1 + l_2 \cos \theta_B) - x l_2 \sin \theta_B}{x(l_1 + l_2 \cos \theta_B) + y l_2 \sin \theta_B} \tag{12.69}$$

or

$$\theta_A = \tan^{-1}\left[\frac{y(l_1 + l_2 \cos \theta_B) - x l_2 \sin \theta_B}{x(l_1 + l_2 \cos \theta_B) + y l_2 \sin \theta_B}\right] \tag{12.70}$$

The signs of the numerator and denominator in Eq. 12.70 can be used to establish the specific quadrant of angle θ_A. Thus there are two solutions for θ_A, corresponding to the two solutions for θ_B as represented by the two signs in Eq. 12.66. Once again, the corresponding configurations of the robot are shown in Figure 12.17. Eqs. 12.63 and 12.70 are the desired relationships for θ_B and θ_A as functions of x and y.

Alternatively, expressions for θ_A and θ_B can be derived directly from the geometry of Figure 12.17. From that figure, it can be seen that angle θ_A can be expressed in terms of angles α and β as

$$\theta_A = \alpha \pm \beta \tag{12.71}$$

where α is obtained from the x and y coordinates as

$$\alpha = \tan^{-1}\left(\frac{y}{x}\right) \tag{12.72}$$

where the signs of x and y will identify the proper quadrant. Angle β can be obtained from the law of cosines as follows:

$$l_2^2 = x^2 + y^2 + l_1^2 - 2 l_1 \sqrt{x^2 + y^2} \cos \beta$$

or

$$\beta = \cos^{-1}\left(\frac{x^2 + y^2 + l_1^2 - l_2^2}{2 l_1 \sqrt{x^2 + y^2}}\right) \tag{12.73}$$

where the positive value of β between $0°$ and $180°$ would be used in Eq.

12.71. Angle θ_B can also be determined from the law of cosines. In this case,

$$x^2 + y^2 = l_1^2 + l_2^2 - 2l_1l_2 \cos(\pi - \theta_B) = l_1^2 + l_2^2 + 2l_1l_2 \cos \theta_B$$

which yields

$$\theta_B = \cos^{-1}\left(\frac{x^2 + y^2 - l_1^2 - l_2^2}{2l_1l_2}\right) \tag{12.74}$$

This is the same as Eq. 12.63, found by the other approach. Care must be exercised in combining the angles from Eqs. 12.71 and 12.74. In particular, inspection of Figure 12.17 reveals that the negative value of θ_B from Eq. 12.74 should be used in combination with the positive sign in Eq. 12.71, while the positive value of θ_B corresponds to the case of the negative sign in Eq. 12.71.

12.6.2 RP Manipulator

Inverse kinematics for mechanisms containing prismatic joints tends to be somewhat easier than for those with revolute joints. This is true for the manipulator of Section 12.5.4 and Figure 12.14, which is a two-degree-of-free-dom manipulator with a combination of a revolute pair and a prismatic pair. Referring to Eq. 12.53, the relations are

$$x = -(s_B + l_2)\sin \theta_A \tag{12.75}$$

$$y = (s_B + l_2)\cos \theta_A \tag{12.76}$$

$$z = s_A \tag{12.77}$$

where, again, subscripts and superscripts have been deleted. In this mechanism, s_A is a fixed dimension and, therefore, coordinate z is constant. Joint rotation variable θ_A is obtained by dividing Eq. 12.75 by Eq. 12.76, leading to

$$\theta_A = \tan^{-1}\left(\frac{(-x)}{y}\right) \tag{12.78}$$

where the combination of the numerator and denominator signs can be used to identify the quadrant. Squaring Eqs. 12.75 and 12.76, and adding yields the following relation for translational variable s_B:

$$s_B = \sqrt{x^2 + y^2} - l_2 \tag{12.79}$$

Note that multiple solutions do not exist in this case.

12.6.3 RRR Spatial Manipulator

A three-degree-of-freedom manipulator with three revolute pairs was considered in Section 12.5.5. See Figure 12.15. The forward kinematics equations that were developed, Eq. 12.59, are

$$x = l_2 \cos \theta_A \cos \theta_B + l_3 \cos \theta_A \cos(\theta_B + \theta_C) \tag{12.80}$$

$$y = l_2 \sin \theta_A \cos \theta_B + l_3 \sin \theta_A \cos(\theta_B + \theta_C) \tag{12.81}$$

$$z - s_A = l_2 \sin \theta_B + l_3 \sin(\theta_B + \theta_C) \tag{12.82}$$

The solution for angles θ_A, θ_B, and θ_C can be patterned after the first approach described in Section 12.6.1. Squaring each of the equations above, and then adding,

$$x^2 + y^2 + (z - s_A)^2 = l_2^2 + l_3^2 + 2l_2 l_3 \cos \theta_C$$

from which

$$\theta_C = \cos^{-1}\left(\frac{x^2 + y^2 + (z - s_A)^2 - l_2^2 - l_3^2}{2l_2 l_3} \right) \tag{12.83}$$

As before, there are two possible solutions for θ_C.

Next, θ_A is obtained by dividing Eq. 12.81 by Eq. 12.80, yielding

$$\frac{y}{x} = \tan \theta_A$$

or

$$\theta_A = \tan^{-1}\left(\frac{y}{x} \right) \tag{12.84}$$

Finally, rewriting Eqs. 12.80 and 12.82,

$$\frac{x}{\cos \theta_A} = (l_2 + l_3 \cos \theta_C) \cos \theta_B - l_3 \sin \theta_C \sin \theta_B$$

and

$$z - s_A = l_3 \sin \theta_C \cos \theta_B + (l_2 + l_3 \cos \theta_C) \sin \theta_B$$

Solving for $\sin \theta_B$ and $\cos \theta_B$, and combining,

$$\theta_B = \tan^{-1}\left(\frac{(z - s_A)(l_2 + l_3 \cos \theta_C) \cos \theta_A - x l_3 \sin \theta_C}{x(l_2 + l_3 \cos \theta_C) + (z - s_A) l_3 \sin \theta_C \cos \theta_A} \right) \tag{12.85}$$

where

$$\sin \theta_C = \pm \sqrt{1 - \cos^2 \theta_C} \tag{12.86}$$

Equations 12.83, 12.84, and 12.85 are the required equations for carrying out the inverse kinematic analysis of this manipulator.

References

1. Craig, J. J., *Introduction to Robotics: Mechanics and Control*, Addison-Wesley, Reading, MA, 1986.
2. Critchlow, A. J., *Introduction to Robotics*, Macmillan, New York, 1985.

3. Denavit, J., and R. S. Hartenberg, "A Kinematic Notation for Lower-Pair Mechanisms Based on Matrices," *Journal of Applied Mechanics*, vol. 77, 1955, pp. 215–221.
4. Fu, K. S., R. C. Gonzalez, and C. S. G. Lee, *Robotics: Control, Sensing, Vision, and Intelligence*, McGraw-Hill, New York, 1987.
5. Hill, R. O., Jr., *Elementary Linear Algebra*, Academic Press, New York, 1986.
6. Paul, R. P., *Robot Manipulators: Mathematics, Programming and Control*, MIT Press, Cambridge, MA, 1981.

PROBLEMS

12.1 Determine the number of degrees of freedom of each of the spatial mechanisms shown in Figure P12.1, where, in each case, the mechanism is a closed chain where the first and last members are part of the ground link.

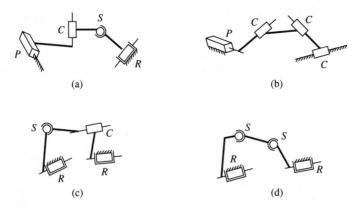

Figure P12.1

12.2 Determine the number of degrees of freedom of each of the spatial mechanisms shown in Figure P12.2, where, in each case, the mechanism is an open chain with the first member part of ground and the last member free to move.

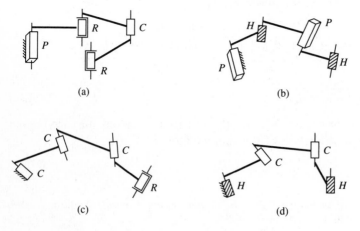

Figure P12.2

12.3 Derive the 3×3 rotational transformation matrix $^i_j[R]$ for the following sequence of rotations: rotation β about the y_i axis, followed by rotation γ about the z_i axis, and then rotation α about the x_i axis.

12.4 Show that finite rotations about perpendicular axes are not commutative by comparing the sequence of an x-axis rotation followed by a y-axis rotation with the sequence of a y-axis rotation followed by an x-axis rotation. Use transformation matrices.

12.5 Reference frames i and j have the same origin. The orientation of frame j relative to frame i is given by a rotation of $-90°$ about the x_i axis, followed by a rotation of $45°$ about the y_i axis, and then $-45°$ about the z_i axis. A point has coordinates $\{1, 1, 1\}^T$ in the j frame. Determine the coordinates of the point in the i frame.

12.6 Reference frames i and j have the same origin. The orientation of frame j relative to frame i is given by a rotation of $90°$ about the y_i axis, followed by a rotation of $90°$ about the z_i axis, and then $90°$ about the x_i axis. Determine the rotational transformation matrices $^i_j[R]$ and $^j_i[R]$.

12.7 Reference frames i and j have the same origin. The orientation of frame j relative to frame i is given by a rotation of $-90°$ about an axis $i_{(u)} = \{1/\sqrt{2}, 1/\sqrt{2}, 0\}^T$. A point has coordinates $\{1, 1, 1\}^T$ in the j frame. Determine the coordinates of the point in the i frame.

12.8 Derive an expression for the rotational transformation matrix $^i_j[R]$, where $^j_i[R]$ is given by Eq. 12.17.

12.9 Derive Eq. 12.12 from Eq. 12.20.

12.10 Frame j has the same origin as frame i and an orientation obtained from the following sequence: a $90°$ rotation about x_i, followed by a $90°$ rotation about y_i, followed by a $90°$ rotation about z_i. Determine the axis of rotation and angle of rotation for an equivalent single axis rotation from frame i to frame j.

12.11 The orientation of frame 2 relative to frame 1 is given by a rotation of $90°$ about the y_1 axis, followed by a rotation of $90°$ about the x_1 axis. The position of the origin of frame 2 relative to the origin of frame 1 is $^1\{Q\} = \{2\ 3\ 0\}^T$. The coordinates of a point in frame 2 are $^2\{P\} = \{3\ 4\ 1\}^T$. Determine the coordinates of the point in frame 1. Draw a sketch of the frames and point to verify your results.

12.12 The orientation of frame 2 relative to frame 1 is given by a rotation of $-45°$ about the x_1 axis, followed by a rotation of $+45°$ about the y_1 axis, followed by a rotation of $-90°$ about the z_1 axis. The position of the origin of frame 2 relative to the origin of frame 1 is $^1\{Q\} = \{1\ 0\ -2\}^T$. The coordinates of a point measured in frame 2 are $^2\{P\} = \{1\ 2\ -1\}^T$. Determine the coordinates of the point in frame 1.

12.13 An RRR planar manipulator is shown in Figure P12.3, where the axes of the three revolute joints are parallel. Assume that the link offsets are zero; that is,

$s_A = s_B = s_C = 0$. Point D is a point on member 3. The home position of the robot is shown in Figure P12.3a, and joint rotations $\theta_A, \theta_B, \theta_C$ are relative to this position. Derive a relationship for $^0\{P\}_D$ as a function of $^3\{P\}_D$, and evaluate these equations for the case of $l_1 = l_2 = l_3 = 10$ in. and $\theta_A = \theta_B = \theta_C = 30°$.

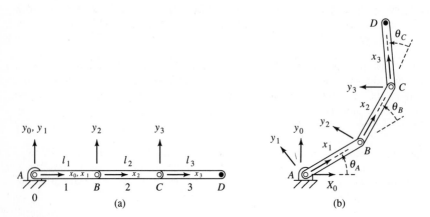

Figure P12.3

12.14 Determine the transformation matrix $^0_3[T]$ for the RRR manipulator in Figure P12.4 as a function of the dimensions shown and the joint angles θ_A, θ_B, and θ_C. The position shown corresponds to $\theta_A = \theta_B = \theta_C = 0$. Compute the position vector $^0\{P\}_D$ for point D on member 3 when $\theta_A = \theta_B = \theta_C = 30°$ for dimensions $l_1 = l_2 = l_3 = 10$ in.

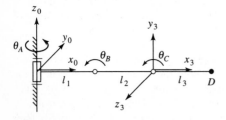

Figure P12.4

12.15 Determine the transformation matrix $^0_2[T]$ for the RR manipulator in Figure P12.5 as a function of the dimensions shown and the joint angles θ_A and θ_B. The position shown corresponds to $\theta_A = \theta_B = 0$. Compute the position vector $^0\{P\}_C$ for point C on member 2 when $\theta_A = \theta_B = 45°$ for dimensions $l_1 = l_2 = 10$ in and $b_1 = b_2 = 5$ in.

12.16 Determine the transformation matrix $^0_2[T]$ for the PP manipulator in Figure P12.6 as a function of joint translations s_A and s_B, and dimensions l_2 and b_2. Compute the position vector $^0\{P\}_C$ for point C on member 2 when $s_A = s_B = 5$ in, and $l_2 = b_2 = 10$ in.

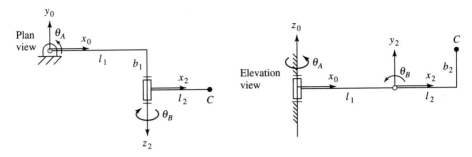

Figure P12.5

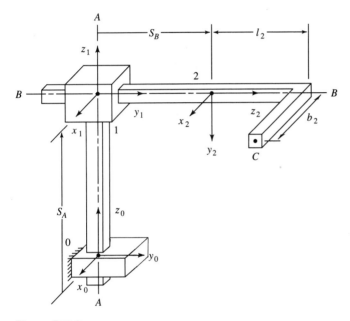

Figure P12.6

12.17 The RR planar manipulator of Figure 12.13 has dimensions $l_1 = l_2 = 10$ in. Carry out the inverse kinematics calculations for joint angles θ_A and θ_B when point C is at the location $^0\{P\}_C = \{-6 \ 12 \ 0\}^T$.

12.18 The manipulator of Figure 12.15 has dimensions $s_A = l_2 = l_3 = 10$ in. Carry out the inverse kinematic analysis to determine joint angles θ_A, θ_B, and θ_C when point D is at the location $^0\{P\}_D = \{12 \ 6 \ 18\}^T$.

12.19 The RP manipulator of Figure 12.14 has dimension $l_2 = 10$ in. Carry out the inverse kinematics calculations for joint coordinates θ_A, s_A, and s_B when point C is at the location $^0\{P\}_C = \{12 \ 6 \ 8\}^T$.

12.20 Derive the inverse kinematics relations for determining the joint angles θ_A, θ_B, and θ_C in terms of the end effector coordinates $^0\{P\}_D$ for the manipulator of Figure P12.4.

Partial Answers to Selected Problems

CHAPTER 1

1.1 **(a)** $DF_{spatial} = 6$
   (b) $DF_{spatial} \geq -5$
   (c) $DF_{planar} = 2$
   (d) Motion occurs in a plane or a set of parallel planes.

1.3 $v_{av} = 400$ in/s

1.5 378 and 553 in/s

1.7 For $R = 1$, piston position is given as follows:

Crank Angle T_1°	Piston Position X_2
0	2.487
20	2.433
40	2.199
60	1.844

1.9 For $R = 1$, piston position is given as follows:

Crank Angle T_1°	Piston Position X_2
0	2.755
20	2.739
40	2.550
60	2.239

1.11 **(b)** 19.99 in/s to the left
    (c) 22.83 in/s to the left
1.13 Triple rocker (non-Grashof)
1.15 Change-point mechanism
1.17 Triple rocker (non-Grashof)
1.19 Change-point mechanism
1.21 Drag link
1.23 $120 \leq L_2 \leq 480$ mm
1.25 **(a)** $0 < L_1 < 20$
    (b) None
    (c) None
    (d) $L_1 = 20$ and 60
    (e) $0 < L_1 \leq 20$ and $L_1 = 60$
    (f) $60 < L_1 < 140$ and $20 < L_1 < 60$
1.27 **(a)** None
    (b) None
    (c) $180 < L_3 < 260$
    (d) $L_3 = 180$ and 260
    (e) $180 \leq L_3 \leq 260$
    (f) $20 < L_3 < 180$ and $260 < L_3 < 420$
1.29 **(a)** $1.5 < L_2/L_1 < 2.9$
    (b) None
    (c) $0 < L_2/L_1 < 0.5$
    (d) $L_2/L_1 = 0.5$ or 1.5 or 2.9
    (e) $0 < L_2/L_1 \leq 0.5$ and $1.5 \leq L_2/L_1 \leq 2.9$
    (f) $0.5 < L_2/L_1 < 1.5$ and $2.9 < L_2/L_1 < 4.9$
1.31 Permissible values for the crank-rocker mechanism lie within the region bounded by the inequalities $L_3/L_1 < L_2/L_1 + 1$, $L_3/L_1 > 3 - L_2/L_1$, and $L_3/L_1 > L_2/L_1 - 1$
1.33 Enter link lengths; sort for L_{max}, etc. Test: $L_{max} \geq L_{min} + L_a + L_b$. If YES, print "NOT A MECHANISM." If NO, test for triple rocker. If YES, print "TRIPLE ROCKER." If NO, print "GRASHOF MECHANISM"; test for change-point, crank rocker, and double-rocker mechanisms. (Steps in the solution may vary.)
1.35 No value of L_3 satisfies the criteria.
1.37 No value of L_3 satisfies the criteria.
1.39 $L_2/L_1 = 3.045$. Other solutions are possible.
1.41 For $L_0/L_1 = 4$, the permissible region is bounded by the following points:

L_2/L_1	L_3/L_1
1.5	3.867
2	4.060
3	4.243
4	3.828
4.22	3.294
1.5	3.826
2	3.382
3	2.406
3.5	1.870

1.43 848.5 mm

1.45 494.97 mm

1.47 0.4571 L

1.51 Set $d\phi/d\theta = R \cos \theta/(L \cos \phi) = 0$.

1.53 In a double-rocker linkage (of the first kind), the coupler must be free to rotate through 360°.

1.55 In Figure 1.25 a, let $DA/DF = 2$. F is the tracing point.

1.57 In Figure 1.25a, let $AB/AC = 0.40$. Point A traces the pattern.

1.59 Many solutions are possible. Let $\phi = 30°$, $N = 6$ cylinders, $A = (d/6)^2$. Then, $d = 129.6$ mm.

1.61 $L_2 - L_1 = 0.005$ in

1.63 One possible solution: Use $L = 15$ mm, $N = 15$ teeth, $r_3/r_0 = 0.5$, and $\phi_{min} = 45°$, from which $r_{1(max)} = 0.287r_0$ and $r_2 = 1.002r_0$. A minimum crank length $r_{1(min)} = 0.090r_0$ is satisfactory.

1.65 $n_2 = 965.9$ to 1035.3 rev/min

1.67 13.86°

1.69 One possible solution: Refer to Figures 1.19 and 1.20; let $L_0 = 2$, $L_1 = 4.5$, and $L_3 = 5$. $L_2 = 7.25$ results in a ratio of 2.5 to 1 (approx).

1.71 One possible solution: $L_3 = 75$ mm. Let $L_0 = 20$ mm, $L_1 = 45$ mm, and $L_4 = 80$ mm. $L_2 = 67.5$ is satisfactory.

1.73 One possible solution: $O_1O_2 = 7$, $L_2 = 8.03$, $L_3 = 4$, and $L_{1(max)} = 4.36$.

1.75 $O_2C = 453$ mm. Selecting $O_1O_2 = 250$ mm, $O_1B_{max} = 77.25$ mm and $O_1B_{min} = 43.67$ mm.

1.77 The stroke is 3.14 in.

1.79 **(a)** 1.05   **(b)** 4.016

1.85 $100 < L_2 < 300$ mm

1.87 $95 < L_2 < 305$ mm

1.89 $\phi_{max} = 85.5°$; $L_2 = 212.13$ mm

1.91 $L_2 = 219.08$ mm; $\phi_{max} = 80.83°$

1.93 Away from crank: $v_{av} = 26,960$ mm/s; toward crank: $v_{av} = 21,940$ mm/s

1.95 $DF_{spatial} \geq 4$

1.97 $DF_{spatial} = 6$

1.101 The conditions are met with an eccentricity of 1.163 times the crank length.

CHAPTER 2

2.1 **(a)** Piston, rack, reciprocating cam follower, straight edge on a drafting machine, the straight section of a belt or chain that is in tension, etc.

2.3 **(a)** 9.425 m/s
   (b) 2961 m/s^2
   (c) 930, 190 m/s^3

2.5 $D = S = 303$ mm

2.7 $D = S = 2.323$ in

2.9 $x = 20[1 - \cos(100t)]$; $v = 4000 \sin(200t)$; $a = 800,000 \cos(200t)$

2.11 44.27 rad/s

2.13 $a_{max} = 31.42$ in/s; $v_{max} = 3948$ in/s^2

2.15 2.909 $\angle 50.10°$

2.17 1.759 $\angle 176.39°$

2.19 $A \times B = k1$

2.21 $C \times (D \times E) = k7.954$

2.23 $A \cdot B = 1.732$

2.25 $A \cdot (B \times C) = 0$

2.27 $C \cdot (A \times B) = 0$

2.29 $-i[3r_y + (5 + 3t)dr_y/dt] + j[3r_x + (5 + 3t)dr_x/dt]$

2.31 In this case, reversing the order of the rotations affects only the altitude change. In general, the commutative law of vector addition does not apply to finite rotations.

2.33 $r_1 = -75.54$; $r_2 = -18.63$

2.35 $T = 3672$ $N \cdot$ mm cw

2.37 $r_0 = i0.5893r_1$ or $-i2.1213r_1$; $r_2 = r_1(-0.6428j \mp 1.3553i)$

2.41 and 2.43 $r_2 = i139.6 + j10.8$; $r_3 = -i49.6 - j62.8$

2.45 $r_2 = i90.56 - j106.76$; $r_3 = -i22.52 + j76.77$

2.47 $dR_1/dt = -20.78 + j12$

2.49 $R_2 = 261.85 + j159.75 = 306.7e^{j0.548}$

2.51 $r_0 = -0.5892r_1$ or $2.1212r_1$; $r_2 = (\pm 1.3552 - j0.6428)r_1$

2.53 $r_3 = 180.4$ mm at $\theta = 45°$

2.55 $D \times C = \begin{bmatrix} 106 \\ 22 \\ 152 \end{bmatrix}$ $E \cdot (E \times D) = 0$

2.57 $B \times D = \begin{bmatrix} -32 \\ 31 \\ -19 \end{bmatrix}$

2.59 $r_{2x} = [-r_2^2 + c_y^2 + r_3^2 - c_x^2]/[2 c_x]$ where $c_x = -r_0 + r_1 \cos \theta$ and $c_y = r_1 \sin \theta$

2.61 $\theta_2 = 0.296$, $\theta_3 = 4.132$ rad after four iterations.

2.63 $\theta_2 = 0.293$, $\theta_3 = 4.698$, $r_D = 4$, $r_E = 4.944$

2.65 $\theta_2 = 0.16$, $\theta_3 = 5.476$, $r_D = 61.39$, $r_E = 158.79$

CHAPTER 3

3.1 $v = -i100 - j2600 + k2000$

3.3 $v = 1260$ in/s

3.5 $n = 12s/(\pi d)$

3.7 $\dot{R} = I572.4 - J82.6$

3.9 $v_{CB} = 670.2$ in/s opposite the direction of rolling

3.11 $v_B = 37.7 \angle -10.8°$

3.13 $\theta_2 = 16.35°$, $\theta_3 = -122.19°$; $\omega_2 = -2.87$ rad/s, $\omega_3 = 10.86$ rad/s; $v_C = 217.25$ mm/s $\angle 147.8°$

3.15

θ_1(degrees):	15	30	45	60	75
ω_3/ω_1:	-0.129	0.173	0.362	0.467	0.519

3.17 $\omega_2 = 5.03$ rad/s ccw

3.19 $O_2B = 79.36$; $\theta_2 = 5.61°$; $\omega_2 = 3.73$ rad/s

3.21 $O_2B = 74.31$; $\theta_2 = 16.59$; $\omega_2 = 3.55$ rad/s

3.23

θ_1 (degrees):	0	15	30	45
ω_2/ω_1:	0.25	0.247	0.237	0.219

3.27 $\theta = 54.6°$; $v = 1.13R\omega$

3.29 $\omega_2 = 3.1$ rad/s

3.31 $\omega_2 = 60$ rad/s

3.33 (a) 20 in/s

(b) 20 in/s

3.35 Sliding velocity = 32 in/s

3.37 (a) and (b) $v_{CB} = \omega R$

3.39 Follower velocity = 7.1 in/s

3.41 $v_D = 17.5$; $v_E = 30$ in/s; $\omega_3 = 12.9$ rad/s cw

3.43 Sliding velocity = 163 in/s

3.45 $v_D = 62$ in/s

3.47 Sliding velocity = 17 in/s

3.49 $v_C = 210$ in/s

3.51 $\omega_4 = 433$ rad/s ccw

3.53 $v_C = 367$ mm/s

3.57

θ_1 (Degrees)	ω_2(rad/s)	ω_3(rad/s)
30	-0.633	-4.106
60	-0.696	-3.927
90	-0.809	-3.599
120	-0.977	-3.077

3.59 $v_{B_2} = v_D = \omega_1 = 0$; $\alpha_{1(av)} = 181{,}000$ rad/s^2 cw; $a_{D(av)} = 76{,}000$ m/s^2 to the right

3.61 The angular velocities of links 2 and 3 are 1.06 rad/s cc and 32.3 rad/s cw, respectively.

3.63 and 3.65 $v_c = 7.37$ in/s to the left.

3.67 At $\theta_1 = 45°$, the angular velocities of links 2 and 3 are 0.205 and -0.238 rad/s, respectively.

3.69 At $\theta_1 = 45°$, $v_D = 490.2$ at 99.4°.

3.71 At $\theta_1 = 60°$, normalized slider velocity is -0.96.

3.73 At $\theta_1 = 30°$, the angular velocities of links 2 and 3 are 0.25 and -0.22 rad/s, respectively.

CHAPTER 4 4.1 a = $-i\,100{,}800 - j\,88{,}600 - k\,30{,}0\,0\,0$

4.1 $a = -i\,100{,}800 - j\,132{,}600 - k\,54{,}000$

4.3 $a_{av} = 2000$ in/s^2

4.5 (a) $\alpha_{av} = 5.236$ rad/s^2

(b) $\alpha_{av} - 2.09$ rad/s^2

4.7 $a^n = 19.29$ m/s^2

4.9 $a_B' = 12{,}500$ mm/s^2; $a_B^n = 72{,}000$ mm/s^2; $a_B = 73{,}077$ mm/s^2

4.11 $R = i20 + j60$; $\dot{R} = -i8995 + j2985 + k10$; $\ddot{R} = -i499{,}790 - j1{,}346{,}870 + k20$

4.13 $a_C = 31{,}381i - 48{,}131j = 57{,}457$ mm/s^2 $\angle -56.9°$

4.15

θ_1 (Degrees)	ω_3/ω_1	α_3/ω_1^2
30	1.333	−0.246
60	1.205	−0.220
90	1.106	−0.164

4.19 $\alpha_2 = 45.7$ rad/s^2

4.21 $a_D = 23{,}000$ in/s^2; $\alpha_2 = 2230$ rad/s^2

4.23 $a_D = 5040$ in/s^2; $\alpha_2 = 147$ rad/s^2; $\alpha_3 = 267$ rad/s^2

4.25 $\alpha_2 = 138.5$ rad/s^2; $\alpha_3 = 440$ rad/s^2

4.27 $\alpha_1 = 63$ rad/s^2

4.29 (b) $a_C = 33{,}300$ in/s^2 to the left when $\theta = 0$

4.31 $a_E = 40{,}000$ in/s^2

4.33 $a_{C_1}{}^N = 100$ in/s^2; $a_{C_1}{}^T = 0$; $a_{C_2} = 70.7$ in/s^2

4.35 $a_E = 1100$ in/s^2

4.37 $a_D = 700$ in/s^2

4.39 $\alpha_2 = 120$ rad/s^2

4.41 Crank speed $= 903$ rev/min

4.43 $\alpha_1 = 5780$ rad/s^2; $\alpha_3 = 5150$ rad/s^2

4.53 $a_C = 625{,}000$ mm/s^2; $\alpha_2 = 1233$ rad/s^2

4.55 $a_C = 1615$ mm/s^2

4.57 $a_C = 900$ m/s^2

4.63 $\alpha_2/\omega_1^2 = -0.05197$ at $\theta_1 = 30°$

4.65 $\alpha_2 = 391.5$ rad/s^2; $\alpha_3 = 541.9$ rad/s^2

4.67 $a_3 = a_{s_2} = -8117$ in/s^2

4.69

θ (Degrees)	$a_3/\omega_1^2 r_1$
0	−0.802
30	−0.727
60	−0.501

4.71 At $\theta_1 = 35°$, $\alpha_2 = 0.2938^*$ and $\alpha_3 = 0.6791^*$. *: Multiply by angular velocity squared of link 1.

4.73 At $\theta_1 = 35°$, $\alpha_2 = 0.0847^*$ and $\alpha_3 = 0.4205^*$. *: Multiply by angular velocity squared of link 1.

4.75 At $\theta_1 = 160°$, $\alpha_2 = 0.119^*$. *: Multiply by angular velocity squared of link 1.

CHAPTER 5

5.27 (b) $v = 10.7$ in/s from $\theta = \pi/4$ to $3\pi/4$; $v = -10.7$ in/s from $\theta = 5\pi/4$ to $7\pi/4$; $a = 85.3$ in/s^2 for θ from 0 to $\pi/4$ and $7\pi/4$ to 2π; $a = -85.3$ in/s^2 for θ from $3\pi/4$ to $5\pi/4$

(c) $s = 2.89$ in, $v = 10.7$ in/s, $a = 0$, $j = 0$

(d) $s = 3.0$ in, $v = 10.7$ in/s, $a = 0$, $j = 0$

5.29 **(b)** $v = 16$ in/s at $\theta = \pi/2$, $v = -16$ in/s at $\theta = 3\pi/2$; $a = 32\pi$ in/s^2 at $\theta = \pi/4$ and $7\pi/4$, $a = -32\pi$ in/s^2 at $\theta = 3\pi/4$ and $5\pi/4$

(c) $s = 3.22$ in, $v = 12$ in/s, $a = -87.1$ in/s^2, $j = -632$ in/s^3

(d) $s = 3.0$ in, $v = 13.4$ in/s, $a = -74.7$ in/s^2, $j = -845$ in/s^3

5.31 **(b)** $v = 15$ in/s at $\theta = \pi/2$, $v = -15$ in/s at $\theta = 3\pi/2$; $a = 92.4$ in/s^2 at $\theta = 0.211\pi$ and 1.789π, $a = -92.4$ in/s^2 at $\theta = 0.789\pi$ and 1.211π

(c) $s = 3.16$ in, $v = 11.8$ in/s, $a = -71.1$ in/s^2, $j = -640$ in/s^3

(d) $s = 3.0$ in, $v = 12.7$ in/s, $a = -61.9$ in/s^2, $j = -734$ in/s^3

5.33 **(b)** $v = 75\pi$ mm/s at $\theta = 60°$, $v = -75\pi$ mm/s at $\theta = 240°$; $a = 225\pi^2$ mm/s^2 at $\theta = 0$ and $300°$, $a = -225\pi^2$ mm/s^2 at $\theta = 120$ and $180°$

5.35 **(b)** $v = 300$ mm/s at $\theta = 60°$, $v = -300$ mm/s at $\theta = 240°$; $a = 1800$ mm/s^2 for θ from 0 to 60° and 240 to 300°, $a = -1800$ mm/s^2 for θ from 60 to 120° and 180 to 240°

5.37 $v_{max} = 9900$ mm/s

5.39 $a = 1190$ in/s^2, $v_{max} = 42.2$ in/s

5.47 **(a)** $y = (-2 + \sqrt{3})x$ and $y = (-2 - \sqrt{3})x$

(b) $x = (3/4 \pm \sqrt{3}/4)\lambda$, $y = (-3/4 \pm \sqrt{3}/4)\lambda$

5.49 $x = -(r_b + h\theta/\pi)\sin\theta - (h/\pi)\cos\theta$, $y = (r_b + h\theta/\pi)\cos\theta - (h/\pi)\sin\theta$; minimum follower face width = 31.8 mm

5.51 $x = -25\sin\theta\{(6 - \cos\theta) \pm [(6 - 2\cos\theta)/\sqrt{37 - 12\cos\theta}]\}$, $y = 25\cos\theta\{(6 - \cos\theta) \pm [(6 - \cos\theta + \sin\theta\tan\theta)/\sqrt{37 - 12\cos\theta}]\}$; $\phi = 8.9°$ at $\theta = 60°$

5.53 $\rho = r_b + h\theta/\pi = 100 + 15.9\theta$; neglecting possible discontinuities at ends of travel, $\rho_{min} = 100$ mm and $\rho_{max} = 150$ mm

5.55 $\rho_p = 25(37 - 12\cos\theta)^{3/2}/(38 - 18\cos\theta)$; $\rho_{p(max)} = 156.2$ mm, $\rho_{p(min)} = 147.9$ mm

---

CHAPTER 6

6.5 $d = 20.0$ in

6.7 $n = 596.8$ rev/min

6.9 $c = 8.33$ in

6.11 $v_p = 78.5$ in/s

6.13 $N_1 = 40$, $N_2 = 100$, $v_p = 2618$ mm/s

6.15 $c = 16$ in, $r_{b2} = 11.3$ in

6.17 $N_2 = 115$, $c = 15.43$ in

6.19 Contact ratio = 1.67

6.21 $c = 8.0$ in

6.23 Contact ratio = 1.66

6.25 17.5 in/s at beginning of contact; 16.6 in/s at end of contact

6.27 $\phi = 21.9°$

6.31 Interference; remove 0.25 mm from teeth of larger gear

6.33 $N_2 = 45$

6.35 Load moves 897 in.

6.39 $N_1 = 20$, $N_2 = 100$, $v_p = 65.4$ in/s

6.41 (a) $c_{AB} = c_{BC} = 240$ mm

(b) $\omega_1 = 104.7$ rad/s cw; $\omega_2 = 52.4$ rad/s ccw; $\omega_3 = 20.9$ rad/s ccw

(c) $T_A = 191$ N · m, $T_B = 0$, $T_C = 955$ N · m

(d) $F_t = 2388$ N, $F_r = 869$ N

6.43 $n_A = 1200$ rev/min ccw, $n_B = 480$ rev/min cw, $n_C = 230$ rev/min ccw; $T_A = 239$ N · m, $T_B = 597$ N · m, $T_C = 1243$ N · m; $F_A = 3970$ N, $F_B = 12,280$ N, $F_C = 9144$ N

CHAPTER 7

7.1 $p'' = 0.481$ in, $N_{f1} = 44.5$, $N_{f2} = 66.7$

7.3 $N_1 = 20$, $N_2 = 38$, $\phi = 20.6°$

7.7 $p'' = 16.32$ mm, $p = 18.85$ mm, $N_1 = 24$, $N_2 = 60$, $N_{f1} = 37.0$, $N_{f2} = 92.4$

7.9 $c = 275$ mm, $N_{f1} = 54.9$, $N_{f2} = 109.8$

7.11 $F_t = 5980$ N, $F_r = 2410$ N, $F_a = 2890$ N

7.13 $c = 14.2$ in

7.15 $N_1 = 24$, $N_2 = 60$

7.17 $d_w = 1.82$ in, $d_g = 3.18$ in

7.19 $c = 16.22$ in

7.21 $d_w = 2.41$ in, $d_g = 17.59$ in

7.23 $T_A = 9.55$ N · m, $T_B = 19.1$ N · m, $T_C = 764$ N · m, $F_{aA} = 87$ N, $F_{aB} = 6077$ N, $F_{aC} = 844$ N

7.25 (a) $\gamma_p = 16.2°$, $\Gamma_g = 43.8°$

(b) $r_{bp} = 1.56$ in, $r_{bg} = 5.18$ in

(c) $N_{fg} = 62.2$

7.27 $F_{tp} = F_{tg} = 1236$ lb, $F_{rp} = F_{ag} = 402$ lb, $F_{ap} = F_{rg} = 201$ lb

CHAPTER 8

8.1 Output speeds may range from 500 to 667 rev/min if all gears have the same module.

8.3 One possible solution: Let $N_1 = 24$. Then, $N_3 = 30$, $N_4 = 32$, $N_5 = 40$, $N_6 = 48$, $N_7 = 60$, $N_2 = 26$.

8.5 One possible solution: $N_1 = 29$, $N_2 = 40$, $N_3 = 39$, $N_4 = 50$

8.7 $n_R = 200(1 - N_S/N_R)$

8.9 (a) $n_{R_2} = 0.0915$ rev/min cw

(b) Impossible due to friction

8.11 $n_{S_2} = 31$ rev/min cw

8.13 S_4 makes one revolution in 10^8 s (about 3 years and 2 months)

8.15 $n_{S_2} = n_C + (n_{S_1} - n_C)(N_{S_1}N_{P_2})/(N_{P_1}N_{S_2})$

8.17 $n_C = (N_{S_1}N_{P_2}n_{S_1})/(N_{S_1}N_{P_2} - N_{P_1}N_{S_2})$

8.19 $n_C = (N_{R_1}N_{P_2}N_{P_4}n_{R_1})/(N_{R_1}N_{P_2}N_{P_4} + N_{P_1}N_{P_3}N_{R_2})$

8.21 $n_C = (N_{R_1}N_{P_2}n_{R_1})/(N_{P_1}N_{R_2} + N_{R_1}N_{P_2})$

8.23 $n_C = (N_S N_{P_2} n_S)/(N_S N_{P_2} - N_{P_1}N_R)$

8.25 $n_C = (n_S N_{P_2}N_{P_4}N_S)/(N_S N_{P_2}N_{P_4} - N_{P_1}N_{P_3}N_R)$

8.27 $n_C = (n_{S_1}N_{S_1}N_{P_3})/(N_{S_1}N_{P_3} + N_{P_2}N_{S_2})$

8.29 $n_C = n_{R_2}/\{1 + [N_{R_1}N_{P_2}/(N_{P_1}N_{R_2})]\}$

8.31 $n_C = n_R/\{1 - [N_S N_{P_2}/(N_{P_1}N_R)]\}$

8.33 $n_C = n_R/\{1 - [N_S N_{P_2}N_{P_4}/(N_{P_1}N_{P_3}N_R)]\}$

8.35 $n_C = (n_{S_1}N_{S_1}N_{P_2})/(N_{P_2}N_{S_1} - N_{S_2}N_{P_1})$

8.37 $n_C = (n_{R_1}N_{R_1}N_{P_2}N_{P_4})/(N_{R_1}N_{P_2}N_{P_4} + N_{R_2}N_{P_1}N_{P_3})$

8.39 $n_C = (N_{P_2}N_{R_1}n_{R_1})/(N_{P_2}N_{R_1} + N_{R_2}N_{P_1})$

8.41 $n_C = n_S N_S N_{P_2}/(N_S N_{P_2} - N_R N_{P_1})$

8.43 $n_C = (N_S n_S N_{P_2}N_{P_4})/(N_S N_{P_2}N_{P_4} - N_R N_{P_1}N_{P_3})$

8.45 $n_C = (n_{R_2}N_{R_2}N_{P_1})/(N_{P_1}N_{R_2} + N_{R_1}N_{P_2})$

8.47 $n_C = (n_R N_R N_{P_1})/(N_R N_{P_1} - N_S N_{P_2})$

8.49 $n_C = (n_R N_R N_{P_1}N_{P_3})/(N_{P_1}N_{P_3}N_R - N_{P_2}N_{P_4}N_S)$

8.51 $N_S = 20$; $N_R = 80$; $d_R = 320$ mm; $15 \le N_p \le 37$

8.53 $n_{S_2}/n_{S_1} = 0.92795$

8.57 $\omega_P = 225$ rad/s cw; $\omega_C = 105.88$ rad/s cw

8.59 $\omega_{S_2} = 25.14$ rad/s

8.61 **(a)** 17.083 carrier rotations; $n_{PC} = 36.458$ rotations
   (b) Carrier rotations $= (7/12)x + (5/12)y$

8.63 $5 \le d_2 \le 9$ in; $5 \le d_3 \le 7.92$ in

8.65 $r_1 = 0.3019$ in for 16 rev/min, 0.5406 for $33\frac{1}{3}$ rev/min, 0.667 for 45 rev/min, and 0.9286 for 78 rev/min

8.67 The drive may be similar to Figure 8.31 except that disk 2 drives and the diameter of disk 2 exceeds twice the diameter of disk 1.

8.71 There are 36 combinations producing 19 different ratios ranging from 0.64 to 1.5625.

8.73 $n_c/n_P = 3$

8.75 $N_{P1} = 20$

8.77 The desired reduction is produced (approximately) if the sun, planet, and ring gears have 19, 21, and 61 teeth, respectively. The train may be balanced with four planets.

8.79 The desired reduction is produced (approximately) if the sun, planet, and ring gears have 26, 18, and 62 teeth, respectively. The train may be balanced with four planets.

8.81 Torques are 30,000, 0, 82,500, and $-112,500$ N $\cdot$ mm on the sun, planet, ring, and carrier shafts, respectively.

8.83 Torques are 4375, 0, 13,125, and $-17,500$ N $\cdot$ mm on the sun, planet, ring, and carrier shafts, respectively.

CHAPTER 9

9.1 (a) $T_1 = 18.3$ N · m cw

9.3 $T_1 = 2.84$ N · m cw

9.5 $T_1 = 230$ N · mm cw

9.7 $T_1 = 4.92$ N · m cw

9.9 $F_s = 168$ N

9.11 $T_2 = 1340$ lb · in ccw

9.13 $T_1 = 8.22$ N · m cw

9.15 $T_1 = 121$ N · m cw

9.17 $P = 13$ lb

9.18 (a) $T_1 = 18.3$ N · m cw

9.20 $T_1 = 2.74$ N · m cw

9.22 $T_1 = 234$ N · mm cw

9.24 $T_1 = 5.09$ N · m cw

9.26 $F_s = 168$ N

9.28 $T_2 = 1340$ lb · in ccw

9.30 $T_1 = 8.44$ N · m cw

9.32 $T_1 = 121$ N · m cw

9.34 $P = 13$ lb

9.35 (a) $T_1 = 18.3$ N · m cw

9.37 $T_1 = 2.72$ N · m cw

9.39 $T_1 = 233$ N · mm cw

9.41 $T_1 = 5.09$ N · m cw

9.44 $T_2 = 1340$ lb · in ccw

9.45 $T_1 = 121$ N · m cw

9.47 (a) $T_1 = 22.8$ N · m cw

(b) $e = 0.80$

9.48 (a) $T_1 = 22.8$ N · m cw

(b) $e = 0.80$

9.51 $T_1 = 4.51$ N · m cw

9.53 $T_1 = 6.90$ lb · in cw

9.55 $T_1 = 10.7$ lb · in cw

9.57 $T_1 = 14.1$ lb · in ccw

CHAPTER 10

10.1 $|m_2 a_{G2}| = 64$ N, $h_2 = 12.6$ mm

10.3 $T_1 = 4.4$ N · m ccw

10.5 (a) $F_2 = 49.5 \angle 45°$ N, $h_2 = 15.3$ mm

10.7 $T_1 = 26.9$ lb · in ccw

10.9 $T_1 = 34.4$ lb · in cw

10.13 $T = 8.93$ N · m when $\phi = 30°$

10.15 $T = 375$ lb · ft when $\phi = 30°$

10.17 $T = 6.94$ N · m when $\phi = 30°$

10.19 $T = 5.94$ N · m cw

10.21 $w_P r_P = 5.59$ oz · in, $\theta_P = 253°$, $w_Q r_Q = 4.27$ oz · in, $\theta_Q = 219°$

10.23 $w_P r_P = 6.36$ oz · in, $\theta_P = 35°$, $w_Q r_Q = 9.37$ oz · in, $\theta_Q = 268°$

10.25 $m_P r_P = m_Q r_Q = 141$ kg · mm, $\theta_P = 90°$, $\theta_Q = 270°$

10.27 At $\phi = 0$, $F_{32x} = 158$ N, $F_{32y} = F_{03y} = F_{12y} = F_{01y} = 0$, $F_{12x} = -251$ N, $F_{01x} = -97$ N

10.29 $F_s = 2000(\cos \omega t - \sin \omega t)$, $a = (4 \sin \omega t + \cos 2\omega t)/(\sin \omega t - \cos \omega t)$; secondary shaking forces balance

10.31 Primary forces do not balance; secondary forces balance.

10.33 Primary moments do not balance.

10.35 Neither primary nor secondary forces balance.

10.37 Correction magnitude = 2 mr, counterweight location = 180° from crank

CHAPTER 11

11.1 The coordinates of the fixed pivot are 125, 344.

11.3 Fixed center O_B must lie along a line through point 4.5, 5.5 at angle 135°.

11.5 The coordinates of the fixed pivot are 124.999, 344.

11.7 Fixed center O_c must lie along a line through point 7.23, 8.23 at an angle of 151°.

11.9 Locate fixed pivot O_B at 2.5, −7.

11.11 Locate fixed pivot O_B at 25.96, 12.02.

11.13 Locate fixed pivot O_B at 26.18, −18.53.

11.15 $R_1 = j4$, $R_2 = j2$, $R_3 = 6 − j2$

11.17 $r_1 = −9.48 − j1.68$, $r_2 = −168.10 + j10.08$, $r_3 = 77.58 − j8.40$

11.19 $r_1 = −0.308 − j\,0.462$, $r_2 = −0.154 − j\,0.231$, $r_3 = −0.538 + j\,0.692$

11.21 $r_1/r_0 = 2.32$, $r_2/r_0 = 0.62$, $r_3/r_0 = 2.43$

CHAPTER 12

12.1 (a) $DF = 1$ **(b)** $DF = 1$ **(c)** $DF = 1$ **(d)** $DF = 2$

12.3 $\,_j^i[R] = \begin{bmatrix} (\cos \beta \cos \gamma) & (−\sin \gamma) & (\sin \beta \cos \gamma) \\ (\cos \alpha \cos \beta \sin \gamma + \sin \alpha \sin \beta) & (\cos \alpha \cos \gamma) & (\cos \alpha \sin \beta \sin \gamma − \sin \alpha \cos \beta) \\ (\sin \alpha \cos \beta \sin \gamma − \cos \alpha \sin \beta) & (\sin \alpha \cos \gamma) & (\sin \alpha \sin \beta \sin \gamma + \cos \alpha \cos \beta) \end{bmatrix}$

12.5 $\,^i\{P\} = \begin{Bmatrix} \dfrac{\sqrt{2}}{2} \\ \dfrac{\sqrt{2}}{2} \\ −\sqrt{2} \end{Bmatrix}$

12.7 $\,^i\{P\} = \begin{Bmatrix} 1 − \dfrac{1}{\sqrt{2}} \\ 1 + \dfrac{1}{\sqrt{2}} \\ 0 \end{Bmatrix}$

12.11 $\,^1\{P\} = \begin{Bmatrix} 3 \\ 6 \\ 4 \\ 1 \end{Bmatrix}$

12.13 $\,^0\{P\}_D = \begin{Bmatrix} 5(1 + \sqrt{3}) \\ 5(3 + \sqrt{3}) \\ 0 \\ 1 \end{Bmatrix}$

12.15 $\,^0\{P\}_C = \begin{Bmatrix} 13.11 \\ 6.04 \\ 10.61 \\ 1 \end{Bmatrix}$

12.17 $\theta_A = 68.7°$, $164.5°$; $\theta_B = 95.7°$, $−95.7°$

12.19 $\theta_A = −63.44°$, $s_A = 8$ in, $s_B = 3.42$ in

Index